Evaluating Oral Probiotics and Prebiotics

Evaluating Oral Probiotics and Prebiotics

Edited by

Amir M. Mortazavian
Department of Food Science and Technology, Faculty of Nutrition Sciences and Food Technology, National Nutrition and Food Technology Research Institute, Shahid Beheshti University of Medical Sciences, Tehran, Iran

Adriano Gomes da Cruz
Science and Technology of Rio de Janeiro, Rio de Janeiro, Brazil

Zahra Yari
Department of Nutrition Research, National Nutrition and Food Technology Research Institute, Shahid Beheshti University of Medical Sciences, Tehran, Iran

ELSEVIER

ACADEMIC PRESS
An imprint of Elsevier

Academic Press is an imprint of Elsevier
125 London Wall, London EC2Y 5AS, United Kingdom
50 Hampshire Street, 5th Floor, Cambridge, MA 02139, United States

ISBN: 978-0-443-29835-6

For Information on all Academic Press publications
visit our website at https://www.elsevier.com/books-and-journals

Publisher: Mara Conner
Acquisitions Editor: Nicole Denis
Editorial Project Manager: Vaishali Thapa
Production Project Manager: Rajeswari R.
Cover Designer: Vicky Pearson Esser

Typeset by MPS Limited, Chennai, India

Contents

2. Application of postbiotics in medicine and food products

Delara Moradi Mirhesari, Mahshid Bahraminejad, Hossein Hashemi, Behzad Ebrahimi, Zahra Sarlak and Reza Mohammadi

3. Postbiotics and viable probiotic ingestion strategy in supplements: a critical approach on health effectiveness and technological challenges

Vincenzina Fusco, Marco Montemurro, Daniele Chieffi, Michela Verni, Thatyane Albuquerque, Maiara Lima, Carlo Giuseppe Rizzello and Evandro de Souza

4. Viable probiotic ingestion strategy in foods: a critical approach on health effectiveness, technological challenges, and consumer view

Amin Abasi, Sahar Sabahi, Yalda Saadat, Elham Sheykhsaran, Negin Hosseinzadeh, Aidin Azadi and Amir M. Mortazavian

Section II
New trends on oral probiotic

8. Oral probiotics and prebiotics: general definitions and concepts

Sahar A. Khairy and Mervat A. Fouad

9. Nonviable probiotic ingestion strategy: research facts and usage in drugs, supplements, and food industries

Mahmoud Yolmeh and Anderson S. Sant'Ana

List of contributors

Amin Abasi Department of Food Science and Technology, Faculty of Nutrition and Food Sciences, Tabriz University of Medical Sciences, Tabriz, Iran

Hafiz Abdullah Department of Food Science, Government College University, Food Safety & Biotechnology Lab, Faisalabad, Pakistan

Muhammad Afzaal Department of Food Sciences, Government College University Faisalabad, Faisalabad, Pakistan

Baharak Aghapour Department of Health, Sarab Faculty of Medical Sciences, Sarab, East Azerbaijan, Iran

Samreen Ahsan Institute of Food Science and Technology, Khwaja Fareed University of Engineering and Information Technology, Rahim Yar Khan, Pakistan

Thatyane Albuquerque Department of Nutrition, Health Sciences Center, Federal University of Paraíba, João Pessoa, PB, Brazil

Diana Alhaddad Nutrition and Food Technology Department, Faculty of Agriculture, Mutah University, Karak, Jordan

Abid Ali Department of Food Science, Government College University, Food Safety & Biotechnology Lab, Faisalabad, Pakistan

Hossein Alikhah Research Center for Evidence-Based Health Management, Maragheh University of Medical Sciences, Maragheh, Iran

Arantxa Almada-Corral Laboratorio de Química y Biotecnología de Productos Lácteos, Centro de Investigación en Alimentación y Desarrollo, Hermosillo, Sonora, Mexico

Rami M. Althnaibat Nutrition and Food Technology Department, Faculty of Agriculture, Mutah University, Karak, Jordan

Muhammad Alu'datt Department of Nutrition and Food Technology, Faculty of Agriculture, Jordan University of Science and Technology, Irbid, Jordan

Chiraz Amrouche Department of Outpatient Clinic and Functional Explorations, National Institute of Nutrition and Food Technology of Tunis, Tunis, Tunisia

Arturo Anadón Department of Pharmacology and Toxicology, Faculty of Veterinary Medicine, Universidad Complutense de Madrid, Madrid, Spain

María Aránzazu Martínez Department of Pharmacology and Toxicology, Faculty of Veterinary Medicine, Universidad Complutense de Madrid, Madrid, Spain

Irma Ares Department of Pharmacology and Toxicology, Faculty of Veterinary Medicine, Universidad Complutense de Madrid, Madrid, Spain

Huda Ateeq Institute of Food Science and Technology, The University of Lahore, Lahore, Pakistan

Aidin Azadi Department of Food Science and Technology, Faculty of Nutrition Sciences and Food Technology, National Nutrition and Food Technology Research Institute, Shahid Beheshti University of Medical Sciences, Tehran, Iran

Mahshid Bahraminejad Department of Food Science and Technology, Student Research Committee, Kermanshah University of Medical Sciences, Kermanshah, Iran

Sanchita Biswas Murmu ICAR – National Institute of Natural Fibres Engineering and Technology, Kolkata, West Bengal, India

Muhammad Usman Butt Department of Food Science, Government College University, Food Safety & Biotechnology Lab, Faisalabad, Pakistan

S. Chakkaravarthi Interdisciplinary Sciences, National Institute of Food Technology Entrepreneurship and Management-Kundli, Sonipat, Haryana, India

Daniele Chieffi Institute of Sciences of Food Production, National Research Council of Italy, Bari, Italy

Evandro de Souza Department of Nutrition, Health Sciences Center, Federal University of Paraíba, João Pessoa, PB, Brazil

Behzad Ebrahimi Department of Food Science and Technology, Maragheh University of Medical Sciences, Maragheh, Iran

Hanieh-Sadat Ejtahed Obesity and Eating Habits Research Center, Endocrinology and Metabolism Clinical Sciences Institute, Tehran University of Medical Sciences, Tehran, Iran

Danial Fotros Department of Nutritional Sciences, College of Human Sciences, Texas Tech University, Lubbock, TX, United States

Mervat A. Fouad National Nutrition Institute, Cairo, Egypt

Vincenzina Fusco Institute of Sciences of Food Production, National Research Council of Italy, Bari, Italy

Marina Golowczyc Center for Research and Development in Food Science and Technology (CIDCA, CCT-CONICET, La Plata), La Plata, Argentina

Andrea Gomez-Zavaglia Center for Research and Development in Food Science and Technology (CIDCA, CCT-CONICET, La Plata), La Plata, Argentina

Muammer Goncuoglu Department of Food Hygiene and Technology, Faculty of Veterinary Medicine, Ankara University, Ankara, Türkiye

Aarón F. González-Córdova Laboratorio de Química y Biotecnología de Productos Lácteos, Centro de Investigación en Alimentación y Desarrollo, Hermosillo, Sonora, Mexico

Rimsha Gulzar Department of Food Science, Government College University Faisalabad, Faisalabad, Pakistan

Hossein Hashemi Department of Food Science and Technology, Student Research Committee, Kermanshah University of Medical Sciences, Kermanshah, Iran

Adrian Hernández-Mendoza Laboratorio de Química y Biotecnología de Productos Lácteos, Centro de Investigación en Alimentación y Desarrollo, Hermosillo, Sonora, Mexico

Aziz Homayouni-Rad Department of Food Science and Technology, Faculty of Nutrition and Food Sciences, Tabriz University of Medical Sciences, Tabriz, Iran

Hamideh Homayouni-Rad Department of Food Science and Technology, Faculty of Nutrition and Food Sciences, Tabriz University of Medical Sciences, Tabriz, Iran

Zahra Hoseini Tavassol Endocrinology and Metabolism Research Center, Endocrinology and Metabolism Clinical Sciences Institute, Tehran University of Medical Sciences, Tehran, Iran

Fatemeh-Sadat Hosseini Endocrinology and Metabolism Research Center, Endocrinology and Metabolism Clinical Sciences Institute, Tehran University of Medical Sciences, Tehran, Iran

Negin Hosseinzadeh Department of Food Science and Technology, Faculty of Nutrition and Food Sciences, Tabriz University of Medical Sciences, Tabriz, Iran

Jalil Houshyar Endocrine Research Center, Tabriz University of Medical Sciences, Tabriz, Iran

Ali Jafari Student Research Committee, Department of Community Nutrition, Faculty of Nutrition Sciences and Food Technology, National Nutrition and Food Technology Research Institute, Shahid Beheshti University of Medical Sciences, Tehran, Iran

Elaheh Karimzadeh-Soureshjani Student Research Committee, Abadan University of Medical Sciences, Abadan, Khuzestan, Iran

Sonya Keisan Department of Food Science and Technology, Faculty of Nutrition and Food Sciences, Tabriz University of Medical Sciences, Tabriz, Iran

Erhan Keyvan Department of Food Hygiene and Technology, Faculty of Veterinary Medicine, Burdur Mehmet Akif Ersoy University, Burdur, Türkiye

Nilay Keyvan Burdur Representative Office, Turkish Standards Institute, Burdur, Türkiye

Sahar A. Khairy National Nutrition Institute, Cairo, Egypt

Nader Khani Department of Food Science and Technology, Faculty of Nutrition and Food Sciences, Tabriz University of Medical Sciences, Tabriz, Iran

Daryoush Kiani-Shamsabadi Student Research Committee, Abadan University of Medical Sciences, Abadan, Khuzestan, Iran

Maiara Lima Department of Nutrition, Health Sciences Center, Federal University of Paraíba, João Pessoa, PB, Brazil

Shubham Mandliya Department of Food and Nutritional Sciences, University of Reading, Reading, England, United Kingdom

Helia Mardani Students' Scientific Research Center (SSRC), Tehran University of Medical Sciences, Tehran, Iran

Marta Martinez Department of Pharmacology and Toxicology, Faculty of Veterinary Medicine, Universidad Complutense de Madrid, Madrid, Spain

María Rosa Martínez-Larrañaga Department of Pharmacology and Toxicology, Faculty of Veterinary Medicine, Universidad Complutense de Madrid, Madrid, Spain

Sourav Misra ICAR – National Institute of Natural Fibres Engineering and Technology, Kolkata, West Bengal, India

S. Mithul Aravind Faculty of Agro-Industry, Prince of Songkla University, Hat Yai, Thailand

Saba Kamalledin Moghadam Department of Food Science and Technology, Faculty of Nutrition Sciences and Food Technology, National Nutrition and Food Technology Research Institute, Shahid Beheshti University of Medical Sciences, Tehran, Iran

Reza Mohammadi Department of Food Science and Technology, School of Nutrition Sciences and Food Technology, Kermanshah University of Medical Sciences, Kermanshah, Iran

Hooriyeh Mohammadpour Department of Food, Environmental and Nutritional Sciences, Universita Degli Studi di Milano, Milan, Italy

Marco Montemurro Institute of Sciences of Food Production, National Research Council of Italy, Bari, Italy

Delara Moradi Mirhesari Department of Food Science and Technology, Student Research Committee, Kermanshah University of Medical Sciences, Kermanshah, Iran

Amir M. Mortazavian Department of Food Science and Technology, Faculty of Nutrition Sciences and Food Technology, National Nutrition and Food Technology Research Institute, Shahid Beheshti University of Medical Sciences, Tehran, Iran

Basavaprabhu Haranahalli Nataraj Southern Regional Station, ICAR-National Dairy Research Institute, Adugodi, Bengaluru, Karnataka, India

Abhisek Pradhan Agricultural and Food Engineering Department, Indian Institute of Technology Kharagpur, Kharagpur, West Bengal, India

Ali Raza Department of Food Science, Government College University, Food Safety & Biotechnology Lab, Faisalabad, Pakistan

Carlo Giuseppe Rizzello Department of Environmental Biology, Sapienza University of Rome, Rome, Italy

Hugo S. Garcia Unidad de Investigación y Desarrollo de Alimentos, Instituto Tecnológico de Veracruz, Veracruz, México

Yalda Saadat Kidney Research Center, Tabriz University of Medical Sciences, Tabriz, Iran

Sahar Sabahi Department of Nutrition, School of Allied Medical Sciences, Ahvaz Jundishapur University of Medical Sciences, Ahvaz, Iran

Farhan Saeed Food Sciences, Government College University Faisalabad Pakistan, Faisalabad, Pakistan

Sania Saleem Department of Food Science, Government College University, Food Safety & Biotechnology Lab, Faisalabad, Pakistan

Lourdes Santiago-López Laboratorio de Química y Biotecnología de Productos Lácteos, Centro de Investigación en Alimentación y Desarrollo, Hermosillo, Sonora, Mexico

Anderson S. Sant'Ana Department of Food Science and Nutrition, Faculty of Food Engineering, University of Campinas, Campinas, São Paulo, Brazil

Vahideh Sarabi-Aghdam Department of Food Science and Technology, Faculty of Nutrition and Food Sciences, Tabriz University of Medical Sciences, Tabriz, Iran

Zahra Sarlak Department of Food Science and Technology, School of Nutrition Sciences and Food Technology, Kermanshah University of Medical Sciences, Kermanshah, Iran

Elham Sheykhsaran Department of Microbiology, Faculty of Medicine, Tabriz University of Medical Sciences, Tabriz, Iran

P. Sivamma ICAR - National Institute of Seed Science & Technology, Kushmaur, Mau, Uttar Pradesh, India

Belinda Vallejo-Cordoba Laboratorio de Química y Biotecnología de Productos Lácteos, Centro de Investigación en Alimentación y Desarrollo, Hermosillo, Sonora, Mexico

Michela Verni Department of Human Science and Quality of Life Promotion, San Raffaele University, Rome, Italy

Zahra Yari Department of Nutrition Research, National Nutrition and Food Technology Research Institute, Shahid Beheshti University of Medical Sciences, Tehran, Iran

Mahmoud Yolmeh Department of Food Engineering, Faculty of Engineering, Inonu University, Malatya, Türkiye

Erdi Şen Department of Food Hygiene and Technology, Faculty of Veterinary Medicine, Burdur Mehmet Akif Ersoy University, Burdur, Türkiye

Preface

Oral probiotics have been used in various food, pharmaceutical, and supplement industries for years, and their products have gained wide acceptance and have occupied a significant world market. Also, a lot of researches have been conducted all over the world on various dimensions and fields of science related to probiotics, such as microbiology, medicine, genetics, nutrition, biotechnology, product production technology, and analysis methods, and significant, reliable articles and books have been published in this field. Given the dynamic and expanding trend of probiotics in terms of concept, function, and production technology, published sources are being updated and developed rapidly. Therefore, there is a need for new and more comprehensive sources in the fields related to probiotics to be published from time to time. In this regard, the present book has been designed and written.

An important and sensitive point about the concept of probiotics is that initially, this term was applied only to "viable microorganisms," but recently, nonviable (inviable) types, whether in the form of intact cells or ruptured cells, have also been considered, researched, and commercially produced; and currently, the term postbiotic is used in this regard, although a precise, comprehensive, and universal definition has not yet been provided; and I personally prefer to use the terms "viable probiotic" and "inviable probiotic" (intact cell or ruptured cell). The rapid development of the "biotic family" has caused a kind of confusion and lexical chaos in the definitions related to probiotics.

In this book, consisting of 19 chapters, an up-to-date and comprehensive collection in the field of probiotics from the viewpoints of history, classification, definitions and concepts, challenges, production technology and principles of product development, production economics, regulations, application of live and nonlive probiotics, new generation of probiotics, and health-promoting benefits (living and nonliving species) in the area of food, medicine, and supplements are presented. Also, the concepts of "prebiotic" and "synbiotic" have been discussed. We hope that this book will make a small contribution to the advancement and development of the vast and extensive field of probiotics (and related branches) and will be used by scientists, experts, researchers, students, manufacturers, legal authorities, and related experts.

Section I

Traditional trends on oral probiotics

Chapter 1

Probiotic: an overview on definition, trend, concept, and species

Danial Fotros[1], Zahra Yari[2], and Amir M. Mortazavian[3]
[1]*Department of Nutritional Sciences, College of Human Sciences, Texas Tech University, Lubbock, TX, United States,* [2]*Department of Nutrition Research, National Nutrition and Food Technology Research Institute, Shahid Beheshti University of Medical Sciences, Tehran, Iran,* [3]*Department of Food Science and Technology, Faculty of Nutrition Sciences and Food Technology, National Nutrition and Food Technology Research Institute, Shahid Beheshti University of Medical Sciences, Tehran, Iran*

1.1 Introduction

The term "probiotic," derived from the Greek term meaning "for life," (Ji et al., 2023) was first proposed by Kollath (1953), followed by Vergin (1954), and subsequently utilized by Lilly and Stillwell (1965). Since its introduction, the concept and definition of probiotics, as well as the strains used, have evolved. Historically, probiotics were primarily associated with fermented foods such as yogurt, cheese, and wine. However, as scientific understanding of their health benefits has advanced, these microorganisms are now also available in concentrated forms, including capsules, tablets, and powders, making them more accessible to consumers worldwide (Wang et al., 2022a; Ranadheera et al., 2017). This expanded availability reflects the growing recognition of probiotics' important role in health maintenance and disease prevention.

Probiotics are most widely known for their ability to prevent and treat various gastrointestinal disorders, including diarrhea, irritable bowel syndrome (IBS), inflammatory bowel disease (IBD), and Helicobacter pylori infections (Rau et al., 2024). However, emerging evidence suggests that their therapeutic effects may extend beyond the gut. Research indicates that probiotics may also have an impact on conditions such as allergies, eczema, obesity, and even diabetes (Nagpal et al., 2012; Aponte et al., 2020; Fijan, 2014). Additionally,

Evaluating Oral Probiotics and Prebiotics. DOI: https://doi.org/10.1016/B978-0-443-29835-6.00014-9

ongoing studies continue to uncover a wide range of potential applications for probiotics in human health.

Despite these promising developments, the efficacy and safety of probiotics depend on several factors, including the specific strain, dose, formulation, delivery method, and the characteristics of the host (Marco and Tachon, 2013; Wang et al., 2022b; Pot and Vandenplas, 2021). While numerous bacterial and yeast species have been identified as potential probiotics, *Lactobacillus* and *Bifidobacterium* strains remain the most widely studied and used in both clinical and commercial applications (Matera, 2024). Nevertheless, recent scientific findings have challenged the traditional view that only live (viable) microorganisms provide the full benefits of probiotics. Concerns regarding the use of viable probiotics (VPs), especially in immunocompromised individuals, have emerged due to risks such as microbial translocation, inflammation, and antibiotic resistance (Liong, 2008). These concerns have sparked a shift in probiotic research, with growing interest in the health-promoting potential of nonviable cells and their metabolites (Taverniti and Guglielmetti, 2011).

New concepts, such as paraprobiotics (inactivated microbial cells) and postbiotics (bioactive compounds produced by probiotics), have emerged as safer and more stable alternatives to live probiotics (Siciliano et al., 2021; Mehta et al., 2023). These forms are particularly valuable in populations where live microbes may pose risks and are more stable under various processing and storage conditions. In response to these developments, Zendeboodi et al. proposed a broader classification system that includes both viable and non-VP (NVP) forms (Zendeboodi et al., 2020). This system defines true probiotics (TP) as viable and active cells, pseudoprobiotics (PP) as viable but inactive cells, and ghost probiotics (GP) as nonviable, dead cells, either intact (GP_I) or ruptured (GP_R). A simplified version introduced in 2021 categorizes probiotics into two main groups: VPs, subdivided into active (VPA) and inactive (VPI), and NVPs, which include intact cells (NVPI) and ruptured cells or metabolites (NVPR) (Zendeboodi and Mortazavian, 2021). Indeed, this expanded framework more comprehensively encompasses the diverse range of functional probiotic agents, enabling safer and more versatile applications across foods, supplements, and pharmaceuticals.

This chapter will review the foundational and evolving definitions, concepts, and strains associated with probiotics, while also discussing their current and potential applications in human health. It will further explore recent advancements in probiotic terminology and classification, highlighting the increasing relevance of nonviable forms in addressing emerging safety and efficacy concerns.

1.2 Definition

The term "probiotic" originates from the Latin word "pro" and the Greek word "βιος" (biotic), meaning "for life" (Gasbarrini et al., 2016). The term, first coined by German scientist Werner Kollath in 1953, referred to "active substances that are crucial for the healthy development of life" (Kollath, 1953). Ferdinand Vergin further developed the concept in 1954, using the term in his article Anti-und Probiotika to contrast the harmful effects of antibiotics, which kill bacteria in the gut, with the positive effects of beneficial bacteria (Al-Qysi et al., 2020; Vergin, 1954). In 1965, Lilly and Stillwell provided a slightly different interpretation, defining probiotics as "substances secreted by one organism which stimulate the growth of another" (Lilly and Stillwell, 1965). A more specific definition was proposed by Fuller in 1992, who described probiotics as "a live microbial feed supplement that beneficially affects the host animal by improving its intestinal microbial balance" (Fuller, 1992). Similarly, Havenaar and Huis in 't Veld (1992) defined probiotics as "a viable mono or mixed culture of bacteria which, when applied to animals or humans, beneficially affects the host by improving the properties of the indigenous flora" (Havenaar and Huis In't Veld, 1992).

Expanding the scope, Salminen et al. (1999) and Schaafsma (1996) emphasized that the health benefits of probiotics should not be limited to native microflora. Salminen defined a probiotic as "a live culture of microbes or a cultured dairy product that has a positive impact on the host's health and nutrition," while Schaafsma stated that "oral probiotics are living microbes that, when consumed in adequate amounts, have health benefits beyond basic nutrition." In 1998 Guarner and Schaafsma provided a more recent and widely accepted definition, describing probiotics as "live microorganisms, which, when consumed in adequate amounts, confer a health effect on the host" (Guarner and Schaafsma, 1998). This definition formed the basis for the definition put forth by the Food and Agriculture Organization (FAO) and the World Health Organization (WHO), which has since become the standard and most widely accepted description of probiotics (Probiotics in food health and nutritional properties and guidelines for evaluation, 2006).

According to this definition, probiotics are defined as "live microorganisms which, when administered in adequate amounts, confer a health benefit on the host" (Probiotics in food health and nutritional properties and guidelines for evaluation, 2006).

A 2013 consensus statement by an expert panel, including Hill et al. (2014), further refined the terminology, affirming that the FAO/WHO definition is sufficiently broad to support a wide range of probiotic products, yet specific enough to impose essential criteria. The panel emphasized that the term "probiotic" should only apply to live microorganisms that, when consumed in adequate amounts, provide a health benefit to the host. This distinction ensures that probiotics are clearly differentiated from other microbial products, such as metabolic by-products or nonviable microorganisms, which, despite their potential health benefits, do not meet the criteria for classification as probiotics.

The panel also underscored the importance of scientific evidence to support health claims related to probiotics. It was stressed that any health claims beyond the general statement "contains probiotics" must be substantiated by robust, controlled studies demonstrating the specific health benefits of the microbial strain(s) involved. In line with this, the panel recommended a structured approach to categorize probiotic products based on the level of evidence supporting their health claims. This approach aims to ensure that only products with proven efficacy are marketed with specific health benefits.

Despite this, accumulating evidence suggests that nonviable microorganisms and their cellular components or metabolites may also exert significant health effects (Vinderola et al., 2022). Conventional terms such as *paraprobiotics* (nonviable cells) and *postbiotics* (bioactive microbial metabolites) have emerged in the literature to describe these alternatives. Although widely used, these terms reflect fragmented interpretations and have caused inconsistencies in scientific discourse.

To address this, Zendeboodi et al. proposed a comprehensive and unified classification system that eliminates the need for separate terminologies like postbiotic (Zendeboodi et al., 2020). Table 1.1 elucidates the components of this framework.

They also introduced a simplified version in 2021 (Zendeboodi and Mortazavian, 2021), as shown in Table 1.2.

This inclusive terminology, rather than relying on auxiliary concepts like postbiotics, offers a clear and practical framework for categorizing functional microbial agents across food, nutraceutical, and pharmaceutical applications. It also facilitates a better understanding of the safety, stability, and efficacy of these agents, particularly in sensitive populations where live probiotics may pose risks.

TABLE 1.1 **Probiotic classification framework: True, Pseudo, and Ghost Forms (2020).**

Category	Abbreviation	Description
True probiotics	TP	Viable, metabolically active cells
Pseudo-probiotics	PP	Viable but metabolically inactive cells
Ghost probiotics	GP_I/GP_R	Nonviable cells, either structurally intact (GP_I) or ruptured (GP_R)

TABLE 1.2 Novel and streamlined probiotic classification (2021).

Category	Subcategory	Description
Viable (V)	Active (VA)	Viable and metabolically active probiotic cells
	Inactive (VI)	Viable but metabolically inactive probiotic cells
Nonviable (NV)	Intact (NVI)	Nonviable probiotic cells that are structurally intact
	Ruptured (NVR)	Nonviable probiotic cells that are ruptured or contain metabolites, including cell fragments, metabolites, and compounds that are secreted or extracted after cell death

1.3 Concept

The modern concept of probiotics can be traced back to the work of Metchnikoff (1908), a Russian scientist and Nobel laureate at the Pasteur Institute in Paris. In the early 20th century, Metchnikoff observed that the rural population of Bulgaria had a longer life expectancy compared to other groups. He attributed this longevity to their regular consumption of fermented dairy products, such as yogurt. Metchnikoff identified a bacterium, later named *Bacillus bulgaricus*, which he believed could counteract the harmful effects of intestinal bacteria that contributed to disease and aging. These harmful bacteria, which he termed "putrefying bacteria," are now known as *proteolytic clostridia*. He proposed that by introducing beneficial bacteria into the gut, the intestinal flora could be altered in a way that enhanced health and prevented aging. This idea formed the foundation of the modern probiotic concept. Metchnikoff's theory also contributed to the growth of the dairy industry in France, which was the first in Europe to produce fermented milk containing *B. bulgaricus* (Mackowiak, 2013; Metchnikoff, 1908; Gasbarrini et al., 2016).

While Metchnikoff's work was instrumental in shaping the scientific understanding of probiotics, the practice of consuming fermented foods has a much longer history (Liong, 2015; Ranadheera et al., 2017; Pineiro and Stanton, 2007; Ozen and Dinleyici, 2015). Evidence suggests that humans have been producing fermented foods and beverages for over 10,000 years, a practice likely tied to the advent of agriculture and animal husbandry (Gasbarrini et al., 2016). Archeological discoveries support this, such as evidence from Jiahu in southwestern China, where fermented beverages made with microorganisms were found to date back to around 7000 BCE (McGovern et al., 2004). Similarly, the region of Georgia is often cited as the birthplace of wine, with evidence suggesting that wine was being produced there around

6000 BCE (McGovern, 2013). The oldest wine jars, dating to 5500 BCE, have also been discovered in the Zagros Mountains of western Iran (Owen, 2011), with similar findings from other parts of the Fertile Crescent, including Armenia, northern Turkey, and northwest Iran (McGovern et al., 2003).

The origins of fermented dairy products likely stemmed from the domestication of animals and the practice of milking (Kindstedt, 2012). It is believed that early humans may have discovered fermented milk products by accident, possibly when milk was stored in warm, unsanitary conditions, allowing beneficial bacteria to thrive. A popular legend suggests that a shepherd in the Turkish desert left milk in a goatskin bag under the hot sun, and it transformed into a thick, creamy custard. This new product was later named "yogurt," derived from the Turkish verb yoğuşmak, meaning to thicken or curdle (Mcgee, 2007).

Fermented milk products became widely consumed in the Middle East around 10000 BCE, with similar practices emerging in Egypt, Greece, and Italy by 7000–5000 BC. In East Asia, fermented foods such as soy sauce are believed to have originated around 3000 BCE, with fermented rice appearing around 2000 BCE (Talwalkar and Kailasapathy, 2004; Santacroce et al., 2019; Gasbarrini et al., 2016). The earliest evidence of milking practices comes from Ur, an ancient city in Mesopotamia, where milking activities date back to about 3100 BCE (Englund, 1995). Fermented milk also has a long history in ancient civilizations such as Egypt, Phoenicia, and across the East (McCormick, 2012). Nomadic shepherds in the Orient, including the Phrygians, Sarmatians, and Macedonians, are believed to have used animal skins or stomachs to store milk, creating an environment where bacteria, such as *Lactobacillus acidophilus* and *B. bulgaricus*, could grow (Gasbarrini et al., 2016).

The introduction of dairying marked a major innovation in prehistoric agriculture and quickly became a vital component of early human diets. Numerous studies have highlighted the importance of fermented dairy products in the diets of ancient societies. The extended shelf life of fermented dairy provided an important strategy for food preservation, enabling surplus food to be stored and consumed during periods of scarcity. Furthermore, the process of fermenting dairy enhanced its nutritional value by concentrating the fat, protein, and mineral content of the milk, thereby creating a highly energy-dense and nutrient-dense food. Additionally, fermentation effectively removed lactose, making the resulting products accessible to adults, who were predominantly lactose intolerant and unable to digest milk after infancy (Leeuwendaal et al., 2022; Kindstedt, 2012).

While the exact time when humans first began consuming fermented foods remains uncertain, evidence suggests that food fermentation was a widespread practice across various civilizations (Marsh et al., 2014a). The process of fermentation has been utilized by nearly all cultures, applied to a wide range of plant-based materials (such as fruits, seeds, tubers, and other vegetative and nonvegetative substances) as well as animal-derived materials (including meat, milk, fish, and eggs). This widespread practice reflects the diverse base foods available across different regions and highlights the adaptability of fermentation in various cultural contexts (Voidarou et al., 2021; Tamang et al., 2020; Leeuwendaal et al., 2022). In Asia, fermented drinks were traditionally made from rice, while ancient Egypt and Mesopotamia used fruits, honey, and malted grain for fermentation (Sicard and Legras, 2011; Samuel, 1996).

The therapeutic potential of yogurt and other fermented dairy products was recognized long before the microbial mechanisms were understood. The Turks, for example, considered yogurt to be an elixir of life, believing it could improve physical and mental well-being and extend lifespan. The scientific exploration of probiotics began in the early 20th century with the isolation of *Bifidobacterium* by Tissier (1907), who identified the bacterium in the intestines of infants and suggested it could outcompete pathogenic bacteria in the gut. This discovery sparked further interest in the role of beneficial microbes in preventing diseases. In 1922 *L. acidophilus* was one of the first probiotics to be tested on humans, with a study conducted on 30 patients suffering from chronic constipation, diarrhea, or eczema (Rettger and Cheplin, 1922). However, the therapeutic benefits of *L. acidophilus* for these conditions were only confirmed in later studies, such as one in 1932 (Kopeloff et al., 1932; McFarland, 2015).

In the 1930s it was also discovered that yogurt starters (*Lactobacillus delbrueckii* ssp. *bulgaricus* and *Streptococcus thermophilus*) could not colonize the human gut, questioning the assumption that yogurt was the optimal medium for delivering probiotics. Research in the 1940s shifted focus toward identifying pathogenic bacteria, rather than studying beneficial microbes. In the subsequent decades, from the 1950s to the 1980s, scientific attention turned to identifying and testing probiotic strains derived from natural or human sources and to explaining how they worked (McFarland, 2015). This period also saw the development of the concept of "colonization resistance," where beneficial bacteria prevent the invasion of harmful pathogens in the gut (McFarland, 2000). From 1980 to 2020, probiotic research advanced significantly, transitioning from basic studies to a sophisticated, multidisciplinary field. The 1980s and 1990s saw the development of molecular techniques for bacterial identification, large-scale clinical trials, and the establishment of safety guidelines, leading to the commercialization of probiotic products. The 2000s brought breakthroughs in microbiome research, with genomic sequencing revealing the strain-specific effects of probiotics and their role in gut health and beyond. This era also expanded probiotic applications to nondigestive conditions, such as immune modulation and metabolic disorders. From the 2010s onward, research has integrated microbiome science with personalized medicine, leading to the development of next-generation probiotics (NGPs), engineered strains, and advanced delivery systems. Emerging fields such as psychobiotics, which target the gut-brain axis, further underscore the therapeutic potential of probiotics (Kimse et al., 2024). These advancements, supported by robust clinical research and regulatory frameworks, have transformed probiotics into a cornerstone of modern health and precision medicine.

In recent years, the concept of probiotics has expanded to include both viable and nonviable forms. NVP, such as intact cells, cell fragments, and metabolites, are now recognized for their potential health benefits, especially when microorganism viability is compromised (Zendeboodi and Mortazavian, 2021; Zendeboodi et al., 2020). This shift reflects the evolving probiotic science, which is increasingly exploring the therapeutic potential of both VP and NVPs, broadening their applications in human health.

1.4 Traditional probiotic species

The term "probiotic" refers to a broad range of microorganisms, primarily bacteria and yeasts, that confer health benefits to the host. Although many microorganisms may exhibit probiotic properties, most documented benefits are linked to specific strains from genera such as *Lactobacillus*, *Bifidobacterium*, *Lactococcus*, *Streptococcus*, *Enterococcus*, *Bacillus*, *Escherichia*, and *Saccharomyces*. The taxonomy of traditional probiotic species is shown in Table 1.3.

1.4.1 *Lactobacillus*

The *Lactobacillus* is a Gram-positive, nonsporing bacterial genus with rod-shaped cells that can grow in anaerobic or microaerophilic environments (De Angelis and Gobbetti, 2016). Lactobacilli are commonly present in the human vagina, gastrointestinal tract, and colon, and they are among the first bacteria to establish themselves in the infant's gut together with *Bifidobacterium* (Lidbeck and Nord, 1993). Lactobacilli are also involved in the production of fermented foods such as yogurt, cheese, sauerkraut, pickles, sourdough, wine, and more (Bauer et al., 2010; De Vuyst et al., 2014; Siragusa et al., 2014) (National Research Council Panel on the Applications of Biotechnology to Traditional Fermented, 1992). They convert sugars into lactic acid in these foods, which enhances their shelf life and preservation (Zapasnik et al., 2022). Indeed, in the presence of these bacteria, glucose is fermented into lactic acid, which creates an acidic environment that inhibits the growth of many pathogenic organisms (Goldstein et al., 2015). Furthermore, lactobacilli species are commonly used as probiotics because of their tolerance to acid and bile, ability to adhere to intestinal surfaces, resistance to gastric enzymes, inhibition of potentially pathogenic species, resistance to antibiotics, and production of exopolysaccharides (Romero-Luna et al., 2019; Kosin and Rakshit, 2006; Ragavan and Das, 2017; Mainville et al., 2005; Gorbach, 2002; Lee et al., 2014; Tulumoglu et al., 2013). Several species of *Lactobacillus* are

TABLE 1.3 Taxonomy of traditional probiotic species.

Domain	Phylum	Class	Order	Family	Genus
Bacteria	Firmicutes	Bacilli	Lactobacillales	Lactobacillaceae	*Lactobacillus*
Bacteria	Actinobacteria	Actinobacteria	Bifidobacteriales	Bifidobacteriaceae	*Bifidobacterium*
Bacteria	Firmicutes	Bacilli	Lactobacillales	Streptococcaceae	*Lactococcus*
Bacteria	Firmicutes	Bacilli	Lactobacillales	Streptococcaceae	*Streptococcus*
Bacteria	Firmicutes	Bacilli	Lactobacillales	Enterococcaceae	*Enterococcus*
Bacteria	Firmicutes	Bacilli	Bacillales	Bacillaceae	*Bacillus*
Bacteria	Proteobacteria	Gammaproteobacteria	Enterobacterales	Enterobacteriaceae	*Escherichia*
Fungi	Ascomycota	Saccharomycetes	Saccharomycetales	Saccharomycetaceae	*Saccharomyces*

considered probiotics due to their specific properties beneficial to health, such as *L. acidophilus*, *L. rhamnosus*, *L. gasseri*, *L. fermentum*, *L. casei*, *L. johnsonii*, *L. reuteri*, *L. plantarum*, and *L. delbrueckii* subsp. *bulgaricus* (Kim et al., 2019; Giraffa et al., 2010; Fijan, 2014). Several studies showed *Lactobacillus* strains to prevent antibiotic-associated diarrhea (Johnston et al., 2011; Hempel et al., 2012). Moreover, several probiotic strains, including *L. acidophilus* and *Lactobacillus rhamnosus* GG (along with *Saccharomyces boulardii*, *Bifidobacterium bifidum*, and *Bacillus coagulans*), are effective in the treatment of travelers' diarrhea caused primarily by bacterial pathogens (McFarland, 2007; Doron et al., 2008). Lactobacilli and other probiotics can also benefit the treatment of bacterial vaginosis when used in conjunction with antibiotics (McFarland, 2007). In addition to these findings, a metaanalysis found that *Lactobacillus* probiotic strains effectively prevented recurrent urinary tract infections in adult women (Grin et al., 2013). Likewise, a review of randomized controlled trials on probiotics for functional constipation showed that *Bifidobacterium lactis* DN-173 010, *Lactobacillus casei* Shirota, and *Escherichia coli* Nissle 1917 were the most effective probiotics for adults (Chmielewska and Szajewska, 2010). Additionally, it has been demonstrated that *L. rhamnosus* CRL1505, an immunobiotic strain, has been shown to reduce virus-associated pulmonary damage by controlling immune-coagulative responses and removing respiratory viruses, as well as improving the innate immune response and the respiratory and systemic adaptive humoral response (Zelaya et al., 2014; Kolling et al., 2018).

1.4.2 *Bifidobacterium*

The genus *Bifidobacterium* comprises diverse bacteria that are anaerobic, Gram-positive, nonmotile, and nonspore-forming (Pino et al., 2022). They have a rod-shaped morphology with a clubbed end and frequently exhibit bifurcations into Y and V shapes (Lee and O'Sullivan, 2010). These bacteria live in a symbiotic relationship with the gut and vaginal flora of mammals, including humans (Chen et al., 2007). Several strains of *Bifidobacterium* are used as probiotics due to their ability to cope with bile salts (Fijan, 2014). Some of the bifidobacteria that are considered important probiotics are *B. infantis*, *B. adolescentis*, and *B. animalis* subsp. *animalis*, *B. animalis* subsp. *lactis*, *B. bifidum*, *B. longum*, and *B. breve* (Chen et al., 2021; Fijan, 2014). Moreover, the effects of *Bifidobacterium* species, in conjunction with other probiotics, have been demonstrated to help with constipation (Chmielewska and Szajewska, 2010), diarrhea associated with antibiotics (Hempel et al., 2012), travelers' diarrhea (Hempel et al., 2012), maintaining remission of mild ulcerative colitis and intestinal inflammation (Huang et al., 2023), radiation-induced diarrhea (Demers et al., 2014), cholesterol reduction (Bordoni et al., 2013), lowering the risk of developing eczema and food allergies (Isolauri et al., 2012).

Bifidobacterium strains have demonstrated a broad range of beneficial effects through various mechanisms (Chen et al., 2021). For example, *B. longum* ATCC 15708 produces bacteriocins to combat infections (Igbafe et al., 2020), while *B. animalis* AHC7 attenuates pro-inflammatory responses (O'Mahony et al., 2010). Strains like *B. longum* 5 (1 A) activate immune pathways (Vieira et al., 2016), and *B. adolescentis* SPM1605 has been shown to inhibit viral replication (Kim et al., 2014). In cancer research, *B. longum* BCRC 910051 enhances macrophage function (Foo et al., 2011), while *B. Bifidum* BGN4 inhibits cancer cell proliferation (Ku et al., 2009). Many probiotics also play a role in regulating inflammation: *B. breve* M-16V helps balance immune responses (Wong et al., 2019), and *B. lactis* DN-173010 reduces inflammatory markers in gingival fluid (Kuru et al., 2017). For psychological health, *B. adolescentis* 150 produces the inhibitory neurotransmitter GABA (Yunes et al., 2020), and *B. longum* 1714 modulates brain activity (Allen et al., 2016). Some strains, like *B. animalis* subsp. *Lactis* CECT 8145, are associated with reduced fat accumulation (Martorell et al., 2016), while *B. longum* ATCC 15707 supports bone health (Parvaneh et al., 2015). Furthermore, probiotic mixtures have been linked to cardiovascular and cognitive benefits (Mauricio et al., 2017; Al-Sheraji et al., 2012).

1.4.3 *Lactococcus*

Lactococcus, often called "mesophilic lactic streptococci," is a genus of Gram-positive lactic acid bacteria used in the dairy industry to produce fermented products (Teuber, 1995). Several strains of *L. lactis* subsp. *Lactis* exhibit probiotic properties, including adhesion to vaginal epithelium cells and the production of nisin (Gao et al., 2011; Yang et al., 2013), which can also be used in conjunction with other probiotics to treat antibiotic-associated diarrhea (Johnston et al., 2011). Moreover, *Lactococcus lactis* NJ414 shows promise as a probiotic candidate, exhibiting strong antimicrobial effects, particularly against *Listeria monocytogenes* (Alizadeh Behbahani and Noshad, 2024).

L. lactis also demonstrates promising immunomodulatory effects, stimulating cytokine production, enhancing macrophage activity, and modulating immune and inflammatory signaling, with potential therapeutic applications in immune-related disorders like IBD and allergies (Jeong et al., 2023; Cho et al., 2023).

1.4.4 *Streptococcus*

Generally, streptococci are Gram-positive, nonmotile, nonspore-forming, catalase-negative bacteria that occur in chains or pairs. Most streptococci are facultative anaerobes, and some are obligate anaerobes (Patterson, 1996). One strain from this genus with probiotic characteristics is *Streptococcus thermophiles* (Pieniz et al., 2013), which is also used for yogurt production along with

L. delbrueckii subsp. *bulgaricus* (Fijan, 2014). *Streptococcus thermophilus* can also modulate uremic toxins in the gut of chronic kidney disease patients, potentially preventing disease progression (Lin et al., 2019; Vitetta et al., 2019). Similarly, *Streptococcus dentisani* strains 7746 and 7747 may serve as oral probiotics against tooth decay by inhibiting major oral pathogens through bacteriocin production and buffering acidic pH via an arginolytic pathway (López-López et al., 2017).

1.4.5 *Enterococcus*

Enterococci are Gram-positive, facultative anaerobic cocci with short and medium chains. They were discovered in the human gastrointestinal tract for the first time in 1899 (Said et al., 2023). Although *Enterococcus faecium* has long been considered a probiotic, particularly for preventing antibiotic-induced diarrhea (Hempel et al., 2012), certain strains are opportunistic pathogens that may harbor antibiotic resistance (Dirienzo, 2014). As a result, they are generally not considered safe for humans but continue to represent an important probiotic for animals (Cao et al., 2013; Bednorz et al., 2013).

The use of *Enterococcus* in treating various diseases dates back to the 1950s, with *Enterococcus faecalis* first described as a probiotic by Heinz Kolb in 1955 for postantibiotic therapy (Krawczyk et al., 2021; Rusch and Rusch, 2001). Probiotic formulations of *E. faecalis*, occasionally supplemented with *E. coli* and lactobacilli, are advised for the management of conditions, including urinary tract inflammation, sinusitis, bronchitis, and the common cold (Krawczyk et al., 2021). In Germany, *E. faecalis* DSM 16431 is sold as Symbioflor 1 for acute and recurrent sinusitis or bronchitis (Habermann et al., 2002; Franz et al., 2011). Additionally, *E. faecium* M-74, isolated from healthy children, exhibits immunomodulatory and antimutagenic properties (Hlivak et al., 2005; Mikeš et al., 1995).

1.4.6 *Bacillus*

Bacillus is a large, heterogeneous genus of spore-forming, aerobic, or facultative anaerobic bacteria that have undergone considerable reclassification in recent years due to advances in molecular biology (Beveridge, 1990; Maughan and Van der Auwera, 2011). Some strains with claimed probiotic properties include *B. subtilis*, *B. coagulans*, *B. subtilis*, and *B. cereus* (Fijan, 2014). *B. coagulans* has proven to be successful in preventing or treating antibiotic-associated diarrhea (Doron et al., 2008; Hempel et al., 2012). Additionally, *Bacillus* spores have been used safely as probiotics for a long time (Elshaghabee et al., 2017). For instance, *B. subtilis* spores have been considered probiotics for animal consumption (Zokaeifar et al., 2014; Larsen et al., 2014) and proposed for treating diarrhea and eradicating *H. pylori* in humans (Tompkins et al., 2010). Nevertheless, recurrent septicemia occurred in an immunocompromised patient following the consumption of *B. subtilis* spores (Oggioni et al., 1998), suggesting that the ingestion of *B. subtilis* spores is generally safe only in individuals with normal immune function. In addition, *B. cereus* NVH 75/95 shows promise as a probiotic for animals (Trapecar et al., 2011). Overall, certain strains of *B. cereus* are highly toxic and lethal for humans; however, other strains exhibit probiotic properties (Kamar et al., 2013).

1.4.7 *Escherichia*

E. coli is a Gram-negative bacillus, which is primarily known for its highly virulent serotypes (e.g., *E. coli* O157:H7) (Fijan, 2014). Nevertheless, one of its strains is used as a probiotic and therapeutic agent. *E. coli* Nissle 1917 (EcN) is a probiotic strain widely known for its potential therapeutic applications in the context of the lower intestine, particularly in addressing conditions such as IBD (Behnsen et al., 2013) and hepatic encephalopathy (Manzhalii et al., 2022). It is also possible that this strain may be beneficial for other conditions, such as constipation (Chmielewska and Szajewska, 2010) and colon cancer (Behnsen et al., 2013), but further research is needed to be done.

1.4.8 *Saccharomyces*

Saccharomyces are ascomycete fungi in the subphylum Saccharomycotina, which mostly comprises yeasts (Hibbett et al., 2007). The genus is famously responsible for food and beverage fermentations and is estimated to be tens of millions of years old (Paleo-López et al., 2016; Ono et al., 2020; Peris et al., 2014). There are many yeast species within this genus, including *S. cerevisiae*, which produces wine, bread, beer, and other products; *S. bayanus*, which produces wine; and *S. cerevisiae var. boulardii (S. boulardii)*, which is used as a probiotic (Fijan, 2014). Furthermore, *Saccharomyces* yeast forms symbiotic relationships with bacteria to produce kefir (Witthuhn et al., 2005), and it is sometimes employed to produce kombucha (Marsh et al., 2014b). The majority of studies indicate that *S. boulardii* decreases the duration of diarrhea regardless of the cause (Dinleyici et al., 2012; Szajewska and Mrukowicz, 2005; Shan et al., 2013). *S. boulardii* is a beneficial probiotic yeast that can ameliorate the condition of patients with IBS (Choi et al., 2011) as well as inhibit and alleviate the exacerbations of IBD and moderate ulcerative colitis symptoms (Guslandi et al., 2003; Guslandi et al., 2000). It can also diminish the likelihood of recurrent pseudomembranous colitis infection induced by *Clostridium difficile* if administered daily with conventional antibiotics (Fitzpatrick, 2013).

In chronic diseases, *S. boulardii* has exhibited a potential role in inhibiting colorectal cancer metastasis by inducing cancer cell apoptosis and modulating immune responses (Li et al., 2020). In conditions like ulcerative colitis and

Crohn's disease, the probiotic yeast helps maintain gut microbiota balance, reduces inflammation, and improves colonic permeability (Abid et al., 2022; Profir et al., 2015; Sivignon et al., 2015; Guslandi et al., 2003). It has also proven effective against vaginal candidiasis by inhibiting *Candida albicans* adhesion and reducing pathogenicity (Pericolini et al., 2017).

Moreover, *S. boulardii* possesses significant antibacterial and antiviral properties, targeting pathogens like *E. coli*, *Shigella*, and *rotavirus* (Kaźmierczak-Siedlecka et al., 2020; Czerucka and Rampal, 2019). Additionally, it exhibits antioxidant properties, reducing oxidative stress and improving gastrointestinal health (Suryavanshi et al., 2013; Menezes et al., 2020; Abid et al., 2022). Lastly, *S. boulardii* is known for its resistance to both broad- and narrow-spectrum antibiotics, making it a valuable therapeutic tool in restoring gut microbiota balance disrupted by antibiotic treatments, without causing harm to the host (Waitzberg et al., 2024; Abid et al., 2022).

1.5 Next-generation probiotics

NGPs are advanced probiotics that extend beyond traditional strains. These include some genera such as *Faecalibacterium*, *Akkermansia*, *Bacteroides*, *Christensenella*, *Parabacteroides*, and *Prevotella*, which are selected for their potential to provide targeted health benefits and improve gut microbiota balance. The taxonomy of NGP is shown in Table 1.4.

1.5.1 *Faecalibacterium*

Faecalibacterium is a genus of Gram-positive, rod-shaped bacteria that are strictly anaerobic, exhibit extreme sensitivity to oxygen, are nonmotile, and do not produce spores (Duncan et al., 2002; Martín et al., 2023).

This genus plays a crucial role in regulating immune function and gut health, and its reduced abundance has been observed in conditions like IBD, IBS, obesity, type 2 diabetes (T2D), and neurodegenerative diseases (Martín et al., 2023).

In IBD, *Faecalibacterium* depletion is associated with increased inflammation and impaired immune responses (Sokol et al., 2008), suggesting potential therapeutic benefits from its restoration. Similarly, in IBS, reduced *Faecalibacterium* levels correlate with symptoms like bloating and altered motility (Collins et al., 2009), and supplementation with *F. prausnitzii* may improve gut function (Miquel et al., 2016). In metabolic diseases like obesity and T2D, *Faecalibacterium* is negatively correlated with body mass index (BMI), and its supplementation has shown potential in improving metabolic parameters (Navab-Moghadam et al., 2017). *Faecalibacterium* levels also increase after weight loss in T2D patients (Hippe et al., 2016), suggesting its role in metabolic regulation. Indeed, in murine models, *F. duncaniae* A2–165 supplementation has shown promise for addressing obesity and its associated

TABLE 1.4 Taxonomy of next-generation probiotics (NGP).

Domain	Phylum	Class	Order	Family	Genus
Bacteria	Firmicutes	Clostridia	Clostridiales	Oscillospiraceae (Ruminococcaceae)	*Faecalibacterium*
Bacteria	Verrucomicrobia	Verrucomicrobiae	Verrucomicrobiales	Akkermansiaceae	*Akkermansia*
Bacteria	Bacteroidetes	Bacteroidia	Bacteroidales	Bacteroidaceae	*Bacteroides*
Bacteria	Firmicutes	Clostridia	Clostridiales	Christensenellaceae	*Christensenella*
Bacteria	Bacteroidetes	Bacteroidia	Bacteroidales	Tannerellaceae	*Parabacteroides*
Bacteria	Bacteroidetes	Bacteroidia	Bacteroidales	Prevotellaceae	*Prevotella*

complications. Compared to control mice, those fed a high-fat diet and supplemented biweekly with *F. duncaniae* A2–165 displayed smaller adipocytes, reduced hepatic inflammation, less lipid accumulation in the liver, and decreased inflammatory cell infiltration in adipose tissue (Munukka et al., 2017).

Alterations in the gut microbiome, including reduced *Faecalibacterium*, have also been observed in neurological conditions such as Parkinson's disease, Alzheimer's disease, and depression (Martín et al., 2023; Nishiwaki et al., 2020; Xi et al., 2021). Animal studies have indicated potential cognitive benefits from *Faecalibacterium* supplementation (Ueda et al., 2021). In cancer, particularly in response to immune checkpoint inhibitors, *Faecalibacterium* has been linked to better outcomes in melanoma and lung cancer treatments (Chaput et al., 2017; Newsome et al., 2022). Furthermore, *F. duncaniae* supplementation may help mitigate radiation-induced colon damage (Lapiere et al., 2020). Moreover, in dermatitis and allergies, reduced *F. duncaniae* A2–165 levels are associated with conditions like atopic dermatitis and psoriasis (Song et al., 2016). *Faecalibacterium* may also alleviate asthma symptoms by modulating inflammation (Demirci et al., 2019). In conclusion, *Faecalibacterium* is critical for gut health, immune regulation, and metabolic balance. Its altered levels are associated with a wide range of diseases, and while promising results in animal models suggest therapeutic potential, further human studies are needed to explore these interventions (Martín et al., 2023).

1.5.2 *Akkermansia*

Akkermansia muciniphila is an anaerobic, Gram-negative, mucin-degrading bacterium from the phylum Verrucomicrobia, first isolated from human feces in 2004 (Derrien et al., 2004; Iwaza et al., 2022). *A. muciniphila* plays a vital role in maintaining intestinal homeostasis and protecting against a range of disorders (Rodrigues et al., 2022). It supports gut barrier integrity by enhancing tight junctions, which help prevent pathogen translocation (Salim and Söderholm, 2011). Numerous studies have identified variations in the gut microbiota composition between healthy individuals and those with IBD, notably a significant decrease in *A. muciniphila* among patients with ulcerative colitis (Pittayanon et al., 2020; Qu et al., 2021). Studies have shown that *A. muciniphila*, particularly its outer membrane proteins like Amuc_1100, can alleviate colitis by reducing inflammation and promoting tissue repair through the production of short-chain fatty acids (SCFAs) (Bian et al., 2019; Kim et al., 2021; Rodrigues et al., 2022; Wang et al., 2020).

In addition to its role in gut health, *A. muciniphila* has significant benefits in combating obesity and metabolic syndrome. The abundance of *A. muciniphila* in the gut is inversely related to body weight, insulin resistance, and other metabolic disturbances (Dao et al., 2016; Santacruz et al., 2010; Karlsson et al., 2012; Everard et al., 2013). The bacterium helps regulate glucose metabolism by enhancing thermogenesis, reducing carbohydrate absorption, and stimulating the release of beneficial hormones like glucagon-like peptide 1 (GLP-1) (Lee et al., 2022; Yoon et al., 2021; Depommier et al., 2020). Furthermore, clinical studies have confirmed that *A. muciniphila* supplementation can improve metabolic parameters, such as insulin sensitivity and glucose tolerance, in patients with obesity and T2D (McMurdie et al., 2022; Perraudeau et al., 2020; Depommier et al., 2021; Depommier et al., 2019).

1.5.3 *Bacteroides*

Bacteroides species are anaerobic, bile-resistant, nonspore-forming, Gram-negative rods commonly found in the human gastrointestinal tract (Wexler, 2007). These bacteria play a key role in the digestion of complex carbohydrates and the maintenance of gut homeostasis (Hooper et al., 2002). Some species, such as *Bacteroides fragilis*, are also known for their involvement in immune modulation, influencing both pro-inflammatory and antiinflammatory pathways. For instance, *B. fragilis* has been shown to reduce tumor necrosis factor alpha (TNF-α)-induced colonic cell injury, promote colonic epithelial cell migration, and modulate inflammation by downregulating the nuclear factor kappa B (NF-κB) pathway, decreasing pro-inflammatory cytokines like interleukin (IL)-6 and IL-1β, and increasing the production of the antiinflammatory cytokine IL-10 (He et al., 2023). Recent animal studies have provided further support for the potential use of *Bacteroides* species as probiotics (Tufail and Schmitz, 2025; Sofi et al., 2021; Zhou et al., 2022). *B. fragilis* strain ZY-312 was found to mitigate radiation-induced intestinal injury in mice, reducing weight loss, shortening intestinal length, and improving intestinal permeability (Zhou et al., 2022). Additionally, *B. fragilis* has been shown to preserve gut integrity and inhibit inflammation in mouse models (Sofi et al., 2021). The nontoxic *B. fragilis* ZY-312 strain also demonstrated protective effects against antibiotic-associated diarrhea in rats, alleviating symptoms by modulating the gut microbiota, enhancing epithelial organization, and improving barrier function, ultimately reducing inflammation and restoring gut homeostasis (Zhang et al., 2018; Tufail and Schmitz, 2025). These studies suggest that *Bacteroides* species hold promise for promoting intestinal barrier integrity, mitigating inflammation, and supporting gut health.

1.5.4 *Christensenella*

Christensenella is a genus of nonspore-forming, anaerobic, and nonmotile bacteria within the Christensenellaceae family (Morotomi et al., 2012). This family, Christensenellaceae, has been extensively studied in relation to human metabolism and metabolic markers (Ignatyeva et al., 2023). A deficiency of Christensenellaceae has been linked to metabolic syndrome and obesity in both adults (He et al., 2018; Peters et al., 2018) and children (Alcazar et al., 2022). Notably, Christensenellaceae abundance is negatively correlated with body weight and metabolic indicators linked to prediabetes and cardiovascular conditions. Specifically, it shows an inverse association with BMI, visceral adipose tissue (VAT), liver fat, insulin resistance (HOMA-IR), as well as triglyceride and cholesterol levels in the blood (Sowah et al., 2022). Reduced levels of Christensenellaceae are also linked to metabolic syndrome conditions, including hypertension (Gomez-Arango et al., 2016) and prediabetes (Jayanama et al., 2022).

In gastrointestinal disorders such as IBD, particularly Crohn's disease and ulcerative colitis, depletion of Christensenellaceae is common (Zakrzewski et al., 2019; Kennedy et al., 2018; Pascal et al., 2017). Importantly, levels of this bacterium tend to drop in quiescent Crohn's disease patients before flare-ups (Braun et al., 2019). Decreased abundance has also been linked to diarrheal conditions (Brunkwall et al., 2021). Beyond the gut, Christensenellaceae appears to play a role in immune regulation (Ignatyeva et al., 2023). Higher levels have been associated with a reduced incidence of eczema and allergic sensitivities in children (Brunkwall et al., 2021), and lower asthma prevalence in households with greater environmental exposure to the bacterium (Gangneux et al., 2020). Additionally, lower Christensenellaceae abundance has been connected to various noncommunicable conditions, including kidney stones (Stanford et al., 2020), thyroid cancer (Lu et al., 2022), polycystic ovary syndrome (Dong et al., 2021), and recurrent aphthous stomatitis (Wang et al., 2022b). However, some studies have found that elevated levels of Christensenellaceae may be associated with neurodegenerative and psychiatric conditions such as Parkinson's disease (Barichella et al., 2019; Shen et al., 2021), Alzheimer's disease (Kaiyrlykyzy et al., 2022; Verhaar et al., 2022), multiple sclerosis (Galluzzo et al., 2021), and anorexia nervosa (Prochazkova et al., 2021), often associated with poor outcomes.

In summary, while Christensenellaceae plays a key role in metabolic and gastrointestinal health, its involvement in neurocognitive diseases warrants further investigation to clarify its broader impact on human health.

1.5.5 *Parabacteroides*

The genus *Parabacteroides* consists of Gram-negative, nonspore-forming, rod-shaped, strictly anaerobic bacteria commonly found in the gastrointestinal tract of both humans and animals (Ezeji et al., 2021; Liu et al., 2018; Sakamoto and Benno, 2006). *Parabacteroides* may promote host health through several mechanisms: (1) immune modulation, (2) inflammation reduction, (3) metabolism regulation, and (4) metabolite secretion (Cui et al., 2022). Among the most studied species are *Parabacteroides distasonis* and *P. goldsteinii*. *P. distasonis* can modulate innate inflammatory responses by inhibiting the release of TNF-α, IL-6, IL-17, IL-12, and interferon-gamma (IFN-γ) (Kverka et al., 2011) and has demonstrated antitumor effects on HT-29/SW480 cells and AOM-induced tumors (Koh et al., 2018). Both species have also been shown to alleviate obesity (Cui et al., 2022). In ob/ob mice, *P. distasonis* promoted the production of succinate and secondary bile acids, which enhanced intestinal gluconeogenesis and improved gut permeability (Wang et al., 2019). Similarly, *P. goldsteinii* reduced obesity-related metabolic disorders in mice by lowering endotoxins, IL-1β, TNF-α, insulin resistance, and protecting gut permeability (Wu et al., 2019). Additionally, some *Parabacteroides* species produce beneficial SCFAs, such as acetate, propionate, and butyrate, which improve gut health (Fu et al., 2019). For example, acetate alleviates intestinal inflammation by inhibiting mitogen-activated protein kinase (MAPK) and NF-kB pathways (Fu et al., 2019), while sodium propionate mitigates tissue damage in rats (Filippone et al., 2020).

1.5.6 *Prevotella*

Prevotella species are Gram-negative, nonspore-forming, rod-shaped anaerobic bacteria (Sharma et al., 2022). Prevotella *copri* is a notable member of this microbial genus, influencing host health in various ways (Zang et al., 2024). *P. copri* has been shown to improve glucose tolerance (Kovatcheva-Datchary et al., 2015) and is negatively associated with allergies (Vuillermin et al., 2020), cholestasis, and liver fibrosis (Jiang et al., 2022), yet positively correlated with conditions like rheumatoid arthritis (Scher et al., 2013) and hypertension (Li et al., 2017).

Changes in *P. copri* abundance have been observed in conditions such as allergies, severe aplastic anemia, and diarrhea (Shao et al., 2021; Noh and Lee, 2020; Goldberg et al., 2020; Zang et al., 2024), though its role as either a cause or consequence of these conditions remains uncertain. *P. copri* appears beneficial in conditions like allergies, cholestasis, and hepatic fibrosis (Jiang et al., 2022) but may be harmful in diarrhea (Gilchrist et al., 2016) and neurological diseases (Zou et al., 2020; Zang et al., 2024). Moreover, its impact on inflammatory diseases and glucose tolerance is complex, influenced by factors such as diet, age, geography, and strain variations (Zang et al., 2024).

1.6 The health effects of viable probiotics and nonviable probiotics

1.6.1 Viable probiotics-intact cells

Despite being nonliving, VPI provides a wide range of health benefits. They modulate the immune system by enhancing cytokine production (e.g., IL-12 and IFN-γ), promoting Th1 responses, and increasing macrophage activity (Fujiki et al., 2012; Miyazawa et al., 2011). VPI also protects against pathogens like *Salmonella*, *Listeria*, and H1N1 virus by inhibiting adhesion and invasion, and strengthening mucosal immunity (Kawase et al., 2012; Ishikawa et al., 2010). Additionally, they help restore gut health by balancing microbiota, reducing inflammation, preserving the intestinal barrier, and aiding in colitis recovery (de Almada et al., 2016; Generoso et al., 2011; Ueno et al., 2011; Kimoto-Nira et al., 2009).

Beyond gastrointestinal effects, VPI supports metabolic and systemic health. They have been shown to alleviate diarrhea, lactose intolerance, and IBS (Rampengan et al., 2010; Xiao et al., 2003), lower cholesterol levels, reduce alcohol-induced liver damage (Segawa et al., 2008; Shin et al., 2010; de Almada et al., 2016), and improve respiratory conditions like allergic rhinitis and pneumonia (Tanaka et al., 2011; Peng and Hsu, 2005). Furthermore, VPI may exert anticancer effects by inhibiting tumor growth and promoting apoptosis (Orlando et al., 2012), and it can reduce symptoms of atopic dermatitis and visceral pain through immune regulation (Kamiya et al., 2006; Kim et al., 2013). These findings highlight the broad therapeutic potential of VPI across multiple body systems (de Almada et al., 2016).

1.6.2 Nonviable probiotics

NVP plays a crucial role in several physiological processes, including pathogen protection, strengthening of the epithelial barrier, and immune system modulation (He et al., 2023). Key components of NVP, such as lactic acid, bacteriocins, and exopolysaccharides, possess antimicrobial properties that are particularly effective against Gram-negative bacteria (He et al., 2023). For instance, lactic acid bacteria and bifidobacteria produce bacteriocins and other antimicrobial substances that can inhibit biofilm formation and prevent pathogen adhesion, thereby offering a natural defense mechanism against harmful microorganisms (Schaefer et al., 2010; Wu et al., 2010; Sarkar and Mandal, 2016; Lukic et al., 2017). In addition to their antimicrobial effects, NVP contribute to immune modulation by activating immune cells like macrophages and influencing cytokine production (He et al., 2023). Metabolites such as lactic acid, SCFAs, and indole derivatives have antiinflammatory effects and enhance immune responses (Ranjbar et al., 2021; Meng et al., 2020), which can be beneficial in conditions like IBD. Moreover, structural components of NVP such as peptidoglycan (PGN) and lipopolysaccharide (LPS) also play a key role in immune stimulation (He et al., 2023). These molecules are recognized by pattern recognition receptors (PRRs) on host cells, particularly Toll-like receptors (TLRs) and NOD-like receptors (NLRs), which identify peptidoglycan (PGN) and lipopolysaccharide (LPS) as pathogen-associated molecular patterns (PAMPs), thereby triggering receptor-dependent immune responses (Philpott and Girardin, 2004). Overall, these findings highlight the role of NVP in modulating microbiota, maintaining gut health, and enhancing immune functions, all without introducing new viable organisms to the host.

1.7 Conclusion

The concept of probiotics has undergone a significant transformation since its inception, evolving from a narrow focus on live microorganisms in fermented foods to a comprehensive understanding that includes both viable and nonviable forms. This progression reflects growing scientific insights into the diverse mechanisms through which probiotics, regardless of viability, can promote health. While VPs have long been valued for their benefits in gastrointestinal and systemic health, recent research underscores the therapeutic potential of NVP, including inactivated cells, cellular fragments, and metabolites. These nonviable forms offer distinct advantages in terms of safety, stability, and applicability, particularly in vulnerable populations where the use of live microorganisms may pose risks. The emergence of new classification systems that incorporate NVP represents an important step toward a more inclusive and practical framework for probiotic use. As knowledge in this area expands, NVPs are increasingly recognized as effective agents in modulating the microbiota, supporting immune function, and enhancing overall health. This expanded view paves the way for broader clinical and commercial applications, positioning NVPs as a promising frontier in microbiome-based therapeutics.

References

Abid, R., Waseem, H., Ali, J., Ghazanfar, S., Ali, G.M., Elasbali, A.M., et al., 2022. Probiotic yeast Saccharomyces: back to nature to improve human health. Journal of Fungi 8 (5) Available from. Available at https://doi.org/10.3390/jof8050444, https://www.mdpi.com/2309-608X/8/5/444/pdf.

Al-Qysi, L., Mohammad, M., Al-iedani, A., AbuKhader, M.M., 2020. Investigating the characteristics of probiotics marketed in the Middle East and pharmacists' perception of use in Muscat, Oman. PharmaNutrition 13, 100202. Available at https://doi.org/10.1016/j.phanu.2020.100202.

Al-Sheraji, S.H., Ismail, A., Manap, M.Y., Mustafa, S., Yusof, R.M., Hassan, F.A., 2012. Hypocholesterolaemic effect of yoghurt containing *Bifidobacterium pseudocatenulatum* G4 or *Bifidobacterium longum* BB536. Food Chemistry 135 (2), 356–361. Available at https://doi.org/10.1016/j.foodchem.2012.04.120.

Alcazar, M., Escribano, J., Ferré, N., Closa-Monasterolo, R., Selma-Royo, M., Feliu, A., et al., 2022. Gut microbiota is associated with metabolic health in children with obesity. Clinical Nutrition 41 (8), 1680–1688. Available at https://doi.org/10.1016/j.clnu.2022.06.007.

Alizadeh Behbahani, B., Noshad, M., 2024. An investigation into the probiotic, anti-bacterial, anti-biofilm, and safety properties of *Lactococcus lactis* NJ414. LWT 201, 116251. Available at https://doi.org/10.1016/j.lwt.2024.116251.

Allen, A.P., Hutch, W., Borre, Y.E., Kennedy, P.J., Temko, A., Boylan, G., et al., 2016. Bifidobacterium longum 1714 as a translational psychobiotic: modulation of stress, electrophysiology and neurocognition in healthy volunteers. Translational Psychiatry 6 (11) Available from. Available at https://doi.org/10.1038/tp.2016.191, http://www.nature.com/tp/index.html.

Aponte, M., Murru, N., Shoukat, M., 2020. Therapeutic, prophylactic, and functional use of probiotics: a current perspective. Frontiers in Microbiology 11. Available at https://doi.org/10.3389/fmicb.2020.562048.

Barichella, M., Severgnini, M., Cilia, R., Cassani, E., Bolliri, C., Caronni, S., et al., 2019. Unraveling gut microbiota in Parkinson's disease and atypical parkinsonism. Movement Disorders 34 (3), 396–405. Available at https://doi.org/10.1002/mds.27581.

Bauer, R., du Toit, M., Kossmann, J., 2010. Influence of environmental parameters on production of the acrolein precursor 3-hydroxypropionaldehyde by *Lactobacillus reuteri* DSMZ 20016 and its accumulation by wine lactobacilli. International Journal of Food Microbiology 137 (1), 28–31. Available at https://doi.org/10.1016/j.ijfoodmicro.2009.10.012.

Bednorz, C., Guenther, S., Oelgeschläger, K., Kinnemann, B., Pieper, R., Hartmann, S., et al., 2013. Feeding the probiotic *Enterococcus faecium* strain NCIMB 10415 to piglets specifically reduces the number of *Escherichia coli* pathotypes that adhere to the gut mucosa. Applied and Environmental Microbiology 79 (24), 7896–7904. Available from. Available at https://doi.org/10.1128/AEM.03138-13Germany, http://aem.asm.org/content/79/24/7896.full.pdf.

Behnsen, J., Deriu, E., Sassone-Corsi, M., Raffatellu, M., 2013. Probiotics: properties, examples, and specific applications. Cold Spring Harbor Perspectives in Medicine 3 (3), a010074. Available at https://doi.org/10.1101/cshperspect.a010074.

Beveridge, T.J., 1990. Mechanism of gram variability in select bacteria. Journal of Bacteriology 172 (3), 1609–1620. Available at https://doi.org/10.1128/jb.172.3.1609-1620.1990.

Bian, X., Wu, W., Yang, L., Lv, L., Wang, Q., Li, Y., et al., 2019. Administration of akkermansia muciniphila ameliorates dextran sulfate sodium-induced ulcerative colitis in mice. Frontiers in Microbiology 10. Available at https://doi.org/10.3389/fmicb.2019.02259.

Bordoni, A., Amaretti, A., Leonardi, A., Boschetti, E., Danesi, F., Matteuzzi, D., et al., 2013. Cholesterol-lowering probiotics: in vitro selection and in vivo testing of bifidobacteria. Applied Microbiology and Biotechnology 97 (18), 8273–8281. Available at https://doi.org/10.1007/s00253-013-5088-2.

Braun, T., Di Segni, A., BenShoshan, M., Neuman, S., Levhar, N., Bubis, M., et al., 2019. Individualized dynamics in the gut microbiota precede Crohn's disease flares. American Journal of Gastroenterology 114 (7), 1142–1151. Available at https://doi.org/10.14309/ajg.0000000000000136.

Brunkwall, L., Ericson, U., Nilsson, P.M., Orho-Melander, M., Ohlsson, B., 2021. Self-reported bowel symptoms are associated with differences in overall gut microbiota composition and enrichment of *Blautia* in a population-based cohort. Journal of Gastroenterology and Hepatology (Australia) 36 (1), 174–180. Available from. Available at https://doi.org/10.1111/jgh.15104, http://onlinelibrary.wiley.com/journal/10.1111/(ISSN)1440-1746.

Cao, G.T., Zeng, X.F., Chen, A.G., Zhou, L., Zhang, L., Xiao, Y.P., et al., 2013. Effects of a probiotic, *Enterococcus faecium*, on growth performance, intestinal morphology, immune response, and cecal microflora in broiler chickens challenged with *Escherichia coli* K88. Poultry Science 92 (11), 2949–2955. Available at https://doi.org/10.3382/ps.2013-03366.

Chaput, N., Lepage, P., Coutzac, C., Soularue, E., Le Roux, K., Monot, C., et al., 2017. Baseline gut microbiota predicts clinical response and colitis in metastatic melanoma patients treated with ipilimumab. Annals of Oncology 28 (6), 1368–1379. Available at https://doi.org/10.1093/annonc/mdx108.

Chen, J., Chen, X., Ho, C.L., 2021. Recent development of probiotic Bifidobacteria for treating human diseases. Frontiers in Bioengineering and Biotechnology 9 Available from. Available at https://doi.org/10.3389/fbioe.2021.770248, http://journal.frontiersin.org/journal/bioengineering-and-biotechnology#archive.

Chen, J., Cai, W., Feng, Y., 2007. Development of intestinal bifidobacteria and lactobacilli in breast-fed neonates. Clinical Nutrition 26 (5), 559–566. Available at https://doi.org/10.1016/j.clnu.2007.03.003.

Chmielewska, A., Szajewska, H., 2010. Systematic review of randomised controlled trials: probiotics for functional constipation. World Journal of Gastroenterology 16 (1), 69–75. Available from. Available at https://doi.org/10.3748/wjg.v16.i1.69, http://www.wjgnet.com/1007-9327/16/69.pdf.

Cho, Y., Han, H.T., Kim, Tr, Sohn, M., Park, Y.S., 2023. Immunostimulatory activity of *Lactococcus lactis* LM1185 isolated from *Hydrangea macrophylla*. Food Science and Biotechnology 32 (4), 497–506. Available from. Available at https://doi.org/10.1007/s10068-022-01199-5, https://www.springer.com/journal/10068.

Choi, C.H., Jo, S.Y., Park, H.J., Chang, S.K., Byeon, J.S., Myung, S.J., 2011. A randomized, double-blind, placebo-controlled multicenter trial of saccharomyces boulardii in irritable bowel syndrome: effect on quality of Life. Journal of Clinical Gastroenterology 45 (8), 679–683. Available at https://doi.org/10.1097/MCG.0b013e318204593e.

Collins, S.M., Denou, E., Verdu, E.F., Bercik, P., 2009. The putative role of the intestinal microbiota in the irritable bowel syndrome. Digestive and Liver Disease 41 (12), 850–853. Available at https://doi.org/10.1016/j.dld.2009.07.023.

Cui, Y., Zhang, L., Wang, X., Yi, Y., Shan, Y., Liu, B., et al., 2022. Roles of intestinal Parabacteroides in human health and diseases. FEMS Microbiology Letters 369 (1). Available at https://doi.org/10.1093/femsle/fnac072.

Czerucka, D., Rampal, P., 2019. Diversity of *Saccharomyces boulardii* CNCM I-745 mechanisms of action against intestinal infections. World Journal of Gastroenterology 25 (18), 2188–2203. Available at https://doi.org/10.3748/wjg.v25.i18.2188.

Dao, M.C., Everard, A., Aron-Wisnewsky, J., Sokolovska, N., Prifti, E., Verger, E.O., et al., 2016. Akkermansia muciniphila and improved metabolic health during a dietary intervention in obesity: relationship with gut microbiome richness and ecology. Gut 65 (3), 426–436. Available from. Available at https://doi.org/10.1136/gutjnl-2014-308778, http://gut.bmj.com/content/65/3/426.full.pdf+html.

de Almada, C.N., Almada, C.N., Martinez, R.C.R., Sant'Ana, A.S., 2016. Paraprobiotics: evidences on their ability to modify biological responses, inactivation methods and perspectives on their application in foods. Trends in Food Science and Technology 58, 96–114. Available from. Available at https://doi.org/10.1016/j.tifs.2016.09.011, http://www.elsevier.com/wps/find/journaldescription.cws_home/601278/description#description.

De Angelis, M., Gobbetti, M., 2016. *Lactobacillus* SPP.: General Characteristics☆. Reference Module in Food Science. Elsevier BV. Available at https://doi.org/10.1016/B978-0-08-100596-5.00851-9.

De Vuyst, L., Van Kerrebroeck, S., Harth, H., Huys, G., Daniel, H.-M., Weckx, S., 2014. Microbial ecology of sourdough fermentations: diverse or uniform? Food Microbiology 37, 11–29. Available at https://doi.org/10.1016/j.fm.2013.06.002.

Demers, M., Dagnault, A., Desjardins, J., 2014. A randomized double-blind controlled trial: Impact of probiotics on diarrhea in patients treated with pelvic radiation. Clinical Nutrition 33 (5), 761–767. Available at https://doi.org/10.1016/j.clnu.2013.10.015.

Demirci, M., Tokman, H.B., Uysal, H.K., Demiryas, S., Karakullukcu, A., Saribas, S., et al., 2019. Reduced Akkermansia muciniphila and Faecalibacterium prausnitzii levels in the gut microbiota of children with allergic asthma. Allergologia et Immunopathologia 47 (4), 365–371. Available at https://doi.org/10.1016/j.aller.2018.12.009.

Depommier, C., Everard, A., Druart, C., Plovier, H., Van Hul, M., Vieira-Silva, S., et al., 2019. Supplementation with Akkermansia muciniphila in overweight and obese human volunteers: a proof-of-concept exploratory study. Nature Medicine 25 (7), 1096–1103. Available from. Available at https://doi.org/10.1038/s41591-019-0495-2, http://www.nature.com/nm/index.html.

Depommier, C., Van Hul, M., Everard, A., Delzenne, N.M., De Vos, W.M., Cani, P.D., 2020. Pasteurized *Akkermansia muciniphila* increases whole-body energy expenditure and fecal energy excretion in diet-induced obese mice. Gut Microbes 11 (5), 1231–1245. Available from. Available at https://doi.org/10.1080/19490976.2020.1737307, http://www.tandfonline.com/toc/kgmi20/current.

Depommier, C., Vitale, R.M., Iannotti, F.A., Silvestri, C., Flamand, N., Druart, C., et al., 2021. Beneficial effects of *Akkermansia muciniphila* are not associated with major changes in the circulating endocannabinoidome but linked to higher mono-palmitoyl-glycerol levels as new pparα agonists. Cells 10 (1), 1–22. Available from. Available at https://doi.org/10.3390/cells10010185, https://www.mdpi.com/2073-4409/10/1/185/pdf.

Derrien, M., Vaughan, E.E., Plugge, C.M., de Vos, W.M., 2004. *Akkermansia municiphila* gen. nov., sp. nov., a human intestinal mucin-degrading bacterium. International Journal of Systematic and Evolutionary Microbiology 54 (5), 1469–1476. Available at https://doi.org/10.1099/ijs.0.02873-0.

Dinleyici, E.C., Eren, M., Ozen, M., Yargic, Z.A., Vandenplas, Y., 2012. Effectiveness and safety of *Saccharomyces boulardii* for acute infectious diarrhea. Expert Opinion on Biological Therapy 12 (4), 395–410. Available at https://doi.org/10.1517/14712598.2012.664129.

Dirienzo, D.B., 2014. Effect of probiotics on biomarkers of cardiovascular disease: implications for heart-healthy diets. Nutrition Reviews 72 (1), 18–29. Available at https://doi.org/10.1111/nure.12084.

Dong, S., jiao, J., Jia, S., Li, G., Zhang, W., Yang, K., et al., 2021. 16S rDNA full-length assembly sequencing technology analysis of intestinal microbiome in polycystic ovary syndrome. Frontiers in Cellular and Infection Microbiology 11 Available from. Available at https://doi.org/10.3389/fcimb.2021.634981, http://www.frontiersin.org/Cellular_and_Infection_Microbiology/archive.

Doron, S.I., Hibberd, P.L., Gorbach, S.L., 2008. Probiotics for prevention of antibiotic-associated diarrhea. Journal of Clinical Gastroenterology 42.

Duncan, S.H., Hold, G.L., Harmsen, H.J.M., Stewart, C.S., Flint, H.J., 2002. Growth requirements and fermentation products of *Fusobacterium prausnitzii*, and a proposal to reclassify it as *Faecalibacterium prausnitzii* gen. nov., comb. nov. International Journal of Systematic and Evolutionary Microbiology 52 (6), 2141–2146. Available at https://doi.org/10.1099/ijs.0.02241-0.

Elshaghabee, F.M.F., Rokana, N., Gulhane, R.D., Sharma, C., Panwar, H., 2017. *Bacillus* as potential probiotics: status, concerns, and future perspectives. Frontiers in Microbiology 8 Available from. Available at https://doi.org/10.3389/fmicb.2017.01490, https://fjfsdata01prod.blob.core.windows.net/articles/files/271541/pubmed-zip/.versions/1/.package-entries/fmicb-08-01490/fmicb-08-01490.pdf.

Englund, R.K., 1995. Late Uruk period cattle and dairy products: evidence from proto-cuneiform sources. Bulletin of Sumerian Agriculture 8, 33–48.

Everard, A., Belzer, C., Geurts, L., Ouwerkerk, J.P., Druart, C., Bindels, L.B., et al., 2013. Cross-talk between *Akkermansia muciniphila* and intestinal epithelium controls diet-induced obesity. Proceedings of the National Academy of Sciences 110 (22), 9066–9071. Available at https://doi.org/10.1073/pnas.1219451110.

Ezeji, J.C., Sarikonda, D.K., Hopperton, A., Erkkila, H.L., Cohen, D.E., Martinez, S.P., et al., 2021. *Parabacteroides distasonis*: intriguing aerotolerant gut anaerobe with emerging antimicrobial resistance and pathogenic and probiotic roles in human health. Gut Microbes 13 (1) Available from. Available at https://doi.org/10.1080/19490976.2021.1922241, http://www.tandfonline.com/toc/kgmi20/current.

Fijan, S., 2014. Microorganisms with claimed probiotic properties: an overview of recent literature. International Journal of Environmental Research and Public Health 11 (5), 4745–4767. Available at https://doi.org/10.3390/ijerph110504745.

Filippone, A., Lanza, M., Campolo, M., Casili, G., Paterniti, I., Cuzzocrea, S., et al., 2020. The Anti-inflammatory and antioxidant effects of sodium propionate. International Journal of Molecular Sciences 21 (8), 3026. Available at https://doi.org/10.3390/ijms21083026.

Fitzpatrick, L.R., 2013. Probiotics for the treatment of *Clostridium difficile* associated disease. World Journal of Gastrointestinal Pathophysiology 4 (3), 47. Available at https://doi.org/10.4291/wjgp.v4.i3.47.

Foo, N.P., Yang, H.O., Chiu, H.H., Chan, H.Y., Liao, C.C., Yu, C.K., et al., 2011. Probiotics prevent the development of 1,2-dimethylhydrazine (DMH)-induced colonic tumorigenesis through suppressed colonic mucosa cellular proliferation and increased stimulation of macrophages. Journal of Agricultural and Food Chemistry 59 (24), 13337–13345. Available at https://doi.org/10.1021/jf203444d.

Franz, C.M.A.P., Huch, M., Abriouel, H., Holzapfel, W., Gálvez, A., 2011. Enterococci as probiotics and their implications in food safety. International Journal of Food Microbiology 151 (2), 125–140. Available at https://doi.org/10.1016/j.ijfoodmicro.2011.08.014.

Fu, X., Liu, Z., Zhu, C., Mou, H., Kong, Q., 2019. Nondigestible carbohydrates, butyrate, and butyrate-producing bacteria. Critical Reviews in Food Science and Nutrition 59 (sup1), S130. Available at https://doi.org/10.1080/10408398.2018.1542587.

Fujiki, T., Hirose, Y., Yamamoto, Y., Murosaki, S., 2012. Enhanced immunomodulatory activity and stability in simulated digestive juices of *Lactobacillus plantarum* L-137 by heat treatment. Bioscience, Biotechnology, and Biochemistry 76 (5), 918–922. Available from. Available at https://doi.org/10.1271/bbb.110919Japan, https://www.jstage.jst.go.jp/article/bbb/76/5/76-110919/-pdf.

Fuller, R., 1992. The Effect of Probiotics on the Gut Micro-ecology of Farm Animals. Springer Science and Business Media LLC, pp. 171–192. Available at 10.1007/978-1-4615-3522-5_7.

Galluzzo, P., Capri, F.C., Vecchioni, L., Realmuto, S., Scalisi, L., Cottone, S., et al., 2021. Comparison of the intestinal microbiome of Italian patients with multiple sclerosis and their household relatives. Life (Chicago, Ill. 1978) 11 (7) Available from. Available at https://doi.org/10.3390/life11070620, https://www.mdpi.com/2075-1729/11/7/620/pdf.

Gangneux, J.P., Sassi, M., Lemire, P., Le Cann, P., 2020. Metagenomic characterization of indoor dust bacterial and fungal microbiota in homes of asthma and non-asthma patients using next generation sequencing. Frontiers in Microbiology 11 Available from. Available at https://doi.org/10.3389/fmicb.2020.01671, https://www.frontiersin.org/journals/microbiology#.

Gao, Y., Lu, Y., Teng, K.L., Chen, M.L., Zheng, H.J., Zhu, Y.Q., et al., 2011. Complete genome sequence of *Lactococcus lactis* subsp. lactis CV56, a probiotic strain isolated from the vaginas of healthy women. Journal of Bacteriology 193 (11), 2886–2887. Available from. Available at https://doi.org/10.1128/JB.00358-11China, http://jb.asm.org/cgi/reprint/193/11/2886.pdf.

Gasbarrini, G., Bonvicini, F., Gramenzi, A., 2016. Probiotics history. Journal of Clinical Gastroenterology 50 (Supplement 2), S116. Available at https://doi.org/10.1097/mcg.0000000000000697.

Generoso, S.V., Viana, M.L., Santos, R.G., Arantes, R.M.E., Martins, F.S., Nicoli, J.R., et al., 2011. Protection against increased intestinal permeability and bacterial translocation induced by intestinal obstruction in mice treated with viable and heat-killed *Saccharomyces boulardii*. European Journal of Nutrition 50 (4), 261–269. Available at https://doi.org/10.1007/s00394-010-0134-7.

Gilchrist, C.A., Petri, S.E., Schneider, B.N., Reichman, D.J., Jiang, N., Begum, S., et al., 2016. Role of the gut microbiota of children in diarrhea due to the protozoan parasite *Entamoeba histolytica*. Journal of Infectious Diseases 213 (10), 1579–1585. Available from. Available at https://doi.org/10.1093/infdis/jiv772, http://jid.oxfordjournals.org/content/current.

Giraffa, G., Chanishvili, N., Widyastuti, Y., 2010. Importance of lactobacilli in food and feed biotechnology. Research in Microbiology 161 (6), 480–487. Available at https://doi.org/10.1016/j.resmic.2010.03.001.

Goldberg, M.R., Mor, H., Magid Neriya, D., Magzal, F., Muller, E., Appel, M.Y., et al., 2020. Microbial signature in IgE-mediated food allergies. Genome Medicine 12 (1) Available from. Available at https://doi.org/10.1186/s13073-020-00789-4, http://www.genomemedicine.com/.

Goldstein, E.J.C., Tyrrell, K.L., Citron, D.M., 2015. *Lactobacillus* species: taxonomic complexity and controversial susceptibilities. Clinical Infectious Diseases 60, S98. Available from. Available at https://doi.org/10.1093/cid/civ072, http://cid.oxfordjournals.org/content/by/year.

Gomez-Arango, L.F., Barrett, H.L., McIntyre, H.D., Callaway, L.K., Morrison, M., Dekker Nitert, M., 2016. Increased systolic and diastolic blood pressure is associated with altered gut microbiota composition and butyrate production in early pregnancy. Hypertension 68 (4), 974–981. Available from. Available at https://doi.org/10.1161/HYPERTENSIONAHA.116.07910, http://hyper.ahajournals.org/.

Gorbach, S.L., 2002. Probiotics in the third millennium. Digestive and Liver Disease 34 (2), S2. Available at https://doi.org/10.1016/s1590-8658(02)80155-4.

Grin, P.M., Kowalewska, P.M., Alhazzani, W., Fox-Robichaud, A.E., 2013. *Lactobacillus* for preventing recurrent urinary tract infections in women: meta-analysis. Canadian Journal of Urology 20 (1), 6607–6614. Available from. Available at http://www.canjurol.com/html/subscriber/Spdf/V20I01/V20I1_05_DrGrin.pdf.

Guarner, F., Schaafsma, G.J., 1998. Probiotics. International Journal of Food Microbiology 39 (3), 237–238. Available at https://doi.org/10.1016/S0168-1605(97)00136-0.

Guslandi, M., Mezzi, G., Sorghi, M., Testoni, P.A., 2000. *Saccharomyces boulardii* in maintenance treatment of Crohn's disease. Digestive Diseases and Sciences 45 (7), 1462–1464. Available at https://doi.org/10.1023/A:1005588911207.

Guslandi, M., Giollo, P., Testoni, P.A., 2003. A pilot trial of *Saccharomyces boulardii* in ulcerative colitis. European Journal of Gastroenterology and Hepatology 15 (6), 697–698. Available at https://doi.org/10.1097/00042737-200306000-00017.

Habermann, W., Zimmermann, K., Skarabis, H., Kunze, R., Rusch, V., 2002. Verminderung der Rezidivhäufigkeit bei Patienten mit chronisch rezidivierender hypertrophischer Sinusitis unter Behandlung mit einem bakteriellen Immunstimulans (*Enterococcus faecalis*-Bakterien humaner Herkunft). Arzneimittel-Forschung 52 (08), 622–627. Available at https://doi.org/10.1055/s-0031-1299941.

Havenaar, R., Huis In't Veld, J.H.J., 1992. Probiotics: A General View. Springer Science and Business Media LLC, pp. 151–170. Available at 10.1007/978-1-4615-3522-5_6.

He, Q., Niu, M., Bi, J., Du, N., Liu, S., Yang, K., et al., 2023. Protective effects of a new generation of probiotic *Bacteroides fragilis* against colitis in vivo and in vitro. Scientific Reports 13 (1). Available at https://doi.org/10.1038/s41598-023-42481-8.

He, Y., Wu, W., Wu, S., Zheng, H.M., Li, P., Sheng, H.F., et al., 2018. Linking gut microbiota, metabolic syndrome and economic status based on a population-level analysis. Microbiome 6 (1) Available from. Available at https://doi.org/10.1186/s40168-018-0557-6, http://www.microbiomejournal.com/.

Hempel, S., Newberry, S.J., Maher, A.R., Wang, Z., Miles, J.N.V., Shanman, R., Shekelle, P.G., et al., 2012. Probiotics for the prevention and treatment of antibiotic-associated diarrhea: a systematic review and meta-analysis. JAMA the Journal of the American Medical Association 307 (18), 1959–1969. Available from. Available at https://doi.org/10.1001/jama.2012.3507, http://jama.ama-assn.org/content/307/18/1959.full.pdf+html.

Hibbett, D.S., Binder, M., Bischoff, J.F., Blackwell, M., Cannon, P.F., Eriksson, O.E., et al., 2007. A higher-level phylogenetic classification of the fungi. Mycological Research 111 (5), 509–547. Available at https://doi.org/10.1016/j.mycres.2007.03.004.

Hill, C., Guarner, F., Reid, G., Gibson, G.R., Merenstein, D.J., Pot, B., et al., 2014. Expert consensus document: the international scientific association for probiotics and prebiotics consensus statement on the scope and appropriate use of the term probiotic. Nature Reviews Gastroenterology and Hepatology 11 (8), 506–514. Available from. Available at https://doi.org/10.1038/nrgastro.2014.66, http://www.nature.com/nrgastro/index.html.

Hippe, B., Remely, M., Aumueller, E., Pointner, A., Magnet, U., Haslberger, A.G., 2016. *Faecalibacterium prausnitzii* phylotypes in type two diabetic, obese, and lean control subjects. Beneficial Microbes 7 (4), 511–517. Available from. Available at https://doi.org/10.3920/BM2015.0075, http://www.wageningenacademic.com/doi/pdf/10.3920/BM2015.0075.

Hlivak, P., Odraska, J., Ferencik, M., Ebringer, L., Jahnova, E., Mikes, Z., 2005. One-year application of probiotic strain *Enterococcus faecium* M-74 decreases serum cholesterol levels. Bratislavské lekárske listy 106 (2), 67–72.

Hooper, L.V., Midtvedt, T., Gordon, J.I., 2002. Host microbial interactions shape the nutrient environment of the mammalian intestine. Annual Review of Nutrition 22 (1), 283–307. Available at https://doi.org/10.1146/annurev.nutr.22.011602.092259.

Huang, C., Hao, W., Wang, X., Zhou, R., Lin, Q., 2023. Probiotics for the treatment of ulcerative colitis: a review of experimental research from 2018 to 2022. Frontiers in Microbiology 14. Available at https://doi.org/10.3389/fmicb.2023.1211271.

Igbafe, J., Kilonzo-Nthenge, A., Nahashon, S.N., Mafiz, A.I., Nzomo, M., 2020. Probiotics and antimicrobial effect of *Lactiplantibacillus plantarum*, *Saccharomyces cerevisiae*, and *Bifidobacterium longum* against common foodborne pathogens in poultry. Agriculture 10 (9), 368. Available at https://doi.org/10.3390/agriculture10090368.

Ignatyeva, O., Tolyneva, D., Kovalyov, A., Matkava, L., Terekhov, M., Kashtanova, D., et al., 2023. *Christensenella minuta*, a new candidate next-generation probiotic: current evidence and future trajectories. Frontiers in Microbiology 14 Available from. Available at https://doi.org/10.3389/fmicb.2023.1241259, https://www.frontiersin.org/journals/microbiology#.

Ishikawa, H., Kutsukake, E., Fukui, T., Sato, I., Shirai, T., Kurihara, T., et al., 2010. Oral administration of heat-killed *Lactobacillus plantarum* strain b240 protected mice against *Salmonella enterica* serovar Typhimurium. Bioscience, Biotechnology, and Biochemistry 74 (7), 1338–1342. Available from. Available at https://doi.org/10.1271/bbb.90871Japan, http://www.jstage.jst.go.jp/article/bbb/74/7/1338/_pdf.

Isolauri, E., Rautava, S., Salminen, S., 2012. Probiotics in the development and treatment of allergic disease. Gastroenterology Clinics of North America 41 (4), 747–762. Available at https://doi.org/10.1016/j.gtc.2012.08.007.

Iwaza, R., Wasfy, R.M., Dubourg, G., Raoult, D., Lagier, J.C., 2022. *Akkermansia muciniphila*: the state of the art, 18 years after its first discovery. Frontiers in Gastroenterology 1 Available from. Available at https://doi.org/10.3389/fgstr.2022.1024393, https://www.frontiersin.org/journals/gastroenterology.

Jayanama, K., Phuphuakrat, A., Pongchaikul, P., Prombutara, P., Nimitphong, H., Reutrakul, S., et al., 2022. Association between gut microbiota and prediabetes in people living with HIV. Current Research in Microbial Sciences 3, 100143. Available at https://doi.org/10.1016/j.crmicr.2022.100143.

Jeong, H., Kim, S., Hwang, U.S., Choi, H., Park, Y.S., 2023. Immunostimulatory activity of *Lactococcus lactis* subsp. lactis CAB701 lsolated from Jeju Cabbage. Microorganisms 11 (7) Available from. Available at https://doi.org/10.3390/microorganisms11071718, www.mdpi.com/journal/microorganisms.

Ji, J., Jin, W., Liu, S.J., Jiao, Z., Li, X., 2023. Probiotics, prebiotics, and postbiotics in health and disease. MedComm 4 (6). Available at https://doi.org/10.1002/mco2.420 Available from: onlinelibrary.wiley.com/journal/26882663.

Jiang, B., Yuan, G., Wu, J., Wu, Q., Li, L., Jiang, P., 2022. *Prevotella copri* ameliorates cholestasis and liver fibrosis in primary sclerosing cholangitis by enhancing the FXR signalling pathway. Biochimica et Biophysica Acta (BBA) - Molecular Basis of Disease 1868 (3), 166320. Available at https://doi.org/10.1016/j.bbadis.2021.166320.

Johnston, B.C., Goldenberg, J.Z., Vandvik, P.O., Sun, X., Guyatt, G.H., 2011. Probiotics for the prevention of pediatric antibiotic-associated diarrhea. Cochrane Database of Systematic Reviews (Online) (11), CD004827.

Kaiyrlykyzy, A., Kozhakhmetov, S., Babenko, D., Zholdasbekova, G., Alzhanova, D., Olzhayev, F., et al., 2022. Study of gut microbiota alterations in Alzheimer's dementia patients from Kazakhstan. Scientific Reports 12 (1) Available from. Available at https://doi.org/10.1038/s41598-022-19393-0, www.nature.com/srep/index.html.

Kamar, R., Gohar, M., Jéhanno, I., Réjasse, A., Kallassy, M., Lereclus, D., et al., 2013. Pathogenic potential of *Bacillus cereus* strains as revealed by phenotypic analysis. Journal of Clinical Microbiology 51 (1), 320–323. Available at https://doi.org/10.1128/jcm.02848-12.

Kamiya, T., Wang, L., Forsythe, P., Goettsche, G., Mao, Y., Wang, Y., et al., 2006. Inhibitory effects of *Lactobocillus reuteri* on visceral pain induced by colorectal distension in Sprague-Dawley rats. Gut 55 (2), 191–196. Available at https://doi.org/10.1136/gut.2005.070987.

Karlsson, C.L.J., Önnerfält, J., Xu, J., Molin, G., Ahrné, S., Thorngren-Jerneck, K., 2012. The microbiota of the gut in preschool children with normal and excessive body weight. Obesity 20 (11), 2257–2261. Available at https://doi.org/10.1038/oby.2012.110.

Kawase, M., He, F., Kubota, A., Yoda, K., Miyazawa, K., Hiramatsu, M., 2012. Heat-killed *Lactobacillus gasseri* TMC0356 protects mice against influenza virus infection by stimulating gut and respiratory immune responses. FEMS Immunology & Medical Microbiology 64 (2), 280–288. Available at https://doi.org/10.1111/j.1574-695x.2011.00903.x.

Kaźmierczak-Siedlecka, K., Ruszkowski, J., Fic, M., Folwarski, M., Makarewicz, W., 2020. *Saccharomyces boulardii* CNCM I-745: a non-bacterial microorganism used as probiotic agent in supporting treatment of selected diseases. Current Microbiology 77 (9), 1987–1996. Available at https://doi.org/10.1007/s00284-020-02053-9.

Kennedy, N.A., Lamb, C.A., Berry, S.H., Walker, A.W., Mansfield, J., Parkes, M., et al., 2018. The impact of NOD2 variants on fecal microbiota in Crohn's disease and controls without gastrointestinal disease. Inflammatory Bowel Diseases 24 (3), 583–592. Available from. Available at https://doi.org/10.1093/ibd/izx061, https://academic.oup.com/ibdjournal/issue.

Ķimse, L., Reinis, A., Miķelsone-Jansone, L., Gintere, S., Krūmiņa, A., 2024. A narrative review of psychobiotics: probiotics that influence the gut–brain axis. Medicina 60 (4), 601. Available at https://doi.org/10.3390/medicina60040601.

Kim, J.-Y., Park, B.-K., Park, H.-J., Park, Y.-H., Kim, B.-O., Pyo, S., 2013. Atopic dermatitis-mitigating effects of new *Lactobacillus strain*, *Lactobacillus sakei* probio 65 isolated from Kimchi. Journal of Applied Microbiology 115 (2), 517–526. Available at https://doi.org/10.1111/jam.12229.

Kim, J.-A., Bayo, J., Cha, J., Choi, Y.J., Jung, M.Y., Kim, D.-H., et al., 2019. Investigating the probiotic characteristics of four microbial strains with potential application in feed industry. PLoS One 14 (6), e0218922. Available at https://doi.org/10.1371/journal.pone.0218922.

Kim, M.J., Lee, D.K., Park, J.E., Park, I.H., Seo, J.G., Ha, N.J., 2014. Antiviral activity of *Bifidobacterium adolescentis* SPM1605 against coxsackievirus B3. Biotechnology and Biotechnological Equipment 28 (4), 681–688. Available from. Available at https://doi.org/10.1080/13102818.2014.945237, http://www.tandfonline.com/doi/pdf/10.1080/13102818.2014.945237.

Kim, S., Shin, Y.C., Kim, T.Y., Kim, Y., Lee, Y.S., Lee, S.H., et al., 2021. Mucin degrader *Akkermansia muciniphila* accelerates intestinal stem cell-mediated epithelial development. Gut Microbes 13 (1), 1–20. Available from. Available at https://doi.org/10.1080/19490976.2021.1892441, http://www.tandfonline.com/toc/kgmi20/current.

Kimoto-Nira, H., Mizumachi, K., Okamoto, T., Sasaki, K., Kurisaki, J.I., 2009. Influence of long-term consumption of a *Lactococcus lactis* strain on the intestinal immunity and intestinal flora of the senescence-accelerated mouse. British Journal of Nutrition 102 (2), 181–185. Available at https://doi.org/10.1017/S0007114508143574.

Kindstedt, P., 2012. Cheese and culture: a history of cheese and its place in western civilization.

Koh, G.Y., Kane, A., Lee, K., Xu, Q., Wu, X., Roper, J., et al., 2018. *Parabacteroides distasonis* attenuates toll-like receptor 4 signaling and Akt activation and blocks colon tumor formation in high-fat diet-fed azoxymethane-treated mice. International Journal of Cancer 143 (7), 1797–1805. Available from. Available at https://doi.org/10.1002/ijc.31559, http://onlinelibrary.wiley.com/journal/10.1002/(ISSN)1097-0215.

Kollath, W., 1953. Die Zunahme der Zivilisationskrankheiten und ihre Verhütung. Münchener medizinische Wochenschrift (1950) 95 (47), 1260–1262.

Kolling, Y., Salva, S., Villena, J., Alvarez, S., Forestier, C., 2018. Are the immunomodulatory properties of *Lactobacillus rhamnosus* CRL1505 peptidoglycan common for all Lactobacilli during respiratory infection in malnourished mice. PLoS One 13 (3), e0194034. Available at https://doi.org/10.1371/journal.pone.0194034.

Kopeloff, N., Blackman, N., McGinn, B., 1932. The incidence of *Lactobacillus acidophilus* in adults. Journal of Infectious Diseases 50 (5-6), 426–429. Available at https://doi.org/10.1093/infdis/50.5-6.426.

Kosin, B., Rakshit, S.K., 2006. Microbial and processing criteria for production of probiotics: a review. Food Technology and Biotechnology 44 (3), 371–379. Available from. Available at http://public.carnet.hr/ftbrfd/44-371.pdf.

Kovatcheva-Datchary, P., Nilsson, A., Akrami, R., Lee, Y.S., De Vadder, F., Arora, T., et al., 2015. Dietary fiber-induced improvement in glucose metabolism is associated with increased abundance of Prevotella. Cell Metabolism 22 (6), 971–982. Available from. Available at https://doi.org/10.1016/j.cmet.2015.10.001, http://www.cellmetabolism.org/.

Krawczyk, Wityk, Gałęcka, Michalik, 2021. The many faces of Enterococcus spp.-commensal, probiotic and opportunistic pathogen. Microorganisms 9.

Ku, S.-M, You, H.-J., Ji, G.-E., 2009. Enhancement of anti-tumorigenic polysaccharide production, adhesion, and branch formation of Bifidobacterium bifidum BGN4 by phytic acid. Food Science and Biotechnology 18, 749–754.

Kuru, B.E., Laleman, I., Yalnizoğlu, T., Kuru, L., Teughels, W., 2017. The influence of a bifidobacterium animalis probiotic on gingival health: a randomized controlled clinical trial. Journal of Periodontology 88 (11), 1115–1123. Available from. Available at https://doi.org/10.1902/jop.2017.170213, http://www.joponline.org/doi/pdf/10.1902/jop.2017.170213.

Kverka, M., Zakostelska, Z., Klimesova, K., Sokol, D., Hudcovic, T., Hrncir, T., et al., 2011. Oral administration of *Parabacteroides distasonis* antigens attenuates experimental murine colitis through modulation of immunity and microbiota composition. Clinical and Experimental Immunology 163 (2), 250–259. Available at https://doi.org/10.1111/j.1365-2249.2010.04286.x.

Lapiere, A., Geiger, M., Robert, V., Demarquay, C., Auger, S., Chadi, S., et al., 2020. Prophylactic *Faecalibacterium prausnitzii* treatment prevents the acute breakdown of colonic epithelial barrier in a preclinical model of pelvic radiation disease. Gut Microbes 12 (1), 1–15. Available from. Available at https://doi.org/10.1080/19490976.2020.1812867, http://www.tandfonline.com/toc/kgmi20/current.

Larsen, N., Thorsen, L., Kpikpi, E.N., Stuer-Lauridsen, B., Cantor, M.D., Nielsen, B., et al., 2014. Characterization of *Bacillus* spp. strains for use as probiotic additives in pig feed. Applied Microbiology and Biotechnology 98 (3), 1105–1118. Available at https://doi.org/10.1007/s00253-013-5343-6 Available from: link.springer.de/link/service/journals/00253/index.htm.

Lee, J.H., O'Sullivan, D.J., 2010. Genomic insights into bifidobacteria. Microbiology and Molecular Biology Reviews 74 (3), 378–416. Available from. Available at https://doi.org/10.1128/MMBR.00004-10, http://mmbr.asm.org/cgi/reprint/74/3/378.

Lee, J.S., Song, W.S., Lim, J.W., Choi, T.R., Jo, S.H., Jeon, H.J., et al., 2022. An integrative multiomics approach to characterize anti-adipogenic and anti-lipogenic effects of *Akkermansia muciniphila* in adipocytes. Biotechnology Journal 17 (2) Available from. Available at https://doi.org/10.1002/biot.202100397, http://onlinelibrary.wiley.com/journal/10.1002/(ISSN)1860-7314.

Lee, S.J., Bose, S., Seo, J.G., Chung, W.S., Lim, C.Y., Kim, H., 2014. The effects of co-administration of probiotics with herbal medicine on obesity, metabolic endotoxemia and dysbiosis: a randomized double-blind controlled clinical trial. Clinical Nutrition 33 (6), 973–981. Available from. Available at https://doi.org/10.1016/j.clnu.2013.12.006, http://www.elsevier-international.com/journals/clnu/.

Leeuwendaal, N.K., Stanton, C., O'toole, P.W., Beresford, T.P., 2022. Fermented foods, health and the gut microbiome. Nutrients 14 (7) Available from. Available at https://doi.org/10.3390/nu14071527, https://www.mdpi.com/2072-6643/14/7/1527/pdf.

Li, J.Q., Li, J.L., Xie, Y.H., Wang, Y., Shen, X.N., Qian, Y., et al., 2020. *Saccharomyces cerevisiae* may serve as a probiotic in colorectal cancer by promoting cancer cell apoptosis. Journal of Digestive Diseases 21 (10), 571–582. Available from. Available at https://doi.org/10.1111/1751-2980.12930, http://onlinelibrary.wiley.com/journal/10.1111/(ISSN)1751-2980.

Li, J., Zhao, F., Wang, Y., Chen, J., Tao, J., Tian, G., et al., 2017. Gut microbiota dysbiosis contributes to the development of hypertension. Microbiome 5 (1). Available at https://doi.org/10.1186/s40168-016-0222-x.

Lidbeck, A., Nord, C.E., 1993. Lactobacilli and the normal human anaerobic microflora. Clinical Infectious Diseases 16 (Supplement_4), S181. Available at https://doi.org/10.1093/clinids/16.Supplement_4.S181.

Lilly, D.M., Stillwell, R.H., 1965. Probiotics: growth-promoting factors produced by microorganisms. Science (New York, N.Y.) 147 (3659), 747–748. Available at https://doi.org/10.1126/science.147.3659.747.

Lin, T.-L., Shu, C.-C., Lai, W.-F., Tzeng, C.-M., Lai, H.-C., Lu, C.-C., 2019. Investiture of next generation probiotics on amelioration of diseases – strains do matter. Medicine in Microecology 1-2, 100139. Available at https://doi.org/10.1016/j.medmic.2019.100002.

Liong, M.-T., 2015. Beneficial microorganisms in medical and health applications. Microbiology Monographs 28. Available at https://doi.org/10.1007/978-3-319-23213-3.

Liong, M.T., 2008. Safety of probiotics: translocation and infection. Nutrition Reviews 66 (4), 192–202. Available at https://doi.org/10.1111/j.1753-4887.2008.00024.x.

Liu, H.Y., Hou, R., Yang, G.Q., Zhao, F., Dong, W.G., 2018. In vitro effects of inulin and soya bean oligosaccharide on skatole production and the intestinal microbiota in broilers. Journal of Animal Physiology and Animal Nutrition 102 (3), 706–716. Available at https://doi.org/10.1111/jpn.12830.

Lu, G., Yu, X., Jiang, W., Luo, Q., Tong, J., Fan, S., et al., 2022. Alterations of gut microbiome and metabolite profiles associated with anabatic lipid dysmetabolism in thyroid cancer. Frontiers in Endocrinology 13. Available at https://doi.org/10.3389/fendo.2022.893164.

Lukic, J., Chen, V., Strahinic, I., Begovic, J., Lev-Tov, H., Davis, S.C., et al., 2017. Probiotics or pro-healers: the role of beneficial bacteria in tissue repair. Wound Repair and Regeneration 25 (6), 912–922. Available from. Available at https://doi.org/10.1111/wrr.12607, http://onlinelibrary.wiley.com/journal/10.1111/(ISSN)1524-475X.

López-López, A., Camelo-Castillo, A., Ferrer, M.D., Simon-Soro, Á., Mira, A., 2017. Health-associated niche inhabitants as oral probiotics: the case of *Streptococcus dentisani*. Frontiers in Microbiology 8 Available from. Available at https://doi.org/10.3389/fmicb.2017.00379, http://journal.frontiersin.org/article/10.3389/fmicb.2017.00379/full.

Mackowiak, P.A., 2013. Recycling Metchnikoff: probiotics, the intestinal microbiome and the quest for long life. Frontiers in Public Health 1 Available from. Available at https://doi.org/10.3389/fpubh.2013.00052, http://journal.frontiersin.org/journal/public-health/section/public-health-education-and-promotion#archive.

Mainville, I., Arcand, Y., Farnworth, E.R., 2005. A dynamic model that simulates the human upper gastrointestinal tract for the study of probiotics. International Journal of Food Microbiology 99 (3), 287–296. Available at https://doi.org/10.1016/j.ijfoodmicro.2004.08.020.

Manzhalii, E., Moyseyenko, V., Kondratiuk, V., Molochek, N., Falalyeyeva, T., Kobyliak, N., 2022. Effect of a specific *Escherichia coli* Nissle 1917 strain on minimal/mild hepatic encephalopathy treatment. World Journal of Hepatology 14 (3), 634–646. Available from. Available at https://doi.org/10.4254/wjh.v14.i3.634, http://www.wjgnet.com/1948-5182/index.htm.

Marco, M.L., Tachon, S., 2013. Environmental factors influencing the efficacy of probiotic bacteria. Current Opinion in Biotechnology 24 (2), 207–213. Available at https://doi.org/10.1016/j.copbio.2012.10.002.

Marsh, A.J., Hill, C., Ross, R.P., Cotter, P.D., 2014a. Fermented beverages with health-promoting potential: past and future perspectives. Trends in Food Science and Technology 38 (2), 113–124. Available from. Available at https://doi.org/10.1016/j.tifs.2014.05.002, http://www.elsevier.com/wps/find/journaldescription.cws_home/601278/description#description.

Marsh, A.J., O'Sullivan, O., Hill, C., Ross, R.P., Cotter, P.D., 2014b. Sequence-based analysis of the bacterial and fungal compositions of multiple kombucha (tea fungus) samples. Food Microbiology 38, 171–178. Available at https://doi.org/10.1016/j.fm.2013.09.003.

Martorell, P., Llopis, S., González, N., Chenoll, E., López-Carreras, N., Aleixandre, A., et al., 2016. Probiotic strain Bifidobacterium animalis subsp. lactis CECT 8145 reduces fat content and modulates lipid metabolism and antioxidant response in *Caenorhabditis elegans*. Journal of Agricultural and Food Chemistry 64 (17), 3462–3472. Available from. Available at https://doi.org/10.1021/acs.jafc.5b05934, http://pubs.acs.org/journal/jafcau.

Martín, R., Rios-Covian, D., Huillet, E., Auger, S., Khazaal, S., Bermúdez-Humarán, L.G., et al., 2023. Faecalibacterium: a bacterial genus with promising human health applications. FEMS Microbiology Reviews 47 (4) Available from. Available at https://doi.org/10.1093/femsre/fuad039, https://academic.oup.com/femsre/pages/About.

Matera, M., 2024. Bifidobacteria, Lactobacilli. when, how and why to use them. Global Pediatrics 8, 100139. Available at https://doi.org/10.1016/j.gpeds.2024.100139.

Maughan, H., Van der Auwera, G., 2011. Bacillus taxonomy in the genomic era finds phenotypes to be essential though often misleading. Infection, Genetics and Evolution 11 (5), 789–797. Available at https://doi.org/10.1016/j.meegid.2011.02.001.

Mauricio, M.D., Serna, E., Fernández-Murga, M.L., Portero, J., Aldasoro, M., Valles, S.L., et al., 2017. *Bifidobacterium pseudocatenulatum* CECT 7765 supplementation restores altered vascular function in an experimental model of obese mice. International Journal of Medical Sciences 14 (5), 444–451. Available from. Available at https://doi.org/10.7150/ijms.18354, http://www.medsci.org/v14p0444.pdf.

McCormick, F., 2012. Cows, milk and religion: the use of dairy produce in early societies. Anthropozoologica 47 (2), 101–113. Available at https://doi.org/10.5252/az2012n2a7.

McFarland, L.V., 2000. Normal flora: diversity and functions. Microbial Ecology in Health and Disease 12 (4), 193–207. Available at https://doi.org/10.1080/08910600050216183.

McFarland, L.V., 2007. Meta-analysis of probiotics for the prevention of traveler's diarrhea. Travel Medicine and Infectious Disease 5 (2), 97–105. Available from. Available at https://doi.org/10.1016/j.tmaid.2005.10.003, http://www.elsevier.com/wps/find/journaldescription.cws_home/643125/description#description.

McFarland, L.V., 2015. From yaks to yogurt: the history, development, and current use of probiotics. Clinical Infectious Diseases 60 (suppl 2), S85. Available at https://doi.org/10.1093/cid/civ054.

McGovern, P.E., 2013. Ancient Wine: The Search for the Origins of Viniculture. Princeton University Press, pp. 1–365. undefined. Available at http://press.princeton.edu/titles/7591.html.

McGovern, P.E., Katz, S.H., Fleming, S.J., 2003. The Origins and Ancient History of Wine: Food and Nutrition in History and Antropology. Gordon and Breach Publishers, United States, pp. 1–409. Available from. Available at http://www.taylorandfrancis.com/books/details/9780203392836, 10.4324/9780203392836.

McGovern, P.E., Zhang, J., Tang, J., Zhang, Z., Hall, G.R., Moreau, R.A., et al., 2004. Fermented beverages of pre- and proto-historic China. Proceedings of the National Academy of Sciences of the United States of America 101 (51), 17593–17598. Available at https://doi.org/10.1073/pnas.0407921102.

McMurdie, P.J., Stoeva, M.K., Justice, N., Nemchek, M., Sieber, C.M.K., Tyagi, S., et al., 2022. Increased circulating butyrate and ursodeoxycholate during probiotic intervention in humans with type 2 diabetes. BMC Microbiology 22 (1) Available from. Available at https://doi.org/10.1186/s12866-021-02415-8, http://www.biomedcentral.com/bmcmicrobiol/.

Mcgee, H., 2007. On food and cooking: the science and lore of the kitchen.

Mehta, J.P., Ayakar, S., Singhal, R.S., 2023. The potential of paraprobiotics and postbiotics to modulate the immune system: a Review. Microbiological Research 275 Available from. Available at https://doi.org/10.1016/j.micres.2023.127449, www.urbanfischer.de/journals/microbiolres/microbio.htm.

Menezes, A.G.T., Ramos, C.L., Cenzi, G., Melo, D.S., Dias, D.R., Schwan, R.F., 2020. Probiotic potential, antioxidant activity, and phytase production of indigenous yeasts isolated from indigenous fermented foods. Probiotics and Antimicrobial Proteins 12 (1), 280–288. Available from. Available at https://doi.org/10.1007/s12602-019-9518-z, http://www.springer.com/new+%26+forthcoming+titles+%28default%29/journal/12602.

Meng, D., Sommella, E., Salviati, E., Campiglia, P., Ganguli, K., Djebali, K., et al., 2020. Indole-3-lactic acid, a metabolite of tryptophan, secreted by *Bifidobacterium longum* subspecies infantis is anti-inflammatory in the immature intestine. Pediatric Research 88 (2), 209–217. Available from. Available at https://doi.org/10.1038/s41390-019-0740-x, http://www.nature.com/pr/index.html.

Metchnikoff, E., 1908. The prolongation of life.

Mikeš, Z., Ferenčík, M., Jahnová, E., Ebringer, L., Čižnár, I., 1995. Hypocholesterolemic and immunostimulatory effects of orally applied *Enterococcus fæcium* M-74 in man. Folia Microbiologica 40 (6), 639–646. Available at https://doi.org/10.1007/bf02818522.

Miquel, S., Martín, R., Lashermes, A., Gillet, M., Meleine, M., Gelot, A., et al., 2016. Anti-nociceptive effect of *Faecalibacterium prausnitzii* in non-inflammatory IBS-like models. Scientific Reports 6 (1). Available at https://doi.org/10.1038/srep19399.

Miyazawa, K., He, F., Kawase, M., Kubota, A., Yoda, K., Hiramatsu, M., 2011. Enhancement of immunoregulatory effects of *Lactobacillus gasseri* TMC0356 by heat treatment and culture medium. Letters in Applied Microbiology 53 (2), 210–216. Available at https://doi.org/10.1111/j.1472-765x.2011.03093.x.

Morotomi, M., Nagai, F., Watanabe, Y., 2012. Description of *Christensenella minuta* gen. nov., sp. nov., isolated from human faeces, which forms a distinct branch in the order clostridiales, and proposal of christensenellaceae fam. nov. International Journal of Systematic and Evolutionary Microbiology 62 (1), 144–149. Available at https://doi.org/10.1099/ijs.0.026989-0.

Munukka, E., Rintala, A., Toivonen, R., Nylund, M., Yang, B., Takanen, A., et al., 2017. *Faecalibacterium prausnitzii* treatment improves hepatic health and reduces adipose tissue inflammation in high-fat fed mice. The ISME Journal 11 (7), 1667–1679. Available at https://doi.org/10.1038/ismej.2017.24.

Nagpal, R., Kumar, A., Kumar, M., Behare, P.V., Jain, S., Yadav, H., 2012. Probiotics, their health benefits and applications for developing healthier foods: a review. FEMS Microbiology Letters 334 (1), 1–15. Available at https://doi.org/10.1111/j.1574-6968.2012.02593.x.

Navab-Moghadam, F., Sedighi, M., Khamseh, M.E., Alaei-Shahmiri, F., Talebi, M., Razavi, S., et al., 2017. The association of type II diabetes with gut microbiota composition. Microbial Pathogenesis 110, 630–636. Available from. Available at https://doi.org/10.1016/j.micpath.2017.07.034, http://www.elsevier.com/inca/publications/store/6/2/2/9/1/5/index.htt.

Newsome, R.C., Gharaibeh, R.Z., Pierce, C.M., da Silva, W.V., Paul, S., Hogue, S.R., et al., 2022. Interaction of bacterial genera associated with therapeutic response to immune checkpoint PD-1 blockade in a United States cohort. Genome Medicine 14 (1) Available from. Available at https://doi.org/10.1186/s13073-022-01037-7, http://www.genomemedicine.com/.

Nishiwaki, H., Ito, M., Ishida, T., Hamaguchi, T., Maeda, T., Kashihara, K., et al., 2020. Meta-analysis of gut dysbiosis in Parkinson's disease. Movement Disorders 35 (9), 1626–1635. Available at https://doi.org/10.1002/mds.28119.

Noh, C.K., Lee, K.J., 2020. Fecal microbiota alterations and small intestinal bacterial overgrowth in functional abdominal bloating/distention. Journal of Neurogastroenterology and Motility 26 (4), 539–549. Available from. Available at https://doi.org/10.5056/jnm20080, http://www.jnmjournal.org/journal/list.html?pn=vol&TG=vol&sm=&s_v=26&s_n=4&year=2020.

O'Mahony, D., Murphy, S., Boileau, T., Park, J.S., O'Brien, F., Groeger, D., et al., 2010. Bifidobacterium animalis AHC7 protects against pathogen-induced NF-κB activation in vivo. BMC Immunology 11 (1). Available at https://doi.org/10.1186/1471-2172-11-63.

Oggioni, M.R., Pozzi, G., Galieni, P., Valensin, P.E., Bigazzi, C., 1998. Recurrent septicemia in an immunocompromised patient due to probiotic strains of *Bacillus subtilis*. Journal of Clinical Microbiology 36 (1), 325–326. Available at https://doi.org/10.1128/jcm.36.1.325-326.1998 Available from: jcm.asm.org.

Ono, J., Greig, D., Boynton, P.J., 2020. Defining and disrupting species boundaries in *Saccharomyces*. Annual Review of Microbiology 74, 477–495. Available from. Available at https://doi.org/10.1146/annurev-micro-021320-014036, http://www.annualreviews.org/journal/micro.

Orlando, A., Refolo, M.G., Messa, C., Amati, L., Lavermicocca, P., Guerra, V., et al., 2012. Antiproliferative and proapoptotic effects of viable or heat-killed *Lactobacillus paracasei* IMPC2.1 and *Lactobacillus rhamnosus* GG in HGC-27 gastric and DLD-1 colon cell lines. Nutrition and Cancer 64 (7), 1103–1111. Available at https://doi.org/10.1080/01635581.2012.717676.

Owen, J., 2011. Earliest known winery found in Armenian cave. National Geographic News 10.

Ozen, M., Dinleyici, E.C., 2015. The history of probiotics: the untold story. Beneficial Microbes 6 (2), 159–165. Available at https://doi.org/10.3920/bm2014.0103.

Paleo-López, R., Quintero-Galvis, J.F., Solano-Iguaran, J.J., Sanchez-Salazar, A.M., Gaitan-Espitia, J.D., Nespolo, R.F., 2016. A phylogenetic analysis of macroevolutionary patterns in fermentative yeasts. Ecology and Evolution 6 (12), 3851–3861. Available from. Available at https://doi.org/10.1002/ece3.2097, http://onlinelibrary.wiley.com/journal/10.1002/(ISSN)2045-7758.

Parvaneh, K., Ebrahimi, M., Sabran, M.R., Karimi, G., Hwei, A.N.M., Abdul-Majeed, S., et al., 2015. Probiotics (*Bifidobacterium longum*) increase bone mass density and upregulate Sparc and Bmp-2 genes in rats with bone loss resulting from ovariectomy. BioMed Research International 2015 Available from. Available at https://doi.org/10.1155/2015/897639, http://www.hindawi.com/journals/biomed/.

Pascal, V., Pozuelo, M., Borruel, N., Casellas, F., Campos, D., Santiago, A., et al., 2017. A microbial signature for Crohn's disease. Gut 66 (5), 813–822. Available at https://doi.org/10.1136/gutjnl-2016-313235.

Patterson, M.J., 1996. Streptococcus. Medical Microbiology.

Peng, G.C., Hsu, C.H., 2005. The efficacy and safety of heat-killed *Lactobacillus paracasei* for treatment of perennial allergic rhinitis induced by house-dust mite. Pediatric Allergy and Immunology 16 (5), 433–438. Available at https://doi.org/10.1111/j.1399-3038.2005.00284.x.

Pericolini, E., Gabrielli, E., Ballet, N., Sabbatini, S., Roselletti, E., Cayzeele Decherf, A., et al., 2017. Therapeutic activity of a *Saccharomyces* cerevisiae-based probiotic and inactivated whole yeast on vaginal candidiasis. Virulence 8 (1), 74–90. Available from. Available at https://doi.org/10.1080/21505594.2016.1213937, http://www.tandfonline.com/loi/kvir20.

Peris, D., Sylvester, K., Libkind, D., Gonçalves, P., Sampaio, J.P., Alexander, W.G., et al., 2014. Population structure and reticulate evolution of *Saccharomyces eubayanus* and its lager-brewing hybrids. Molecular Ecology 23 (8), 2031–2045. Available from. Available at https://doi.org/10.1111/mec.12702, http://www.blackwellpublishing.com/journals/MEC.

Perraudeau, F., McMurdie, P., Bullard, J., Cheng, A., Cutcliffe, C., Deo, A., et al., 2020. Improvements to postprandial glucose control in subjects with type 2 diabetes: a multicenter, double blind, randomized placebo-controlled trial of a novel probiotic formulation. BMJ Open Diabetes Research and Care 8 (1) Available from. Available at https://doi.org/10.1136/bmjdrc-2020-001319, http://drc.bmj.com/.

Peters, B.A., Shapiro, J.A., Church, T.R., Miller, G., Trinh-Shevrin, C., Yuen, E., et al., 2018. A taxonomic signature of obesity in a large study of American adults. Scientific Reports 8 (1) Available from. Available at https://doi.org/10.1038/s41598-018-28126-1, www.nature.com/srep/index.html.

Philpott, D., Girardin, S., 2004. The role of Toll-like receptors and Nod proteins in bacterial infection. Molecular Immunology 41 (11), 1099–1108. Available at https://doi.org/10.1016/j.molimm.2004.06.012.

Pieniz, S., Andreazza, R., Pereira, J.Q., De Oliveira Camargo, F.A., Brandelli, A., 2013. Production of selenium-enriched biomass by *Enterococcus durans*. Biological Trace Element Research 155 (3), 447–454. Available at https://doi.org/10.1007/s12011-013-9818-1.

Pineiro, M., Stanton, C., 2007. Probiotic bacteria: legislative framework— requirements to evidence basis. The Journal of Nutrition 137 (3), 850S. Available at https://doi.org/10.1093/jn/137.3.850S.

Pino, A., Benkaddour, B., Inturri, R., Amico, P., Vaccaro, S.C., Russo, N., et al., 2022. Characterization of *Bifidobacterium asteroides* Isolates. Microorganisms 10 (3) Available from. Available at https://doi.org/10.3390/microorganisms10030655, https://www.mdpi.com/2076-2607/10/3/655/pdf.

Pittayanon, R., Lau, J.T., Leontiadis, G.I., Tse, F., Yuan, Y., Surette, M., et al., 2020. Differences in gut microbiota in patients with vs without inflammatory bowel diseases: a systematic review. Gastroenterology 158 (4), 930. Available from. Available at https://doi.org/10.1053/j.gastro.2019.11.294, http://www.journals.elsevier.com/gastroenterology/.

Pot, B., Vandenplas, Y., 2021. Factors that influence clinical efficacy of live biotherapeutic products. European Journal of Medical Research 26 (1). Available at https://doi.org/10.1186/s40001-021-00509-7.

Prochazkova, P., Roubalova, R., Dvorak, J., Kreisinger, J., Hill, M., Tlaskalova-Hogenova, H., et al., 2021. The intestinal microbiota and metabolites in patients with anorexia nervosa. Gut Microbes 13 (1), 1–25. Available from. Available at https://doi.org/10.1080/19490976.2021.1902771, http://www.tandfonline.com/toc/kgmi20/current.

Probiotics in food health and nutritional properties and guidelines for evaluation. 85 (2006).

Profir, A.-G., Buruiana, C.-T., Vizireanu, C., 2015. Effects of Saccharomyces cerevisiae var. boulardii in gastrointestinal disorders. Journal of Agroalimentary Processes and Technologies 21, 148–155.

Qu, S., Fan, L., Qi, Y., Xu, C., Hu, Y., Chen, S., et al., 2021. *Akkermansia muciniphila* alleviates dextran sulfate sodium (DSS)-induced acute colitis by NLRP3 activation. Microbiology Spectrum 9 (2). Available at https://doi.org/10.1128/spectrum.00730-21.

Ragavan, M.L., Das, N., 2017. Isolation and characterization of potential probiotic yeasts from different sources. Asian Journal of Pharmaceutical and Clinical Research 10 (4), 451–455. Available from. Available at https://doi.org/10.22159/ajpcr.2017.v10i4.17067, http://innovareacademics.in/journals/index.php/ajpcr/article/download/17067/10643.

Rampengan, N.H., Manoppo, J., Warouw, S.M., 2010. Comparison of efficacies between live and killed probiotics in children with lactose malabsorption. Southeast Asian Journal of Tropical Medicine and Public Health 41 (2), 474–481. Available from. Available at https://seameotropmednetwork.org/publication_current_issue.html.

Ranadheera, C.S., Vidanarachchi, J.K., Rocha, R.S., Cruz, A.G., Ajlouni, S., 2017. Probiotic delivery through fermentation: dairy vs. non-dairy beverages. Fermentation 3 (4) Available from. Available at https://doi.org/10.3390/fermentation3040067, https://www.mdpi.com/2311-5637/3/4/67/pdf.

Ranjbar, R., Vahdati, S.N., Tavakoli, S., Khodaie, R., Behboudi, H., 2021. Immunomodulatory roles of microbiota-derived short-chain fatty acids in bacterial infections. Biomedicine and Pharmacotherapy 141, 111817. Available at https://doi.org/10.1016/j.biopha.2021.111817.

Rau, S., Gregg, A., Yaceczko, S., Limketkai, B., 2024. Prebiotics and probiotics for gastrointestinal disorders. Nutrients 16 (6), 778. Available at https://doi.org/10.3390/nu16060778.

Rettger, L.F., Cheplin, H.A., 1922. *Bacillus acidophilus* and its therapeutic application. Archives of Internal Medicine 29 (3), 357–367. Available at https://doi.org/10.1001/archinte.1922.00110030082005.

Rodrigues, V.F., Elias-Oliveira, J., Pereira, Í.S., Pereira, J.A., Barbosa, S.C., Machado, M.S.G., et al., 2022. *Akkermansia muciniphila* and gut immune system: a good friendship that attenuates inflammatory bowel disease, obesity, and diabetes. Frontiers in Immunology 13 Available from. Available at https://doi.org/10.3389/fimmu.2022.934695, https://www.frontiersin.org/journals/immunology#.

Romero-Luna, H.E., Hernández-Sánchez, H., Ribas-Aparicio, R.M., Cauich-Sánchez, P.I., Dávila-Ortiz, G., 2019. Evaluation of the probiotic potential of *Saccharomyces cerevisiae* strain (C41) isolated from tibicos by in vitro studies. Probiotics and Antimicrobial Proteins 11 (3), 794–800. Available from. Available at https://doi.org/10.1007/s12602-018-9471-2, http://www.springer.com/new+%26+forthcoming+titles+%28default%29/journal/12602.

Rusch, K., Rusch, V., 2001. Mikrobiologische Therapie: Grundlagen und Praxis; mit57. Tabellen.

Said, M.S., Tirthani, E., Lesho, E., 2023. Enterococcus Infections.

Sakamoto, M., Benno, Y., 2006. Reclassification of *Bacteroides distasonis*, *Bacteroides goldsteinii* and *Bacteroides merdae* as *Parabacteroides distasonis* gen. nov., comb. nov., *Parabacteroides goldsteinii* comb. nov. and *Parabacteroides merdae* comb. nov. International Journal of Systematic and Evolutionary Microbiology 56 (7), 1599–1605. Available at https://doi.org/10.1099/ijs.0.64192-0.

Salim, S.Y., Söderholm, J.D., 2011. Importance of disrupted intestinal barrier in inflammatory bowel diseases. Inflammatory Bowel Diseases 17 (1), 362–381. Available at https://doi.org/10.1002/ibd.21403.

Salminen, S., Ouwehand, A., Benno, Y., Lee, Y.K., 1999. Probiotics: how should they be defined? Trends in Food Science and Technology 10 (3), 107–110. Available at https://doi.org/10.1016/s0924-2244(99)00027-8.

Samuel, D., 1996. Investigation of ancient egyptian baking and brewing methods by correlative microscopy. Science (New York, N.Y.) 273 (5274), 488–490. Available at https://doi.org/10.1126/science.273.5274.488.

Santacroce, L., Charitos, I.A., Bottalico, L., 2019. A successful history: probiotics and their potential as antimicrobials. Expert Review of Anti-Infective Therapy 17 (8), 635–645. Available from. Available at https://doi.org/10.1080/14787210.2019.1645597, http://www.tandfonline.com/loi/ierz20.

Santacruz, A., Collado, M.C., García-Valdés, L., Segura, M.T., Martín-Lagos, J.A., Anjos, T., et al., 2010. Gut microbiota composition is associated with body weight, weight gain and biochemical parameters in pregnant women. British Journal of Nutrition 104 (1), 83–92. Available at https://doi.org/10.1017/s0007114510000176.

Sarkar, A., Mandal, S., 2016. Bifidobacteria—Insight into clinical outcomes and mechanisms of its probiotic action. Microbiological Research 192, 159–171. Available at https://doi.org/10.1016/j.micres.2016.07.001.

Schaafsma, G., 1996. State of the art concerning probiotic strains in milk products. IDF Nutrition Newsletter 5, 23–24.

Schaefer, L., Auchtung, T.A., Hermans, K.E., Whitehead, D., Borhan, B., Britton, R.A., 2010. The antimicrobial compound reuterin (3-hydroxypropionaldehyde) induces oxidative stress via interaction with thiol groups. Microbiology (Reading, England) 156 (6), 1589–1599. Available from. Available at https://doi.org/10.1099/mic.0.035642-0, http://mic.sgmjournals.org/cgi/reprint/156/6/1589.

Scher, J.U., Sczesnak, A., Longman, R.S., Segata, N., Ubeda, C., Bielski, C., et al., 2013. Expansion of intestinal *Prevotella copri* correlates with enhanced susceptibility to arthritis. eLife 2013 (2) Available from. Available at https://doi.org/10.7554/eLife.01202.001, http://elife.elifesciences.org/content/elife/2/e01202.full.pdf.

Segawa, S., Wakita, Y., Hirata, H., Watari, J., 2008. Oral administration of heat-killed *Lactobacillus brevis* SBC8803 ameliorates alcoholic liver disease in ethanol-containing diet-fed C57BL/6N mice. International Journal of Food Microbiology 128 (2), 371–377. Available at https://doi.org/10.1016/j.ijfoodmicro.2008.09.023.

Shan, L.S., Hou, P., Wang, Z.J., Liu, F.R., Chen, N., Shu, L.H., et al., 2013. Prevention and treatment of diarrhoea with *Saccharomyces boulardii* in children with acute lower respiratory tract infections. Beneficial Microbes 4 (4), 329–334. Available at https://doi.org/10.3920/bm2013.0008.

Shao, Y., Qi, W., Zhang, X., Ran, N., Liu, C., Fu, R., et al., 2021. Plasma metabolomic and intestinal microbial analyses of patients with severe aplastic anemia. Frontiers in Cell and Developmental Biology 9. Available at https://doi.org/10.3389/fcell.2021.669887.

Sharma, G., Garg, N., Hasan, S., Shirodkar, S., 2022. Prevotella: an insight into its characteristics and associated virulence factors. Microbial Pathogenesis 169, 105673. Available at https://doi.org/10.1016/j.micpath.2022.105673.

Shen, T., Yue, Y., He, T., Huang, C., Qu, B., Lv, W., et al., 2021. The association between the gut microbiota and parkinson's disease, a meta-analysis. Frontiers in Aging Neuroscience 13 Available from. Available at https://doi.org/10.3389/fnagi.2021.636545, https://www.frontiersin.org/journals/aging-neuroscience#.

Shin, H.S., Park, S.Y., Lee, D.K., Kim, S.A., An, H.M., Kim, J.R., et al., 2010. Hypocholesterolemic effect of sonication-killed *Bifidobacterium longum* isolated from healthy adult Koreans in high cholesterol fed rats. Archives of Pharmacal Research 33 (9), 1425–1431. Available at https://doi.org/10.1007/s12272-010-0917-7.

Sicard, D., Legras, J.L., 2011. Bread, beer and wine: yeast domestication in the *Saccharomyces sensu* stricto complex. Comptes Rendus - Biologies 334 (3), 229–236. Available from. Available at https://doi.org/10.1016/j.crvi.2010.12.016, http://www.elsevier.com/journals/comptes-rendus-biologies/.

Siciliano, R.A., Reale, A., Mazzeo, M.F., Morandi, S., Silvetti, T., Brasca, M., 2021. Paraprobiotics: a new perspective for functional foods and nutraceuticals. Nutrients 13 (4) Available from. Available at https://doi.org/10.3390/nu13041225, https://www.mdpi.com/2072-6643/13/4/1225/pdf.

Siragusa, S., De Angelis, M., Calasso, M., Campanella, D., Minervini, F., Di Cagno, R., et al., 2014. Fermentation and proteome profiles of *Lactobacillus plantarum* strains during growth under food-like conditions. Journal of Proteomics 96, 366–380. Available at https://doi.org/10.1016/j.jprot.2013.11.003.

Sivignon, A., de Vallée, A., Barnich, N., Denizot, J., Darcha, C., Pignède, G., et al., 2015. *Saccharomyces cerevisiae* CNCM I-3856 prevents colitis induced by AIEC bacteria in the transgenic mouse model mimicking Crohn's disease. Inflammatory Bowel Diseases 21 (2), 276–286. Available at https://doi.org/10.1097/mib.0000000000000280.

Sofi, M.H., Wu, Y., Ticer, T., Schutt, S., Bastian, D., Choi, H.J., et al., 2021. A single strain of *Bacteroides fragilis* protects gut integrity and reduces GVHD. JCI Insight 6 (3) Available from. Available at https://doi.org/10.1172/JCI.INSIGHT.136841, https://insight.jci.org/articles/view/136841.

Sokol, H., Pigneur, B., Watterlot, L., Lakhdari, O., Bermúdez-Humarán, L.G., Gratadoux, J.J., et al., 2008. *Faecalibacterium prausnitzii* is an anti-inflammatory commensal bacterium identified by gut microbiota analysis of Crohn disease patients. Proceedings of the National Academy of Sciences of the United States of America 105 (43), 16731–16736. Available from. Available at https://doi.org/10.1073/pnas.0804812105France, http://www.pnas.org/content/105/43/16731.full.pdf.

Song, H., Yoo, Y., Hwang, J., Na, Y.C., Kim, H.S., 2016. *Faecalibacterium prausnitzii* subspecies-level dysbiosis in the human gut microbiome underlying atopic dermatitis. Journal of Allergy and Clinical Immunology 137 (3), 852–860. Available from. Available at https://doi.org/10.1016/j.jaci.2015.08.021, http://www.elsevier.com/inca/publications/store/6/2/3/3/6/8/index.htt.

Sowah, S.A., Milanese, A., Schübel, R., Wirbel, J., Kartal, E., Johnson, T.S., et al., 2022. Calorie restriction improves metabolic state independently of gut microbiome composition: a randomized dietary intervention trial. Genome Medicine 14 (1) Available from. Available at https://doi.org/10.1186/s13073-022-01030-0, http://www.genomemedicine.com/.

Stanford, J., Charlton, K., Stefoska-Needham, A., Ibrahim, R., Lambert, K., 2020. The gut microbiota profile of adults with kidney disease and kidney stones: a systematic review of the literature. BMC Nephrology 21 (1). Available at https://doi.org/10.1186/s12882-020-01805-w.

Suryavanshi, A., Agarwal, A., Kaler, A., Bihade, U., Kaur, J., Tikoo, K., et al., 2013. Comparative studies on the antioxidant potential of vanillin-producing *Saccharomyces boulardii* extracts. Oxidants and Antioxidants in Medical Science 2 (3), 201. Available at https://doi.org/10.5455/oams.190413.or.040.

Szajewska, H., Mrukowicz, J., 2005. Meta-analysis: non-pathogenic yeast *Saccharomyces boulardii* in the prevention of antibiotic-associated diarrhoea. Alimentary Pharmacology and Therapeutics 22 (5), 365–372. Available at https://doi.org/10.1111/j.1365-2036.2005.02624.x.

Talwalkar, A., Kailasapathy, K., 2004. A review of oxygen toxicity in probiotic yogurts: influence on the survival of probiotic bacteria and protective techniques. Comprehensive Reviews in Food Science and Food Safety 3 (3), 117–124. Available at https://doi.org/10.1111/j.1541-4337.2004.tb00061.x.

Tamang, J.P., Cotter, P.D., Endo, A., Han, N.S., Kort, R., Liu, S.Q., et al., 2020. Fermented foods in a global age: East meets West. Comprehensive Reviews in Food Science and Food Safety 19 (1), 184–217. Available from. Available at https://doi.org/10.1111/1541-4337.12520, http://onlinelibrary.wiley.com/journal/10.1111/(ISSN)1541-4337.

Tanaka, A., Seki, M., Yamahira, S., Noguchi, H., Kosai, K., Toba, M., et al., 2011. *Lactobacillus pentosus* strain b240 suppresses pneumonia induced by *Streptococcus pneumoniae* in mice. Letters in Applied Microbiology 53 (1), 35–43. Available at https://doi.org/10.1111/j.1472-765x.2011.03079.x.

Taverniti, V., Guglielmetti, S., 2011. The immunomodulatory properties of probiotic microorganisms beyond their viability (ghost probiotics: proposal of paraprobiotic concept). Genes and Nutrition 6 (3), 261–274. Available at https://doi.org/10.1007/s12263-011-0218-x.

Teuber, M., 1995. The genus *Lactococcus*. In: Wood, B.J.B., Holzapfel, W.H. (Eds.), The Genera of Lactic Acid Bacteria. The Lactic Acid Bacteria, vol 2. Springer, Boston, MA, pp. 173–234. Available at https://doi.org/10.1007/978-1-4615-5817-0_6.

Tissier, H., 1907. Traitement des infections intestinales par la methode de transformation de la flore bacterienne de l'intestin. Available at https://books.google.co.in/books?id=MydomgEACAAJ.

Tompkins, T., Xu, X., Ahmarani, J., 2010. A comprehensive review of post-market clinical studies performed in adults with an Asian probiotic formulation. Beneficial Microbes 1 (1), 93–106. Available at https://doi.org/10.3920/bm2008.1005.

Trapecar, M., Leouffre, T., Faure, M., Jensen, H.E., Granum, P.E., Cencic, A., et al., 2011. The use of a porcine intestinal cell model system for evaluating the food safety risk of *Bacillus cereus* probiotics and the implications for assessing enterotoxigenicity. APMIS Acta Pathologica, Microbiologica, et Immunologica Scandinavica 119 (12), 877–884. Available at https://doi.org/10.1111/j.1600-0463.2011.02797.x.

Tufail, M.A., Schmitz, R.A., 2025. Exploring the probiotic potential of Bacteroides spp. within one health paradigm. Probiotics and Antimicrobial Proteins 17 (2), 681–704. Available from. Available at https://doi.org/10.1007/s12602-024-10370-9, https://www.springer.com/journal/12602.

Tulumoglu, S., Yuksekdag, Z.N., Beyatli, Y., Simsek, O., Cinar, B., Yaşar, E., 2013. Probiotic properties of *Lactobacilli species* isolated from children's feces. Anaerobe 24, 36–42. Available at https://doi.org/10.1016/j.anaerobe.2013.09.006.

Ueda, A., Shinkai, S., Shiroma, H., Taniguchi, Y., Tsuchida, S., Kariya, T., et al., 2021. Identification of *Faecalibacterium prausnitzii* strains for gut microbiome-based intervention in Alzheimer's-type dementia. Cell Reports Medicine 2 (9), 100398. Available at https://doi.org/10.1016/j.xcrm.2021.100398.

Ueno, N., Fujiya, M., Segawa, S., Nata, T., Moriichi, K., Tanabe, H., et al., 2011. Heat-killed body of *Lactobacillus brevis* SBC8803 ameliorates intestinal injury in a murine model of colitis by enhancing the intestinal barrier function. Inflammatory Bowel Diseases 17 (11), 2235–2250. Available at https://doi.org/10.1002/ibd.21597.

Vergin, F., 1954. Anti- und probiotika. Hippokrates 25 (4), 116–119.

Verhaar, B.J.H., Hendriksen, H.M.A., de Leeuw, F.A., Doorduijn, A.S., van Leeuwenstijn, M., Teunissen, C.E., et al., 2022. Gut microbiota composition is related to AD pathology. Frontiers in Immunology 12 Available from. Available at https://doi.org/10.3389/fimmu.2021.794519, https://www.frontiersin.org/journals/immunology#.

Vieira, A.T., Rocha, V.M., Tavares, L., Garcia, C.C., Teixeira, M.M., Oliveira, S.C., et al., 2016. Control of *Klebsiella pneumoniae* pulmonary infection and immunomodulation by oral treatment with the commensal probiotic *Bifidobacterium longum* 51A. Microbes and Infection 18 (3), 180–189. Available from. Available at https://doi.org/10.1016/j.micinf.2015.10.008, http://www.journals.elsevier.com/microbes-and-infection/.

Vinderola, G., Sanders, M.E., Salminen, S., 2022. The concept of postbiotics. Foods 11 (8) Available from. Available at https://doi.org/10.3390/foods11081077, https://www.mdpi.com/2304-8158/11/8/1077/pdf.

Vitetta, L., Llewellyn, H., Oldfield, D., 2019. Gut dysbiosis and the intestinal microbiome: *Streptococcus thermophilus* a key probiotic for reducing uremia. Microorganisms 7 (8), 228. Available at https://doi.org/10.3390/microorganisms7080228.

Voidarou, C., Antoniadou, M., Rozos, G., Tzora, A., Skoufos, I., Varzakas, T., et al., 2021. Fermentative foods: microbiology, biochemistry, potential human health benefits and public health issues. Foods 10 (1) Available from. Available at https://doi.org/10.3390/foods10010069, https://www.mdpi.com/2304-8158/10/1/69/pdf.

Vuillermin, P.J., O'Hely, M., Collier, F., Allen, K.J., Tang, M.L.K., Harrison, L.C., et al., 2020. Maternal carriage of prevotella during pregnancy associates with protection against food allergy in the offspring. Nature Communications 11 (1) Available from. Available at https://doi.org/10.1038/s41467-020-14552-1, http://www.nature.com/ncomms/index.html.

Waitzberg, D., Guarner, F., Hojsak, I., Ianiro, G., Polk, D.B., Sokol, H., 2024. Can the evidence-based use of probiotics (notably *Saccharomyces boulardii* CNCM I-745 and *Lactobacillus rhamnosus* GG) mitigate the clinical effects of antibiotic-associated dysbiosis? Advances in Therapy 41 (3), 901–914. Available from. Available at https://doi.org/10.1007/s12325-024-02783-3, https://www.springer.com/journal/12325.

Wang, G., Chen, Y., Xia, Y., Song, X., Ai, L., 2022a. Characteristics of probiotic preparations and their applications. Foods 11 (16), 2472. Available at https://doi.org/10.3390/foods11162472.

Wang, X., Luo, N., Mi, Q., Kong, W., Zhang, W., Li, X., et al., 2022b. Influence of cigarette smoking on oral microbiota in patients with recurrent aphthous stomatitis. Journal of Investigative Medicine 70 (3), 805–813. Available at https://doi.org/10.1136/jim-2021-002119.

Wang, K., Liao, M., Zhou, N., Bao, L., Ma, K., Zheng, Z., et al., 2019. Parabacteroides distasonis alleviates obesity and metabolic dysfunctions via production of succinate and secondary bile acids. Cell Reports 26, 222–235.

Wang, L., Tang, L., Feng, Y., Zhao, S., Han, M., Zhang, C., et al., 2020. A purified membrane protein from *Akkermansia muciniphila* or the pasteurised bacterium blunts colitis associated tumourigenesis by modulation of CD8 + T cells in mice. Gut 69 (11), 1988–1997. Available from. Available at https://doi.org/10.1136/gutjnl-2019-320105, http://gut.bmj.com/content/by/year.

Wexler, H.M., 2007. Bacteroides: the good, the bad, and the nitty-gritty. Clinical Microbiology Reviews 20 (4), 593–621. Available at https://doi.org/10.1128/CMR.00008-07.

Witthuhn, R.C., Schoeman, T., Britz, T.J., 2005. Characterisation of the microbial population at different stages of Kefir production and Kefir grain mass cultivation. International Dairy Journal 15 (4), 383–389. Available at https://doi.org/10.1016/j.idairyj.2004.07.016.

Wong, C.B., Iwabuchi, N., Xiao, J.Z., 2019. Exploring the science behind *Bifidobacterium breve* M-16V in infant health. Nutrients 11 (8) Available from. Available at https://doi.org/10.3390/nu11081724, https://www.mdpi.com/2072-6643/11/8/1724/pdf.

Wu, M.H., Pan, T.M., Wu, Y.J., Chang, S.J., Chang, M.S., Hu, C.Y., 2010. Exopolysaccharide activities from probiotic bifidobacterium: immunomodulatory effects (on J774A.1 macrophages) and antimicrobial properties. International Journal of Food Microbiology 144 (1), 104–110. Available at https://doi.org/10.1016/j.ijfoodmicro.2010.09.003.

Wu, T.R., Lin, C.S., Chang, C.J., Lin, T.L., Martel, J., Ko, Y.F., et al., 2019. Gut commensal *Parabacteroides goldsteinii* plays a predominant role in the anti-obesity effects of polysaccharides isolated from *Hirsutella sinensis*. Gut 68 (2), 248–262. Available from. Available at https://doi.org/10.1136/gutjnl-2017-315458, http://gut.bmj.com/content/by/year.

Xi, J., Ding, D., Zhu, H., Wang, R., Su, F., Wu, W., et al., 2021. Disturbed microbial ecology in Alzheimer's disease: evidence from the gut microbiota and fecal metabolome. BMC

Microbiology 21 (1) Available from. Available at https://doi.org/10.1186/s12866-021-02286-z, http://www.biomedcentral.com/bmcmicrobiol/.

Xiao, S.D., Zhang, D.Z., Lu, H., Jiang, S.H., Liu, H.Y., Wang, G.S., et al., 2003. Multicenter, randomized, controlled trial of heat-killed *Lactobacillus acidophilus* LB in patients with chronic diarrhea. Advances in Therapy 20 (5), 253–260. Available from. Available at https://doi.org/10.1007/BF02849854, http://www.springer.com/springer+healthcare/journal/12325.

Yang, X., Wang, Y., Huo, G., 2013. Complete genome sequence of *Lactococcus lactis* subsp. lactis KLDS4.0325. Genome Announcements 1 (6). Available at https://doi.org/10.1128/genomea.00962-13.

Yoon, H.S., Cho, C.H., Yun, M.S., Jang, S.J., You, H.J., Kim, Jh, et al., 2021. Akkermansia muciniphila secretes a glucagon-like peptide-1-inducing protein that improves glucose homeostasis and ameliorates metabolic disease in mice. Nature Microbiology 6 (5), 563–573. Available from. Available at https://doi.org/10.1038/s41564-021-00880-5, www.nature.com/nmicrobiol/.

Yunes, R.A., Poluektova, E.U., Vasileva, E.V., Odorskaya, M.V., Marsova, M.V., Kovalev, G.I., et al., 2020. A multi-strain potential probiotic formulation of GABA-producing *Lactobacillus plantarum* 90sk and *Bifidobacterium adolescentis* 150 with antidepressant effects. Probiotics and Antimicrobial Proteins 12 (3), 973–979. Available at https://doi.org/10.1007/s12602-019-09601-1.

Zakrzewski, M., Simms, L.A., Brown, A., Appleyard, M., Irwin, J., Waddell, N., et al., 2019. IL23R-protective coding variant promotes beneficial bacteria and diversity in the ileal microbiome in healthy individuals without inflammatory bowel disease. Journal of Crohn's and Colitis 13 (4), 451–461. Available from. Available at https://doi.org/10.1093/ecco-jcc/jjy188, https://academic.oup.com/ecco-jcc/issue.

Zang, X., Xiao, M., Yu, L., Chen, Y., Duan, H., Zhang, C., et al., 2024. Prevotella copri —a potential next-generation probiotic. Food Frontiers 5 (4), 1391–1409. Available at https://doi.org/10.1002/fft2.417.

Zapasnik, A., Sokolowska, B., Bryla, M., 2022. Role of lactic acid bacteria in food preservation and safety. Foods 11.

Zelaya, H., Tsukida, K., Chiba, E., Marranzino, G., Alvarez, S., Kitazawa, H., et al., 2014. Immunobiotic lactobacilli reduce viral-associated pulmonary damage through the modulation of inflammation–coagulation interactions. International Immunopharmacology 19 (1), 161–173. Available at https://doi.org/10.1016/j.intimp.2013.12.020.

Zendeboodi, F., Mortazavian, A.M., 2021. Probiotic concept: an overview. Vitro Functionality of Probiotics in Foods. Nova Science Publishers, Inc., Iran, pp. 1–14. Available from. Available at https://novapublishers.com/shop/in-vitro-functionality-of-probiotics-in-foods/.

Zendeboodi, F., Khorshidian, N., Mortazavian, A.M., da Cruz, A.G., 2020. Probiotic: conceptualization from a new approach. Current Opinion in Food Science 32, 103–123. Available from. Available at https://doi.org/10.1016/j.cofs.2020.03.009, http://www.journals.elsevier.com/current-opinion-in-food-science/.

Zhang, W., Zhu, B., Xu, J., Liu, Y., Qiu, E., Li, Z., et al., 2018. *Bacteroides fragilis* protects against antibiotic-associated diarrhea in rats by modulating intestinal defenses. Frontiers in Immunology 9 (MAY). Available at https://doi.org/10.3389/fimmu.2018.01040.

Zhou, Q., Shen, B., Huang, R., Liu, H., Zhang, W., Song, M., et al., 2022. *Bacteroides fragilis* strain ZY-312 promotes intestinal barrier integrity via upregulating the STAT3 pathway in a radiation-induced intestinal injury mouse model. Frontiers in Nutrition 9. Available at https://doi.org/10.3389/fnut.2022.1063699.

Zokaeifar, H., Babaei, N., Saad, C.R., Kamarudin, M.S., Sijam, K., Balcazar, J.L., 2014. Administration of *Bacillus subtilis* strains in the rearing water enhances the water quality, growth performance, immune response, and resistance against *Vibrio harveyi* infection in juvenile white shrimp, *Litopenaeus vannamei*. Fish and Shellfish Immunology 36 (1), 68–74. Available from. Available at https://doi.org/10.1016/j.fsi.2013.10.007, http://www.elsevier.com/inca/publications/store/6/2/2/8/3/2/index.htt.

Zou, R., Xu, F., Wang, Y., Duan, M., Guo, M., Zhang, Q., et al., 2020. Changes in the gut microbiota of children with autism spectrum disorder. Autism Research 13 (9), 1614–1625. Available at https://doi.org/10.1002/aur.2358.

Chapter 2

Application of postbiotics in medicine and food products

Delara Moradi Mirhesari[1], Mahshid Bahraminejad[1], Hossein Hashemi[1], Behzad Ebrahimi[2], Zahra Sarlak[3], and Reza Mohammadi[3]

[1]*Department of Food Science and Technology, Student Research Committee, Kermanshah University of Medical Sciences, Kermanshah, Iran,* [2]*Department of Food Science and Technology, Maragheh University of Medical Sciences, Maragheh, Iran,* [3]*Department of Food Science and Technology, School of Nutrition Sciences and Food Technology, Kermanshah University of Medical Sciences, Kermanshah, Iran*

2.1 Application of postbiotics in food products

2.1.1 Introduction

Today, probiotics, prebiotics, and synbiotics are increasingly important for gut health and offer several advantages. These health-boosting, wellness-supporting, and microbe-containing supplements are promising and have captured attention. However, while these additives show potential, certain limitations prevent them from reaching their peak effectiveness. Introducing live microorganisms into a biological system inevitably carries some risk that needs careful consideration, especially when dealing with a compromised gut microbiome. Individuals with weakened gut microbiomes may face the risk of fungal or bacterial overgrowth, a possibility that must be carefully evaluated. Moreover, the effectiveness of prebiotics and probiotics relies on the presence of beneficial microbes that can effectively utilize prebiotic substrates, such as fructans and galactans. These beneficial microbes need to thrive and outcompete potential harmful invaders in the gut. This approach might prove less successful for those with compromised gut barriers, individuals recovering postsurgery, and notably, very young infants experiencing an imbalance in their gut flora (dysbiosis) and lacking enough helpful bacteria like *Lacticaseibacillus* and *Bifidobacterium*. This understanding prompted the exploration and development of postbiotics (Tarique et al., 2022). Crucially, the updated definition of postbiotics removes the requirement that they originate exclusively from probiotics. This expands the selection of bacteria that can be utilized for their creation. This is especially

Evaluating Oral Probiotics and Prebiotics. DOI: https://doi.org/10.1016/B978-0-443-29835-6.00008-3

pertinent considering the potential negative effects associated with probiotics, including the risk of infections, the generation of harmful compounds, and the potential transfer of antibiotic resistance genes from probiotic bacteria to other microorganisms residing in the gut (Álvarez et al., 2023; Ansari et al., 2024). The advantages of postbiotics arise from the fact that they eliminate the need to maintain a viable population of live microorganisms, which leads to enhanced stability and safety. Postbiotics can be easily added to foods or animal feeds, thus extending their shelf life. The improved stability and other logistical advantages of using postbiotics in transport, storage, and handling are significantly better compared to other biotics. A key area of focus in postbiotic research is their inherent stability, both during manufacturing and storage. Because postbiotics cannot reproduce, they present a safer option compared to probiotics (Kim et al., 2024; Serter et al, 2024). Proper formulation of postbiotics plays an important role in maintaining their biological activity, stability, and effective delivery. Components such as cell walls, lipoteichoic acid, and metabolites such as short-chain fatty acids (SCFAs) play a key role in the body's immune and metabolic activity and must be protected during the manufacturing and storage process. Methods such as microencapsulation can protect these compounds from gastric acid and high temperatures and increase bioavailability. Ultimately, such formulation strategies improve the safety, therapeutic efficacy, and quality control of postbiotics (Rouhi et al., 2024; Teame et al., 2020; Zhao et al., 2024). Another definition describes postbiotics as bioactive substances generated when beneficial gut bacteria (probiotic bacteria) consume prebiotic nutrients, such as fiber, within the colon. Even though these bioactive compounds are considered by-products of probiotic metabolism, they offer several health benefits to the human body (Álvarez et al., 2023). In fact, numerous studies suggest that the positive effects of probiotics on health may not necessarily depend on the microorganisms' survival. It appears that many of the health benefits attributed to prebiotics, probiotics, and synbiotics are derived from various metabolic products, cellular and intracellular components, and intact or broken-down dead microbes. Postbiotics are structural and metabolic microbial products that include cell-free supernatants, destroyed bacterial components, cell wall fragments, teichoic acid, SCFAs, vitamins, enzymes, exopolysaccharides, various peptides, amino acids, and fermentation by-products. Therefore many of the health benefits associated with prebiotics and probiotics actually result from the production of postbiotics (Álvarez et al., 2023; Tarique et al., 2022). Fig. 2.1 illustrates the history and evolution of postbiotics from 1986 to 2021 (Serter et al, 2024).

Key environmental factors during fermentation, including initial pH, temperature, incubation time, and nutrient composition (e.g., yeast extract concentration), significantly influence postbiotic yield and quality. A typical production protocol involves culturing microorganisms in MRS media at 38°C

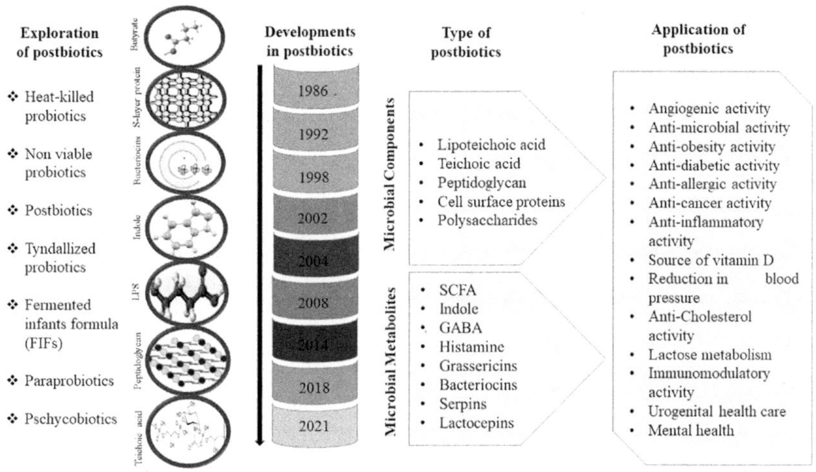

FIGURE 2.1 History and early definition of postbiotics.

for 48 hours, followed by centrifugation at 6000 rpm for 10 minutes and filtration through a 0.2 μm filter (Álvarez et al., 2023).

2.1.2 The origin and applications of postbiotics in the dairy industry

The growing interest in postbiotics has encouraged the dairy industry to explore new strategies for improving food safety, extending shelf life, and enhancing the functional value of conventional products. Dairy matrices such as whey, yogurt, pasteurized milk, powdered milk, kefir, and various cheeses provide ideal environments for the growth of probiotic bacteria and the subsequent formation of beneficial postbiotic compounds. These metabolites, including antimicrobial peptides, organic acids, exopolysaccharides, and phenolic derivatives, can improve the nutritional, microbial, and sensory quality of dairy products. As a result, postbiotics are increasingly being investigated not only as natural preservatives but also as health-promoting additives. The following sections outline key findings regarding the production and application of different postbiotics derived from probiotic strains within dairy-based systems. (Sharafi et al., 2024; Isaac-Bamgboye et al., 2024). Table 2.1 lists postbiotics used in various food matrices and their functional benefits.

2.1.2.1 Whey proteins

Whey, a byproduct of cheese production, is increasingly recognized as a valuable resource for postbiotic production (Vera-Santander et al., 2024). Bacteriocins (BACs), antimicrobial peptides produced by probiotic bacteria

TABLE 2.1 Postbiotics used in different food matrices and their functional benefits.

Postbiotic strain(s)	Food matrix	Reason for use/benefits
Lactobacillus acidophilus and *Bifidobacterium lactis*	Whey	Production of Antimicrobial Peptides (AMPs), preservation, inhibition of various microbes
L. acidophilus LA-5 and *Bifidobacterium animalis* BB-12	Whey/ Mozzarella cheese	Improves shelf life (up to 8 days), maintains texture and pH, protects against mesophile and psychrophile bacteria
B. animalis subspecies lactis BB-12	Low-fat yogurt	Enhances antioxidant activity without negative effect on viscosity, viscoelastic properties, shear thinning behavior, water retention, and yogurt structure
Escherichia coli Nissle 1917	Yogurt	Broad antibacterial and antifungal activity due to organic acids, peptides, SCFAs, phenolic compounds
Lactobacillus species	Pasteurized milk	Antioxidant activity, inhibition of *Staphylococcus aureus*, preservation
Lactobacillus plantarum RM1 and *Lactobacillus paracasei* KC39	Milk powder	Reduction of aflatoxin B1 and inhibiting the growth of *Fusarium moniliforme* and *Aspergillus flavus*
Lactobacillus kefiri LK1 and *Enterococcus faecium* EFM2	Kefir/Raw milk	Prevent mastitis pathogens, antibiofilm, strong lactic acid production
E.faecium BGZLM1–5	Raw milk	Bacteriocin production effective against *Listeria monocytogenes*
Leuconostoc spp.	Cheese	Production of GABA, ornithine, tryptophan, contributes to flavor, stress reduction, and shelf life
L. plantarum TW57-4	Meat (general)	Antibiofilm effect against *Listeria monocytogenes*
Lacticaseibacillus paracasei DTA-83	Pork sausage	Natural nitrite substitute, improves color, reduces nitrite use, and antibiofilm
Lactiplantibacillus plantarum and *Lacticaseibacillus sakei*	Chicken breast meat	Extends shelf life by 14 days, reduces Salmonella and Listeria growth

(*Continued*)

TABLE 2.1 (Continued)

Postbiotic strain(s)	Food matrix	Reason for use/benefits
Bifidobacterium bifidum DSM 20456 and *B. bifidum* BB-12	Turkey meat (coated with alginate)	Antioxidant effect, reduction of spoilage microorganisms, maintains quality
L. plantarum	Fermented sausage	Partial replacement of sodium nitrite, color improvement, and reduction of biogenic amines
Molds (*Monascus, Aspergillus* spp., *Chrysoniliacrassa, Rhizopus*)	Fermented products (rice, tea, Tempeh)	Antioxidant, antiinflammatory, hypocholesterolemic, antidiabetic effects
L. plantarum K8	Kimchi	Antiobesity, antioxidant, immunostimulatory, and beneficial even after cooking

like *Lactobacillus* and *Bifidobacterium*, have gained attention as natural food preservatives due to their broad-spectrum activity against pathogens and spoilage organisms (Bisht et al., 2024; Kim et al., 2024; Serter et al, 2024). Maximum production of AMPs in whey from *L. acidophilus* and *B*ifidobacterium *lactis* occurred under optimized fermentation conditions (temperature, time, and yeast extract concentration). The produced AMPs exhibited inhibitory effects against *L. acidophilus*, *B. lactis*, and Listeria monocytogenes, with high stability to temperature (40°C–100°C) and across a wide pH range. These broad-spectrum bacteriocin-like AMPs are effective against both Gram-positive and Gram-negative bacteria, as well as yeasts (da Silva et al., 2023; Kim et al., 2024; Serter et al, 2024). Postbiotics derived from *L. acidophilus* LA-5 and *B. animalis* BB-12, when added to high-moisture mozzarella cheese, extended shelf life up to 8 days by reducing mesophilic and psychrophilic bacteria while maintaining sensory and textural properties. The presence of exopolysaccharides in postbiotic solutions contributed to the desired texture, and pH remained stable after 4 days (Brasili et al., 2013; Jungersen et al., 2014). In recent years, adding *Bifidobacterium animalis* subsp. *Lactis* BB-12 postbiotics (B12-CW and B12-SM) in low-fat yogurt improved antioxidant activity without affecting starter bacteria counts. While B12-CW slightly impacted taste and aroma, other postbiotic combinations enhanced antioxidant properties while maintaining viscosity, water retention, and overall acceptability, offering consumers improved nutritional value (Jungersen et al., 2014; Shakirova et al., 2013).

2.1.2.2 Yogurt

Yogurt is one of the most widely consumed fermented dairy products, traditionally produced through the metabolic activity of lactic acid bacteria, which impart desirable texture, flavor, and functional properties. Beyond conventional probiotics, yogurt also provides an effective medium for generating postbiotics, as different beneficial microorganisms can grow, metabolize nutrients, and release bioactive compounds during fermentation. Within this context, *Escherichia coli* Nissle 1917 has attracted attention for its ability to produce postbiotics with notable antimicrobial and functional characteristics (Sadighbathi et al., 2023; Isaac-Bamgboye et al., 2024). Within the Enterobacteriaceae family, *E. coli* Nissle 1917 stands out for its nonpathogenic nature. As a probiotic bacterium, it offers numerous benefits for digestive health, particularly in restoring balance to the intestinal microbiota. Crucially, this strain of *E. coli* does not produce any pathological factors; thus it does not harm intestinal epithelial cells (Pradhan and Weiss, 2020; Zhao et al., 2022). To create yogurt containing postbiotics, skim milk is inoculated with *E. coli* Nissle 1917, and the mixture is then incubated at 37°C for 24 hours. The yogurt aqueous extract is prepared by mixing samples with water, adjusting the pH, heat-inactivating any replicating *E. coli* Nissle 1917, and centrifuging the mixture. Subsequently, the precipitate is acidified, centrifuged again, and the residue is adjusted to pH 7 and filtered. Finally, the postbiotics are lyophilized at −50°C for 48 hours (Pradhan and Weiss, 2020; Sonnenborn and Robertson, 2016). The postbiotics derived from *E. coli* Nissle 1917 exhibit broad-spectrum antibacterial activity, capable of inhibiting both Gram-negative and Gram-positive pathogenic bacteria. They also possess moderate "in vitro" antifungal activity. The antimicrobial effect of these postbiotics is thought to be mediated by antimicrobial agents, including organic acids, biopeptides, SCFAs, and phenolic compounds (Falzone et al., 2024; Rebai et al., 2023).

Based on their molecular weight, the polyphenolic compounds found in the resulting postbiotics are grouped as follows:

- 11 phenolic compounds with a molecular weight lower than 200 g/mol
- 7 phenolic compounds with a molecular weight ranging from 200 to 400 g/mol
- 1 phenolic compound with a high molecular weight (above 400 g/mol)

There is a positive correlation between the amount of low molecular weight phenolic compounds and antimicrobial activity. These low-mass phenolic compounds might be responsible for the higher antimicrobial effect of postbiotics after storage (7 and 15 days) compared to their initial activity. Another notable point is the increase in the concentration of salicylic acid in the postbiotic with storage time, even though no trace of this compound was present initially. Salicylic acid possesses antimicrobial properties. Two other phenolic compounds, pyrogallol and caffeic acid, also exhibit antimicrobial properties in this postbiotic, and their mechanism of action involves the inhibition of enzymes by oxidized compounds (Kumar et al, 2024; Rafique et al., 2023).

2.1.2.3 Pasteurized milk

Pasteurized milk serves as a widely consumed dairy product and an important substrate for studying the behavior of probiotic bacteria and their derived postbiotics. Because milk contains lactose, proteins, minerals, and key micronutrients, it offers a favorable environment for microbial metabolism and the production of bioactive compounds with antimicrobial and antioxidant properties. In this context, postbiotics produced by Lactobacillus species have gained considerable attention for their potential to enhance milk safety and quality (Khakpour et al, 2024; Tarique et al., 2022). The antibacterial effect of postbiotics is mainly attributed to compounds involved in the production of lactic acid, laurostearic acid, and isopropylidene-3,3-dimethyl. Additionally, postbiotics exhibit significant antioxidant effects, which are linked to the development of cell surface proteins and molecules such as exopolysaccharides. *L. plantarum* has strong resistance to hydrogen peroxide and exhibits stronger antioxidant ability compared to others. It also demonstrates a significant ability to neutralize DPPH and superoxide free radicals. Given that lactic acid is a key component of postbiotics and contains hydroxyl groups (which can release electrons), its ability to inhibit free radicals is enhanced. Among the postbiotics derived from three *Lactobacillus* strains (*acidophilus, paracasei*, and *plantarum*), the type derived from *L. plantarum* possesses the best antioxidant potential (Chang et al., 2021; Khakpour et al, 2024). The minimum effective concentration of antibacterials refers to the lowest amount of antibacterial agents required to inhibit bacterial growth in food. This concentration varies significantly depending on the target ecosystem or desired outcome. In MRS culture medium, the postbiotic derived from *L. plantarum* demonstrates the highest effectiveness. However, when applied to pasteurized milk, the minimum effective concentration for all postbiotics tends to be quite high. This is likely due to the interaction of postbiotic constituents with the food matrix, leading to subsequent absorption into the food. Postbiotics derived from *Lactobacillus* species can be safely used as additives in the food industry. For instance, postbiotics from *L. plantarum* can serve as an effective preservative to inhibit the growth of *Staphylococcus aureus* in milk. Therefore postbiotics show promise as preservatives in various food products (Ooi et al., 2021; Tong et al., 2025).

2.1.2.4 Powdered milk

Powdered milk is widely used not only as a direct nutritional source but also as a key ingredient in the manufacturing of numerous dairy-based and processed foods. The high vulnerability of powdered milk to aflatoxin contamination highlights the importance of identifying natural strategies, such as postbiotics, to mitigate these risks and enhance product safety (Fahim et al., 2021; Oglio et al., 2023). Milk and dairy products are crucial components of the diets of vulnerable age groups, including infants, children, and the elderly. However, these foods are prone to contamination with aflatoxins (AFs), which are considered among the most potent mycotoxins in terms of poisoning and carcinogenicity. In cows, liver enzymes convert AFB_1 into AFM_1, which is then excreted in the milk. AFM_1 contamination is not only found in milk but also in powdered milk. In addition to direct consumption, powdered milk serves as a raw material in the production of numerous dairy products such as ice cream, cheese, evaporated milk, and condensed milk, as well as in bakery products, meat products, and soups. Therefore AFM_1 contamination in the formula means that this contamination is carried over into all finished products. The issue of AFM_1 contamination is more serious than it appears because common methods, such as heating or the use of weak acids, are not very effective in reducing AFM_1 levels (Maggira et al., 2021; Vaz et al., 2020). Postbiotics derived from *L. plantarum* RM1 and *L. paracasei* KC39 bacteria produce growth inhibition zones against *Fusarium moniliforme*, with diameters of 16.7 ± 1.3 mm and 23.7 ± 1.5 mm, respectively. Additionally, these postbiotics not only inhibit the growth of *Aspergillus flavus* but also reduce the amount of AFB_1. The fungal growth inhibitory (antifungal) activity of compounds derived from the RM1 bacterial strain is less potent compared to that obtained from the KC39 strain (Fahim et al., 2021). The postbiotics extracted from the KC39 strain are the most effective agents, demonstrating an 89.8% reduction in AFB_1 levels in experimental models. It is thought that bioactive compounds in the bacterial culture medium degrade aflatoxins into less harmful substances. The reduction of AFB_1 in the liquid culture medium containing RM1 postbiotics is most likely due to the presence of a novel protein with a specific molecular weight within these postbiotics. The antifungal activity of KC39 postbiotics is attributed to organic acids (lactic, phenyllactic, hydroxyphenyllactic, and indole lactic) and a novel bacteriocin (bacteriocin KC39) (Chaisakul et al., 2020; Fahim et al., 2021).

2.1.2.5 Kefir and raw milk

Bovine mastitis is an inflammatory disease affecting the udder of lactating cows, causing swelling and potentially stemming from various bacteria, including *S. aureus*, *Enterococcus faecalis*, *Pseudomonas aeruginosa*, and *E. coli*. This condition poses a significant concern for the dairy industry due to its potential to compromise both milk quality and the health of the cows. Postbiotics generated by *Lactobacillus kefiri* LK1 from kefir exhibit greater potency against bacteria that cause bovine mastitis compared to postbiotics produced by *Enterococcus faecium* EFM2 from raw milk. This difference is likely attributable to the more robust lactic acid fermentation process of *Lactobacillus kefiri* LK1. At their minimum effective concentrations, both types of postbiotics can inhibit infection and biofilm formation in the udder by modulating certain characteristics of the pathogenic bacteria (Kim et al., 2024). LK1 generates more lactic acid than EFM2, which is attributed to its ability to breakdown a specific sugar called melibiose, facilitated by the presence of the enzyme alpha-galactosidase. Additionally, LK1 can utilize arabinose as an energy source, whereas EFM2 uses maltose. The breakdown of arabinose by LK1 leads to increased lactic acid production and a more rapid acidification of the surrounding environment. Among the various antimicrobial compounds, lactic acids create an unfavorable environment for pathogen growth by lowering the pH and increasing the binding affinity of antimicrobial proteinase molecules. Lactic acids release protons through the bacterial cell membrane; the loss of these protons inhibits energy production, leading to antimicrobial and antibiofilm activities (Amobonye et al., 2025; Wilm et al., 2024). Lactic acid disrupts the outer membrane of pathogens, facilitating the penetration of other antimicrobial agents and altering the integrity and hydrophobicity of the cell surface. Considering these biochemical activities, LK1, which is positive for arbutin, effectively inhibits the accumulation and production of exopolysaccharides by *Enterococcus faecalis*. Arbutin is degraded into glucose and hydroquinone. Hydroquinone reduces the formation of biofilms of pathogens by disrupting the cell membrane and regulating virulence based on phenotypic and genotypic expression. In conclusion, the use of LK1 postbiotics represents a promising new strategy for alternative or complementary therapy in the management of bovine mastitis, offering a safe and effective solution for the prevention of this disease in the dairy industry (Amobonye et al., 2025; Kim et al., 2024). The bacteriocin postbiotic produced by *E. faecium* BGZLM1–5, originating from raw milk, can mitigate cross-contamination by the foodborne pathogen *Listeria monocytogenes*. There is a potential application of utilizing the bacteriocin-producing bacteria themselves (rendered inactive, of course) in milk, which could lead to a substantial reduction in *Listeria monocytogenes* bacteria (Popović et al., 2023).

2.1.2.6 Cheese

Cheese is one of the most diverse and widely consumed fermented dairy products, produced through complex microbial interactions that shape its texture, flavor, and functional properties. Among the microorganisms involved in cheese fermentation, certain lactic acid bacteria play a key role not only in product quality but also in generating valuable postbiotics with potential health benefits. Within this context, the Leuconostoc genus has received significant attention due to its ability to synthesize a range of bioactive compounds during cheese fermentation (Coelho et al, 2022; Ramos et al., 2022).

The *Leuconostoc* genus comprises heterofermentative lactic acid bacteria cultures crucial for the fermentation of various foods, including cheese and vegetables. Their capacity to produce technically important gases through fermentation, which contributes to cheese formation, the generation of flavor compounds, and the development of dextran, has garnered significant research interest. These bacteria can produce postbiotics such as gamma-aminobutyric acid (GABA), as well as ornithine (Orn), tryptophan (Trp), and certain organic acids. GABA is formed through the irreversible decarboxylation of L-glutamic acid. In recent years, the discovery of novel physiological and pharmacological attributes of GABA has sparked considerable investigation. GABA acts as a neurotransmitter that promotes relaxation, sleep, and anxiety reduction. Additionally, it aids in the treatment of epilepsy, diabetes, and cancer. Consequently, the availability of GABA-producing bacteria has increased in recent years. Tryptophan and ornithine are also noteworthy amino acids due to their health benefits for consumers, which include anxiety and stress relief, fatigue reduction, bioactive metabolite production, and the modulation of various hormones, including growth hormone stimulation (Ahn and Park, 2023; Dhakal et al, 2012; Cui et al., 2020). Organic acids generated during lactic acid bacteria fermentation reside in the metabolites and perform several functions in fermented foods. These acids work concurrently and paradoxically to lower pH, thereby limiting or inhibiting undesirable microorganisms for preservation purposes, and also to enhance the food's aroma and flavor to improve consumer appeal. Postingestion, they are also known for their bacteriogenic properties, targeting potentially hazardous gut bacteria for suppression or elimination (Tang et al, 2023; Punia Bangar et al., 2022).

2.2 Application of postbiotics in medicine

2.2.1 Identification of active postbiotic components

The therapeutic efficacy of postbiotics stems from diverse bioactive components derived from microbial cells. These components are broadly classified into cellular structures, metabolites, and enzymes, as summarized in Table 2.2 and illustrated in Fig. 2.2 (Álvarez et al., 2023).

2.2.1.1 Cellular components

Cellular components of postbiotics primarily include structural molecules derived from microbial cell walls and membranes. These molecules are recognized by host immune receptors and can modulate immune responses, gut barrier function, and inflammation (Prajapati et al., 2023). Their fragments, generated during bacterial lysis or enzymatic processing, are known to activate pattern recognition receptors such as NOD-like receptors (NOD1 and NOD2), leading to immune modulation. The stability and bioavailability of these fragments are critical factors influencing their efficacy in postbiotic

TABLE 2.2 Classification of active postbiotic components, cellular elements, metabolites, and enzymes with their primary mechanisms and delivery forms.

Category	Main component(s)	Microbial source and structure	Mechanisms and functions	Pharmaceutical/ production forms	Key references
Cellular components	Peptidoglycan fragments (e.g., muramyl dipeptide)	Bacterial cell wall	Activation of NOD1/NOD2 receptors, enhanced macrophage activity, cytokine production	Mainly injectable/oral (adjuvant, capsules); chemical synthesis	Mashayekh et al. (2021), Vacariu and Tanner (2022)
	Lipoteichoic acids (LTAs)	Plasma membrane of Gram-positive bacteria (e.g., *Staphylococcus aureus*)	TLR2 stimulation with context- and dose-dependent anti or proinflammatory effects; regulation of NF-κB, TNF-α, IL-6, IL-10	Mainly injectable (adjuvants, inhibitors) and nasal (immunomodulator); microbial extraction	Fang et al. (2025), Percy and Gründling (2014)
	Exopolysaccharides (EPS) and capsular polysaccharides	Extracellular/ surface polysaccharides of bacteria (e.g., *Lactobacillus rhamnosus*)	Strain- and context-dependent prebiotic effects; adhesion to gut mucosa; biofilm formation; increased IL-10, decreased TNF-α	Oral (capsules, prebiotics); LAB fermentation	Fanning et al. (2012), Paik et al. (2019)
	S-layer proteins	Self-assembled crystalline surface proteins of Gram-positive bacteria	Adhesion to intestinal epithelium; stimulation of TLR2, DC-SIGN; combined pro- and antiinflammatory effects in immune and epithelial cells	Vaccine adjuvants and oral drug delivery systems	Assandri et al. (2023), Taverniti et al. (2013)

(*Continued*)

TABLE 2.2 (Continued)

Category	Main component(s)	Microbial source and structure	Mechanisms and functions	Pharmaceutical/ production forms	Key references
Metabolites and enzymes	SCFAs	Fermentation of dietary fibers by gut microbiota	Energy source for colonocytes; regulation of immune responses; lipid and glucose metabolism	Oral (capsules) and rectal (suppositories)	Deleu et al. (2021), Mukhopadhya and Louis (2025)
	Organic acids (lactic, succinic, citric)	Fermentation by *Lactobacillus, Aspergillus,* Rhizopus	Antimicrobial activity; pH regulation; metabolic pathway involvement	Oral/injectable (tablets, ampoules)	Hijová (2024), Jahedi and Pashangeh (2025)
	Bacteriocins	Ribosomally synthesized peptides from Lactobacillus and Bifidobacterium spp.	Inhibition of pathogenic bacteria; maintenance of gut microbial balance	Oral/topical (tablets, ointments)	Aljohani et al. (2023), Egan et al. (2016)
	Bioactive peptides	Produced by certain *Lactobacillus strains*	Angiotensin-converting enzyme (ACE) inhibition; antihypertensive and immunomodulatory effects	Oral supplements	Raveschot et al. (2020), Vera-Santander et al. (2023)
	GABA	Lactobacillus brevis and other LAB	Inhibitory neurotransmitter; may improve sleep and reduce anxiety; neuroprotective support; blood pressure regulation; clinical evidence limited	Oral supplements; nanoencapsulation technologies	Lyu et al. (2017), Wu et al. (2017)

	B and K vitamins	Synthesized by Bacteroidota, Bacillota, Corynebacterium	Cofactors in energy metabolism, blood coagulation, antioxidant defense	Probiotic supplements, functional foods	Fang et al. (2017), Tarracchini et al. (2024)
	Polyphenolic metabolites	Microbial biotransformation of polyphenols by Lactobacillus and others	Antioxidant and antiinflammatory effects; antibacterial and antiviral; cardiovascular benefits	Nutraceuticals and dietary supplements	Alqudah and Claesen (2024), Kumar et al. (2025)
	Bile salt hydrolases (BSHs)	Enzymes mainly from *Lactobacillus* and *Bifidobacterium*	Deconjugation of bile acids may reduce serum cholesterol and LDL by 20%–50% in models	Oral supplements and probiotic products	Dong et al. (2025), Zhu et al. (2022)
	Lactase (β-galactosidase)	Various bacteria, including *Lactobacillus* and *Bifidobacterium*	Hydrolysis of lactose improves lactose intolerance symptoms	Oral supplements and fermented dairy products	Dong et al. (2025), Zhu et al. (2022)

GABA, Gamma-aminobutyric acid; *SCFAs*, short-chain fatty acids.

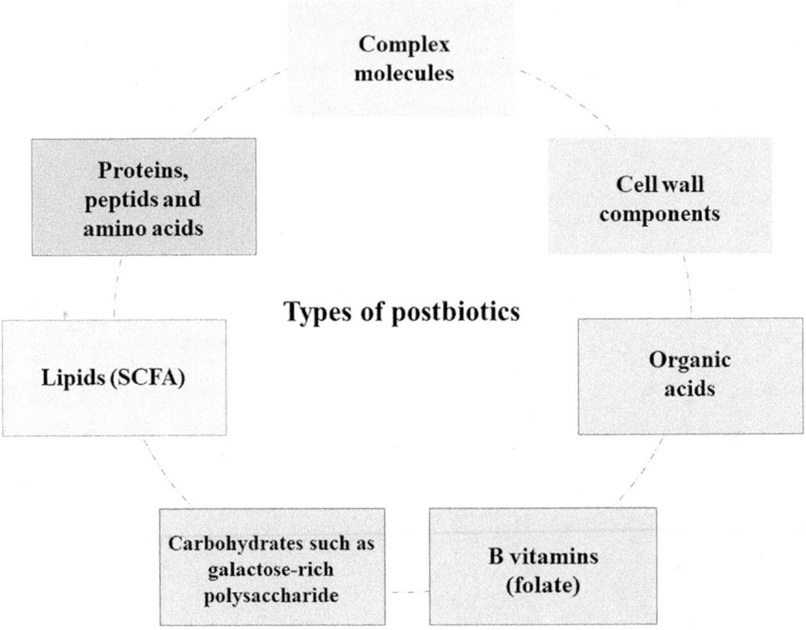

FIGURE 2.2 Types of postbiotics.

formulations (Vacariu and Tanner, 2022; Yilmaz, 2024). Lipoteichoic acids (LTAs) are amphiphilic polymers in the plasma membrane of Gram-positive bacteria that play key roles in immune interaction through TLR2 stimulation, inducing context-dependent antiinflammatory or proinflammatory responses (Fang et al., 2025; Percy and Gründling, 2014). Exopolysaccharides (EPS) and capsular polysaccharides are complex extracellular/surface polymers that mediate adhesion to gut mucosa, biofilm formation, and immune modulation (TLR2/4, DC-SIGN modulation), exerting strain-dependent prebiotic and immunomodulatory effects (Fanning et al., 2012; Paik et al, 2019).

2.2.1.2 *Microbial metabolites*

Metabolites produced by microbial fermentation are key active components in postbiotics, mediating various health benefits, including antiinflammatory, antimicrobial, and metabolic effects. SCFAs, primarily acetate, propionate, and butyrate, are produced during the fermentation of dietary fibers by gut microbiota. They serve as energy sources for colonocytes, regulate immune responses, and influence lipid and glucose metabolism (Deleu et al., 2021; Mukhopadhya and Louis, 2025). Organic acids such as lactic acid, succinic acid, and citric acid contribute to antimicrobial activity and pH regulation in the gut, produced from strains of Lactobacillus, Aspergillus, or Rhizopus during fermentation (Hijová, 2024; Jahedi and Pashangeh, 2025; Vandenberghe et al., 2017). Bacteriocins are ribosomally synthesized antimicrobial peptides produced by bacteria such as Lactobacillus and Bifidobacterium species. They inhibit pathogenic bacteria, contributing to gut microbial balance (Aljohani et al., 2023; Egan et al., 2016). Microbial-derived peptides with specific bioactivities, such as antihypertensive or immunomodulatory effects, are increasingly recognized as postbiotic components. For example, certain Lactobacillus strains produce peptides that inhibit angiotensin-converting enzyme (ACE), contributing to blood pressure regulation. Their stability and activity depend on preservation during processing (Raveschot et al., 2020; Vera-Santander et al., 2023). Gamma-aminobutyric acid (GABA), produced by Lactobacillus brevis via glutamic acid decarboxylase (GAD), acts as an inhibitory neurotransmitter with potential benefits for sleep, anxiety, and blood pressure, though clinical evidence remains limited (Lyu et al., 2017; Wu et al., 2017). B vitamins (B1, B2, B6, B12, folate) and vitamin K2 are synthesized by gut microbiota (Bacteroidota, Bacillota, Corynebacterium), serving as essential cofactors in energy metabolism, blood coagulation, and antioxidant defense (Tarracchini et al., 2024; Yu et al., 2025; Fang et al, 2017). Polyphenolic metabolites, generated through bacterial biotransformation by Lactobacillus and other gut bacteria, exhibit enhanced antioxidant and antiinflammatory properties compared to their parent compounds. These metabolites (e.g., hydroxyphenylacetic acids) also demonstrate antibacterial and cardiovascular benefits (Alqudah and Claesen, 2024; Bouarab-Chibane et al., 2019; Makarewicz et al., 2021). Bile salt hydrolases (BSHs) deconjugate bile acids, reducing their reabsorption and promoting hepatic cholesterol conversion to bile acids, potentially lowering serum cholesterol by 20–50% in animal models. Lactase (β-galactosidase) hydrolyzes lactose, alleviating intolerance symptoms (Dong et al., 2025; Sridevi et al, 2009; Zhu et al., 2022). S-layer proteins are self-assembled crystalline surface structures in Gram-positive bacteria (Lactobacillus, Bifidobacterium) that mediate adhesion to intestinal epithelium and immune modulation via TLR2, DC-SIGN, and CLRs, inducing both pro- and antiinflammatory responses depending on cell type (Assandri et al., 2023; Taverniti et al., 2013).

2.2.2 Criteria for selecting postbiotics

The selection of active postbiotic components is guided by several important criteria that ensure efficacy, safety, and stability. First, the components must show observable biological activity with respect to an intended health benefit. This must be supported by evidence from in vitro assays, animal models, or clinical trials (Kumar et al, 2024; Salminen et al., 2021). Bioactive molecules must endure processing onslaughts like heat for inactivation or drying (Salminen et al., 2021). Stability testing under different conditions informs the choice of formulations. There should also be compatibility of the components selected with the delivery system, whether capsule, hydrogel, or nanocarrier. Compatibility ensures that, during formulation, storage, or when traversing the GI tract, bioactives are either not degraded or inactivated (Vinderola et al., 2022). All the components selected should meet safety standards like Generally Recognized As Safe (GRAS) or Qualified Presumption of Safety (QPS). This is all the more critical for regulatory approval and consumer acceptance (Salminen et al., 2021). In conclusion, the identification of active postbiotic components is the bedrock of developing effective and safe postbiotic products. This demands a multidisciplinary approach that will merge microbiology, biochemistry, immunology, and pharmaceutical sciences to select, validate, and optimize bioactive constituents for therapeutic applications.

2.2.3 Selection of microbial strain and production strategy

2.2.3.1 Strain selection criteria

Selecting an appropriate microbial strain is a fundamental step in the development of postbiotic formulations, as it directly influences the quality, consistency, and efficacy of the final product. The criteria for strain selection are multifaceted, encompassing genetic, metabolic, safety, and functional considerations (Binda et al., 2020). A primary criterion is the genomic stability of the strain. Strains with stable genomes are less prone to genetic drift, which ensures consistent production of bioactive components. Whole-genome sequencing (WGS) has become a standard method for assessing genetic stability, identifying genes related to metabolite production, and detecting potential virulence factors or antibiotic resistance genes (Khakpour et al, 2024). The strain must possess the genetic and enzymatic machinery to produce the desired bioactive metabolites, such as SCFAs, bacteriocins, or bioactive peptides. Screening methods such as metabolomics and transcriptomics assist in identifying strains with high-yield production capabilities (Chang et al., 2021). Safety is paramount; strains must be nonpathogenic and free from transferable antibiotic resistance genes. Regulatory agencies such as EFSA and FDA require comprehensive safety assessments, including in vitro and in vivo toxicity studies. Strains designated as GRAS (Generally Recognized As Safe) or QPS (Qualified Presumption of Safety) are preferred (Binda et al., 2020). Strains with documented health benefits in clinical or preclinical studies are prioritized. Evidence of immunomodulatory, antiinflammatory, or gut barrier-enhancing effects supports their selection (Chang et al., 2022; Conway, 1996). The strain must thrive under industrial fermentation conditions, including tolerance to pH, temperature, oxygen levels, and osmotic stress. Strains with robust growth kinetics and resilience to processing stresses are preferred to ensure high-yield and cost-effective production (Binda et al., 2020; Khakpour et al, 2024). In summary, the ideal microbial strain for postbiotic production should possess a stable genome, produce targeted bioactives efficiently, demonstrate safety in humans, and be compatible with industrial processes. A comprehensive screening combining genomic, metabolic, safety, and functional assessments ensures the selection of optimal strains for subsequent development stages.

2.2.3.2 Culture optimization

Once the suitable strain is selected, optimizing the cultivation conditions is crucial to maximize biomass yield and the production of desired bioactive metabolites. Culture optimization involves fine-tuning parameters such as pH, nutrient composition, oxygen levels, and fermentation strategy. Maintaining optimal pH (usually between 5.5 and 6.5 for Lactobacilli) enhances growth rate and metabolite production. For example, a study demonstrated that controlling pH during fermentation increased bacteriocin yield by 30% (Tong et al., 2023). Adjusting Carbon/Nitrogen (C/N) ratios influences primary metabolism and secondary metabolite synthesis (Prajapati et al., 2023). While many probiotic strains are microaerophilic or anaerobic, oxygen availability affects growth and metabolite profiles. For example, *Lactobacillus plantarum* exhibits enhanced SCFA production under microaerophilic conditions (Tong et al., 2023). The choice of substrate impacts biomass and metabolite yields. Using agro-industrial by-products like wheat bran or fruit extracts can enhance cost-effectiveness and sustainability (Prajapati et al., 2023). In fed-batch fermentation, the gradual addition of nutrients prevents substrate inhibition and increases cell density and metabolite yield. Continuous cultivation provides steady-state conditions that are ideal for large-scale, continuous production (Tong et al., 2023). Real-time monitoring of parameters such as pH, dissolved oxygen, and metabolite concentrations is vital (Prajapati et al., 2023).

2.2.3.3 Yield optimization for desired bioactive compounds

Maximizing the yield of specific bioactive compounds requires targeted manipulation of microbial metabolic pathways, harvesting strategies, and process parameters (Prajapati et al., 2023). Engineering strains to overexpress key biosynthetic genes can significantly increase production of desired metabolites. For example, overexpression of butyrate-producing pathways in Clostridium species has enhanced SCFA yields (Prajapati et al., 2023; Tong et al., 2023). This can be achieved through process conditions or genetic modifications. The timing of biomass collection significantly influences bioactive content. Many metabolites peak at specific growth phases; for example, bacteriocin production often correlates with late exponential or early stationary phases. Monitoring metabolite levels during fermentation guides optimal harvest points (Zhang et al., 2025). Adjusting parameters such as temperature, pH, and oxygen to favor biosynthesis pathways enhances metabolite accumulation. For example, maintaining slightly acidic pH during fermentation can increase bacteriocin stability and yield (Prajapati et al., 2023;

Tong et al., 2023). Postfermentation concentration and purification steps, such as ultrafiltration or chromatography, can increase the concentration of bioactives, improving their efficacy in formulations. Yield optimization must align with formulation needs. For example, producing higher concentrations of stable bacteriocins allows for lower doses and more efficient encapsulation (Alavi and Ciftci, 2023; Xin et al., 2019). In general, achieving high yields of target bioactive compounds involves a combination of metabolic engineering, process timing, environmental control, and downstream processing. These strategies ensure that postbiotic products contain sufficient active constituents to exert their intended health benefits effectively.

2.2.4 Inactivation and biomass processing methods

Inactivation and biomass processing are critical steps that define the core characteristics of postbiotics. The goal is to produce a stable, safe, and bioactive product by effectively inactivating the microbial cells while preserving their structural components and metabolites. This section discusses various inactivation techniques, their impact on bioactivity, and subsequent biomass processing strategies.

2.2.4.1 Inactivation techniques

The choice of inactivation method influences the stability, safety, and biological activity of the final postbiotic product. The techniques are broadly categorized into thermal, physical, and nonthermal methods (Yan et al., 2024; Zhu et al., 2025). Thermal inactivation includes dry heat and moist heat. Dry heat involves exposure to high temperatures (above 160°C). It effectively inactivates microbes but can denature sensitive bioactive molecules. Moist heat (autoclaving) utilizes saturated steam at 121°C under pressure. It ensures complete inactivation but may degrade heat-labile metabolites like bacteriocins and bioactive peptides (Almahbashi and Gunes Altuntas, 2025; Yan et al., 2024). The high-pressure processing (HPP) uses hydrostatic pressures (100–600 MPa) at ambient or moderate temperatures. HPP effectively inactivates microbes while maintaining the structural integrity of cell wall components and metabolites. Advantages of this method include minimal thermal degradation, preservation of biological activity, and suitability for sensitive compounds (Yan et al., 2024; Zhu et al., 2025). The other two methods are ultraviolet (UV) and pulsed electric field (PEF). PEF applies high voltage for short periods of time, disrupting cell membranes (Ashrafudoulla et al., 2024; Yemmireddy et al., 2022). In general, the selection of inactivation techniques depends on the nature of the bioactive components, processing scale, and intended application. Combining methods, such as mild heat with PEF, can optimize microbial inactivation while preserving bioactivity

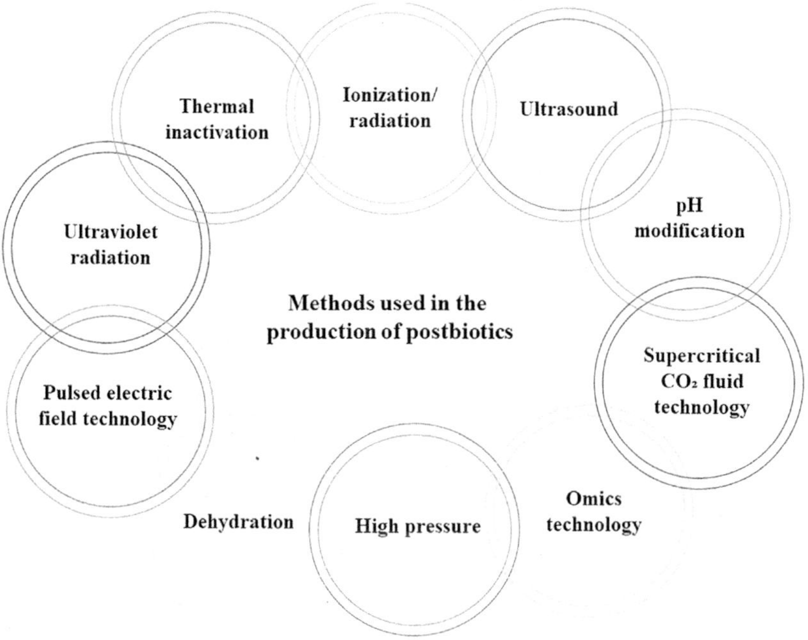

FIGURE 2.3 Methods used in the production of postbiotics.

(Almahbashi and Gunes Altuntas, 2025). In addition, the methods for producing postbiotics are shown in Fig. 2.3 (Álvarez et al., 2023).

2.2.4.2 Effect of inactivation methods on bioactivity

Different inactivation methods induce specific structural and molecular changes in biomass, directly impacting bioactivity. Thermal methods can cause cell lysis or denaturation of surface molecules, affecting immunostimulatory capacity. HPP preserves cell wall architecture better, maintaining the integrity of peptidoglycans and lipoteichoic acids essential for immune modulation. PEF causes transient pore formation, leading to inactivation while maintaining overall cell morphology. Heat treatments may degrade labile metabolites like bacteriocins and SCFAs, reducing antimicrobial activity (Almahbashi and Gunes Altuntas, 2025; Ma et al., 2023; Zhu et al., 2025). Nonthermal techniques tend to retain these molecules, enhancing the functional properties of the postbiotic (Pujato et al., 2024; Putri et al., 2024). The method chosen influences not only safety but also the therapeutic efficacy of postbiotics.

2.2.4.3 Primary biomass processing

Postinactivation biomass undergoes processing steps to produce a stable, uniform, and bioactive product suitable for formulation (Almahbashi and

Gunes Altuntas, 2025). Centrifugation is used to concentrate biomass and remove residual media components. High-speed centrifugation (e.g., 10,000–20,000 g) effectively separates biomass from supernatant. Membrane filtration (microfiltration or ultrafiltration) further purifies biomass and removes debris, ensuring clarity and consistency. Ultrafiltration can also concentrate bioactives like bacteriocins (Almahbashi and Gunes Altuntas, 2025; Hernández et al., 2025). The quality of inactivated biomass is assessed based on residual viability (should be nonviable), structural integrity, and retention of bioactive molecules (Almahbashi and Gunes Altuntas, 2025; Jahedi and Pashangeh, 2025). Lyophilization (freeze-drying), often employed to stabilize biomass for storage, maintains cell structures and metabolites. Spray drying, suitable for scaling, rapid drying minimizes thermal damage if parameters are optimized. Washing and buffer exchange are critical to remove residual nutrients or inhibitory substances that may affect stability or safety. The final quality features ultimately include consistent particle size and moisture content, absence of viable microbes and contaminants, preservation of structural molecules and metabolites at levels sufficient to confer desired bioactivities (Tong et al., 2023; Aggarwal et al., 2022). So, effective biomass processing involves centrifugation, filtration, washing, and drying techniques tailored to preserve bioactive integrity and ensure product safety. Rigorous quality assessment ensures the final biomass meets regulatory and functional standards for postbiotic applications (Almahbashi and Gunes Altuntas, 2025; Jahedi and Pashangeh, 2025).

2.2.5 Drying technologies

Drying technologies are fundamental in stabilizing postbiotic biomass, extending shelf life, and facilitating storage, transport, and formulation (Almahbashi and Gunes Altuntas, 2025; Jahedi and Pashangeh, 2025). The choice of drying method significantly influences the preservation of structural integrity, bioactivity, and safety of the final product. Several drying techniques are employed, each with distinct advantages and limitations, depending on the nature of the bioactive components and the intended application. Freeze-drying (lyophilization) involves freezing the biomass followed by sublimation of water under vacuum at low temperatures. This process effectively preserves cellular structures, metabolites, and bioactive molecules such as bacteriocins and cell wall polysaccharides, making it the gold standard for sensitive postbiotic materials. Its advantages include maintaining bioactivity, structural integrity, and stability over long periods; minimal thermal stress; and suitability for delicate compounds, and its limitations include high operational costs, longer processing times, and equipment complexity (Almahbashi and Gunes Altuntas, 2025; Ganchev, 2024). The vacuum drying method reduces water activity by applying vacuum conditions, often at moderate temperatures.

This method has less thermal stress than conventional drying and better preserves bioactive substances. However, the process is slow and requires higher equipment costs. Emerging technologies, including fluid bed drying (uniform drying with minimal loss of bioactive materials) and microwave-assisted drying, have been developed, which, in the case of microwave-assisted drying, offer rapid and energy-efficient drying, although it is still under investigation for postbiotics (Panigrahi et al., 2025; Vega-Gálvez et al., 2022). It should be noted that drying-induced stresses, such as moisture content, temperature, and oxidative conditions, can affect the stability of bioactives. For this reason, incorporating stabilizers and optimizing process parameters are essential to mitigate degradation (Jahedi and Pashangeh, 2025).

2.2.6 Storage and sustainability challenges

Postbiotic products face several stability challenges during storage that can compromise their efficacy, safety, and shelf life. Understanding these challenges is critical for designing robust formulations and storage conditions (Ma et al., 2023; Salminen et al., 2021). Residual moisture facilitates microbial growth (if viability remains), hydrolytic reactions, and microbial degradation of sensitive compounds like bacteriocins or peptides. Control measures to prevent it can include the use of desiccants, moisture-proof packaging, and low-moisture formulations (Salminen et al., 2021). Labile compounds such as organic acids, bioactive peptides, and cell wall polysaccharides are susceptible to oxidative degradation, leading to loss of activity and structural integrity. Ways to prevent this include adding antioxidants, oxygen-impermeable packaging, and neutral atmospheres during storage (Ma et al., 2023; Prajapati et al., 2023). Over time, bioactive molecules can undergo hydrolysis, racemization, or aggregation, diminishing their functional properties. Therefore stabilizers such as trehalose or glycerol, controlled temperature, and protection from light should be used. Elevated temperatures accelerate degradation processes, while UV exposure can induce structural modifications. Therefore they should be stored in cool, dark environments, and UV-protective packaging should be used (Jannah et al., 2022; Ma et al., 2023; Salminen et al., 2021). The stability of cell wall components and metabolites can be pH-dependent. Maintaining optimal pH during storage minimizes hydrolysis or structural alterations (Ma et al., 2023; Prajapati et al., 2023). In general, addressing these stability challenges involves a combination of formulation strategies (e.g., stabilizers and antioxidants), appropriate packaging (UV-protective, oxygen-barrier films), and controlled storage conditions (low temperature, humidity). Regular stability testing under accelerated and real-time conditions guides shelf life determination and ensures product efficacy throughout its lifecycle.

2.2.7 Encapsulation and delivery system design

Efficient delivery of postbiotics to the target site hinges on advanced encapsulation techniques and delivery system design. The primary objectives are to protect bioactive components from environmental and physiological stresses, control release kinetics, and enhance bioavailability. This section explores various encapsulation methods, nanotechnology approaches, and route-specific considerations.

2.2.7.1 Microencapsulation techniques

Microencapsulation involves enclosing bioactive compounds within a protective matrix at the micron scale, enabling controlled-release and improved stability. Alginate, a natural polysaccharide, forms gels with calcium ions, creating beads that encapsulate bioactives. It maintains bioactivity, is scalable, and has applications in oral delivery targeting intestinal release (e.g., bacteriocins and cell wall fragments) (Lai et al., 2024; Rahimi et al., 2024a; Wei et al., 2024). Chitosan, a cationic polymer with mucoadhesive properties, provides pH-responsive release in the intestine, protecting bioactives from stomach acidity and enabling colon-targeted delivery (Kulkarni et al., 2022b). Double emulsions (W/O/W) encapsulate aqueous bioactives within a lipid phase, providing protection against oxidation and precise control over release profiles. They are useful for unstable peptides or metabolites in oral or topical formulations (Kumar et al., 2022). Other techniques, including spray drying with protective excipients, coacervation, and extrusión, are also employed, each tailored to specific bioactives and formulations. Spray drying converts emulsions or suspensions into stable dry powders for oral tablets, capsules, inhalation formulations, or injectable suspensions using protective carriers (maltodextrin, whey protein, gum arabic). It protects heat-sensitive postbiotics (bacteriocins, SCFAs, peptidoglycans) during rapid dehydration at 120°C–160°C, achieving >80% encapsulation efficiency and up to 12 months shelf life. Careful control of outlet temperature (<80°C) and atomization pressure is essential (Barbosa et al., 2022; Rodrigues et al., 2020; Ziaee et al., 2019). Coacervation utilizes the phase separation of oppositely charged polymers (gelatin-chitosan or alginate-chitosan) under controlled pH (4.5–6.0) or temperature changes to create multilayer microcapsules (5–50 μm) without heat or organic solvents. Complex coacervation has shown superior gastroprotection (>90% survival at pH 2.0) and colon-specific release of immunomodulatory peptides and EPS. This solvent-free approach is particularly suitable for biologics requiring precise pharmacokinetics in inflammatory bowel disease or metabolic syndrome treatments (Muhoza et al., 2023; Timilsena et al., 2019). Extrusion produces hydrogel beads (0.5–3 mm) by extruding polymer-bioactive slurries into divalent cation baths ($CaCl_2$, $BaCl_2$). It offers high loading capacity (20%–40% w/w), sustained release (8–24 hours), and compatibility with tableting, ideal for temperature-sensitive compounds (GABA, LTAs, vitamins) in chronic disease applications (Timilsena et al., 2019; Voo et al., 2015). In general, when using these microencapsulation techniques to encapsulate postbiotics, considerations should be taken into account, including the compatibility of the encapsulating materials with the bioactive materials, the particle size affecting release and bioavailability, and the stability of the encapsulating matrix during storage.

2.2.7.2 Nanoformulation approaches

Nanotechnology offers innovative strategies to enhance the delivery, targeting, and efficacy of postbiotics (Mohammed and Çon, 2024). Nanoliposomes are spherical vesicles composed of phospholipid bilayers capable of encapsulating hydrophilic and lipophilic compounds. They have high biocompatibility, the ability to cross biological barriers, and targeted delivery. They are used to encapsulate bacteriocins or bioactive peptides for systemic or local delivery (Lopes et al., 2023; Mohammed and Çon, 2024). Solid lipid nanoparticles (SLNs), nanocarriers made from solid lipids stabilized by surfactants, protect labile compounds. SLNs have the advantages of controlled-release, improved stability, and scalable production and are useful in delivering bacteriocins and SCFAs to the gastrointestinal tract with increased bioavailability (Mohammed and Çon, 2024; Arratia-Quijada et al., 2024). Nanofibers (electrospinning) are made from nanofibrous mats containing bioactive materials for topical or mucosal applications. The advantages of this method include high surface area, sustained release, potential for targeted therapy, and dermal applications for delivering postbiotic fractions with antiinflammatory or antimicrobial effects (Chen et al., 2025a; Mohammed and Çon, 2024). In general, nanoformulation considerations include particle size (<200 nm for cellular uptake), surface modifications (targeting ligands for specificity), and stability during storage and administration. Nanocarriers improve bioavailability, protect sensitive molecules, enable site-specific delivery, and reduce required doses, minimizing side effects (Abbasi et al., 2022; Mohammed and Çon, 2024).

2.2.8 Selection of delivery route

The choice of delivery route for postbiotic formulations critically affects bioavailability, stability, safety, and therapeutic efficacy (Prajapati et al., 2023). It determines how bioactive components are protected against physiological stresses, interact with target tissues, and ultimately confer health benefits (Govender et al., 2014). An evidence-based selection process considers physicochemical properties of bioactives, target site, patient compliance, and regulatory considerations (Yang et al., 2022). This section discusses delivery routes, technological strategies, and the scientific rationale behind their choice, summarized in Table 2.3.

TABLE 2.3 Delivery routes for postbiotics, challenges, strategies, and applications.

Delivery route	Key challenges/considerations	Formulation strategies	Therapeutic applications
Oral delivery	– Acidic gastric environment – Enzymatic degradation – Limited intestinal absorption	– Acid-resistant coatings – pH – responsive carriers – Nanocarriers to protect bioactives	– Gut microbiota modulation – Immune regulation – Metabolic health improvement
Dermal/topical delivery	– Skin barrier (stratum corneum) – Need for local penetration	– Nanofibers- Hydrogels – Nanoemulsions enabling transdermal transport	– Antiinflammatory effects – Wound healing – Skin microbiome modulation
Systemic (injectable) delivery	– Requires sterility – Need for high purity – Biocompatibility concerns	– Controlled-release injectable systems – Sterile formulations for purified peptides or bacteriocins	– Precise dosing of purified fractions – Systemic immune or metabolic modulation

2.2.8.1 Oral delivery

Oral administration remains the most favored route for postbiotics, owing to its convenience, high patient compliance, and direct access to the gastrointestinal (GI) tract, which is often the primary target for postbiotic effects such as modulation of microbiota, immune regulation, and metabolic benefits (Yang et al., 2022). However, the GI environment presents significant barriers, including acidic pH in the stomach, digestive enzymes, bile salts, and variable transit times, which can degrade or inactivate bioactive compounds (Prajapati et al., 2023). Various encapsulation strategies have been developed to address these challenges (Homayouni Rad et al., 2023). For instance, alginate-based microcapsules coated with pH-sensitive polymers can resist gastric acidity and release their payload in the higher pH of the small intestine (Prajapati et al., 2023). Similarly, nanocarriers such as liposomes, solid lipid nanoparticles, and nanoemulsions have been employed to improve the stability and absorption of labile metabolites like SCFAs and bioactive peptides (Yang et al., 2023). Oral mucosal routes (buccal, sublingual) offer noninvasive alternatives that bypass the gastric environment and first-pass metabolism (Enck et al., 2020). Adhesive dosage forms (gels, films, pills, patches) provide localized delivery, extended retention time, and controlled-release of sensitive peptides and metabolites (Homayouni Rad et al., 2023). Recent formulations of oral mucoadhesive pills and films containing postbiotic metabolites have demonstrated sustained release profiles and increased local therapeutic effects, underscoring their potential in oral health applications and beyond (Kumar et al., 2024). In clinical and preclinical studies, postbiotic lysates are often administered in oral capsules due to their stability and patient acceptance. For example, a randomized trial in type 2 diabetes used *L. rhamnosus* DV lysate capsules (100 mg twice daily), improving glycemic parameters and beta-cell function (Mosca et al., 2022; Savytska et al., 2024; Sichel et al., 2013). Tablets are increasingly being recognized as suitable platforms for "pharmabiotics" and are considered a category that includes probiotics, paraprobiotics, and postbiotics. A review of oral pharmabiotic tablet formulations highlights uncoated tablets, effervescent tablets, chewable tablets, orally disintegrating tablets, and multilayer designs as promising technologies for delivering microbiome-targeted active ingredients, citing their low cost, high stability, ease of transport, and flexibility in modifying their release profiles. Concepts developed for probiotic tablets (e.g., use of protective excipients, moisture barrier coatings, and pH-dependent polymers) can be adapted for postbiotics to protect sensitive peptides or bacteriocins and target their release to the gut, even if viability is no longer required (Baral et al., 2021; Bogdanović et al., 2024). Beyond solids, postbiotic active ingredients can be formulated as liquid suspensions, sirups, or emulsions, especially for children or the elderly. Their nonliving nature simplifies cold chain requirements and allows for combination with flavorings, sweeteners, and stabilizers without compromising viability, although chemical stability and interactions with excipients still need to be evaluated. In functional food contexts, postbiotics have been incorporated into fermented beverages and dairy products. When these experiences translate into drug development, they inform the design of oral solutions or suspensions with controlled concentrations and shelf lives (Thorakkattu et al., 2022).

2.2.8.2 *Dermal and mucosal delivery*

Topical and mucosal routes allow direct delivery of postbiotics to the skin, oral, nasal, vaginal, or rectal mucosa, thereby bypassing the corrosive environment of the gastrointestinal tract and allowing for local or, if appropriately designed, systemic effects. This approach is particularly attractive when the goal is to treat skin diseases, enhance mucosal immunity, or modulate systemic inflammation with minimal systemic side effects, as the effective dose accumulates at the site and the burden of initial hepatic metabolism is reduced (De Almeida et al., 2023; Prajapati et al., 2025; Scott et al., 2022). To overcome protective barriers (stratum corneum, mucous membranes), advanced delivery systems (hydrogels, nanofibers, nanogels, liposomes, nanoliposomes, nanoemulsions) are often required. These carriers enhance penetration, targeting, and local persistence, enabling controlled and localized release critical for managing skin disorders, accelerating wound closure, and modulating inflammation (De Almeida et al., 2023; Machado et al., 2025; Prajapati et al., 2025). For instance, nanofibrous mats loaded with postbiotic fractions or antimicrobial peptides can form an adhesive scaffold over the wound, gradually releasing the active substances while at the same time providing antimicrobial protection and supporting tissue regeneration (Kulkarni et al., 2022a). In comparison, conventional gels or lotions are easily washed away with sweat or cleansing routines and usually have a shorter effective residence time on the skin (Prajapati et al., 2025). Likewise, liposomal gels or creams and nanoemulsions designed for dermal application can transport antiinflammatory and barrier-restoring agents across the superficial layers and promote their accumulation in the epidermis, rather than just on top of the stratum corneum (De Almeida et al., 2023). This approach has been explored, with varying degrees of success, in conditions such as atopic dermatitis, acne lesions, and photo-damaged skin, where both barrier repair and local immune modulation are clinically relevant (Machado et al., 2025). Carrier type, excipients, and particle size are crucial for topical success. Particles <200 nm penetrate skin layers better, while mucoadhesive polymers (chitosan, carbopol) increase mucosal residence time and absorption (Subramanian, 2021; Jawadi et al., 2022). For example, chitosan-coated nanoliposomes adhere better to oral, nasal, or vaginal mucosa, releasing immunomodulatory postbiotics gradually—an approach proposed for enhancing mucosal immune responses (Nair et al., 2023). Formulations must balance adhesion, persistence, and release profile to maintain effective concentrations without irritation or systemic leakage (de Lima et al., 2022).

In the field of drug delivery to the skin, a pilot clinical trial was carried out to evaluate an 8% *L. plantarum* VHProbi E15 fermented lysate lotion containing emollient, humectant, emulsifier, and preservative. After 4 weeks of treatment, the lotion was shown to enhance skin hydration, reduce epidermal water loss, and help maintain skin pH, with reported excellent safety and tolerability, indicating the potential application of postbiotic fermented lysates into skin emulsions to modulate the skin microbiome and its barrier function (Cui et al., 2022). Industrial and research formulations have also used postbiotic compounds derived from various probiotic strains (e.g., Lactobacillus fermentum, Lactobacillus reuteri, and Bacillus subtilis) in creams, serums, and hair care products, demonstrating the flexibility of oil-in-water and water-in-oil emulsions as carriers. Studies suggest that the nonbiodegradability of postbiotics reduces the risk of infection, especially in damaged skin, and allows the use of conventional cosmetic preservatives, but accurate characterization of endotoxin or proinflammatory components remains essential when targeting medicinal claims (Duarte et al., 2022). Oral, nasal, rectal, and vaginal mucosa are important sites for postbiotic administration, allowing rapid absorption and bypassing first-pass metabolism (Scott et al., 2022). Adhesive films on oral mucosa target dysbiosis and local infections (Rebelo et al., 2023). Nasal sprays and thermoresponsive gels can have local or systemic effects, though safety and irritation must be assessed (Thwala et al., 2017). Rectal and vaginal mucoadhesive gels, suppositories, or films are attractive for distal IBD or vaginal infections, bypassing the upper GI tract and enabling controlled-release (Song et al., 2023). Overall, topical and mucosal routes combined with nanocarriers, hydrogels, and mucoadhesive systems provide a flexible platform for targeted local treatment and, potentially, systemic effects by bypassing GI limitations. The key challenge is balancing penetration, persistence, safety, and formulation complexity, which varies with postbiotic type, target tissue, and therapeutic objective (Subramanian, 2021).

2.2.8.3 Parenteral and systemic delivery

Researchers believe that the routes of intravenous, intramuscular, or subcutaneous administration are reserved for highly purified and potent postbiotic fractions such as specific bacteriocins, certain bioactive peptides, or cell wall components aimed at systemic targets (Prajapati et al., 2023). This route is helpful in achieving specific drug targets because it is the fastest route, allows for more precise dosing, avoids first-pass hepatic and gastrointestinal metabolism, and helps achieve maximum bioavailability. Formulation strategies for systemic drug delivery often include encapsulation within liposomes, polymeric nanoparticles, or biodegradable carriers to protect the bioactive

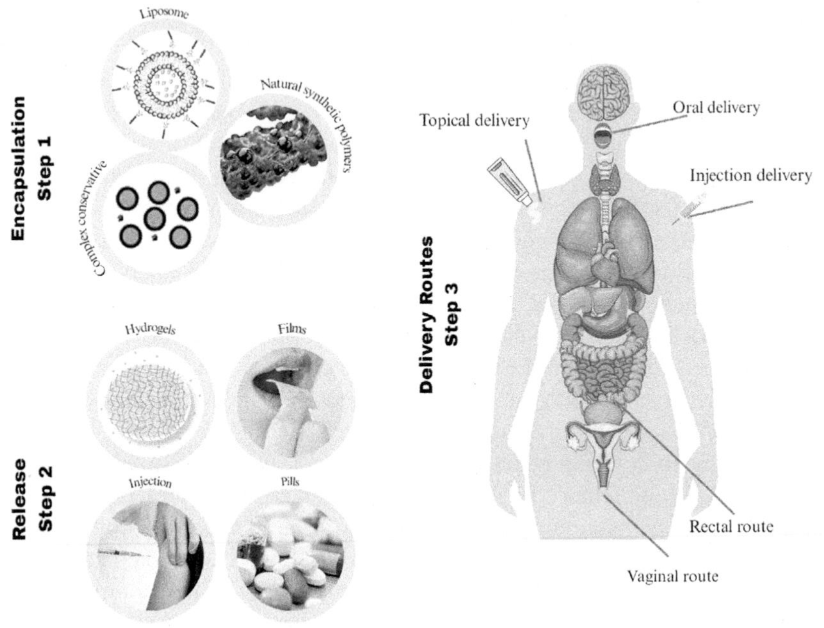

FIGURE 2.4 **Formulation and delivery of postbiotic-based drugs.** An example of the formulation and drug delivery pathway for postbiotic-based drugs.

agents from enzymatic degradation and facilitate controlled-release. For example, bacteriocins formulated within liposomes have been shown to increase stability and exhibit targeted antimicrobial activity when administered intravenously in animal models. (Prajapati et al., 2023). Systemic delivery, however, is generally limited to those clinical applications where high purity and potency are required, such as targeted antimicrobial therapy or immune modulation, due to the high invasiveness and complexity of the process (Kumar et al., 2024). An example of the formulation and drug delivery pathway for postbiotic-based drugs is shown in Fig. 2.4.

2.2.9 Factors influencing route selection

The choice of route of administration for postbiotics should be made as a planned and integrated decision that simultaneously considers both the characteristics of the molecule itself, clinical conditions, and safety and regulatory requirements (Ragavan and Hemalatha, 2024; Smolinska et al., 2025). The first step in design is to precisely define the target tissue; for example, in intestinal diseases, the oral route with colon-targeted formulations (pH-dependent or sensitive to microbial enzymes) is more logical than systemic administration, while for skin disorders, topical formulations or a "gut–skin axis" approach via oral administration can also be efficient (De Almeida et al., 2023; Smolinska et al., 2025). For mucosal applications (oral, nasal, vaginal), formulations should maximize contact time and local concentration without increasing systemic absorption (Bogdanović et al., 2024). pH-sensitive, oxidation-prone postbiotics (e.g., unstable peptides) require encapsulation in polysaccharides, lipid matrices, or enteric nanostructures to prevent premature degradation (Fang et al., 2025; Smolinska et al., 2025). In contrast, stable small molecules like SCFAs can use controlled-release systems or acid-resistant carriers for sustained colonic delivery (Pérez-Reytor et al., 2021; Ragavan and Hemalatha, 2024). The pattern of absorption, distribution, metabolism, and excretion of each postbiotic determines whether achieving effective concentrations in the systemic circulation is essential or whether focusing on local and paracrine effects is sufficient (Pérez-Reytor et al., 2021). For metabolites that act primarily through local surface receptors (such as SCFA receptors on intestinal epithelium and immune cells), formulation design should focus on increasing local contact and residence time at the epithelial surface, while for more distant targets, optimizing passage across biological barriers and tailoring the systemic release profile becomes important (Hays et al., 2024; Ragavan and Hemalatha, 2024). In chronic applications, noninvasive routes (oral, topical, mucosal) are preferred for ease, safety, and patient acceptance (De Almeida et al., 2023). In acute cases requiring rapid action or tight systemic control, invasive routes may be considered with careful clinical evaluation and pharmacokinetic justification (Chambers et al., 2018; Shete et al., 2022). Thus route selection is a strategic decision considering physicochemical properties, target access, patient acceptance, and safety. Advances in nanotechnology and encapsulation continue to expand optimization possibilities (Fang et al., 2025). Since postbiotics are often presented as "nonliving" compounds with a better safety profile than probiotics, oral and topical routes within the context of supplements, functional foods, and cosmeceuticals are subject to relatively less stringent regulations but still require toxicological data, DOA/NOAEL assessment, and clinical trial evidence (Homayouni Rad et al., 2023; Smolinska et al., 2025). In contrast, nanoformulations or injectable systems carrying postbiotics face stricter requirements for nanomaterial safety, formulation stability, long-term safety, and immunological risks (Chen et al., 2025b; Chen et al., 2025a; Fang et al., 2025) (Y; Z).

2.2.10 Health effects of postbiotics

Postbiotics have demonstrated potential in the prevention and treatment of chronic diseases such as inflammatory bowel disease (IBD), metabolic syndrome, chronic diarrhea, and allergies. Their more recently proposed action mechanisms of immune modulation, inflammation inhibition, gut–barrier strengthening, and metabolic control have now been substantiated. Postbiotics, made up of nonviable microbial metabolites, cell wall components, and soluble factors, are establishing themselves as potential candidates across many health care sectors. Animal studies have demonstrated mechanisms and safety; however, human effectiveness is still under investigation (Asefa et al., 2025; Scott et al., 2022).

2.2.10.1 Mechanisms and scientific evidence

Postbiotics, with their various mechanisms, can exert significant immune-modulating effects. In vitro and in animal models, postbiotics derived from Lactobacillus strains have been shown to inhibit proinflammatory cytokines IL-6 and TNF-α, whilst elevating antiinflammatory cytokines such as IL-10. Bifidobacterium longum heat-inactivated postbiotics also demonstrated anti-oxidant effects and inhibited pathogen adhesion in experimental settings (Asefa et al., 2025; Ma et al., 2023). In-clinic, in a 2024 trial, it was shown that postbiotic supplementation has diminished serum levels of IL-1β and high-sensitivity CRP, indicating antiinflammatory benefits in stroke recovery (Rahimi et al., 2024a). Postbiotics influence host metabolism through both direct and indirect mechanisms. For example, microbial metabolites, such as SCFAs including acetate, propionate, and butyrate, originate from postbiotic fermentation and play a crucial role in regulating lipid profiles, improving insulin sensitivity, and reducing the risks of obesity, type 2 diabetes, and cardiovascular diseases (Asefa et al., 2025; Calvanese et al., 2025). SCFAs promote fat oxidation and elevated energy expenditure, thus contributing to overall metabolic health (Calvanese et al., 2025; Yilmaz, 2024). Postbiotics enhance the integrity of this barrier in many ways by increasing the expression of tight junction proteins that anchor epithelial cell–cell junctions and decrease permeability to gut-derived toxins. Liu et al. showed in a liver-disease model the upregulation of barrier-protective proteins by postbiotics. They also induce mucin production via upregulation of the MUC2 gene expression, which lays down a protective mucus layer for shielding against pathogens (Asefa et al., 2025). Postbiotics are reported to have considerable antimicrobial activities by varied mechanisms, including production of bacteriocins against pathogens, including Listeria and *E. coli*, and organic acids (Asadi et al., 2024; Ma et al., 2023). Postbiotics exert much therapeutic benefit through increased antiinflammatory and antioxidant measures. For instance, they inhibit the NF-kB pathway by reducing nuclear translocation and limiting the expression of key inflammatory genes (Asefa et al., 2025). Consumption of postbiotics was also shown to decrease inflammatory and oxidative markers in people at increased risk for poststroke pneumonia, as per research (Rahimi et al., 2024b). Moreover, those postbiotics containing phenolic compounds and flavonoids show good scavenging capacity on the free radicals, acting to reduce oxidative damage and render protection against cardiovascular and neurodegenerative conditions, as per Asadi et al. (2024). Finally, this combined antiinflammatory and antioxidative activity reinforces the defense mechanisms both in the intestine and the systemic tissues against different categories of injury and stress (Asadi et al., 2024; Asefa et al., 2025; Rahimi et al., 2024b) (a; D).

2.2.10.2 Specific health effects

Compounds such as SCFAs, bacteriocins, and cell wall components exert their effects by modulating immune activity, strengthening tight junctions in the intestinal lining, and restoring microbial balance. Wang et al. (2025b) demonstrated that postbiotics not only suppress inflammatory processes but also maintain microbial stability and promote mucosal repair, positioning them as a valuable adjunctive option for IBD management. Furthermore, bacterial lysates from *Lactobacillus casei* have proven effective in preventing IBD by ameliorating dysbiosis and promoting intestinal tissue repair (Om et al., 2024; Wang et al., 2025a). Postbiotics offer significant advantages in addressing key features of metabolic syndrome, including obesity, dyslipidemia, and insulin resistance. Their antiobesity and antidiabetic mechanisms encompass altering gut microbiota composition, reducing lipid absorption, and improving insulin sensitivity (Asefa et al., 2025; Mishra et al., 2024). Furthermore, gut bacteria-derived elements, such as cell wall components and SCFAs, provide antiinflammatory and metabolic benefits, underscoring the potential of postbiotics in mitigating diabetes and cardiovascular risks (Asefa et al., 2025). Clinical trials have demonstrated the effectiveness of postbiotics in managing chronic diarrhea. A randomized, controlled study involving young adults observed significant improvements in key symptoms, including stool consistency, defecation frequency, and urgency (Guo et al., 2024). Growing research highlights the efficacy of postbiotics in preventing and managing allergies. These compounds bolster immune tolerance by modulating the Th1/Th2 balance, promoting the development of regulatory T cells, and optimizing immunoglobulin profiles, thereby enhancing allergy control. Clinical evidence shows reduced IgE, increased IgA, and symptom alleviation in allergic conditions (Capponi et al., 2022; Homayouni Rad et al., 2021).

2.3 Regulatory hurdles and considerations

The regulatory landscape for postbiotic-based products is complex and evolving, exhibiting global variations in classification and approval pathways depending on the intended use. These range from medicinal products requiring rigorous clinical trials and GMP compliance to food supplements with safety-focused regulations (Guglielmetti et al., 2025; Salminen et al., 2021). Extensive characterization of the heterogeneous postbiotic compositions, adherence to manufacturing practices that ensure batch stability and sterility, and comprehensive safety evaluations, including toxicity and immunogenicity testing, are crucial (Salminen et al., 2021). Clinical development requires well-designed trials with relevant endpoints under regulatory scrutiny, while labeling and claims must be strictly aligned with regulatory categories to avoid misleading information. Challenges include a lack of harmonized global guidelines, variability in regulatory interpretations across agencies such as the FDA, EMA, and TGA, and regulatory gaps for late-stage expansion and commercialization (Guglielmetti et al., 2025; Vinderola et al., 2023). Emerging efforts by international bodies aim to standardize definitions and criteria, exemplified by clinical trials of IBD treatments adhering to stringent drug standards. Continuous cooperation between industries, regulators, and academics is needed to refine regulatory pathways and accelerate safe market access. Overcoming these obstacles through early dialog and rigorous quality control will facilitate the translation of postbiotics into effective, safe treatments, enabling broader clinical and commercial adoption in the near future. Optimizing the delivery and formulation of postbiotic-based products is crucial to realizing their clinical potential. Postbiotic products can be delivered orally or topically, with advanced encapsulation technologies and custom release mechanisms representing promising delivery methods. Despite the enhanced safety and effectiveness afforded by customized release mechanisms, regulatory challenges persist. Ongoing collaboration is required between scientists, industry, and regulators to realize the full potential of postbiotics and translate research findings into reliable, effective medical interventions and clinical nutrition strategies (Guglielmetti et al., 2025; Vinderola et al., 2023).

2.4 Safety profile

Postbiotics, as nonviable microbial components, present a safer alternative to probiotics, which can sometimes trigger translocation, infection, or unexpected immune responses, particularly in immunocompromised or critically ill individuals. Emerging reviews and clinical trials indicate that postbiotics are suitable for high-risk groups, such as newborns, the elderly, and those with weakened immune systems, populations for whom probiotics are often avoided due to potential risks. For instance, studies on postbiotics derived from *Lactobacillus* and *Bifidobacterium* species in hospitalized children report no instances of bacteremia or infections, contrasting with sporadic risks associated with live probiotics (Calvanese et al., 2025; Mosca et al., 2022). Compared to prebiotics, postbiotics generally exhibit less severe adverse effects and can even reduce gastrointestinal issues such as inflammation or excessive gas (Żółkiewicz et al., 2020). Furthermore, postbiotics demonstrate greater resilience to heat, pH variations, and storage conditions, mitigating contamination risks and ensuring reliable dosing, unlike probiotics, which can be compromised during processing or in the stomach (Calvanese et al., 2025; Srivastava et al., 2024; Żółkiewicz et al., 2020). Their nonviable nature also streamlines standardization and simplifies regulatory approval by eliminating concerns about transmission or genetic alterations (Vinderola et al., 2023). Studies indicate clinical equivalence in health outcomes, such as improved intestinal function, reduced inflammation, and balanced gut microbiota, between postbiotics and probiotics in pregnancy, with postbiotics offering the added benefit of avoiding the risks associated with live organisms (Żółkiewicz et al., 2020; Zhang et al., 2020). While prebiotics are proven to support beneficial bacteria and SCFA production, they can trigger gastrointestinal distress such as gas and inflammation. Postbiotics circumvent these deficiencies by directly delivering beneficial compounds without relying on microbial fermentation, thereby addressing major needs and providing significant benefits, including antiinflammatory, antioxidant, and intestinal effects (Smolinska et al., 2025). Postbiotics confer diverse health benefits by modulating immune responses, enhancing gut barrier integrity, improving metabolic processes, and providing antimicrobial and antiinflammatory functions. They hold promise in the prevention and treatment of inflammatory bowel disease (IBD), metabolic syndrome, certain types of diarrhea, and even allergies and may offer a safer profile than probiotics. Initial preclinical and clinical evidence is encouraging, but more large-scale studies are needed. Future research aims to refine their applications and develop comprehensive health strategies incorporating postbiotics.

2.5 Future trends and research directions

The area of postbiotics is relatively new, resting at the crossroads of innovative biotechnology, individualized health care, and growing market opportunities. Future progress looks promising due to three important factors: need-based medicine and nutrition, innovations in delivery systems led by nanotechnology, and ever-changing market and regulatory environments (Prajapati et al., 2023; Aggarwal et al., 2022). The inclusion of postbiotics within precision medicine paradigms defines an evolving shift. Nutrition plans tailored to an individual's biology, like microbiome data, metabolic indicators, and personal health history, have proven more effective than traditional generalized approaches. For instance, studying lipid profiles, body composition, and gut microbiota composition through multilevel personalization highlighted the possibility of customizing postbiotic interventions. With further progress on the interplay between host and microbiome, developing actionable and identifiable mechanisms to target the microbiome will facilitate refined categorization of patients and more tailored therapeutics, thus situating postbiotics as central elements in personalized medicine (Bermingham et al., 2024; Yu et al., 2024). Nanotechnology is a promising strategy to overcome the limitations of bioavailability, stability, and targeted release of postbiotics. In functional and nutraceutical products, their application can be revolutionary due to the postbiotic nanoparticles (postbiotics-NPs) enhancing the absorption and safeguarding microbial metabolites, while also controlling their release. Research that has created chitosan nanogels encapsulating probiotic lysates demonstrates stable nanoscale formulations with in vivo wound-healing effects. However, there are still gaps in the field, including safety evaluations, clarity in regulatory guidance, a comprehensive nanoscale carrier model of action evaluations, and the mechanistic understanding of the nanocarrier's behavior. Regardless of these challenges, the possible use cases of postbiotics nanoencapsulation for scaling the clinical utility of postbiotics encapsulation are immense, including but not limited to the intestines and immune system, the brain, and even metabolic disorders (Huang et al., 2024; Mohammed and Çon, 2024). Recent evaluations of the postbiotics market estimate it at roughly USD 147 million in 2025 and expect it to exceed USD 225 million by 2030, which indicates an annual growth rate of nearly 9%. This growth is largely attributed to the increasing consumer insight into the microbiome and probiotics' challenges of viability and stability. The increasing use of postbiotics in supplements, functional foods, and personal care products is attributed to their scientifically proven benefits and, more importantly, their shelf stability as compared to probiotics. Globally, regulatory frameworks are still in their infancy, with varying definitions of products as medicines, novel foods, or supplements. This creates a barrier for standardization. There is progress toward streamlined frameworks and clearer pathways for approvals, which is important for turning scientific breakthroughs into marketable therapies (Meena et al., 2025). As postbiotics gain traction, their integration with advances in personalized medicine, nanotechnology, and the evolving market and regulatory landscape will undoubtedly shape future innovations. Further investigation into individualized applications, designing safe and effective nanodelivery systems, and consistent regulatory frameworks will enhance the adoption of postbiotics in clinical and commercial settings. This heralds a paradigm shift in the therapeutic arsenal of the microbiome, where medicine is precisely tailored to the individual and publicly beneficial health outcomes are achieved (Asefa et al., 2025; Puccetti et al., 2024).

2.6 Conclusion

This chapter focuses on the flourishing role of postbiotics in food science and human health. Being postbiotics, nonliving entities, they obviate bacterial survival issues and are capable of providing extensive health benefits. Technologically, the development of food products with postbiotics has turned out to be a promising option through advances in formulation and delivery techniques for keeping these substances effective in functional foods. Therefore applications can now be extended to the various categories, such as dairy, beverages, bakery items, and meat alternatives. There are also interesting health effects associated with postbiotic ingestion. Studies suggest that they assist in modulating immune function, reducing antioxidative stress, and maintaining metabolic balance. This evidence places them not just on the table for preventive nutrition but also as candidates for therapeutic use. However, many aspects remain unresolved. The fact that postbiotics represent a hugely heterogeneous group of compounds makes their mechanism of action incompletely understood. Also, the variability in production processes complicates establishing uniform rules for dosage, labeling, and safety. An answer to these challenges would require carefully designed clinical trials and clearer regulatory standards. In the future, the field would lean toward precision nutrition, where postbiotic constituents could be formulated to suit specific health concerns or, indeed, tailor individual ones to individual microbiomes. To realize tangible benefits, there will have to be a synergy of effort among researchers, clinicians, and industry to transmute laboratory discoveries into reliable consumer products. Postbiota are, in essence, an intersection of generic functional foods and newly emerged types of biomedicine applications. Their durability, safety, and generalized effects on health make them different from most options that are currently available. Nevertheless, ongoing research, proper regulation, and more public awareness would need to be established for their successful incorporation into future foods worldwide. If all these measures succeed, postbiotics might become a normal aspect in future diets instead of remaining a good idea.

References

Abbasi, A., Hajipour, N., Hasannezhad, P., Baghbanzadeh, A., Aghebati-Maleki, L., 2022. Potential in vivo delivery routes of postbiotics. Critical Reviews in Food Science and Nutrition 62 (12), 3345–3369. Available at https://doi.org/10.1080/10408398.2020.1865260.

Aggarwal, S., Sabharwal, V., Kaushik, P., Joshi, A., Aayushi, A., Suri, M., 2022. Postbiotics: from emerging concept to application. Frontiers in Sustainable Food Systems 6. Available at https://doi.org/10.3389/fsufs.2022.887642, www.frontiersin.org/journals/sustainable-food-systems#.

Ahn, J., Park, J.Y., 2023. Potential of γ-aminobutyric acid-producing leuconostoc mesenteroides strains isolated from kimchi as a starter for high-γ-aminobutyric acid kimchi fermentation. Preventive Nutrition and Food Science 28 (4), 492–501. Available at https://doi.org/10.3746/pnf.2023.28.4.492, https://www.pnfs.or.kr/journal/view.html?doi=10.3746/pnf.2023.28.4.492.

Álvarez, M., Andrade, M.J., Cebrián, E., Roncero, E., Delgado, J., 2023. Perspectives on the probiotic potential of indigenous moulds and yeasts in dry-fermented sausages. Microorganisms 11 (7), 1746. Available at https://doi.org/10.3390/microorganisms11071746.

Alavi, F., Ciftci, O.N., 2023. Purification and fractionation of bioactive peptides through membrane filtration: a critical and application review. Trends in Food Science and Technology 131, 118–128. Available at https://doi.org/10.1016/j.tifs.2022.11.024, http://www.elsevier.com/wps/find/journaldescription.cws_home/601278/description#description.

Aljohani, A.B., Al-Hejin, A.M., Shori, A.B., 2023. Bacteriocins as promising antimicrobial peptides, definition, classification, and their potential applications in cheeses. Food Science and Technology 43. Available at https://doi.org/10.1590/fst.118021.

Almahbashi, A., Gunes Altuntas, E., 2025. From preparation to bioactivity: a comparative study on preparation methods and characterization of postbiotics. Food Science and Nutrition 13 (5). Available at https://doi.org/10.1002/fsn3.70294.

Alqudah, S., Claesen, J., 2024. Mechanisms of gut bacterial metabolism of dietary polyphenols into bioactive compounds. Gut Microbes 16 (1). Available at https://doi.org/10.1080/19490976.2024.2426614.

Amobonye, A., Pillay, B., Hlope, F., Asong, S.T., Pillai, S., 2025. Postbiotics: an insightful review of the latest category in functional biotics. World Journal of Microbiology and Biotechnology 41 (8). Available at https://doi.org/10.1007/s11274-025-04483-8, https://www.springer.com/journal/11274.

Ansari, F., Lee, C.C., Rashidimehr, A., Eskandari, S., Ashaolu, T.J., Mirzakhani, E., et al., 2024. The role of probiotics in improving food safety; detoxification of heavy metals and chemicals. Toxin Reviews 43 (1), 63–91. Available at https://doi.org/10.1080/15569543.2023.2283768, http://www.tandfonline.com/loi/itxr20.

Arratia-Quijada, J., Nuño, K., Ruíz-Santoyo, V., Andrade-Espinoza, B.A., 2024. Nano-encapsulation of probiotics: need and critical considerations to design new non-dairy probiotic products. Journal of Functional Foods 116, 106192. Available at https://doi.org/10.1016/j.jff.2024.106192.

Asadi, Abbasi, Ghaemi, Montazeri, Akrami, 2024. Investigating the properties and antibacterial, antioxidant, and cytotoxicity activity of postbiotics derived from *Lacticaseibacillus casei* on various gastrointestinal pathogens in vitro and in food models. GMS Hygiene and Infection Control 19.

Asefa, Z., Belay, A., Welelaw, E., Haile, M., 2025. Postbiotics and their biotherapeutic potential for chronic disease and their feature perspective: a review. Frontiers in Microbiomes 4. Available at https://doi.org/10.3389/frmbi.2025.1489339, https://www.frontiersin.org/journals/microbiomes.

Ashrafudoulla, M., Park, J., Toushik, S.H., Shaila, S., Ha, A.J.W., Rahman, M.A., et al., 2024. Synergistic mechanism of UV-C and postbiotic of Leuconostoc mesenteroides (J.27) combination to eradicate Salmonella Thompson biofilm in the poultry industry. Food Control 164. Available at https://doi.org/10.1016/j.foodcont.2024.110607, https://www.sciencedirect.com/science/journal/09567135.

Assandri, M.H., Malamud, M., Trejo, F.M., Serradell, MdlA., 2023. S-layer proteins as immune players: tales from pathogenic and non-pathogenic bacteria. Current Research in Microbial Sciences 4. Available at https://doi.org/10.1016/j.crmicr.2023.100187, www.journals.elsevier.com/current-research-in-microbial-sciences.

Baral, K.C., Bajracharya, R., Lee, S.H., Han, H.K., 2021. Advancements in the pharmaceutical applications of probiotics: dosage forms and formulation technology. International Journal of Nanomedicine 16, 7535–7556. Available at https://doi.org/10.2147/IJN.S337427, https://www.dovepress.com/getfile.php?fileID=75762.

Barbosa, J.C., Almeida, D., Machado, D., Sousa, S., Freitas, A.C., Andrade, J.C., et al., 2022. Spray-drying encapsulation of the live biotherapeutic candidate *Akkermansia muciniphila* DSM 22959 to survive aerobic storage. Pharmaceuticals 15 (5). Available at https://doi.org/10.3390/ph15050628, https://www.mdpi.com/1424-8247/15/5/628/pdf?version=1653027635.

Bermingham, K.M., Linenberg, I., Polidori, L., Asnicar, F., Arrè, A., Wolf, J., et al., 2024. Effects of a personalized nutrition program on cardiometabolic health: a randomized controlled trial. Nature Medicine 30 (7), 1888–1897. Available at https://doi.org/10.1038/s41591-024-02951-6, https://www.nature.com/nm/.

Binda, S., Hill, C., Johansen, E., Obis, D., Pot, B., Sanders, M.E., et al., 2020. Criteria to qualify microorganisms as "probiotic" in foods and dietary supplements. Frontiers in Microbiology 11. Available at https://doi.org/10.3389/fmicb.2020.01662, https://www.frontiersin.org/journals/microbiology#.

Bisht, V., Das, B., Hussain, A., Kumar, V., Navani, N.K., 2024. Understanding of probiotic origin antimicrobial peptides: a sustainable approach ensuring food safety. npj Science of Food 8 (1). Available at https://doi.org/10.1038/s41538-024-00304-8, https://www.nature.com/npjscifood/about.

Bogdanović, M., Mladenović, D., Mojović, L., Djuriš, J., Djukić-Vuković, A., 2024. Intraoral administration of probiotics and postbiotics: an overview of microorganisms and formulation strategies. Brazilian Journal of Pharmaceutical Sciences 60. Available at https://doi.org/10.1590/s2175-97902024e23272, https://www.scielo.br/j/bjps/a/S64FSmfrbLWT8yJbPxZH6pC/?lang=en&format=pdf.

Bouarab-Chibane, L., Forquet, V., Lantéri, P., Clément, Y., Léonard-Akkari, L., Oulahal, N., et al., 2019. Antibacterial properties of polyphenols: characterization and QSAR (quantitative structure–activity relationship) models. Frontiers in Microbiology 10. Available at https://doi.org/10.3389/fmicb.2019.00829.

Brasili, E., Mengheri, E., Tomassini, A., Capuani, G., Roselli, M., Finamore, A., et al., 2013. Lactobacillus acidophilus La5 and Bifidobacterium lactis Bb12 induce different age-related metabolic profiles revealed by 1H-NMR spectroscopy in urine and feces of mice. The Journal of Nutrition 143 (10), 1549–1557. Available at https://doi.org/10.3945/jn.113.177105.

Calvanese, C.M., Villani, F., Ercolini, D., De Filippis, F., 2025. Postbiotics versus probiotics: possible new allies for human health. Food Research International 217, 116869. Available at https://doi.org/10.1016/j.foodres.2025.116869.

Capponi, M., Gori, A., De Castro, G., Ciprandi, G., Anania, C., Brindisi, G., et al., 2022. (R) Evolution in allergic rhinitis add-on therapy: from probiotics to postbiotics and parabiotics. Journal of Clinical Medicine 11 (17), 5154.

Chaisakul, J., Rusmili, M.R.A., Alsolaiss, J., Albulescu, L.O., Harrison, R.A., Othman, I., et al., 2020. In vitro immunological cross-reactivity of thai polyvalent and monovalent antivenoms with asian viper venoms. Toxins 12 (12). Available at https://doi.org/10.3390/toxins12120766, https://www.mdpi.com/2072-6651/12/12/766/pdf.

Chambers, E.S., Preston, T., Frost, G., Morrison, D.J., 2018. Role of gut microbiota-generated short-chain fatty acids in metabolic and cardiovascular health. Current Nutrition Reports 7 (4), 198–206. Available at https://doi.org/10.1007/s13668-018-0248-8, http://www.springer.com/new+%26+forthcoming+titles+%28default%29/journal/13668?changeHeader.

Chang, H.M., Foo, H.L., Loh, T.C., Lim, E.T.C., Abdul Mutalib, N.E., 2021. Comparative studies of inhibitory and antioxidant activities, and organic acids compositions of postbiotics produced by probiotic *Lactiplantibacillus plantarum* strains isolated from Malaysian foods. Frontiers in Veterinary Science 7. Available at https://doi.org/10.3389/fvets.2020.602280, journal.frontiersin.org/journal/veterinary-science.

Chang, H.M., Loh, T.C., Foo, H.L., Lim, E.T.C., 2022. *Lactiplantibacillus plantarum* postbiotics: alternative of antibiotic growth promoter to ameliorate gut health in broiler chickens. Frontiers in Veterinary Science 9. Available at https://doi.org/10.3389/fvets.2022.883324, journal.frontiersin.org/journal/veterinary-science.

Chen, A., Gong, Y., Wu, S., Du, Y., Liu, Z., Jiang, Y., et al., 2025a. Navigating a challenging path: precision disease treatment with tailored oral nano-armor-probiotics. Journal of Nanobiotechnology 23 (1). Available at https://doi.org/10.1186/s12951-025-03141-3, https://jnanobiotechnology.biomedcentral.com.

Chen, Y., Zhao, N., Han, S., Zhao, G., Jiang, Y., Li, C., et al., 2025b. Tricobalt tetraselenide nanoparticles improve intestinal barrier function by reshaping the gut microbiota and fortifying epithelial tight junctions. Biomaterials Science 13 (24), 6775–6790. Available at https://doi.org/10.1039/d5bm00712g.

Coelho, M.C., Malcata, F.X., Silva, C.C.G., 2022. Lactic acid bacteria in raw-milk cheeses: from starter cultures to probiotic functions. Foods 11 (15). Available at https://doi.org/10.3390/foods11152276, www.mdpi.com/journal/foods.

Conway, P.L., 1996. Selection criteria for probiotic microorganisms. Asia Pacific Journal of Clinical Nutrition 5 (1), 10–14.

Cui, H., Guo, C., Wang, Q., Feng, C., Duan, Z., 2022. A pilot study on the efficacy of topical lotion containing anti-acne postbiotic in subjects with mild -to -moderate acne. Frontiers in Medicine 9. Available at https://doi.org/10.3389/fmed.2022.1064460.

Cui, Y., Miao, K., Niyaphorn, S., Qu, X., 2020. Production of gamma-aminobutyric acid from lactic acid bacteria: a systematic review. International Journal of Molecular Sciences 21 (3), 995. Available at https://doi.org/10.3390/ijms21030995.

da Silva, B.S., Díaz-Roa, A., Yamane, E.S., Hayashi, M.A.F., Silva Junior, P.I., 2023. Doderlin: isolation and characterization of a broad-spectrum antimicrobial peptide from *Lactobacillus acidophilus*. Research in Microbiology 174 (3). Available at https://doi.org/10.1016/j.resmic.2022.103995, http://www.elsevier.com/wps/find/journaldescription.cws_home/522493/description#description.

de Lima, C.S.A., Varca, J.P.R.O., Alves, V.M., Nogueira, K.M., Cruz, C.P.C., Rial-Hermida, M.I., et al., 2022. Mucoadhesive polymers and their applications in drug delivery systems for the treatment of bladder cancer. Gels 8 (9). Available at https://doi.org/10.3390/gels8090587, www.mdpi.com/journal/gels.

De Almeida, C.V., Antiga, E., Lulli, M., 2023. Oral and topical probiotics and postbiotics in skincare and dermatological therapy: a concise review. Microorganisms 11 (6). Available at https://doi.org/10.3390/microorganisms11061420, www.mdpi.com/journal/microorganisms.

Deleu, S., Machiels, K., Raes, J., Verbeke, K., Vermeire, S., 2021. Short chain fatty acids and its producing organisms: an overlooked therapy for IBD? eBioMedicine 66, 103293. Available at https://doi.org/10.1016/j.ebiom.2021.103293.

Dhakal, R., Bajpai, V.K., Baek, K.H., 2012. Production of GABA (γ-aminobutyric acid) by microorganisms: a review. Brazilian Journal of Microbiology 43 (4), 1230–1241. Available at https://doi.org/10.1590/S1517-83822012000400001, http://www.scielo.br/pdf/bjm/v43n4/01.pdf.

Dong, Z., Yang, S., Tang, C., Li, D., Kan, Y., Yao, L., 2025. New insights into microbial bile salt hydrolases: from physiological roles to potential applications. Frontiers in Microbiology 16. Available at https://doi.org/10.3389/fmicb.2025.1513541.

Duarte, M., Oliveira, A.L., Oliveira, C., Pintado, M., Amaro, A., Madureira, A.R., 2022. Current postbiotics in the cosmetic market—an update and development opportunities. Applied Microbiology and Biotechnology 106, 5879–5891. Available at https://doi.org/10.1007/s00253-022-12116-5, https://www.springer.com/journal/253.

Egan, K., Field, D., Rea, M.C., Ross, R.P., Hill, C., Cotter, P.D., 2016. Bacteriocins: novel solutions to age old spore-related problems? Frontiers in Microbiology 7. Available at https://doi.org/10.3389/fmicb.2016.00461, http://journal.frontiersin.org/article/10.3389/fmicb.2016.00461/full.

Enck, K., Banks, S., Yadav, H., Welker, M.E., Opara, E.C., 2020. Development of a novel oral delivery vehicle for probiotics. Current Pharmaceutical Design 26 (26), 3134–3140. Available at https://doi.org/10.2174/1381612826666200210111925, https://www.eurekaselect.com/179171/article.

Fahim, K.M., Badr, A.N., Shehata, M.G., Hassanen, E.I., Ahmed, L.I., 2021. Innovative application of postbiotics, parabiotics and encapsulated *Lactobacillus plantarum* RM1 and *Lactobacillus paracasei* KC39 for detoxification of aflatoxin M1 in milk powder. Journal of Dairy Research 88 (4), 429–435. Available at https://doi.org/10.1017/S002202992100090X, http://journals.cambridge.org/action/displayJournal?jid=DAR.

Falzone, L., Lavoro, A., Candido, S., Salmeri, M., Zanghì, A., Libra, M., 2024. Benefits and concerns of probiotics: an overview of the potential genotoxicity of the colibactin-producing

Escherichia coli Nissle 1917 strain. Gut Microbes 16 (1). Available at https://doi.org/10.1080/19490976.2024.2397874.

Fang, H., Rodrigues E-Lacerda, R., Barra, N.G., Kukje Zada, D., Robin, N., Mehra, A., et al., 2025. Postbiotic impact on host metabolism and immunity provides therapeutic potential in metabolic disease. Endocrine Reviews 46 (1), 60–79. Available at https://doi.org/10.1210/endrev/bnae025, https://academic.oup.com/edrv/issue.

Fang, H., Kang, J., Zhang, D., 2017. Microbial production of vitamin B12: a review and future perspectives. Microbial Cell Factories 16 (1). Available at https://doi.org/10.1186/s12934-017-0631-y.

Fang, Z., Yang, X., Shang, L., 2025. Microfluidic-derived montmorillonite composite microparticles for oral codelivery of probiotic biofilm and postbiotics. Science Advances 11 (12). Available at https://doi.org/10.1126/sciadv.adt2131.

Fanning, S., Hall, L.J., Cronin, M., Zomer, A., MacSharry, J., Goulding, D., et al., 2012. Bifidobacterial surface-exopolysaccharide facilitates commensal-host interaction through immune modulation and pathogen protection. Proceedings of the National Academy of Sciences 109 (6), 2108–2113. Available at https://doi.org/10.1073/pnas.1115621109.

Ganchev, I., 2024. Impact of *Spirulina platensis* biomass on the viability of *Lactobacillus delbrueckii* subsp. bulgaricus strain during the freeze-drying process. BioTechnologia 105 (2), 109–119. Available at https://doi.org/10.5114/bta.2024.139751.

Govender, M., Choonara, Y.E., Kumar, P., Du Toit, L.C., Van Vuuren, S., Pillay, V., 2014. A review of the advancements in probiotic delivery: conventional vs. non-conventional formulations for intestinal flora supplementation. AAPS PharmSciTech 15 (1), 29–43. Available at https://doi.org/10.1208/s12249-013-0027-1.

Guglielmetti, S., Boyte, M.E., Smith, C.L., Ouwehand, A.C., Paraskevakos, G., Younes, J.A., 2025. Commercial and regulatory frameworks for postbiotics: an industry-oriented scientific perspective for non-viable microbial ingredients conferring beneficial physiological effects. Trends in Food Science and Technology 163. Available at https://doi.org/10.1016/j.tifs.2025.105130, https://www.sciencedirect.com/science/journal/09242244.

Guo, S., Ma, T., Kwok, L.Y., Quan, K., Li, B., Wang, H., et al., 2024. Effects of postbiotics on chronic diarrhea in young adults: a randomized, double-blind, placebo-controlled crossover trial assessing clinical symptoms, gut microbiota, and metabolite profiles. Gut Microbes 16 (1). Available at https://doi.org/10.1080/19490976.2024.2395092, http://www.tandfonline.com/toc/kgmi20/current.

Hays, K.E., Pfaffinger, J.M., Ryznar, R., 2024. The interplay between gut microbiota, short-chain fatty acids, and implications for host health and disease. Gut Microbes 16 (1). Available at https://doi.org/10.1080/19490976.2024.2393270, http://www.tandfonline.com/toc/kgmi20/current.

Hernández, M., Quevedo, B., Cabrera, M., Ulloa, J., 2025. Lab-scale production of postbiotic proteins from Bifidobacterium adolescentis with antiviral and epithelial-protective properties. Frontiers in Microbiology 16. Available at https://doi.org/10.3389/fmicb.2025.1646082.

Hijová, E., 2024. Postbiotics as metabolites and their biotherapeutic potential. International Journal of Molecular Sciences 25 (10), 5441. Available at https://doi.org/10.3390/ijms25105441.

Homayouni Rad, A., Aghebati Maleki, L., Samadi Kafil, H., Abbasi, A., 2021. Postbiotics: a novel strategy in food allergy treatment. Critical Reviews in Food Science and Nutrition 61 (3), 492–499. Available at https://doi.org/10.1080/10408398.2020.1738333.

Homayouni Rad, A., Pourjafar, H., Mirzakhani, E., 2023. A comprehensive review of the application of probiotics and postbiotics in oral health. Frontiers in Cellular and Infection Microbiology 13. Available at https://doi.org/10.3389/fcimb.2023.1120995.

Huang, H.L., Lai, C.H., Tsai, W.H., Chen, K.W., Peng, S.L., Lin, J.H., et al., 2024. Nanoparticle-enhanced postbiotics: revolutionizing cancer therapy through effective delivery. Life Sciences 337. Available at https://doi.org/10.1016/j.lfs.2023.122379, https://www.sciencedirect.com/science/journal/00243205.

Isaac-Bamgboye, F.J., Mgbechidinma, C.L., Onyeaka, H., Isaac-Bamgboye, I.T., Chukwugozie, D.C., 2024. Exploring the potential of postbiotics for food safety and human health improvement. Journal of Nutrition and Metabolism 2024. Available at https://doi.org/10.1155/2024/1868161, www.hindawi.com/journals/jnume/.

Jahedi, S., Pashangeh, S., 2025. Bioactivities of postbiotics in food applications: a review. Iranian Journal of Microbiology 17 (3), 348–357. Available at https://doi.org/10.18502/ijm.v17i3.18816, https://ijm.tums.ac.ir/index.php/ijm/article/download/4740/1775.

Jannah, S.R., Rahayu, E.S., Yanti, R., Suroto, D.A., Wikandari, R., 2022. Study of viability, storage stability, and shelf life of probiotic instant coffee *Lactiplantibacillus plantarum* subsp. plantarum Dad-13 in vacuum and nonvacuum packaging at different storage temperatures. International Journal of Food Science 2022. Available at https://doi.org/10.1155/2022/1663772, http://www.hindawi.com/journals/ijfs/.

Jawadi, Z., Yang, C., Haidar, Z.S., Santa Maria, P.L., Massa, S., 2022. Bio-inspired muco-adhesive polymers for drug delivery applications. Polymers 14 (24). Available at https://doi.org/10.3390/polym14245459, http://www.mdpi.com/journal/polymers.

Jungersen, M., Wind, A., Johansen, E., Christensen, J.E., Stuer-Lauridsen, B., Eskesen, D., 2014. The science behind the probiotic strain bifidobacterium animalis subsp. Lactis bb-12® Microorganisms 2 (2), 92–110. Available at https://doi.org/10.3390/microorganisms 2020092, https://www.mdpi.com/2076-2607/2/2/92/pdf.

Khakpour, M., Mohsenzadeh, M., Salari, A., 2024. Feasibility of *Lactiplantibacillus plantarum* postbiotics production in challenging media by different techniques. AMB Express 14 (1). Available at https://doi.org/10.1186/s13568-024-01704-5.

Kim, H.J., Youn, H.Y., Moon, J.S., Kim, H., Seo, K.H., 2024. Comparative anti-microbial and anti-biofilm activities of postbiotics derived from kefir and normal raw milk lactic acid bacteria against bovine mastitis pathogens. LWT 191. Available at https://doi.org/10.1016/j.lwt.2023.115699, https://www.sciencedirect.com/science/journal/00236438.

Kulkarni, M.A., Gangal, A.R., Khare, S.S., Mundada, H.G., Gawade, A.R., R. Pujari, R., 2022a. Electrospun collagen based nanofibrous mats for wound healing: an integrative review. Biosciences Biotechnology Research Asia 19 (2), 515–528. Available at https://doi.org/10.13005/bbra/3006.

Kulkarni, N., Jain, P., Shindikar, A., Suryawanshi, P., Thorat, N., 2022b. Advances in the colon-targeted chitosan based multiunit drug delivery systems for the treatment of inflammatory bowel disease. Carbohydrate Polymers 288, 119351. Available at https://doi.org/10.1016/j.carbpol.2022.119351.

Kumar, A., Green, K.M., Rawat, M., 2024. A comprehensive overview of postbiotics with a special focus on discovery techniques and clinical applications. Foods 13 (18). Available at https://doi.org/10.3390/foods13182937, www.mdpi.com/journal/foods.

Kumar, A., Saranyadevi, S., Thirumalaisamy, S.K., Dapana Durage, T.T., Jaiswal, S.G., Kavitake, D., et al., 2025. Phenolic acids in fermented foods: microbial biotransformation, antioxidant mechanisms, and functional health implications. Frontiers in Molecular Biosciences 12. Available at https://doi.org/10.3389/fmolb.2025.1678673, journal.frontiersin.org/journal/molecular-biosciences.

Kumar, A., Kaur, R., Kumar, V., Kumar, S., Gehlot, R., Aggarwal, P., 2022. New insights into water-in-oil-in-water (W/O/W) double emulsions: properties, fabrication, instability mechanism, and food applications. Trends in Food Science and Technology 128, 22–37. Available at https://doi.org/10.1016/j.tifs.2022.07.016.

Lai, J., Azad, A.K., Sulaiman, W.M.A.W., Kumarasamy, V., Subramaniyan, V., Alshehade, S.A., 2024. Alginate-based encapsulation fabrication technique for drug delivery: an updated review of particle type, formulation technique, pharmaceutical ingredient, and targeted delivery system. Pharmaceutics 16 (3). Available at https://doi.org/10.3390/pharmaceutics16030370, http://www.mdpi.com/journal/pharmaceutics.

Lopes, S.A., Roque-Borda, C.A., Duarte, J.L., Di Filippo, L.D., Borges Cardoso, V.M., Pavan, F.R., et al., 2023. Delivery strategies of probiotics from nano- and microparticles: trends in the treatment of inflammatory bowel disease—an overview. Pharmaceutics 15 (11). Available at https://doi.org/10.3390/pharmaceutics15112600, http://www.mdpi.com/journal/pharmaceutics.

Lyu, C.J., Zhao, W.R., Hu, S., Huang, J., Lu, T., Jin, Z.H., et al., 2017. Physiology-oriented engineering strategy to improve gamma-aminobutyrate production in Lactobacillus brevis. Journal of Agricultural and Food Chemistry 65 (4), 858–866. Available at https://doi.org/10.1021/acs.jafc.6b04442, http://pubs.acs.org/journal/jafcau.

Ma, L., Tu, H., Chen, T., 2023. Postbiotics in human health: a narrative review. Nutrients 15 (2), 291. Available at https://doi.org/10.3390/nu15020291.

Machado, P., Ribeiro, F.N., Giublin, F.C.W., Mieres, N.G., Tonin, F.S., Pontarolo, R., et al., 2025. Next-generation wound care: a scoping review on probiotic, prebiotic, synbiotic, and postbiotic cutaneous formulations. Pharmaceuticals 18 (5). Available at https://doi.org/10.3390/ph18050704, https://www.mdpi.com/journal/pharmaceuticals.

Maggira, M., Ioannidou, M., Sakaridis, I., Samouris, G., 2021. Determination of Aflatoxin M1 in raw milk using an HPLC-FL method in comparison with commercial ELISA Kits—application in raw milk samples from various regions of Greece. Veterinary Sciences 8 (3), 46. Available at https://doi.org/10.3390/vetsci8030046.

Makarewicz, M., Drożdż, I., Tarko, T., Duda-Chodak, A., 2021. The interactions between polyphenols and microorganisms, especially gut microbiota. Antioxidants 10 (2), 188. Available at https://doi.org/10.3390/antiox10020188.

Mashayekh, S., Bersch, K.L., Ramsey, J., Harmon, T., Prather, B., Genova, L.A., et al., 2021. Synthesis of bacterial-derived peptidoglycan cross-linked fragments. Journal of Organic Chemistry 85 (24), 16243–16253. Available at https://doi.org/10.1021/acs.joc.0c01852, http://pubs.acs.org/joc.

Meena, K.K., Joshi, M., Gupta, L., Meena, S., 2025. Comprehensive insights into postbiotics: Bridging the gap to real-world application. Food Nutrition 1 (2), 100024. Available at https://doi.org/10.1016/j.fnutr.2025.100024.

Mishra, N., Garg, A., Ashique, S., Bhatt, S., 2024. Potential of postbiotics for the treatment of metabolic disorders. Drug Discovery Today 29 (4), 103921. Available at https://doi.org/10.1016/j.drudis.2024.103921.

Mohammed, S., Çon, A.H., 2024. Postbiotic nanoparticles (postbiotics-NPs): a novel strategy for providing probiotics' health advantages through food consumption. Food Science and Biotechnology 33 (12), 2729–2736. Available at https://doi.org/10.1007/s10068-024-01629-6, https://www.springer.com/journal/10068.

Mosca, A., Abreu Y Abreu, A.T., Gwee, K.A., Ianiro, G., Tack, J., Nguyen, T.V.H., et al., 2022. The clinical evidence for postbiotics as microbial therapeutics. Gut Microbes 14 (1). Available at https://doi.org/10.1080/19490976.2022.2117508, http://www.tandfonline.com/toc/kgmi20/current.

Muhoza, B., Yuyang, H., Uriho, A., Harindintwali, J.D., Liu, Q., Li, Y., 2023. Spray-and freeze-drying of microcapsules prepared by complex coacervation method: a review. Food Hydrocolloids 140, 108650. Available at https://doi.org/10.1016/j.foodhyd.2023.108650.

Mukhopadhya, I., Louis, P., 2025. Gut microbiota-derived short-chain fatty acids and their role in human health and disease. Nature Reviews Microbiology 23 (10), 635–651. Available at https://doi.org/10.1038/s41579-025-01183-w.

Nair, V.V., Cabrera, P., Ramírez-Lecaros, C., Jara, M.O., Brayden, D.J., Morales, J.O., 2023. Buccal delivery of small molecules and biologics: of mucoadhesive polymers, films, and nanoparticles – an update. International Journal of Pharmaceutics 636. Available at https://doi.org/10.1016/j.ijpharm.2023.122789, www.elsevier.com/locate/ijpharm.

Oglio, F., Bruno, C., Coppola, S., De Michele, R., Masino, A., Carucci, L., 2023. Evidence on the preventive effects of the postbiotic derived from cow's milk fermentation with *Lacticaseibacillus paracasei* CBA L74 against pediatric gastrointestinal infections. Microorganisms 11 (1). Available at https://doi.org/10.3390/microorganisms11010010, www.mdpi.com/journal/microorganisms.

Om, H., Chand, U., Kushawaha, P.K., 2024. Postbiotics: an alternative and innovative intervention for the therapy of inflammatory bowel disease. Microbiological Research 279, 127550.

Ooi, M.F., Foo, H.L., Loh, T.C., Mohamad, R., Rahim, R.A., Ariff, A., 2021. A refined medium to enhance the antimicrobial activity of postbiotic produced by *Lactiplantibacillus plantarum* RS5. Scientific Reports 11 (1). Available at https://doi.org/10.1038/s41598-021-87081-6, www.nature.com/srep/index.html.

Paik, W., Alonzo, F., Knight, K.L., 2019. Probiotic exopolysaccharide protects against systemic *Staphylococcus aureus* infection, inducing dual-functioning macrophages that restrict bacterial growth and limit inflammation. Infection and Immunity 87 (1). Available at https://doi.org/10.1128/IAI.00791-18, http://iai.asm.org/content/by/year.

Panigrahi, S.S., Mishra, G., Kumar, V., Panigrahi, C., Kaur, G., Shaikh, A.E.Y., et al., 2025. Next-generation food drying: specialized and smart approaches to boost efficiency and quality. Comprehensive Reviews in Food Science and Food Safety 24 (5). Available at https://doi.org/10.1111/1541-4337.70269, http://onlinelibrary.wiley.com/journal/10.1111/(ISSN)1541-4337.

Percy, M.G., Gründling, A., 2014. Lipoteichoic acid synthesis and function in gram-positive bacteria. Annual Review of Microbiology 68, 81–100. Available at https://doi.org/10.1146/annurev-micro-091213-112949, http://arjournals.annualreviews.org/loi/micro.

Popović, N., Stevanović, D., Radojević, D., Veljović, K., Đokić, J., Golić, N., et al., 2023. Insight into the postbiotic potential of the autochthonous bacteriocin-producing *Enterococcus faecium* BGZLM1-5 in the reduction in the abundance of *Listeria monocytogenes* ATCC19111 in a milk model. Microorganisms 11 (12), 2844. Available at https://doi.org/10.3390/microorganisms11122844.

Pradhan, S., Weiss, A.A., 2020. Probiotic properties of *Escherichia coli* nissle in human intestinal organoids. mBio 11 (4), 1–16. Available at https://doi.org/10.1128/mBio.01470-20, https://mbio.asm.org/content/mbio/11/4/e01470-20.full.pdf.

Prajapati, N., Patel, J., Singh, S., Yadav, V.K., Joshi, C., Patani, A., et al., 2023. Postbiotic production: harnessing the power of microbial metabolites for health applications. Frontiers in Microbiology 14. Available at https://doi.org/10.3389/fmicb.2023.1306192, https://www.frontiersin.org/journals/microbiology#.

Prajapati, S.K., Lekkala, L., Yadav, D., Jain, S., Yadav, H., 2025. Microbiome and postbiotics in skin health. Biomedicines 13 (4). Available at https://doi.org/10.3390/biomedicines13040791, www.mdpi.com/journal/biomedicines.

Puccetti, M., Pariano, M., Schoubben, A., Giovagnoli, S., Ricci, M., 2024. Biologics, theranostics, and personalized medicine in drug delivery systems. Pharmacological Research 201, 107086. Available at https://doi.org/10.1016/j.phrs.2024.107086.

Pujato, S.A., Mercanti, D.J., Briggiler Marcó, M., Capra, M.L., Quiberoni, A., Guglielmotti, D.M., 2024. Bacteriocins from lactic acid bacteria: strategies for the bioprotection of dairy foods. Frontiers in Food Science and Technology 4. Available at https://doi.org/10.3389/frfst.2024.1439891, www.frontiersin.org/journals/food-science-and-technology.

Punia Bangar, S., Suri, S., Trif, M., Ozogul, F., 2022. Organic acids production from lactic acid bacteria: a preservation approach. Food Bioscience 46, 101615. Available at https://doi.org/10.1016/j.fbio.2022.101615.

Putri, D.A., Lei, J., Rossiana, N., Syaputri, Y., 2024. Biopreservation of food using bacteriocins from lactic acid bacteria: classification, mechanisms, and commercial applications. International Journal of Microbiology 2024 (1). Available at https://doi.org/10.1155/ijm/8723968, https://onlinelibrary.wiley.com/journal/8472.

Pérez-Reytor, D., Puebla, C., Karahanian, E., García, K., 2021. Use of short-chain fatty acids for the recovery of the intestinal epithelial barrier affected by bacterial toxins. Frontiers in Physiology 12. Available at https://doi.org/10.3389/fphys.2021.650313.

Rafique, N., Jan, S.Y., Dar, A.H., Dash, K.K., Sarkar, A., Shams, R., et al., 2023. Promising bioactivities of postbiotics: a comprehensive review. Journal of Agriculture and Food Research 14. Available at https://doi.org/10.1016/j.jafr.2023.100708, www.journals.elsevier.com/journal-of-agriculture-and-food-research.

Ragavan, M.L., Hemalatha, S., 2024. The functional roles of short chain fatty acids as postbiotics in human gut: future perspectives. Food Science and Biotechnology 33 (2), 275–285. Available at https://doi.org/10.1007/s10068-023-01414-x, https://www.springer.com/journal/10068.

Rahimi, A., Qaisar, S.A., Janeh, T., Karimpour, H., Darbandi, M., Moludi, J., 2024a. Clinical trial of the effects of postbiotic supplementation on inflammation, oxidative stress, and clinical outcomes in patients with CVA. Scientific Reports 14 (1). Available at https://doi.org/10.1038/s41598-024-76153-y, https://www.nature.com/srep/.

Rahimi, D., Sadeghi, A., Kashaninejad, M., Ebrahimi, M., 2024b. Postbiotic characterization of a potential probiotic yeast isolate, and its microencapsulation in alginate beads coated layer-by-layer with chitosan. Heliyon 10 (7), e28452. Available at https://doi.org/10.1016/j.heliyon.2024.e28452.

Ramos, I.M., Rodríguez-Sánchez, S., Seseña, S., Palop, M.L., Poveda, J.M., 2022. Assessment of safety characteristics, postbiotic potential, and technological stress response of *Leuconostoc strains* from different origins for their use in the production of functional dairy foods. LWT 165. Available at https://doi.org/10.1016/j.lwt.2022.113722, https://www.journals.elsevier.com/lwt.

Raveschot, C., Deracinois, B., Bertrand, E., Flahaut, C., Frémont, M., Drider, D., et al., 2020. Integrated continuous bioprocess development for ace-inhibitory peptide production by *Lactobacillus helveticus* strains in membrane bioreactor. Frontiers in Bioengineering and Biotechnology 8. Available at https://doi.org/10.3389/fbioe.2020.585815.

Rebai, Y., Wagner, L., Gnaien, M., Hammer, M.L., Kapitan, M., Niemiec, M.J., et al., 2023. *Escherichia coli* Nissle 1917 antagonizes *Candida albicans* growth and protects intestinal cells from *Candida albicans*-mediated damage. Microorganisms 11 (8). Available at https://doi.org/10.3390/microorganisms11081929, www.mdpi.com/journal/microorganisms.

Rebelo, M.B., Oliveira, C.S., Tavaria, F.K., 2023. Novel strategies for preventing dysbiosis in the oral cavity. Frontiers in Bioscience - Elite 15 (4). Available at https://doi.org/10.31083/j.fbe1504023, https://www.imrpress.com/journal/FBE/15/4/10.31083/j.fbe1504023.

Rodrigues, F.J., Cedran, M.F., Bicas, J.L., Sato, H.H., 2020. Encapsulated probiotic cells: relevant techniques, natural sources as encapsulating materials and food applications – a narrative review. Food Research International 137, 109682. Available at https://doi.org/10.1016/j.foodres.2020.109682.

Rouhi, A., Falah, F., Azghandi, M., Alizadeh Behbahani, B., Mortazavi, S.A., Tabatabaei-Yazdi, F., et al., 2024. Investigating the effect of *Lactiplantibacillus plantarum* TW57-4 in preventing biofilm formation and expression of virulence genes in *Listeria monocytogenes* ATCC 19115. LWT 191. Available at https://doi.org/10.1016/j.lwt.2023.115669, https://www.sciencedirect.com/science/journal/00236438.

Sadighbathi, S., Amiri, S., Saris, P.E.J., Yousefvand, A., 2023. Development and properties of functional yoghurt enriched with postbiotic produced by yoghurt cultures using cheese whey and skim milk. Frontiers in Microbiology 14. Available at https://doi.org/10.3389/fmicb.2023.1276268, https://www.frontiersin.org/journals/microbiology#.

Salminen, S., Collado, M.C., Endo, A., Hill, C., Lebeer, S., Quigley, E.M.M., et al., 2021. The International Scientific Association of Probiotics and Prebiotics (ISAPP) consensus statement on the definition and scope of postbiotics. Nature Reviews Gastroenterology and Hepatology 18 (9), 649–667. Available at https://doi.org/10.1038/s41575-021-00440-6, http://www.nature.com/nrgastro/index.html.

Savytska, M., Baka, O., Manzhalii, E., Kyriienko, D., Falalyeyeva, T., Sichel, L., et al., 2024. Efficacy and safety of postbiotic supplementation for type 2 diabetes patients: randomized clinical Ttrial (DELI Diab study). Endocrine Abstracts.

Scott, E., De Paepe, K., Van de Wiele, T., 2022. Postbiotics and their health modulatory biomolecules. Biomolecules 12 (11). Available at https://doi.org/10.3390/biom12111640, http://www.mdpi.com/journal/biomolecules.

Serter, B., Önen, A., Ilhak, O.I., 2024. Antimicrobial efficacy of postbiotics of lactic acid bacteria and their effects on food safety and shelf life of chicken meat. Annals of Animal Science 24 (1), 277–287. Available at https://doi.org/10.2478/aoas-2023-0081, https://sciendo.com/journal/aoas.

Shakirova, L., Grube, M., Gavare, M., Auzina, L., Zikmanis, P., 2013. *Lactobacillus acidophilus* La5 and *Bifidobacterium lactis* Bb12 cell surface hydrophobicity and survival of the cells under adverse environmental conditions. Journal of Industrial Microbiology and Biotechnology 40 (1), 85–93. Available at https://doi.org/10.1007/s10295-012-1204-z.

Sharafi, H., Divsalar, E., Rezaei, Z., Liu, S.Q., Moradi, M., 2024. The potential of postbiotics as a novel approach in food packaging and biopreservation: a systematic review of the latest developments. Critical Reviews in Food Science and Nutrition 64 (33), 12524–12554. Available at https://doi.org/10.1080/10408398.2023.2253909, https://www.tandfonline.com/loi/bfsn20.

Shete, M.R., Popere, S.V., John, J., 2022. Pharmacokinetic availability of proteolytic enzymes after oral administration: a narrative review of the literature. International Journal of Basic and Clinical Pharmacology 11 (5), 515. Available at https://doi.org/10.18203/2319-2003.ijbcp20222046.

Sichel, L., Timoshok, N., Pidgorskyy, V., Spivak, N., 2013. Study of interferonogenous activity of the new probiotic formulation Del-Immune V® Journal of Probiotics and Health 1, 1000107.

Smolinska, S., Popescu, F.D., Zemelka-Wiacek, M., 2025. a review of the influence of prebiotics, probiotics, synbiotics, and postbiotics on the human gut microbiome and intestinal integrity.

Journal of Clinical Medicine 14 (11). Available at https://doi.org/10.3390/jcm14113673, https://www.mdpi.com/journal/jcm.

Song, D., Wang, X., Ma, Y., Liu, N.N., Wang, H., 2023. Beneficial insights into postbiotics against colorectal cancer. Frontiers in Nutrition 10. Available at https://doi.org/10.3389/fnut.2023.1111872, journal.frontiersin.org/journal/veterinary-science, journal.frontiersin.org/journal/nutrition.

Sonnenborn, U., Robertson, L., 2016. Escherichia coli strain Nissle 1917—from bench to bedside and back: history of a special Escherichia coli strain with probiotic properties. FEMS Microbiology Letters 363 (19), fnw212. Available at https://doi.org/10.1093/femsle/fnw212.

Sridevi, N., Vishwe, P., Prabhune, A., 2009. Hypocholesteremic effect of bile salt hydrolase from *Lactobacillus buchneri* ATCC 4005. Food Research International 42 (4), 516–520. Available at https://doi.org/10.1016/j.foodres.2009.02.016.

Srivastava, S., Basak, U., Naghibi, M., Vijayakumar, V., Parihar, R., Patel, J., et al., 2024. A randomized double-blind, placebo-controlled trial to evaluate the safety and efficacy of live *Bifidobacterium longum* CECT 7347 (ES1) and heat-treated *Bifidobacterium longum* CECT 7347 (HT-ES1) in participants with diarrhea-predominant irritable bowel syndrome. Gut Microbes 16 (1). Available at https://doi.org/10.1080/19490976.2024.2338322.

Subramanian, P., 2021. Mucoadhesive delivery system: a smart way to improve bioavailability of nutraceuticals. Foods 10 (6), 1362. Available at https://doi.org/10.3390/foods10061362.

Tang, H., Huang, W., Yao, Y.F., 2023. The metabolites of lactic acid bacteria: classification, biosynthesis and modulation of gut microbiota. Microbial Cell 10 (3), 49–62. Available at https://doi.org/10.15698/mic2023.03.792, http://microbialcell.com/researcharticles/2023a-tang-microbial-cell/.

Tarique, M., Abdalla, A., Masad, R., Al-Sbiei, A., Kizhakkayil, J., Osaili, T., et al., 2022. Potential probiotics and postbiotic characteristics including immunomodulatory effects of lactic acid bacteria isolated from traditional yogurt-like products. LWT 159, 113207. Available at https://doi.org/10.1016/j.lwt.2022.113207.

Tarracchini, C., Lugli, G.A., Mancabelli, L., van Sinderen, D., Turroni, F., Ventura, M., et al., 2024. Exploring the vitamin biosynthesis landscape of the human gut microbiota. mSystems 9 (10). Available at https://doi.org/10.1128/msystems.00929-24, https://journals.asm.org/doi/10.1128/msystems.00929-24.

Taverniti, V., Stuknyte, M., Minuzzo, M., Arioli, S., De Noni, I., Scabiosi, C., et al., 2013. S-Layer protein mediates the stimulatory effect of *Lactobacillus helveticus* MIMLH5 on innate immunity. Applied and Environmental Microbiology 79 (4), 1221–1231. Available at https://doi.org/10.1128/AEM.03056-12Italy, http://aem.asm.org/content/79/4/1221.full.pdf.

Teame, T., Wang, A., Xie, M., Zhang, Z., Yang, Y., Ding, Q., et al., 2020. Paraprobiotics and postbiotics of probiotic Lactobacilli, their positive effects on the host and action mechanisms: a review. Frontiers in Nutrition 7. Available at https://doi.org/10.3389/fnut.2020.570344, https://www.frontiersin.org/journals/nutrition.

Thorakkattu, P., Khanashyam, A.C., Shah, K., Babu, K.S., Mundanat, A.S., Deliephan, A., et al., 2022. Postbiotics: current trends in food and pharmaceutical industry. Foods 11 (19). Available at https://doi.org/10.3390/foods11193094, www.mdpi.com/journal/foods.

Thwala, L.N., Préat, V., Csaba, N.S., 2017. Emerging delivery platforms for mucosal administration of biopharmaceuticals: a critical update on nasal, pulmonary and oral routes. Expert Opinion on Drug Delivery 14 (1), 23–36. Available at https://doi.org/10.1080/17425247.2016.1206074.

Timilsena, Y.P., Akanbi, T.O., Khalid, N., Adhikari, B., Barrow, C.J., 2019. Complex coacervation: principles, mechanisms and applications in microencapsulation. International

Journal of Biological Macromolecules 121, 1276–1286. Available at https://doi.org/10.1016/j.ijbiomac.2018.10.144, www.elsevier.com/locate/ijbiomac.

Tong, Y., Guo, H., Abbas, Z., Zhang, J., Wang, J., Cheng, Q., et al., 2023. Optimizing postbiotic production through solid-state fermentation with *Bacillus amyloliquefaciens* J and *Lactiplantibacillus plantarum* SN4 enhances antibacterial, antioxidant, and anti-inflammatory activities. Frontiers in Microbiology 14. Available at https://doi.org/10.3389/fmicb.2023.1229952, https://www.frontiersin.org/journals/microbiology#.

Tong, Y., Abbas, Z., Zhang, J., Wang, J., Zhou, Y., Si, D., et al., 2025. Antimicrobial activity and mechanism of novel postbiotics against foodborne pathogens. LWT 217, 117464. Available at https://doi.org/10.1016/j.lwt.2025.117464.

Vacariu, C.M., Tanner, M.E., 2022. Recent advances in the synthesis and biological applications of peptidoglycan fragments. Chemistry - A European Journal 28 (43). Available at https://doi.org/10.1002/chem.202200788, http://onlinelibrary.wiley.com/journal/10.1002/(ISSN)1521-3765.

Vandenberghe, L.P.S., Karp, S.G., de Oliveira, P.Z., de Carvalho, J.C., Rodrigues, C., Soccol, C.R., 2017. Solid-state fermentation for the production of organic acids. Current Developments in Biotechnology and Bioengineering: Current Advances in Solid-State Fermentation. Elsevier, Brazil, pp. 415–434. Available at https://www.sciencedirect.com/book/9780444639905/current-developments-in-biotechnology-and-bioengineering#:~:text=Description-,Current%20Developments%20in%20Biotechnology%20and%20Bioengineering%3A%20Current%20Advances%20in%20Solid,genetic%20engineering%2C%20metabolic%20engineering%20and, 10.1016/B978-0-444-63990-5.00018-9.

Vaz, A., Cabral Silva, A.C., Rodrigues, P., Venâncio, A., 2020. Detection methods for aflatoxin M1 in dairy products. Microorganisms 8 (2), 246. Available at https://doi.org/10.3390/microorganisms8020246.

Vega-Gálvez, A., Uribe, E., Gómez-Pérez, L.S., García, V., Mejias, N., Pastén, A., 2022. Drying kinetic modeling and assessment of mineral content, antimicrobial activity, and potential α-glucosidase activity inhibition of a green seaweed (Ulva spp.) subjected to different drying methods. ACS Omega 7 (38), 34230–34238. Available at https://doi.org/10.1021/acsomega.2c03617, https://pubs.acs.org/journal/acsodf.

Vera-Santander, V.E., Hernández-Figueroa, R.H., Jiménez-Munguía, M.T., Mani-López, E., López-Malo, A., 2023. Health benefits of consuming foods with bacterial probiotics, postbiotics, and their metabolites: a review. Molecules (Basel, Switzerland) 28 (3). Available at https://doi.org/10.3390/molecules28031230, http://www.mdpi.com/journal/molecules.

Vera-Santander, V.E., Mani-López, E., López-Malo, A., Jiménez-Munguía, M.T., 2024. Use of whey for a sustainable production of postbiotics with potential bioactive metabolites. Sustainable Food Technology 2 (4), 1101–1112. Available at https://doi.org/10.1039/d4fb00061g, https://www.rsc.org/journals-books-databases/about-journals/sustainable-food-technology/.

Vinderola, G., Sanders, M.E., Salminen, S., Szajewska, H., 2022. Postbiotics: the concept and their use in healthy populations. Frontiers in Nutrition 9. Available at https://doi.org/10.3389/fnut.2022.1002213, https://www.frontiersin.org/journals/nutrition.

Vinderola, G., Druart, C., Gosálbez, L., Salminen, S., Vinot, N., Lebeer, S., 2023. Postbiotics in the medical field under the perspective of the ISAPP definition: scientific, regulatory, and marketing considerations. Frontiers in Pharmacology 14. Available at https://doi.org/10.3389/fphar.2023.1239745.

Voo, W.P., Lee, B.B., Idris, A., Islam, A., Tey, B.T., Chan, E.S., 2015. Production of ultra-high concentration calcium alginate beads with prolonged dissolution profile. RSC Advances 5

(46), 36687–36695. Available at https://doi.org/10.1039/c5ra03862f, http://pubs.rsc.org/en/journals/journalissues.

Wang, S., Wang, P., Wang, D., Shen, S., Wang, S., Li, Y., et al., 2025a. Postbiotics in inflammatory bowel disease: efficacy, mechanism, and therapeutic implications. Journal of the Science of Food and Agriculture 105 (2), 721–734.

Wang, T., Xue, H., Liu, H., Yuan, H., Huang, D., Jiang, Y., 2025b. Advancements in metabolic engineering: unlocking the potential of key organic acids for sustainable industrial applications. Frontiers in Bioengineering and Biotechnology 13, 1556516.

Wei, L., Wong, D., Jeoh, T., Marco, M.L., 2024. Intestinal delivery of encapsulated bacteriocin peptides in cross-linked alginate microcapsules. Food Research International 188, 114473. Available at https://doi.org/10.1016/j.foodres.2024.114473.

Wilm, J., Svennesen, L., Kirkeby, C., Krömker, V., 2024. Treatment of clinically severe bovine mastitis – a scoping review. Frontiers in Veterinary Science 11. Available at https://doi.org/10.3389/fvets.2024.1286461.

Wu, Q., Tun, H.M., Law, Y.S., Khafipour, E., Shah, N.P., 2017. Common distribution of gad operon in *Lactobacillus brevis* and its GadA contributes to efficient GABA synthesis toward cytosolic near-neutral pH. Frontiers in Microbiology 8. Available at https://doi.org/10.3389/fmicb.2017.00206, http://journal.frontiersin.org/article/10.3389/fmicb.2017.00206/full.

Xin, X., Ambati, R.R., Cai, Z., Lei, B., 2019. Development of universal purification protocols for fibrinolytic enzyme-producing bacilli. CYTA - Journal of Food 17 (1), 112–120. Available at https://doi.org/10.1080/19476337.2018.1561521, http://www.tandfonline.com/toc/tcyt20/current.

Yan, R., Zeng, X., Shen, J., Wu, Z., Guo, Y., Du, Q., et al., 2024. New clues for postbiotics to improve host health: a review from the perspective of function and mechanisms. Journal of the Science of Food and Agriculture 104 (11), 6376–6387. Available at https://doi.org/10.1002/jsfa.13444.

Yang, K., Wang, X., Huang, R., Wang, H., Lan, P., Zhao, Y., 2022. Prebiotics and postbiotics synergistic delivery microcapsules from microfluidics for treating colitis. Advanced Science 9 (16). Available at https://doi.org/10.1002/advs.202104089.

Yang, X., Wang, C., Wang, Q., Zhang, Z., Nie, W., Shang, L., 2023. Armored probiotics for oral delivery. Smart Medicine 2 (4). Available at https://doi.org/10.1002/smmd.20230019.

Yemmireddy, V., Adhikari, A., Moreira, J., 2022. Effect of ultraviolet light treatment on microbiological safety and quality of fresh produce: an overview. Frontiers in Nutrition 9. Available at https://doi.org/10.3389/fnut.2022.871243.

Yilmaz, Y., 2024. Postbiotics as antiinflammatory and immune-modulating bioactive compounds in metabolic dysfunction-associated steatotic liver disease. Molecular Nutrition and Food Research 68 (23). Available at https://doi.org/10.1002/mnfr.202400754.

Yu, H.L., Hou, X.W., Zhao, J.X., Liu, G.H., Meng, J.X., Wei, Y.J., et al., 2025. Insights from metagenomics on microbial biosynthesis of vitamins B and K2 in chicken gut microbiota. Frontiers in Veterinary Science 12. Available at https://doi.org/10.3389/fvets.2025.1646825, journal.frontiersin.org/journal/veterinary-science.

Yu, T., Bai, R., Wang, Z., Qin, Y., Wang, J., Wei, Y., et al., 2024. Colon-targeted engineered postbiotics nanoparticles alleviate osteoporosis through the gut-bone axis. Nature Communications 15 (1). Available at https://doi.org/10.1038/s41467-024-55263-1.

Zhang, H., Lang, J., Lan, P., Yang, H., Lu, J., Wang, Z., 2020. Study on the dissolution mechanism of cellulose by ChCl-based deep eutectic solvents. Materials 13 (2), 278. Available at https://doi.org/10.3390/ma13020278.

Zhang, J., Zhang, R., Wang, J., Abbas, Z., Tong, Y., Fang, Y., et al., 2025. Efficient production strategy of a novel postbiotic produced by *Bacillus subtilis* and its antioxidant and anti-inflammatory effects. Molecules (Basel, Switzerland) 30 (10), 2089.

Zhao, X., Liu, S., Li, S., Jiang, W., Wang, J., Xiao, J., et al., 2024. Unlocking the power of postbiotics: a revolutionary approach to nutrition for humans and animals. Cell Metabolism 36 (4), 725–744. Available at https://doi.org/10.1016/j.cmet.2024.03.004, https://www.sciencedirect.com/science/journal/15504131.

Zhao, Z., Xu, S., Zhang, W., Wu, D., Yang, G., 2022. Probiotic *Escherichia coli* NISSLE 1917 for inflammatory bowel disease applications. Food and Function 13 (11), 5914–5924. Available at https://doi.org/10.1039/d2fo00226d.

Zhu, H., Zhao, F., Zhang, W., Xia, W., Chen, Y., Liu, Y., et al., 2022. Cholesterol-lowering effect of bile salt hydrolase from a Lactobacillus johnsonii strain mediated by FXR pathway regulation. Food was Function 13 (2), 725–736. Available at https://doi.org/10.1039/d1fo03143k.

Zhu, Y., Xiao, M., Kang, T., He, Y., Zhang, J., Zhao, Y., et al., 2025. The role of inactivation methods in shaping postbiotic composition and modulating bioactivity: a review. Foods 14 (13), 2358. Available at https://doi.org/10.3390/foods14132358.

Ziaee, A., Albadarin, A.B., Padrela, L., Femmer, T., O'Reilly, E., Walker, G., 2019. Spray drying of pharmaceuticals and biopharmaceuticals: critical parameters and experimental process optimization approaches. European Journal of Pharmaceutical Sciences 127, 300–318. Available at https://doi.org/10.1016/j.ejps.2018.10.026, www.elsevier.com/locate/ejps.

Żółkiewicz, J., Marzec, A., Ruszczyński, M., Feleszko, W., 2020. Postbiotics—a step beyond pre- and probiotics. Nutrients 12 (8), 2189. Available at https://doi.org/10.3390/nu12082189.

Chapter 3

Postbiotics and viable probiotic ingestion strategy in supplements: a critical approach on health effectiveness and technological challenges

Vincenzina Fusco[1], Marco Montemurro[1], Daniele Chieffi[1], Michela Verni[2], Thatyane Albuquerque[3], Maiara Lima[3], Carlo Giuseppe Rizzello[4], and Evandro de Souza[3]
[1]*Institute of Sciences of Food Production, National Research Council of Italy, Bari, Italy,* [2]*Department of Human Science and Quality of Life Promotion, San Raffaele University, Rome, Italy,* [3]*Department of Nutrition, Health Sciences Center, Federal University of Paraíba, João Pessoa, PB, Brazil,* [4]*Department of Environmental Biology, Sapienza University of Rome, Rome, Italy*

3.1 Introduction

According to the latest widely accepted definition, probiotics are "live microorganisms that, when administered in adequate amounts, confer a health benefit on the host" (Hill et al., 2014). In the last few decades, mainly due to the rise in the number of health-conscious consumers, the increase in the aging population, the rise in probiotics awareness, and the increase in the number of gut-related disorders, the market for probiotic products has increased worldwide. The global probiotic dietary supplement market size reached US$7.3 billion in 2023 and is expected to reach US$ 12.6 billion by 2032, with a growth rate (CAGR) of 6% during 2024–2032 (https://www.imarcgroup.com/probiotic-dietary-supplement-market). These dietary supplements have been extensively studied and applied in infant formula, food processing, agriculture, livestock feeding, medical, and aquaculture industries (Amenyogbe, 2023; Arsène et al., 2021; Latif et al., 2023; Leistikow et al., 2022; Liu et al., 2024; Raman et al., 2022; Todorov et al., 2024; de Souza et al., 2022; Fusco et al., 2022a). However, challenges occurring during the lifetime of probiotics, which include processing, storage, transportation, and, finally, passage through the

Evaluating Oral Probiotics and Prebiotics. DOI: https://doi.org/10.1016/B978-0-443-29835-6.00015-0

oro-gastrointestinal tract, might affect their viability so that these microorganisms may not reach the intestine in sufficient numbers to exert the beneficial effect on the host (Fusco et al., 2022c). To overcome these challenges, various formulations have been developed (Brito Sampaio et al., 2022) so that from the planktonic form of the first-generation probiotics, they evolved into the dual-coated fourth-generation probiotics, whose survival has been assessed only by in vitro studies.

However, in the last decades, nonviable cells, cellular components, and metabolites have been demonstrated to exert functional properties that improve human health. According to the International Scientific Association of Probiotics and Prebiotics (ISAPP) (Salminen et al., 2021), a postbiotic is a "preparation of inanimate microorganisms and/or their components that confers a health benefit on the host." Postbiotics overcome many drawbacks of probiotics. Indeed postbiotics could have a longer shelf life and higher stability during manufacturing, storage, transportation, and passage through the oro-gastrointestinal tract. Moreover, prostbiotics can be easily produced in high quantities and are less risky than probiotics for immunocompromised hosts. Although they are still neophytes among healthy products, the postbiotic supplements market is estimated to be valued at USD 10.8 million in 2024. It is expected to grow notably, reaching a valuation of USD 30.5 million by 2034, with a CAGR of 10.9% from 2024 to 2034 (https://www.futuremarketinsights.com/reports/postbiotic-supplements-market).

In light of the above, the foremost technological steps involved in the production of probiotic supplements and postbiotics, as well as the challenges encountered during their production and passage through the oro-gastrointestinal tract, will be reviewed. Moreover the pros and cons of postbiotic and probiotic supplement consumption will be highlighted.

3.2 Physiological challenges in the ingestion and passage through the gastrointestinal tract of probiotic supplements

Probiotics have been incorporated into different food matrices as part of the dietary components with health-related functional properties (Flach et al., 2018; Nguyen et al., 2020; Pimentel et al., 2021). Traditionally, fermented dairy foods have been used to carry probiotics (Abdi-Moghadam et al., 2023; Hadjimbei et al., 2022; Quintieri et al., 2024); however, recently, several studies have reported the use of probiotics in vegetal matrices (Oguntoyinbo et al., 2016; Afzaal et al., 2020; Akman et al., 2019; de Almeida Bianchini Campos et al., 2019; Feitoza et al., 2023; Rodrigues et al., 2022) and especially in supplements and nutraceuticals for oral administration (de Araújo Henriques Ferreira et al., 2022; Sampaio et al., 2024; da Silva et al., 2024).

Using plant-based products contributes to the increasing consumer demand for functional, nutrient-rich, and environmentally sustainable alternatives (Pimentel et al., 2021). The high abundance of soluble and insoluble fiber, macronutrients, and phenolic compounds in vegetables contributes to increasing nutritional and improving probiotic stability during the processing and shelf life of probiotic products, in addition to improving functional product effects (Lacerda Massa et al., 2020; Rodrigues et al., 2022).

Probiotic products, including supplements and nutraceuticals, are usually presented as concentrated dried cells after centrifugation of probiotic biomass from cultivation in broth media, incorporated in capsules, pills, or powders (Lacerda Massa et al., 2020; Zhong et al., 2024; da Silva et al., 2024; Wang and Zhong, 2024). Freeze-drying is one of the most used conservation methods, reducing water activity and improving the shelf life of the powdered products. Nevertheless, probiotic cells could be exposed to several stress factors, such as extreme dryness and oxidative stress, so the use of cryoprotectants could reduce these detrimental effects (Araújo et al., 2024; Kiepś and Dembczyński, 2022).

Nutraceuticals are described as nutritional and pharmaceutical appeal combinations, usually including a functional ingredient, containing integral foods or their constituents, byproducts, or isolated functional substances, such as probiotics, phenolic compounds, dietary fiber, terpenoids, unsaturated fatty acids, or vitamins (Damián et al., 2022; Sampaio et al., 2024; da Silva et al., 2024).

Probiotic viability during the storage of probiotic-carrying products does not necessarily prove the robustness of a culture when exposed to the challenging conditions found in the gastrointestinal tract (GIT) and the presence of a diverse resident microbiota (da Silva et al., 2024). The cell membrane integrity, intracellular pH, and functional enzymes of probiotic cells may be constantly challenged by stressors during their journey through the GIT (Rodrigues et al., 2019).

Fig. 3.1 shows the primary challenging conditions that probiotic cells encounter during passage through the GIT and the physiological alterations resulting from exposure of probiotic supplements and nutraceuticals to distinct GIT compartments. It also highlights the complex interactions of probiotic cells when exposed to the harsh environment of the GIT and the subsequent physiological cellular adaptive mechanisms.

After oral ingestion, the possible stressors that probiotic cells could face through the GIT are chewing, digestive enzymes, pH fluctuations, mucin layer, substrate deprivation, bile salts, osmolarity changes, and immunological factors produced in the different GIT compartments (Han et al., 2021; Bustos et al., 2024), before colonizing the gut to produce beneficial effects in the host (Han et al., 2021). However, probiotic microorganisms have strategies to manage environmental stress (Muscariello et al., 2020; Bustos et al., 2024), developing adaptive responses for each type of stress to survive. Changes in the coordination of gene suppression or expression can culminate in changes in their cell surface architecture (Sengupta et al., 2013). In addition, other modifications, including the cell surface proteins, peptidoglycan, and

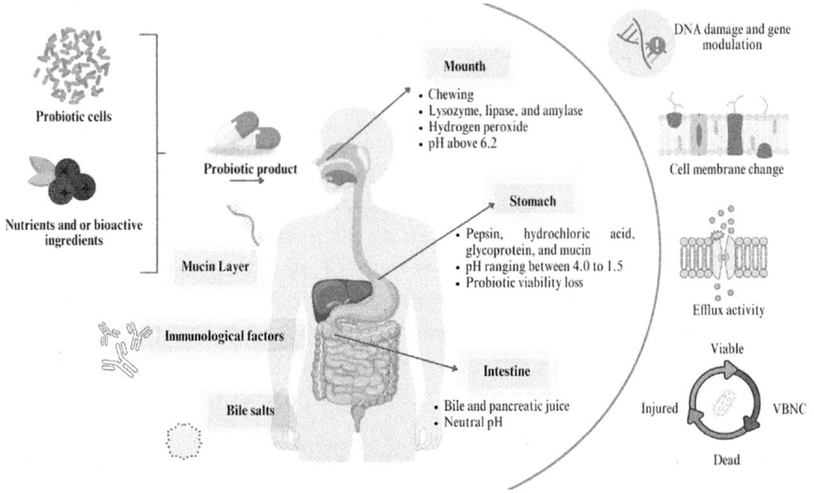

FIGURE 3.1 The primary challenging conditions of probiotic cells in gastrointestinal tract (GIT).

fatty acid biosynthesis, redox balance changes, modulation of energy production pathways, nitrogen metabolism, and alkali generation, have been reported (Cotter and Hill, 2003; Sánchez et al., 2007; Sengupta et al., 2013).

The oral cavity is the first section in which probiotic products are exposed after ingestion (Bustos et al., 2024). Saliva, an extracellular secretion, is the key to homeostasis in the mouth and contains enzymes, such as lysozyme, lipase, and amylase, and antimicrobial compounds, such as hydrogen peroxide, with a pH of around 6.2 and 7.4. In addition, chewing, a relevant process during digestion, aims to disrupt food particles during digestion (Vila et al., 2019). However, studies showing the effects of saliva or chewing on probiotic viability are still scarce (Haukioja et al., 2006; Horz et al., 2007). However, a few studies have reported no significant impacts of oral digestion on probiotic viability (de Oliveira et al., 2023; Sampaio et al., 2024), probably because of the short time spent in the oral cavity.

The stomach has been an essential step during digestion since gastric fluid secretion comprises water, pepsin, hydrochloric acid, glycoprotein, and mucin. The extreme pH, ranging between 4.0 and 1.5, harshly challenges probiotic cells since it could lead to DNA damage, peptidoglycan membrane impairments, and desaturation of essential enzymes (Wendel, 2022). During gastrointestinal digestion, the stomach step exposure typically negatively affects probiotic cell viability (Araújo et al., 2024; Athayde et al., 2024; de Oliveira et al., 2023).

A previous study investigated the impacts of simulated gastrointestinal digestion of a nutraceutical composed of *Limosilactobacillus fermentum* 296 and quercetin and or resveratrol on viable cell counts and physiological status of probiotic cells. Results showed that, considering the whole digestive model, exposure to gastric conditions (pH 2.0 and pepsin solution 25 mg/mL) was the step with significant impairments in viable cell count and physiological status of *L. fermentum* cells. However, samples without the tested phenolic compounds had more damage, suggesting a protective effect from quercetin and resveratrol during the digestive process (Sampaio et al., 2024).

In the intestine, another pH variation occurs (6.0–7.0), that is, a sudden neutralization caused by intestinal fluid. In addition, exposure to trypsin, lipase, amylase, and bile (Han et al., 2021) could cause probiotic cell denaturation and DNA impairment, affecting probiotic cell viability and survival (Castro-López et al., 2023). Physiological tools of probiotic cells to survive drastic fluctuations of the pH arise in the modulation of plasmatic membrane fatty acid composition, decreasing the permeability and increasing the proton pump efflux activity by chemiosmotic gradients (Guan and Liu, 2020).

A previous study investigated the survival rates of *Lacticaseibacillus rhamnosus* GG encapsulated with alginate-chitosan when exposed to different pH conditions (2.0, 3.0, 5.0, and 7.2) and bile salt concentrations (0.1%, 0.15%, and 0.2%). The results showed that encapsulation could protect the cells and reduce the drop in the viable cell counts throughout the tested conditions, suggesting that a protective barrier was formed when cells were encapsulated with alginate-chitosan (Athayde et al., 2024).

The effects of peristaltic movement through the GIT, a physical condition caused by a digestive physiological process, on probiotic cells have been unexplored. It could be considered a challenge for probiotic cells to colonize the GIT since contraction may modulate the colonizing microorganisms, avoiding overgrowing (Castro-López et al., 2023). The physical barrier through the GIT includes a layer of epithelial cells, their intercellular tight junctions, and the mucus covering the epithelial surface (Bustos et al., 2024).

From the esophagus to the colon, the GIT is covered by a layer of mucus consisting of water, electrolytes, and proteins. Mucin, the predominant protein in mucus, is a glycoprotein and the mucus's main structural and functional component. In the intestine, mucus is crucial in maintaining intestinal integrity, providing physical protection against chemical and biological agents. The capability of intestinal microorganisms to adhere to mucus is important since it promotes more significant colonization (Paone and Cani, 2020).

Probiotic cells require adhesion mechanisms to colonize the host persistently. The mucus-binding proteins (MUB) are bacterial surface proteins that bind to mucin proteins, enabling adherence of probiotics to host surfaces in GIT (Muscariello et al., 2020). Probiotics via MUB action could (1) promote the aggregation of cells, facilitating colonization (Muscariello et al., 2020), (2) enable bacteria to compete effectively with microbial populations for adhesion sites (Schroeder, 2019), and (3) modulate immune regulatory response, increasing intestinal protection to the host, via the Toll-like receptor 4 (TLR4) signaling pathway and inhibiting mitogen-activated protein kinase (MAPK) signaling pathway (Singh et al., 2018; Xiong et al., 2022), contributing to keeping bacterial viability under GIT conditions.

Probiotic defense mechanisms could be analogous among the same genera and species (Castro-López et al., 2023). However, investigations have suggested strain-dependent results (De Albuquerque et al., 2018; Araújo et al., 2024). The need to develop different defense mechanisms against stressful GIT conditions could result in physiological cell changes, which could reduce the viability and/or efficacy of probiotic products (Rodrigues et al., 2019).

As a consequence, the probiotic cells may be in different physiological states as alive (viable and culturable, with active and intact cells), dormant (viable but not culturable—VBNC), injured (viable and culturable in conditions of undergrowth; may be reparably damaged when causing sublethal cell damage or irreparably damaged when no grow with vigor under any conditions due to progressive metabolic decline), or dead (inactive and not culturable, no metabolic activity) (Davis, 2014). The most notable changes in the physiological status of cells during exposure to simulated gastrointestinal conditions occur after exposure to gastric conditions, causing an increase in damaged or dead subpopulations of probiotic cells even in the presence of food products with protective effects (Araújo et al., 2024).

Among these states, VBNC stands out as a protective response in which the cell has low metabolic activity, besides being unable to divide or to form a colony on an agar plate without a preceding resuscitation phase, that is, cannot replicate but still possesses the capability to do this (Foglia et al., 2020). Cells enter the VBNC state as a critical survival strategy during passage through the GIT when threatened by the stressful environment with disturbance of physiological functions (such as starvation, high osmotic, or oxygen concentrations) or have suffered some damage that can later be repaired (Ahire et al., 2022).

Nevertheless, VBNC cells could maintain metabolic activity, biosynthesis, continuous gene expression, and organic acid concentration like exponentially growing cells (Liu et al., 2017). Studies involving probiotic cells in the VBNC physiological state demonstrated that these cells can still benefit health (Davis, 2014; Pazos-Rojas et al., 2024). Microorganisms in this state can reestablish broad functioning and replicate in a more hospitable environment (Sampaio et al., 2024). This happens when VBNC reaches the colon and can resuscitate to a cultivable state due to the favorable conditions of this compartment (Ahire et al., 2022).

Additionally, irreparable damage during passage through the GIT results in probiotic cell death (Sampaio et al., 2024). These nonviable cells or parts of their structure can continue to confer health benefits as postbiotics (Vinderola et al., 2024), such as suppressing the growth of colorectal cancer, modulating immune response, having an antiinflammatory and antioxidant effect, and reducing fat deposition (Wei et al., 2024).

Supplements containing only probiotic cells are more susceptible to damage during passage through the GIT than when combined with dietary ingredients (Rodrigues et al., 2019). Obtaining information about all individual bacteria and their physiological status is relevant since many probiotic effects may depend on their metabolic activity and subsequent byproducts rather than culturability/replicating status (Davis, 2014). Therefore environmental stress, with a consequent decrease in probiotic cell survival during the passage through the GIT, is one of the reasons for the recommendation to constantly consume probiotics to promote health (Douglas and Sanders, 2008).

3.3 Technological challenges in the manufacturing and storage of probiotic supplements

Before starting large-scale manufacturing of a probiotic supplement, some preventive lab-scale steps are carried out to first isolate, characterize, and then select the probiotic strain that will undergo industrial production.

Briefly, probiotics are isolated from a starting material that is typically a human-derived specimen (e.g., feces and breast milk), although also foods, especially spontaneously fermented ones, may represent interesting sources of autochthonous probiotic microorganisms (Fentie et al., 2024; Fusco et al., 2022b). Microbial isolates (such as bacteria usually belonging to *Lactobacillus* sensu lato and *Bifidobacterium* genera, or yeasts such as *Saccharomyces cerevisiae* var. *boulardii*) are then cultivated using laboratory culture media and purified to obtain pure cultures that could be long-term stored (e.g., as frozen glycerol stock cultures), as well as further characterized to evaluate their probiotic properties.

Typically, a DNA-based characterization could be performed in silico on genomic data obtained after whole-genome sequencing of the candidate probiotic strain, to assess the absence of virulence and antibiotic-resistance-related genes, as well as the presence of genes related to beneficial effects in the human host (e.g., bacterial production of short-chain fatty acids or vitamins) (Fanelli et al., 2023; Grumet et al., 2020). Probiotic-related phenotypic abilities could also be assessed in vitro, such as the ability to adhere to the intestinal epithelial cells, modulate immune cell response, resist simulated gastrointestinal conditions, and possess antimicrobial capabilities towards pathogenic microorganisms (Fanelli et al., 2023; Forssten et al., 2020; Maldonado Galdeano et al., 2019). Then, the efficacy of the probiotic strain may be tested in vivo using animal models (typically murine, but also nematode and insect models) (Wolfe et al., 2023) and lastly verified in human clinical trials. Here, a specific dose of the probiotic microorganism, delivered through a selected administration route, is related to specific health effects (Forssten et al., 2020).

Once the probiotic candidate is chosen, the first challenge it faces from a manufacturing standpoint is its culturability under industrial conditions. Based on the laboratory stage, much information on the growth conditions, such as temperature, oxygen/carbon dioxide demand, and nutrient requirements, is known. Nevertheless, such conditions need to be tested and standardized on an industrial scale to achieve the highest possible yield of viable probiotic cells, while limiting the number of microbial generations during the production that could represent a potential risk for mutations and genetic drift with respect to the original mother culture (Fenster et al., 2019; Jackson et al., 2019). Interestingly, the presence of multiple genotypes has been reported when *Lactobacillus rhamnosus* GG-containing probiotic products were analyzed, suggesting the possibility of genetic drift (although also the presence of a mixed population in the original seed stock culture could not be excluded) (Jackson et al., 2019; Sybesma et al., 2013). To avoid this problem, usually, the frozen seed culture is intermediately inoculated into the culture liquid medium and briefly allowed to replicate for a low number of generations before being transferred into the main fermenter vessel. Alternatively, a direct vat inoculation is performed, thus utilizing a greater amount of the frozen seed culture that is employed to directly inoculate the microbial medium in the main fermenter vessel (Fenster et al., 2019).

It is important to note that the microbial medium in the fermenter vessel should contain all the nutrients to support the probiotic growth, such as water, proteins and other nitrogen sources, carbohydrates, salts, and micronutrients (Fenster et al., 2019). Obviously, food-grade ingredients should be used, and particular attention must be paid to the needs of the end-consumer. The presence of ingredients containing allergens such as those derived from soybean, gluten-containing cereals, nuts, and others (European Parliament and Council, 2011; Food Allergen Labeling and Consumer Protection Act of 2004, 2004; Food Allergy Safety, Treatment, Education, and Research Act of 2021, 2021) should be considered for possible carryover. Moreover, the presence of ingredients derived from animal products (e.g., milk and meat) should be evaluated if a probiotic product suitable for vegan and vegetarian consumers is to be manufactured (Fenster et al., 2019).

Once the desired industrial scale growth of the probiotic microorganism has been achieved, the cells are then collected by a centrifugation process that separates them from the spent liquid medium. At this stage of production, the harvested cells should undergo drying and eventually encapsulation processes that will improve their stability in the product and in the gastrointestinal tract (GIT), as well as their release in the host after intake. It should be underlined that, depending on the application of such technological processes, different probiotic products can be obtained by the manufacturer. These could be classified as first-, second-, third-, or fourth-generation probiotics (Salas-Jara et al., 2016; Trush et al., 2020), as briefly explained below.

The first-generation probiotics are defined as noncoated microorganisms that may be in a planktonic state (i.e., free-floating cells) or dried state (lyophilized) (Salas-Jara et al., 2016). Although there may be a strain-, species-, and genus-dependent resistance (Castro-López et al., 2023), first-generation probiotics may be easily damaged during transit through the GIT due to the action, among others, of enzymes and harsh pH conditions. To overcome this problem, the second-generation probiotics were conceived, consisting of lyophilized microorganisms enclosed in polymeric capsules or compressed together with natural or synthetic excipients (e.g., in tablets), that may improve their survivability in the GIT (Salas-Jara et al., 2016; Trush et al., 2020). Although some capsules are able to achieve a controlled probiotic release (Wang et al., 2022), the main drawback of second-generation probiotics may be the rapid release in the GIT (Trush et al., 2020), therefore encapsulation technologies (e.g., micro- or nanoencapsulation) were exploited in third-generation probiotics as a way for their protection and controlled GIT release (Agriopoulou et al., 2023; Trush et al., 2020). Lastly, the fourth-generation probiotics recently leveraged the possibility to deliver probiotics in the form of biofilm (Gao et al., 2022; Trush et al., 2020). Indeed, some probiotic yeast and bacterial strains can form biofilm (Salas-Jara et al., 2016; Speranza et al., 2020) that could be obtained by further growing on various food-grade materials such as polysaccharide-based materials (e.g., dextranomer microspheres and calcium alginate beads), food fibers (e.g., grape seeds and starch fiber), and minerals (e.g., smectite) (Gao et al., 2022). Being a consortium of adhered microbial cells surrounded by self-produced extracellular polymeric substances (EPS), the biofilm is therefore less susceptible to the GIT adverse conditions and has better adhesion properties compared to the planktonic counterpart. Additionally, encapsulation techniques (e.g., using polymers and minerals) can make it further resistant against environmental stresses (Gao et al., 2022).

As pointed out, probiotics in supplements should be stabilized to ensure high viability over time. Specifically, freeze-drying (also known as lyophilization) is a desiccation process that involves freezing the substance followed by sublimation of ice in a vacuum. This dehydrates sensitive products while preserving their integrity (Ward and Matejtschuk, 2021). Such a method is widely used for probiotics due to its ability to ensure high microbial viability, extended shelf life, and ease of transportation (Ge et al., 2024). However, the freezing stage of freeze-drying can cause mechanical and solute damage to probiotics, affecting cell membranes, sensitive proteins, enzymes, and DNA, potentially leading to cell death. Research emphasizes that adjusting culture conditions or using cryoprotectants can enhance strain survival. However, differences in cell size and structure among microbial genera result in varying resistance to freeze-drying across strains (Ge et al., 2024). Cryoprotectants like glycerol are used for storing bacteria in frozen form but are unsuitable as lyoprotectants for freeze-drying due to high unfrozen water content and low glass transition temperature (Tg), leading to matrix collapse (Haseley and Oetjen, 2017). Lyoprotectants must meet specific criteria, including removing intracellular water, forming a protective amorphous glass matrix, increasing Tg, and stabilizing membrane proteins through hydrogen bonding. Nonreducing disaccharides, such as sucrose, are commonly used as effective lyoprotectants (Bodzen et al., 2021).

Considering the importance of third-generation probiotics, a discussion about the techniques and materials used is necessary. Microencapsulation is the process of enclosing microparticles or droplets in a coating, or embedding them in a polymeric matrix to create tiny capsules with useful properties (Safeer Abbas et al., 2023). Successful microencapsulation methods include spray freezing, spray drying, layer-by-layer drying, spray chilling, extrusion, fluidized bed drying, electro-spraying, coacervation, and emulsification. While probiotic capsules are classified by size as macro (>5000 μm), micro (0.2–5000 μm), and nano (<0.2 μm), the survival of probiotics in the GIT decreases as the size of their encapsulation increases (Ayoub et al., 2019; Safeer Abbas et al., 2023). However, overall, these techniques facilitate the probiotic release into the target portion of the intestine in a viable and functional state (Picot and Lacroix, 2004). Moreover, it should be noted that the viability of the encapsulated probiotics is influenced by the physicochemical properties of the encapsulating materials (Chen and Chen, 2007), the most important of which are summarized in Table 3.1.

Chewable tablets and mixing probiotics with different active ingredients are other growing trends in dietary supplements, but they introduce challenges for probiotic viability. Compression during tablet production can significantly reduce viable cell count, influenced by the formulation and strain, while active ingredients may increase water activity or contain antimicrobial polyphenols. Strategies like optimizing compression pressure, selecting low water activity ingredients, and using protective packaging can help preserve probiotics, while gastric simulators can be used to predict the effects of such formulations after consumption (Fenster et al., 2019).

In the end, the final product to be delivered to the end-consumer must contain viable probiotic microorganisms that should remain stable throughout the shelf life of the probiotic product.

Usually, probiotic dietary supplements have a shelf life of up to 24 months (Grumet et al., 2020) and are transported and stored at room temperature and humidity. Since an effective dose, as assessed during the clinical human trials (Fusco et al., 2022b), has to be administered to the consumer, it is a common strategy to include in the final product an overage of microorganisms that could compensate for any possible reduction of the probiotic amount during storage (Fenster et al., 2019). Indeed, probiotics in the final products can be damaged by several factors, such as large temperature variations, pH fluctuations, light, humidity, and oxygen exposure, especially for strictly anaerobic strains (Ahire et al., 2024; Grumet et al., 2020; Torp et al., 2022). For these reasons, appropriate packaging solutions may prevent both contamination and impairment of the probiotic product (Ahire et al., 2024).

TABLE 3.1 List of the main encapsulating materials usually chosen in probiotic production.

Material	Composition	Strengths	Weaknesses
Alginate	Natural polysaccharides from algae or bacteria with ionic gelation with Ca^{2+}	Easy and scalable production method using the emulsion technique. Possibility to produce chitosan-alginate coating to improve protective effect	Uncontrollable swelling behavior and susceptibility to acid pH. Low mechanical durability
Gums	Xanthan, acacia, guar, and locust bean gums are the most important	Resistance to a wide range of pH and thermal stresses. Possibility to combine with other materials	No significant disadvantages
Carrageenan	Sulfated polysaccharide from red seaweed Gelation with K^+ and Ca^{2+} among other cations	Solubilization in water at different temperatures according to the type	No significant disadvantages
Chitosan	Chitosan produced when chitin is partially deacetylated or extracted from insects	Possibility to combine it with other polysaccharides or proteins	No significant disadvantages
Proteins	Different proteins used by achieving enzymatic or chemical cross-linking or temperature-dependent gelation	Amphiphilic nature and possibility to couple with other coating mediator. Gelatine protein has a higher oxygen resistance	Possibility to carry out allergenic compounds and need to select plant-based products for vegetarians/vegans
Lipids	Different classes with different properties, including also waxes, phospholipids, liposomes	They act well as moisture barrier. They can be combined with polysaccharides or proteins	Use of solvents can make their use expensive

(Continued)

TABLE 3.1 (Continued)

Material	Composition	Strengths	Weaknesses
Synthetic polymers	Different polymers used in a water-in-oil-in-water double emulsion method with solvent evaporation	Possibility to obtain a time-dependent release	Involvement of organic solvents during fabrication

Fenster et al. (2019) reviewed materials that may be used for probiotic product packaging, especially to prevent moisture damage. Polyethylene terephthalate (PET) should be avoided as it allows high moisture levels to enter the package. High-density polyethylene (HDPE) may represent a better solution, while glass could be the best packaging material since it has the lowest moisture vapor transmission rate (MVTR). Moreover, desiccant sachets may be used to further contribute to the maintenance of the low moisture content (Fenster et al., 2019).

In addition, oxygen scavengers may be employed to lower the oxidative stress for probiotics. For this purpose, food-grade iron-based scavengers in the form of sachets may be added to the probiotic product package, as previously proposed to preserve probiotic dairy products (Norouzbeigi et al., 2021). However, independently of the technology used, the efficacy of probiotic supplements depends on maintaining a sufficient density of live microorganisms (typically 10^8–10^9 CFU per dose) to ensure health benefits (Binda et al., 2020). Therefore it is worth mentioning that, in 2017, the International Probiotics Association, in collaboration with the Council for Responsible Nutrition, expanded the FAO/WHO guidelines by incorporating aspects of US regulations and proposed new recommendations for probiotic labeling. Product labels should specify the quantitative amount of live microorganisms, expressed as colony-forming units (CFUs), and include the expiration date. For formulations containing multiple species and/or strains, the total amount of microorganisms should be reported, and, if technically feasible, it should be specified also the amount of each contained strain/species (Mazzantini et al., 2021).

3.4 Technological challenges in the production and storage of postbiotics

Postbiotics are bioactive compounds produced by microorganisms and include short-chain fatty acids (SCFAs), exopolysaccharides (EPS), secreted proteins, peptides, and other low-molecular-weight functional compounds. Postbiotics can be associated with inactivated cells, bacterial lysates, extracellular vesicles, and cell wall fragments (Suthar et al., 2023).

The advent of the postbiotics concept has prompted the development of diverse categorization systems for postbiotics. These include classifications based on the biomolecules' chemical group (e.g., proteins, peptides, carbohydrates, and lipids), their physiological effects (e.g., antimicrobial, antiinflammatory, and immunomodulatory actions), or the microorganism involved (e.g., *Lactobacillus* spp., *Bifidobacterium* spp., and *Streptococcus* spp.) (Barros et al., 2020).

The inherent complexity of postbiotics, which are mixtures of metabolites derived from a multitude of bacterial strains with a vast array of effects and compositions, presents a significant challenge in classification. For instance, EPS, which are secreted outside bacterial cell walls, are well-established in industrial food applications due to their emulsifying, stabilizing, and water-binding properties. *Bifidobacterium* spp., for example, produces heteropolysaccharides containing D-glucose, D-galactose, and L-rhamnose. EPS composition, molecular mass, and properties depend on the strain and environmental conditions of growth (Pyclik et al., 2020; Suthar et al., 2023). EPS health benefits, such as antioxidant and antiinflammatory effects, modulated not only by the composition of monosaccharides like galactose, rhamnose, and glucosamine but also by their arrangement into the molecule, have been largely demonstrated (Chen et al., 2019). EPS produced by strains belonging to *Lactiplantibacillus plantarum* and *Lacticaseibacillus paracasei* have been successfully applied in food and pharmaceutical industries, thanks to their multifunctionality, including cholesterol-lowering activity and antiinflammatory effects (Bhat and Bajaj, 2019; Silva et al., 2019).

SCFAs (e.g., acetate, butyrate, and propionate), key metabolites of the gut microbiota, derive from the fermentation of dietary fibers. They reduce the pH in the gut, suppress pathogenic microorganisms, and contribute to immune modulation, energy metabolism, and the management of obesity (Peluzio et al., 2021). Several studies have demonstrated that different lactic acid bacteria and *Bifidobacterium* strains are prolific producers of SCFAs, with strain-specific differences in the types and amounts of SCFAs synthesized (Hati et al., 2019; Han et al., 2021). It was also reported that cross-feeding mechanisms among microorganisms enable the conversion of acetate to butyrate, thereby enhancing the metabolic benefits (da Silva et al., 2021). Additionally, SCFAs have been shown to effectively improve gut barrier integrity, regulate blood pressure, and manage lipid metabolism (Nogal et al., 2021).

Bacterial cell walls, comprising 15%–30% of total cell dry weight, include components like peptidoglycan, lipoteichoic acids (LTA), and teichoic acids. These structures modulate immune responses by interacting with host immune system receptors. For instance, lipoteichoic acids and muramyl dipeptides have been demonstrated to elicit both immune-stimulatory and antiinflammatory effects (Żółkiewicz et al., 2020). These components also demonstrate therapeutic potential, including antioxidant effects and modulation of oxidative stress (Chorawala et al., 2021).

Certain probiotics, such as *Limosilactobacillus reuteri*, produce vital vitamins like B_{12}, rendering them valuable to fortify food and address deficiencies. Vitamin B complex, particularly B_9 (folic acid) and B_{12}, is crucial for various metabolic functions and can be synthesized by gut bacteria like *Escherichia coli* and bifidobacteria (Zakrzewska et al., 2022). Probiotic-derived vitamin K_2 has also been noted for its protective roles against osteoporosis and cardiovascular diseases (Hernández-Granados and Franco-Robles, 2020).

Microbial extracellular metabolites, including bacteriocins, organic acids, and peptides, have been demonstrated to exhibit a range of biological activities, including antimicrobial, antioxidant, and immunomodulatory properties. Some of these compounds can modulate physiologic functions and are worthy of significant attention as postbiotics. For example, lactosporin, derived from *Bacillus coagulans*, has shown strong antimicrobial activity and potential applications in food preservation (Majeed et al., 2020); γ-aminobutyric acid (GABA), which can be produced by lactic acid bacteria (Di Cagno et al., 2010; Rizzello et al., 2008; Verni et al., 2022), acts as the principal inhibitory neurotransmitter in the mammalian central nervous system and plays a critical role in reducing neuronal excitability and modulating hypertensive state.

The production and storage of postbiotics face significant technological challenges, impacting their scalability, efficacy, and commercial viability. Techniques such as heat inactivation, sonication, and UV radiation are used to obtain nonviable microbes and/or to recover their metabolites. The production/stabilization process strongly affects the composition and bioactivity of postbiotics, leading to large variability in functionality; thus the selection of appropriate methods depends on the intended use, which complicates process standardization. Guided fermentation, enzymatic treatments, and synthetic biology are currently considered the most efficient tools for probiotic production (Prajapati et al., 2023).

Strains belonging to the genera *Lactobacillus* (reclassified into 25 genera), *Streptococcus, Bifidobacterium, Eubacterium, Faecalibacterium*, and *Saccharomyces* are the most widespread microorganisms employed to obtain postbiotics. The production of several fermented foods involves the activities of these microorganisms. Spontaneous or not properly assisted fermentation processes can theoretically produce postbiotics but are often characterized by unpredictable compositions that may lack sufficient concentrations to elicit meaningful physiological effects in vivo. To support their application in pharmaceutical and nutraceutical industries, controlled and efficient bioprocessing methods to standardize postbiotic production were developed (Prajapati et al., 2023; Thorakkattu et al., 2022). Guided fermentation with *Bifidobacterium* and lactic acid bacteria-selected strains was successfully adopted at the industrial level for postbiotics production (e.g., organic acids, peptides, and exopolysaccharides with antiinflammatory, antioxidant, and immunomodulatory activities) (Rafique et al., 2023; Thorakkattu et al., 2022).

Enzymatic conversion could lead to postbiotics release/synthesis from precursors, as reported for nitrite generated by using nitrate reductase on celery powder (then used as a preservative in cured meat) (Oliveira et al., 2021), urolithins obtained from ellagic acid by catechol-dehydroxylases, identified in *Gordonibacter urolithinfaciens* and *Ellagibacter isourolithinifaciens*, two bacterial species identified within the human gastrointestinal tract (García-Villalba et al., 2020).

The development and rapid analysis of extensive genetic component libraries, enabled by next-generation sequencing techniques, and the advances in genomic engineering (Marchand and Collins, 2016), significantly propelled research in the biosynthesis of postbiotics (Bober et al., 2018). Synthetic biology plays a crucial role in postbiotic production by facilitating the efficient and sustainable synthesis of valuable compounds using recombinant living organisms integrated with synthetic or biological genetic components. This approach has already enabled the production of pharmacologically important molecules, such as insulin and vaccines, through more cost-effective and scalable methods (Fang et al., 2020). Despite the potential of synthetic biology, the development of reliable microbial platforms remains resource-intensive, and regulatory hurdles regarding genetically modified organisms (GMOs) limit their large-scale employment.

Optimizing yield in postbiotic production is a further complex challenge that requires balancing multiple intricate factors, including the regulation of metabolic pathways, strain-specific variability, and precise adjustment of fermentation conditions to maximize productivity (Thorakkattu et al., 2022). Optimization involves the selection of suitable probiotic strains, engineering their metabolic pathways, refining cultivation conditions, and employing efficient downstream processing techniques, all while ensuring economic feasibility and regulatory compliance (Fenster et al., 2019). The complexity intensifies during scale-up from laboratory settings to industrial production, where additional variables such as bioreactor design, process control, and scalability of purification methods must be addressed (Du et al., 2022). Achieving successful yield optimization is essential to unlocking the full potential of postbiotics for health and industrial applications, driving innovation, and ensuring cost-effective production while maintaining quality and adherence to regulatory standards (Prajapati et al., 2023; Wegh et al., 2019).

Thermal treatments represent one of the most prevalent methods for preparing postbiotics, encompassing techniques such as pasteurization, sterilization, ohmic heating, radio-frequency heating, and microwave heating. These methods primarily rely on thermal effects to inactivate microorganisms, although their efficiency varies depending on the specific heat generation mechanism employed.

Pasteurization, a moderate heat treatment, is widely applied to inactivate microorganisms belonging to the former *Lactobacillus* genus (Yoshitake et al., 2021), *Bifidobacterium* (Sugahara et al., 2017), *Streptococcus thermophilus* (Kang et al., 2021), and *Akkermansia muciniphila* (Ashrafian et al., 2021). In contrast, sterilization involves the application of higher temperatures, typically exceeding 100°C, for shorter durations (5–30 minutes). The method is employed for the preparation of postbiotics derived from species, including *Lactobacillus* (Hsieh et al., 2016), *Enterococcus faecium* (Popović et al., 2019), and *Bifidobacterium* (Martorell et al., 2021) at temperatures ranging between 100°C and 121°C. Both methods disrupt cell membranes, degrade nucleic acids, inactivate essential enzymes, denature structural proteins, and polymerize ribosomes.

The selection of the appropriate temperature and duration for postbiotic preparation is dependent upon a number of factors, including the type of microorganism, culture conditions, growth stage, pH, and water activity. Microorganisms inactivated through pasteurization or sterilization can retain a number of beneficial health effects, including the modulation of the intestinal microbiota and the improvement of metabolic syndrome (Ashrafian et al., 2021; Nam et al., 2022).

Ohmic heating employs resistance-generated heat for sterilization processes, effectively inactivating microorganisms primarily through thermal effects, though nonthermal mechanisms also contribute (de Alwis and Fryer, 1990). During this process, an accumulation of electrical charges on cell membranes occurs, which results in the formation of pores (Knirsch et al., 2010). While nonthermal damage does occur, heat remains the principal agent of microbial inactivation (Sun et al., 2008). Compared to conventional heating, ohmic heating offers several advantages, including faster and more uniform heat distribution, which preserves the nutritional value, color, and flavor of foods (Alkanan et al., 2021).

Research into the use of ohmic heating for the preparation of inactivated probiotics has garnered significant attention. For example, *Lactobacillus acidophilus* LA-05, *Bifidobacterium animalis* Bb-12, and *Lacticaseibacillus casei* 01 were effectively inactivated at 95°C for 5–7 minutes under an 8 V/cm field at 60 Hz. These inactivated probiotics demonstrated functional benefits, such as a reduction in postprandial glycemia, when incorporated into grape-flavored whey drinks (Barros et al., 2021a,b).

Similarly, microwave and radio-frequency heating methods can inactivate microorganisms through a combination of thermal and nonthermal mechanisms. Microwave irradiation results in alterations of cell membranes, characterized by an increase in ion and electron concentrations, which in turn leads to enhanced permeability and DNA damage (Hong et al., 2004). Radio-frequency heating, on the other hand, disrupts protein folding and membrane integrity through electromagnetic effects. Nonetheless, the thermal effects of both methods are more pronounced and efficacious than their nonthermal components.

The use of nonthermal sterilization techniques is becoming more prevalent in postbiotic preparation to mitigate the loss of heat-sensitive bioactive compounds occurring during thermal processing. These methods include ionizing radiation, ultrasound, high-pressure, ultraviolet (UV) light, extreme pH, pulsed light, magnetic field heating, high-voltage electrical discharge, and gliding arc discharge. Among these techniques, ionizing radiation, UV light, ultrasound, supercritical CO_2, and extreme pH have been effectively employed in postbiotic production.

Ionizing radiation, which encompasses gamma rays, X-rays, and high-speed electrons, induces damage to nucleic acids and oxidative stress in microbial cells through water ionization. Gamma irradiation (below 10 kGy) has been demonstrated to be safe and nutritionally nondestructive, thereby enabling its application in postbiotics derived from lactic acid bacteria and *Bifidobacterium* strains (Almada et al., 2021; Kamiya et al., 2006; Zhong et al., 2024).

UV radiation, an economical and widely adopted technique, suppresses DNA transcription and translation while damaging cellular structures. UV light has minimal impact on food quality and is therefore an ideal method for microbial inactivation in juices and dairy products. Research indicates its effectiveness in postbiotic preparation, such as UV-inactivated *L. rhamnosus* GG (Lopez et al., 2008) and *Bacillus* strains (Kamilya et al., 2015).

Ultrasound treatment employs sound waves with frequencies above 20 kHz to disrupt microbial membranes and generate free radicals. It is valued for its capacity to preserve food quality while extending shelf life. Though less explored for probiotics, the ultrasonic inactivation of *L. casei* has shown benefits in modulating gut microbiota and mitigating cardiovascular risks (Brandão et al., 2021).

The use of supercritical $CO_2$for microbial inactivation is based on the alteration of intracellular pH, disruption of cell membranes, and deactivation of key enzymes. Studies have shown that *Bifidobacterium lactis* inactivated by supercritical CO_2 can influence lipid profiles and kidney function markers (Almada et al., 2021).

Additional nonthermal methods, including high-pressure treatment, pulsed electric fields, and pulsed light, also show promise for microbial inactivation. These techniques minimally impact food qualities while disrupting microbial integrity and homeostasis. However, their application to postbiotic production remains unexplored. Future research could focus on combining sterilization methods, potentially enhancing efficiency and preserving the functionality of postbiotic preparations as the field advances.

Postbiotics, which encompass a range of metabolic byproducts and cellular components derived from microorganisms, hold great potential in the food industry. Their diverse applications include the use as functional food supplements, food quality improvers, and biological preservatives. As functional supplements, postbiotics enhance health by addressing metabolic syndrome, inflammatory diseases, and microbial infections, and they can be incorporated into various foods, including fermented milk, juices, and bread, to improve nutritional benefits and metabolic health (Zhong et al., 2024). In the domain of food quality, postbiotic components like EPS improve texture, stability, and sensory attributes, particularly in dairy and bakery products. However, the low EPS yields currently limit their broader commercial use. As food preservatives, postbiotics such as bacteriocins and organic acids demonstrate bacteriostatic and antimicrobial properties. Innovations include the use of nisin-based antimicrobial packaging and organic acid-infused films, which enhance food safety, quality, and shelf life. Despite the challenges associated with commercial scalability, postbiotics offer a transformative opportunity to enhance food functionality, quality, and preservation while addressing health and safety concerns (Zhong et al., 2024). However, safety concerns and the lack of in vivo and clinical trials prevent the use of postbiotics in food technology, thus limiting commercially available postbiotics for human use (Moradi et al., 2021).

At present, regulatory agencies have yet to establish a comprehensive regulatory framework that specifically addresses the use of postbiotics in foods or dietary supplements. Although certain regulations governing the use of postbiotic formulations in medical or pharmaceutical contexts exist, a dedicated framework for their broader applications remains absent. In Europe, the European Food Safety Authority (EFSA) oversees food safety standards, frequently updating guidelines to ensure compliance, whereas the European Pharmacopoeia specifies permissible levels of live microorganisms in pharmaceutical and therapeutic products (Thorakkattu et al., 2022). In the United States, the Food and Drug Administration (FDA) has yet to issue formal guidelines specific to postbiotics. Instead, the regulatory classification chosen for a product in development determines the FDA requirements concerning its safety, efficacy, and intended use (Yelin et al., 2019). It is noteworthy that the existing probiotic regulatory frameworks from EFSA and FDA, which rely on the "generally recognized as safe" (GRAS) and "qualified presumption of safety" (QPS) lists, do not apply to postbiotics because they lack live microorganisms. This regulatory gap not only creates flexibility in the development and commercialization of postbiotic products but also raises concerns about the absence of standardized safety measures. It is therefore imperative that dedicated regulatory frameworks tailored to postbiotics are established. Until such guidelines are provided by the relevant authorities, research must focus on defining appropriate safety and regulatory standards for these preparations to ensure public health and consumer trust (Scott et al., 2022).

Postbiotics are typically more stable than live probiotics, making them a preferred choice in health and industrial applications. However, the use of postbiotics is hindered by significant stability-related challenges.

Postbiotics possess intrinsic stability across a broad range of temperatures and pH, allowing for the incorporation into foods with high acidity and before thermal treatment (Siciliano et al., 2021). Postbiotics can also be incorporated into nondairy beverages that are unsuitable for probiotics' survival due to their low pH and water activity (Barros et al., 2021b). Notably, the majority of postbiotics exhibit minimal or no interaction with other food components (Barros et al., 2020). Moreover, there is no need to consider the maintenance of postbiotic activity during the storage and transportation of postbiotic products. Furthermore, postbiotics are expected to be safer than probiotics because they have lost the ability to replicate and migrate, which effectively eliminates the risk of bacteraemia or fungemia. Thus this approach can be applied to infants with compromised immune systems, individuals with abnormal gastrointestinal mucosal barriers, and patients undergoing surgical procedures (Kumar et al., 2023).

Ensuring proper storage is therefore critical to preserve the bioactivity, effectiveness, and safety of postbiotics. Components such as supernatants, cell wall fragments, and extracellular metabolites are particularly susceptible to environmental factors like temperature, humidity, and light. To mitigate the risks associated with a decline in functionality, postbiotics are typically stored under controlled conditions, often requiring refrigeration at 2–8°C or freezing at −20°C or below to prevent microbial degradation and maintain their functionality (Vinderola et al., 2022).

Furthermore, the packaging of postbiotics is of great consequence with regard to their stability. The degradation of bioactive components can be accelerated by exposure to oxygen and moisture; therefore the use of impermeable materials and vacuum-sealed containers is necessary to limit these influences. To prevent light-induced deterioration, opaque or UV-resistant packaging is frequently employed. Moreover, regular monitoring during storage is essential to ensure the long-term stability of postbiotic products. Stability testing entails the assessment of critical parameters, including the concentration of active ingredients, pH levels, microbial contamination, and overall quality. Such tests assist manufacturers in establishing suitable shelf life and storage guidelines, thereby ensuring the efficacy of postbiotic products throughout their intended usage period (De Almeida et al., 2023).

3.5 Probiotics versus postbiotics: pros and cons of two health-promoting agents

Being live microorganisms, probiotics may be completely or partially damaged under the biotic and abiotic stresses they are subjected to during the production, storage, and ingestion, thus resulting in a population of viable, dead, and viable but not culturable (VBNC) cells (Fusco et al., 2022a; Di Lena et al., 2015). As happens for VBNC cells of pathogenic microorganisms

(Fusco et al., 2012; Fusco and Quero, 2014; Quero et al., 2014), probiotic cells in a VBNC state can still maintain some metabolic activity and, in adequate conditions, may "resuscitate" from their dormant state, reacquiring the ability to reproduce (Di Lena et al., 2015). For this reason, it should be mandatory to provide information about the number of dead, viable (and intact), and VBNC probiotic cells on the label of the probiotic products and during a clinical trial. Fiore et al. (2020) suggested using plating and flow cytometry to enumerate probiotic cells and determine their proportions of viable, dead, and VBNC cells.

Although less challenging than probiotics, the production process of postbiotics may affect their integrity and, therefore, their biological activities. Indeed, the structural components of the cells may be affected by different methods of inactivation, influencing their biological activities (Piqué et al., 2019; Ma et al., 2023). Postbiotics have a longer shelf life and easier transport and storage than probiotics. Nevertheless, to preserve their efficacy and safety, the storage and packaging of postbiotics should be appropriate. Indeed, environmental elements such as humidity, temperature, and light may influence postbiotics' stability (Prajapati et al., 2023). However, while the viability and effectiveness of probiotics are affected during the passage through the oro-gastrointestinal tract, certain postbiotics, once ingested, may be metabolized by the microbiota before reaching the large intestine, determining a reduction of the relevant health benefits (Liang and Xing, 2023).

From a safety point of view, both probiotics and postbiotics may exert potential adverse effects in specific situations. It has been reported that the main side effects of probiotics are the risk of translocation and the subsequent bacteremia and sepsis (systemic infections), excessive immune stimulations in susceptible individuals, harmful metabolic activities, as well as (antibiotic-resistance and virulence) gene transfer. (Doron and Snydman, 2015; Wang et al., 2020). Such adverse reactions mainly occur in immunocompromised individuals, such as premature newborns, pediatric patients, the elderly, pregnant women, patients in intensive care units assuming broad-spectrum antibiotics on central venous catheters, or patients critically ill with severe comorbidities (Ma et al., 2023).

Postbiotics are primarily safer than probiotics since they derive from inactivated bacteria, which cannot multiply and thus pose no theoretical risk of infection. Conversely, unlike probiotics, postbiotics do not grow or proliferate in the intestine and, therefore, do not provide a sustained effect when supplementation is ceased (Liang and Xing, 2023). However, apart from a few studies, clinical and epidemiological evidence regarding any potential risks associated with the consumption of postbiotics is lacking. Certain *in vitro* and *in vivo* studies demonstrated the nontoxic effects of postbiotics (Sabahi et al., 2023). Malagón-Rojas et al. (2020) systematically reviewed evidence on postbiotics for preventing and treating common infectious diseases among children younger than 5 years. Among seven randomized controlled trials

(RCTs) comparing postbiotics with placebo or no intervention, involving 1740 children younger than 5 years, three clinical trials reported pediatric D-lactic acidosis, a higher rate of vomiting and abdominal distention, and severe dehydration (Embleton et al., 2016; Łukasik et al., 2018; Sharieff et al., 2006). Moreover, Łukasik et al. (2018) suggested that infants fed with postbiotic acidified and fermented infant formula may develop a possible subclinical accumulation of D-lactate. No side effects were reported for the remaining RCTs (Malagón-Rojas et al., 2020). Clinical trials recently carried out by Kim et al. (2024), and other clinical trials reviewed by Sabahi et al. (2023), do not report side effects of postbiotics. Recently, Liang et al. (2024) investigated through a systematic review and metaanalysis the safety and efficacy of adding postbiotics in infant formula, concluding that the postbiotics added to infant formula are safe and not ineffective. However, Liang and Xing (2023) reviewed the most recent preclinical and clinical studies, highlighting their crucial beneficial role in treating or preventing many diseases, including those that do not have effective cures, such as multiple sclerosis and Alzheimer's disease (Aggarwal et al., 2022), or necrotizing enterocolitis in premature neonates, or newborns with very low birth weights (Sajankila et al., 2023).

These findings allow to conclude that, with equal beneficial effects on the health of the host, postbiotics represent a valuable alternative to probiotics and can be fruitfully used in human and veterinary medicine and applied in the pharmacological, food, and cosmetic industries.

References

Abdi-Moghadam, Z., Darroudi, M., Mahmoudzadeh, M., Mohtashami, M., Jamal, A.M., Shamloo, E., et al., 2023. Functional yogurt, enriched and probiotic: a focus on human health. Clinical Nutrition ESPEN 57, 575–586. Available at https://doi.org/10.1016/j.clnesp.2023.08.005, http://www.journals.elsevier.com/clinical-nutrition-espen.

Afzaal, M., Saeed, F., Saeed, M., Ahmed, A., Ateeq, H., Nadeem, M.T., et al., 2020. Survival and stability of free and encapsulated probiotic bacteria under simulated gastrointestinal conditions and in pasteurized grape juice. Journal of Food Processing and Preservation 44 (3). Available at https://doi.org/10.1111/jfpp.14346, http://onlinelibrary.wiley.com/journal/10.1111/(ISSN)1745-4549.

Aggarwal, S., Sabharwal, V., Kaushik, P., Joshi, A., Aayushi, A., Suri, M., 2022. Postbiotics: from emerging concept to application. Frontiers in Sustainable Food Systems 6. Available at https://doi.org/10.3389/fsufs.2022.887642, www.frontiersin.org/journals/sustainable-food-systems#.

Agriopoulou, S., Tarapoulouzi, M., Varzakas, T., Jafari, S.M., 2023. Application of encapsulation strategies for probiotics: from individual loading to co-encapsulation. Microorganisms 11 (12), 2896. Available at https://doi.org/10.3390/microorganisms11122896.

Ahire, J.J., Mokashe, N.U., Kashikar, M.S., Madempudi, R.S., 2022. Survival of *Limosilactobacillus reuteri* UBLRu-87 during passage through the in vitro gut model system. LWT 164, 113652. Available at https://doi.org/10.1016/j.lwt.2022.113652.

Ahire, J.J., Rohilla, A., Kumar, V., Tiwari, A., 2024. Quality management of probiotics: ensuring safety and maximizing health benefits. Current Microbiology 81 (1). Available at https://doi.org/10.1007/s00284-023-03526-3, https://www.springer.com/journal/284.

Akman, P.K., Uysal, E., Ozkaya, G.U., Tornuk, F., Durak, M.Z., 2019. Development of probiotic carrier dried apples for consumption as snack food with the impregnation of *Lactobacillus paracasei*. LWT 103, 60–68. Available at https://doi.org/10.1016/j.lwt.2018.12.070, https://www.journals.elsevier.com/lwt.

Alkanan, Z.T., Altemimi, A.B., Al-Hilphy, A.R.S., Watson, D.G., Pratap-Singh, A., 2021. Ohmic heating in the food industry: developments in concepts and applications during 2013–2020. Applied Sciences 11 (6), 2507. Available at https://doi.org/10.3390/app11062507.

Almada, C.N., Almada-Érix, C.N., Roquetto, A.R., Santos-Junior, V.A., Cabral, L., Noronha, M.F., et al., 2021. Paraprobiotics obtained by six different inactivation processes: impacts on the biochemical parameters and intestinal microbiota of Wistar male rats. International Journal of Food Sciences and Nutrition 72 (8), 1057–1070. Available at https://doi.org/10.1080/09637486.2021.1906211, http://www.tandfonline.com/loi/iijf20.

de Almeida Bianchini Campos, R.C., Martins, E.M.F., de Andrade Pires, B., do Carmo Gouveia Peluzio, M., da Rocha Campos, A.N., Ramos, A.M., et al., 2019. In vitro and in vivo resistance of *Lactobacillus rhamnosus* GG carried by a mixed pineapple (*Ananas comosus* L. Merril) and jussara (Euterpe edulis Martius) juice to the gastrointestinal tract. Food Research International 116, 1247–1257. Available at https://doi.org/10.1016/j.foodres.2018.10.012, www.elsevier.com/inca/publications/store/4/2/2/9/7/0.

de Alwis, A.A.P., Fryer, P.J., 1990. The use of network theory in the finite-element analysis of coupled thermo-electric fields. Communications in Applied Numerical Methods 6 (1), 61–66. Available at https://doi.org/10.1002/cnm.1630060110.

Amenyogbe, E., 2023. Application of probiotics for sustainable and environment-friendly aquaculture management - a review. Cogent Food and Agriculture 9 (1). Available at https://doi.org/10.1080/23311932.2023.2226425.

Araújo, C.M., Sampaio, K.B., da Silva, J.Y.P., de Oliveira, J.N., de Albuquerque, T.M.R., Lima, M.D.C., et al., 2024. Exploiting tropical fruit processing coproducts as circular resources to promote the growth and maintain the culturability and functionality of probiotic lactobacilli. Food Microbiology 123. Available at https://doi.org/10.1016/j.fm.2024.104596, https://www.sciencedirect.com/science/journal/07400020.

de Araújo Henriques Ferreira, G., Magnani, M., Cabral, L., Brandão, L.R., Noronha, M.F., de Campos Cruz, J., et al., 2022. Potentially probiotic Limosilactobacillus fermentum fruit-derived strains alleviate cardiometabolic disorders and gut microbiota impairment in male rats fed a high-fat diet. Probiotics and Antimicrobial Proteins 14 (2), 349–359. Available at https://doi.org/10.1007/s12602-021-09889-y, http://www.springer.com/new+%26+forthcoming+titles+%28default%29/journal/12602.

Arsène, M.M.J., Davares, A.K.L., Andreevna, S.L., Vladimirovich, E.A., Carime, B.Z., Marouf, R., et al., 2021. The use of probiotics in animal feeding for safe production and as potential alternatives to antibiotics. Veterinary World 14 (2), 319–328. Available at https://doi.org/10.14202/VETWORLD.2021.319-328, http://www.veterinaryworld.org/Vol.14/February-2021/3.html.

Ashrafian, F., Keshavarz Azizi Raftar, S., Lari, A., Shahryari, A., Abdollahiyan, S., Moradi, H.R., et al., 2021. Extracellular vesicles and pasteurized cells derived from Akkermansia muciniphila protect against high-fat induced obesity in mice. Microbial Cell Factories 20 (1). Available at https://doi.org/10.1186/s12934-021-01709-w, http://www.microbialcellfactories.com/home/.

Athayde, A.J.A.A., Berger, L.R.R., de Albuquerque, T.M.R., Sampaio, K.B., Fernandes, K.F.D., do Nascimento, H.M.A., et al., 2024. Physiological and technological properties of probiotic *Lacticaseibacillus rhamnosus* GG encapsulated with alginate-chitosan mixture and its incorporation in whole milk. Probiotics and Antimicrobial Proteins. Available at https://doi.org/10.1007/s12602-024-10345-w, https://www.springer.com/journal/12602.

Ayoub, A., Sood, M., Singh, J., Bandral, J.D., Gupta, N., Bhat, A., 2019. Microencapsulation and its applications in food industry. Journal of Pharmacognosy and Phytochemistry 8 (3), 32–37.

Barros, C.P., Guimarães, J.T., Esmerino, E.A., Duarte, M.C.K., Silva, M.C., Silva, R., et al., 2020. Paraprobiotics and postbiotics: concepts and potential applications in dairy products. Current Opinion in Food Science 32, 1–8. Available at https://doi.org/10.1016/j.cofs.2019.12.003, http://www.journals.elsevier.com/current-opinion-in-food-science/.

Barros, C.P., Pires, R.P.S., Guimarães, J.T., Abud, Y.K.D., Almada, C.N., Pimentel, T.C., et al., 2021a. Ohmic heating as a method of obtaining paraprobiotics: impacts on cell structure and viability by flow cytometry. Food Research International 140. Available at https://doi.org/10.1016/j.foodres.2020.110061, www.elsevier.com/inca/publications/store/4/2/2/9/7/0.

Barros, C.P., Grom, L.C., Guimarães, J.T., Balthazar, C.F., Rocha, R.S., Silva, R., et al., 2021b. Paraprobiotic obtained by ohmic heating added in whey-grape juice drink is effective to control postprandial glycemia in healthy adults. Food Research International 140. Available at https://doi.org/10.1016/j.foodres.2020.109905, www.elsevier.com/inca/publications/store/4/2/2/9/7/0.

Bhat, B., Bajaj, B.K., 2019. Hypocholesterolemic potential and bioactivity spectrum of an exopolysaccharide from a probiotic isolate *Lactobacillus paracasei* M7. Bioactive Carbohydrates and Dietary Fibre 19, 100191. Available at https://doi.org/10.1016/j.bcdf.2019.100191.

Binda, S., Hill, C., Johansen, E., Obis, D., Pot, B., Sanders, M.E., et al., 2020. Criteria to qualify microorganisms as "probiotic" in foods and dietary supplements. Frontiers in Microbiology 11. Available at https://doi.org/10.3389/fmicb.2020.01662, https://www.frontiersin.org/journals/microbiology#.

Bober, J.R., Beisel, C.L., Nair, N.U., 2018. Synthetic biology approaches to engineer probiotics and members of the human microbiota for biomedical applications. Annual Review of Biomedical Engineering 20, 277–300. Available at https://doi.org/10.1146/annurev-bioeng-062117-121019, http://www.annualreviews.org/journal/bioeng.

Bodzen, A., Jossier, A., Dupont, S., Mousset, P.Y., Beney, L., Lafay, S., et al., 2021. Design of a new lyoprotectant increasing freeze-dried *Lactobacillus strain* survival to long-term storage. BMC Biotechnology 21 (1). Available at https://doi.org/10.1186/s12896-021-00726-2, http://www.biomedcentral.com/bmcbiotechnol/.

Brandão, L.R., De Brito Alves, J.L., Da Costa, W.K.A., Ferreira, G.D.A.H., De Oliveira, M.P., Gomes Da Cruz, A., et al., 2021. Live and ultrasound-inactivated: *Lacticaseibacillus casei* modulate the intestinal microbiota and improve biochemical and cardiovascular parameters in male rats fed a high-fat diet. Food and Function 12 (12), 5287–5300. Available at https://doi.org/10.1039/d1fo01064f, http://pubs.rsc.org/en/journals/journal/fo.

Bustos, A.Y., Taranto, M.P., Gerez, C.L., Agriopoulou, S., Smaoui, S., Varzakas, T., et al., 2024. Recent advances in the understanding of stress resistance mechanisms in probiotics: relevance for the design of functional food systems. Probiotics and Antimicrobial Proteins. Available at https://doi.org/10.1007/s12602-024-10273-9, https://www.springer.com/journal/12602.

Castro-López, C., Romero-Luna, H.E., García, H.S., Vallejo-Cordoba, B., González-Córdova, A.F., Hernández-Mendoza, A., 2023. Key stress response mechanisms of probiotics during their journey through the digestive system: a review. Probiotics and Antimicrobial Proteins 15 (5), 1250–1270. Available at https://doi.org/10.1007/s12602-022-09981-x, https://www.springer.com/journal/12602.

Chen, M.J., Chen, K.N., 2007. Applications of Probiotic Encapsulation in Dairy Products Encapsulation and Controlled Release Technologies in Food Systems. Wiley Blackwell, Taiwan, pp. 83–112. Available at http://onlinelibrary.wiley.com/book/10.1002/9780470277881, https://doi.org/10.1002/9780470277881.ch4.

Chen, Y.C., Wu, Y.J., Hu, C.Y., 2019. Monosaccharide composition influence and immunomodulatory effects of probiotic exopolysaccharides. International Journal of Biological Macromolecules 133, 575–582. Available at https://doi.org/10.1016/j.ijbiomac.2019.04.109, http://www.elsevier.com/locate/ijbiomac.

Chorawala, M.R., Chauhan, S., Patel, R., Shah, G., 2021. Cell wall contents of probiotics (*Lactobacillus species*) protect against lipopolysaccharide (LPS)-induced murine colitis by limiting immuno-inflammation and oxidative stress. Probiotics and Antimicrobial Proteins 13 (4), 1005–1017. Available at https://doi.org/10.1007/s12602-020-09738-4, http://www.springer.com/new+%26+forthcoming+titles+%28default%29/journal/12602.

Cotter, P.D., Hill, C., 2003. Surviving the acid test: responses of gram-positive bacteria to low pH. Microbiology and Molecular Biology Reviews 67 (3), 429–453. Available at https://doi.org/10.1128/MMBR.67.3.429-453.2003.

Damián, M.R., Cortes-Perez, N.G., Quintana, E.T., Ortiz-Moreno, A., Noguez, C.G., Cruceño-Casarrubias, C.E., et al., 2022. Functional foods, nutraceuticals and probiotics: a focus on human health. Microorganisms 10 (5). Available at https://doi.org/10.3390/microorganisms10051065, https://www.mdpi.com/2076-2607/10/5/1065/pdf?version=1653129062.

Davis, C., 2014. Enumeration of probiotic strains: review of culture-dependent and alternative techniques to quantify viable bacteria. Journal of Microbiological Methods 103, 9–17. Available at https://doi.org/10.1016/j.mimet.2014.04.012.

De Albuquerque, T.M.R., Garcia, E.F., De Oliveira Araújo, A., Magnani, M., Saarela, M., De Souza, E.L., 2018. In vitro characterization of *Lactobacillus strains* isolated from fruit processing by-products as potential probiotics. Probiotics and Antimicrobial Proteins 10 (4), 704–716. Available at https://doi.org/10.1007/s12602-017-9318-2, http://www.springer.com/new+%26+forthcoming+titles+%28default%29/journal/12602.

De Almeida, C.V., Antiga, E., Lulli, M., 2023. Oral and topical probiotics and postbiotics in skincare and dermatological therapy: a concise review. Microorganisms 11 (6), 1420. Available at https://doi.org/10.3390/microorganisms11061420.

Di Cagno, R., Mazzacane, F., Rizzello, C.G., De Angelis, M., Giuliani, G., Meloni, M., et al., 2010. Synthesis of γ-aminobutyric acid (GABA) by *Lactobacillus plantarum* DSM19463: functional grape must beverage and dermatological applications. Applied Microbiology and Biotechnology 86 (2), 731–741. Available at https://doi.org/10.1007/s00253-009-2370-4.

Di Lena, M., Quero, G.M., Santovito, E., Verran, J., De Angelis, M., Fusco, V., 2015. A selective medium for isolation and accurate enumeration of *Lactobacillus casei*-group members in probiotic milks and dairy products. International Dairy Journal 47, 27–36. Available at https://doi.org/10.1016/j.idairyj.2015.01.018, www.elsevier.com/inca/publications/store/4/0/5/8/6/0/index.htt.

Doron, S., Snydman, D.R., 2015. Risk and safety of probiotics. Clinical Infectious Diseases 60, S129. Available at https://doi.org/10.1093/cid/civ085, http://cid.oxfordjournals.org/content/by/year.

Douglas, L.C., Sanders, M.E., 2008. Probiotics and prebiotics in dietetics practice. Journal of the American Dietetic Association 108 (3), 510–521. Available at https://doi.org/10.1016/j.jada.2007.12.009, http://www.elsevier.com/inca/publications/store/6/6/2/1/7/3/index.htt.

Du, Y.H., Wang, M.Y., Yang, L.H., Tong, L.L., Guo, D.S., Ji, X.J., 2022. Optimization and scale-up of fermentation processes driven by models. Bioengineering 9 (9). Available at https://doi.org/10.3390/bioengineering9090473, www.mdpi.com/journal/bioengineering.

Embleton, N.D., Zalewski, S., Berrington, J.E., 2016. Probiotics for prevention of necrotizing enterocolitis and sepsis in preterm infants. Current Opinion in Infectious Diseases 29 (3), 256–261. Available at https://doi.org/10.1097/QCO.0000000000000269, http://journals.lww.com/co-infectiousdiseases/pages/default.aspx.

Fanelli, F., Montemurro, M., Verni, M., Garbetta, A., Bavaro, A.R., Chieffi, D., et al., 2023. Probiotic potential and safety assessment of type strains of *Weissella* and *Periweissella* species. Microbiology Spectrum 11 (2). Available at https://doi.org/10.1128/spectrum.03047-22, https://journals.asm.org/doi/10.1128/spectrum.03047-22.

Fang, K., Park, O.J., Hong, S.H., 2020. Controlling biofilms using synthetic biology approaches. Biotechnology Advances 40. Available at https://doi.org/10.1016/j.biotechadv.2020.107518, www.elsevier.com/inca/publications/store/5/2/5/4/5/5/index.htt.

Feitoza, T.G., de Lima Ponciano Costa, B., Sampaio, K.B., dos Santos Lima, M., Garcia, E.F., de Albuquerque, T.M.R., et al., 2023. An in vitro study of the impacts of sweet potato chips with potentially probiotic *Levilactobacillus brevis* and *Lactiplantibacillus plantarum* on human intestinal microbiota: impacts of potato chips with probiotics on intestinal microbiota. Probiotics and Antimicrobial Proteins. Available at https://doi.org/10.1007/s12602-023-10168-1, https://www.springer.com/journal/12602.

Fenster, K., Freeburg, B., Hollard, C., Wong, C., Rønhave Laursen, R., Ouwehand, A.C., 2019. The production and delivery of probiotics: a review of a practical approach. Microorganisms 7 (3), 83. Available at https://doi.org/10.3390/microorganisms7030083.

Fentie, E.G., Lim, K., Jeong, M., Shin, J.H., 2024. A comprehensive review of the characterization, host interactions, and stabilization advancements on probiotics: addressing the challenges in functional food diversification. Comprehensive Reviews in Food Science and Food Safety 23 (5). Available at https://doi.org/10.1111/1541-4337.13424, http://onlinelibrary.wiley.com/journal/10.1111/(ISSN)1541-4337.

Fiore, W., Arioli, S., Guglielmetti, S., 2020. The neglected microbial components of commercial probiotic formulations. Microorganisms 8 (8), 1177. Available at https://doi.org/10.3390/microorganisms8081177.

Flach, J., van der Waal, M.B., van den Nieuwboer, M., Claassen, E., Larsen, O.F.A., 2018. The underexposed role of food matrices in probiotic products: reviewing the relationship between carrier matrices and product parameters. Critical Reviews in Food Science and Nutrition 58 (15), 2570–2584. Available at https://doi.org/10.1080/10408398.2017.1334624, https://www.tandfonline.com/loi/bfsn20.

Foglia, C., Allesina, S., Amoruso, A., De Prisco, A., Pane, M., 2020. New insights in enumeration methodologies of probiotic cells in finished products. Journal of Microbiological Methods 175, 105993. Available at https://doi.org/10.1016/j.mimet.2020.105993.

Food Allergen Labeling and Consumer Protection Act of 2004.

Food Allergy Safety, Treatment, Education, and Research Act of 2021.

Forssten, S.D., Laitila, A., Maukonen, J., Ouwehand, A.C., 2020. Probiotic triangle of success; strain production, clinical studies and product development. FEMS Microbiology Letters 367 (19). Available at https://doi.org/10.1093/femsle/fnaa167, http://femsle.oxfordjournals.org/.

Fusco, V., Quero, G.M., 2014. Culture-dependent and culture-independent nucleic-acid-based methods used in the microbial safety assessment of milk and dairy products. Comprehensive Reviews in Food Science and Food Safety 13 (4), 493–537. Available at https://doi.org/10.1111/1541-4337.12074, http://www.interscience.wiley.com/jpages/1541-4337.

Fusco, V., Riccardi, M., Quero, G.M., 2012. Thin agar layer- versus most probable number-PCR to enumerate viable and stressed *Escherichia coli* O157: H7 and application in a traditional raw milk pasta filata cheese. International Journal of Food Microbiology 159 (1), 1–8. Available at https://doi.org/10.1016/j.ijfoodmicro.2012.07.013.

Fusco, V., Chieffi, D., Benomar, N., Abriouel, H., 2022a. Indigenous probiotic microorganisms in fermented foods. Probiotics for Human Nutrition in Health and Disease. Elsevier, Italy, pp. 75–114. Available at https://www.sciencedirect.com/book/9780323899086, https://doi.org/10.1016/B978-0-323-89908-6.00014-5.

Fusco, V., Fanelli, F., Chieffi, D., 2022b. Authenticity of probiotic foods and supplements: Up-to-date situation and methods to assess it. Probiotics for Human Nutrition in Health and Disease. Elsevier, Italy, pp. 45–74. Available at https://www.sciencedirect.com/book/9780323899086, https://doi.org/10.1016/B978-0-323-89908-6.00015-7.

Fusco, V., Fanelli, F., Chieffi, D., 2022c. Authenticity of probiotic foods and dietary supplements: a pivotal issue to address. Critical Reviews in Food Science and Nutrition 62 (25), 6854–6871. Available at https://doi.org/10.1080/10408398.2021.1907300.

Gao, J., Sadiq, F.A., Zheng, Y., Zhao, J., He, G., Sang, Y., 2022. Biofilm-based delivery approaches and specific enrichment strategies of probiotics in the human gut. Gut Microbes 14 (1). Available at https://doi.org/10.1080/19490976.2022.2126274, http://www.tandfonline.com/toc/kgmi20/current.

García-Villalba, R., Beltrán, D., Frutos, M.D., Selma, M.V., Espín, J.C., Tomás-Barberán, F.A., 2020. Metabolism of different dietary phenolic compounds by the urolithin-producing human-gut bacteria: *Gordonibacter urolithinfaciens* and *Ellagibacter isourolithinifaciens*. Food and Function 11 (8), 7012–7022. Available at https://doi.org/10.1039/d0fo01649g, http://pubs.rsc.org/en/journals/journal/fo.

Ge, S., Han, J., Sun, Q., Zhou, Q., Ye, Z., Li, P., et al., 2024. Research progress on improving the freeze-drying resistance of probiotics: a review. Trends in Food Science and Technology 147, 104425. Available at https://doi.org/10.1016/j.tifs.2024.104425.

Grumet, L., Tromp, Y., Stiegelbauer, V., 2020. The development of high-quality multispecies probiotic formulations: from bench to market. Nutrients 12 (8), 2453. Available at https://doi.org/10.3390/nu12082453.

Guan, N., Liu, L., 2020. Microbial response to acid stress: mechanisms and applications. Applied Microbiology and Biotechnology 104 (1), 51–65. Available at https://doi.org/10.1007/s00253-019-10226-1.

Hadjimbei, E., Botsaris, G., Chrysostomou, S., 2022. Beneficial effects of yoghurts and probiotic fermented milks and their functional food potential. Foods 11 (17), 2691. Available at https://doi.org/10.3390/foods11172691.

Han, S., Lu, Y., Xie, J., Fei, Y., Zheng, G., Wang, Z., et al., 2021. Probiotic gastrointestinal transit and colonization after oral administration: a long journey. Frontiers in Cellular and Infection Microbiology 11. Available at https://doi.org/10.3389/fcimb.2021.609722.

Haseley, P., Oetjen, G.-W., 2017. Foundations and Process Engineering. Wiley, pp. 1–176. Available at https://doi.org/10.1002/9783527808946.ch1.

Hati, S., Patel, M., Mishra, B.K., Das, S., 2019. Short-chain fatty acid and vitamin production potentials of *Lactobacillus* isolated from fermented foods of Khasi Tribes, Meghalaya, India. Annals of Microbiology 69 (11), 1191–1199. Available at https://doi.org/10.1007/s13213-019-01500-8, https://rd.springer.com/journal/13213.

Haukioja, A., Yli-Knuuttila, H., Loimaranta, V., Kari, K., Ouwehand, A.C., Meurman, J.H., et al., 2006. Oral adhesion and survival of probiotic and other lactobacilli and bifidobacteria in vitro. Oral Microbiology and Immunology 21 (5), 326–332. Available at https://doi.org/10.1111/j.1399-302x.2006.00299.x.

Hernández-Granados, M.J., Franco-Robles, E., 2020. Postbiotics in human health: possible new functional ingredients? Food Research International 137, 109660. Available at https://doi.org/10.1016/j.foodres.2020.109660.

Hill, C., Guarner, F., Reid, G., Gibson, G.R., Merenstein, D.J., Pot, B., et al., 2014. Expert consensus document: the international scientific association for probiotics and prebiotics consensus statement on the scope and appropriate use of the term probiotic. Nature Reviews Gastroenterology and Hepatology 11 (8), 506–514. Available at https://doi.org/10.1038/nrgastro.2014.66, http://www.nature.com/nrgastro/index.html.

Hong, S.M., Park, J.K., Lee, Y.O., 2004. Mechanisms of microwave irradiation involved in the destruction of fecal coliforms from biosolids. Water Research 38 (6), 1615–1625. Available at https://doi.org/10.1016/j.watres.2003.12.011, www.elsevier.com/locate/watres.

Horz, H.-P., Meinelt, A., Houben, B., Conrads, G., 2007. Distribution and persistence of probiotic *Streptococcus salivarius* K12 in the human oral cavity as determined by real-time quantitative polymerase chain reaction. Oral Microbiology and Immunology 22 (2), 126–130. Available at https://doi.org/10.1111/j.1399-302x.2007.00334.x.

Hsieh, F.C., Lan, C.C.E., Huang, T.Y., Chen, K.W., Chai, C.Y., Chen, W.T., et al., 2016. Heat-killed and live: *Lactobacillus reuteri* GMNL-263 exhibit similar effects on improving metabolic functions in high-fat diet-induced obese rats. Food and Function 7 (5), 2374–2388. Available at https://doi.org/10.1039/c5fo01396h, http://www.rsc.org/Publishing/Journals/FO/about.asp.

Jackson, S.A., Schoeni, J.L., Vegge, C., Pane, M., Stahl, B., Bradley, M., et al., 2019. Improving end-user trust in the quality of commercial probiotic products. Frontiers in Microbiology 10. Available at https://doi.org/10.3389/fmicb.2019.00739, https://www.frontiersin.org/journals/microbiology#.

Kamiya, T., Wang, L., Forsythe, P., Goettsche, G., Mao, Y., Wang, Y., et al., 2006. Inhibitory effects of *Lactobocillus reuteri* on visceral pain induced by colorectal distension in Sprague-Dawley rats. Gut 55 (2), 191–196. Available at https://doi.org/10.1136/gut.2005.070987.

Kamilya, D., Baruah, A., Sangma, T., Chowdhury, S., Pal, P., 2015. Inactivated probiotic bacteria stimulate cellular immune responses of Catla, Catla catla (Hamilton) in vitro. Probiotics and Antimicrobial Proteins 7 (2), 101–106. Available at https://doi.org/10.1007/s12602-015-9191-9.

Kang, X., Liang, H., Luo, Y., Li, Z., He, F., Han, X., et al., 2021. Anti-adipogenesis and metabolism-regulating effects of heat-inactivated *Streptococcus thermophilus* MN-ZLW-002. Letters in Applied Microbiology 72 (6), 677–687. Available at https://doi.org/10.1111/lam.13398.

Kiepś, J., Dembczyński, R., 2022. Current trends in the production of probiotic formulations. Foods 11 (15), 2330. Available at https://doi.org/10.3390/foods11152330.

Kim, M.S., Kim, H.J., Kang, S.M., Heo, Y.M., Kang, J., Ryu, T.K., et al., 2024. Efficacy and safety of topical *Streptococcus postbiotic* emollient in adolescents and adults with mild-to-moderate atopic dermatitis: a randomized, double-blind, vehicle-controlled trial. Allergy: European Journal of Allergy and Clinical Immunology 79 (6), 1612–1616. Available at https://doi.org/10.1111/all.16077, http://onlinelibrary.wiley.com/journal/10.1111/(ISSN)1398-9995.

Knirsch, M.C., Alves dos Santos, C., Martins de Oliveira Soares Vicente, A.A., Vessoni Penna, T.C., 2010. Ohmic heating - a review. Trends in Food Science and Technology 21 (9), 436–441. Available at https://doi.org/10.1016/j.tifs.2010.06.003.

Kumar, H., Schütz, F., Bhardwaj, K., Sharma, R., Nepovimova, E., Dhanjal, D.S., et al., 2023. Recent advances in the concept of paraprobiotics: nutraceutical/functional properties for promoting children health. Critical Reviews in Food Science and Nutrition 63 (19),

3943–3958. Available at https://doi.org/10.1080/10408398.2021.1996327, https://www.tandfonline.com/loi/bfsn20.

Lacerda Massa, N.M., Dantas Duarte Menezes, F.N., de Albuquerque, T.M.R., de Oliveira, S.P.A., Lima, Md.S., Magnani, M., et al., 2020. Effects of digested jabuticaba (*Myrciaria jaboticaba* (Vell.) Berg) by-product on growth and metabolism of *Lactobacillus* and *Bifidobacterium* indicate prebiotic properties. LWT 131. Available at https://doi.org/10.1016/j.lwt.2020.109766, https://www.journals.elsevier.com/lwt.

Latif, A., Shehzad, A., Niazi, S., Zahid, A., Ashraf, W., Iqbal, M.W., et al., 2023. Probiotics: mechanism of action, health benefits and their application in food industries. Frontiers in Microbiology 14. Available at https://doi.org/10.3389/fmicb.2023.1216674, https://www.frontiersin.org/journals/microbiology#.

Leistikow, K.R., Beattie, R.E., Hristova, K.R., 2022. Probiotics beyond the farm: benefits, costs, and considerations of using antibiotic alternatives in livestock. Frontiers in Antibiotics 1. Available at https://doi.org/10.3389/frabi.2022.1003912, https://www.frontiersin.org/journals/antibiotics.

Liang, B., Xing, D., 2023. The current and future perspectives of postbiotics. Probiotics and Antimicrobial Proteins 15 (6), 1626–1643. Available at https://doi.org/10.1007/s12602-023-10045-x.

Liang, X., Li, Y., Zhao, Z., Ding, R., Sun, J., Chi, C., 2024. Safety and efficacy of adding postbiotics in infant formula: a systematic review and meta-analysis. Pediatric Research 95, 43–51. Available at https://doi.org/10.1038/s41390-023-02813-w.

Liu, J., Li, L., Li, B., Peters, B.M., Deng, Y., Xu, Z., et al., 2017. Study on spoilage capability and VBNC state formation and recovery of *Lactobacillus plantarum*. Microbial Pathogenesis 110, 257–261. Available at https://doi.org/10.1016/j.micpath.2017.06.044, http://www.elsevier.com/inca/publications/store/6/2/2/9/1/5/index.htt.

Liu, M., Chen, J., Dharmasiddhi, I.P.W., Chen, S., Liu, Y., Liu, H., 2024. Review of the potential of probiotics in disease treatment: mechanisms, engineering, and applications. Processes 12 (2). Available at https://doi.org/10.3390/pr12020316, http://www.mdpi.com/journal/processes.

Lopez, M., Li, N., Kataria, J., Russell, M., Neu, J., 2008. Live and ultraviolet-inactivated *Lactobacillus Rhamnosus* GG decrease flagellin-induced interleukin-8 production in Caco-2 cells. The Journal of Nutrition 138 (11), 2264–2268. Available at https://doi.org/10.3945/jn.108.093658.

Łukasik, J., Salminen, S., Szajewska, H., 2018. Rapid review shows that probiotics and fermented infant formulas do not cause d -lactic acidosis in healthy children. Acta Paediatrica 107 (8), 1322–1326. Available at https://doi.org/10.1111/apa.14338.

Ma, L., Tu, H., Chen, T., 2023. Postbiotics in human health: a narrative review. Nutrients 15 (2), 291. Available at https://doi.org/10.3390/nu15020291.

Majeed, M., Majeed, S., Nagabhushanam, K., Lawrence, L., Arumugam, S., Mundkur, L., 2020. Skin protective activity of lactosporin-the extracellular metabolite from *Bacillus coagulans* MTCC 5856. Cosmetics 7 (4), 76. Available at https://doi.org/10.3390/cosmetics7040076.

Malagón-Rojas, J.N., Mantziari, A., Salminen, S., Szajewska, H., 2020. Postbiotics for preventing and treating common infectious diseases in children: a systematic review. Nutrients 12 (2), 389. Available at https://doi.org/10.3390/nu12020389.

Maldonado Galdeano, C., Cazorla, S.I., Lemme Dumit, J.M., Vélez, E., Perdigón, G., 2019. Beneficial effects of probiotic consumption on the immune system. Annals of Nutrition and Metabolism 74 (2), 115–124. Available at https://doi.org/10.1159/000496426, www.karger.ch/journals/anm/anm_jh.htm.

Marchand, N., Collins, C.H., 2016. Synthetic quorum sensing and cell-cell communication in gram-positive bacillus megaterium. ACS Synthetic Biology 5 (7), 597–606. Available at https://doi.org/10.1021/acssynbio.5b00099, http://pubs.acs.org/journal/asbcd6.

Martorell, P., Alvarez, B., Llopis, S., Navarro, V., Ortiz, P., Gonzalez, N., et al., 2021. Heat-treated *Bifidobacterium longum* CECT-7347: a whole-cell postbiotic with antioxidant, anti-inflammatory, and gut-barrier protection properties. Antioxidants 10 (4), 536. Available at https://doi.org/10.3390/antiox10040536.

Mazzantini, D., Calvigioni, M., Celandroni, F., Lupetti, A., Ghelardi, E., 2021. Spotlight on the compositional quality of probiotic formulations marketed worldwide. Frontiers in Microbiology 12. Available at https://doi.org/10.3389/fmicb.2021.693973.

Moradi, M., Molaei, R., Guimarães, J.T., 2021. A review on preparation and chemical analysis of postbiotics from lactic acid bacteria. Enzyme and Microbial Technology 143, 109722. Available at https://doi.org/10.1016/j.enzmictec.2020.109722.

Muscariello, L., De Siena, B., Marasco, R., 2020. *Lactobacillus* cell surface proteins involved in interaction with mucus and extracellular matrix components. Current Microbiology 77 (12), 3831–3841. Available at https://doi.org/10.1007/s00284-020-02243-5.

Nam, Y.H., Yoon, S., Baek, J., Kim, J.-H., Park, M., Hwang, K.W., Kim, W., 2022. Heat-killed *Lactiplantibacillus plantarum* LRCC5314 mitigates the effects of stress-related type 2 diabetes in mice via gut microbiome modulation. Journal of Microbiology and Biotechnology 32, 324–332. Available at https://doi.org/10.4014/jmb.2111.11008.

Nguyen, M., Ferge, K.K., Vaughn, A.R., Burney, W., Teng, L.H., Pan, A., et al., 2020. Probiotic supplementation and food intake and knowledge among patients and consumers. Probiotics and Antimicrobial Proteins 12 (3), 824–833. Available at https://doi.org/10.1007/s12602-019-09602-0, http://www.springer.com/new+%26+forthcoming+titles+%28default%29/journal/12602.

Nogal, A., Valdes, A.M., Menni, C., 2021. The role of short-chain fatty acids in the interplay between gut microbiota and diet in cardio-metabolic health. Gut Microbes 13 (1), 1–24. Available at https://doi.org/10.1080/19490976.2021.1897212, http://www.tandfonline.com/toc/kgmi20/current.

Norouzbeigi, S., Vahid-Dastjerdi, L., Yekta, R., Farhoodi, M., Mortazavian, A.M., 2021. Effects of using different O_2 scavengers on the qualitative attributes of bifidus yogurt during refrigerated storage. Food Research International 140, 109953. Available at https://doi.org/10.1016/j.foodres.2020.109953.

Oguntoyinbo, F.A., Fusco, V., Cho, G.S., Kabisch, J., Neve, H., Bockelmann, W., et al., 2016. Produce from Africa's gardens: potential for leafy vegetable and fruit fermentations. Frontiers in Microbiology 7. Available at https://doi.org/10.3389/fmicb.2016.00981, http://journal.frontiersin.org/article/10.3389/fmicb.2016.00981/full.

de Oliveira, S.P.A., do Nascimento, H.M.A., Rodrigues, N.P.A., Sampaio, K.B., Lima, M.D.S., da Conceição, M.L., et al., 2023. Different parts from the whole red beet (*Beta vulgaris* L.) valorization with stimulatory effects on probiotic lactobacilli and protection against gastrointestinal conditions. Food Bioscience 52. Available at https://doi.org/10.1016/j.fbio.2023.102439, http://www.journals.elsevier.com/food-bioscience/.

Oliveira, W.A., Rodrigues, A.R.P., Oliveira, F.A., Oliveira, V.S., Laureano-Melo, R., Stutz, E.T.G., et al., 2021. Potentially probiotic or postbiotic pre-converted nitrite from celery produced by an axenic culture system with probiotic lacticaseibacilli strain. Meat Science 174. Available at https://doi.org/10.1016/j.meatsci.2020.108408, http://www.elsevier.com/locate/meatsci.

Paone, P., Cani, P.D., 2020. Mucus barrier, mucins and gut microbiota: the expected slimy partners? Gut 69 (12), 2232–2243. Available at https://doi.org/10.1136/gutjnl-2020-322260, http://gut.bmj.com/content/by/year.

Pazos-Rojas, L.A., Cuellar-Sánchez, A., Romero-Cerón, A.L., Rivera-Urbalejo, A., Van Dillewijn, P., Luna-Vital, D.A., et al., 2024. The viable but non-culturable (VBNC) state, a poorly explored aspect of beneficial bacteria. Microorganisms 12 (1). Available at https://doi.org/10.3390/microorganisms12010039, http://www.mdpi.com/journal/microorganisms.

Peluzio, M.D.C.G., Martinez, J.A., Milagro, F.I., 2021. Postbiotics: metabolites and mechanisms involved in microbiota-host interactions. Trends in Food Science and Technology 108, 11–26. Available at https://doi.org/10.1016/j.tifs.2020.12.004, http://www.elsevier.com/wps/find/journaldescription.cws_home/601278/description#description.

Picot, A., Lacroix, C., 2004. Encapsulation of bifidobacteria in whey protein-based microcapsules and survival in simulated gastrointestinal conditions and in yoghurt. International Dairy Journal 14 (6), 505–515. Available at https://doi.org/10.1016/j.idairyj.2003.10.008.

Pimentel, T.C., Costa, W.K.A.D., Barão, C.E., Rosset, M., Magnani, M., 2021. Vegan probiotic products: a modern tendency or the newest challenge in functional foods. Food Research International 140. Available at https://doi.org/10.1016/j.foodres.2020.110033, http://www.elsevier.com/inca/publications/store/4/2/2/9/7/0.

Piqué, N., Berlanga, M., Miñana-Galbis, D., 2019. Health benefits of heat-killed (tyndallized) probiotics: an overview. International Journal of Molecular Sciences 20 (10), 2534. Available at https://doi.org/10.3390/ijms20102534.

Popović, N., Djokić, J., Brdarić, E., Dinić, M., Terzić-Vidojević, A., Golić, N., et al., 2019. The influence of heat-killed Enterococcus faecium BGPAS1-3 on the tight junction protein expression and immune function in differentiated Caco-2 cells infected with *Listeria monocytogenes* ATCC 19111. Frontiers in Microbiology 10. Available at https://doi.org/10.3389/fmicb.2019.00412.

Prajapati, N., Patel, J., Singh, S., Yadav, V.K., Joshi, C., Patani, A., et al., 2023. Postbiotic production: harnessing the power of microbial metabolites for health applications. Frontiers in Microbiology 14. Available at https://doi.org/10.3389/fmicb.2023.1306192, https://www.frontiersin.org/journals/microbiology#.

Pyclik, M., Srutkova, D., Schwarzer, M., Górska, S., 2020. Bifidobacteria cell wall-derived exo-polysaccharides, lipoteichoic acids, peptidoglycans, polar lipids and proteins – their chemical structure and biological attributes. International Journal of Biological Macromolecules 147, 333–349. Available at https://doi.org/10.1016/j.ijbiomac.2019.12.227, http://www.elsevier.com/locate/ijbiomac.

Quero, G.M., Santovito, E., Visconti, A., Fusco, V., 2014. Quantitative detection of *Listeria monocytogenes* in raw milk and soft cheeses: culture-independent versus liquid- and solid-based culture-dependent real time PCR approaches. LWT - Food Science and Technology 58 (1), 11–20. Available at https://doi.org/10.1016/j.lwt.2014.03.005.

Quintieri, L., Fanelli, F., Monaci, L., Fusco, V., 2024. Milk and its derivatives as sources of components and microorganisms with health-promoting properties: probiotics and bioactive peptides. Foods 13 (4), 601. Available at https://doi.org/10.3390/foods13040601.

Rafique, N., Jan, S.Y., Dar, A.H., Dash, K.K., Sarkar, A., Shams, R., et al., 2023. Promising bioactivities of postbiotics: a comprehensive review. Journal of Agriculture and Food Research 14. Available at https://doi.org/10.1016/j.jafr.2023.100708, http://www.journals.elsevier.com/journal-of-agriculture-and-food-research.

Raman, J., Kim, J.S., Choi, K.R., Eun, H., Yang, D., Ko, Y.J., et al., 2022. Application of lactic acid bacteria (LAB) in sustainable agriculture: advantages and limitations. International

Journal of Molecular Sciences 23 (14). Available at https://doi.org/10.3390/ijms23147784, http://www.mdpi.com/journal/ijms.

Rizzello, C.G., Cassone, A., Di Cagno, R., Gobbetti, M., 2008. Synthesis of angiotensin i-converting enzyme (ACE)-inhibitory peptides and γ-aminobutyric acid (GABA) during sourdough fermentation by selected lactic acid bacteria. Journal of Agricultural and Food Chemistry 56 (16), 6936–6943. Available at https://doi.org/10.1021/jf800512u.

Rodrigues, V.Cd.C., Silva, L.G.Sd, Simabuco, F.M., Venema, K., Antunes, A.E.C., 2019. Survival, metabolic status and cellular morphology of probiotics in dairy products and dietary supplement after simulated digestion. Journal of Functional Foods 55, 126–134. Available at https://doi.org/10.1016/j.jff.2019.01.046, http://www.elsevier.com/wps/find/journaldescription.cws_home/717426/description#description.

Rodrigues, N.P.A., Garcia, E.F., Sampaio, K.B., do Nascimento, H.M.A., de Sousa Guedes, J.P., de Souza, E.L., 2022. Dynamics of physiological responses of potentially probiotic fruit-derived *Limosilactobacillus fermentum* in apple and orange juices during refrigeration storage and exposure to simulated gastrointestinal conditions. Archives of Microbiology 204 (1). Available at https://doi.org/10.1007/s00203-021-02672-1, http://link.springer.de/link/service/journals/00203/index.htm.

Sabahi, S., Homayouni Rad, A., Aghebati-Maleki, L., Sangtarash, N., Ozma, M.A., Karimi, A., et al., 2023. Postbiotics as the new frontier in food and pharmaceutical research. Critical Reviews in Food Science and Nutrition 63 (26), 8375–8402. Available at https://doi.org/10.1080/10408398.2022.2056727, https://www.tandfonline.com/loi/bfsn20.

Safeer Abbas, M., Afzaal, M., Saeed, F., Asghar, A., Jianfeng, L., Ahmad, A., et al., 2023. Probiotic viability as affected by encapsulation materials: recent updates and perspectives. International Journal of Food Properties 26 (1), 1324–1350. Available at https://doi.org/10.1080/10942912.2023.2213408, https://www.tandfonline.com/loi/ljfp20.

Sajankila, N., Wala, S.J., Ragan, M.V., Volpe, S.G., Dumbauld, Z., Purayil, N., et al., 2023. Current and future methods of probiotic therapy for necrotizing enterocolitis. Frontiers in Pediatrics 11. Available at https://doi.org/10.3389/fped.2023.1120459, https://www.frontiersin.org/journals/pediatrics.

Salas-Jara, M., Ilabaca, A., Vega, M., García, A., 2016. Biofilm forming lactobacillus: new challenges for the development of probiotics. Microorganisms 4 (3), 35. Available at https://doi.org/10.3390/microorganisms4030035.

Salminen, S., Collado, M.C., Endo, A., Hill, C., Lebeer, S., Quigley, E.M.M., et al., 2021. The International Scientific Association of Probiotics and Prebiotics (ISAPP) consensus statement on the definition and scope of postbiotics. Nature Reviews Gastroenterology and Hepatology 18 (9), 649–667. Available at https://doi.org/10.1038/s41575-021-00440-6, http://www.nature.com/nrgastro/index.html.

Sampaio, K.B., de Brito Alves, J.L., do Nascimento, Y.M., Tavares, J.F., da Silva, M.S., dos Santos Nascimento, D., et al., 2024. Effects of simulated gastrointestinal conditions on combined potentially probiotic *Limosilactobacillus fermentum* 296, quercetin, and/or resveratrol as bioactive components of novel nutraceuticals. Probiotics and Antimicrobial Proteins 16 (1), 308–319. Available at https://doi.org/10.1007/s12602-023-10046-w, https://www.springer.com/journal/12602.

Sampaio, K.B., Fusco, V., Alves, J.Ld.B., de Souza, E.L., 2022. Probiotics: Concepts, evolution, and Applications Probiotics for Human Nutrition in Health and Disease. Elsevier, Brazil, pp. 3–24. Available at https://www.sciencedirect.com/book/9780323899086, https://doi.org/10.1016/B978-0-323-89908-6.00019-4.

Sánchez, B., Champomier-Vergès, M.C., Stuer-Lauridsen, B., Ruas-Madiedo, P., Anglade, P., Baraige, F., et al., 2007. Adaptation and response of *Bifidobacterium animalis* subsp. lactis to bile: a proteomic and physiological approach. Applied and Environmental Microbiology 73 (21), 6757–6767. Available at https://doi.org/10.1128/AEM.00637-07.

Schroeder, B.O., 2019. Fight them or feed them: how the intestinal mucus layer manages the gut microbiota. Gastroenterology Report 7 (1), 3–12. Available at https://doi.org/10.1093/gastro/goy052, http://gastro.oxfordjournals.org/?code=gastro&homepage=&buy=&.cgifields=code.

Scott, E., De Paepe, K., Van de Wiele, T., 2022. Postbiotics and their health modulatory biomolecules. Biomolecules 12 (11), 1640. Available at https://doi.org/10.3390/biom12111640.

Sengupta, R., Altermann, E., Anderson, R.C., McNabb, W.C., Moughan, P.J., Roy, N.C., 2013. The role of cell surface architecture of lactobacilli in host-microbe interactions in the gastrointestinal tract. Mediators of Inflammation 2013. Available at https://doi.org/10.1155/2013/237921.

Sharieff, W., Bhutta, Z., Schauer, C., Tomlinson, G., 2006. Micronutrients (including zinc) reduce diarrhoea in children: the Pakistan sprinkles diarrhoea study. Archives of Disease in Childhood 91 (7), 573–579. Available at https://doi.org/10.1136/adc.2005.086199.

Siciliano, R.A., Reale, A., Mazzeo, M.F., Morandi, S., Silvetti, T., Brasca, M., 2021. Paraprobiotics: a new perspective for functional foods and nutraceuticals. Nutrients 13 (4). Available at https://doi.org/10.3390/nu13041225, https://www.mdpi.com/2072-6643/13/4/1225/pdf.

da Silva, T.F., Casarotti, S.N., de Oliveira, G.L.V., Penna, A.L.B., 2021. The impact of probiotics, prebiotics, and synbiotics on the biochemical, clinical, and immunological markers, as well as on the gut microbiota of obese hosts. Critical Reviews in Food Science and Nutrition 61 (2), 337–355. Available at https://doi.org/10.1080/10408398.2020.1733483, https://www.tandfonline.com/loi/bfsn20.

da Silva, J.Y.P., do Nascimento, H.M.A., de Albuquerque, T.M.R., Sampaio, K.B., dos Santos Lima, M., Monteiro, M., et al., 2024. Revealing the potential impacts of nutraceuticals formulated with freeze-dried jabuticaba peel and *Limosilactobacillus fermentum* strains candidates for probiotic use on human intestinal microbiota. Probiotics and Antimicrobial Proteins 16 (5), 1773–1789. Available at https://doi.org/10.1007/s12602-023-10134-x, https://www.springer.com/journal/12602.

Silva, L.A., Lopes Neto, J.H.P., Cardarelli, H.R., 2019. Exopolysaccharides produced by *Lactobacillus plantarum*: technological properties, biological activity, and potential application in the food industry. Annals of Microbiology 69 (4), 321–328. Available at https://doi.org/10.1007/s13213-019-01456-9, http://springerlink.com/content/121707/.

Singh, K.S., Kumar, S., Mohanty, A.K., Grover, S., Kaushik, J.K., 2018. Mechanistic insights into the host-microbe interaction and pathogen exclusion mediated by the Mucus-binding protein of *Lactobacillus plantarum*. Scientific Reports 8 (1). Available at https://doi.org/10.1038/s41598-018-32417-y, www.nature.com/srep/index.html.

de Souza, E.L., Alves, J.Ld.B., Fusco, V., 2022. Probiotics for Human Nutrition in Health and Disease Probiotics for Human Nutrition in Health and Disease. Elsevier, Brazil, pp. 1–594. Available at https://www.sciencedirect.com/book/9780323899086, 10.1016/C2020-0-03501-2.

Speranza, B., Corbo, M.R., Campaniello, D., Altieri, C., Sinigaglia, M., Bevilacqua, A., 2020. Biofilm formation by potentially probiotic *Saccharomyces cerevisiae* strains. Food Microbiology 87, 103393. Available at https://doi.org/10.1016/j.fm.2019.103393.

Sugahara, H., Yao, R., Odamaki, T., Xiao, J.Z., 2017. Differences between live and heat-killed bifidobacteria in the regulation of immune function and the intestinal environment. Beneficial Microbes 8 (3), 463–472. Available at https://doi.org/10.3920/bm2016.0158.

Sun, H., Kawamura, S., Himoto, J.I., Itoh, K., Wada, T., Kimura, T., 2008. Effects of ohmic heating on microbial counts and denaturation of proteins in milk. Food Science and Technology Research 14 (2), 117–123. Available at https://doi.org/10.3136/fstr.14.117.

Suthar, P., Kumar, S., Kumar, V., Sharma, V., Dhiman, A., 2023. Postbiotics: an exposition on next generation functional food compounds-opportunities and challenges. Critical Reviews in Food Science and Nutrition 8, 1–20. Available at https://doi.org/10.1080/10408398.2023.2289646, https://doi.org/10.1080/10408398.2023.2289646.

Sybesma, W., Molenaar, D., van IJcken, W., Venema, K., Kort, R., 2013. Genome instability in *Lactobacillus rhamnosus* GG. Applied and Environmental Microbiology 79 (7), 2233–2239. Available at https://doi.org/10.1128/aem.03566-12.

Thorakkattu, P., Khanashyam, A.C., Shah, K., Babu, K.S., Mundanat, A.S., Deliephan, A., et al., 2022. Postbiotics: current trends in food and pharmaceutical industry. Foods 11 (19). Available at https://doi.org/10.3390/foods11193094, http://www.mdpi.com/journal/foods.

Todorov, S.D., Lima, J.M.S., Bucheli, J.E.V., Popov, I.V., Tiwari, S.K., Chikindas, M.L., 2024. Probiotics for aquaculture: hope, truth, and reality. Probiotics and Antimicrobial Proteins 16 (6), 2007–2020. Available at https://doi.org/10.1007/s12602-024-10290-8, https://www.springer.com/journal/12602.

Torp, A.M., Bahl, M.I., Boisen, A., Licht, T.R., 2022. Optimizing oral delivery of next generation probiotics. Trends in Food Science and Technology 119, 101–109. Available at https://doi.org/10.1016/j.tifs.2021.11.034, http://www.elsevier.com/wps/find/journaldescription.cws_home/601278/description#description.

Trush, E.A., Poluektova, E.A., Beniashvilli, A.G., Shifrin, O.S., Poluektov, Y.M., Ivashkin, V.T., 2020. The evolution of human probiotics: challenges and prospects. Probiotics and Antimicrobial Proteins 12 (4), 1291–1299. Available at https://doi.org/10.1007/s12602-019-09628-4, http://www.springer.com/new+%26+forthcoming+titles+%28default%29/journal/12602.

Verni, M., Vekka, A., Immonen, M., Katina, K., Rizzello, C.G., Coda, R., 2022. Biosynthesis of γ-aminobutyric acid by lactic acid bacteria in surplus bread and its use in bread making. Journal of Applied Microbiology 133 (1), 76–90. Available at https://doi.org/10.1111/jam.15332, https://academic.oup.com/jambio.

Vila, T., Rizk, A.M., Sultan, A.S., Jabra-Rizk, M.A., 2019. The power of saliva: antimicrobial and beyond. PLoS Pathogens 15 (11). Available at https://doi.org/10.1371/journal.ppat.1008058, https://journals.plos.org/plospathogens/article/file?id=10.1371/journal.ppat.1008058&type=printable.

Vinderola, G., Sanders, M.E., Salminen, S., 2022. The concept of postbiotics. Foods 11 (8), 1077. Available at https://doi.org/10.3390/foods11081077.

Vinderola, G., Sanders, M.E., Cunningham, M., Hill, C., 2024. Frequently asked questions about the ISAPP postbiotic definition. Frontiers in Microbiology 14. Available at https://doi.org/10.3389/fmicb.2023.1324565.

Wang, A., Zhong, Q., 2024. Drying of probiotics to enhance the viability during preparation, storage, food application, and digestion: a review. Comprehensive Reviews in Food Science and Food Safety 23 (1). Available at https://doi.org/10.1111/1541-4337.13287.

Wang, Y., Jiang, Y., Deng, Y., Yi, C., Wang, Y., Ding, M., et al., 2020. Probiotic supplements: hope or hype? Frontiers in Microbiology 11. Available at https://doi.org/10.3389/fmicb.2020.00160.

Wang, G., Chen, Y., Xia, Y., Song, X., Ai, L., 2022. Characteristics of probiotic preparations and their applications. Foods 11 (16), 2472. Available at https://doi.org/10.3390/foods11162472.

Ward, K.R., Matejtschuk, P., 2021. The principles of freeze-drying and application of analytical technologies. Methods in Molecular Biology 2180, 99–127. Available at https://doi.org/10.1007/978-1-0716-0783-1_3, http://www.springer.com/series/7651.

Wegh, C.A.M., Geerlings, S.Y., Knol, J., Roeselers, G., Belzer, C., 2019. Postbiotics and their potential applications in early life nutrition and beyond. International Journal of Molecular Sciences 20 (19), 4673. Available at https://doi.org/10.3390/ijms20194673.

Wei, L., Wang, B., Bai, J., Zhang, Y., Liu, C., Suo, H., et al., 2024. Postbiotics are a candidate for new functional foods. Food Chemistry: X 23, 101650. Available at https://doi.org/10.1016/j.fochx.2024.101650.

Wendel, U., 2022. Assessing viability and stress tolerance of probiotics—a review. Frontiers in Microbiology 12. Available at https://doi.org/10.3389/fmicb.2021.818468.

Wolfe, W., Xiang, Z., Yu, X., Li, P., Chen, H., Yao, M., et al., 2023. The challenge of applications of probiotics in gastrointestinal diseases. Advanced Gut and Microbiome Research 2023, 1–10. Available at https://doi.org/10.1155/2023/1984200.

Xiong, R.G., Zhou, D.D., Wu, S.X., Huang, S.Y., Saimaiti, A., Yang, Z.J., et al., 2022. Health benefits and side effects of short-chain fatty acids. Foods 11 (18). Available at https://doi.org/10.3390/foods11182863, http://www.mdpi.com/journal/foods.

Yelin, I., Flett, K.B., Merakou, C., Mehrotra, P., Stam, J., Snesrud, E., et al., 2019. Genomic and epidemiological evidence of bacterial transmission from probiotic capsule to blood in ICU patients. Nature Medicine 25 (11), 1728–1732. Available at https://doi.org/10.1038/s41591-019-0626-9, http://www.nature.com/nm/index.html.

Yoshitake, R., Hirose, Y., Murosaki, S., Matsuzaki, G., 2021. Heat-killed *Lactobacillus plantarum* L-137 attenuates obesity and associated metabolic abnormalities in C57BL/6 J mice on a high-fat diet. Bioscience of Microbiota, Food and Health 40 (2), 84–91. Available at https://doi.org/10.12938/bmfh.2020-040.

Zakrzewska, Z., Zawartka, A., Schab, M., Martyniak, A., Skoczeń, S., Tomasik, P.J., et al., 2022. Prebiotics, probiotics, and postbiotics in the prevention and treatment of anemia. Microorganisms 10 (7), 1330. Available at https://doi.org/10.3390/microorganisms10071330.

Zhong, Y., Wang, T., Luo, R., Liu, J., Jin, R., Peng, X., 2024. Recent advances and potentiality of postbiotics in the food industry: composition, inactivation methods, current applications in metabolic syndrome, and future trends. Critical Reviews in Food Science and Nutrition 64 (17), 5768–5792. Available at https://doi.org/10.1080/10408398.2022.2158174.

Żółkiewicz, J., Marzec, A., Ruszczyński, M., Feleszko, W., 2020. Postbiotics—a step beyond pre- and probiotics. Nutrients 12 (8), 2189. Available at https://doi.org/10.3390/nu12082189.

European Parliament and Council. (2011). Regulation (EU) No 1169/2011 of the European Parliament and of the Council of 25 October 2011 on the provision of food information to consumers, amending Regulations (EC) No 1924/2006 and (EC) No 1925/2006 of the European Parliament and of the Council, and repealing Commission Directive 87/250/EEC, Council Directive 90/496/EEC, Commission Directive 1999/10/EC, Directive 2000/13/EC of the European Parliament and of the Council, Commission Directives 2002/67/EC and 2008/5/EC and Commission Regulation (EC) No 608/2004. Official Journal of the European Union, L 304/18.

Chapter 4

Viable probiotic ingestion strategy in foods: a critical approach on health effectiveness, technological challenges, and consumer view

Amin Abasi[1], Sahar Sabahi[2], Yalda Saadat[3], Elham Sheykhsaran[4], Negin Hosseinzadeh[1], Aidin Azadi[5], and Amir M. Mortazavian[5]

[1]*Department of Food Science and Technology, Faculty of Nutrition and Food Sciences, Tabriz University of Medical Sciences, Tabriz, Iran,* [2]*Department of Nutrition, School of Allied Medical Sciences, Ahvaz Jundishapur University of Medical Sciences, Ahvaz, Iran,* [3]*Kidney Research Center, Tabriz University of Medical Sciences, Tabriz, Iran,* [4]*Department of Microbiology, Faculty of Medicine, Tabriz University of Medical Sciences, Tabriz, Iran,* [5]*Department of Food Science and Technology, Faculty of Nutrition Sciences and Food Technology, National Nutrition and Food Technology Research Institute, Shahid Beheshti University of Medical Sciences, Tehran, Iran*

4.1 Introduction

Probiotics are viable microorganisms that, when consumed inappropriate quantities, provide health benefits to the host. These benefits are mostly associated with gastrointestinal health, including the enhancement of intestinal barrier functions, regulation of the immune system, and reduction of the risk of gastrointestinal infections (Abbasi et al., 2025a). Current studies have broadened the potential health advantages of probiotics to encompass immunological modulation, reduction of systemic inflammation, and enhancements in mental health through the gut–brain axis (Abbasi et al., 2024b,c; Asadi et al., 2024). The incorporation of probiotics into food systems enables consumers to easily and accessibly reap the advantages of these beneficial bacteria. Probiotics have been incorporated into yogurt, kefir, fermented vegetables, and various functional beverages. These foods offer a practical method for individuals to easily incorporate probiotics into their meals rather than relying on dietary supplements (Ozma et al., 2024; Sabahi et al., 2024). Still, including probiotics in food items presents significant difficulties.

Evaluating Oral Probiotics and Prebiotics. DOI: https://doi.org/10.1016/B978-0-443-29835-6.00019-8

Temperature, pH, and oxygen levels are among the many environmental factors that significantly impact the survival and efficacy of probiotics. Furthermore, probiotics must remain viable throughout food preparation, which includes storage and transportation (Ozma et al., 2023b; Abbasi et al., 2024a). Most importantly, they have to make it to the intestines, where they may show their healing powers, from the hostile acidic environment of the stomach. An important factor affecting the overall performance of food items is their efficiency in probiotics (Moradi et al., 2024). The effectiveness of meals enriched with probiotics largely depends on how well they are received by consumers. Elements such as flavor, texture, perceived health benefits, and convenience significantly impact consumers' willingness to incorporate these meals into their daily diets (Hosseini et al., 2024; Abbasi et al., 2024d). To promote the broad acceptance of probiotic-based functional foods, it is essential to identify the obstacles to consumer approval and develop approaches to address them. Hence, this chapter aims to explore the health effectiveness and technological challenges of probiotic-enriched foods, analyze consumer behavior models to enhance the adoption of probiotic products, and discuss strategies for improving the stability and viability of probiotics in food systems.

4.2 Health effectiveness of probiotics

4.2.1 Effect on oral health

A recent emphasis has led to the development of prebiotics as well as probiotics aimed at providing significant therapeutic health advantages. Recent studies have highlighted the potential for utilizing pre and probiotic types beyond gut health, specifically in oral health goods aimed at preventing dental caries. The primary emphasis of these elements in oral health functions is to manage cariogenic streptococci that colonize the oral cavity (Robin et al., 2025).

The probiotic type that is frequently used in individuals who have persistent periodontitis is *Lactobacillus reuteri*. Administered as a pill at least once daily, it has demonstrated effectiveness as a supportive measure alongside root planing and scaling. The therapeutic findings seen encompassed reductions in plaque index, gingival index, and hemorrhaging probing pocket depth. Furthermore, it was observed that the administration of a probiotic capsule containing Lactobacillus *salivarius* to healthy people resulted in a rise in salivary buffering abilities. The researchers believed that probiotic management may lead to improved resistance to caries based on these findings (Kaur et al., 2018; Xu et al., 2023). The study focused on the effects of chewing gums that contain probiotic organisms and xylitol on the levels of salivary *Streptococcus mutans* in kids aged from seven to twelve years old. Additionally, it aimed to evaluate any changes in plaque and gingival outcomes following 3 weeks of chewing these gums. Chewing gum with xylitol is commonly utilized in commerce. Several preclinical studies have demonstrated the efficacy of xylitol in decreasing the population of *S. mutans* in both saliva and plaque. Xylitol, a five-carbon polyol, undergoes metabolism through the phospho-enolpyruvate-phospho-transferase (PEP-PTS) route in *S. mutans*, in contrast to sucrose, which is processed via the glycolysis path. The PEP-PTS route transforms xylitol into xylitol-5-phosphate, which competes with phosphofructokinase, subsequently halting glycolysis through the intracellular buildup of glucose 6-phosphate. Multiple studies have demonstrated that the use of sorbitol or xylitol alone markedly reduced the formation of *S. mutans* biofilms. Conversely, approaches for quantifying and visualizing biofilms indicate that the addition of sucrose significantly diminished the inhibitory effects of polyols, irrespective of whether xylitol, sorbitol, or their mixtures were present at a level of 10%. Consequently, it can be concluded that the consistent presence of sucrose in the oral cavity will hinder the effectiveness of xylitol-containing healthcare goods, such as chewing gum, in preventing the formation of dental plaque that includes *S. mutans*. Substituting fermentable carbohydrates in the meals with alternative carbohydrates has been proposed as a strategy to mitigate dental caries, as evidenced by the influential findings of the Turku sugar research (Chan et al., 2020).

A recent in vitro investigation revealed that continuous management of arginine enhances the oral microbiota's capacity to resist acidification and suppresses the development of pathogenic infections (Rosier et al., 2018; Higuchi et al., 2019). Research indicates that probiotics can serve as a substitute for chemical approaches for biological plaque management. The biological basis of this treatment encompasses antibacterial actions alongside various processes, including alterations in immune system function, nutritional rivals, and alterations to the oral setting, without engaging in bactericidal action (Nadelman et al., 2019). Probiotics have demonstrated greater resilience in dairy goods derived from sheep, where they can synthesize bioactive peptides. For probiotics to effectively address oral conditions, they must adhere to dental areas, generate antimicrobial substances toward oral infections, modify the oral setting, and reduce the inflammatory reaction from the patient (Twetman et al., 2018).

Specific strains of *L. reuteri* may enhance gingival wound recovery through the upregulation of the neuropeptide hormone oxytocin. Research conducted in vitro has shown that metabolites from probiotic lactobacilli can enhance the production of prostaglandin E2 (PGE2) by human gingival fibroblasts when interleukin (IL)-1β is present, which may accelerate the healing process of oral wounds (Keller et al., 2018). An active intervention lasting 4 weeks, which involved administering probiotic tablets with *L. curvatus* and *L. rhamnosus*, demonstrated positive outcomes on gingival crevicular fluid flow, gingival inflammatory processes, and the development of supragingival plaque (Dassi et al., 2018). The taxonomic composition of the saliva microbiota was changed in a number of immediate ways by a commercial probiotic supplement (fermented milk). In contrast to people who did not partake in the probiotic beverage, those who did generally exhibited more intricate salivary microbial communities (Alamoudi et al., 2018). The three primary domains of contemporary caries prevention techniques encompass the elimination of plaque biofilm, the influence of food factors, and the characteristics of hosts. Probiotic medication represents a novel approach to avoiding tooth decay that has recently been evaluated in a cohort of preschoolers in Jeddah. Employing nonpathogenic commensal or endogenous bacteria to substitute pathogenic bacteria, probiotics have emerged as a practical and organic alternative for addressing infectious diseases (Klaenhammer et al., 2012).

4.2.2 Effect on intestinal health

Probiotics have been utilized for an extended period to enhance intestinal health. They play a crucial role in maintaining the epithelial barrier, promoting the longevity of cells, increasing the generation of antimicrobial substances and protective proteins, enhancing immune responses, and inhibiting the release of proinflammatory cytokines like IL-13, IL-4, TNF-α, TNF-α, and IFN-γ, thereby regulating the activities of the intestinal epithelium. Pathogenic microbes, including enteropathogenic Escherichia *coli*, *Vibrio cholerae*, and *Clostridium perfringens*, along with their toxic substances, can disrupt the digestive system. Furthermore, it has been observed that the administration of alcohol and nonsteroidal antiinflammatory drugs can significantly impair intestinal barrier activities (Yan and Polk, 2020).

Probiotics release short-chain fatty acids (SCFAs), which have been shown to positively impact the integrity of the intestinal barrier. Research indicates that SCFAs, including acetate, propionate, and butyrate, exhibit a protective role in countering the ethanol-triggered changes in barrier structure (Tripathy et al., 2021). The presence of ethanol increases metabolic stress and leads to the disruption of tight junctions (TJs) as well as the epithelial cytoskeletons. SCFAs mitigate oxidative stress and enhance TJs by activating AMP-activated protein kinase (AMPK) in Caco-II cells. Butyrate derived from bacteria enhanced the functionality of the epithelial barrier and maintained the hypoxia-inducible factor (HIF) through decreasing the level of oxygen (Tripathy et al., 2021; Abbasi et al., 2023a). An investigation utilizing an animal model of colitis provoked by dextran sulfate sodium (DSS) demonstrated that propionate enhanced barrier function while reducing oxidative damage and inflammatory conditions. Several dietary fibers yield different amounts of SCFAs through fermentation, which play a role in both enhancing and safeguarding the integrity of the epithelial barrier. Acetate, generated by *Bifidobacterium longum*, supports the resistance of host intestinal epithelial cells toward the spread of *E. coli* O157. H7's Shiga toxin translocates from the intestinal lumen into the bloodstream (Tripathy et al., 2021).

In addition to their direct influence on cells in the intestines via probiotics or probiotic-derived effective elements, there is evidence that probiotics can enhance intestinal epithelial strength by restoring the balance of the gut microbiota population. A controlled randomized trial demonstrated that supplementing overweight individuals with *Bifidobacterium animalis* resulted in higher amounts of *Akkermansia* and *Lactobacillus*, thereby promoting a lean metabolism. The importance of probiotics' regulating effects on gut microbiota in maintaining intestinal epithelial homeostasis is shown by these studies (Tripathy et al., 2021).

A notable relationship between beneficial bacteria and warfarin has been established. Antibiotics have the potential to modify gut flora, which can lead to a decrease in vitamin K production. It is well established that intestinal bacteria play a significant role in the production of vitamin K (Nath et al., 2018). This indicates a typical enhancement of warfarin impacts among individuals receiving broad-spectrum antibiotics. The gut microbiota in adulthood exhibits a significant degree of stability, yet it can be influenced by numerous external factors such as anxiety, diseases, chemotherapy, and medications (Frazier et al., 2011). Probiotics contribute significantly to the well-being of the upper respiratory tract, help in the prevention of tooth decay and tonsillitis, support urogenital well-being, and assist in the treatment of wounds and throat-related infections. Currently, probiotics can be found in the form of nutritional supplements and have demonstrated effectiveness comparable to that of conventional medication (Ranjan, 2022).

4.2.3 Irritable bowel syndrome

Intestinal homeostasis disorders are associated with irritable bowel syndrome. The identified abnormalities result in an unregulated response of intestinal immune and epithelial cells to the gut microbiota, potentially leading to damaging outcomes like fibrosis and ulcers. Prebiotics serve as essential components that promote the proliferation of beneficial bacteria, correlating with alterations in the intestinal microbiota. In the context of irritable bowel syndrome, evidence suggests that both probiotics and prebiotics can offer significant benefits (Benjak Horvat et al., 2021).

4.2.4 Effect on the immune system

The immune system of individuals is influenced by probiotics and prebiotics. Their impact extends to cellular metabolism, development, and the operation of the epithelial barrier, alongside numerous other positive impacts on well-being (Benjak Horvat et al., 2021). The initial colonization through *Bifidobacterium* and *Bacteroides* species could play a vital role in establishing immunological regulation, given that the gut microbiome is an evolving system. The diet of a mother, antibiotic usage, the method of birth, environmental factors, household shifts, and various influences on infants can all significantly affect early life expansion (EFSA Panel on Dietetic Products, Nutrition and Allergies (NDA), 2010). The surface of immune cells is equipped with toll-like receptors (TLR), which undergo stimulation modifications in response to immune system activities, allowing them to differentiate between the microbiota of the gut and pathogenic germs. Probiotics and their effector compounds impact the gut barrier through several mechanisms, such as modifying the generation of mucus, reducing adhesion of bacteria, enhancing tight connections, promoting survival of cells, and stimulating the generation of IgA or defensins (Kerry et al., 2018). The quantity of probiotics in an average meal can vary significantly, typically falling between 2 and 20 grams per day, depending on the specific ingredient and the intended outcome. The increasing understanding of the beneficial effects of prebiotics and probiotics on gut health, their potential to reduce disease risk, and their application in therapeutic contexts has generated significant interest in the development of functional foods that incorporate these biological elements (Dahiya and Nigam, 2023).

4.2.5 Allergy prevention

Allergy, recognized as a form of hypersensitivity, is a condition marked by an immune system reaction to an antigen, reflecting the complexities of hypersensitive immune responses. The prevalence of allergies has reached nearly 50% of people in Europe and North America, accompanied by a rising incidence rate. These allergic reactions are triggered by one or more prevalent environmental substances, or antigens. Common allergic responses include asthma, edema, urticaria, dermatitis, rhinitis, hay fever, and hypersensitivity to various medicines, foods, and pests. The gut microbiome plays a crucial role in modulating the immune and inflammatory responses, which subsequently influence the progression of sensitization and allergies. Therefore it presents a promising avenue for therapeutic intervention in the treatment of allergic conditions (Latif et al., 2023). Significant antiallergy benefits are achieved by *Lactobacillus plantarum*, which decreases the generation of inducible nitric oxide synthase, cyclooxygenase-2, T-helper 2-related cytokines, tumor necrosis factor-α, and nitric oxide (Latif et al., 2023; Lee et al., 2014). A separate investigation indicated that oral administration of *L. reuteri* contributed to a reduction in allergic diarrhea and enhanced the compromised profile of the cinonic microflora. Furthermore, it decreased the expression of GATA3, increased the levels of Foxp3, IL-10, and TGF-b, diminished the generation of T-helper 1 and 2 inflammatory mediators, and enhanced the stimulation of mast cells as well as IgE. Research results indicated that probiotics exhibited antiallergic characteristics by modifying intestinal flora and enhancing tolerogenic immune reactions (Huang et al., 2017).

4.2.6 *Helicobacter pylori* infection

Over 4.4 billion individuals are estimated to be infected with *Helicobacter pylori* globally (De Brito et al., 2019). In contrast to industrialized nations, there is a higher prevalence of *H. pylori* infections in developing countries, with infection rates reaching as high as 80% of people. Probiotics demonstrate the ability to inhibit *H. pylori* through competitive mechanisms, acting as bacteriostatic drivers while also promoting the health of the gut microbiota. Probiotics, including *Saccharomyces*, *Bacilluslicheniformis*, *Bifidobacterium*, and *Lactobacilli*, are presently utilized and have shown effectiveness in addressing digestive issues linked to *H. pylori* (Mestre et al., 2022). Numerous clinical studies have indicated a reduction in antibiotic-related side effects when compared to probiotics, highlighting an improvement in *H. pylori* eradication levels. The enzyme activity of *H. pylori* urease is subject to suppression by lactic acid. Furthermore, the generation of reactive oxygen species by probiotics inflicts destruction on the bacteria's cell wall and its membranes (Lionetti et al., 2010).

The bacterial surface comprises several components that enable *H. pylori* to adhere to the intestinal epithelium. Investigations indicate that probiotics can enhance IgA production, thereby strengthening the intestinal barrier's protection against infections. Probiotics engage in competition with *H. pylori* at microbial attachment regions, thereby enhancing the immune system's reaction and mitigating its pathogenic effects. Current investigations are focused on the glycolipid-binding function of probiotics and *H. pylori*, aiming to explore the potential application of probiotics as antiadhesion agents in the treatment of *H. pylori*-triggered ulcers in the stomach (Bai et al., 2022; Sjomina et al., 2024).

4.2.7 On nutritional status

Individuals' apprehensions about their well-being hold considerable importance in today's context. Approximately one hundred years ago, the exploration of probiotic supplementation and its possible advantages for health began, with the objective of disease prevention (Shojaee-Aliabadi et al., 2023). Fermented foods and drinks possess a rich and significant cultural and culinary heritage. Recently, fermented foods have been acknowledged for their nutritional benefits following the incorporation of probiotics. These meals are generally crafted from nutrient-rich foundational components such as legumes, grains, meat, and milk, which inherently possess high levels of protein, minerals, and vitamins (Abbasi et al., 2023a).

Recent studies highlight that the live microorganisms found in fermented foods contribute positively to the wellness of the gastrointestinal tract and overall well-being (Abbasi et al., 2022b; Abbasi et al., 2023d).

The addition of particular probiotic microbes to fermented foods has become a prevalent practice. Consequently, a significant number of manufactured yogurt and cultured milk goods currently incorporate probiotic strains of *Bifidobacteria* and *Lactobacilli*. In such instances, the fermented food serves as the medium for the probiotic delivery. Furthermore, the majority of cheeses and yogurts are created through the earlier method, which incorporates the incorporation of carefully selected strains of lactic acid-producing bacteria. Nonetheless, the creation of various fermented foods, such as sauerkraut and kimchi, relies on the presence of wild or natural bacteria (Marco et al., 2017; Abbasi et al., 2023c).

Research indicates that the yogurt bacteria, *Lactobacillus delbrueckii* subsp., *Lactobacillus bulgaricus,* and *Streptococcus thermophilus,* can enhance lactose digestion in individuals experiencing lactose maldigestion. This finding is supported by the European Food Safety Authority (EFSA Panel on Dietetic Products, Nutrition and Allergies (NDA), 2010), which has recognized a nutritional claim for yogurt (EFSA Panel on Dietetic Products and Allergies) (EFSA Panel on Dietetic Products, Nutrition and Allergies (NDA), 2010). Several epidemiological investigations have provided evidence for the nutritional advantages of probiotics. A large cohort investigation of older adults found an association between yogurt-rich diets and a decreased likelihood of metabolic syndrome (Babio et al., 2015; Abbasi et al., 2025b). In a comparable extensive cohort research, a daily intake of yogurt was linked to reduced permanent obesity (An et al., 2011). The findings from the Malmo nutrition and cancer cohort research indicated that the intake of fermented dairy goods, particularly sour milk and yogurt, was oppositely related to the risk of heart disease (An et al., 2011). The effect of consumption of cheese was observed, albeit exclusively in women. Two extensive cross-sectional studies involving adults in Korea indicated that a high intake (2–4 servings daily) of kimchi, along with other fermented foods and drinks, correlated with a lower incidence of atopic dermatitis (Park and Bae, 2016).

The intake of natto, miso, and fermented soy goods demonstrated an inverse relationship with the risk of developing high blood pressure (Beltrán-Barrientos et al., 2018). Furthermore, findings from an investigation carried out on a group of people with type 2 diabetes indicate that probiotic yogurt may offer benefits to these patients. The participants who ingested probiotic yogurt with *Bifidobacterium lactis* Bb12 and *L. acidophilus* La5 exhibited enhancements in their fasting blood sugar levels and capacity for antioxidants (Ejtahed et al., 2012). Fermented milk may prevent infectious illness in children (Nocerino et al., 2017), lower blood pressure in individuals with hypertension (Nozue et al., 2017), and temporarily enhance bone health indicators in individuals with osteoporosis, according to recent clinical studies. Lastly, similar to cultured dairy goods as well as other fermented foods, the presence of fermentation-associated bacteria has the potential to enhance the nutritional value of food (Chamlagain et al., 2018).

4.2.8 Obesity

Obesity has emerged as a widespread clinical issue, primarily driven by increased food consumption and absorption coupled with diminished energy expenditure (Sheykhsaran et al., 2023). Recent studies have confirmed that human intestinal microbes play a crucial role in the progression of obesity through effective energy generation and nutrient absorption. Furthermore, it remains recognized that the duodenal microbial community in individuals with obesity exhibits greater diversity compared to that of those who are lean. Obesity is associated with the proliferation of certain gut microbial species, such as *Staphylococcus aureus*, *E. coli*, and various generic bacterial species like *Bifidobacterium* (Ejtahed et al., 2019).

The initial mechanism believed to be responsible for this increase in body fat was attributed to the capacity of microbes to obtain energy from food components and regulate the energy status of the host. The alteration of nutritional polysaccharides and fibers via *Bacteroides* and *Firmicutes* in the gastrointestinal tract leads to the production of SCFAs. Propionate serves as an important energy source for the host, facilitating the synthesis of lipids and glucose in the liver (Vallianou et al., 2023). A new study evaluated 61 initial investigations. Variations were observed in the gut bacteria composition between individuals classified as overweight and those identified as nonobese. Changes were noted in energy balance, with gut microbiota influencing dietary intake and lipid storage (Sankararaman et al., 2023) (Fig. 4.1).

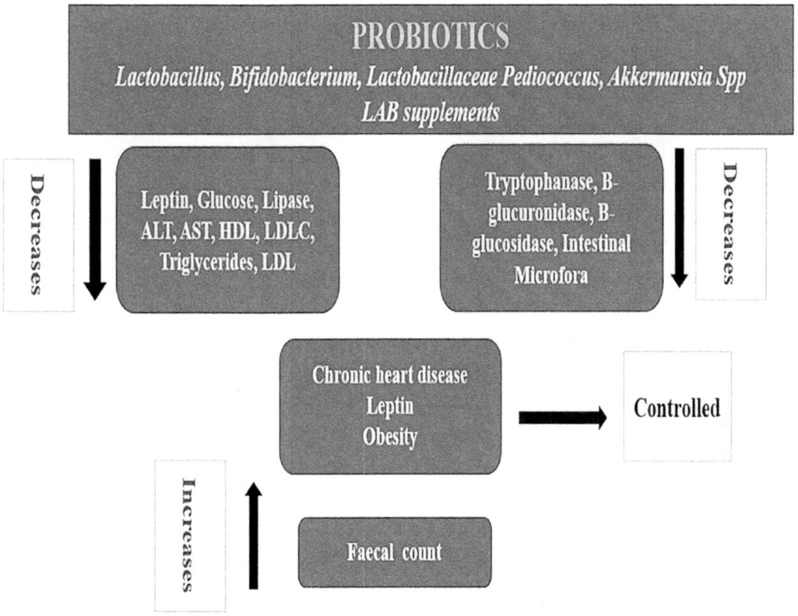

FIGURE 4.1 Illustrates the mechanisms through which probiotic dietary supplements exert their antiobesity effects.

These probiotics function by decreasing total cholesterol and low-density lipoprotein (LDL) levels, aiding in the upkeep of a normal weight with reduced leptin levels, and diminishing the occurrence of chronic coronary artery disease. Despite an observed rise in fecal count, there is a concurrent reduction in triglycerides, lipase, glucose, HDL, ALT, AST, and LDL-C. The intestinal microflora, along with the generation of β-glucuronidase, β-glucosidase, and tryptophanase, is also diminished by these probiotics (Barathikannan et al., 2019; An et al., 2011).

4.2.9 Diabetes mellitus

A serious global health concern, diabetes mellitus is marked by insufficient insulin production and/or function. A medical condition characterized by an autoimmune breakdown of pancreatic beta cells leads to type 1 diabetes, commonly known as diabetes with insulin dependence. In type 2 diabetes mellitus, the body fails to secrete or utilize insulin properly, often resulting in a state of insulin resistance. People with diabetes exhibit changes in their gut microbiota (Zhang et al., 2025). A comprehensive analysis of the human metagenome revealed a notable association among gut microbes, bacterial DNA, and metabolic processes in individuals with type 2 diabetes. The quantity of *Lactobacillus* species showed a significant variation when contrasted with patients with no diabetes. The percentages of glycated hemoglobin (HbA1c) and fasting glucose showed a positive association with the *Lactobacillus* species. *Clostridium* species exhibited an inverse correlation with plasma triglycerides, C-peptide, insulin, HbA1c, and fasting glucose, while demonstrating a positive correlation with high-density lipoprotein (HDL) cholesterol and adiponectin (Pavlidou et al., 2025).

The prevalence of diabetic retinopathy in individuals with type 2 diabetes mellitus is 60%. In diabetic individuals, a higher prevalence of gram-positive microbes along with coagulase-negative staphylococci was identified, particularly in cases of retinopathy. The prevalence of *S. aureus* detected in the subject's eyes exceeded that observed in healthy individuals and patients with T1DM (Zhang and Zhang, 2013). The gut microbiota plays a significant role in the pathophysiology of T2D. Alterations in the structure and function of gut microbiota are linked to the onset of T2D. Research on stool samples reveals various mechanisms by which microbiota impacts the progression of T2DM, such as the onset of insulin resistance, production of SCFAs, metabolic endotoxemia, and changes in incretin secretion. A variety of cytokines play a crucial role in the development of T2DM, and within the intestinal immune system, there are essential signals that influence these processes.

A variety of preclinical investigations have elucidated the effects of probiotics on the metabolism of glucose. A significant number of studies have shown that the onset of diabetes is linked to both oxidative stress and the capacity for antioxidation. The suppression of lipid peroxidation and the enhanced synthesis of glutathione stimulated by probiotics contribute to a reduction in oxidative damage observed in diabetic rats. Probiotics possess antidiabetic properties, which may aid in the prevention of insulin resistance. This capability additionally leads to a rise in natural killer-T cells within the liver. Furthermore, probiotics have the potential to decrease inflammation through the modulation of TNF-alpha expression, the maintenance of insulin stability, and the inhibition of NF-kB binding activity. Probiotics have the potential to enhance glucose metabolism through improving the bioavailability of certain antidiabetic medications, including gliclazide. They also contribute to the inhibition of the absorption of glucose in the gut and the regulation of the autonomic nervous system action (An et al., 2021). The diversity of microbial colonies contributes to the strength of the gastrointestinal integrity, enhances glucose homeostasis, reduces inflammation, supports the generation of insulin, and improves the absorption of nutrients (Taylor et al., 2017).

4.2.10 Chronic kidney disease

As the number of chronic kidney disease (CKD) cases increased, so did the concern regarding the addition of probiotics and prebiotics (Pei et al., 2018). The development of chronic renal failure can result in severe uremia at different rates, depending on the underlying cause. CKD can result in the necessity for peritoneal dialysis, kidney transplantation, and hemodialysis (Pei et al., 2018). Studies have indicated that uremic toxins could facilitate the advancement of kidney disease through the disruption of tubular cells (Saadat et al., 2025). In chronic kidney disease, inflammation represents a complex and multifaceted phenotype. During the course of the disease, probiotics have surfaced as a potential solution, modulating the gut microbiota to decrease uremic retention solutes and enhance cardiovascular health. The primary objective of administering probiotics in the context of CKD is the elimination of uremic retention solutes. Hyperoxaluria and hypercalciuria represent significant risk factors for the formation of stones in the kidneys, which can severely impact kidney health.

The majority of oxalate production takes place in the gastrointestinal tract. Lactobacilli are provided as dietary supplements in nephrolithiasis and might help in preventing stone formation and reducing the risk of urolithiasis (Fig. 4.2). A study conducted by Cheol Kwak indicates that *L. casei* can inhibit the formation of oxalate through several mechanisms (Kwak et al., 2006). The enzymatic breakdown of oxalate involves particular bacterial enzymes, including oxalyl-CoA decarboxylase and formyl-CoA transferase, which facilitate the transformation of oxalate into benign byproducts like carbon dioxide and formate. Furthermore, these probiotic strains attach right away to oxalate molecules, inhibiting their absorption in the intestines and facilitating their excretion through feces, thereby lowering systemic oxalate amounts. Furthermore, they have the potential to inhibit oxalate absorption effectively by filling binding spots on epithelial cells in the intestine (Di Cerbo et al.,

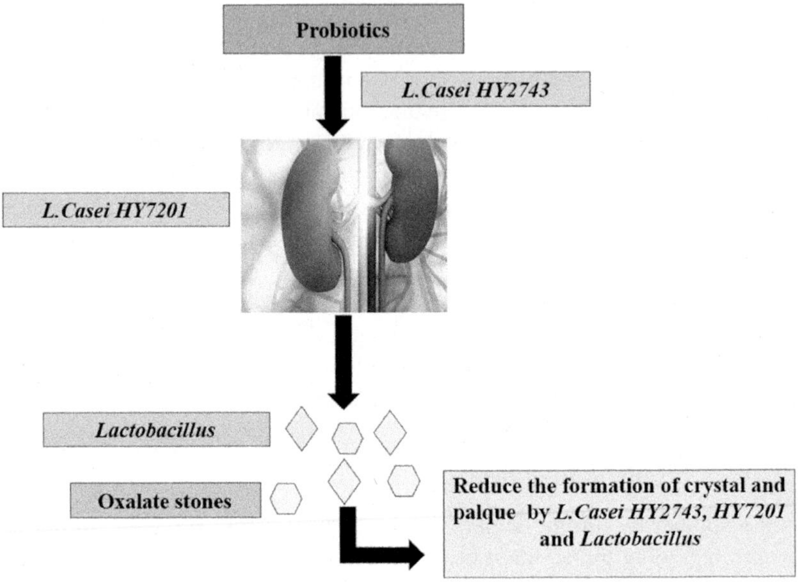

FIGURE 4.2 Illustrates the mechanism by which *Lactobacillus* species acts toward oxalate stones.

2013). Moreover, by adjusting the composition and functionality of the gut microbiota, *L. casei* may have an indirect effect on oxalate metabolism. With increasing understanding of the complex relationship between gut microbiota and the development of kidney stones, interventions based on lactobacilli may prove to be significant assets in the holistic management of urolithiasis (Turroni et al., 2010; Wigner et al., 2022).

4.2.11 Osteoporosis

Deterioration of the skeletal system, low bone mass density, and heightened vulnerability to fractures are all consequences of osteoporosis. Fractures are most frequently observed in the spine, femur, and distal forearm. Women who have gone through menopause are at a higher risk of encountering these issues. Bone loss is observed during menopause, as estrogens are essential for the formation and upkeep of bone health (Zeng et al., 2021). Osteoporosis is a prevalent condition, and numerous individuals are at significant risk of experiencing inadequate bone mass. Both females and males can experience osteoporosis at various points in their lives; however, older females exhibit a higher susceptibility to the condition (Ohlsson and Sjögren, 2015). The hormonal control of bone mass is notably affected by gut bacteria. Shifts in the immune system play a crucial role in how gut microbiota affects bone mass, while in an optimal condition, it governs the process of osteoclastogenesis. The ongoing reconstruction process involving bone-resorbing osteoclasts and bone-making osteoblasts leads to the development of osteoporosis when an imbalance occurs (Yousf et al., 2015; Saadat et al., 2025).

Rheumatoid arthritis and osteoporosis are two challenging bone and joint conditions that may be addressed through the use of probiotics as a form of treatment (Collins et al., 2017). Probiotics play a crucial role in generating antimicrobial peptides, maintaining the luminal pH in the intestines, and facilitating mucus development. They also modulate changes in the stomach's microbiota influenced by the host's immunity (Parvaneh et al., 2014). Probiotics exhibit a range of mechanisms influencing bone health, with their most nuanced effects stemming from their interaction with vitamins. The metabolism of calcium is associated with vitamins K, C, D, and folate, which are crucial for the development of bones (Mccabe et al., 2013).

In addition to limiting bone uptake and preserving the condition of the bones, *L. reuteri* significantly lowers the host's tumor necrosis factor. *L. reuteri* plays a crucial role in influencing bone density and the development of osteoporosis (Parvaneh et al., 2015). Following the consumption of *B. longum*, an increase in bone resorption variables was observed, alongside a reduction in osteoclasts and serum C-terminal telopeptide (CTX). Additionally, researchers noticed an increase in blood levels of osteocalcin (OC) and osteoblast action. The consumption of *B. longum* resulted in modifications to the femoral bone's architecture (Xu et al., 2017). Furthermore, various *Lactobacillus* types have shown significant promise as probiotics for enhancing bone wellness. For instance, *L. brevis* and *L. paracasei* have been found to mitigate cortical fractures in ovariectomized rats by decreasing the generation of two inflammatory mediators (IL-1β and TNF-α) and preventing osteoclastogenesis (Duffuler et al., 2024). Furthermore, oral consumption of *L. rhamnosus* GG led to a reduction in gut permeability by enhancing the generation of TJ proteins and inhibiting the generation of proinflammatory cytokines associated with bone loss. The synthesis of SCFAs was also enhanced, which positively influences the growth of bones (Li et al., 2016). Conversely, the administration of *L. reuteri*, *Bacillus coagulans*, *Bifidobacterium*, *L. casei*, and *L. acidophilus* notably elevated serum calcium levels and 1,25-$(OH)_2$-vitamin D, while also safeguarding against bone loss in an animal model (Tyagi et al., 2018). Oral supplemental intake with *L. reuteri* resulted in an increase in the blood level of 25-hydroxyvitamin D in humans (Duffuler et al., 2024).

In addition to lowering TNF-α expression and downregulating the microbial lipopolysaccharide synthesis, *B. longum* and *L. plantarum* increase blood levels of osteocalcin, phosphorus, and calcium and improve the development of bones (Kim et al., 2019). Consumption of *B. vulgatus* has been shown to minimize dysbiosis, downregulate markers of inflammation like TNF-α, and decrease microbial generation of lipopolysaccharide, which contributes to enhanced bone strength and structure (Yuan and Shen, 2021).

Bacterial antigens penetrate the intestinal epithelial barrier and initiate immune system reactions, particularly in the context of postmenopausal osteoporosis, where bone loss is associated with diminished levels of estrogen. Furthermore, probiotics play a crucial role in supporting intestinal calcium absorption, regulating abnormal immune responses, enhancing the integrity of the intestinal epithelial barrier, restoring diversity of bacteria to prevent bone resorption, and facilitating the latent production of an estrogen-like compound (Dar et al., 2018).

L. acidophilus facilitates raised mineral content and diversity in bones, along with improved cortical and trabecular microstructure. The immune system of the host is influenced by a characteristic of *Lactobacillus*. *L. acidophilus* plays a role in reducing the expression of osteoclastogenic variables (RANKL, TNF-α, IL-17, and IL-6) while enhancing the expression of antiosteoclastogenic elements. Furthermore, *L. acidophilus* exhibits therapeutic properties that contribute to osteoprotection in postmenopausal osteoporosis by modulating the physiological equilibrium of Treg-Th17 cells (Lamsal, 2012).

4.2.12 Wound healing

Probiotics have been demonstrated to accelerate the healing process of skin-infected wounds and ulcers in the gut. The skin microbial community in healthy skin acts as a barrier of protection, regulating the inflammatory reaction to minor epidermal damage by modulating the generation of cytokines, both enhancing and diminishing it (Parhizkar et al., 2025). Probiotics demonstrate advantageous effects via several mechanisms, such as direct pathogen elimination, improved epithelial barrier activity, selective eradication of harmful bacteria, and stimulating fibroblasts. Probiotics have the potential to reduce the number of bacteria in ulcerated regions, which is particularly beneficial for individuals with burn injuries (Tsiouris and Tsiouri, 2017). A skin injury alters the microbiota and elevates the levels of bacteria that adversely affect the healing process of wounds. Wounds represent complex physiological challenges that result in alterations in neuroendocrine activity, susceptibility to infection, and compromised healing processes (Johnson et al., 2018).

Diabetic foot ulcers, decubitus ulcers, and venous leg ulcers are instances of persistent wounds that present significant healing challenges and can impose considerable burdens on both patients and healthcare providers. The development of impaired wounds is primarily attributed to the existence of microbial biofilms in persistent wounds, which promote the proliferation of detrimental microorganisms and hinder the process of healing (Habeebuddin et al., 2022). Probiotics not only positively influence gut health but also demonstrate efficacy in addressing skin-related concerns such as infections, scars, and burns. They enhance the skin's innate defenses and support the renewal of healthy skin (Habeebuddin et al., 2022).

4.2.13 Autism

Gastrointestinal issues frequently occur in children with autism. Autism spectrum disorder (ASD) represents a significant growth-oriented and neurobehavioral condition. ASD is linked to alterations in the microbiota of the gut. Observational findings indicate that a gluten-free and free-of-casein diet may lead to a reduction in gastrointestinal issues and/or enhancements in behavioral characteristics. Tomova and coworkers assessed the effects of four months of mixed probiotic management on the composition of the gut microbiota in children with ASD, demonstrating a modulation of the Bacteroides proportion and a rise in bifidobacterial counts. A reduction in the intake of particular food elements, such as casein or gluten, was associated with a decrease in various digestive discomforts. On the other hand, a wealth of studies has shown that minerals, vitamins, omega-3 fatty acids, and probiotic supplements, in conjunction with medicinal and behavioral therapies, can address dietary deficiencies in people with autism (Cristiano et al., 2018).

Probiotics offered a promising approach for addressing *Clostridium* infections, while also leading to a reduction in autistic characteristics. These results could provide valuable insights into therapeutic approaches related to microbial health and may reduce the likelihood of autism development. Oral probiotics effectively mitigated social deficits created by maternal immune system activation, which are characteristic symptoms associated with ASD. Maltodextrin and fructo-oligosaccharides (FOS) were utilized for the oral ingestion of prebiotics (Wang et al., 2019). An investigation was carried out on the supplementation of a mixed probiotic–bovine colostrum goods (BCG) among kids with ASD. The main indicators of outcome were the changes in the gut microbiota and the tolerability of the supplements. The effectiveness of the research was evaluated by monitoring gastrointestinal indications and any behavioral alterations via metabolomic analysis of feces, urine, and plasma. Additionally, secondary indicators of outcome were investigated, including alterations in the level of cytokines in peripheral blood mononuclear cells (PBMCs), intestinal microbiota, and both host and microbial metabolism (Sanctuary et al., 2019).

Probiotics generate vitamins and enzymes, modify the body's acidic setting, and suppress the proliferation of pathogenic microbes. Ultimately, their goal was to restore the gut bacteria to their original, optimal state, which could have been influenced by ecological or dietary variables such as excessive consumption of antibiotics. Most probiotics intended for use by humans are predominantly composed of gram-positive bacteria. Probiotic therapy constantly led to improvements in the intestinal microbiota and urine substances regardless of deviations in quantity, species, strains, and length of time.

4.2.14 Synergistic effect of probiotics with vitamins and minerals in host health

The interplay of vitamins, minerals, and probiotics demonstrates significant advantages for human health, especially in enhancing immune system performance and gastrointestinal function. By outcompeting harmful bacteria and producing beneficial metabolites like SCFAs, probiotics play a crucial role in sustaining a healthy gut flora, which significantly impacts our overall health (Huang et al., 2021). The integrity of the intestinal barrier and immune responses—both of which are primarily regulated by SCFAs—are essential for the development of general gut health. Intestinal permeability is increased by probiotics, which lowers inflammation and improves the body's absorption of vital vitamins and minerals (Latif et al., 2023; Luu and Visekruna, 2019). Together, vitamins and probiotics improve gut health and immune system functioning. Important immunological regulators that change the composition of the gut bacteria and hence influence antimicrobial peptide generation and disease resistance include vitamins A, D, B6, and C (Degnan et al., 2014; Ly et al., 2011) (Fig. 4.3). Furthermore, helping to regulate immune response and maintain intestinal barrier integrity are iron, magnesium, and zinc. Probiotics also boost immunological response and nutrient absorption (Yamamoto and Jørgensen, 2020; Steinert et al., 2020). The complex and fascinating interactions, alongside vitamins and minerals, highlight the therapeutic potential of probiotics. Additional clinical studies are essential to fully comprehend the mechanisms that contribute to these synergistic benefits, despite the promising results observed in current research.

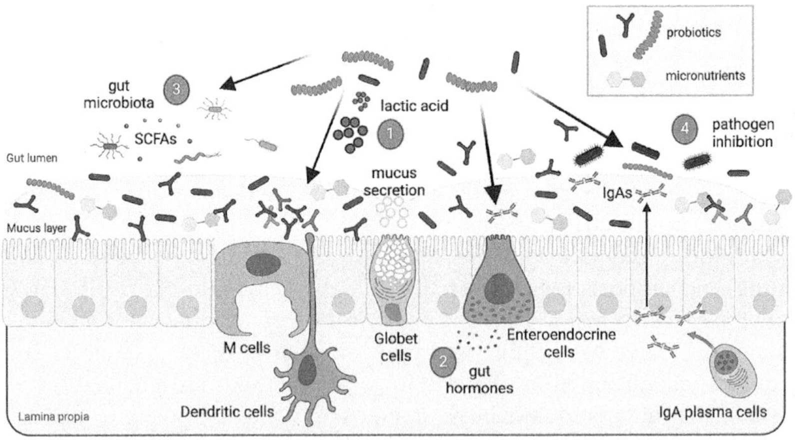

FIGURE 4.3 Suggested probiotics mechanisms in optimizing the intestinal milieu for improved nutrient absorption.

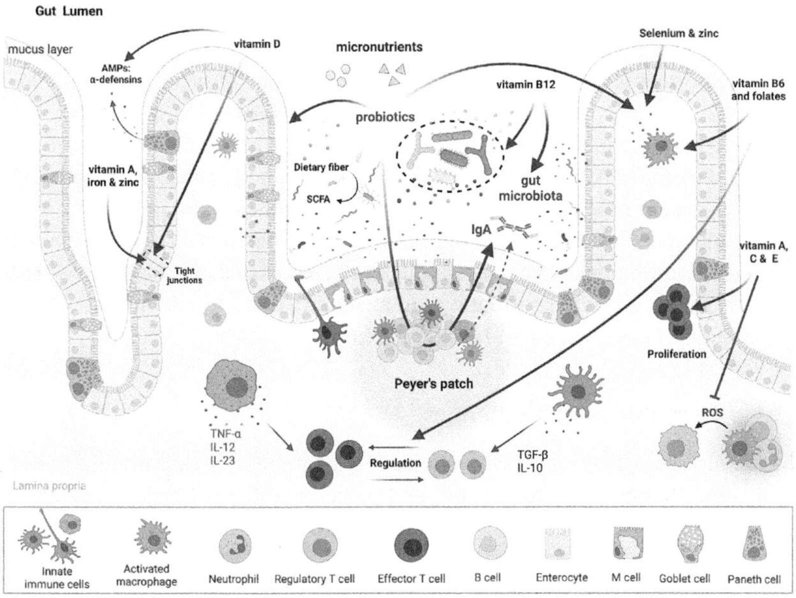

FIGURE 4.4 Suggested probiotics mechanisms in the synergistic stimulation of intestinal immunity with micronutrients (vitamins and minerals).

By employing improved treatment strategies that leverage the combined benefits of probiotics, vitamins, and minerals, individuals can achieve superior gut health, enhanced immunity, and overall well-being (Fig. 4.4).

4.3 Technological challenges in probiotic food systems

The fragility of probiotic bacteria presents various technological challenges that limit their widespread application in food products. The influence of environmental variables such as pH, temperature, and humidity on probiotics can be significant. Examine food production as your case study. The viability of probiotics often diminishes considerably when exposed to heat, acidity, or oxidative stress. Modern preservation techniques ensure that probiotics stay viable and functional much into the product's expiration date (Abbasi et al., 2022a; Bazzaz et al., 2025). The primary issue in the industry is maintaining the efficacy of probiotics throughout food processing and storage activities. While heat processing, including pasteurization, may adversely affect several probiotics, high-pressure processing and freezing may also inflict cellular damage on many of them. Consequently, it is essential to produce functional food products containing probiotics that can endure these conditions.

As important as their resistance to food processing is their capacity to survive stomach acidity and bile salts in the small intestine. Before their possible improvement of intestinal condition, the acidity in your stomach removes most probiotics. Other delivery strategies have surfaced to solve this issue: microencapsulation protects probiotics from stomach acid (Abbasi et al., 2024a). Often using biopolymeric materials, including alginate, chitosan, or gelatin, these delivery systems capture probiotics and release them into the intestines in a controlled manner (Abbasi et al., 2022c). Eating synbiotic foods—which include prebiotics and probiotics—helps to improve the efficacy of probiotics in yet another manner. Prebiotics, or indigestible food components, provide the ideal habitat for probiotics—that is, good bacteria—inside the digestive tract. Research shows that adding prebiotics to meals enhanced by probiotics improves both customer acceptance and health benefits of the meal.

4.3.1 Technological challenges

The technologies that have been extensively examined, including rehydration, drying, encapsulation, fermentation, and preservation, have been invented and effectively utilized to safeguard certain probiotics from adverse effects linked to various foods. However, numerous technological challenges remain in the production and preservation of probiotic foods, and these issues must be addressed.

Clearly, the presence of living bacteria in a food item, particularly probiotics, makes the retention of adequate viable bacteria a crucial aspect of quality. The mere selection of bacterial species does not ensure a high viable count in fermented goods postfermentation and throughout storage (Holko et al., 2013). Consequently, selecting suitable probiotic bacteria and their cultures, along with examining the interactions between microbial elements and food ingredients under various circumstances, is crucial. Furthermore, the safety of probiotic goods warrants careful consideration. A significant drop in pH may hinder the development of *Enterobacteriaceae* throughout sausage fermentation; however, adverse impacts on the viability of probiotic organisms and the sensory attributes of fermented meat goods have been observed (Rubio et al., 2013).

The amination and transamination of ketones and aldehydes or the decarboxylation of free amino acids in aged fermented meats are problematic, as these biogenic amines adversely impact health (Kołożyn-Krajewska and Dolatowski, 2012). Additionally, the authors observed that the oxidation of lipids and proteins led to a decline in nutrient content, hue, and other sensory properties. For instance, raw-fermented meat sausages containing *L. casei* exhibited a bitter taste along with other unfavorable flavors, a fatty flavor, an acrid odor, and noticeable fat in the final goods (Trząskowska et al., 2014). Various prevalent technological obstacles, including manufacturing, temperature of preservation, oxygen levels, pH, and outside stresses, can influence the use of probiotics in food items (Kouam et al., 2024). Additionally, water activity serves as a crucial factor that can influence the survival of the organisms in these types of foods. A variety of food components, including salts and sugars, have the ability to bind water, leading to "dry" conditions with low water activity. This phenomenon enhances the survival rates of bacteria. Conversely, elevated water activity, such as that found in fruit juice, can diminish bacterial viability throughout preservation.

4.3.2 Technological advancements

The fermentation process has undergone significant enhancements to achieve high sustainability and generate probiotic cells as well as nutrient-wise active biomass (Sabahi et al., 2022). A variety of probiotic bacteria have demonstrated effective survival rates, with viable cell counts ranging from 10^8 to 10^9 CFU/ml in fermented smoothies (Sharma and Mishra, 2013). Nevertheless, certain limitations associated with the implementation of these techniques for the production of probiotic foods require careful consideration on an individual basis (Rad et al., 2021a,b,c,d).

According to Racioppo et al. (2017), sonication was used to lessen acidity caused by probiotics without compromising their viability, since it may be a significant issue during fermentation. The findings indicated that the acidification levels caused by four probiotic organisms subjected to ultrasound were considerably reduced compared to the control group after 14 days of treatment at 15°C. Nonetheless, the authors found no differences in viability between bacteria that were treated with ultrasound and those that were not (Racioppo et al., 2017). Calcium lactate was utilized to maintain the pH within the range of four to five in fermented soy meals treated with *L. rhamnosus* and a combination of *L. paracasei* and *L. acidophilus*, achieving a viability of 9 to 11 logs CFU/mL. In contrast, the inoculation with *B. longum* resulted in a viability of 8 to 9 logs CFU/mL following an incubation period of 30 days at 4°C (Mondragón-Bernal et al., 2017). Four strains of *L. plantarum* demonstrated commendable tolerance to acidic environments such as fermented oat meals, maintaining high sustainability of 8 logs CFU/g for all strains following preservation at 4°C. However, viscosity decreased owing to the breakdown of oat β-glucan and the generation of microbial exopolysaccharides (Russo et al., 2016). Yeast demonstrated significant antioxidant capacity throughout fermentation via rice and cassava, thereby mitigating oxidative stress impacts on probiotic survival throughout the fermentation process (Freire et al., 2017).

To obtain high survivability of probiotic microorganisms in fruits and vegetables, microencapsulation was given careful consideration (Antunes et al., 2013). Incorporating protectants into the culture media can enhance the number of viable types of probiotics in the finished item. Various lyoprotectants have been employed to enhance the growth of *B. infantis* following freeze-drying (Basholli-Salihu et al., 2014). The study revealed that trehalose, sucrose, lactose, and cellobiose can improve the survival of *B. infantis*, with cellobiose demonstrating superior effectiveness by preserving the membranes of cells and inhibiting intracellular accumulation of ice.

The use of air drying combined with electromagnetic vacuum drying for dehydrating slices of apples, along with the addition of *L. rhamnosus*, resulted in an acceptable count of 10^5 CFU/g after being stored at 25°C for 60 days (Mestry et al., 2011). A further investigation by Borges et al. (2016) revealed a significant correlation between the procedure of drying and formulation with the survival rate of probiotics in the fruit powder throughout preservation at 4°C and at ambient temperatures (Borges et al., 2016).

The inclusion of protectants in the preparation of microencapsulated probiotics demonstrated the potential for maintaining high survival of probiotics postdrying. Alginate, as a fundamental coating substance, regularly garners significant interest, particularly when paired with other innovative substances. After 5 weeks of preservation at 4°C, gelatin-alginate wrapped *L. salivarius* exhibited a loss of 1.7 logs CFU, while alginate alone showed a decreased survival of 2.4 logs CFU, and free cells demonstrated a decline of 3.5 logs CFU after only 4 weeks. Nonetheless, under mimicked small intestine circumstances, gelatin-alginate microgels exhibited either erosion or swelling (Yao et al., 2017). Muhammad et al. (2017) found that the survival rate of *L. plantarum*, when wrapped with potato resistant starches or potassium alginate, was 5.7 logs CFU and 4.5 logs CFU, respectively, after being stored at 25°C for 4 days. In contrast, free cells experienced a viability loss of 3.1 logs CFU. The amorphous transparent condition and low moisture content of coated organisms were proposed as factors contributing to high bacterial survivals. Nonetheless, the study revealed that cells wrapped in potassium alginate were not as favorable for bacterial growth in mimicked gastric circumstances (Muhammad et al., 2017).

To mitigate the decomposition of alginate in mimicked digestive juice, the cell wall fragment of *S. cerevisiae* was employed for covering alginate elements of *B. bifidum*, or *L. acidophilus*, which were further enveloped in a further covering of alginate wrapping (Mokhtari et al., 2017). While the sensitivity of multilayer wrapped *L. acidophilus* to digestive juice showed an important rise contrasted to free as well as single-layer alginate wrapped cells, it is noteworthy that the dimension of the multilayer microcapsules (103 mm) was twice that of the one-layer setup. Furthermore, the analysis revealed no notable variation in growth rates of wrapped *B. bifidum* when comparing single-layer to multilayer coatings in mimicked digestive juice. Research has indicated that the addition of cellulose nanocrystal (CNC), lecithin, or starch can effectively minimize the expansion of alginate capsules in solutions with pH levels of 1.5 and 5. The study revealed that incorporating lecithin and CNC into alginate microbeads resulted in reduced gastric fluid uptake while prolonging the dissolution period in comparison to standard alginate microbeads (Huq et al., 2017; Abbasi et al., 2023b).

Alternative wrapping substances, including resistant rice starch, were employed for covering three Lactobacillus strains through emulsification. These exhibited impressive viability, exceeding seven logs CFU/g for a duration of 60 days at 4°C or for 12 hours within mimicked gastrointestinal circumstances (Ashwar et al., 2018; Rad et al., 2020a). Nonetheless, an inadequate efficiency of encapsulation, ranging from 43% to 48%, and an average microcapsule dimension ranging from 45.5 to 49.3 mm in length were observed, which could complicate the integration of this wrapped probiotic into food items.

Given that instances like the ones mentioned above have shown sustained probiotic viabilities throughout preservation and/or with artificial gastrointestinal fluids, the encapsulation method may seem as a comprehensive way to preserve the sustainability of probiotic cells in the enveloped particles (Ozma et al., 2022). Nonetheless, the generation and dimensions of encapsulated probiotics often fall short of expectations, and the viability of these encapsulated probiotics within various food products is also contingent upon the specific strain used. In summary, a formulation that effectively incorporates probiotic A may not be appropriate for probiotic B when it comes to the development of innovative probiotic meals. It is noteworthy that the volume of research focused on encapsulation for the development of probiotic cell formulations significantly exceeds the investigations into encapsulated probiotic cells within actual food systems. It has been noted that there is a scarcity of investigations focusing on wrapped probiotics in food, a finding that has been documented in other sources (Sabahi et al., 2022). The limited availability of commercial goods featuring covered probiotic cells can be attributed to several factors, including inadequate attention to the structural impacts of encapsulating substances, inappropriate encapsulation techniques, challenges in methods for assessment, and shortcomings in risk assessments for usage (Chen et al., 2017). The material used in microcapsules is a crucial element influencing the performance of the microencapsulation process (Dianawati et al., 2016).

In the assessment of food probiotics, embedded probiotics have not lived up to expectations. The authors highlighted the significance of steering this science regarding the effective generation of beneficial probiotics in food and proposed novel polymers for probiotics that merit attention (Abbasi et al., 2021a). Huq et al. (2013) conducted a review on the impact of biopolymeric systems on the packaging of probiotics. They highlighted that when incorporating probiotic encapsulation into novel food products, several factors warrant further exploration. These include the selection of suitable encapsulation approaches, the use of reliable and efficient encapsulating substances, the identification of effective probiotic varieties, and the potential benefits of increasing layers of biopolymers. Nevertheless, the researchers overlooked the consideration of the food chain in relation to the preservation of living cells throughout preservation (Huq et al., 2013).

Probiotic microcapsules' creation and use within foods may suggest that the food structure contains a complex ecosystem that is not conducive to bacterial life. The application of microencapsulation innovation for the straightforward production of probiotic capsules does not guarantee the viability and stability of probiotics within a food item (Ozma et al., 2023a). The exploration of various types of coatings and protectants, as well as the improvement of the procedure for drying, may not adequately address the technological challenges associated with enhancing the persistence of probiotics in meals, unless the interactions between probiotics and food ingredients are thoroughly examined (Ozma et al., 2022). Additional investigation into the use of probiotics in food should focus on understanding the fundamental connection between free probiotic bacteria and the food structure throughout preparation, preservation, and within the host, aligning with the results of Flach and collaborators (Flach et al., 2018). The significance of examining the connection between carrier matrix structures and the overall quality of probiotic goods was emphasized. Additionally, investigating the formation of biofilms from either pure or mixed probiotic species within specific circumstances could provide significant insights into bacterial behavior when faced with different stressors.

4.4 Consumer perspectives on probiotic foods

The adoption of probiotic-enriched meals by consumers is crucial for their continued popularity in the market. Despite the growing awareness of probiotics in recent decades, a significant number of consumers remain uninformed regarding particular strains and their related health advantages. Moreover, the scientific data backing the health assertions of probiotics frequently lacks clarity or consistency, leading to potential confusion and doubt among customers. The perceived health advantages significantly influence consumer attitudes regarding probiotic foods. Probiotics are frequently linked to digestive health, with a common belief that they enhance the digestive process and help prevent digestive issues (Gere et al., 2019). Nonetheless, the true health advantages of probiotics might not be readily apparent to consumers, particularly when these goods are promoted without transparent, evidence-supported assertions (Abbasi et al., 2021b). Consequently, initiatives aimed at education and promoting consciousness are crucial for enhancing customer confidence in probiotic-rich foods.

The sensory characteristics of probiotic foods, including flavor, feel, and appearance, play a crucial role in shaping the acceptance of consumers (Abbasi et al., 2022c). A variety of probiotic-enriched meals exhibit unique tastes or textures, which might not resonate with every consumer's preferences. For instance, kefir and yogurt, often utilized as vehicles for probiotics, can possess a tangy or sour taste that certain individuals may find unappealing (Rad et al., 2021c). Furthermore, the texture of these goods might be unappealing to some consumers. Consequently, it is essential to create probiotic meals that are not only functional but also enjoyable to eat to promote their widespread adoption. The aspect of convenience significantly influences the acceptance of consumers (Aldred Cheek and Wansink, 2017). A significant number of consumers are seeking convenient probiotic goods that seamlessly integrate into their hectic routines. The creation of convenient probiotic foods, including functional snack bars and drinks, can effectively meet this demand (Karimi et al., 2020). Furthermore, advanced packaging ways that safeguard the probiotics from external influences and offer straightforward usage guidelines can improve the usability and attractiveness of these goods.

4.4.1 Consumer buying behavior models

At present, various models are employed to forecast consumer purchasing behavior. A straightforward conceptual model includes the subsequent steps illustrated in Fig. 4.5: (1) an individual recognizes a need that has not been met; (2) following this recognition, the individual seeks information from multiple sources, including society, private, or business; (3) the individual encounters a range of alternative options and assesses them based on various factors like cost, effectiveness, company, and availability, ultimately choosing the most suitable option. Following choosing the best option, consumers proceed to the next phase, which involves purchasing the products. The last phase is the postpurchase selections, where the consumer's continued purchases are influenced by their own fulfillment with the good.

The models of consumer purchasing behavior serve to illustrate and clarify the different decision-making processes of consumers, offering a framework

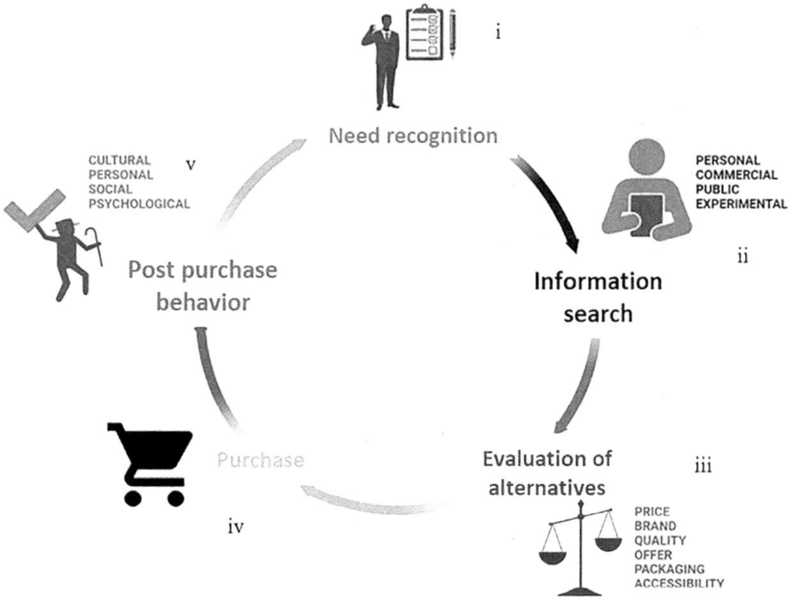

FIGURE 4.5 A conceptual model of consumer buying behavior.

for adapting to shifting parameters and marketplace circumstances. A notable model applicable for examining consumer purchasing behavior regarding biotic-family is the Health Belief Model (HBM) (Precup et al., 2022). The objective behind the creation of the HBM was to explore variations in personal decision-making regarding the utilization of healthcare facilities in the United States, particularly in relation to the motivation for avoiding and relieving illness (Rausch and Kopplin, 2021). This model serves to forecast people who engage in health behaviors, driven by the belief that such actions will enhance or alleviate the severity of their health conditions.

The HBM concentrated on two aspects of how individuals perceive health and health-related behaviors: the evaluation of behaviors and their assessment of threats. Perception of threat encompasses two primary beliefs: a sense of vulnerability to diseases or health problems and the anticipated severity of the consequences of illness. In contrast, behavioral evaluation involves two distinct sets of opinions: one regarding the advantages of the recommended health behavior and the other concerning the financial implications or barriers to implementing that behavior.

According to the latest research by Chong and Teh on the purchasing habits of Malaysians regarding probiotic and nutraceutical goods, consumers' probiotic item beliefs—rather than their understanding of the nutraceutical properties—have a major impact on their intentions to purchase (Le Chong and Teh, 2020). A different model of consumer behavior that may be utilized to forecast the adoption of synbiotic goods is the Theory of Planned Behavior (TPB). The TPB predicts a person's intention to engage in a behavior at a particular time and place. Rezai and colleagues employed a combination of TPB and HBM to investigate the purchasing behavior of Malaysian consumers regarding natural functional food. The study identified the perceived advantages (the belief that a specific action will reduce the probability or degree of an influence), after the subjective norm (social influence), and perceived vulnerability (the likelihood of being diagnosed with an illness or condition) as the main variables influencing consumer intentions throughout the purchasing process (Rezai et al., 2017).

An additional significant factor influencing customer buying habits is the willingness to pay (WTP). Pappalardo and Lusk carried out a social study to examine consumers' willingness to pay a greater utility cost for functional food in contrast to conventional products, utilizing the Expected Utility Theory (EUT). Their findings indicated that consumers' willingness to pay was influenced not only by the health benefits of the food but also by more variables such as cost, naturalness, and safety (Pappalardo and Lusk, 2016). Furthermore, another study indicated that the extent of consumer knowledge influences the acceptance of probiotic foods and the decision-making process regarding purchases. Informed consumers tend to seek out and buy food items that feature probiotic labels (Bogueva and Danova, 2024).

4.5 Regulatory and safety considerations

Probiotics in food are regulated differently depending on the country. Public health professionals have long maintained that any probiotics included in food products must be absolutely safe for consumption and improve people's health. Like other food ingredients, probiotics are governed in the United States by the Food and Drug Administration (FDA) (Abbasi et al., 2022b). Probiotics' safety is assessed in part by their capacity to generate toxins, induce medication resistance, or cause infections (Das et al., 2025). Strong testing is done on probiotic strains, proving possible for food consumption to avoid consumer damage. Probiotics must be included in food products using good manufacturing practices (GMP) if one wants to ensure the safety and high quality of the finished product. In terms of health claims, regulatory authorities such as the European Food Safety Authority (EFSA) require that health claims made about probiotics be substantiated by scientific evidence (Rad et al., 2020b). Health claims related to probiotics must be based on rigorous clinical studies that demonstrate the health benefits of the specific strains used in the product (Szajewska and Vinderola, 2024).

4.6 Future directions and challenges

Advances in personalized nutrition, sustainability, and consumer education might shape probiotic-enriched foods going forward. Customizing dietary recommendations based on an individual's unique genetic makeup, microbiota, and metabolic profile will help personalized nutrition provide more targeted probiotic treatment (Rad et al., 2021a,b,c,d). Recent developments in microbiomics and genetics might finally point to the most beneficial probiotic strains for certain individuals, therefore enabling possibly more effective probiotic-based diets generally. One of the most important things to think about is the direction of probiotic meals. For probiotic manufacture, the growing functional food industry depends on environmentally friendly and reasonably priced technology (Rad et al., 2021b,c). Using modern fermenting technologies, alternative substrates, and improved manufacturing techniques will make probiotics more affordable and easily available for a larger population. Customer knowledge and confidence play a significant role in the widespread adoption of products such as probiotics. The company's success is dependent on delivering consistent, evidence-based information on the benefits of probiotics and developing products that suit customer preferences for taste, usability, and health benefits (Rad et al., 2020c).

References

Abbasi, A., Sheykhsaran, E., Kafil, H.S., 2021. Postbiotics: Science, Technology and Applications. Bentham Science Publishers. Available at https://doi.org/10.2174/97816810883891210101.

Abbasi, A., Aghebati-Maleki, A., Yousefi, M., Aghebati-Maleki, L., 2021. Probiotic intervention as a potential therapeutic for managing gestational disorders and improving pregnancy outcomes. Journal of Reproductive Immunology 143, 103244. Available at https://doi.org/10.1016/j.jri.2020.103244.

Abbasi, A., Hajipour, N., Hasannezhad, P., Baghbanzadeh, A., Aghebati-Maleki, L., 2022a. Potential in vivo delivery routes of postbiotics. Critical Reviews in Food Science and Nutrition 62 (12), 3345–3369. Available at https://doi.org/10.1080/10408398.2020.1865260, https://www.tandfonline.com/loi/bfsn20.

Abbasi, A., Rad, A.H., Ghasempour, Z., Sabahi, S., Kafil, H.S., Hasannezhad, P., et al., 2022b. The biological activities of postbiotics in gastrointestinal disorders. Critical Reviews in Food Science and Nutrition 62 (22), 5983–6004. Available at https://doi.org/10.1080/10408398.2021.1895061, https://www.tandfonline.com/loi/bfsn20.

Abbasi, A., Aghebati-Maleki, L., Homayouni-Rad, A., 2022c. The promising biological role of postbiotics derived from probiotic *Lactobacillus* species in reproductive health. Critical Reviews in Food Science and Nutrition 62 (32), 8829–8841. Available at https://doi.org/10.1080/10408398.2021.1935701.

Abbasi, A., Rad, A.H., Maleki, L.A., Kafil, H.S., Baghbanzadeh, A., 2023a. Antigenotoxicity and cytotoxic potentials of cell-free supernatants derived from *Saccharomyces cerevisiae* var. boulardii on HT-29 human colon cancer cell lines. Probiotics and Antimicrobial Proteins 15 (6), 1583–1595. Available at https://doi.org/10.1007/s12602-022-10039-1, https://www.springer.com/journal/12602.

Abbasi, A., Sabahi, S., Shahbazi, N., Karimi, A., Barkhordari, H., 2023b. Nano lipid carrier in augmenting the bioavailability of functional bio-compounds. Biointerface Research in Applied Chemistry 13 (1). Available at https://doi.org/10.33263/BRIAC131.051, https://biointerfaceresearch.com/wp-content/uploads/2022/02/BRIAC131.051.pdf.

Abbasi, A., Sanej, K.D., Mohammadi, A., Moradi, S., Nayeri, B.D., Bazzaz, S., et al., 2023c. The role of postbiotics in food safety promotion. Electronic Journal of Food Processing and Preservation 15 (1), 75–108. Available at https://www.magiran.com/p2647435.

Abbasi, A., Sabahi, S., Bazzaz, S., Tajani, A.G., Lahouty, M., Aslani, R., et al., 2023d. An edible coating utilizing *Malva sylvestris* seed polysaccharide mucilage and postbiotic from Saccharomyces cerevisiae var. boulardii for the preservation of lamb meat. International Journal of Biological Macromolecules 246, 125660. Available at https://doi.org/10.1016/j.ijbiomac.2023.125660.

Abbasi, A., Bazzaz, S., Da Cruz, A.G., Khorshidian, N., Saadat, Y.R., Sabahi, S., et al., 2024a. A critical review on *Akkermansia muciniphila*: functional mechanisms, technological challenges, and safety issues. Probiotics and Antimicrobial Proteins 16 (4), 1376–1398. Available at https://doi.org/10.1007/s12602-023-10118-x, https://www.springer.com/journal/12602.

Abbasi, A., Hashemi, M., Pourjafar, H., Hosseini, H., 2024b. Malva neglecta seed polysaccharide mucilage coating enriched by the *Lactobacillus brevis* TD4 postbiotics: a promising strategy to promote the shelf life of fresh beef. International Journal of Biological Macromolecules 280 (Pt 2), 135789. Available at https://doi.org/10.1016/j.ijbiomac.2024.135789.

Abbasi, A., Hashemi, M., Pourjafar, H., Hosseini, S.M., Kafil, H.S., Rad, A.H., et al., 2024c. Chemical characterization, cell-based safety, and antioxidant assessments of *Lactobacillus helveticus* postbiotics and their potential antibacterial effects and mode of action against food-borne multidrug-resistant *Staphylococcus aureus* and enterohaemorrhagic *Escherichia coli* O157: H7. Journal of Food Safety 44 (6). Available at https://doi.org/10.1111/jfs.13174.

Abbasi, A., Hashemi, M., Samadi Kafil, H., Abbasi Astamal, M., Lahouty, M., Ghorbani Tajani, A., et al., 2024d. A critical review on the bioavailability promotion of the food bioactive compounds: nano lipid carriers perspective. Pharmaceutical Sciences 30 (3). Available at https://doi.org/10.34172/ps.2024.11.

Abbasi, A., Pourjafar, H., Moradi, S., Sabahi, S., Hosseini, H., 2025a. Anticancer and antioxidant activity of postbiotics. Postbiotics. pp. 459–468. Available at 10.1016/B978-0-443-22188-0.00025-5.

Abbasi, A., Hosseini, H., Pourjafar, H., Moradi, S., Shojaee-Aliabadi, S., 2025b. *Malva neglecta* essential oil: a promising approach for bio-preserving mayonnaise and assessing variable correlations through principal component analysis and heat map. Heliyon 11 (3), e41971. Available at https://doi.org/10.1016/j.heliyon.2025.e41971.

Alamoudi, N.M., Almabadi, E.S., El Ashiry, E.A., El Derwi, D.A., 2018. Effect of probiotic *Lactobacillus reuteri* on salivary cariogenic bacterial counts among groups of preschool children in Jeddah, Saudi Arabia: a randomized clinical trial. Journal of Clinical Pediatric Dentistry 42 (5), 331–337. Available at https://doi.org/10.17796/1053-4625-42.5.2, http://jocpd.org/?code=clpd-site.

Aldred Cheek, K., Wansink, B., 2017. Making it part of the package: edible packaging is more acceptable to young consumers when it is integrated with food. Journal of Food Products Marketing 23 (6), 723–732. Available at https://doi.org/10.1080/10454446.2017.1244793, http://www.tandfonline.com/toc/wfpm20/current.

An, H.M., Park, S.Y., Lee, D.K., Kim, J.R., Cha, M.K., Lee, S.W., et al., 2011. Antiobesity and lipid-lowering effects of Bifidobacterium spp. in high fat diet-induced obese rats. Lipids in Health and Disease 10. Available at https://doi.org/10.1186/1476-511X-10-116.

An, Z., Wang, H., Mokadem, M., 2021. Role of the autonomic nervous system in mechanism of energy and glucose regulation post bariatric surgery. Frontiers in Neuroscience 15. Available at https://doi.org/10.3389/fnins.2021.770690.

Antunes, A.E.C., Liserre, A.M., Coelho, A.L.A., Menezes, C.R., Moreno, I., Yotsuyanagi, K., et al., 2013. Acerola nectar with added microencapsulated probiotic. LWT - Food Science and Technology 54 (1), 125–131. Available at https://doi.org/10.1016/j.lwt.2013.04.018.

Asadi, Z., Abbasi, A., Ghaemi, A., Montazeri, E.A., Akrami, S., 2024. Investigating the properties and antibacterial, antioxidant, and cytotoxicity activity of postbiotics derived from *Lacticaseibacillus casei* on various gastrointestinal pathogens in vitro and in food models. GMS Hygiene and Infection Control 19, Doc60. Available at 10.3205/dgkh000515.

Ashwar, B.A., Gani, A., Gani, A., Shah, A., Masoodi, F.A., 2018. Production of RS4 from rice starch and its utilization as an encapsulating agent for targeted delivery of probiotics. Food Chemistry 239, 287–294. Available at https://doi.org/10.1016/j.foodchem.2017.06.110, www.elsevier.com/locate/foodchem.

Babio, N., Becerra-Tomás, N., Martínez-González, M.Á., Corella, D., Estruch, R., Ros, E., et al., 2015. Consumption of yogurt, low-fat milk, and other low-fat dairy products is associated with lower risk of metabolic syndrome incidence in an elderly mediterranean population. The Journal of Nutrition 145 (10), 2308–2316. Available at https://doi.org/10.3945/jn.115.214593.

Bai, X., Zhu, M., He, Y., Wang, T., Tian, D., Shu, J., 2022. The impacts of probiotics in eradication therapy of *Helicobacter pylori*. Archives of Microbiology 204 (12). Available at https://doi.org/10.1007/s00203-022-03314-w.

Barathikannan, K., Chelliah, R., Rubab, M., Daliri, E.B.M., Elahi, F., Kim, D.H., et al., 2019. Gut microbiome modulation based on probiotic application for anti-obesity: a review on efficacy and validation. Microorganisms 7 (10). Available at https://doi.org/10.3390/microorganisms7100456, https://www.mdpi.com/2076-2607/7/10/456/pdf.

Basholli-Salihu, M., Mueller, M., Salar-Behzadi, S., Unger, F.M., Viernstein, H., 2014. Effect of lyoprotectants on β-glucosidase activity and viability of *Bifidobacterium infantis* after freeze-drying and storage in milk and low pH juices. LWT - Food Science and Technology 57 (1), 276–282. Available at https://doi.org/10.1016/j.lwt.2014.01.011.

Bazzaz, S., Abbasi, A., Ghotbabad, A.G., Pourjafar, H., Hosseini, H., 2025. Novel encapsulation approaches in the functional food industry: with a focus on probiotic cells and bioactive compounds. Probiotics and Antimicrobial Proteins 17 (3), 1132–1170. Available at https://doi.org/10.1007/s12602-024-10364-7, https://www.springer.com/journal/12602.

Beltrán-Barrientos, L.M., González-Córdova, A.F., Hernández-Mendoza, A., Torres-Inguanzo, E.H., Astiazarán-García, H., Esparza-Romero, J., et al., 2018. Randomized double-blind controlled clinical trial of the blood pressure–lowering effect of fermented milk with *Lactococcus lactis*: a pilot study2. Journal of Dairy Science 101 (4), 2819–2825. Available at https://doi.org/10.3168/jds.2017-13189, http://www.elsevier.com/wps/find/journaldescription.cws_home/721317/description#description, http://www.elsevier.com/wps/find/journaldescription.cws_home/721317/description#description.

Benjak Horvat, I., Gobin, I., Kresović, A., Hauser, G., 2021. How can probiotic improve irritable bowel syndrome symptoms? World Journal of Gastrointestinal Surgery 13 (9), 923–940. Available at https://doi.org/10.4240/wjgs.v13.i9.923.

Bogueva, D., Danova, S., 2024. Comparing precision fermentation and traditional fermentations: consumer views. Consumer Perceptions and Food. Springer Nature, Australia, pp. 563–588. Available at 10.1007/978-981-97-7870-6_27.

Borges, S., Barbosa, J., Silva, J., Gomes, A.M., Pintado, M., Silva, C.L.M., et al., 2016. A feasibility study of *Lactobacillus plantarum* in fruit powders after processing and storage. International Journal of Food Science and Technology 51 (2), 381–388. Available at https://doi.org/10.1111/ijfs.12975.

De Brito, B.B., Da Silva, F.A.F., Soares, A.S., Pereira, V.A., Cordeiro Santos, M.L., Sampaio, M.M., et al., 2019. Pathogenesis and clinical management of *Helicobacter pylori* gastric infection. World Journal of Gastroenterology 25 (37), 5578–5589. Available at https://doi.org/10.3748/wjg.v25.i37.5578, https://www.wjgnet.com/1007-9327/full/v25/i37/5578.htm.

Di Cerbo, A., Pezzuto, F., Palmieri, L., Rottigni, V., Iannitti, T., Palmieri, B., 2013. Clinical and experimental use of probiotic formulations for management of end-stage renal disease: an update. International Urology and Nephrology 45 (6), 1569–1576. Available at https://doi.org/10.1007/s11255-012-0335-5.

Chamlagain, B., Sugito, T.A., Deptula, P., Edelmann, M., Kariluoto, S., Varmanen, P., et al., 2018. In situ production of active vitamin B12 in cereal matrices using *Propionibacterium freudenreichii*. Food Science and Nutrition 6 (1), 67–76. Available at https://doi.org/10.1002/fsn3.528.

Chan, A., Ellepola, K., Truong, T., Balan, P., Koo, H., Seneviratne, C.J., 2020. Inhibitory effects of xylitol and sorbitol on *Streptococcus mutans* and *Candida albicans* biofilms are repressed by the presence of sucrose. Archives of Oral Biology 119, 104886. Available at https://doi.org/10.1016/j.archoralbio.2020.104886.

Chen, J., Wang, Q., Liu, C.M., Gong, J., 2017. Issues deserve attention in encapsulating probiotics: critical review of existing literature. Critical Reviews in Food Science and Nutrition 57 (6), 1228–1238. Available at https://doi.org/10.1080/10408398.2014.977991, www.tandf.co.uk/journals/titles/10408398.asp.

Le Chong, C., Teh, P.H., 2020. Consumer buying behavior towards probiotics nutraceutical products in Malaysia. Asian Journal of Research in Business and Management 2, 1–22.

Collins, F.L., Rios-Arce, N.D., Schepper, J.D., Parameswaran, N., McCabe, L.R., 2017. The potential of probiotics as a therapy for osteoporosis. Microbiology Spectrum 5 (4). Available at https://doi.org/10.1128/microbiolspec.BAD-0015-2016.

Cristiano, C., Lama, A., Lembo, F., Mollica, M.P., Calignano, A., Mattace Raso, G., 2018. Interplay between peripheral and central inflammation in autism spectrum disorders: possible nutritional and therapeutic strategies. Frontiers in Physiology 9. Available at https://doi.org/10.3389/fphys.2018.00184.

Dahiya, D., Nigam, P.S., 2023. Nutraceuticals prepared with specific strains of probiotics for supplementing gut microbiota in hosts allergic to certain foods or their additives. Nutrients 15 (13), 2979. Available at https://doi.org/10.3390/nu15132979.

Dar, H.Y., Shukla, P., Mishra, P.K., Anupam, R., Mondal, R.K., Tomar, G.B., et al., 2018. *Lactobacillus acidophilus* inhibits bone loss and increases bone heterogeneity in osteoporotic mice via modulating Treg-Th17 cell balance. Bone Reports 8, 46–56. Available at https://doi.org/10.1016/j.bonr.2018.02.001, http://www.journals.elsevier.com/bone-reports/.

Das, U., Mehra, R.K., Mandal, S., Ashique, S., Kumar, S., Farid, A., et al., 2025. Toxicology of Probiotics: Challenges and Future Prospects The Role of Probiotics in Cancer Management. Apple Academic Press, India, pp. 379–428. Available at https://appleacademicpress.com/the-role-of-probiotics-in-cancer-management-/9781774919125.

Dassi, E., Ferretti, P., Covello, G., Speccher, A., Migazzi, A., Bosco, B., et al., 2018. The short-term impact of probiotic consumption on the oral cavity microbiome. Scientific Reports 8 (1). Available at https://doi.org/10.1038/s41598-018-28491-x, www.nature.com/srep/index.html.

Degnan, P.H., Taga, M.E., Goodman, A.L., 2014. Vitamin B12 as a modulator of gut microbial ecology. Cell Metabolism 20 (5), 769–778. Available at https://doi.org/10.1016/j.cmet.2014.10.002, http://www.cellmetabolism.org/.

Dianawati, D., Mishra, V., Shah, N.P., 2016. Survival of microencapsulated probiotic bacteria after processing and during storage: a review. Critical Reviews in Food Science and Nutrition 56 (10), 1685–1716. Available at https://doi.org/10.1080/10408398.2013.798779, www.tandf.co.uk/journals/titles/10408398.asp.

Duffuler, P., Bhullar, K.S., Wu, J., 2024. Targeting gut microbiota in osteoporosis: impact of the microbial based functional food ingredients. Food Science and Human Wellness 13 (1), 1–15. Available at https://doi.org/10.26599/FSHW.2022.9250001, https://www.sciopen.com/article/pdf/10.26599/FSHW.2022.9250001.pdf.

EFSA Panel on Dietetic Products, Nutrition and Allergies (NDA), 2010. Scientific opinion on the substantiation of health claims related to live yoghurt cultures and improved lactose digestion (ID 1143, 2976) pursuant to Article 13 (1) of Regulation (EC) No 1924/2006. EFSA Journal 8, 1763.

Ejtahed, H.S., Mohtadi-Nia, J., Homayouni-Rad, A., Niafar, M., Asghari-Jafarabadi, M., Mofid, V., 2012. Probiotic yogurt improves antioxidant status in type 2 diabetic patients. Nutrition (Burbank, Los Angeles County, Calif.) 28 (5), 539–543. Available at https://doi.org/10.1016/j.nut.2011.08.013.

Ejtahed, H.S., Angoorani, P., Soroush, A.R., Atlasi, R., Hasani-Ranjbar, S., Mortazavian, A.M., et al., 2019. Probiotics supplementation for the obesity management; a systematic review of animal studies and clinical trials. Journal of Functional Foods 52, 228–242. Available at https://doi.org/10.1016/j.jff.2018.10.039, http://www.elsevier.com/wps/find/journaldescription.cws_home/717426/description#description.

Flach, J., van der Waal, M.B., van den Nieuwboer, M., Claassen, E., Larsen, O.F.A., 2018. The underexposed role of food matrices in probiotic products: reviewing the relationship between carrier matrices and product parameters. Critical Reviews in Food Science and Nutrition 58 (15), 2570–2584. Available at https://doi.org/10.1080/10408398.2017.1334624, https://www.tandfonline.com/loi/bfsn20.

Frazier, T.H., DiBaise, J.K., McClain, C.J., 2011. Gut microbiota, intestinal permeability, obesity-induced inflammation, and liver injury. Journal of Parenteral and Enteral Nutrition 35 (5). Available at https://doi.org/10.1177/0148607111413772.

Freire, A.L., Ramos, C.L., da Costa Souza, P.N., Cardoso, M.G.B., Schwan, R.F., 2017. Nondairy beverage produced by controlled fermentation with potential probiotic starter cultures of lactic acid bacteria and yeast. International Journal of Food Microbiology 248, 39–46. Available at https://doi.org/10.1016/j.ijfoodmicro.2017.02.011, www.elsevier.com/locate/ijfoodmicro.

Gere, A., Radványi, D., Moskowitz, H., 2019. Consumer Perspectives about Innovations in Traditional Foods. Elsevier BV, pp. 53–84. Available at 10.1016/b978-0-12-814887-7.00003-4.

Habeebuddin, M., Karnati, R.K., Shiroorkar, P.N., Nagaraja, S., Asdaq, S.M.B., Anwer, M.K., et al., 2022. Topical probiotics: more than a skin deep. Pharmaceutics 14 (3). Available at https://doi.org/10.3390/pharmaceutics14030557, https://www.mdpi.com/1999-4923/14/3/557/pdf.

Higuchi, T., Suzuki, N., Nakaya, S., Omagari, S., Yoneda, M., Hanioka, T., et al., 2019. Effects of *Lactobacillus salivarius* WB21 combined with green tea catechins on dental caries, periodontitis, and oral malodor. Archives of Oral Biology 98, 243–247. Available at https://doi.org/10.1016/j.archoralbio.2018.11.027.

Holko, I., Hrabě, J., Šalaková, A., Rada, V., 2013. The substitution of a traditional starter culture in mutton fermented sausages by *Lactobacillus acidophilus* and Bifidobacterium animalis. Meat Science 94 (3), 275–279. Available at https://doi.org/10.1016/j.meatsci.2013.03.005.

Hosseini, H., Abbasi, A., Sabahi, S., Akrami, S., Yousefi-Avarvand, A., 2024. Assessing the potential biological activities of postbiotics derived from S*accharomyces cerevisiae*: an in vitro study. Probiotics and Antimicrobial Proteins 16 (4), 1348–1364. Available at https://doi.org/10.1007/s12602-023-10117-y.

Huang, C.H., Lin, Y.C., Jan, T.R., 2017. *Lactobacillus reuteri* induces intestinal immune tolerance against food allergy in mice. Journal of Functional Foods 31, 44–51. Available at https://doi.org/10.1016/j.jff.2017.01.034, http://www.elsevier.com/wps/find/journaldescription.cws_home/717426/description#description.

Huang, C.Y., Araujo, K., Sánchez, J.N., Kund, G., Trumble, J., Roper, C., et al., 2021. A stable antimicrobial peptide with dual functions of treating and preventing citrus Huanglongbing. Proceedings of the National Academy of Sciences of the United States of America 118 (6). Available at https://doi.org/10.1073/pnas.2019628118, https://www.pnas.org/content/pnas/118/6/e2019628118.full.pdf.

Huq, T., Khan, A., Khan, R.A., Riedl, B., Lacroix, M., 2013. Encapsulation of probiotic bacteria in biopolymeric system. Critical Reviews in Food Science and Nutrition 53 (9), 909–916. Available at https://doi.org/10.1080/10408398.2011.573152.

Huq, T., Fraschini, C., Khan, A., Riedl, B., Bouchard, J., Lacroix, M., 2017. Alginate based nanocomposite for microencapsulation of probiotic: effect of cellulose nanocrystal (CNC) and lecithin. Carbohydrate Polymers 168, 61–69. Available at https://doi.org/10.1016/j.carbpol.2017.03.032.

Johnson, T.R., Gómez, B.I., McIntyre, M.K., Dubick, M.A., Christy, R.J., Nicholson, S.E., et al., 2018. The cutaneous microbiome and wounds: new molecular targets to promote wound healing. International Journal of Molecular Sciences 19 (9). Available at https://doi.org/10.3390/ijms19092699, http://www.mdpi.com/1422-0067/19/9/2699/pdf.

Karimi, N., Jabbari, V., Nazemi, A., Ganbarov, K., Karimi, N., Tanomand, A., et al., 2020. Thymol, cardamom and *Lactobacillus plantarum* nanoparticles as a functional candy with high protection against *Streptococcus mutans* and tooth decay. Microbial Pathogenesis 148, 104481. Available at https://doi.org/10.1016/j.micpath.2020.104481.

Kaur, K., Nekkanti, S., Madiyal, M., Choudhary, P., 2018. Effect of chewing gums containing probiotics and xylitol on oral health in children: a randomized controlled trial. Journal of

International Oral Health 10 (5), 237–243. Available at https://doi.org/10.4103/jioh.jioh_170_18, http://www.jioh.org/.

Keller, M.K., Brandsborg, E., Holmstrøm, K., Twetman, S., 2018. Effect of tablets containing probiotic candidate strains on gingival inflammation and composition of the salivary microbiome: a randomised controlled trial. Beneficial Microbes 9 (3), 487–494. Available at https://doi.org/10.3920/bm2017.0104.

Kerry, R.G., Patra, J.K., Gouda, S., Park, Y., Shin, H.S., Das, G., 2018. Benefaction of probiotics for human health: a review. Journal of Food and Drug Analysis 26 (3), 927–939. Available at https://doi.org/10.1016/j.jfda.2018.01.002, http://www.elsevier.com/journals/journal-of-food-and-drug-analysis/1021-9498#.

Kim, D.E., Kim, J.K., Han, S.K., Jang, S.E., Han, M.J., Kim, D.H., 2019. *Lactobacillus plantarum* NK3 and *Bifidobacterium longum* NK49 alleviate bacterial vaginosis and osteoporosis in mice by suppressing NF-κB-Linked TNF-α expression. Journal of Medicinal Food 22 (10), 1022–1031. Available at https://doi.org/10.1089/jmf.2019.4419, www.liebertonline.com/jmf.

Klaenhammer, T.R., Kleerebezem, M., Kopp, M.V., Rescigno, M., 2012. The impact of probiotics and prebiotics on the immune system. Nature Reviews Immunology 12 (10), 728–734. Available at https://doi.org/10.1038/nri3312.

Kołożyn-Krajewska, D., Dolatowski, Z.J., 2012. Probiotic meat products and human nutrition. Process Biochemistry 47 (12), 1761–1772. Available at https://doi.org/10.1016/j.procbio.2012.09.017.

Kouam, E.M.F., Piame, L.T., Kouteu, S.S., Temgoua, J.B., Ngoufack, F.Z., Kaktcham, P.M., 2024. Probiotic characterisation of *Lactiplantibacillus plantarum* LO3 and use in the development of a golden apple-based non-dairy probiotic beverage. Systems Microbiology and Biomanufacturing 4 (3), 869–881. Available at https://doi.org/10.1007/s43393-024-00251-1, https://www.springer.com/journal/43393.

Kwak, C., Jeong, B.C., Ku, J.H., Kim, H.H., Lee, J.J., Huh, C.S., et al., 2006. Prevention of nephrolithiasis by *Lactobacillus* in stone-forming rats: a preliminary study. Urological Research 34 (4), 265–270. Available at https://doi.org/10.1007/s00240-006-0054-4.

Lamsal, B.P., 2012. Production, health aspects and potential food uses of dairy prebiotic galactooligosaccharides. Journal of the Science of Food and Agriculture 92 (10), 2020–2028. Available at https://doi.org/10.1002/jsfa.5712.

Latif, A., Shehzad, A., Niazi, S., Zahid, A., Ashraf, W., Iqbal, M.W., et al., 2023. Probiotics: mechanism of action, health benefits and their application in food industries. Frontiers in Microbiology 14. Available at https://doi.org/10.3389/fmicb.2023.1216674, https://www.frontiersin.org/journals/microbiology#.

Lee, N.K., Kim, S.Y., Han, K.J., Eom, S.J., Paik, H.D., 2014. Probiotic potential of *Lactobacillus strains* with anti-allergic effects from kimchi for yogurt starters. LWT 58 (1), 130–134. Available at https://doi.org/10.1016/j.lwt.2014.02.028, https://www.journals.elsevier.com/lwt.

Li, J.Y., Chassaing, B., Tyagi, A.M., Vaccaro, C., Luo, T., Adams, J., et al., 2016. Sex steroid deficiency-associated bone loss is microbiota dependent and prevented by probiotics. Journal of Clinical Investigation 126 (6), 2049–2063. Available at https://doi.org/10.1172/JCI86062, http://www.jci.org/articles/view/86062/pdf.

Lionetti, E., Indrio, F., Pavone, L., Borrelli, G., Cavallo, L., Francavilla, R., 2010. Role of probiotics in pediatric patients with *Helicobacter pylori* infection: a comprehensive review of the literature. Helicobacter 15 (2), 79–87. Available at https://doi.org/10.1111/j.1523-5378.2009.00743.x.

Luu, M., Visekruna, A., 2019. Short-chain fatty acids: bacterial messengers modulating the immunometabolism of T cells. European Journal of Immunology 49 (6), 842–848. Available at https://doi.org/10.1002/eji.201848009.

Ly, N.P., Litonjua, A., Gold, D.R., Celedón, J.C., 2011. Gut microbiota, probiotics, and vitamin D: interrelated exposures influencing allergy, asthma, and obesity? Journal of Allergy and Clinical Immunology 127 (5), 1087–1094. Available at https://doi.org/10.1016/j.jaci.2011.02.015.

Marco, M.L., Heeney, D., Binda, S., Cifelli, C.J., Cotter, P.D., Foligné, B., et al., 2017. Health benefits of fermented foods: microbiota and beyond. Current Opinion in Biotechnology 44, 94–102. Available at https://doi.org/10.1016/j.copbio.2016.11.010, http://www.elsevier.com/locate/copbio.

Mccabe, L.R., Irwin, R., Schaefer, L., Britton, R.A., 2013. Probiotic use decreases intestinal inflammation and increases bone density in healthy male but not female mice. Journal of Cellular Physiology 228 (8), 1793–1798. Available at https://doi.org/10.1002/jcp.24340.

Mestre, A., Sathiya Narayanan, R., Rivas, D., John, J., Abdulqader, M.A., Khanna, T., et al., 2022. Role of probiotics in the management of *Helicobacter pylori*. Cureus 14 (6). Available at https://doi.org/10.7759/cureus.26463.

Mestry, A.P., Mujumdar, A.S., Thorat, B.N., 2011. Optimization of spray drying of an innovative functional food: fermented mixed juice of carrot and watermelon. Drying Technology 29 (10), 1121–1131. Available at https://doi.org/10.1080/07373937.2011.566968.

Mokhtari, S., Jafari, S.M., Khomeiri, M., Maghsoudlou, Y., Ghorbani, M., 2017. The cell wall compound of *Saccharomyces cerevisiae* as a novel wall material for encapsulation of probiotics. Food Research International 96, 19–26. Available at https://doi.org/10.1016/j.foodres.2017.03.014, www.elsevier.com/inca/publications/store/4/2/2/9/7/0.

Mondragón-Bernal, O.L., Alves, J.G.L.F., Teixeira, M.A., Ferreira, M.F.P., Maugeri Filho, F., 2017. Stability and functionality of synbiotic soy food during shelf-life. Journal of Functional Foods 35, 134–145. Available at https://doi.org/10.1016/j.jff.2017.05.021, http://www.elsevier.com/wps/find/journaldescription.cws_home/717426/description#description.

Moradi, S., Moslehi Shad, M., Hosseini, H., Mortazavian, A., Shojaei Aliabadi, S., Abbasi, A., 2024. Characterizing chemical profiles and functional attributes of postbiotics derived from *Lactobacillus casei* ATCC 39392 and assessing their antimicrobial activities against *Staphylococcus aureus* under in-vitro conditions. Iranian Journal of Nutrition Sciences and Food Technology 19 (2), 43–58. Available at https://doi.org/10.61186/nsft.19.2.43, http://nsft.sbmu.ac.ir/article-1-3788-en.pdf.

Muhammad, Z., Ramzan, R., Huo, G.C., Tian, H., Bian, X., 2017. Integration of polysaccharide-thermoprotectant formulations for microencapsulation of *Lactobacillus plantarum*, appraisal of survivability and physico-biochemical properties during storage of spray dried powders. Food Hydrocolloids 66, 286–295. Available at https://doi.org/10.1016/j.foodhyd.2016.11.040, https://www.journals.elsevier.com/food-hydrocolloids.

Nadelman, P., Monteiro, A., Balthazar, C.F., Silva, H.L.A., Cruz, A.G., de Almeida Neves, A., et al., 2019. Probiotic fermented sheep's milk containing *Lactobacillus casei* 01: effects on enamel mineral loss and *Streptococcus* counts in a dental biofilm model. Journal of Functional Foods 54, 241–248. Available at https://doi.org/10.1016/j.jff.2019.01.025, http://www.elsevier.com/wps/find/journaldescription.cws_home/717426/description#description.

Nath, A., Haktanirlar, G., Varga, Á., Molnár, M.A., Albert, K., Galambos, I., et al., 2018. Biological activities of lactose-derived prebiotics and symbiotic with probiotics on gastrointestinal system. Medicina 54 (2), 18. Available at https://doi.org/10.3390/medicina54020018.

Nocerino, R., Paparo, L., Terrin, G., Pezzella, V., Amoroso, A., Cosenza, L., et al., 2017. Cow's milk and rice fermented with *Lactobacillus paracasei* CBA L74 prevent infectious diseases in children: a randomized controlled trial. Clinical Nutrition 36 (1), 118–125. Available at https://doi.org/10.1016/j.clnu.2015.12.004.

Nozue, M., Shimazu, T., Sasazuki, S., Charvat, H., Mori, N., Mutoh, M., et al., 2017. Fermented soy product intake is inversely associated with the development of high blood pressure: the Japan public health center-based prospective study. The Journal of Nutrition 147 (9), 1749–1756. Available at https://doi.org/10.3945/jn.117.250282.

Ohlsson, C., Sjögren, K., 2015. Effects of the gut microbiota on bone mass. Trends in Endocrinology and Metabolism 26 (2), 69–74. Available at https://doi.org/10.1016/j.tem.2014.11.004.

Ozma, M.A., Abbasi, A., Akrami, S., Lahouty, M., Shahbazi, N., Ganbarov, K., et al., 2022. Postbiotics as the key mediators of the gut microbiota-host interactions. Infezioni in Medicina 30 (2), 180–193. Available at https://doi.org/10.53854/liim-3002-3, www.infezmed.it.

Ozma, M.A., Abbasi, A., Ahangarzadeh Rezaee, M., Hosseini, H., Hosseinzadeh, N., Sabahi, S., et al., 2023a. A critical review on the nutritional and medicinal profiles of garlic's (*Allium sativum* L.) bioactive compounds. Food Reviews International 39 (9), 6324–6361. Available at https://doi.org/10.1080/87559129.2022.2100417, www.tandf.co.uk/journals/titles/87559129.asp.

Ozma, M.A., Abbasi, A., Sabahi, S., 2023b. Characterization of postbiotics derived from *Lactobacillus paracasei* ATCC 55544 and its application in *Malva sylvestris* seed mucilage edible coating to the improvement of the microbiological, and sensory properties of lamb meat during storage. Biointerface Research in Applied Chemistry 13 (3). Available at https://doi.org/10.33263/BRIAC133.267, https://biointerfaceresearch.com/wp-content/uploads/2022/07/BRIAC133.267.pdf.

Ozma, M.A., Moaddab, S.R., Hosseini, H., Khodadadi, E., Ghotaslou, R., Asgharzadeh, M., et al., 2024. A critical review of novel antibiotic resistance prevention approaches with a focus on postbiotics. Critical Reviews in Food Science and Nutrition 64 (26), 9637–9655. Available at https://doi.org/10.1080/10408398.2023.2214818, https://www.tandfonline.com/loi/bfsn20.

Pappalardo, G., Lusk, J.L., 2016. The role of beliefs in purchasing process of functional foods. Food Quality and Preference 53, 151–158. Available at https://doi.org/10.1016/j.foodqual.2016.06.009.

Parhizkar, E., Vosough, P., Baneshi, M., Keshavarzi, A., Lohrasbi, P., Taghizadeh, S., et al., 2025. Probiotics and gut microbiota modulation: implications for skin health and disease management. Archives of Microbiology 207 (3). Available at https://doi.org/10.1007/s00203-025-04267-6.

Park, S., Bae, J.H., 2016. Fermented food intake is associated with a reduced likelihood of atopic dermatitis in an adult population (Korean National Health and Nutrition Examination Survey 2012-2013). Nutrition Research 36 (2), 125–133. Available at https://doi.org/10.1016/j.nutres.2015.11.011, www.elsevier.com/locate/nutres.

Parvaneh, K., Jamaluddin, R., Karimi, G., Erfani, R., 2014. Effect of probiotics supplementation on bone mineral content and bone mass density. The Scientific World Journal 2014, 1–6. Available at https://doi.org/10.1155/2014/595962.

Parvaneh, K., Ebrahimi, M., Sabran, M.R., Karimi, G., Hwei, A.N.M., Abdul-Majeed, S., et al., 2015. Probiotics (*Bifidobacterium longum*) increase bone mass density and upregulate Sparc and Bmp-2 genes in rats with bone loss resulting from ovariectomy. BioMed Research International 2015. Available at https://doi.org/10.1155/2015/897639, http://www.hindawi.com/journals/biomed/.

Pavlidou, E., Alexatou, O., Tsourouflis, G., Antasouras, G., Papadopoulou, S.K., Papandreou, D., et al., 2025. Probiotic supplementation during pregnancy: evaluating the current clinical evidence against gestational diabetes mellitus. Current Diabetes Reviews 21 (5). Available at https://doi.org/10.2174/0115733998284749240417052006, www.eurekaselect.com/587/journal/current-diabetes-reviews.

Pei, M., Wei, L., Hu, S., Yang, B., Si, J., Yang, H., et al., 2018. Probiotics, prebiotics and synbiotics for chronic kidney disease: protocol for a systematic review and meta-analysis. BMJ Open 8 (7), e020863. Available at https://doi.org/10.1136/bmjopen-2017-020863.

Precup, G., Pocol, C.B., Teleky, B.E., Vodnar, D.C., 2022. Awareness, knowledge, and interest about prebiotics— a study among romanian consumers. International Journal of Environmental Research and Public Health 19 (3). Available at https://doi.org/10.3390/ijerph19031208, https://www.mdpi.com/1660-4601/19/3/1208/pdf.

Racioppo, A., Corbo, M.R., Piccoli, C., Sinigaglia, M., Speranza, B., Bevilacqua, A., 2017. Ultrasound attenuation of lactobacilli and bifidobacteria: effect on some technological and probiotic properties. International Journal of Food Microbiology 243, 78–83. Available at https://doi.org/10.1016/j.ijfoodmicro.2016.12.011, www.elsevier.com/locate/ijfoodmicro.

Rad, A.H., Fathi-zavoshti, H., Douroud, N., Shahbazi, N., Abbasi, A., 2020a. Evaluating the role of postbiotics as a new generation of probiotics in health and diseases. Journal of Ardabil University of Medical Sciences 19 (4), 381–399. Available at https://doi.org/10.29252/jarums.19.4.381.

Rad, A.H., Abbasi, A., Kafil, H.S., Ganbarov, K., 2020b. Potential pharmaceutical and food applications of postbiotics: a review. Current Pharmaceutical Biotechnology 21 (15), 1576–1587. Available at https://doi.org/10.2174/1389201021666200516154833, https://www.eurekaselect.com/182004/article.

Rad, A.H., Maleki, L.A., Kafil, H.S., Zavoshti, H.F., Abbasi, A., 2020c. Postbiotics as novel health-promoting ingredients in functional foods. Health Promotion Perspectives 10 (1), 3–4. Available at https://doi.org/10.15171/hpp.2020.02, http://hpp.tbzmed.ac.ir/PDF/hpp-10-3.pdf.

Rad, A.H., Maleki, L.A., Kafil, H.S., Abbasi, A., 2021a. Postbiotics: a novel strategy in food allergy treatment. Critical Reviews in Food Science and Nutrition 61 (3), 492–499. Available at https://doi.org/10.1080/10408398.2020.1738333.

Rad, A.H., Maleki, L.A., Kafil, H.S., Abbasi, A., Khani, N., 2021b. Postbiotics as a safe alternative to live probiotic bacteria in the food and pharmaceutical industries. Scientific Journal of Kurdistan University of Medical Sciences 26 (4), 132–157. Available at https://doi.org/10.52547/sjku.26.4.132.

Rad, A.H., Maleki, L.A., Kafil, H.S., Zavoshti, H.F., Abbasi, A., 2021c. Postbiotics as promising tools for cancer adjuvant therapy. Advanced Pharmaceutical Bulletin 11 (1), 1–5. Available at https://doi.org/10.34172/apb.2021.007.

Rad, A.H., Aghebati-Maleki, L., Kafil, H.S., Abbasi, A., 2021d. Molecular mechanisms of postbiotics in colorectal cancer prevention and treatment. Critical Reviews in Food Science and Nutrition 61 (11), 1787–1803. Available at https://doi.org/10.1080/10408398.2020.1765310, https://www.tandfonline.com/loi/bfsn20.

Ranjan, A., 2022. The use of probiotics, prebiotics, and synbiotics as an alternative to antibiotics. Alternatives to Antibiotics: Recent Trends and Future Prospects. Springer Nature, United States, pp. 449–465. Available at https://doi.org/10.1007/978-981-19-1854-4, 10.1007/978-981-19-1854-4_18.

Rausch, T.M., Kopplin, C.S., 2021. Bridge the gap: consumers' purchase intention and behavior regarding sustainable clothing. Journal of Cleaner Production 278. Available at https://doi.

org/10.1016/j.jclepro.2020.123882, https://www.journals.elsevier.com/journal-of-cleaner-production.

Rezai, G., Teng, P.K., Shamsudin, M.N., Mohamed, Z., Stanton, J.L., 2017. Effect of perceptual differences on consumer purchase intention of natural functional food. Journal of Agribusiness in Developing and Emerging Economies 7 (2), 153–173. Available at https://doi.org/10.1108/JADEE-02-2015-0014, http://www.emeraldgrouppublishing.com/jadee.htm.

Robin, V., Wim, T., Maria, C.D.L.P., Isabelle, L., 2025. Probiotics for maintaining oral health during fixed orthodontic treatment: a systematic review and meta-analysis. International Journal of Dental Hygiene 23 (1), 100–113. Available at https://doi.org/10.1111/idh.12821, http://onlinelibrary.wiley.com/journal/10.1111/(ISSN)1601-5037.

Rosier, B.T., Marsh, P.D., Mira, A., 2018. Resilience of the oral microbiota in health: mechanisms that prevent dysbiosis. Journal of Dental Research 97 (4), 371–380. Available at https://doi.org/10.1177/0022034517742139.

Rubio, R., Aymerich, T., Bover-Cid, S., Guàrdia, M.D., Arnau, J., Garriga, M., 2013. Probiotic strains *Lactobacillus plantarum* 299V and *Lactobacillus rhamnosus* GG as starter cultures for fermented sausages. LWT - Food Science and Technology 54 (1), 51–56. Available at https://doi.org/10.1016/j.lwt.2013.05.014.

Russo, P., de Chiara, M.L.V., Capozzi, V., Arena, M.P., Amodio, M.L., Rascón, A., et al., 2016. *Lactobacillus plantarum* strains for multifunctional oat-based foods. LWT - Food Science and Technology 68, 288–294. Available at https://doi.org/10.1016/j.lwt.2015.12.040.

Saadat, Y.R., Abbasi, A., Hejazian, S.S., Hekmatshoar, Y., Ardalan, M., Farnood, F., et al., 2025. Combating chronic kidney disease-associated cachexia: a literature review of recent therapeutic approaches. BMC Nephrology 26 (1). Available at https://doi.org/10.1186/s12882-025-04057-8, https://bmcnephrol.biomedcentral.com/.

Sabahi, S., Abbasi, A., Mortazavi, S.A., 2022. Characterization of cinnamon essential oil and its application in *Malva sylvestris* seed mucilage edible coating to the enhancement of the microbiological, physicochemical and sensory properties of lamb meat during storage. Journal of Applied Microbiology 133 (2), 488–502. Available at https://doi.org/10.1111/jam.15578, https://academic.oup.com/jambio.

Sabahi, S., Abbasi, A., Mortazavi, S.A., 2024. Phenolic components from carrot (*Daucus carota* L.) pomace: optimizing the extraction and assessing its potential antioxidant and antimicrobial activities. Heliyon 10 (17). Available at https://doi.org/10.1016/j.heliyon.2024.e36971, https://www.sciencedirect.com/science/journal/24058440.

Sanctuary, M.R., Kain, J.N., Chen, S.Y., Kalanetra, K., Lemay, D.G., Rose, D.R., et al., 2019. Pilot study of probiotic/colostrum supplementation on gut function in children with autism and gastrointestinal symptoms. PLoS One 14 (1). Available at https://doi.org/10.1371/journal.pone.0210064, https://journals.plos.org/plosone/article/file?id=10.1371/journal.pone.0210064&type=printable.

Sankararaman, S., Noriega, K., Velayuthan, S., Sferra, T., Martindale, R., 2023. Gut microbiome and its impact on obesity and obesity-related disorders. Current Gastroenterology Reports 25 (2), 31–44. Available at https://doi.org/10.1007/s11894-022-00859-0.

Sharma, V., Mishra, H.N., 2013. Fermentation of vegetable juice mixture by probiotic lactic acid bacteria. Nutrafoods 12 (1), 17–22. Available at https://doi.org/10.1007/s13749-012-0050-y.

Sheykhsaran, E., Abbasi, A., Ebrahimzadeh Leylabadlo, H., Sadeghi, J., Mehri, S., Naeimi Mazraeh, F., et al., 2023. Gut microbiota and obesity: an overview of microbiota to microbial-based therapies. Postgraduate Medical Journal 99 (1171), 384–402. Available at https://doi.org/10.1136/postgradmedj-2021-141311.

Shojaee-Aliabadi, S., Abbasi, A., Moradi, S., Sanej, K.D., Bazzaz, S., Nayeri, B.D., et al., 2023. Bioactive compounds and biological activities of madonna lily (*Lilium candidum* L.).

Reference Series in Phytochemistry 2023. Available at https://doi.org/10.1007/978-3-031-29006-0_17-1, https://link.springer.com/content/pdf/10.1007/978-3-031-29006-0_17-1?pdf=chapter%20toc.

Sjomina, O., Poļaka, I., Suhorukova, J., Vangravs, R., Paršutins, S., Knaze, V., et al., 2024. Randomised clinical trial: efficacy and safety of H. pylori eradication treatment with and without *Saccharomyces boulardii* supplementation. European Journal of Cancer Prevention 33 (3), 217–222. Available at https://doi.org/10.1097/CEJ.0000000000000858, http://journals.lww.com/eurjcancerprev/pages/default.aspx.

Steinert, R.E., Lee, Y.K., Sybesma, W., 2020. Vitamins for the gut microbiome. Trends in Molecular Medicine 26 (2), 137–140. Available at https://doi.org/10.1016/j.molmed.2019.11.005, www.elsevier.com/locate/molmed.

Szajewska, H., Vinderola, G., 2024. Current regulatory issues for the use of probiotics. Advances in Experimental Medicine and Biology 1449, 187–193. Available at https://doi.org/10.1007/978-3-031-58572-2_12, https://www.springer.com/series/5584.

Taylor, B.L., Woodfall, G.E., Sheedy, K.E., O'Riley, M.L., Rainbow, K.A., Bramwell, E.L., et al., 2017. Effect of probiotics on metabolic outcomes in pregnant women with gestational diabetes: a systematic review and meta-analysis of randomized controlled trials. Nutrients 9 (5). Available at https://doi.org/10.3390/nu9050461, http://www.mdpi.com/2072-6643/9/5/461/pdf.

Tripathy, A., Dash, J., Kancharla, S., Kolli, P., Mahajan, D., Senapati, S., et al., 2021. Probiotics: a promising candidate for management of colorectal cancer. Cancers 13 (13), 3178. Available at https://doi.org/10.3390/cancers13133178.

Trząskowska, M., Kołożyn-Krajewska, D., Wójciak, K., Dolatowski, Z., 2014. Microbiological quality of raw-fermented sausages with *Lactobacillus casei* LOCK 0900 probiotic strain. Food Control 35 (1), 184–191. Available at https://doi.org/10.1016/j.foodcont.2013.07.002.

Tsiouris, C.G., Tsiouri, M.G., 2017. Human microflora, probiotics and wound healing. Wound Medicine 19, 33–38. Available at https://doi.org/10.1016/j.wndm.2017.09.006, http://www.elsevier.com/journals/wound-medicine/2213-9095.

Turroni, S., Bendazzoli, C., Dipalo, S.C.F., Candela, M., Vitali, B., Gotti, R., et al., 2010. Oxalate-degrading activity in bifidobacterium animalis subsp. lactis: Impact of acidic conditions on the transcriptional levels of the oxalyl coenzyme A (CoA) decarboxylase and formyl-CoA ransferase genes. Applied and Environmental Microbiology 76 (16), 5609–5620. Available at https://doi.org/10.1128/AEM.00844-10Italy, http://aem.asm.org/cgi/reprint/76/16/5609.

Twetman, S., Keller, M.K., Lee, L., Yucel-Lindberg, T., Pedersen, A.M.L., 2018. Effect of probiotic lozenges containing *Lactobacillus reuteri* on oral wound healing: a pilot study. Beneficial Microbes 9 (5), 691–696. Available at https://doi.org/10.3920/bm2018.0003.

Tyagi, A.M., Yu, M., Darby, T.M., Vaccaro, C., Li, J.Y., Owens, J.A., et al., 2018. The microbial metabolite butyrate stimulates bone formation via T regulatory cell-mediated regulation of WNT10B expression. Immunity 49 (6), 1116. Available at https://doi.org/10.1016/j.immuni.2018.10.013, www.immunity.com.

Vallianou, N.G., Kounatidis, D., Tsilingiris, D., Panagopoulos, F., Christodoulatos, G.S., Evangelopoulos, A., et al., 2023. The role of next-generation probiotics in obesity and obesity-associated disorders: current knowledge and future perspectives. International Journal of Molecular Sciences 24 (7). Available at https://doi.org/10.3390/ijms24076755, http://www.mdpi.com/journal/ijms.

Wang, X., Yang, J., Zhang, H., Yu, J., Yao, Z., 2019. Oral probiotic administration during pregnancy prevents autism-related behaviors in offspring induced by maternal immune activation via anti-inflammation in mice. Autism Research 12 (4), 576–588. Available at https://doi.org/10.1002/aur.2079.

Wigner, P., Bijak, M., Saluk-Bijak, J., 2022. Probiotics in the prevention of the calcium oxalate urolithiasis. Cells 11 (2), 284. Available at https://doi.org/10.3390/cells11020284.

Xu, X., Jia, X., Mo, L., Liu, C., Zheng, L., Yuan, Q., et al., 2017. Intestinal microbiota: a potential target for the treatment of postmenopausal osteoporosis. Bone Research 5 (1). Available at https://doi.org/10.1038/boneres.2017.46.

Xu, J., Chen, C., Gan, S., Liao, Y., Fu, R., Hou, C., et al., 2023. The potential value of probiotics after dental implant placement. Microorganisms 11 (7). Available at https://doi.org/10.3390/microorganisms11071845, www.mdpi.com/journal/microorganisms.

Yamamoto, E.A., Jørgensen, T.N., 2020. Relationships between vitamin D, gut microbiome, and systemic autoimmunity. Frontiers in Immunology 10. Available at https://doi.org/10.3389/fimmu.2019.03141, https://www.frontiersin.org/journals/immunology#.

Yan, F., Polk, D.B., 2020. Probiotics and probiotic-derived functional factors—mechanistic insights into applications for intestinal homeostasis. Frontiers in Immunology 11. Available at https://doi.org/10.3389/fimmu.2020.01428, https://www.frontiersin.org/journals/immunology#.

Yao, M., Wu, J., Li, B., Xiao, H., McClements, D.J., Li, L., 2017. Microencapsulation of *Lactobacillus salivarious* Li01 for enhanced storage viability and targeted delivery to gut microbiota. Food Hydrocolloids 72, 228–236. Available at https://doi.org/10.1016/j.foodhyd.2017.05.033, https://www.journals.elsevier.com/food-hydrocolloids.

Yousf, H., Tomar, G.B., Srivastava, R.K., 2015. Probiotics and bone health: it takes GUTS to improve bone density. International Journal of Immunotherapy and Cancer Research 1, 018–020.

Yuan, S., Shen, J., 2021. Bacteroides vulgatus diminishes colonic microbiota dysbiosis ameliorating lumbar bone loss in ovariectomized mice. Bone 142, 115710. Available at https://doi.org/10.1016/j.bone.2020.115710.

Zeng, L., Yu, G., Yang, K., Hao, W., Chen, H., Durazzo, A., 2021. the improving effect and safety of probiotic supplements on patients with osteoporosis and osteopenia: a systematic review and meta-analysis of 10 randomized controlled trials. Evidence-Based Complementary and Alternative Medicine 2021, 1–13. Available at https://doi.org/10.1155/2021/9924410.

Zhang, Y., Zhang, H., 2013. Microbiota associated with type 2 diabetes and its related complications. Food Science and Human Wellness 2 (3-4), 167–172. Available at https://doi.org/10.1016/j.fshw.2013.09.002.

Zhang, C., Zhang, Q., Zhang, X., Du, S., Zhang, Y., Wang, X., et al., 2025. Effects of synbiotics surpass probiotics alone in improving type 2 diabetes mellitus: a randomized, double-blind, placebo-controlled trial. Clinical Nutrition 44, 248–258. Available at https://doi.org/10.1016/j.clnu.2024.11.042.

Chapter 5

The economy of probiotics in drugs, supplements, and foods, and the future trend

Marina Golowczyc and Andrea Gomez-Zavaglia
Center for Research and Development in Food Science and Technology (CIDCA, CCT-CONICET, La Plata), La Plata, Argentina

5.1 Introduction

The economy of probiotics in drugs, supplements, and foods represents a dynamic and rapidly evolving sector within the broader landscape of healthcare and nutrition. Probiotics are defined as live microorganisms that, when administered in adequate amounts, confer a health benefit on the host (Gibson et al., 2017). They have garnered substantial attention because of their capacity to promote digestive health, immunomodulation, and even influence mental well-being (Cassani et al., 2020, 2022; Kumar et al., 2015a; Min et al., 2019).

The awareness among consumers to promote healthcare and wellness, which is stimulated by a continuously growing scientific support, has continuously enhanced the demand for probiotic-containing products, which strongly contributed to enlarging the market size and investment. Currently, the industry covers a wide diversity of probiotic-containing products, ranging from pharmaceutical formulations to dietary supplements and a constantly growing functional food market. Dairy matrices have traditionally been preferred for probiotic delivery, yet with the rising demand for health-enhancing products, alternative food matrices such as fruits, vegetables, cereals, meats, soy-based foods, and others are increasingly utilized as effective probiotic carriers (Kumar et al., 2015b). However, both liquid and solid matrices pose challenges for probiotics, exposing them to various biotic and abiotic stresses that can lead to inactivation or damage, resulting in reduced viability and integrity of probiotic cells in the final product. Indeed, current EFSA and FAO/WHO recommendations state that the probiotic food/supplement must have a dose of viable probiotic cells at least equal to that used

Evaluating Oral Probiotics and Prebiotics. DOI: https://doi.org/10.1016/B978-0-443-29835-6.00013-7

in efficacy trials up to the end of its shelf life (Regulation EC No 178/2002 of the European Parliament and of the Council of 28 January 2002 laying down the general principles and requirements of food law, establishing the European Food Safety Authority and laying down procedures in matters of food safety, (EC) No 178/2002 of the European Parliament and of the Council of 28 January 2002 laying down the general principles and requirements of food law, establishing the European Food Safety Authority and laying down procedures in matters of food safety, 2002).

This chapter delves into the key market trends, regulatory challenges, and emerging innovations involving probiotics. The incorporation of probiotics in drugs, supplements, and foods, from personalized probiotic therapies to advancements in delivery systems, is also explored.

5.2 Probiotics market and regulation

The probiotics market has increasingly grown in recent years and is foreseen to continue its expansion (2024). The global probiotics market was valued at 61.15 billion USD in 2021, achieved a value of 71.2 billion USD in 2024, and is projected to reach 91.1 billion USD in 2026 and 105.7 billion USD by 2029, which means a CAGR of 8.2% (2024) (Liang et al., 2024). This growth results from different factors, including the increase in consumer awareness regarding the link between gut health and overall well-being, the growing preference for preventive healthcare, and the innovation in delivery systems and targeted strains.

The increased demand for food and beverages, dietary supplements, and animal feed is the main factor explaining the expected growth of the probiotic market in the coming years. For a market analysis, there are five regions usually considered, and within such regions, some countries are leading the market. These regions are **Asia-Pacific** (China, Japan, Korea, India, Australia, Indonesia, Thailand, Philippines, Malaysia, and Vietnam), **Europe** (Germany, the United Kingdom, France, Italy, Russia, and Turkey), **North America** (the United States, Canada, and Mexico), **South America** (Brazil, Argentina, and Columbia), and **Middle East and Africa** (Saudi Arabia, UAE, Egypt, Nigeria, and South Africa) (2024).

Understanding the evolution of the probiotics market requires a thorough analysis considering the type of probiotics (bacteria, yeasts), their applications (food, feed, supplements, drugs), key players (major producing companies, which occupy 12%–24% of the market share), regionalism, and sometimes end-users, channels, and technology. All these factors are critical to evaluate the market perspectives in the coming years.

The regulatory scenery governing their production, marketing, and sale varies considerably from one region to another around the world. In **North America, in the United States**, the Food and Drug Administration (FDA) regulates the use of probiotics depending on whether they are used as a dietary supplement or a drug. If they are administered as drugs, their use must be submitted and authorized by the FDA Development and approval process [www.fda.gov/drugs/developmentapprovalprocess/default.htm], following the same regulatory process as any therapeutic agent. **In Europe**, the EFSA is the first responsible for the assessment of the probiotics health claim made on food that is presented by various manufacturers (Salvetti and O'Toole, 2017). Probiotics are regulated by the Food Products Directive and Regulation (Regulation EC No 178/2002 of the European Parliament and of the Council of 28 January 2002 laying down the general principles and requirements of food law, establishing the European Food Safety Authority and laying down procedures in matters of food safety, (EC) No 178/2002 of the European Parliament and of the Council of 28 January 2002 laying down the general principles and requirements of food law, establishing the European Food Safety Authority and laying down procedures in matters of food safety, 2002) on food supplements, which established specific criteria regarding composition, production, labeling, marketing, and additional regulation on food safety, hygiene, and sanitary regulation for food production. In the **Asia-Pacific** region, the situation is complex, and the Association of Southeast Asian Nations (ASEAN) is constantly working on the harmonization of the regulations of the different countries. Japan is a global leader in the probiotics market, incorporating them both in food and pharmaceutical products. When products containing probiotics are categorized as foods, they fall under the regulation of the Food for Specific Health Use agency (FOSHU). This regulatory framework permits the labeling of health claims on foods or ingredients only after receiving special permission from the Consumer Affairs Agency. The claims are classified as special dietary uses, specific health applications, and food with nutrient function (Kumar et al., 2015a). China, heavily influenced by traditional dietary culture and habits, has granted health food registration approval to 137 products that feature probiotics as their primary raw material. The Chinese State Food and Drug Administration (SFDA) regulates and supervises functional foods and nutraceuticals. In 2019 the China State Administration for Market Regulation (2019) issued the *Application and Review of Probiotic Health Food*, outlining probiotic definitions and emphasizing safety and functionality at the strain level. It stipulates that probiotic health foods must maintain viable counts of no less than 10^6 CFU/mL or g throughout their shelf life. Since 2019, the FAO/WHO [FAO/WHO] has addressed the existing lack of harmonization by initiating the development of a Codex Standard or Guidelines aimed at establishing "admissibility criteria" and a structured framework for probiotic products. This initiative aims to guide global regulatory agencies in developing regulations focused on probiotics. In addition to complying with the General Standard for Labeling of Prepackaged Foods (CXS 1–1985), it has been proposed that specific provisions should be added. They should include listing the genus, species, and strains of the microorganisms in the ingredients, the quantity of viable cells of each microorganism (CFU/g), the name of the food, serving size, and storage conditions.

In **South America**, Brazil and Argentina are the fastest emerging markets for probiotic products. Brazil has been the most active country in addressing probiotics and incorporating them in their regulations, publishing the first guidelines for their evaluation in 1999 (Salminen et al., 2021). The Brazilian Ministry of Health has a regulatory body, Brazil's National Health Surveillance Agency (Anvisa), that is responsible for approving and enforcing regulations for functional foods (Danielski et al., 2020). In 2018 Brazil implemented a comprehensive regulatory framework for probiotics and prebiotics, aiming to safeguard consumer safety and ensure scientific validity of product claims (www.mordorintelligence.com/industry-reports/south-america-probiotics-market-industry). This process utilizes the same online petitioning system employed for other food products. Proposed claims related to probiotic substances necessitate the provision of specific information, including the character of the microorganism, the antimicrobial resistance profile, determination of hemolytic activity and potential, proofs of potential adverse effect, efficacy, and viability (Koirala and Anal, 2021). In Argentina, the National Administration of Drugs, Food, and Medical Technology (ANMAT) is responsible for ensuring the safety, efficacy, and quality of probiotic products sold in the country, as well as regulating the probiotic market. The expansion of the probiotic market in Argentina is primarily determined by heightened health awareness and a growing preference for natural health remedies among the population. The Argentine government is anticipated to implement regulatory changes designed to foster a favorable environment for market growth.

In summary, establishing a globally recognized definition of probiotics and developing standardized criteria and a gold standard for quality are crucial for regulating the production, marketing, and sale of probiotic products. This initiative seeks to ensure the authenticity and reliability of the probiotics market. By implementing accurate labeling and transparent communication practices, consumers will be empowered to make well-informed decisions tailored to their specific needs.

5.2.1 Factors determining the market behavior

The main **types of probiotics** present in the market are bacteria and yeasts. Lactobacilli, streptococci, and bifidobacteria are the main bacteria, and *Saccharomyces boulardii* is the yeast with the highest market share. Although the probiotic properties are strain-specific, several microorganisms from the species *Lacticaseibacillus rhamnosus*, *Lactobacillus acidophilus*, *Lactiplantibacillus plantarum*, *Lacticaseibacillus casei*, and *Limosilactobacillus reuteri* have a well-established reputation, with sound scientific support regarding their efficacy in promoting gut health. Bacteria are the microorganisms majorly employed as they have a broader diversity than yeasts, are easier to manage at an industrial level, and the associated processes have greater affordability. This consolidates broader consumer accessibility, enabling the development of targeted formulations tailored to address the personal needs of consumers and leading to a market dominance in the next years (2024–32). Examples of that are the spore-forming bacteria with probiotic properties, which improve shelf life and delivery properties, providing versatility and broader applicability of probiotic microorganisms (Soares et al., 2023).

Regarding **applications**, for a matter of systematic classification, food, supplements, and feed represent the main groups for probiotic incorporation. However, very often, the innovative techniques are shared and are not specific to a given group. Food and beverages (e.g., yogurts, fermented and fortified beverages, nondairy beverages, infant formula, and cereals) offer suitable matrices to incorporate probiotics in the daily diets (Cassani et al., 2020), accounting for the highest market share of probiotics during the 2021–28 period. This segment benefits from continuous innovation and diversification, as producers always introduce probiotics in a wide range of products (e.g., probiotic-infused waters, shots, snacks, chocolates, catering products, and products for vegans and lactose-free diets) to meet diverse taste preferences and dietary needs and are increasingly providing personalized solutions. Applications in the probiotics market also extend to the exploration of skin health benefits or the commercialization of psychobiotic strains for their potential mental health benefits. Researching the potential of psychobiotic strains that target the gut–brain axis and promote emotional well-being requires interdisciplinary teams, including mental health professionals, to reach a new customer base (Cassani et al., 2022).

In supplements, probiotics can be incorporated into tablets, capsules, powders, liquids, soft gels, or gel caps. This segment is particularly suitable for applying innovation regarding delivery systems or modified matrices. For example, chewable tablets can be easily manufactured and serve to control the release of probiotics. They are particularly appealing for those who have difficulty swallowing pills or prefer a more palatable option that easily dissolves in the mouth. Using stable matrices when exposed to gastrointestinal conditions (fructo-oligosaccharides—FOS), the safe arrival of bacteria to the gut can be ensured. FOS, being a soluble fiber, can impact the texture and hardness of the tablet. As the hardness of the tablet is defined by the thermophysical properties of FOS, which, in turn, are defined by the composition and degree of polymerization of the employed FOS, the challenge for manufacturers is to properly define the FOS composition that ensures both the tablet's palatability and the maintenance of the chewable quality, and an appropriate bacterial protection during gastrointestinal digestion. Microencapsulation is another technology widely employed for producing capsules containing probiotics, which is also often employed to incorporate probiotics into food products and thus protect bacteria from the harsh environment of the gastrointestinal tract (Cassani et al., 2022).

Feed products constitute another segment in continuous growth in which probiotics provide sustainable alternatives to replace antibiotics and promote animal healthcare. A particularly interesting market is that of pets. As pet owners increasingly seek out natural alternatives to pharmaceuticals, probiotics offer a viable option to enhance the well-being of their animals, addressing concerns ranging from digestive problems to skin allergies and behavioral issues like separation anxiety [MarketsandMarkets (Mali, 2024)]. Probiotic formulations can be tailored to different animal species (dogs, cats, etc.) and dietary needs, thereby improving their efficacy. Flavored powders or chewable tablets are also suitable options to facilitate the administration, showing that a given technological strategy applies to more than one segment. Regarding allergies, probiotics can not only contribute to health improvement, but they also act as immunomodulators. Incorporating strains of probiotics known for their antiinflammatory properties can potentially alleviate symptoms of allergic reactions and skin irritations. These could be administered through topical products like shampoos or oral supplements. Considering the connection between gut and animals' behavior (e.g., anxiety), the formulation of products containing bacteria that generate neurotransmitters can be useful to enhance the gut–brain axis activity and manage the stress and anxiety in pets.

Regarding the **end-users**, the human segment accounts for the highest market share in the probiotics market, either incorporated in functional foods or as food supplements. This can be easily explained considering the awareness of human consumers, who prioritize the consumption of healthy and sustainable products promoting digestive support, immunomodulation, or mental well-being, whose demand is increasing. In a recent survey carried out by the International Food Information Council (IFIC) on the US population, about 32% of surveyed people affirmed their active search for probiotics, and about 60% reported the routine incorporation in their diet (Food and Health Survey, 2023). Another continuously growing segment is that of animals, including both production animals (pigs, poultry, ruminants) and pets. As previously mentioned, international regulations that prohibit the use of antibiotics as growth promoters provide a strong incentive to use alternative products that serve a similar function, such as probiotics. In this regard, the global Animal Probiotics market size is estimated to achieve 7400 Million by 2030, growing at a CAGR of 8.0% during the period 2024–30 (https://www.cognitivemarketresearch.com/animal-probiotics-market-report) (Animal Probiotics Market Size Research Report, 2026). North America is the largest consumer (market share 40%), followed by Europe (market share 28%). The main manufacturers are Chr. Hansen, DuPont, Novozymes, and Elanco. The probiotic market for pets has generated particular interest, considering the important status pets have acquired as members of modern families. In this context, owners are willing to spend whatever is necessary to promote healthy nutrition, which has led to significant growth in this market segment, with an expected strengthening in the coming years. The evidence supporting this assertion is compelling. In 2022 the global market for pet probiotics was valued at 4.301 billion USD and is projected to grow to approximately 6.926 billion USD by 2030, with a CAGR of 7% for the period from 2023 to 2030 (Pet probiotics market report).

Distribution channels for human probiotics encompass a wide array of outlets, including hypermarkets/supermarkets, pharmacies/drugstores, specialty stores, online platforms, and convenience stores/small retail formats. Hypermarkets serve as critical centers for probiotic distribution, providing a convenient one-stop shopping solution for a diverse range of consumers, particularly those who do not frequent specialty stores. Additionally, these distribution channels have the capability to offer an extensive variety of probiotic products (e.g., food, feed, sweets, and dietary supplements), making them particularly effective for launching new products into the market. Regarding pets' probiotics, the main distribution channels include veterinary hospitals and clinics, pharmacies, and drugstores.

5.2.2 Market dynamism

The awareness of consumers regarding well-being and consumption of health products is the main cause of market dynamism, as validated by FAO (2019). Asia-Pacific is a particularly relevant market as it exhibits the highest CAGR for the probiotics market (9.6% for the 2021–2028 period). The aging population (particularly in South Korea) and the positive effects on the gut microbiome are the main factors supporting this growth, which also entails the market a suitable one to test new products. Certain governments (e.g., in India) have been increasingly promoting nutraceutical and probiotics' consumption, also contributing to the high CAGR in the Asia-Pacific (Nori et al., 2023). The European probiotics market is expected to be worth USD 19.4 billion by 2029 (CAGR 5.3%) [MarketsandMarkets], the dairy sector being the most developed segment of the market. Yogurts and fermented milks are the most demanded products, with many different products launched in the market that generally achieve a fast acceptance by consumers. Cheese-related products exemplify this dynamism, with the 2.8 million tons recorded in 2020 systematically increasing each year by 4.1%, driven by a consistent demand from countries such as China, the Republic of Korea, the Russian Federation, and Iraq.

This context provides great opportunities for companies to capitalize on and develop culturally relevant products and distribution strategies. These opportunities require well-consolidated Research and Development (R&D) skills, knowledge, and disruptive thinking, which collectively validate the success of new products and represent the primary barrier to entry in the probiotics market. Innovations are not only addressed to the product itself but also to packaging and distribution channels, as they also contribute to the final cost of the probiotic-containing products. For this reason, counting on R&D departments in the enterprises facilitates the process of launching new products. This, along with the continually increasing collaboration between enterprises and academia, contributes to the implementation of innovative ideas that disrupt the market. The success of these implementations depends on the interdisciplinary collaboration of research teams, which acts as the primary catalyst for introducing disruptive ideas. These ideas often emerge from the application of methodologies and processes well-established in other disciplines (e.g., physics and medicine) to the field of probiotics. For example, the ongoing innovation in the human probiotics market results from adapting matrices or delivery systems typically used in drug delivery to those utilized in food products.

All these approaches showcase the market dynamism in which the main key players include Probi (Sweden), Nestlé (Switzerland), ADM (USA), Danone (France), International Flavors and Fragances Inc (USA), Yakult Honsha Co. Ltd (Japan), BioGaia (Sweden), Morinaga Milk Industry Co, Ltd (Japan), Mejii Holdings Co, Ltd (Japan), Lifeway Foods Inc (USA), Adisseo (France), Winclove Probiotics (USA), AB Biotics S.A. (Belgium), Apsen Farmacêutica (Brazil), Lallemand (Canada), and Arla Foods amba.

5.3 Challenges in integrating probiotics into functional foods

Incorporating probiotic strains into functional products involves a multifaceted network in the probiotic market ecosystem, involving different stakeholders in the selection of microorganisms, characterization, production, distribution, regulation, and consumption of probiotic products by end consumers (Fig. 5.1).

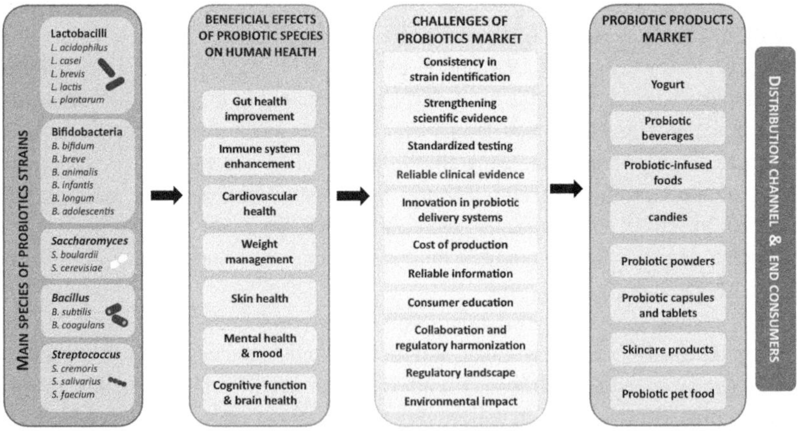

FIGURE 5.1 Overview of probiotic species, benefits, market challenges, and product offerings.

Although most probiotics employed in commercial products belong to a restricted number of lactobacilli, bifidobacteria, and yeast species, their beneficial properties are strain-dependent. Whenever a novel strain is incorporated, it must undergo very strict in vitro and in vivo studies, including animal, preclinical, and clinical studies, to support both its safety and the claimed health benefits (e.g., immunomodulation, cancer prevention, decrease of cholesterol levels, antihypertensive properties, improvement of lactose metabolism, improvement of gut health and microbiome, and production of certain vitamins) (Salvetti and O'Toole, 2017). In addition, certain technological characteristics are desirable, as probiotic strains are often exposed to low pH, temperature changes, and the gastrointestinal environment, and the strains should sustain them when integrated into functional products. Researchers and companies conduct investigations to identify effective strains of bacteria, cultivate them under controlled conditions, and formulate them into products such as capsules, tablets, powders, and beverages.

The basic requirement for probiotics is that products should contain sufficient numbers of microorganisms up to the expiry date. Insufficient and variable data exist to determine the optimal concentration of live probiotics in food products at the point of consumption, and there is no consensus on the matter. In general, the food industry has embraced a concentration of 10^6 colony-forming units (CFU) per milliliter or gram, a value consistent with the recommendations of the FDA (Tripathi and Giri, 2014). Furthermore, the International Scientific Association for Probiotics and Prebiotics (ISAPP) has reached a consensus that the daily intake of probiotics should fall within the range of 10^8 to 10^9 viable cells, equivalent to the daily consumption of approximately 100 grams of probiotic-containing food (Hill et al., 2014). Hence, probiotics must contain specific strains and maintain certain numbers of live cells to produce health benefits in the host. Different countries have decided on the minimum number of viable cells required in the probiotic product for it to be beneficial. The preservation of these probiotic microorganisms presents a challenge as they are affected by exposure to temperature, oxygen, and light. Besides these factors, the Nagoya Protocol shall also be considered for strains isolated after 2008, which also regulates the incorporation of strains isolated in different geographical regions (Greiber et al., 2019). In this context, innovation plays a key role in overcoming these challenges.

Integrating probiotics into functional foods requires innovation, scientific support, evolving consumer preferences, and education, not forgetting the environmental impact of the production processes. As research continues to uncover the potential health benefits of probiotics and technology enables the development of new formulations and delivery methods, the probiotic market is poised for continued growth and diversification (Fig. 5.1).

5.3.1 Emerging applications for innovation in probiotics

5.3.1.1 Next generation of probiotics (personalized probiotics)

The scientifically supported evidence regarding the health benefits of probiotics explains the increasing demand for products containing them. This demand has stimulated the potential of probiotic products as pharmaceutical or nutraceutical agents, which are strain- and dosage-dependent. Different clinical trials support the capacity of probiotics to treat gastrointestinal disorders, including diarrhea associated with antibiotics or rotavirus, inflammatory bowel disease, irritable bowel syndrome, allergic diseases, cancer, *Helicobacter pylori* infection, and lactose intolerance by incorporating fermented dairy products into the diet (Petrariu et al., 2023).

Despite the supporting evidence, the interindividual variability in probiotic efficacy has led to the emergence of a novel concept, "next-generation of probiotics" or "personalized probiotics." Increasing information demonstrates that variations in gut microbiome diversity, microbial composition, and metabolic profiles lead to differences in responses to probiotic interventions (Zmora et al., 2018; Min et al., 2024). Gut bacteria imbalance is now recognized as a crucial factor in the development of many diseases, including chronic diseases, obesity, diabetes, liver diseases, cardiovascular diseases, cancer, neurodegenerative diseases, and irritable bowel syndrome. Other individual-related characteristics, such as ethnicity, age, diet, and lifestyle, also lead to differences in gut microbiome profiles (Asnicar et al., 2021; Gupta et al., 2017; Rosas-Plaza et al., 2022).

Next-generation probiotics promise benefits in both food and therapeutic applications, offering the possibility of personalized therapies, combination therapies, and targeted delivery systems. Advances in microbiome research and molecular techniques are paving the way for implementing probiotic interventions tailored to individual gut microbiome characteristics (Veiga et al., 2020). Analyzing individuals' gut microbiome composition enables the development of targeted probiotic formulations specifically addressing individual needs, thereby overcoming the problem of interindividual variability.

The potential therapeutic applications of next-generation probiotics involve their identification, characterization, and targeted delivery, elucidating their effects on gut health, immune function, and broader health outcomes. Mechanistic insights include the production of bioactive compounds, competitive interactions with pathogenic bacteria, modulation of immune cell activity, and reinforcement of the gut barrier (Abouelela and Helmy, 2024).

The global market offers different opportunities, including personalized probiotic solutions that could involve targeted strains for specific health concerns, genetic testing to determine optimal strains, or even custom formulations based on individual needs. For example, **MyBiotics** is an Israeli microbiome company that developed novel technology for very efficient growth, delivery, and colonization of needed bacteria types in the gut (https://mybiotics-pharma.com/technology/). The technology uses innovative fermentation methods to grow complex bacterial communities on microparticles, dry it to powder, and place them into capsules administered like any probiotic supplement. Using the bacteria without any genetic modification, the final material allows them to survive under extreme conditions related to the production, formulation, and gastrointestinal conditions and demonstrates much better colonization in the target site. The enterprise portfolio includes tablets and pills for antibiotic-associated diarrhea and a combined treatment for vaginal dysbiosis, obtained by advanced fermentation and formulation technologies, new delivery systems, as well as discovery and coculture capabilities.

5.3.1.2 Vegan probiotic-containing products

Globally, millions of individuals avoid animal-derived ingredients for a multitude of reasons, including health, animal cruelty, dietary preference, environmental concerns, and religious or personal beliefs. Recently, this topic has garnered greater attention due to the increasing popularity of veganism, with an estimated 79 million individuals around the world adhering to this lifestyle (Tieu et al., 2024).

The growing vegan market has created a significant demand for plant-based probiotic options, leading to the emergence of a diverse range of vegan probiotic products (Pimentel et al., 2021). These factors have triggered product developers and scientists to design and develop dairy-free vegan foods (Jayarathna et al., 2021; Udayarajan et al., 2022; Küçükgöz and Trząskowska, 2022). Consumers now have access to a diverse range of fermented plant-based foods, like kombucha, coconut kefir, tempeh, and kimchi, which are naturally rich in probiotics (Cassani et al., 2020). However, due to the complex species diversity and relative composition of species/strains in the inoculum of these products, standardizing the production of naturally occurring probiotic properties can pose challenges when pursuing commercial purposes.

The main vegan probiotic products include beverages made from cereals, pseudocereals, and grains such as barley, oats, millet, corn, sorghum, rice, quinoa, and chia (Pimentel et al., 2021). Tea and fruit juices or fermented beverages, which contain various bioactive compounds such as vitamins, minerals, and polyphenols, also serve as interesting matrices for adding probiotics. Although there is a need to evaluate the impact of probiotics on product quality, the presence of phenolic compounds, flavonoids, and betacyanins with recognized prebiotic properties adds additional value to vegan probiotic products. Moreover, the metabolism of probiotics in fruit-derived matrices can enhance the bioaccessibility of phenolic compounds and increase functionality (Pimentel et al., 2021).

Other vegan probiotic products include vegan-friendly capsules, tablets, and powders containing different probiotic strains. Furthermore, vegan yogurt alternatives, granola, and bars are increasingly incorporating probiotics for enhanced gut health benefits. This expanding market, valued at 11.8 billion USD in 2032, offers a wide selection for consumers seeking both ethical and health-conscious probiotic options, while also demonstrating the growing trend towards plant-based lifestyles and sustainable consumption [Vegan Probiotics Market Size, Demand, Share & Trends to 2032 (futuremarketinsights.com)]. Different companies (Christian Hansen, Lallemand, DuPont Nutrition Biosciences ApS, Morinaga Milk Industry Co., Ltd, Probi AB Next Foods, Ganeden Biotech Inc., Lacteol Laboratory, Isabella´s Healthy bakery, Zena´s Brownie, NKD Pizza, Happy Baby, Nestlé, Bio K Plus International Inc, Velle, Skane Dairy, Bioferme Ou, Avenly Oy Ltd, AGM Foods Pvt Ltd, Nancy´s Yogurt, or Puravida) are the main companies producing vegan products with probiotics under the forms of chewable tablets, oat and rice, flavored cereal breakfast, cockies, muffins, cultured coconut, oatmeal, among the most important ones (Pimentel et al., 2021).

The main challenges that vegan probiotics face include limited strain availability and the need for novel production methods, research, and innovation. The demand for transparent labeling, clear ingredient sourcing, and sustainable production practices is crucial for building consumer trust and ensuring the continued growth of this promising sector. For example, the origin of the strains used in commercial products can be problematic, as many available and approved strains for human consumption are isolated from animals or animal-derived products. In the case of supplements, if the strain was isolated from animals or dairy products, the supplements may contain animal ingredients. Naturally fermented vegetable foods (such as kombucha, tempeh, and miso) include a natural microbiota of lactic acid bacteria with probiotic properties. Isolating strains from these products can be a smart strategy to overcome the animal origin of several strains.

5.3.1.3 Advancements in delivery systems

New and **innovative delivery methods** present innovative opportunities (next-generation delivery systems). Innovation refers to systems aiming at enhancing the efficacy, stability, and targeted release of probiotic bacteria. These three pillars are of great importance to overcome the main problems when administering probiotics, namely, the challenge represented by processing conditions and also by the gastrointestinal environment. Next-generation delivery systems involve innovative encapsulation technologies, belonging to the so-called smart delivery systems. They include pH-, enzyme-, and immune-responsive systems as well as targeted delivery of disease markers and signaling molecules (Xu et al., 2024).

pH-responsive systems are generally bio-responsive biopolymers (e.g., hydrogels) that destabilize upon pH changes in response to specific physiological conditions, ensuring targeted and controlled release of probiotics at the desired site of action (Quintana et al., 2017; Quintana et al., 2021). These innovations not only improve the viability and efficacy of probiotics but also open new avenues for their application in personalized medicine, where probiotic treatments can be tailored to the unique microbiome and health conditions of individual patients (next-generation of probiotics). For example, pH-responsive microspheres and calcium alginate hydrogels protect probiotics in acidic environments and release them in neutral or alkaline conditions. Combining alginate with other materials improves mechanical strength and retention (Wang et al., 2020). Commercial polymers like Eudragit are also employed for targeted delivery, creating protective shells that dissolve at specific pH levels, ensuring probiotics reach their intended site (Quintana et al., 2017; Quintana et al., 2021). Variations in individual gastrointestinal pH must be considered when developing these systems (Xu et al., 2024).

Enzyme-specific responsive systems release encapsulated substances at targeted sites using specific enzymes, making them effective for delivering drugs and probiotics to multiple locations. These systems are particularly advantageous for colon-targeted probiotic release, as they are more reliable than pH-responsive systems, which can be influenced by diet and physiology. Enzyme-responsive systems protect probiotics until they encounter specific enzymes that trigger their release. For example, starch walls are hydrolyzed by oral amylases, protein walls by stomach pepsin, and dietary fiber walls by colonic bacterial enzymes. Encapsulating probiotics in materials like chitosan or phosphocholine-balsalazide improves their resistance to gastric conditions and enhances colonization rates in the colon, offering therapeutic benefits for conditions like ulcerative colitis (Xu et al., 2024). Despite their benefits, these microgels can prematurely release probiotics due to swelling and absorption of digestive fluids, a challenge that can be mitigated by using composite materials or biopolymer coatings (Xu et al., 2024).

Concerning immune delivery systems, yeast cell membranes coated with β-glucan can enhance phagocytosis by M cells in the intestinal epithelium. This method improves the mucosal immune response and regulates intestinal microbiota by releasing encapsulated probiotics to lymphatic follicles, promoting strong immune responses and maintaining high levels of sIgA and probiotic aggregation in mesenteric lymph nodes (Lin et al., 2021).

Finally, regarding targeted delivery systems, when the body suffers from diseases like colitis, significant physiological changes, such as decreased pH in the colon, affect the efficacy of pH-responsive delivery systems. Disease states also produce an excess of disease markers and signaling molecules, which can be harnessed to trigger the release of encapsulated probiotics. For instance, elevated levels of nitric oxide (NO) in colitis can be targeted by NO-responsive delivery systems, ensuring that probiotics are released in response to NO presence, thus alleviating inflammation. Wang et al. (2022) developed such a system using benzotriazole groups grafted onto poly-γ-glutamic acid, which protected probiotics in gastric conditions and released them in the colon. Similarly, hydrogen sulfide-responsive systems using sulfhydryl-modified hyaluronic acid release probiotics in the colon by breaking disulfide bonds in the presence of hydrogen sulfide, enhancing both the resistance to gastric acids and adhesion to intestinal sites. These strategies offer promising approaches for targeted probiotic delivery in treating gastrointestinal diseases. Delivery systems can be designed to respond to reactive oxygen species (ROS) and glutathione, which are elevated in tumor microenvironments and inflamed tissues, such as those seen in inflammatory bowel disease (Zhong et al., 2021). These ROS-responsive systems can target and release encapsulated probiotics at sites of inflammation. For example, hydrogels formed from methacrylated-hyaluronic acid and thiolated-thioketal chemical cross-linking release probiotics like *L. reuteri* in response to ROS, helping treat inflammatory bowel diseases (Kaur et al., 2022). Although still in early stages, developing probiotic delivery systems based on disease markers and signaling molecules offers potential for more accurate and personalized treatments.

The ability of probiotics to effectively colonize specific regions within the human gastrointestinal tract can be significantly enhanced by encapsulating them in microgels made from adhesive materials. Natural polysaccharides like anionic alginate and cationic chitosan are commonly used for this purpose, with chitosan particularly effective due to its electrostatic attraction to the negatively charged mucus layer. Studies have shown that chitosan, when conjugated with polydopamine, significantly increases probiotic colonization in the colon, improving their therapeutic effects. Similarly, anionic polysaccharides like sodium alginate and hyaluronic acid, as well as various proteins, have been used to create delivery systems that protect and enhance the adhesion of probiotics within the gastrointestinal tract. Recent advancements in encapsulation technologies, such as electrospinning and 3D printing, utilize specialized polymers like poly-γ-glutamic acid to further improve adhesion and protection, demonstrating promising results in enhancing probiotic effectiveness and stability in harsh gastrointestinal conditions (Ma et al., 2023).

Inspired by mussels' strong adhesion to surfaces, biomimetic approaches have been developed for probiotic delivery systems. Liu et al. (2022) created a poly-norepinephrine coating for probiotics that resists acidic gastric fluids and bile acids, enhancing gut retention. Polydopamine, another mussel-inspired material, improves probiotic adhesion but is limited by cost and color. Plant polyphenols, rich in catechol or pyrogallol groups, offer an alternative with adhesive properties suitable for encapsulation (Centurion et al., 2022). Metal-polyphenol networks, formed through coordination with tannic acid, have been used to enhance probiotic adhesion and viability, though they disassemble under acidic conditions, necessitating combination with other technologies for optimal protection and adhesion in the gastrointestinal tract (Yang et al., 2022).

Traditionally, biofilm formation has been seen as problematic, causing issues like increased fuel usage on ships, food spoilage, and dental decay. However, biofilms also help bacteria withstand harsh environments and improve their adhesion to surfaces. Biofilm formation can protect probiotics within the gastrointestinal tract. In this regard, encapsulating probiotics in microcapsules or cultivating them on scaffolds promotes biofilm formation and enhances their resistance and adhesion. Moreover, combining biofilms with other technologies could further improve the efficacy of probiotic delivery systems (Xu et al., 2024).

Bioadhesive materials also allow probiotics to adhere to the oral mucosa. For example, **ProBiora Health** (Chr. Hansen) has developed a blend of three probiotic strains naturally found in the mouth (*Streptococcus oralis* KJ3, *Streptococcus uberis* KJ2, *Streptococcus rattus* JH145) that employs a patented bioadhesive mechanism. This allows the probiotics to adhere to the oral and gastrointestinal mucosa (Shirbhate et al., 2023). The product is dissolved in the mouth nightly after brushing and flossing. Saliva hydrates the probiotics, allowing them to migrate to their natural habitats on the tooth surfaces, gums, gum pockets, and tongue. Therefore they compete with pathogens, promoting a healthier oral environment. This enhances the colonization and effectiveness of the probiotics, particularly in targeting oral health issues like gum disease and bad breath.

Another product in the market is "PerioBalance" (Gum) formulated to control periodontal disease. It is a patented advanced food supplement that helps to maintain the benefits of professional dental cleaning. It results from a combination of *L. reuteri* Prodentis, one of the very few species uniquely adapted to reside in the oral cavity and in the gastrointestinal tract. It has the ability to bind to saliva and oral mucosa, giving anticariogenic activity. Including this strain in chewing gums has demonstrated beneficial effects in clinical markers (Krasse et al., 2006). Other systems for delivering probiotics addressed to the oral cavity include tablets, lozenges, toothpaste, and mouthwash (Kharwar et al., 2022).

In addition, the next-generation of probiotics themselves can be designed to have tailored functionalities, including the response to different signals, offering solutions for a wide range of applications.

5.4 Present opportunities and potential obstacles for implementing innovations in the near future

The available technological approaches and equipment for obtaining probiotics facilitate the thorough analysis of bacterial mechanisms of action, enabling the engineering of targeted products fulfilling the necessary health claims for their introduction in the market. At the same level of importance, the development of innovative delivery systems, very often resulting from the adaptation of drug delivery ones, contributes to overcoming problems associated with the stabilization of probiotics in food matrices, as well as their viability in the gastrointestinal environment. Altogether, this also strengthens the market innovation in probiotics. This scientific evidence, combined with new marketing and commercialization tools, is critical for the innovation that the probiotic market is experiencing lately. Examples of that synergism are different products recently launched to the market. Shape 100 was launched in February 2024 as a result of a partnership between AB Biotics SA (Netherlands) and Wonderlab, a leader in the Chinese probiotic market. This collaboration aims to expand AB Biotics' presence in Asia, the largest probiotic market in the world (Bolatti, 2024). "AB Life" is a probiotic blend focusing on cardiometabolic health by lowering cholesterol levels for the Chinese market. The probiotic strains included in the product have been recognized as Generally Recognized as Safe (GRAS) by the FDA and received cardiometabolic health claims approval from several global regulatory agencies, including Anvisa (Brazilian Health Regulatory Agency) and South Korea's Ministry of Food and Drug Safety.

Another recent product is **N3 milk**, launched by Nestlé last November 2023 for the Chinese market (Nestlé develops N3 milk with new nutritional benefits, launches first in China, 2023). It consists of cow milk having all the essential nutrients found in milk (e.g., proteins, vitamins, and minerals), enriched with prebiotic fiber and low in lactose, with over 15% fewer calories. The decrease in the lactose content was achieved by enzymatic treatments promoting the conversion of lactose into prebiotic fiber (presumably in galacto-oligosaccharides) (Martins et al., 2019).

"Yakult Gohonmaru Cafe & Gallery" is another innovation launched by Yakult Honsha Co Ltd in Utsunomiya (Japan) in September 2023. It is part of the Yakult service center, and customers can treat themselves to a range of meals, including ice cream and tiramisu made from the Yakult probiotic beverage. Aside from the cafe, the center houses a gallery, a salon, and a daycare center (Woon, 2023). The strategic investment highlights the commitment of Yakult to expand its consumer base and foster brand awareness.

These products were mentioned just as examples of innovation in probiotics, targeted to well-defined markets on the basis of regional variations in terms of availability, consumer preferences, and regulatory frameworks.

The future of probiotics holds significant opportunities, driven by advancements in biotechnology, personalized medicine, and increasing consumer awareness of gut health. Opportunities include the development of highly targeted probiotic strains designed to address specific diseases (e.g., inflammatory bowel disease, obesity, and mental health issues). The integration of probiotics into personalized nutrition plans, guided by individual microbiome analysis, can offer customized solutions for optimal gut health. The potential to incorporate probiotics into a wider range of functional foods, beverages, and dietary supplements also presents substantial market growth opportunities.

5.5 Conclusions

This chapter has provided a comprehensive overview of the economic landscape surrounding probiotics in drugs, supplements, and foods. The evidence suggests a robust and growing market driven by increasing consumer awareness of the potential health benefits of probiotics. From pharmaceutical formulations targeting specific medical conditions to dietary supplements and functional foods promoting general wellness, probiotics have carved out a significant presence in various sectors. The economic impact of probiotics extends beyond income generation for manufacturers to encompass cost savings in healthcare expenditures associated with the prevention and management of certain diseases. As research continues to elucidate the mechanisms of action and therapeutic potential of probiotics, the market is poised for further expansion and diversification. The use of nondairy matrices as a vehicle for probiotics is a promising field for the food industry that can benefit both the general consumer and specific groups who have milk allergies or prefer a vegan diet. However, further research is needed in different fields to develop successful products, both in terms of their acceptance and functionality, with functionality defined as promoting a beneficial health effect. It is necessary to better understand the physiological changes of probiotics subjected to different stress conditions associated with plant-based matrices, and how they respond (sensorially, technologically, and functionally). It is also important to improve knowledge about the most suitable materials and technologies (e.g., immobilization in edible coatings, microencapsulation, 3D printing, and delivery systems) to protect probiotics and conduct quality studies that ensure the stability of probiotics during their shelf life and their beneficial effect on health, without forgetting the urgent need for appropriate legislation for their use in foods. Moreover, the convergence of probiotics with other emerging fields such as microbiology, nutrigenomics, and digital health technologies presents new opportunities for innovation and market growth. Collaborative efforts between industry, academia, and regulatory agencies will be essential to navigate the evolving landscape and realize the full potential of probiotics in improving human health and security.

In summary, in spite of the remaining challenges and uncertainties, the outlook for the probiotic economy is undeniably promising. By leveraging scientific advancements, market insights, and strategic partnerships, stakeholders can drive continued innovation and capitalize on the growing demand for probiotic-based products and interventions in the years to come.

Acknowledgment

This study was supported by the European NeoGiANT project funded by the Horizon 2020 (Grant no. 101036768). M.G. and A.G.-Z. are members of the research career CONICET.

References

Abouelela, M.E., Helmy, Y.A., 2024. Next-generation probiotics as novel therapeutics for improving human health: current trends and future perspectives. Microorganisms 12 (3). Available at https://doi.org/10.3390/microorganisms12030430, www.mdpi.com/journal/microorganisms.

Animal Probiotics Market Size Research Report. 2026. Available at https://www.cognitivemarketresearch.com/animal-probiotics-market-report.

Asnicar, F., Leeming, E.R., Dimidi, E., Mazidi, M., Franks, P.W., Al Khatib, H., et al., 2021. Blue poo: impact of gut transit time on the gut microbiome using a novel marker. Gut 70 (9), 1665–1674. Available from. Available at https://doi.org/10.1136/gutjnl-2020-323877, http://gut.bmj.com/content/by/year.

B. Bolatti, 2024. AB-LIFE® Broadens its Reach in Asia through Partnership with Wonderlab. Available at https://www.ab-biotics.com/ab-life-broadens-its-reach-in-china-with-partnership-with-wonderlab/.

Cassani, L., Gomez-Zavaglia, A., Simal-Gandara, J., 2020. Technological strategies ensuring the safe arrival of beneficial microorganisms to the gut: from food processing and storage to their passage through the gastrointestinal tract. Food Research International 129, 108852. Available at https://doi.org/10.1016/j.foodres.2019.108852.

Cassani, L., Gerbino, E., Gómez-Zavaglia, A., 2022. Technology aspects of probiotic production and live biotherapeutics. Probiotics for Human Nutrition in Health and Disease. Elsevier, Argentina, pp. 143–170. Available at https://www.sciencedirect.com/book/9780323899086, 10.1016/B978-0-323-89908-6.00007-8.

Centurion, F., Merhebi, S., Baharfar, M., Abbasi, R., Zhang, C., Mousavi, M., et al., 2022. Cell-mediated biointerfacial phenolic assembly for probiotic nano encapsulation. Advanced Functional Materials 32 (26). Available at https://doi.org/10.1002/adfm.202200775, http://onlinelibrary.wiley.com/journal/10.1002/(ISSN)1616-3028.

Danielski, R., Silva, C.P.da, Camargo, A.C.de, 2020. The gap between scientific evidence and food regulation: analyzing the updated Brazilian normative on the use of bioactives in food supplements. Journal of Food Bioactives 87–96. Available at https://doi.org/10.31665/jfb.2020.12247.

Food & Health Survey. 2023, Available at https://foodinsight.org/2023-food-health-survey/.

Gibson, G.R., Hutkins, R., Sanders, M.E., Prescott, S.L., Reimer, R.A., Salminen, S.J., et al., 2017. Expert consensus document: the International Scientific Association for Probiotics and Prebiotics (ISAPP) consensus statement on the definition and scope of prebiotics. Nature Reviews Gastroenterology and Hepatology 14 (8), 491–502. Available at https://doi.org/10.1038/nrgastro.2017.75, http://www.nature.com/nrgastro/index.html.

T. Greiber, S.P. Moreno, M. Åhrén, J.N. Carrasco, E.C. Kamau, J.C. Medaglia, et al., 2019. An Explanatory Guide to the Nagoya Protocol on Access and Benefit-Sharing.

Gupta, V.K., Paul, S., Dutta, C., 2017. Geography, ethnicity or subsistence-specific variations in human microbiome composition and diversity. Frontiers in Microbiology 8. Available at https://doi.org/10.3389/fmicb.2017.01162, http://journal.frontiersin.org/article/10.3389/fmicb.2017.01162/full.

Hill, C., Guarner, F., Reid, G., Gibson, G.R., Merenstein, D.J., Pot, B., et al., 2014. Expert consensus document: the international scientific association for probiotics and prebiotics consensus statement on the scope and appropriate use of the term probiotic. Nature Reviews Gastroenterology and Hepatology 11 (8), 506–514. Available at https://doi.org/10.1038/nrgastro.2014.66, http://www.nature.com/nrgastro/index.html.

Jayarathna, S., Priyashantha, H., Johansson, M., Vidanarachchi, J.K., Jayawardana, B.C., Liyanage, R., 2021. Probiotic enriched fermented soy-gel as a vegan substitute for dairy yoghurt. Journal of Food Processing and Preservation 45 (1). Available at https://doi.org/10.1111/jfpp.15092, http://onlinelibrary.wiley.com/journal/10.1111/(ISSN)1745-4549.

Kaur, P., Ghosh, S., Bhowmick, A., Gadhave, K., Datta, S., Ghosh, A., et al., 2022. Bacterioboat—a novel tool to increase the half-life period of the orally administered drug. Science Advances 8 (10). Available at https://doi.org/10.1126/sciadv.abh1419, https://www.science.org/doi/10.1126/sciadv.abh1419.

Kharwar, A., Bazaz, M.R., Dandekar, M.P., 2022. Quantitative and qualitative characterization of commercially available oral suspension of probiotic products containing *Bacillus Clausii* spores. BMC Microbiology 22 (1). Available at https://doi.org/10.1186/s12866-022-02631-w, https://bmcmicrobiol.biomedcentral.com/.

Koirala, S., Anal, A.K., 2021. Probiotics-based foods and beverages as future foods and their overall safety and regulatory claims. Future Foods 3, 100013. Available at https://doi.org/10.1016/j.fufo.2021.100013.

Krasse, P., Carlsson, B., Dahl, C., Paulsson, A., Nilsson, A., Sinkiewicz, G., 2006. Decreased gum bleeding and reduced gingivitis by the probiotic *Lactobacillus reuteri*. Swedish Dental Journal 30 (2), 55–60.

Küçükgöz, K., Trząskowska, M., 2022. Nondairy probiotic products: functional foods that require more attention. Nutrients 14 (4), 753. Available at https://doi.org/10.3390/nu14040753.

Kumar, H., Salminen, S., Verhagen, H., Rowland, I., Heimbach, J., Bañares, S., et al., 2015a. Novel probiotics and prebiotics: road to the market. Current Opinion in Biotechnology 32, 99–103. Available at https://doi.org/10.1016/j.copbio.2014.11.021.

Kumar, B.V., Vijayendra, S.V.N., Reddy, O.V.S., 2015b. Trends in dairy and non-dairy probiotic products - a review. Journal of Food Science and Technology 52 (10), 6112–6124. Available at https://doi.org/10.1007/s13197-015-1795-2, http://www.springerlink.com/content/121580/.

Liang, D., Wu, F., Zhou, D., Tan, B., Chen, T., 2024. Commercial probiotic products in public health: current status and potential limitations. Critical Reviews in Food Science and Nutrition 64 (19), 6455–6476. Available at https://doi.org/10.1080/10408398.2023.2169858.

Lin, S., Mukherjee, S., Li, J., Hou, W., Pan, C., Liu, J., 2021. Mucosal immunity–mediated modulation of the gut microbiome by oral delivery of probiotics into Peyer's patches. Science Advances 7 (20). Available at https://doi.org/10.1126/sciadv.abf0677.

Liu, J., Wang, Y., Heelan, W.J., Chen, Y., Li, Z., Hu, Q., 2022. Mucoadhesive probiotic backpacks with ROS nanoscavengers enhance the bacteriotherapy for inflammatory bowel diseases. Science Advances 8 (45). Available at https://doi.org/10.1126/sciadv.abp8798, https://www.science.org/doi/10.1126/sciadv.abp8798.

Ma, J., Li, T., Wang, Q., Xu, C., Yu, W., Yu, H., et al., 2023. Enhanced viability of probiotics encapsulated within synthetic/natural biopolymers by the addition of gum arabic via electrohydrodynamic processing. Food Chemistry 413, 135680. Available at https://doi.org/10.1016/j.foodchem.2023.135680.

S. Mali, 2024. Pet Probiotics Market Report 2024 (Global Edition). Available at https://www.cognitivemarketresearch.com/pet-probiotics-market-report?gad_source=1&gclid=Cj0KCQjw_qexBhCoARIsAFgBlevzz_MHkcTwcs2biAaGftxpRVpzuL2AVB3Ym3JpMvOM_t1y70TgBcAaAinyEALw_wcB.

Martins, G.N., Ureta, M.M., Tymczyszyn, E.E., Castilho, P.C., Gomez-Zavaglia, A., 2019. Technological aspects of the production of fructo and galacto-oligosaccharides. Enzymatic synthesis and hydrolysis. Frontiers in Nutrition 6 Available from: Available at https://doi.org/10.3389/fnut.2019.00078, journal.frontiersin.org/journal/nutrition.

Min, M., Bunt, C.R., Mason, S.L., Hussain, M.A., 2019. Non-dairy probiotic food products: an emerging group of functional foods. Critical Reviews in Food Science and Nutrition 59 (16), 2626–2641. Available at https://doi.org/10.1080/10408398.2018.1462760, https://www.tandfonline.com/loi/bfsn20.

Min, U., Jin, Y.J., Jang, Y.J., Lim, J., Kim, B.Y., 2024. Personalized probiotic strategy considering bowel habits: impacts on gut microbiota composition and alleviation of gastrointestinal symptoms via Consti-Biome and Sensi-Biome. Frontiers in Nutrition 11 Available from: Available at https://doi.org/10.3389/fnut.2024.1302093, journal.frontiersin.org/journal/nutrition.

Nestlé Develops N3 Milk with New Nutritional Benefits, Launches First in China. 2023, Available at https://www.nestle.com/media/news/nutrition-n3-milk-product-launch.

Nori, L.P., Sowmya, P., Madhusree, T., Raju, K.V., 2023. Nutraceuticals: regulatory landscape and market dynamics in the Indian scenario. Journal of Young Pharmacists 15 (4), 595–603. Available at https://doi.org/10.5530/jyp.2023.15.84.

Petrariu, O.A., Barbu, I.C., Niculescu, A.G., Constantin, M., Grigore, G.A., Cristian, R.E., et al., 2023. Role of probiotics in managing various human diseases, from oral pathology to cancer and gastrointestinal diseases. Frontiers in Microbiology 14. Available at https://doi.org/10.3389/fmicb.2023.1296447, https://www.frontiersin.org/journals/microbiology#.

Pimentel, T.C., Costa, W.K.A.D., Barão, C.E., Rosset, M., Magnani, M., 2021. Vegan probiotic products: a modern tendency or the newest challenge in functional foods. Food Research International 140 Available from: Available at https://doi.org/10.1016/j.foodres.2020.110033, www.elsevier.com/inca/publications/store/4/2/2/9/7/0.

Quintana, G., Simões, M.G., Hugo, A., Alves, P., Ferreira, P., Gerbino, E., et al., 2017. Layer-by-layer encapsulation of *Lactobacillus delbrueckii* subsp. *Bulgaricus* using block-copolymers of poly(acrylic acid) and pluronic for safe release in gastro-intestinal conditions. Journal of Functional Foods 35, 408–417. Available at https://doi.org/10.1016/j.jff.2017.06.007.

Quintana, G., Gerbino, E., Alves, P., Simões, P.N., Rúa, M.L., Fuciños, C., et al., 2021. Microencapsulation of *Lactobacillus plantarum* in W/O emulsions of okara oil and block-copolymers of poly(acrylic acid) and pluronic using microfluidic devices. Food Research International 140 Available from: Available at https://doi.org/10.1016/j.foodres.2020.110053, www.elsevier.com/inca/publications/store/4/2/2/9/7/0.

Regulation (EC) No 178/2002 of the European Parliament and of the Council of 28 January 2002 Laying Down the General Principles and Requirements of Food Law, Establishing the European Food Safety Authority and Laying Down Procedures in Matters of Food Safety. 2002.

Rosas-Plaza, S., Hernández-Terán, A., Navarro-Díaz, M., Escalante, A.E., Morales-Espinosa, R., Cerritos, R., 2022. Human gut microbiome across different lifestyles: from hunter-gatherers to urban populations. Frontiers in Microbiology 13. Available at https://doi.org/10.3389/fmicb.2022.843170, https://www.frontiersin.org/journals/microbiology#.

Salminen, S., Collado, M.C., Endo, A., Hill, C., Lebeer, S., Quigley, E.M.M., et al., 2021. The International Scientific Association of Probiotics and Prebiotics (ISAPP) consensus statement on the definition and scope of postbiotics. Nature Reviews Gastroenterology and Hepatology 18 (9), 649–667. Available at https://doi.org/10.1038/s41575-021-00440-6, http://www.nature.com/nrgastro/index.html.

Salvetti, E., O'Toole, P.W., 2017. When regulation challenges innovation: the case of the genus *Lactobacillus*. Trends in Food Science and Technology 66, 187–194. Available at https://doi.org/10.1016/j.tifs.2017.05.009. Available from: http://www.elsevier.com/wps/find/journaldescription.cws_home/601278/description#description.

Shirbhate, U., Bajaj, P., Chandak, M., Jaiswal, P., Sarangi, S., Suchak, D., et al., 2023. Clinical implications of probiotics in oral and periodontal health: a comprehensive review. Cureus 15 (12). Available at https://doi.org/10.7759/cureus.51177.

Soares, M.B., Almada, C.N., Pereira, E.P.R., Ferreira, B.M., Balthazar, C.F., Khorshidian, N., et al., 2023. Review - sporeforming probiotic bacteria: characteristics, health benefits, and technological aspects for their applications in foods and beverages. Trends in Food Science and Technology 138, 453–469. Available at https://doi.org/10.1016/j.tifs.2023.06.029, http://www.elsevier.com/wps/find/journaldescription.cws_home/601278/description#description.

Tieu, L., Uchi, J., Patel, N., Meghani, M., Patel, P., Nguyen, Y., 2024. Embracing medication needs of patients based on ethical, dietary, and religious preferences. American Journal of Lifestyle Medicine 18 (3), 351–363. Available at https://doi.org/10.1177/15598276221135538, https://journals.sagepub.com/home/AJL.

Tripathi, M.K., Giri, S.K., 2014. Probiotic functional foods: survival of probiotics during processing and storage. Journal of Functional Foods 9 (1), 225–241. Available at https://doi.org/10.1016/j.jff.2014.04.030.

Udayarajan, C.T., Mohan, K., Nisha, P., 2022. Tree nuts: treasure mine for prebiotic and probiotic dairy free vegan products. Trends in Food Science and Technology 124, 208–218. Available at https://doi.org/10.1016/j.tifs.2022.04.022, http://www.elsevier.com/wps/find/journaldescription.cws_home/601278/description#description.

Veiga, P., Suez, J., Derrien, M., Elinav, E., 2020. Moving from probiotics to precision probiotics. Nature Microbiology 5 (7), 878–880. Available at https://doi.org/10.1038/s41564-020-0721-1.

Wang, L., Song, M., Zhao, Z., Chen, X., Cai, J., Cao, Y., et al., 2020. *Lactobacillus acidophilus* loaded pickering double emulsion with enhanced viability and colon-adhesion efficiency. LWT 121, 108928. Available at https://doi.org/10.1016/j.lwt.2019.108928.

Wang, R., Guo, K., Zhang, W., He, Y., Yang, K., Chen, Q., et al., 2022. Poly-γ-glutamic acid microgel-encapsulated probiotics with gastric acid resistance and smart inflammatory factor targeted delivery performance to ameliorate colitis. Advanced Functional Materials 32 (26). Available at https://doi.org/10.1002/adfm.202113034.

Woon, W., 2023. Ice-cream, pastries and facial at first-ever. Yakult Cafe in Japan Available from: Available at www.thestar.com.my/aseanplus/aseanplus-news/2023/11/08/ice-cream-pastries-and-facial-at-first-ever-yakult-cafe-in-japan.

Xu, C., Gantumur, M.A., Sun, J., Guo, J., Ma, J., Jiang, Z., et al., 2024. Design of probiotic delivery systems for targeted release. Food Hydrocolloids 149. Available at https://doi.org/10.1016/j.foodhyd.2023.109588, https://www.sciencedirect.com/science/journal/0268005X.

Yang, X., Yang, J., Ye, Z., Zhang, G., Nie, W., Cheng, H., et al., 2022. Physiologically inspired mucin coated *Escherichia coli* Nissle 1917 enhances biotherapy by regulating the pathological microenvironment to improve intestinal colonization. ACS Nano 16 (3), 4041–4058. Available at https://doi.org/10.1021/acsnano.1c09681.

Zhong, X., Wang, X., Li, J., Hu, J., Cheng, L., Yang, X., 2021. ROS-based dynamic therapy synergy with modulating tumor cell-microenvironment mediated by inorganic nanomedicine. Coordination Chemistry Reviews 437, 213828. Available at https://doi.org/10.1016/j.ccr.2021.213828.

Zmora, N., Zilberman-Schapira, G., Suez, J., Mor, U., Dori-Bachash, M., Bashiardes, S., et al., 2018. Personalized gut mucosal colonization resistance to empiric probiotics is associated with unique host and microbiome features. Cell 174, 1388–1405. Available at https://doi.org/10.1016/j.cell.2018.08.041 e21.

Chapter 6

The European Union's regulations associated to the viable probiotics in industrial products

Arturo Anadón, María Aránzazu Martínez, Marta Martinez, María Rosa Martínez-Larrañaga, and Irma Ares
Department of Pharmacology and Toxicology, Faculty of Veterinary Medicine, Universidad Complutense de Madrid, Madrid, Spain

6.1 Introduction

In general, citizens need access to safe and healthy food of the highest quality. Therefore European Food legislation aims to guarantee the highest standards of quality, efficacy, and food safety. This complexity makes food law challenging to approach from a single perspective. Over the years, various probiotic products known as probiotic foods and supplements, both of which are beneficial to health, have come onto the market. Small amounts of probiotics can also be consumed from fermented foods and beverages such as yogurt, kéfir, sauerkraut, kimchi, tempeh, miso, pickles and other pickles, raw cheese, and nattō.

6.2 Intestinal microbiota

The term "Intestinal microbiota" is often used interchangeably and erroneously with the term "intestinal microbiome" (the genome of intestinal microbes, which is frequently considered a human organ with distinct physiology and pathology), is an ecosystem of commensal, symbiotic, and pathogenic microorganisms that outnumber their host's genes by more than 100 times (Baquero and Nombela, 2012; Collins et al., 2012), and is associated with disease development (Duvallet et al., 2017). It is well known that the composition and metabolic effects of the gastrointestinal microbiota are of key importance for health. The gastrointestinal microbiota with metabolic

Evaluating Oral Probiotics and Prebiotics. DOI: https://doi.org/10.1016/B978-0-443-29835-6.00012-5

capacity is also referred to as a "metabolic organ." In addition to promoting normal gastrointestinal functions and protecting from infections, the microbiota also exerts important effects on systemic metabolism and immune functions. The trillions of microorganisms that inhabit the intestines (i.e., bacteria, viruses, archaea, and fungi) play a fundamental role in human health. The imbalance of microbial composition and heterogeneity at the expense of pathogenic microorganisms, together with the loss of beneficial tax species from the gut microbiota, is due to factors such as age, diet, and the use of antimicrobials to treat various infections, among others. The gut microbiota is highly sensitive to information coming not only from the body's various organs but also from external factors such as diet, physical activity, psychological and physical stress, sleep restrictions, drugs, antibiotics, noise, and temperature, among others (Ortega et al., 2023). The gut microbiota can help us to understand various pathophysiological processes that regulate a large number of metabolic pathways and interact closely with the immune system. Dysregulation of the gut microbiota may be directly associated with low-grade inflammation and a wide range of pathological conditions (i.e., obesity, type 2 diabetes, inflammatory bowel disease, gastrointestinal cancer, and psychiatric disorders such as bipolar depression) (Ortega et al., 2023).

The microbiota is considered a superorganism and an integral part of the gastrointestinal tract (GIT). While intestinal bacteria are useful in promoting human health, certain components of the microbiota can contribute to various pathological disorders in genetically susceptible individuals, including Crohn's disease and ulcerative colitis, the two main types of inflammatory bowel disease (IBD). These diseases are characterized by persistent mucosal inflammation at different levels of the GIT. Clinical and experimental observations indicate an imbalance between beneficial and harmful microbial components in these disorders. Probiotic bacteria are currently studied as a potential treatment for irritable bowel syndrome (Anadón et al., 2021).

The human GIT is an important source of probiotics, mainly found in the intestines and stomach. They can also be found in human feces and colostrum. Nonhuman sources of probiotics include the GIT of animals (cows, pigs, goats, chickens, shrimps, fish, and bees). Milk from animals is also an important source of probiotics (Parichat and Pongsak, 2023). Food is also another important source of probiotics. They are usually found in: (1) raw foods (milk, dairy products, meat), (2) fermented foods (kimchi, pickles, fermented sausages), and (3) a variety of fruits and vegetables. In the environment, probiotics can be found in water, air, and soil (Parichat and Pongsak, 2023).

The term "intestinal dysbiosis" refers to change(s) of the intestinal microbiota rather than an unhealthy microbiome (Shanahan and Hill, 2019). Even though various approaches have been proposed for managing alterations in the intestinal microbiota, their effectiveness in the general population has not yet been confirmed. Drug-induced intestinal dysbiosis refers predominantly to antibiotic-specific intestinal dysbiosis.

When probiotics enter the gut, they help balance the intestinal microbiota and have beneficial effects on the health of the digestive tract. Probiotics are known for their capacity to improve health by stimulating the gut microbiota, enhancing host immunity and resistance to infection, helping digestion, and carrying out various other functions.

Concurrently, the metabolites produced by these microorganisms, termed postbiotics, which include compounds like bacteriocins, lactic acid, and hydrogen peroxide, contribute to inhibiting a wide range of pathogenic bacteria (Petrariu et al., 2024).

6.3 Probiotics

Probiotics are defined as "live microorganisms which when administered in adequate amounts confer a health benefit on the host" (FAO) (Guidelines for the evaluating probiotics in food, 2002). These microorganisms can remain viable for extended periods of storage and survive under intestinal conditions. Probiotics are live microorganisms complying with the following criteria: (1) They must undergo proper classification to genus, species, and strain level, (2) they must be safe for consumers for the intended use, (3) they must be scientifically proven to provide health beneficial, and (4) they must be alive in products at the time of consumption in an amount sufficient to exert health benefits (Binda et al., 2020).

Microorganisms with probiotic potential can be isolated from three main sources in nature: (1) humans and animals, (2) food, and (3) the environment. Although most probiotics are bacteria, some yeast and mold have also been reported to exhibit probiotic properties. Commercially available probiotic foods can be both fermented and nonfermented foods, and can be divided into two categories: dairy-based foods and nondairy-based foods. Dairy-based foods are foods made from milk (i.e., fermented milks, cheese, ice cream, buttermilk, milk powder, and yogurt), and nondairy foods are foods made from nonmilk sources (i.e., foods prepared from soy bean, soy curd, soy yogurt, and soy milk), cereals, vegetables, fruit, and meat (Vijaya Kumar et al., 2015).

Microorganisms from many different genera have been used as probiotics. Most research has focused on probiotics, lactic acid bacteria (LAB), *Lactobacillus*, and *Bifidobacterium* genera. Other probiotics in this group belong to the genera *Lactococcus*, *Enterococcus*, *Streptococcus*, *Pediococcus,* and *Leuconostoc*. Probiotics from the nonlactic acid bacteria group are *Bacillus*, *Escherichia*, and *Propionibacterium*. The most commonly reported probiotic yeast is *Saccharomyces*, particularly *Saccharomyces boulardii*, and fungi such as *Aspergillus* have also received much attention. Today, the food market offers products containing different combinations of probiotics.

Probiotics microorganisms used for humans belong to group of LAB: *Lactobacillus* (*L. rhamnosus, L. acidophilus, L. casei, L. fermentum, L. gasseri, L. johnsonii, L. reuteri, L. paracasei, L. plantarum, L. salivarius, L.*

brevis, L. johnsonii), *Bifidobacterium* (*B. infantis, B. animalis* subsp. *Lactis, B. adolescentis, B. animalis, B. bifidum, B. breve, B. longum*), *Streptococcus* (*S. thermophilus*), *Lactococcus* (*L. lactis* subsp. *Lactis*), *Enterococcus* (*E. durans, E. faecium, E. faecalis*), *Pediocuccus (P. acidilactici), and Leuconostoc (L. mesenteroides);* group of nonlactic acid bacteria:gram-positive bacteria of genus *Bacillus (B. coagulans, B. subtilis, B. cereus), Escherichia (Escherichia coli nissli), and* propionilbacterium *(P. freudenreichii)*; and yeast and fungi group: *Saccharomyces* genus (*S. cerevisiae, S. boulardii) and Aspergillus* (*A. oryzae, A. niger)* (Anadón et al., 2006, 2021). A mixture comprising live microorganisms and substrate(s) selectively utilized by host microorganisms that confers a health benefit on the host (Swanson et al., 2020).

The majority of probiotic microorganisms belong to the LAB group (*Lactobacillus, Enterococcus, Streptococcus)*, being important components of the normal intestinal microflora. Additionally, *Bifidobacteria* constitute an important part of the normal intestinal microflora throughout the life span, and also the yeast *Saccharomyces boulardii*. The gram-positive, nonspore-forming microorganisms have been used in food production, in particular for fermentation processes that generate lactic acid by homo- and heterofermentative pathways (Fijan, 2023). It is necessary to explore the functional effects and health benefits of *Bifidobacterium* species and its metabolic and beneficial properties throughout life (Derrien et al., 2022).

A general recommendation is to choose probiotic products with at least 1 billion colony-forming units and containing the genus *Lactobacillus, Bifidobacterium, Bacillus*, or *Saccharomyces boulardii*, some of the most researched probiotics.

Probiotics have a number of beneficial effects on the health of the host body which can be summarized as follows: (a) Improvement of intestinal health: (1) improving immunity in the GIT tract by stimulating or inhibiting the secretion of certain cytokines and reducing inflammation, (2) reducing the number of harmful bacteria in the GIT, (3) inhibiting the adhesion to the intestinal wall of pathogenic bacteria, (4) altering the number and type of normal microbiota in the gut, (5) increasing the strength of epithelial cells in the intestinal wall, and (6) stimulating the production of receptors on epithelial cells (e.g., mu opioid and cannabinoid receptors); (b) enhancement of immune response:common immune system disorders that have been found to be alleviated by probiotics (allergic diseases such as atopic dermatitis and allergic rinitis and diseases caused by inflammatory response such as rheumatoid arthritis and inflammatory intestinal disease); (c) prevention and treatment of infectious diseases such as (1) infectious diseases in digestive system caused by *Salmonella enterica* serotype Typhimurium, *Escherichia coli*, *Shigella sonnei*, *Clostridium difficile*, *Helicobacter pylori*, *Staphylococcus aureus* and rotavirus; (2) infectious diseases in respiratory system caused by rhinovirus, influenza virus, respiratory syncytial virus, *Streptococcus pneumonia* and group A streptococci; (3) infectious diseases in urinary system caused by uropathogenic *Escherichia coli* (UPEC), *Proteus mirabilis,* and *Klebsiella pneumonia;* and (4) infectious diseases in reproductive system caused by *Candida albicans*, *Guardnerella vaginalis*, *Mycoplasma hominis,* and *Bacteroides* spp.; (d) prevention and treatment of cancer; and (e) other health benefits in the body (reducing cholesterol, reducing diabetes (antidiabetic), controlling overweight/obesity, modulation of brain activity, and prevention of caries) (Andersson et al., 2016).

6.4 Postbiotics

Postbiotics are *produced by bacterial and fungal species like Lactobacillus, Bifidobacterium*, *Streptococcus*, *Eubacterium*, *Faecalibacterium*, and *Saccharomyces*, which are naturally found in fermented foods. The term postbiotic appropriately refers to substances derived after the microorganisms are no longer alive, or, in other words, inanimate, dead, or inactivated (Vinderola et al., 2022). The metabolites produced by these microorganisms, also termed postbiotics, include compounds such as bacteriocins, lactic acid, and hydrogen peroxide, which contribute to inhibiting a wide range of pathogenic bacteria. Postbiotics refer to nonviable microbial cells, metabolic byproducts, and their microbial components released after lysis. Postbiotics modulate the immune response, regulate gene expression, exclude pathogenic microbes, maintain intestinal barrier function, and help control carcinogénesis, pathogen infections, and the synthesis of nutrients (e.g., amino acids, enzymes, vitamins, and carbohydrates) (Aggarwal et al., 2022). Consequently, they have been associated with several health benefits (including the prevention and management of diarrhea, relief from clinical symptoms of irritable bowel syndrome (IBS), modulation of the immune system, and potential roles in metabolic health).

6.5 European regulations of probiotics

Current EU legislation covers substances with a physiological effect, such as probiotic bacteria, but they are not governed under specific EU regulatory frameworks. The new food probiotics are regulated by the Novel Foods Regulation [Regulation (EC) No. 258/97] (EC, 1997a) and the Nutrition and Health Claims regulation [Regulation (EC) No. 1924/2006] (Regulation (EC) No. 1924/2006 of the European Parliament and of the Council of 20 December 2006 on nutrition and health claims made on foods, 2006). Microbial strains must comply with one or both of these regulations. Any health claims proposed for probiotic bacteria to be used in food must be based on, and substantiated by, the generally accepted scientific data (Guidance on the scientific requirements for health claims related to gut and immune function, 2011).

The EU regulations will prohibit any claims referring to the prevention, treatment, or cure of a human disease by a food, which is in contrast to that proposed by other countries, such as Canada and the United States (Sanders et al., 2005). In the United States, probiotic products may be marketed and regulated to a generally healthy population as food, dietary supplement, or biological product (i.e., drugs) depending on the intended use of a probiotic.

Probiotic substantiation domains of efficacy are needed to support claims of health benefits. Probiotic products that claim specific nutritional, functional, or therapeutic characteristics blur the borderlines between food, nutrition, and health claims, and dietary supplements or human medicines are ambiguous, thereby producing questions for regulators (Guidance on the scientific requirements for health claims related to gut and immune function, 2011).

Depending on the intended use of a probiotic, the legislative framework requirements differ because the probiotic may have different considerations: (1) microorganisms as novel food or novel food Ingredients; (2) nutrition and health claim substantiation; (3) probiotics as genetically modified microorganisms (GMMs); (4) food supplements or dietary supplements and (EC, 1997b) other similar nutritional products; and (5) probiotics used in food or as medicines.

The regulations of these products share many similarities with those for human drugs. In all these cases, EFSA has a very central role as the principal risk assessment body providing the European Commission with safety assessment and, in the case of health claims, also of the efficacy aspects of the products (Von Wright, 2020). Regarding the safety assessment of microorganisms, EFSA takes into consideration the Qualified Perception Of Safety (QPS) Concept. The QPS concept would allow strains with established safety to enter the market without extensive toxicity and efficacy testing requirements. The QPS guidelines regarding GRAS are applied to: (1) microorganisms; (2) a positive list that will be available; (3) the inclusion of the microorganism based on the history of use and adverse effects; and (4) a defined taxonomic unit (e.g., genus, species, or strain). For QPS status, the EFSA plays a primary role in the scientific opinions on probiotics and their respective regulations (Anadón et al., 2021). For microorganisms that do not have the QPS status, a basic set of toxicity studies should be provided, consisting of genotoxicity/mutagenicity tests and a sub-chronic (90-day) oral toxicity study (Anadón et al., 2006; Von Wright, 2020). The QPS guidelines regarding GRAS are applied to: (1) microorganisms; (2) a positive list that will be available; (3) the inclusion of the microorganism based on the history of use and adverse effects; and (4) a defined taxonomic unit (e.g., genus, species, or strain).

6.5.1 Microorganisms as novel food or novel food ingredients

The legal instrument document governing novel foods in the European Union (EU) is Regulation (EU) 2015/2283 (EU, 2015) which repealed Regulation

(EC) No 258/97 (EC, 1997a). Regulation (EC) No 258/97 (EC, 1997b) established rules for novel foods that were not used for human consumption to a significant degree within the Community before May 15, 1997.

Foods consumed before this date are considered if no adverse health effects have been reported. However, for new food products introduced after this date, safety is not automatically assumed. There are four novel food/food ingredient categories: (1) those that present a new or intentionally modified primary molecular structure; (2) those that consist of or isolated from microorganisms, fungi, or algae; (3) those consisting of or isolated from plants or isolated from animals, except for foods and food ingredients obtained by traditional propagating or breeding practices and having a history of safe use; and (4) those that have been applied in a production process caused not currently used, where that causes significant changes in the composition or structure that affect their nutritional value, metabolism, or undesirable substance levels (Anadón et al., 2021).

Functional foods containing novel ingredients or produced using novel processes fall under Regulation (EU) 2015/2283 (Regulation (EU) 2015/2283 of the European Parliament and of the Council of 25 November 2015 on Novel Foods, Amending Regulation (EU) No. 1169/2011 of the European Parliament and of the Council and Repealing Regulation (EC) No 258/97 of the European Parliament and of the Council and Commission Regulation (EC) No 1852/2001, 2015). This Regulation, which has come into force on January 1, 2018, aims to facilitate the authorization of novel foods. The authorization of novel foods and supplements containing novel ingredients to be allowed on the EU market can start on the initiative of either the European Commission or through an application submitted by a competent authority (i.e., Member States). The European Commission will then request the opinion of the European Food Safety Authority (EFSA), responsible for the scientific risk assessment, ensuring the convergence from an individual assessment to a unified assessment evaluation. The relevant EFSA Panel for novel foods is the Panel on Dietetic Products, Nutrition and Allergies (NDA Panel). One authorized, approved food is added to the Union list of novel foods.

The Novel Foods Regulation provides two different authorization procedures: (1) simplified procedure, based on a notification system and followed if the applicant can provide evidence that the novel food/food ingredient is substantially equivalent to an existing food ingredient in terms of composition, nutritional value, metabolism, intended use, and level of undesirable substances, and (2) full assessment procedure, required when the novel food or food ingredient consists of or is isolated from microorganisms, fungi, algae, plants, or animals, except for those obtained by traditional propagating or breeding practices and having a history of safe food use. This procedure is widely used worldwide and constitutes the cornerstone of novel food safety evaluation (Constable et al., 2007).

The proposal for a Regulation on Novel Foods (Proposal for a Regulation of the European Parliament and of the Council on novel foods (Text with EEA relevance), 2013) was adopted in 2013, although the European Commission has adopted three draft laws concerning animal cloning and novel food that will provide legal certainty in this field. The draft law on novel food revises the existing Novel Food Regulation to improve access to new and innovative food products in the EU market while still maintaining high consumer protection standards.

A health claim is any statement used on labels, in marketing, or in advertising that suggests health benefits from consuming a specific food or one of its components, such as vitamins and minerals, fiber, and "probiotic" bacteria. There are different types of health claims; for instance, statements that a food can reinforce the body's natural defences or enhance learning ability are classified as "general function" claims.

To date, the novel food ingredients are not governed under specific EU regulatory frameworks, although the Regulation (EC) No. 258/97 (EC, 1997a) outlines safety requirements for the authorization of novel food/ingredients. The principal objective of the Regulation (EC) No. 258/97 (EC, 1997b) was to guarantee that new foods and food ingredients: (1) protect the functioning of the internal market within the Community and (2) protect public health ensuring that novel foods in that they must not: (1) present a danger for the consumer; (2) mislead the consumer; or (3) differ from foods or food ingredients that are intended to replace to an extent that their normal consumption would be nutritionally disadvantageous for the consumer. Commission Recommendation 97/618/EC of July 29, 1997 (EC, 1997a) concerning the scientific aspects and the presentation of information necessary to support applications for the placing on the market of novel foods and novel food ingredients and the preparation of initial assessment reports under Regulation (EC) No. 258/97 (EC, 1997a) recommends as "Novel Food/Ingredient Guidelines," instruction necessary to sustain an application to fulfill the safety assessment. Examples of authorized substances of microbial origin or new microorganisms for human consumption include dextran produced by *Leuconostoc mesenteroides*, vitamin K produced by *Bacillus subtilis* Natto, *Clostridium butyricum* as a food supplement, and *Bacteroides xylanosolvens* in fermented pasteurized milk products. Other examples of microorganisms candidates' safety evaluation for novel food or novel food ingredient include *Bacteroides xylanisolvens*, *Akkermansia muciniphila*, fructophilic lactic acid bacteria (FLAB), and *Faecalibacterium prausnitzii* (Brodmann et al., n.d.).

To ensure that commercial probiotic products deliver their claimed beneficial health effects to patients with serious disorders, labeling must provide accurate information regarding the presence of live bacteria at a specific concentration and the number of dead bacteria (de Simone, 2019).

6.5.2 Nutrition and health claim substantiation

In the EU, health and nutrition claims are governed by Regulation (EC) No. 1924/2006 (Regulation (EC) No. 1924/2006 of the European Parliament and of the Council of 20 December 2006 on nutrition and health claims made on foods, 2006). Article 6 of this regulation states that health claims must be based on and substantiated by generally accepted scientific evidence, and Article 13(1) (claims) indicates that health claims should be evaluated by EFSA after a scientific assessment of the highest possible standard. Many claims were evaluated on a case-by-case basis (some were clustered into a single opinion) regarding uniform scientific criteria for consistency. This rejection highlights the poor characterization of probiotic strains and the lack of solid human data to demonstrate a cause-and-effect relationship. This is why many probiotic products are sold without any substantiation of health claims, either as foods or as food supplements (Von Wright, 2020).

In 2009 the EFSA published a Scientific Opinion on the substantiation of health claims related to noncharacterized microorganisms pursuant to Article 13(1) of Regulation (EC) No 1924/2006 (Scientific Opinion on the substantiation of health claims related to non-characterised microorganisms pursuant to Article 13(1) of Regulation (EC) No 1924/2006, 2019). The NDA Panel was asked to provide a scientific opinion on a list of health claims pursuant to Article 13 of Regulation 1924/2006. This opinion addresses the scientific substantiation of health claims related to a number of microorganisms, for which the Panel considers the data provided insufficient to characterize the microorganisms in question. Based on these criteria, the Panel considered that the microorganisms/combination of microorganisms covered in this opinion are not sufficiently characterized. As the data available are insufficient to characterize the microorganisms/combination of microorganisms addressed in this opinion, and due to the strain-specificity of the effects, the evidence obtained for one strain cannot be extrapolated to another. Therefore the Panel concluded that a cause-and-effect relationship has not been established between the consumption of the microorganisms/combination of microorganisms discussed in this opinion and their claimed effects (Scientific Opinion on the substantiation of health claims related to non-characterised microorganisms pursuant to Article, 2009).

In the case of probiotics, the main health benefits appear to be associated with maintenance of healthy microbiota or improving the microbiota resilience, for instance, by reducing the numbers or colonization of pathogenic bacteria or viruses and maintaining and improving the intestinal integrity and barrier function. Specific probiotics may also be associated with other health benefits, including desirable modulation of lactose intolerance, bowel function and gastrointestinal comfort, diarrhea prevention and clinical symptom alleviation, and upregulation or downregulation of immune response. A probiotic claim is any statement that or implies that a probiotic food has particular characteristics relating to its origin, nutritional properties, and health (Guidelines for the evaluating probiotics in food, 2002; FAO Technical meeting on prebiotics, 2008; Regulation (EC) No. 1924/2006 of the European Parliament and of the Council of 20 December 2006 on nutrition and health claims made on foods, 2006).

Based on experiences gained to date with the evaluation of health claims in the area related to gut and immune function, the NDA Panel of EFSA published a Guidance to facilitate the submission of an application for the authorization of health claims. The guidance document focuses on two key issues: (1) claimed effects which are considered to be beneficial physiological effects and (2) studies/outcome measures which are considered to be appropriate for the substantiation of health claims. The guidance presents examples drawn from previous evaluations to illustrate the Panel's approach, as well as some examples currently under consideration in ongoing evaluations (Guidance on the scientific requirements for health claims related to gut and immune function, 2011).

The difficulties encountered with probiotic health claims can be summarized as follows: identification and characterization of strains (phenotypic and genetic tests); impact of carrier or vehicle food matrix; stability (processes and storage); and clinical intervention studies often performed in diseased populations. Regarding the characterization of microorganisms as food constituents that are the subject of health claims, the NDA Panel establishes two criteria: (1) species identification by DNA–DNA hybridization or 16S rRNA sequence analysis and (2) strain identification by any internationally accepted genetic typing molecular methods (Scientific Opinion on the substantiation of health claims related to non-characterised microorganisms pursuant to Article, 2009). Regarding the claim application challenges, the main ones are as follows: The component should be well-defined; the basis for the effects should be sufficiently characterized; the clinical studies should be performed in the target population; a clear rationale for extrapolation from specific populations should be presented; the cause-and-effects relationship must be demonstrated; and the proposed health claim hypothesis and evidence should be clearly presented throughout the application. The layout of the submission dossier and the corresponding content of the application must be in accordance with the requirements of Regulation (EC) No. 1924/2006 (Regulation (EC) No. 1924/2006 of the European Parliament and of the Council of 20 December 2006 on nutrition and health claims made on foods, 2006; Anadón et al., 2013).

Recently, the European Ombudsman has concluded an investigation into a complaint from a probiotic producers' association regarding the European Commission's (EC) classification of the term "probiotic" as a health claim (Decision on how the European Commission deals with the labelling of foodstuff that contain probiotics as "health claims" (case 2273/2023/MIK), 2025). The association argued that the negative evaluations of probiotics' health benefits and the EC's rejection of related health claim applications effectively ban the term "probiotic" and create legal uncertainty. They contended that "probiotic" should be viewed merely as a descriptor, not implying specific health benefits, and thus should qualify as a nutrition claim.

Furthermore, they noted that several EU Member States have guidelines permitting the use of the term "probiotic." However, none of the applications for authorization of health claims concerning live microorganisms, "probiotics" and similar terms submitted to date received a favorable scientific opinion of EFSA. This is mainly due to the absence of scientific evidence on the effects of probiotics on a healthy population, insufficient characterization of the strains of microorganisms used in the relevant foods, the poor quality of studies, and undefined claims. As regards the complainant's proposal to consider the term "probiotics" as a "nutrition claim" rather than a "health claim," the Commission argued that there is no link between probiotics and the nutritional qualities of foods. In any case, the Claims Regulation requires scientific evidence to substantiate any claims, including nutrition claims, which is currently lacking.

The Ombudsman upheld the EC's interpretation, asserting that "probiotic" suggests a positive health benefit according to the World Health Organization's definition acknowledged by EU Member States. The WHO defines probiotics as "live microorganisms which when administered in adequate amounts confer a health benefit on the host" (FAO) (Guidelines for the evaluating probiotics in food, 2002). The Ombudsman deemed this interpretation reasonable and protective of consumers against misleading claims, given the current insufficient scientific evidence. Additionally, it was noted that Member States allowing the term's use violate EU law, which the EC must address.

6.5.3 Genetically modified microorganisms

The main pieces of legislation on genetically modified organisms (GMOs) are supplemented by various implementing rules, recommendations, and guidelines on more specific aspects. The term "genetically modified food" describes food containing, consisting of, or produced from "genetically modified organisms" (GMOs). A GMO is defined as an organism, such a plant or animal, whose genetic material has been altered in a way that does not occur naturally through mating and/or natural recombination to give them a new property (Article 2(2) Directive 2001/18/EC) (Directive 2001/18/EC of the European Parliament and of the Council of 12 March 2001 on the deliberate release into the environment of genetically modified organisms and repealing Council Directive 90/220/EEC - Commission Declaration, 2001), amended by the subsequent Directive (EU) 2015/412 (EU, 2015).

Initially, foods and food ingredients containing or consisting of GMOs within the meaning of Directive 90/220/EEC (EEC, 1990) and foods and food ingredients produced from, but not containing them, were also part of the scope of Regulation (EC) No. 258/1997 (EC, 1997a). However, Regulation (EC) No. 1829/2003 on genetically modified food and feed (EC, 2003a) established a specific authorization procedure for GMO food and feed products. Directive (EU) 2015/412 amending Directive 2001/18/EC as regards the possibility for Member States to restrict or prohibit the cultivation of GMOs in their territory (EU, 2015) and Regulation (EC) No. 1830/2003 (EC, 2003b) concerns the traceability and labeling of food and feed products produced from GMOs. The GMO intended for food use would be evaluated according to Regulation (EC, No 1829/2003) and assessed by the EFSA (Panel on Genetically Modified Organisms, GMO-Panel). The Commission Implementing Decision (EU) 2018/1790 of 16 November 2018, repealing Decision 2002/623/EC, established guidance notes on the environmental risk assessment of GMOs (EU, 2018). Overall, the legal framework of GMOs legislation aims to: (1) **protect human and animal health and the environment** by introducing a safety assessment of the highest possible standards at EU level before any GMO is placed on the market; (2) implement **harmonized procedures** for risk assessment and authorization of GMOs that are efficient, time-limited, and transparent; (3) ensure **clear labeling** of GMOs placed on the market to enable consumers as well as professionals (e.g. farmers and food feed chain operators) to make an informed choice; (4) ensure the **traceability** of GMOs placed on the market. GMOs are produced in such a way that the new food/foodstuff develops an increased resistance to environmental pressures and competitive/stressful conditions, such as diseases, droughts, insects, or herbicides. GMOs may be endowed with an improved quality, such as an increased nutritional value or yield.

Directive 2009/41/EC is specifically devoted to the use of genetically modified microorganisms (GMMs) (Directive 2009/41/EC of the European parliament and of the Council of 6 May 2009 on the contained use of genetically modified micro-organisms (Recast) (0 J L 125, 21.5.2009), 2009). The different requirements, namely, the procedures for developing GMOs, their proper usage, the mandatory authorizations, and eventual restrictions or prohibitions of GMO cultivation, are contained in Regulation (EC) No 1829/2003 (EC, 2003a). Articles 3 and 4 state that GMOs may be placed on the market only for food use, and only after being authorized by the EFSA and the European Commission. Regulation (EC) 1946/2003 regulates the transboundary movements of GMOs (Bragazzi et al., 2017)

The Commission Implementing Decision (EU) 2018/1790 of November 16, 2018, repealing Decision 2002/623/EC (EU, 2018) revoked the outdated Guidance Notes from 2002. The repeal reduces the number of guidance documents that operators and competent authorities need to take into account when carrying out an environmental risk assessment under Annex II to Directive 2001/18/EC (Directive 2001/83/EC of the European Parliament and of the Council of 6 November 2001 on the Community code relating to medicinal products for human use lays down harmonized rules for the authorization, supervision and pharmacovigilance of medicinal products for human use within the Union, 2001).

6.5.4 Food supplements or dietary supplements and other similar nutritional products

The concept of functional products could cover functional foods, probiotic food, dietary supplements, and nutraceuticals, the use of which is intended to improve the quality of life by preventing nutrition-related diseases. The term "functional food," was introduced in Japan in the mid-1980s and used as a synonym for nutraceuticals; these include designer food, pharmafood, medifood, vitafood, and the more traditional food/dietary supplements and fortified foods. It is also important to note that nutraceutical is a term often applied to a wide range of products, from functional foods and food supplements to dietary foods that provide health benefits. Functional or probiotic foods, food supplements, and nutraceuticals are at the interface between nutrition and medicines (Anadón et al., 2021).

For many years, food and dietary supplements were simply food items intended to balance the diet or provide for special dietary needs. It is known that healthy nutrition requires an appropriate dietary intake of energy in the form of macronutrients (i.e., protein, fat, carbohydrates, and total calories) as well as the adequate intake of essential nutrients, which include vitamins, trace elements, minerals, essential fatty acids, and essential amino acids. Accompanying the new food products with healthy benefits, new terms have also emerged. New terms such as "nutraceutical," which is made up from the combination of "nutrition" and "pharmaceutical," are used to describe foods that contain a supplement or enhancement to provide a specific health benefit (e.g., a combination of vitamins, amino acids, and plant preparations in a capsule). When it comes to promoting health, food is preferred over dietary supplements. This preference may be due to the buffering properties of foods for probiotics during their passage through the gut, the supply of essential nutrients to maintain the activity and efficacy of probiotic bacteria, the synergistic effects of food ingredients on the growth of probiotics, and the attitude of consumers towards probiotic foods versus supplementation based on pharmaceutical forms, preferably tablets and capsules. The efficacy of probiotic bacteria is mainly based on two factors: (1) viability and (2) survivability as well as activity (del Piano et al., 2011; De Vrese and Schrezenmeir, 2008).

An important element in the Food Supplement Directive is that food supplements are considered foodstuffs. Therefore food supplements are defined in the Directive 2002/46/EC (EC, 2002) as "foodstuffs, the purpose of which is to supplement the normal diet and which are concentrated sources of nutrients or other substances with a nutritional or physiological function, alone or in combination, marketed in dose form, namely, forms such as capsules, pastilles, tablets, pills and other similar forms, sachets of powder, ampules of liquids, drop dispensing bottles, and other similar forms of liquids and powders designed to be taken in measured small unit quantities." Food supplements are marketable finished products that are precisely presented to the public for supplementation of the diet and may or may not exceed the intake of micronutrients present in the consumed diet. Micronutrients (vitamins and minerals) are required by the body throughout life in extremely small amounts. Food supplements cannot be presented as medicines or as a substitute for medicines, and the Food Supplement Directive does not apply to medicinal products as defined by Directive 2001/83/EC (Directive 2001/18/EC of the European Parliament and of the Council of 12 March 2001 on the deliberate release into the environment of genetically modified organisms and repealing Council Directive 90/220/EEC - Commission Declaration, 2001) (latest amendment in 2011). It is important to note that the European Directive applies only to products sold in the EU; it does not apply to retailers in the United States. Therefore retailers operating in the United States could continue to sell products without being affected by the Directive. The Directive only applies to EU Member States; however, manufacturers, distributors, and retailers that have exported dietary supplements containing ingredients not listed in Directive 2002/46/EC (EC, 2002) can no longer sell those products in the EU as of August 1, 2005. There is a wide range of nutrients and other substances that can be added to food supplements. At the EU level, the reference legislation on food supplements is Directive 2002/46/EC (Food Supplements Directive, FSD), which partially harmonizes the rules related to food supplements across the Member States. It establishes only harmonized lists and specific rules for vitamins and minerals that may be used as ingredients in the manufacture of food supplements and sets out the labeling requirements for these products. Directive 2002/46/EC constitutes the regulatory framework at the European level and has been amended several times. It applies to all food supplements, and certain requirements, in particular those relating to the indications appearing on the label, are mandatory for all food supplements, regardless of their composition. These supplements are marketed in dose form (i.e., as pills, tablets, capsules, sachets, and liquids in measured doses, among others) in which bacterial species or combinations of species are used, and product details include dose, duration of intake, combination, and methods used to manufacture the formulation, supported by evidence of general beneficial effects. Food supplements may be used to correct nutritional deficiencies or to maintain an adequate intake of certain nutrients.

Food supplements are defined in the Directive 2002/46/EC (EC, 2002) as "foodstuffs, the purpose of which is to supplement the normal diet and which are concentrated sources of nutrients or other substances with a nutritional or physiological function, alone or in combination, marketed in dose form, namely, forms such as capsules, pastilles, tablets, pills and other similar forms, sachets of powder, ampules of liquids, drop dispensing bottles, and other similar forms of liquids and powders designed to be taken in measured small unit quantities."

However, the Directive only sets rules for the use of vitamins and minerals in the manufacture of food supplements. Consequently, the use of substances other than vitamins or minerals in the manufacture of food supplements remains subject to national legislation, which will be applied within the framework of Articles 28–30 of the EC Treaty, without prejudice to other Community provisions of general application which are also applicable to these substances. Food supplements containing substances other than vitamins or minerals are foodstuffs within the meaning of Article 2 of Regulation (EC) No 178 (Working Group Report on Drafting Guidelines for the Evaluation of Probiotics in Food, 2002; Regulation (EC) No 178/2002 of the European Parliament and of the Council of 28 January 2002 laying down the general principles and requirements of food law, establishing the European Food Safety Authority and laying down procedures in matters of food safety, 2002) of the European Parliament and of the Council, which states that "foodstuff" (or "food") means any substance or product, whether processed, partially processed or unprocessed, intended to be, or reasonably expected to be ingested by humans.

Because the Regulation applies to all foodstuffs, its general principles also cover foods with added functional properties (e.g., functional or probiotic foods, nutraceuticals, dietetic foods, and food supplements). Article 2 of the Regulation also states that foodstuff (or food) means any substance or product, whether processed, partially processed, or unprocessed, intended to be, or reasonably expected to be ingested by humans. This Article 2 excludes from the definition of foodstuffs certain categories of products, including medicinal products with the meaning of Directive 2001/83/EC (Directive 2001/18/EC of the European Parliament and of the Council of 12 March 2001 on the deliberate release into the environment of genetically modified organisms and repealing Council Directive 90/220/EEC - Commission Declaration, 2001).

In application of the principle of mutual recognition, any food supplement that is legally marketed in another Member State of the EU can be marketed in Spain, even if its ingredients are not regulated in that state, unless the suspension or limitation of marketing in another Member State is justified and expressly stated, as established in Directive 2002/46/EC on food supplements, Articles 11 and 12, and Regulation (EC) 764/2008 on technical standards applicable to products legally marketed in another Member State (Regulation (EC) No 764/2008 of the European Parliament and of the Council of 9 July 2008 laying down procedures relating to the application of certain national technical rules to products lawfully marketed in another Member State and repealing Decision No 3052/95/EC, 2008).

However, most food supplements do not make disease-specific claims and are, therefore, classified as food supplements or dietary supplements.

6.5.5 Medical or pharmaceutical products

When probiotics are indicated for the prevention, treatment, or diagnosis of a disease, they are classified as medical or pharmaceutical products and are regulated according to the rules for registration of human drugs (e.g., pouchitis, inflammation, ulcerative colitis, or certain other diseases). Human drugs are regulated by the EU under Directive 2001/83/EC (Directive 2001/83/EC of the European Parliament and of the Council of 6 November 2001 on the Community code relating to medicinal products for human use lays down harmonized rules for the authorization, supervision and pharmacovigilance of medicinal products for human use within the Union, 2001), amended by Directive 2004/27/EC (Directive 2004/27/EC of the European Parliament and of the Council of 31 March 2004 amending Directive 2001/83/EC on the Community code relating to medicinal products for human use, 2004; Regulation (EC) No. 726/2004 (EC, 2004) of the European Parliament and of the Council of 31 March 2004 laying down Community procedures for the authorization and supervision of medicinal products for human and veterinary use and establishing a European Medicines Agency, n.d.). A medicinal product may only be placed on the market in the European Economic Area (EEA) when a marketing authorization has been issued by the competent authority of a Member State (or EEA country) for its own territory (national authorization) or when an authorization has been granted in accordance with Regulation (EC) No. 726/2004 (Directive 2004/27/EC of the European Parliament and of the Council of 31 March 2004 amending Directive 2001/83/EC on the Community code relating to medicinal products for human use, 2004; Regulation (EC) No. 726/2004 (EC, 2004) of the European Parliament and of the Council of 31 March 2004 laying down Community procedures for the authorization and supervision of medicinal products for human and veterinary use and establishing a European Medicines Agency, n.d.). The requirements and procedures for marketing authorization, as well as the rules for monitoring authorized products, are primarily laid down in Directive 2001/83/EC and in Regulation (EC) No 726/2004. They also include harmonized provisions for the manufacture, wholesale, or advertising of medicinal products for human use. On April 26, 2023, the European Commission adopted a proposal for a new Directive (EC, 2003a) and a new Regulation (EC, 2003a) to revise and replace the existing general pharmaceutical legislation. This is the largest reform in over 20 years. Probiotics, when registered according to pharmaceutical legislation, must always comply with quality, efficacy, and safety criteria and must be applied to each probiotic product.

Most probiotics are marketed as foodstuffs or medicines (Mäyrä-Mäkinen and Bigret, 1993; Anadón et al., 2021). Currently, the most widely used probiotics include *lactobacilli*, *bifidobacteria*, and some nonpathogenic strains, mostly of human origin, that confer health benefits on the host and enable prevention of or improvement in some diseases when administered in adequate amounts. *Lactobacillus*, *Leuconostoc*, and *Pediococcus* species have been extensively used in food processing throughout human history, and ingestion of foods containing live bacteria, dead bacteria, and metabolites of these microorganisms has taken place for a long time (Anadón et al., 2021; Mäyrä-Mäkinen and Bigret, 1993). An important fact is that probiotics must retain their viability during storage, the manufacturing process, and transit through the stomach and small intestine. Before being categorized as probiotics, organisms need to follow a process of testing, including strain testing, identification by genotype and phenotype, functional characterization and safety assessment testing, and double-blind, placebo-controlled human trials to verify their health benefits, and so the guidelines for the evaluation of probiotics in food have been proposed (FAO/WHO) (Guidelines for the evaluating probiotics in food, 2002).

The use of probiotics in medical practice is of considerable importance. The effects and applications of probiotics in various types of patological conditions have been summarized by Goldin and Gorbach (2008), who classified probiotics into four groups according to the effectiveness level: first group includes applications in acute gastroenteritis/antibiotic-associated gastroenteritis; second group includes applications to allergic reactions (atopic dermatitis I); third group includes applications in childhood respiratory infections, dental caries, IBD, nasal pathogen control, and prevention of recurrent *Clostridium difficile*-induced gastroenteritis; and fourth group covers potential applications such as rheumatoid arthritis, inflammatory bowel syndrome, cancer, alcohol-induced liver disease, diabetes, and graft-versus-host disease.

Overall, probiotics may also maintain or restore the gut microecology during or after antibiotic treatment; bifidobacteria or lactobacilli will decrease the duration and/or severity of acute diarrheal disease in infants and children (Oberhelman et al., 1999). *Lactobacillus* probiotics, as single species or combination probiotic products, have been tested for the prevention of *C. difficile*-induced gastroenteritis. Treatment with antibiotics results in disturbance of the gastrointestinal microbiota, which is associated with diarrhea and abdominal discomfort in a variable fraction of patients, depending on the age group and antibiotic used; approximately 15%–30% of patients experience recurrence of symptoms after discontinuation of antibiotics. A treatment period with antibiotics only temporarily alters the composition of the microbiota, causing environmental changes. In most cases, the cause of the diarrhea is unknown, but a varying proportion of the cases are caused by *C. difficile*.

The toxins produced by *C. difficile* may cause mild diarrhea, which can be cured simply by stopping the antibiotic treatment or, in more severe cases, lead to the life-threatening disease pseudomembranous colitis (Vanderhoof et al., 1999).

Natural remedies represent a gray area between food and medicinal products. Most products are food supplements, and sales of such products have increased in Scandinavian countries (Norway, Sweden, and Denmark).

Natural remedies are considered complementary and alternative medicine and have very limited use within the conventional health care system in Scandinavia. Probiotic bacteria are listed as an active ingredient and as an indication for gastrointestinal disorders (Waaseth et al., 2007). At present, two products have been classified as "natural remedies" with the indication "traditionally used for the normalization of intestinal flora when temporary disturbances, e.g. in mild diarrhea and constipation." It should be noted, however, that the classification as natural remedies has been made on "traditional use" and not based on any evaluation of the product-specific documentation of the products. Classification as medical foods would seem relevant when the product is used as part of a specific medical treatment, for example, to prevent or treat diarrhea associated with the use of antibiotics, or as a complication to, for instance, radiological treatment. Examples of probiotic products that have been assessed by EMA include Symbioflor 2, containing a specific strain of *Escherichia coli* and used to treat irritable bowel syndrome, but not allowed for other intestinal disorders, and the granting of orphan drug status to a probiotic preparation containing *Lactobacillus acidophilus* and *Bifidobacterium bifidum* intended for the treatment of necrotizing colitis. It should be noted that in the case of Symbioflor 2, the case involves an expert opinion on a product not centrally authorized but approved by national decisions in certain Member States.

6.6 Conclusions

Probiotics are generally accepted as being live microorganisms that, when administered in adequate amounts, confer a health benefit to the host. This review may give a comprehensive insight into the paradigm shift from the conventional view of foods and food-related products to their recognition for medicinal purposes. This review provides a comprehensive overview of the different uses of orally administered probiotics, shifting from their traditional role as food and food-related products to acknowledging their medicinal functions, as well as discussing the complexity of the applicable regulatory framework in the EU. European Food legislation intends to guarantee the highest standards of quality, efficacy, and food safety for customers. In the EU, oral probiotics are extensively regulated. *As applicable*, the authorization or notification of the product may occur centrally or within individual EU Member States, particularly when they are associated with health claims and involve microorganisms classified as novel foods or GMOs. Many oral probiotic products are marketed as foods or food supplements.

Acknowledgments

This work was supported by the Project Ref. PID 2020–15979RR-C33 from the Ministerio de Ciencia e Innovación, Spain (Project/AEI/10.13039/501100011033).

References

Aggarwal, S., Sabharwal, V., Kaushik, P., Joshi, A., Aayushi, A., Suri, M., 2022. Postbiotics: from emerging concept to application. Frontiers in Sustainable Food Systems 6. Available at https://doi.org/10.3389/fsufs.2022.887642, www.frontiersin.org/journals/sustainable-food-systems#.

Anadón, A., Rosa Martínez-Larrañaga, M., Aranzazu Martínez, M., 2006. Probiotics for animal nutrition in the European Union. Regulation and safety assessment. Regulatory Toxicology and Pharmacology 45 (1), 91–95. Available at https://doi.org/10.1016/j.yrtph.2006.02.004.

Anadón, A., Castellano, V., Martínez-Larrañaga, M.R., 2013. Regulation and guidelines of probiotics and prebiotics. Probiotics and Prebiotics in Food, Nutrition and Health. CRC Press, Spain, pp. 91–113. Available at https://doi.org/10.1201/b15561-6, https://www.taylorfrancis.com/books/e/9781466586246.

Anadón, A., Ares, I., Martínez-Larrañaga, M.R., Martínez, M.A., 2021. Probiotics: safety and toxicity considerations. Nutraceuticals: Efficacy, Safety and Toxicity. Elsevier, Spain, pp. 1081–1105. Available at https://doi.org/10.1016/B978-0-12-821038-3.00065-3, https://www.sciencedirect.com/book/9780128210383.

Andersson, H., Asp, N.-G., Bruce, Å., Roos, S., Wadström, T., Wold, A.E., 2016. Health effects of probiotics and prebiotics: a literature review on human studies. Näringsforskning 45 (1), 58–75. Available at https://doi.org/10.3402/fnr.v45i0.1790.

Baquero, F., Nombela, C., 2012. The microbiome as a human organ. Clinical Microbiology and Infection 18 (4), 2–4. Available at https://doi.org/10.1111/j.1469-0691.2012.03916.x.

Binda, S., Hill, C., Johansen, E., Obis, D., Pot, B., Sanders, M.E., et al., 2020. Criteria to qualify microorganisms as "probiotic" in foods and dietary supplements. Frontiers in Microbiology 11. Available at https://doi.org/10.3389/fmicb.2020.01662, https://www.frontiersin.org/journals/microbiology#.

Bragazzi, N.L., Martini, M., Saporita, T.C., Nucci, D., Gianfredi, V., Maddalo, F., et al., 2017. Nutraceutical and Functional Food Regulations in the European Union. Elsevier BV, pp. 309–322. Available at https://doi.org/10.1016/b978-0-12-802780-6.00017-1.

Brodmann, T., Endo, M.A., Gueimonde, G., Vinderola, W., Kneifel, de Vos. n.d.

Collins, S.M., Surette, M., Bercik, P., 2012. The interplay between the intestinal microbiota and the brain. Nature Reviews Microbiology 10 (11), 735–742. Available at https://doi.org/10.1038/nrmicro2876.

Constable, A., Jonas, D., Cockburn, A., Davi, A., Edwards, G., Hepburn, P., et al., 2007. History of safe use as applied to the safety assessment of novel foods and foods derived from genetically modified organisms. Food and Chemical Toxicology 45 (12), 2513–2525. Available at https://doi.org/10.1016/j.fct.2007.05.028.

de Simone, C., 2019. The unregulated probiotic market. Clinical Gastroenterology and Hepatology 17 (5), 809–817. Available at https://doi.org/10.1016/j.cgh.2018.01.018.

De Vrese, M., Schrezenmeir, J., 2008. Probiotics, prebiotics, and synbiotics. Advances in Biochemical Engineering/Biotechnology 111, 1–66. Available at https://doi.org/10.1007/10_2008_097.

Decision on How the European Commission Deals with the Labelling of Foodstuff that Contain Probiotics as 'health Claims' (Case 2273/2023/MIK). 2025.

del Piano, M., Carmagnola, S., Ballarè, M., Sartori, M., Orsello, M., Balzarini, M., et al., 2011. Is microencapsulation the future of probiotic preparations? The increased efficacy of gastro-protected probiotics. Gut Microbes 2 (2), 120–123. Available at https://doi.org/10.4161/gmic.

2.2.15784Italy, http://www.landesbioscience.com/journals/gutmicrobes/DelPianoGMIC2-2.pdf.

Derrien, M., Turroni, F., Ventura, M., van Sinderen, D., 2022. Insights into endogenous *Bifidobacterium* species in the human gut microbiota during adulthood. Trends in Microbiology 30 (10), 940–947. Available at https://doi.org/10.1016/j.tim.2022.04.004, www.elsevier.com/locate/tim.

Directive 2001/18/EC of the European Parliament and of the Council of 12 March 2001 on the Deliberate Release into the Environment of Genetically Modified Organisms and Repealing Council Directive 90/220/EEC - Commission Declaration. 2001.

Directive 2001/83/EC of the European Parliament and of the Council of 6 November 2001 on the Community Code Relating to Medicinal Products for Human Use Lays Down Harmonized Rules for the Authorization, Supervision and Pharmacovigilance of Medicinal Products for Human Use within the Union. 2001.

Directive 2004/27/EC of the European Parliament and of the Council of 31 March 2004 Amending Directive 2001/83/EC on the Community Code Relating to Medicinal Products for Human use. 2004.

Directive 2009/41/EC of the European Parliament and of the Council of 6 May 2009 on the Contained Use of Genetically Modified Micro-organisms (Recast) (0J L 125, 21.5.2009). 2009.

Duvallet, C., Gibbons, S.M., Gurry, T., Irizarry, R.A., Alm, E.J., 2017. Meta-analysis of gut microbiome studies identifies disease-specific and shared responses. Nature Communications 8 (1). Available at https://doi.org/10.1038/s41467-017-01973-8, http://www.nature.com/ncomms/index.html.

EC. 1997a. Commission Recommendation of 29 July 1997 Concerning the Scientific Aspects and the Presentation of Information Necessary to Support Applications for the Placing on the Market of Novel Foods and Novel Food Ingredients and the Preparation of Initial Assessment Reports Under Regulation (EC) No. 258/97 of the European Parliament and of the Council (97/618/EC) (OJ L 253, 16.9.1997).

EC. 1997b. Regulation (EC) No. 258/97 of the European Parliament and of the Council of 27 January 1991 Concerning Novel Foods and Novel Food Ingredients (OJ L 43, 14.2.1997). .

EC. 2002. Directive 2002/46/EC of the European Parliament and of the Council of 10 June 2002 on the Approximation of the Laws of the Member States Relating to Food Supplements. (OJ L 183, 12.7.2002).

EC. 2003a. Regulation (EC) No 1829/2003 of the European Parliament and of the Council of 22 September 2003 on Genetically Modified Food and Feed. (OJ L 268, 18.10.2003, pp. 0024-0028).

EC. 2003b. Regulation (EC) No 1830/2003 of the European Parliament and of the Council of 22 September 2003 Concerning the Traceability and Labelling of Genetically Modified Organisms and the Traceability of Food and Feed Products Produced from Genetically Modified Organisms and Amending Directive 2001/18/EC. 0024.

EEC. 1990. Council Directive 90/220/EEC of 23 April 1990 on the deliberate release into the environment of genetically modified organisms. (OJ L 117, 8.5.1990).

EFSA Panel on Dietetic Products, Nutrition and Allergies (NDA), 2019. Scientific Opinion on the Substantiation of Health Claims Related to Non-characterised Microorganisms Pursuant to Article 13(1) of Regulation (EC) No 1924/2006. EFSA Journal 7 (9), 1247.

EU, 2015. Directive (EU) 2015/412 of the European Parliament and of the Council of 11 March 2015 amending Directive 2001/18/EC as Regards the Possibility for the Member States to Restrict or Prohibit the Cultivation of Genetically Modified Organisms (GMOs) in their Territory territory (OJ L 68, 13.3.2015).

EU. 2018. Commission Implementing Decision (EU) 2018/1790 of 16 November 2018 Repealing Decision 2002/623/EC Establishing Guidance notes on the Environmental Risk Assessment of Genetically Modified Organisms (Notified under Document C(2018) 7513) (L 293, 20.11.2018).

FAO Technical Meeting on Prebiotics. 2008, 1–12.

Fijan, S., 2023. Probiotics and their antimicrobial effect. Microorganisms 11 (2). Available at https://doi.org/10.3390/microorganisms11020528, www.mdpi.com/journal/microorganisms.

Goldin, B.R., Gorbach, S.L., 2008. Clinical indications for probiotics: an overview. Clinical Infectious Diseases 46 (s2), S96. Available at https://doi.org/10.1086/523333.

Guidance on the scientific requirements for health claims related to gut and immune function. EFSA Journal. 9 (4) (2011), 1984.

Guidelines for the Evaluating Probiotics in Food. 2002.

Mäyrä-Mäkinen, M. Bigret, 1993. Industrial Use and Production of Lactic Acid Bacteria. 65–95.

Oberhelman, R.A., Gilman, R.H., Sheen, P., Taylor, D.N., Black, R.E., Cabrera, L., et al., 1999. A placebo-controlled trial of *Lactobacillus* GG to prevent diarrhea in undernourished Peruvian children. Journal of Pediatrics 134 (1), 15–20. Available at https://doi.org/10.1016/S0022-3476(99)70366-5, http://www.elsevier.com/inca/publications/store/6/2/3/3/1/1/index.htt.

Ortega, M.A., Álvarez-Mon, M.A., García-Montero, C., Fraile-Martínez, O., Monserrat, J., Martinez-Rozas, L., et al., 2023. Microbiota–gut–brain axis mechanisms in the complex network of bipolar disorders: potential clinical implications and translational opportunities. Molecular Psychiatry 28 (7), 2645–2673. Available at https://doi.org/10.1038/s41380-023-01964-w, https://www.nature.com/mp/.

Parichat, P., Pongsak, R., 2023. Probiotics: sources, selection and health benefits. Research Journal of Biotechnology 18 (5), 102–113. Available at https://doi.org/10.25303/1805rjbt1020113, https://www.worldresearchersassociations.com/Archives/RJBT/Vol(18)2023/May%202023/Probiotics%20Sources,%20selection%20and%20health%20benefits.aspx.

Petrariu, O.A., Barbu, I.C., Niculescu, A.G., Constantin, M., Grigore, G.A., Cristian, R.E., et al., 2024. Role of probiotics in managing various human diseases, from oral pathology to cancer and gastrointestinal diseases. Frontiers in Microbiology 14. Available at https://doi.org/10.3389/fmicb.2023.1296447.

Proposal for a Regulation of the European Parliament and of the Council on novel foods (Text with EEA relevance). 2013.

Regulation (EC) No 178/2002 of the European Parliament and of the Council of 28 January 2002 laying Down The General Principles and Requirements of Food Law, Establishing the European Food Safety Authority and Laying down Procedures In Matters of Food Safety. 2002.

Regulation (EC) No 764/2008 of the European Parliament and of the Council of 9 July 2008 Laying Down Procedures Relating to the Application of Certain National Technical Rules to Products Lawfully Marketed in Another Member State and repealing Decision No 3052/95/EC. 2008.

Regulation (EC) No. 1924/2006 of the European Parliament and of the Council of 20 December 2006 on Nutrition and Health Claims Made on Foods. 2006.

Regulation (EC) No. 726/2004 (EC, 2004) of the European Parliament and of the Council of 31 March 2004 Laying Down Community Procedures for the Authorization and Supervision of Medicinal Products for Human and Veterinary Use And Establishing a European Medicines Agency. n.d.

Regulation (EU) 2015/2283 of the European Parliament and of the Council of 25 November 2015 on Novel Foods, Amending Regulation (EU) No. 1169/2011 of the European Parliament and of the Council and Repealing Regulation (EC) No 258/97 of the European Parliament and of the Council and Commission Regulation (EC) No 1852/2001. 2015.

Sanders, M.E., Tompkins, T., Heimbach, J.T., Kolida, S., 2005. Weight of evidence needed to substantiate a health effect for probiotics and prebiotics. European Journal of Nutrition 44 (5), 303–310. Available at https://doi.org/10.1007/s00394-004-0527-6.

Scientific Opinion on the Substantiation of Health Claims related to Non-characterised Microorganisms Pursuant to Article. 13 (1). 2009.

Shanahan, F., Hill, C., 2019. Language, numeracy and logic in microbiome science. Nature Reviews Gastroenterology and Hepatology 16 (7), 387–388. Available at https://doi.org/10.1038/s41575-019-0163-5, http://www.nature.com/nrgastro/index.html.

Swanson, K.S., Gibson, G.R., Hutkins, R., Reimer, R.A., Reid, G., Verbeke, K., et al., 2020. The International Scientific Association for Probiotics and Prebiotics (ISAPP) consensus statement on the definition and scope of synbiotics. Nature Reviews Gastroenterology and Hepatology 17 (11), 687–701. Available at https://doi.org/10.1038/s41575-020-0344-2, http://www.nature.com/nrgastro/index.html.

Vanderhoof, J.A., Whitney, D.B., Antonson, D.L., Hanner, T.L., Lupo, J.V., Young, R.J., 1999. *Lactobacillus* GG in the prevention of antibiotic-associated diarrhea in children. Journal of Pediatrics 135 (5), 564–568. Available at https://doi.org/10.1016/S0022-3476(99)70053-3, http://www.elsevier.com/inca/publications/store/6/2/3/3/1/1/index.htt.

Vijaya Kumar, B., Vijayendra, S.V.N., Reddy, O.V.S., 2015. Trends in dairy and non-dairy probiotic products - a review. Journal of Food Science and Technology 52 (10), 6112–6124. Available at https://doi.org/10.1007/s13197-015-1795-2, http://www.springerlink.com/content/121580/.

Vinderola, G., Sanders, M.E., Salminen, S., 2022. The concept of postbiotics. Foods 11 (8), 1077. Available at https://doi.org/10.3390/foods11081077.

Von Wright, A., 2020. Regulatory aspects of probiotics and other microbial products intended for skin care: the European approach. Skin Microbiome Handbook: From Basic Research to Product Development. wiley, Finland, pp. 321–341. Available at https://doi.org/10.1002/9781119593058.ch17, https://onlinelibrary.wiley.com/doi/book/10.1002/9781119593058.

Waaseth, M., Eggen, A.E., Grimsgaard, S., 2007. Natural remedies in Scandinavia - authorization and sales. Pharmacy World and Science 29 (3), 137–145. Available at https://doi.org/10.1007/s11096-006-9033-7.

Working Group Report on Drafting Guidelines for the Evaluation of Probiotics in Food. 2002.

Chapter 7

The failure aspects of viable probiotics in foods, supplements, and drugs

Saba Kamalledin Moghadam[1], Jalil Houshyar[2], Hossein Alikhah[3], Sonya Keisan[4], and Aziz Homayouni-Rad[4]

[1]*Department of Food Science and Technology, Faculty of Nutrition Sciences and Food Technology, National Nutrition and Food Technology Research Institute, Shahid Beheshti University of Medical Sciences, Tehran, Iran,* [2]*Endocrine Research Center, Tabriz University of Medical Sciences, Tabriz, Iran,* [3]*Research Center for Evidence-Based Health Management, Maragheh University of Medical Sciences, Maragheh, Iran,* [4]*Department of Food Science and Technology, Faculty of Nutrition and Food Sciences, Tabriz University of Medical Sciences, Tabriz, Iran*

7.1 Introduction

The Food and Agriculture Organization of the United Nations and the World Health Organization in 2001 defined probiotics as live microorganisms that, when given in sufficient quantities, provide a health benefit to the host. Commercially available probiotics contain various lactic acid bacteria from different genera, including *Limosilactobacillus reuteri, Lacticaseibacillus rhamnosus*, selected strains of the *Lacticaseibacillus casei* and *Lactobacillus acidophilus* group, bifidobacteria, and *Bacillus coagulans*. Additionally, commercially produced probiotics also include lactic acid bacteria from genera like *Propionibacterium*, *Pediococcus*, *Lactococcus*, *Streptococcus*, *Leuconostoc*, *Enterococcus* (specifically *Enterococcus faecium* SF68), and the yeast *Saccharomyces boulardii*, as illustrated in Fig. 7.1 (Zendeboodi et al., 2020). In recent years, numerous health conditions have been linked to variations in microbiota composition and activity, known as dysbiosis. Imbalances in the intestinal microbiota have been related to disorders such as inflammatory bowel disease, diarrhea, irritable bowel syndrome, allergic disorders, colorectal cancer, and metabolic disorders (Shafie et al., 2022; Homayouni-Rad et al., 2023; Homayouni et al., 2020a,b; Bastani et al., 2016b; Homayoni Rad et al., 2012; Maleki et al., 2016; Bastani et al., 2016a; Ranjbar et al., 2016; Homayouni Rad et al., 2013). Similar associations have been

Evaluating Oral Probiotics and Prebiotics. DOI: https://doi.org/10.1016/B978-0-443-29835-6.00011-3

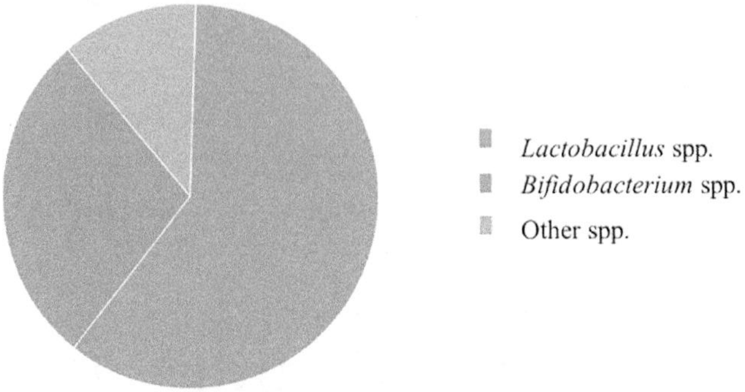

FIGURE 7.1 Classification of probiotic microorganisms. Probiotic microorganisms can be classified into different categories based on their genus, species, and strain;Probiotic microorganisms can be classified into different categories based on their genus, species, and strain; (1) *Lactobacillus,* including *L. acetotolerans, L. acidophilus, L. amylolyticus, L. amylovorus, L. apis, L. bombicola, L. casei, L. colini, L. crispatus, L. delbrueckii, L. delbrueckii subsp. Bulgaricus*a, *L. delbrueckii* subsp. *Delbrueckii, L. delbrueckii* subsp. *Indicus, L. delbrueckii* subsp. *Jakobsenii, L. delbrueckii* subsp. *Lactis, L. delbrueckii* subsp. *Sunkii, L. equicursoris, L. fermentum, L. fornicalis, L. gallinarum, L. jensenii, L. gasseri, L. gigeriorum, L. hamster, L. helsingborgensis, L. helveticus, L. hominis, L. iners, L. intestinalis, L. johnsonii, L. kalixensis, L. kefiranofaciens, L. kefiranofaciens* subsp. *Kefiranofaciens, L. kefiranofaciens* subsp. *Kefirgranum, L. kimbladii, L. kitasatonis, L. kullabergensis, L. lactis, L. melliventris, L. mulieris, L. panisapium, L. paracasei, L. paragasseri, L. pasteurii, L. plantarum, L. porci, L.psittaci, L. reuteri, L. rhamnosus, L. rodentium, L. helveticus, L. sporogenes, L. taiwanensis, L. ultunensis, L. xujianguonis*; (2) *Bifidobacterium,* including *B. animalis, B. breve, B. bifidum, B. essensis, B. infantis, B. lactis, B. laterosporus, B. longum*; (3) Other spp., including *Clostridium butyricum, Enterococcus franciumb*, *Escherichia coli Nissle, Propionibacterium freudenreichii, Kluyveromyces lactis, Leuconostoc lactis* subsp. *Cremoris, Propionibacterium freudenreichii* subsp. *Shermanii, Pediococcus, Saccharomyces boulardii, Saccharomyces cerevisiae, Staphylococcus, Intermedius, Streptococcus salivarius, Streptococcus thermophiles*).

found with imbalances in the vaginal microbiota and vaginal infections. Probiotics have been studied for their potential to prevent or improve these disorders and are increasingly recommended by healthcare professionals. The number of commercially available probiotic products is rising, and their success in development depends on various factors, regardless of their classification as food, dietary supplement, medical food, or drug (Gheorghita et al., 2021; Homayoni Rad et al., 2016; George Kerry et al., 2018). However, the viability and effectiveness of probiotic microorganisms in these products can be influenced by various factors, ultimately impacting their functionality and health-promoting properties. This chapter aims to explore the failure aspects of viable probiotics in food, drugs, and dietary supplements, shedding light on the challenges faced in maintaining the viability and efficacy of probiotic strains throughout the manufacturing, storage, and consumption processes (Table 7.1).

TABLE 7.1 Failure of live probiotics in foods, supplements, and medications.

Survival and viability	One of the key challenges with live probiotics is ensuring their survival and viability throughout processing, storage, and consumption. Studies have shown that many probiotic strains may not survive the harsh conditions of the manufacturing, storage, and gastrointestinal tract, leading to a decrease in their effectiveness.
Tolerate the gastrointestinal environment	The journey of probiotics through the digestive system can be challenging. Many probiotic strains may not survive the acidic environment of the stomach or the bile salts in the small intestine, which can reduce their effectiveness. This can result in a lower number of viable probiotics reaching the large intestine where they are needed.
Storage conditions	Proper storage conditions are crucial for maintaining the viability of live probiotics. Exposure to light, heat, moisture, or oxygen during storage can lead to a decline in the number of viable probiotic cells. Improper storage practices by retailers or consumers can compromise the quality and efficacy of probiotic products.
Strain-specificity	Different probiotic strains have varying abilities to survive and colonize the gut. Some strains may be more resilient and effective than others in providing health benefits. Studies have highlighted the importance of strain-specificity when selecting probiotics for specific health conditions.
Dosage and duration	The dosage and duration of probiotic supplementation can significantly impact their efficacy. Researchers have found that inadequate dosages or short durations of probiotic use may not be sufficient to exert beneficial effects on gut health or immune function.
Quality assurance	The maintenance of quality control standards throughout the manufacturing, storage, and distribution processes is essential to preserve the viability and potency of live probiotics. Scientific investigations have pinpointed cases where inadequate quality control procedures resulted in a reduction in the quantity of viable probiotic cells in products, thereby compromising their efficacy.
Clinical trial methodology	Clinical trial methodology: The structure of clinical trials assessing the effectiveness of probiotics can significantly influence the results. Studies have

(*Continued*)

TABLE 7.1 (Continued)

	underscored the significance of meticulously planned, randomized controlled trials with adequate sample sizes, control groups, and outcome measures to precisely evaluate the impact of live probiotics on health outcomes.
Lack of compatibility with existing gut microbiota	The intricate composition of the gut microbiota, a diverse community of microorganisms, is essential for maintaining gastrointestinal health. Certain probiotic strains may face challenges in establishing themselves within the existing microbial balance in the gut, hindering their ability to effectively exert their beneficial properties.
Microbiological contamination	The presence of harmful microorganisms in probiotic products poses a potential risk to consumer health. Contamination events during various stages of production, including manufacturing, packaging, and storage, can jeopardize the safety and efficacy of probiotics. Vigilant monitoring and strict quality control measures are essential to mitigate the risk of harmful agent presence in probiotic formulations.
Publication bias	Publication bias, where studies with positive results are more likely to be published than those with negative or neutral findings, can also influence the perception of probiotic effectiveness. Scientific researchers have emphasized the importance of considering all available evidence, including unpublished studies, to avoid bias in interpreting the overall efficacy of probiotics.

7.2 Probiotic foods, dietary supplements, and drugs

7.2.1 Probiotic foods

Nowadays, consumers are becoming extra health-conscious about the nutritional value of food, due to the key role of food in keeping healthy and growing, and manufacturers emphasize the production of functional foods. The development of probiotic food formulations is indeed a key area of research for the future functional food market. The global market for probiotics was valued at about 58 billion US dollars in 2022 and was forecast to reach over 85 billion dollars by 2027 (Shahbandeh, 2023). This growth is indicative of the increasing interest and investment in functional foods and their potential health benefits. Probiotics are deliberated more as probiotics-fortified foods such as curd, yogurt, ice cream, and cereals, or health supplements. These are being promoted without adequate mention of formulation, efficacy, and consumer safety, particularly in fermented foods. Probiotics are naturally found in fermented dairy products and other foods like kimchi, kombucha, sourdough, pickles, and sauerkraut. Probiotics are recognized for their ability to promote digestive health by increasing beneficial gut bacteria. Commercial probiotic production commonly involves fermentation. Probiotic products should be standardized based on viable counts since probiotic viability is a practical measure of probiotic activity and is more important than the concentration of probiotics in the diet (Roy, 2005). Industries shall carry out shelf-life tests at the date of sale, and at the same time when the product leaves the factory, till the lowest counts of probiotics (10^7 CFU/mL) are reached or are maintained in the digestive system under different conditions, as shown in Fig. 7.2 (Holzapfel et al., 2001; Yamamura et al., 2023). These foods undergo a fermentation process that encourages the growth of beneficial bacteria, which then act as probiotics when consumed. Commercial probiotic production commonly involves fermentation (Tripathi and Giri, 2014). Examples of commercial probiotic products are represented in Table 7.2 Dairy products, which are effective carriers for delivering probiotics to humans. In countries like Japan, South Korea, Europe, North Africa, North, Central, and South America, the dairy industry utilizes bifidobacteria to produce fermented milk that is milder in taste compared to traditional yogurt. Fermented milk containing *B. longum* or *B. breve* has received "foods for specific health use

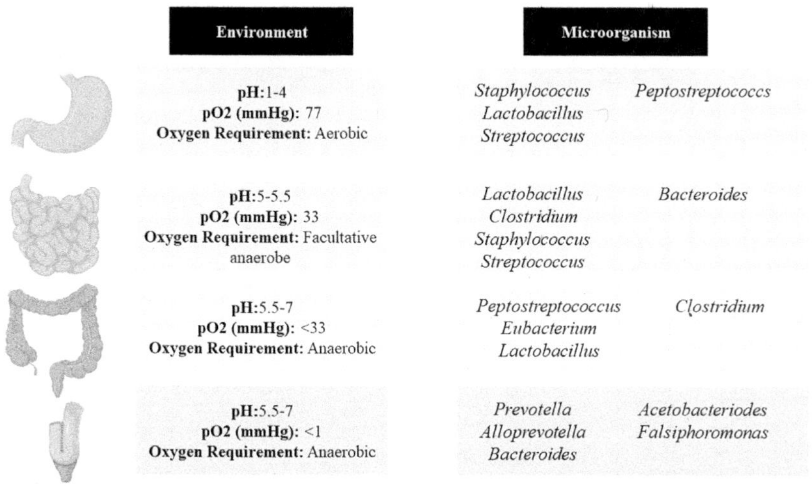

FIGURE 7.2 Various gut microbiota colonize specific regions of the gut (different regions of the gut have distinct pH levels, oxygen levels, and nutrient availability, which can influence the types of probiotics that thrive in each area).

TABLE 7.2 Some probiotic foods in the market.

Type of food	Type of bacteria	Storage	Form of probiotic	Effect	References
Mayonnaise sauce	Lactobacillus *acidophilus*	Storage for 30 days/40°C	Microencapsulated cells/free cells	The microencapsulation significantly increased the survival of probiotic compared to free state	Mohammadi et al. (2012)
Mayonnaise sauce	*L. acidophilus* and Lacticaseibacillus *casei*	91 days/4°C	Encapsulated cells/ free cells	The samples containing free probiotic cells had the highest loosing viability compared with encapsulated cells	Bigdelian and Razavi (2014)
Dry fermented sausage	*L. acidophilus, L. casei, L. sakei* and *Pediococcus pentosaceus*	21 days	Free cells	An autochthonous isolate of *Lactobacillus casei* adapted the best to the final products	Magra et al. (2021)
Sausages	*L. sakei, Staphylococcus equorum* and *Staphylococcus succinus*	–	Free cells	Improved safety compared with the traditional natural fermentation of sausages, by inhibiting the pathogen Listeria monocytogenes, decreasing the level of biogenic amines, and limiting fatty acid and cholesterol oxidation do not affect the typical sensory quality of the traditional sausages	Talon et al. (2008)
Probiotic beet juice	*L. acidophilus, L. casei, L. delbrueckii, L. plantarum*	4 weeks/4°C	Free cells	*L. acidophilus* and *L. plantarum produced* a greater amount of lactic acid than other cultures and reduced the pH of fermented beet juice	Yoon et al. (2005)

Baby food purees such as ahuyama, carrot, and figs	*B. coagulans*	Ahuyama: 90 days/ carrot: 30 days	Free cells	Baby food purees could have a functional effect as seen from the count of *B. coagulans*	Marín et al. (2020)
Gluten-free sorghum bread	*L. acidophilus* and *L. plantarum*	–	Encapsulated cells	Encapsulation increased water absorption (from 56% to 59%), dough development time, and farinograph number, and decreased the softening degree and dough stability time-increased the survival of probiotics during baking and storage-delay staling of bread	Ghasemi et al. (2022)
Pickled vegetable	*L. casei* *L. rhnmnosus* *L. acidophilus, L. plantarum* *Pediococcus acidilactici, L. plantarum* *Lactobacillus peniusus*	63 days/ 4°C	Free	The acid-resistant strains had slower growth rates than the acid-sensitive strains	Fan et al. (2017)
Bread	*L. acidophilus* and *L. rhamnosus*	–	Free cells	Elevated levels of proteins, carbohydrates, and fats. Increased vitamin B_9	Azmi et al. (2023)

(Continued)

TABLE 7.2 (Continued)

Type of food	Type of bacteria	Storage	Form of probiotic	Effect	References
Cashew apple juice	*L. casei* NRRL B-442	42 days/ 4°C	Free cells	Viable cell counts were higher than 8.00 Log CFU/mL throughout the storage. The values of lightness, yellowness, and total color change increased, and the values of redness reduced along the fermentation and refrigerated storage period.	Pereira et al. (2011)
Passion fruit pulp	*L. reuteri*	Incubation at 30, 20 and 10°C/48 h	Encapsulated cells	*Lactobacillus reuteri* survival showed dependency on droplet sizes.	Monteiro et al. (2020)
Minas Frescal cheese	*L. casei Zhang*	21 d/ refrigerated storage	Free cells	*L. casei* Zhang population reaching 8.16 and 9.02 log cfu/g using the direct acidification and enzymatic coagulation protocols, respectively, negatively affected the sensory acceptance	Dantas et al. (2016)
Buffalo Ricotta cheese	*L. acidophilus* La-05	7°C	Free cells	The compositional, color, textural and sensorial parameters of control and probiotic Ricotta cheese were statistically nonsignificant	Sameer et al. (2020)

Wagashi cheese	*L. casei* 39, *L. rhamnosus*, *L. plantarum* and *E. faecium*	30 days/4°C	Free cells	Addition of probiotic strains did not affect protein content in contrast to fat, dry matter, pH, and titratable activity. Cheese containing probiotics had a significantly higher acceptability.	Anihouvi and Kesenkaş (2022)
Low-fat Feta cheese	*Streptococcus thermophilus* and *L. bulgaricus*, *Bifidobacterium* cultures (*Bifidobacterium bifidum* and *Bifidobacterium longum*), and a mix of them	30 days/4°C	Free cells	The mixed cultures improved the flavor, body, and texture of cheese	Hamdy et al. (2021)
Ice creams	*Saccharomyces boulardii* and *L. rhamnosus* GG	6 days/-18°C	Free cells	Both probiotic strains successfully survived during storage period, and their numbers were above 6.18 log CFU/g at final date of storage. The rheological parameters of ice cream products improved with the coinoculation of *S. boulardii* and *L. rhamnosus*.	Goktas et al. (2022)
Chocolate	*Bifidobacterium lactis* HN019, *L. acidophilus* NCFM and *L. rhamnosus* HN001	2°C/6 months	Free cells	The survival of these strains was above 90%, with a viable cell count of about 8.1 log(CFU/g) during storage	Zarić et al. (2016)

(FOSHU)" approval in Japan, defined as processed foods that support specific bodily functions and are nutritious (Ono and Ono, 2015).

Yogurt has long been recognized as a beneficial product for consumers, with traditional fermentation bacteria like *L. bulgaricus* and *S. thermophilus* not being part of the gut flora, lacking bile acid resistance, and not surviving gut passage. However, these bacteria may still have positive effects through fermentation metabolites, inhibiting pathogens, or aiding lactose digestion (Ejtahed et al., 2012; Khalili et al., 2020; Asgharian et al., 2020; Nyanzi et al., 2021; Rad et al., 2019; Homayouni-Rad et al., 2016).

Labneh, also known as laban or strained yogurt, is a popular Middle Eastern dairy product made by straining yogurt to remove the whey, resulting (Homayouni et al., 2018) in a thicker, creamier consistency. It has a tangy flavor and is often enjoyed as a spread, dip, or condiment in Middle Eastern cuisine. It is also seen as a promising carrier for probiotics due to its higher total solids content, making it a suitable matrix for probiotic incorporation. However, various process factors such as pH, redox potential, probiotic inoculation levels, and the use of flavoring agents can impact the viability of probiotics in labneh (Kebary et al., 2021; Mehanna et al., 2021).

Cheese has become a top product in dairy marketing and is increasingly recognized as an effective carrier for probiotic microorganisms. The buffer capacity of cheese helps protect probiotics from the acidic environment of the gastrointestinal tract (GIT), and the dense matrix and high-fat content provide additional protection in the stomach. Fresh cheese, such as cottage cheese, is particularly suitable for probiotic delivery due to its unripened nature, refrigerated storage, and shorter shelf life compared to ripened cheeses (Heller et al., 2003; Ansari et al., 2020; Karimi et al., 2011; Homayouni et al., 2012).

Ice cream, a popular dairy product, is also considered a good carrier for probiotics. Its neutral pH and high-fat and solids content help shield probiotics from the negative effects of freezing. Studies have shown that probiotic strains like *L. acidophilus* and *L. rhamnosus* can maintain stability during cold storage at -20°C for up to two months. However, the viability of probiotics in dairy frozen yogurt and fermented ice cream with a pH below 4.5 may be lower compared to ice cream (Homayouni et al., 2012, 2018; Pimentel et al., 2022).

The survival of probiotic bacteria in fermented dairy products depends on factors such as the strains used, species interactions, culture conditions, medium composition, acidity, solids content, nutrient availability, growth promoters/inhibitors, sugar concentration, oxygen levels (especially for Bifidobacterium sp.), inoculation level, temperature, fermentation time, and storage conditions (Gao et al., 2021; Farahmand et al., 2021). Low pH in fermented milk can negatively impact probiotic viability. Studies have shown that *L. acidophilus* declines rapidly at pH 2.0 but remains stable at pH 4.0 or above. Lactobacillus strains generally exhibit good stability during storage, while Bif. Lactis LAFTI B94 cell counts may decrease (Hood and Zoitola,

1988; Shah et al., 1995). Though dairy products are the primary source of probiotics, nondairy foods offer unique advantages over dairy products. These include catering to the needs of vegetarians, avoiding allergens found in dairy, offering low-cholesterol options for those with cardiovascular or obesity issues, accommodating diverse dietary preferences, and enhancing overall nutritional value (Min et al., 2019; Rivera-Espinoza and Gallardo-Navarro, 2010). Probiotics assist in the digestion and breakdown of lactose by releasing β-galactosidase in the small intestine. Consuming sufficient probiotics may help reduce the symptoms of lactose intolerance, although their effectiveness is influenced by the lactose content and the probiotic cell count in the product (Oak and Jha, 2019). Hence, nondairy probiotic foods could serve as a beneficial treatment option for individuals with lactose intolerance.

The majority of nondairy probiotics found in markets are typically derived from fruit juices. In addition to the high sugar content in fruit juices, the presence of prebiotic fibers in fruits can also support the growth and activity of probiotic bacteria in the gut (Dey, 2018; Luckow and Delahunty, 2004). Prebiotics are nondigestible food components that promote the growth of beneficial bacteria in the gut, which can enhance the overall health benefits of probiotic fruit products. Organic acids can elevate the levels of dissociated forms, leading to a more acidic environment with intrinsic antibacterial properties that can impact the survival of probiotic bacteria. Studies by Patel (2017) and Fernandes Pereira and Rodrigues (2018) have highlighted the detrimental effects of accumulated organic acids on probiotic viability. Furthermore, multiple researchers have noted a decline in probiotic viability when stored at cold temperatures (4°C) (Ferdousi et al., 2013; Călinoiu et al., 2016).

Moreover, certain strains of lactic acid bacteria, such as *L. acidophilus* CICC20710, *Lb. casei, Lb. plantarum, Lb. rhamnosus* GR-1, *Lb. reuteri*, and *Bifidobacterium animalis* ssp. *Lactis*, BB-12, have been used in the production of vegetable-based probiotic foods as an alternative to dairy products. Various vegetables can be either combined or used individually to create probiotic products (Mojikon et al., 2022; Rasika et al., 2021). Similar to fruit-based items, studies have shown a decrease in probiotic viability during cold storage in vegetable-based probiotic products (Do and Fan, 2019; Tamminen, 2013). A study reported that the viable count of the cabbage juice fermented by *Lactobacillus casei* 431 remained at 10^7 CFU/ml during 14 days of cold storage despite initially high cell viability of 109 CFU/ml postfermentation (Dimitrovski et al., 2016). Furthermore, cereal and legume-based probiotic food products are becoming increasingly popular due to their potential health benefits. These products combine the nutritional benefits of cereals and legumes with the beneficial effects of probiotics, providing consumers with a convenient way to incorporate probiotics into their diet. Cereals and legumes are rich sources of vitamins, minerals, fatty acids, amino acids, protein, carbohydrates, resistant starch, water-soluble fiber (arabinoxylan and β-

glucan), and oligosaccharides (fructo- and galacto-oligosaccharides) (Cichońska and Ziarno, 2022; Soltani et al., 2018). Barley, finger millet, maize, millet flour, millet malt, oat, rice, sorghum, soybeans, and wheat flour can be utilized to create cereal and legume-based probiotic products. Additionally, cereal and cereal components offer an effective means to incorporate probiotics, prebiotics, and fiber into the human diet. Whole grains contain antioxidants, phenolic compounds, sterols, phytic acid, and phytoestrogens as phytochemicals that could increase probiotic viability (Oh et al., 2021; Rezende et al., 2021; Arslan-Tontul et al., 2022). In addition, cereal fermentation creates an optimal pH environment for an enzymatic breakdown of phytate, releasing essential minerals like calcium, iron, zinc, and manganese (important for LAB growth) (Fekri et al., 2020). This process can be seen as a cost-effective way to produce a nutrient-rich substrate that supports the proliferation of beneficial microbes. Traditional nondairy fermented beverages, commonly nonalcoholic, are made worldwide using cereal as the main ingredient. Within cereal grains, dietary fiber and phenolic compounds are linked to polysaccharides via ester bonds, which can be hydrolyzed through fermentation to release phenolic acids like ferulic acid (Valero-Cases et al., 2020).

Overall, maintaining high cell viability throughout transit, storage, and shelf life is essential for utilizing probiotics in the food industry. The design of probiotic food formulations is more intricate compared to pharmaceuticals, as it involves ensuring probiotic stability during processing, storage, and reaching appropriate sensory attributes. This is because probiotics can potentially alter the sensory characteristics of foods. Additionally, probiotic foods may face more challenges from a regulatory and legislative perspective compared to supplements. Cell density assessment and contamination checks are performed to monitor probiotic growth during fermentation, typically maintaining a temperature of 37°C. After fermentation, the concentrated cell suspension is spray-dried or freeze-dried to manufacture a dry powder probiotic that may be blended with food or made in capsule or tablet form. High cell density and viability are crucial for commercial viability during the fermentation process. Some materials, such as cryoprotectants and lyoprotectants, are added before freeze- and spray-drying to preserve bacterial viability. Studies have shown that freeze-drying with lactulose as a cryoprotectant can effectively preserve probiotic strains (Tripathi and Giri, 2014; Chiquetti et al., 2016). Furthermore, to overcome the limitation of the application of live probiotics, numerous research studies are underway to enhance the survival of probiotic cultures. These approaches include exposing probiotic cultures to sublethal stresses before use, adding micronutrients like peptides and amino acids, genetic modifications, immobilization, two-step fermentations, utilizing oxygen-impermeable containers, and encapsulation. Encapsulation, defined as "trap, target, and transport," entrapping materials into thin films by chemical or mechanical means, is another strategy to maintain the probiotic viability. The encapsulation process protects the probiotics from detrimental environmental factors (e.g., low pH, process stress, transport status, and digestive enzymes), increasing the survivability and functionality of the final product (Gharsallaoui et al., 2007; Nedovic et al., 2011; Zheng et al., 2020). Encapsulation has been shown to significantly improve the survival of live microorganisms during fermentation, drying, freezing, thawing, and heating (Nedovic et al., 2011; Steenson et al., 1987). Different microencapsulation techniques such as extrusion, freeze-drying, spray-drying, and vacuum-drying have commonly been practiced for the microencapsulation of probiotics based on their physicochemical properties as well as the wall component; however, some studies suggest that spray-drying is equally effective, providing a protective barrier for probiotics against harsh conditions (Rodrigues et al., 2020; Liliana and Vladimir, 2013).

7.2.2 Probiotic dietary supplements

To create an effective probiotic supplement, the probiotics must withstand the strict conditions of the GIT and remain viable at the targeted site. This consideration should guide supplement formulation, production, and storage. Typically, probiotic dosage forms are produced using lyophilized microorganisms, a preservation process commonly used for biological samples. However, the environmental stresses during freeze-drying can harm the cells' structure and viability. To mitigate these effects, protective agents are often added before freezing or freeze-drying. The resilience of probiotics in oral solid dosage forms like tablets, pellets, and capsules has also been studied to develop a probiotic supplement (Homayoni Rad et al., 2016; Abdollahpour et al., 2023; Vaghef-Mehrabany et al., 2014).

7.2.3 Probiotic drugs

Formulated pharmaceutical probiotic products should have the capability to deliver a high concentration of viable probiotics to the GIT without significant losses during storage. Some research suggests that probiotics may have potential therapeutic benefits for certain medical disorders, such as irritable bowel syndrome, inflammatory bowel disease, and antibiotic-associated diarrhea. In these cases, probiotics could be considered a form of drug therapy. It is important to note that the application of probiotics as a drug should be approached with caution and under the guidance of a healthcare professional. More research is needed to comprehend the potential benefits and risks of using probiotics as a drug for specific medical conditions (Nikolaeva et al., 2020).

7.3 Manufacture of the live biotherapeutic products

The manufacturing process involves culture expansion, propagation, harvest, as well as purification, including all critical inputs and outputs controlling the process. Key process inputs for products containing microorganisms include propagation settings and controls during harvest. Product stability is influenced by buffer composition and storage temperatures. Process progress often happens in tandem with preclinical and clinical trials, where clinical data confirms the effectiveness of the established processes. Final process changes in clinical development may necessitate additional studies to ensure that product quality is not adversely affected. Process-output quantities, including growth monitoring, pH change, and purity assessment, ensure consistent product quality. The yield from the harvest can be quantified either by mass or by the count of viable cells per unit volume. Harvest and purification steps may be combined, with yields typically decreasing with successive purification steps. After achieving the desired purity, stabilizers can be included in the final suspension (Cordaillat-Simmons et al., 2020).

7.3.1 Biotherapeutic products formulation

Live biotherapeutics contain the active ingredient of harvested and concentrated live organisms, which are formulated into a convenient dosage form to ensure accurate dosage content and quality. Concentrated harvests may be adjusted by dilution, or dried preparations may be blended with stabilizers or fillers. Liquid preparations for freezing may be stabilized with cryoprotectants. The dosage form's uniformity is guaranteed by monitoring weights and volumes during the filling or compression process. Tablets and capsules can be coated with enteric coatings to ensure that they pass through the stomach and dissolve in the intestine. Additionally, powdered products may come with prepackaged buffered reconstituting solutions (Trinchieri et al., 2017).

7.3.2 Biotherapeutic products quality control and consistency

To ensure minimal adjustments and maintain product quality, the process should be designed with this in mind. Any discrepancies from the design must be documented for assessment during product-release decisions. Master production and control records (MCPRs) are essential in preventing deviations, as they outline each manufacturing step, necessary inputs, expected output ranges, and control of the manufacturing environment. Product integrity and quality are maintained by defining conditions and times for in-process steps and specifying storage conditions. A batch record, a lot- or batch-specific executed MCPR, details all component materials and test results, ensuring compliance with the manufacturer's specifications. It also provides traceability for equipment use, maintenance, cleaning, processing locations, and environmental conditions. Notably, pharmaceutical products should not use potable tap water, and instead, purified water is recommended for sterilized propagation media and formulation. For nonparenteral dosage forms, sterile purified water should be used (Georgieva et al., 2014).

7.3.3 Biotherapeutic products contamination and control

Biological product manufacturing relies on aseptic processing to prevent degradation by contaminating microorganisms, as these products are not typically heat sterilized due to heat sensitivity. The entire process must be aseptic, with controlled bioburden. Water and air are managed for bioburden and nonviable particulates, depending on the manufacturing stage and final product use. Component materials may need sterilization and bioburden monitoring. Sterilized disposable equipment can be used, or equipment that comes into contact with products should be sterilized. Open manufacturing areas are monitored for bioburden to prevent cross-contamination with other products. Cleaning and disinfection processes reduce bioburden and remove previous products, and multiple products should not be manufactured simultaneously in the same areas (Korčok, 2020).

7.4 Specifications for probiotic products

Specifications are essential criteria that active ingredients and drug products must meet before being approved for human use. They depend on the testing methods used to obtain results. These criteria should be set for critical materials used in drug and biological manufacturing, as well as for intermediates and final products. Product-release specifications are selected to ensure product quality, rather than to characterize the product, as shown in Fig. 7.3.

7.4.1 Selection criteria and characteristics of probiotics

The section addresses the standards that should be followed when evaluating bacteria for use as probiotics, as stated in a joint FAO and WHO report from 2002. It underscores the necessity of particular selection standards and inherent physicochemical properties, along with protocols for standardizing processes when evaluating probiotics for food use and verifying health claims through human trials.

The good sensory qualities, phage resistance, viability, and stability in the product and during storage are among the technological factors that must be taken into account when choosing probiotics. Probiotic resistance to bile and gastric acidity, adherence to mucus and/or human epithelial cells, bile salt hydrolase activity, decreased pathogen adherence to surfaces, and

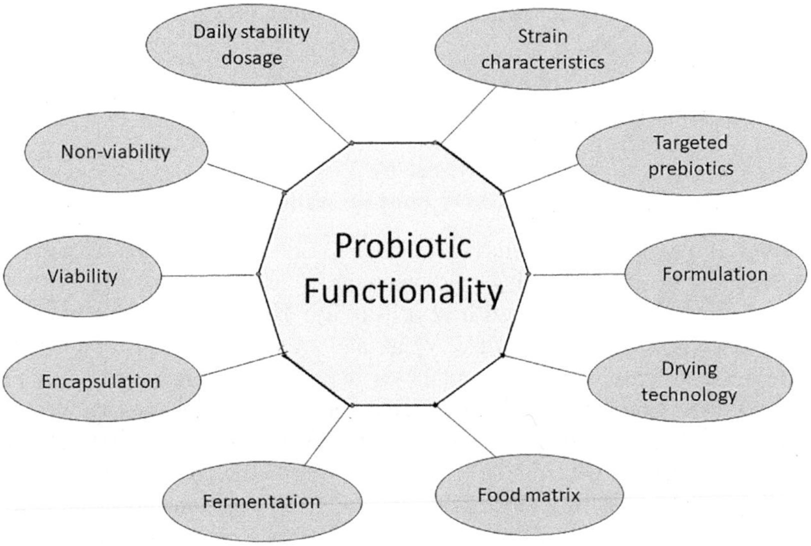

FIGURE 7.3 Technological factors influencing the functionality of probiotics (technological factors can significantly influence the functionality of probiotics, which are live microorganisms that confer health benefits).

antimicrobial activity against particular pathogenic bacteria are other common in vitro tests.

Additionally, the report recommends conducting animal trials, when possible and appropriate, before human studies and highlights the rarity of systemic infections attributed to probiotics, particularly in patients with serious underlying medical conditions. Demonstrating the lack of infectivity in immunocompromised animals is also suggested to emphasize the safety profile of the probiotic strain (Shewale et al., 2014).

7.4.2 Quality control of the harvested live biotherapeutic

The harvested live organism serves as the active pharmaceutical ingredient and is tested against acceptance criteria to confirm the quality of the ultimate product. Specific identity testing is conducted to approve the strain and its critical features, including verification of products of transgene expression for recombinant strains. Microbiological purity is a critical aspect of ensuring the safety and quality of pharmaceutical products. It involves evaluating the absence of harmful microorganisms in the purified active pharmaceutical ingredient and throughout high-risk manufacturing processes. Various testing methods, such as the US Pharmacopoeia microbial limit tests, are used to confirm the absence of foreign organisms. These tests are more precise and targeted compared to those conducted after dilution for the final dosage form.

Techniques are carefully chosen to detect potential contaminants and ensure the product meets stringent purity standards. Monitoring and controlling extraneous bioburden at key stages of the manufacturing process is crucial, and the number of viable organisms is determined through methods like sample dilution and colony counts. This rigorous approach helps to maintain microbiological purity and safeguard the integrity of pharmaceutical products. The yield of the active ingredient is assessed as colony-forming units per product unit, guiding formulation and serving as a monitor of process consistency (Zommiti et al., 2020).

7.4.3 Biotherapeutic products dosage control

Biologics regulations require the evaluation of several product qualities for each lot, including potency, general safety, sterility, purity as well as identity. Potency is measured by the specific ability of the product to produce a given result and may involve a quantitative measure such as colony-forming units per dose unit. General safety testing involves observing the effects of intraperitoneal injection on mice and guinea pigs for 7 days to detect extraneous toxic contaminants. Sterility testing may be modified for live biotherapeutics and can be waived for orally administered products. Purity testing is performed for the dosage form to eliminate extraneous contamination, while identity testing ensures differentiation from similar products produced in the same manufacturing areas (Baral et al., 2021).

7.4.4 Biotherapeutic products stability

Stability studies ensure product safety and effectiveness throughout its shelf life and GIT. Dosage forms are tested in intended containers at recommended temperatures. The FDA reviews data supporting recommended usage and administration. Key points include studying the stability of reconstituted products, presenting stability data for products combined with consumer-supplied matrices, and instructing consumers on storage limits postreconstitution. A stability protocol outlines testing types and schedules, including potency, purity, moisture, and bioburden assays (Gueimonde and Sánchez, 2012).

7.4.5 Biotherapeutic products safety

Safety considerations encompass the determination of antibiotic resistance profiles, evaluation of metabolic activities and side effects in human trials, epidemiological surveillance of adverse effects in consumers, toxin production, and hemolytic activity. The deliberate use of microorganisms in human applications, particularly in food and dietary supplements, necessitates a thorough safety evaluation, both from a moral and legal standpoint. In the

European Union, for instance, regulations such as EU regulation (EC) No 178/2002 oversee the safety of introducing microorganisms into the food chain. To ensure safety, guidelines and recommendations have been developed to assess the safety of microorganisms intended for human use.

One critical aspect of safety evaluation is the taxonomic identification of the probiotic strain, requiring the use of advanced molecular biology, biochemistry, and physiology methods. While classical approaches based on observable characteristics have been historically used, molecular techniques, such as whole-genome sequencing, are now becoming the standard for strain identification (Kim et al., 2020). Safety assessment involves a thorough examination of the strain's properties to ensure it does not produce harmful metabolites, cause infections, trigger excessive immune responses, or transfer genes, such as antimicrobial resistance genes. The safety of the strain is contingent on its designated purpose, method of delivery, extent of exposure, and the health condition of the target population (Sanders et al., 2014). Guidelines provided by agencies like the European Food Safety Authority (EFSA) offer recommendations for genome comparative analyzes and the assessment of antimicrobial resistance. These guidelines include the examination of virulence factors, deleterious metabolic activities, and the potential for antimicrobial resistance transfer (Roe et al., 2022). Furthermore, a probiotic strain with a well-documented history of safe use in the human population is considered to have a strong safety profile (Ebner et al., 2014; Thakur et al., 2016). Organizations like the International Dairy Federation keep records of microorganisms that have been safely used in food. In the European Union, the Qualified Presumption of Safety (QPS) approach streamlines safety evaluations for microorganisms needing premarket approval. Ultimately, the safety assessment of a potential probiotic strain follows a standard procedure, but the exact requirements may vary based on the intended application and target consumer base of the final probiotic product. These assessments are crucial to ensure the safety of microorganisms deliberately introduced into human applications and the food chain.

7.5 Adverse effects related to probiotics

Side effects of probiotics include systemic infections, deleterious metabolic activities, excessive immune stimulation in susceptible individuals, and gene transfer.

7.5.1 Risk of infections

In an ideal scenario, the occurrence of side effects from probiotics should not surpass the placebo effect (Marteau and Lerman, 2001). It was suggested that complete risk elimination is not feasible, and acknowledging the potential for probiotics to have both positive effects and side effects is crucial. GIT discomfort, including symptoms like bloating, constipation, flatulence, dyspepsia, nausea, and abdominal discomfort, is the most commonly reported mild side effect of probiotics (Sotoudegan et al., 2019). However, these symptoms are unlikely to significantly impact the safety of the probiotic. The incidence of diarrhea associated with probiotic treatment varied across studies, as indicated in the AHRQ Safety Report (Hempel et al., 2011). According to the studies, it was found that Lactobacillus spp. was responsible for infectious complications following probiotic use. Among these, 14 cases were attributed to *L. rhamnosus*, 4 to *L. acidophilus*, 3 to *L. paracasei*, 1 to *L. casei*, and 4 cases could not be identified at the species level. Additionally, out of the 12 patients who experienced infectious complications due to *Bifidobacterium* spp., 10 were newborns (9 of whom were preterm). Furthermore, a case series reported 4 instances of bacteremia caused by *B. subtilis* linked to the use of probiotics comprising these microorganisms, which were administered to diminish the incidence of diarrhea associated with enteral feeding (Costa et al., 2018). Clinically significant side effects can be categorized into various groups, including pathogenicity, infectivity, and virulence factors, which encompass toxicity, metabolic activity, and the inherent characteristics of the microorganisms (Anadón et al., 2016).

7.5.2 Probiotics safety in healthy populations

The GIT, the principal target for probiotic activity, is effectively separated from the bloodstream and other organs by a barrier that serves to protect the host from potential infections initiated by ingested or colonizing microbes. The breakdown of the GI barrier or deficiencies in the immune system can increase the likelihood of infection. In healthy populations, probiotics are believed to fortify the GIT barrier and bolster immune responses (Salminen et al., 2004). The risk of probiotic-related infections is exceedingly low, particularly in healthy populations. Salminen et al. (2002) reported data on Lactobacillus bacteremia in Finland, showing no increase in infections despite the growing consumption from 1990 to 2000 (Salminen et al., 2002). A study conducted by the same authors in 2004 found that lactobacilli were detected in just 0.24% of positive blood cultures received by the laboratory (Salminen et al., 2004). Bernardeau et al. (2006) conducted a study on probiotic consumption in France and estimated the risk of Lactobacillus infection as very low, at one case per 10 million people over a century (Bernardeau et al., 2006). Borriello et al. (2003) suggested that the risk of probiotic *Lactobacillus* was "unequivocally negligible" (Borriello et al., 2003). One of the key documents evaluating the safety of probiotics is the report titled "Safety of Probiotics to Lower Risk and Prevent or Treat Disease," commissioned by the Agency for Healthcare Research and Quality (AHRQ) from the US Department of Health and Human Services, which was released in April 2011 (Hempel et al., 2011). Based on reports, randomized controlled trials (RCTs) showed no statistically significant increase in the relative risk of adverse effects of probiotics (Zielińska et al., 2018). However, the authors concluded that rare adverse events are difficult to assess.

7.5.3 Probiotic safety in ill and immunocompromised patients

The risk group people, such as neonates, premature infants, pregnant women, and elderly patients, are at increased risk of infections. Immunocompromised individuals, including those receiving immunosuppressive treatment, undergoing chemotherapy, or postorgan transplantation, are particularly sensitive to contamination (Oggioni et al., 1998). Probiotic supplementation can provide benefits for immunocompromised patients, but it also carries a higher risk of adverse effects in these vulnerable populations due to the disrupted balance between intestinal microflora and the immune system (Happel et al., 2018). Dysfunctional intestinal barriers in vulnerable populations may promote infection, leading to sepsis or inflammation. Central venous catheters, commonly used for parenteral nutrition and postpyloric feeding, have been identified as a potential risk factor for probiotic infection. While certain strains of probiotics have been safely administered to patients who are receiving nutritional support, certain products may increase the risk of complications in high-risk populations, such as transplant patients and those with pancreatitis (Sanders, 2011). There have been worries about the possibility of immune system overstimulation in susceptible individuals, but this hypothetical response has not been verified in human subjects. Risk factors for probiotic use include an immune-compromised state, impaired intestinal barrier function, and the presence of a central venous catheter. Appropriate design, transmission, and continuous supervision of clinical trials are essential to ensure the safety and effectiveness of probiotic use in vulnerable populations (Doron and Snydman, 2015). Probiotic administration concurrent with antibiotic treatment has shown potential preventive effects, especially in reducing antibiotic-associated diarrhea (AAD). However, conflicting results exist regarding the effectiveness of probiotics in preventing *Clostridium difficile* infection (CDI). The use of probiotics in high-risk populations remains a topic of ongoing research and debate due to various and conflicting study results (Goodman et al., 2021). Additionally, there have been case reports of probiotic bacteria leading to endocarditis. While trial results do not point out that the use of probiotics raises the risk of endocarditis, it is important to consider the potential risks associated with probiotic use, especially in vulnerable populations. Hygiene standards, including hand-washing and sterilization, should be strictly followed to prevent contamination in hospitals (Boumis et al., 2018).

7.5.4 Deleterious metabolic activities

Another vital safety consideration is the metabolic activity in the food matrix and the intestine following consumption. Lactobacillus spp. is often associated with d-lactic acidosis, principally in patients suffering from short bowel syndrome (Kowlgi and Chhabra, 2015). Some cases of d-lactic acidosis have been reported in the literature. When the small bowel is heavily colonized with bacteria, high numbers of microorganisms can prompt diarrhea and intestinal lesions, especially through the deconjugation and dehydroxylation of bile salts. Ingesting certain probiotics can lead to high concentrations of microorganisms in the small bowel, similar to those indicated during small bowel bacterial overgrowth. Studies have shown that fermented dairy products containing *L. acidophilus* and *Bifidobacterium* spp. can transform conjugated primary bile salts into toxin-free secondary bile salts during digestion in the small bowel (He et al., 2010; Seifert et al., 2019). Firmicutes, including bacteria such as 7α-dehydroxylate, convert primary bile acids to toxic secondary bile acids. These secondary bile salts can be harmful, potentially exhibiting carcinogenic effects by promoting cell proliferation or acting as promoters of carcinogenesis. Various bacterial enzymes, such as reductases and hydrolases, play a crucial role in metabolizing carcinogens and other toxic substances in the intestines (Ridlon et al., 2014). Lactic bacteria are enzymatically active and display proteolytic, lipolytic, and/or saccharolytic properties. Lactic acid bacteria must not exhibit harmful biochemical activities associated with intestinal diseases and the generation of carcinogens and tumor promoters (Kieliszek et al., 2021). Enterococcus, known for its β-hemolysin/bacteriocin or cytolysin, is considered a confirmed virulence factor. Proteases are also believed to be involved in enterococcal pathology. A sepsis outbreak resembling a toxic shock-like syndrome in humans was linked to a particular strain of *Enterococcus faecium* (Said et al., 2021; Li et al., 2023). Furthermore, Bacillus species, particularly *B. cereus*, are frequently associated with food poisoning incidents. Some Bacillus species are used in the animal feed industry, where they improve zootechnical parameters and livestock health. Nevertheless, in human microorganisms, Lactobacilli and Bifidobacteria are more dominant. Toxin production has been observed among Bacillus species, especially strains of *B. licheniformis* (Tewari and Abdullah, 2015; Oliveira et al., 2023). Various strais of the same probiotic may also produce contrasting effects, and outcomes derived from animal models may not align with those from human studies. Additional research is necessary to more accurately discern the advantages and possible drawbacks of probiotic use.

7.5.5 Gene transfer

LABs are highly adaptable, as evidenced by their relatively small genomes, containing a variety of genes indicating both gene loss and acquisition. Probiotic strains may exhibit antibiotic resistance and possess virulence factors. The potential transfer of resistance genes in probiotics raises concerns about the spread of antibiotic resistance to more virulent bacteria. LAB often harbor plasmids containing antibiotic resistance determinants, and the presence and transfer of resistance in bacteria is critical for controlling antibiotic resistance. Intrinsic and attained resistance in bacteria present different risks of horizontal spread, with acquired resistance mediated by additional genes posing a higher risk. The transfer of resistance genes from beneficial bacteria to pathogenic bacteria is a significant concern (Van Reenen and Dicks, 2011; Shenderov, 2013). Antibiotic resistance is a global health issue, with natural gene transfer occurring through various mechanisms such as transformation, transduction, and conjugation (Gueimonde et al., 2013). Plasmids in LAB play a key role in conferring bacterial characteristics, including antibiotic resistance. The transfer of resistance plasmids has been observed in various bacterial genera. In LAB, conjugative transfer systems are frequently observed, and plasmids encoding antibiotic resistance have been identified in *Enterococcus*, *Staphylococcus*, *Lactobacillus*, and *Bacillus*. Intrinsic resistance in LAB includes vancomycin-resistant phenotypes and resistance to mupirocin and aminoglycosides in *Bifidobacterium* strains. *Lactobacillus* bacteria are generally vulnerable to β-lactamase and penicillin but typically show resistance to cephalosporins (Warnecke et al., 2007; Fatahi-Bafghi et al., 2022; Detection of oxalyl-CoA decarboxylase (oxc) and formyl-CoA transferase (frc) genes in novel probiotic isolates capable of oxalate degradation in vitro, 2024). They are sensitive to macrolides, chloramphenicol, lincosamides, and tetracycline but often resistant to aminoglycosides. Several genes responsible for atypical antibiotic resistance properties have been identified in *L. acidophilus*, *L. delbrueckii* subsp. *bulgaricus*, *L. johnsonii*, *L. reuteri*, and *L. plantarum*. Tetracycline resistance genes are common in lactobacilli, with 11 different genes found so far (Gueimonde et al., 2013; Detection of oxalyl-CoA decarboxylase (oxc) and formyl-CoA transferase (frc) genes in novel probiotic isolates capable of oxalate degradation in vitro, 2024). *Bifidobacterium* species are typically susceptible to macrolides, vancomycin, beta-lactams, and spectinomycin, with only rare strains showing resistance to streptomycin. Some *Bifidobacterium* strains possess mutated genes responsible for low erythromycin susceptibility. Tetracycline resistance genes have been reported in several *Bifidobacterium* species, including *B. longum* subsp. *infantis, B. longum* subsp. *longum, B. breve, B. animalis* subsp. *lactis, B. pseudocatenulatum, B. bifidum*, and *B. thermophilum*, with the tet(W) gene detected at high frequencies in *B. longum* and *B. animalis subsp. Lactis* strains (Egervärn et al., 2010). Some Bacillus species possess macrolide-resistance genes, plasmid-encoded tetracycline resistance genes, and cfr-like genes encoding ribosome methyltransferases (Wash et al., 2022). The spread of vancomycin resistance in enterococcal strains poses a significant clinical challenge, highlighting the need for a thorough evaluation of the safety of *Enterococcus faecium* strains employed as probiotics. Virulence factors in different Enterococcus species found in food vary by strain, with limited research on the functional and safety characteristics of select probiotic strains (Lebreton et al., 2011). The *E. faecalis* strain Symbioflor 1 genome lacks genes associated with pathogenicity, pointing towards gene loss (Bourgogne et al., 2008). Antibiotic resistance has been detected in lactic acid bacteria (LAB) obtained from diverse food sources, suggesting the potential for gene transfer between probiotics and other microorganisms. While there is no clinical proof of antimicrobial-resistant gene transfer, the issue of antibiotic resistance in bacteria poses a potential threat to public health. Bacteria present in food or deliberately introduced into food products may serve as a reservoir of antibiotic-resistant genetic elements (Zavišić et al., 2023). Fermented foods, especially those containing lactic acid bacteria, have high adaptability and produce resistance mechanisms, raising the risk of dynamic progress of antibiotic resistance in the human GIT. Therefore strains used as starter cultures or probiotics must be verified for resistance to antibiotics and their ability to transfer these genes.

7.6 Standards implementation

The adoption of definitions and standards for probiotics is crucial for ensuring the quality and efficacy of probiotic products. Probiotics are living microorganisms that, when consumed in sufficient quantities, offer a health advantage to the host. However, the lack of clear definitions and standards can lead to confusion and inconsistency in the probiotic market. Several organizations and regulatory bodies have established guidelines and criteria for probiotics. For example, the Food and Agriculture Organization of the United Nations (FAO) and the World Health Organization (WHO) have developed a definition of probiotics and guidelines for their evaluation and use. These guidelines emphasize the importance of strain-specificity, viability, and documented health benefits (Pineiro and Stanton, 2007). In addition, some countries have specific regulations and standards for probiotic products. The European Food Safety Authority (EFSA) has recognized a list of approved probiotic strains and health claims. In the United States, the Food and Drug Administration (FDA) regulates probiotic products as dietary supplements, and manufacturers are responsible for confirming the safety and quality of their products. Standardization of probiotics involves establishing criteria for strain identification, viability, stability, and health benefits. This can include methods for strain characterization, quality control, and clinical evaluation. Standardized methods for probiotic enumeration and viability assessment are also important for ensuring consistency and reliability in probiotic products (Degnan, 2008). Adoption of definitions and standards for probiotics can help consumers make informed choices and ensure that they are receiving high-quality, effective products. It also provides a framework for manufacturers to develop and market probiotic products responsibly.

7.7 Principles for clinical research and industrialization

The regulation of probiotics varies internationally. Probiotics can be marketed as ingredients in several regulatory categories, determined by intended use, formulation, route of administration, target consumers, and safety in the United States. Manufacturers are responsible for packaging, labeling, and maintaining scientific support for claims. Clinical studies are often required. When studied clinically, the regulatory category depends on the country's regulations. In the United States, drugs are classified according to their intended purposes, and clinical trials may mandate the submission of an Investigational New Drug (IND) application. Recommendations include open policy discussions, familiarity with regulatory requirements, and transparent guidance for researchers and manufacturers. Regulation of probiotics lacks international harmonization. In the United States, probiotics can be marketed as ingredients in various regulatory categories, including conventional foods, dietary supplements, foods for special dietary uses, drugs, and veterinary products. The product's regulatory category is determined by intended use, formulation, route of administration, target consumers, and safety, which in turn dictates the regulations imposed on clinical research, improvement, manufacturing, and marketing processes. Manufacturers are responsible for packaging, labeling, and marketing their products and ensuring scientific support for claims made on the product label and in promotional materials. Clinical studies are often required to support label and marketing claims. The regulatory classification of a probiotic being studied clinically is determined by the regulations of the country where the product is undergoing testing. In the United States, the legal definition of a drug is based on its intended use, and to conduct a clinical trial, it may be necessary to submit an Investigational New Drug (IND) application to the FDA. This process requires time, human resources, and financial resources. Recommendations from the workshop included the need for regulators to actively participate in policy discussions and the scientific debate, for funding agencies to become more familiar with regulatory requirements, and for federal agencies to provide appropriate guidance to researchers and manufacturers regarding the development process and research study design for probiotics (Arora et al., 2013).

7.8 Novel strategies for overcoming the side effects of probiotics

7.8.1 Application of postbiotics

Administering inactivated microbial cells or cell fractions, referred to as postbiotics, in adequate quantities, can offer health advantages to consumers, akin to probiotics. Postbiotics with distinct chemical compositions, excellent safety records, extended shelf life, resilience to mammalian digestive enzymes, and a rigorous gastrointestinal environment have shown no adverse effects on consumers (Rad et al., 2020b; Tsilingiri et al., 2012; Homayouni Rad et al., 2021). These soluble bioactive metabolites, which are secreted by live microbes or produced after bacterial lysis in the host intestinal environment, confer physiological health benefits by regulating cellular progress and metabolic pathways in the host (Teame et al., 2020; Sabahi et al., 2023; Khani et al., 2022; Abbasi et al., 2022; Rad et al., 2021a, 2021b; Homayouni-Rad et al., 2023). Additionally, by-products of probiotic microorganisms, which are released after bacterial lysis or secreted by live bacteria, such as short-chain fatty acids, bacteriocins, enzymes, ethanol, diacetyl, peptides, hydrogen peroxide, teichoic acids, peptidoglycan-derived peptides, endo- and exo-polysaccharides, cell surface proteins, vitamins, plasmalogens, and organic acids, are also mentioned as postbiotics. These products and by-products have been consistently reported as bioactives with various health-promoting activities (Tsilingiri and Rescigno, 2013; Wegh et al., 2019). Shenderov (2013) demonstrated that postbiotics with effective absorption, metabolism, distribution, and excretion capabilities can signal to numerous organs and tissues in the host and elicit multiple biological responses (Shenderov, 2013).

The balance of viable and nonviable microbes can vary significantly among different probiotic supplements, with some products containing more nonviable cells than viable ones. This suggests that the presumed beneficial effects of probiotics may be attributed to the presence of nonviable components. While probiotics have demonstrated health benefits, certain medical studies have highlighted potential risks associated with microbial food cultures. These risks include the presence of virulence factors in probiotic strains, the transfer of resistance genes within gut microbial populations, viability issues with probiotic species in products, varying colonization patterns, stimulation of inflammatory responses, translocation to tissues and blood, establishment of persistent colonies that disrupt normal microbiota colonization, and the potential to promote opportunistic infections in humans (such as meningitis, pneumonia, sepsis, abdominal abscesses, infective endocarditis, peritonitis, urological infections, and bacteremia), as well as rheumatic vascular diseases, especially in patients undergoing antibiotic treatment or with compromised immune systems (Rad et al., 2020a, 2021a; Homayouni Rad et al., 2021).

7.8.2 Application of prebiotics

The growth and proliferation of gut microbiota rely profoundly on the availability of essential nutrients. These microorganisms secrete low molecular weight metabolites that play a crucial role in promoting the growth, development, and reproduction of both themselves and other beneficial microorganisms. They also facilitate communication between cells and protect against stress factors (Roberfroid et al., 2010). Moreover, the use of prebiotics with similar health effects can help mitigate probiotic-associated risks.

Prebiotics are resistant to gastrointestinal enzymes, environmental stresses, food processing, and storage conditions, unlike probiotics, which are susceptible to these factors (Carlson et al., 2018; Ansari et al., 2020; Homayoni Rad et al., 2012). Additionally, prebiotics can be obtained inexpensively from several sources, making them available to industries and consumers. As crucial components in functional foods, prebiotics can be utilized by probiotic microorganisms in the intestines, leading to an increase in beneficial microbes in the gut. This, in turn, improves the consumer's health, reduces the risk of disease, and has been consistently demonstrated to regulate cholesterol levels, maintain hormonal balance, reduce anxiety, lower the risk of cardiovascular disease and obesity, improve digestion and gut health, enhance immune function, as well as reduce inflammatory and autoimmune responses (Carlson et al., 2018; Yang et al., 2023; Li et al., 2023; Rani et al., 2023). Consequently, foods or supplements containing prebiotic ingredients could enhance the growth, survival, and cultivation of endogenous beneficial microbial cells. Therefore the food and pharmaceutical industries could safely utilize post and prebiotics to mitigate probiotic-associated risks and provide therapeutic effects to consumers. Thus prebiotics and postbiotic components with effects comparable to live probiotic microbes should be deliberated for noncytotoxic physiological health benefits. However, postbiotics need more research to approve safe labeling.

7.9 Conclusion

In conclusion, the viability of probiotics in drugs, food, and supplements presents a complex and multifaceted challenge. While probiotics offer potential health benefits, there are concerns regarding their safety, efficacy, and long-term impact on the host. The failure aspect of viable probiotics highlights the need for a more comprehensive approach to promoting gut health and overall well-being. This includes exploring alternative strategies such as prebiotics and postbiotic components, which offer comparable physiological benefits without the associated risks of live probiotic microbes. Additionally, further research and development in the field of microbiota modulation are necessary to address the limitations and potential drawbacks of viable probiotics. By recognizing and addressing the failure aspect of viable probiotics, It will be easy to modulate the gut microbiota and pave the way for safer and more effective interventions in the realm of drugs, food, and supplements.

Conflict of interest

The authors declared no conflicts of interest.

References

Abbasi, A., Rad, A.H., Ghasempour, Z., Sabahi, S., Kafil, H.S., Hasannezhad, P., et al., 2022. The biological activities of postbiotics in gastrointestinal disorders. Critical Reviews in Food Science and Nutrition 62 (22), 5983–6004. Available at https://doi.org/10.1080/10408398.2021.1895061.

Abdollahpour, D., Homayouni-Rad, A., Nourizadeh, R., Hakimi, S., Mehrabi, E., 2023. The effect of probiotic supplementation on episiotomy wound healing among primiparous women: a triple-blind randomized clinical trial. BMC Complementary Medicine and Therapies 23 (1). Available at https://doi.org/10.1186/s12906-023-03980-3.

Anadón, A., Martínez-Larrañaga, M.R., Ares, I., Martínez, M.A., 2016. Interactions between Nutraceuticals/Nutrients and Therapeutic Drugs. Elsevier BV, pp. 855–874. Available at https://doi.org/10.1016/b978-0-12-802147-7.00060-7.

Anihouvi, E.S., Kesenkaş, H., 2022. Wagashi cheese: probiotic bacteria incorporation and significance on microbiological, physicochemical, functional and sensory properties during storage. LWT 155, 112933. Available at https://doi.org/10.1016/j.lwt.2021.112933.

Ansari, F., Pourjafar, H., Tabrizi, A., Homayouni, A., 2020. The effects of probiotics and prebiotics on mental disorders: a review on depression, anxiety, alzheimer, and autism spectrum disorders. Current Pharmaceutical Biotechnology 21 (7), 555–565. Available at https://doi.org/10.2174/1389201021666200107113812.

Arora, M., Sharma, S., Baldi, A., 2013. Comparative insight of regulatory guidelines for probiotics in USA, India and Malaysia: a critical review. International Journal of Biotechnology for Wellness Industries 2, 51.

Arslan-Tontul, S., Candal Uslu, C., Mutlu, C., Erbaş, M., 2022. Expected glycemic impact and probiotic stimulating effects of whole grain flours of buckwheat, quinoa, amaranth and chia. Journal of Food Science and Technology 59 (4), 1460–1467. Available at https://doi.org/10.1007/s13197-021-05156-8.

Asgharian, H., Homayouni-Rad, A., Mirghafourvand, M., Mohammad-Alizadeh-Charandabi, S., 2020. Effect of probiotic yoghurt on plasma glucose in overweight and obese pregnant women: a randomized controlled clinical trial. European Journal of Nutrition 59 (1), 205–215. Available at https://doi.org/10.1007/s00394-019-01900-1.

Azmi, S., Ahmed, J., Ramalingam, K., 2023. Nutritional enhancement of wheat bread by probiotic *Lactobacillus consortium*. Letters in Applied NanoBioScience 12 (4). Available at https://doi.org/10.33263/LIANBS124.101, http://nanobioletters.com/wp-content/uploads/2022/09/LIANBS124.101.pdf.

Baral, K.C., Bajracharya, R., Lee, S.H., Han, H.-K., 2021. Advancements in the pharmaceutical applications of probiotics: dosage forms and formulation technology. International Journal of Nanomedicine 16, 7535–7556. Available at https://doi.org/10.2147/ijn.s337427.

Bastani, P., Akbarzadeh, F., Homayouni, A., Javadi, M., Khalili, L., 2016a. Health benefits of probiotic consumption. Microbes in Food and Health. Springer International Publishing, Iran, pp. 163–184. Available at http://doi.org/10.1007/978-3-319-25277-3, https://doi.org/10.1007/978-3-319-25277-3_9.

Bastani, P., Homayouni, A., Norouzi-Panahi, L., Tondhoush, A., Norouzi, S., Mehrabany, E.V., et al., 2016b. The mechanisms of immune system regulation by probiotics in immune-related diseases. Journal of Pharmacy and Nutrition Sciences 6 (3), 105–111. Available at https://doi.org/10.6000/1927-5951.2016.06.03.4.

Bernardeau, M., Guguen, M., Vernoux, J.P., 2006. Beneficial lactobacilli in food and feed: long-term use, biodiversity and proposals for specific and realistic safety assessments. FEMS

Microbiology Reviews 30 (4), 487–513. Available at https://doi.org/10.1111/j.1574-6976.2006.00020.x.

Bigdelian, E., Razavi, S.H., 2014. Evaluation of survival rate and physicochemical properties of encapsulated bacteria in alginate and resistant starch in mayonnaise sauce. Journal of Bioprocessing and Biotechniques 4, 1.

Borriello, S.P., Hammes, W.P., Holzapfel, W., Marteau, P., Schrezenmeir, J., Vaara, M., et al., 2003. Safety of probiotics that contain lactobacilli or bifidobacteria. Clinical Infectious Diseases 36 (6), 775–780. Available at https://doi.org/10.1086/368080.

Boumis, E., Capone, A., Galati, V., Venditti, C., Petrosillo, N., 2018. Probiotics and infective endocarditis in patients with hereditary hemorrhagic telangiectasia: a clinical case and a review of the literature. BMC Infectious Diseases 18 (1). Available at https://doi.org/10.1186/s12879-018-2956-5, http://www.biomedcentral.com/bmcinfectdis/.

Bourgogne, A., Garsin, D.A., Qin, X., Singh, K.V., Sillanpaa, J., Yerrapragada, S., et al., 2008. Large scale variation in *Enterococcus faecalis* illustrated by the genome analysis of strain OG1RF. Genome Biology 9 (7). Available at https://doi.org/10.1186/gb-2008-9-7-r110.

Călinoiu, L.F., Vodnar, D.C., Precup, G., 2016. The probiotic bacteria viability under different conditions. Bulletin UASVM Food Science and Technology 73, 55–60.

Carlson, J.L., Erickson, J.M., Lloyd, B.B., Slavin, J.L., 2018. Health effects and sources of prebiotic dietary fiber. Current Developments in Nutrition 2 (3), nzy005. Available at https://doi.org/10.1093/cdn/nzy005.

Chiquetti, R.L., Castro, E.M., Valério, G.D., Bernini, L.J., Suguimoto, H.H., Santana, E.H.W., et al., 2016. Viability of the probiotic *Lactobacillus acidophilus* La-5 in ice cream: effect of lactose hydrolysis and overrun. International Food Research Journal 23 (6), 2631–2637. Available at http://www.ifrj.upm.edu.my/23%20(06)%202016/(45).pdf.

Cichońska, P., Ziarno, M., 2022. Legumes and legume-based beverages fermented with lactic acid bacteria as a potential carrier of probiotics and prebiotics. Microorganisms 10 (1), 91. Available at https://doi.org/10.3390/microorganisms10010091.

Cordaillat-Simmons, M., Rouanet, A., Pot, B., 2020. Live biotherapeutic products: the importance of a defined regulatory framework. Experimental and Molecular Medicine 52 (9), 1397–1406. Available at https://doi.org/10.1038/s12276-020-0437-6.

Costa, R.L., Moreira, J., Lorenzo, A., Lamas, C.C., 2018. Infectious complications following probiotic ingestion: a potentially underestimated problem? A systematic review of reports and case series. BMC Complementary and Alternative Medicine 18.

Dantas, A.B., Jesus, V.F., Silva, R., Almada, C.N., Esmerino, E.A., Cappato, L.P., et al., 2016. Manufacture of probiotic minas frescal cheese with *Lactobacillus casei* Zhang. Journal of Dairy Science 99 (1), 18–30. Available at https://doi.org/10.3168/jds.2015-9880.

Degnan, F.H., 2008. The us food and drug administration and probiotics: regulatory categorization. Clinical Infectious Diseases 46 (s2), S133. Available at https://doi.org/10.1086/523324.

Detection of oxalyl-CoA decarboxylase (oxc) and formyl-CoA transferase (frc) Genes in Novel Probiotic Isolates Capable of Oxalate Degradation in vitro Folia Microbiologica. 2024, 1–10.

Dey, G., 2018. Non-dairy Probiotic Foods: Innovations and Market Trends. Springer Nature, pp. 159–173. Available at https://doi.org/10.1007/978-3-319-74820-7_9.

Dimitrovski, D., Stojanovska, M.C., Simonovska, J., Sapceska, D.D., 2016. Lactic acid fermentation of cabbage juice by probiotic strain III International Congress. Food Technology, Quality and Safety 67–73.

Do, T.V.T., Fan, L., 2019. Probiotic viability, qualitative characteristics, and sensory acceptability of vegetable juice mixture fermented with *Lactobacillus* strains. Food and Nutrition Sciences 10 (04), 412–427. Available at https://doi.org/10.4236/fns.2019.104031.

Doron, S., Snydman, D.R., 2015. Risk and safety of probiotics. Clinical Infectious Diseases 60, S129. Available at https://doi.org/10.1093/cid/civ085, http://cid.oxfordjournals.org/content/by/year.

Ebner, S., Smug, L.N., Kneifel, W., Salminen, S.J., Sanders, M.E., 2014. Probiotics in dietary guidelines and clinical recommendations outside the European Union. World Journal of Gastroenterology 20 (43), 16095. Available at https://doi.org/10.3748/wjg.v20.i43.16095.

Egervärn, M., Lindmark, H., Olsson, J., Roos, S., 2010. Transferability of a tetracycline resistance gene from probiotic *Lactobacillus reuteri* to bacteria in the gastrointestinal tract of humans. Antonie Van Leeuwenhoek 97 (2), 189–200. Available at https://doi.org/10.1007/s10482-009-9401-0.

Ejtahed, H.S., Mohtadi-Nia, J., Homayouni-Rad, A., Niafar, M., Asghari-Jafarabadi, M., Mofid, V., 2012. Probiotic yogurt improves antioxidant status in type 2 diabetic patients. Nutrition (Burbank, Los Angeles County, Calif.) 28 (5), 539–543. Available at https://doi.org/10.1016/j.nut.2011.08.013.

Fan, S., Breidt, F., Price, R., Pérez-Díaz, I., 2017. Survival and growth of probiotic lactic acid bacteria in refrigerated pickle products. Journal of Food Science 82 (1), 167–173. Available at https://doi.org/10.1111/1750-3841.13579.

Farahmand, N., Ouoba, L.I.I., Naghizadeh Raeisi, S., Sutherland, J., Ghoddusi, H.B., 2021. Probiotic lactobacilli in fermented dairy products: selective detection, enumeration and identification scheme. Microorganisms 9 (8), 1600. Available at https://doi.org/10.3390/microorganisms9081600.

Fatahi-Bafghi, M., Naseri, S., Alizehi, A., 2022. Genome analysis of probiotic bacteria for antibiotic resistance genes. Antonie Van Leeuwenhoek 115 (3), 375–389. Available at https://doi.org/10.1007/s10482-021-01703-7.

Fekri, A., Torbati, M., Yari Khosrowshahi, A., Bagherpour Shamloo, H., Azadmard-Damirchi, S., 2020. Functional effects of phytate-degrading, probiotic lactic acid bacteria and yeast strains isolated from Iranian traditional sourdough on the technological and nutritional properties of whole wheat bread. Food Chemistry 306, 125620. Available at https://doi.org/10.1016/j.foodchem.2019.125620.

Ferdousi, R., Rouhi, M., Mohammadi, R., Mortazavian, A.M., Khosravi-Darani, K., Rad, A.H., 2013. Evaluation of probiotic survivability in yogurt exposed to cold chain interruption. Iranian Journal of Pharmaceutical Research 12, 137–142. Available at http://ijpr.sbmu.ac.ir/?_action=showPDF&article=1289&_ob=6d69ec1f11fd226e5b432867eb3846ce&fileName=full_text.pdf.

Fernandes Pereira, A.L., Rodrigues, S., 2018. Turning Fruit Juice Into Probiotic Beverages. Elsevier BV, pp. 279–287. Available at https://doi.org/10.1016/b978-0-12-802230-6.00015-1.

Gao, J., Li, X., Zhang, G., Sadiq, F.A., Simal-Gandara, J., Xiao, J., et al., 2021. Probiotics in the dairy industry—advances and opportunities. Comprehensive Reviews in Food Science and Food Safety 20 (4), 3937–3982. Available at https://doi.org/10.1111/1541-4337.12755.

George Kerry, R., Patra, J.K., Gouda, S., Park, Y., Shin, H.-S., Das, G., 2018. Benefaction of probiotics for human health: a review. Journal of Food and Drug Analysis 26 (3), 927–939. Available at https://doi.org/10.1016/j.jfda.2018.01.002.

Georgieva, M., Andonova, L., Peikova, L., Zlatkov, A., 2014. Probiotics - health benefits, classification, quality assurance and quality control - review. Pharmacia 61 (4), 22–31. Available at https://pharmacia.pensoft.net/.

Gharsallaoui, A., Roudaut, G., Chambin, O., Voilley, A., Saurel, R., 2007. Applications of spray-drying in microencapsulation of food ingredients: an overview. Food Research International 40 (9), 1107–1121. Available at https://doi.org/10.1016/j.foodres.2007.07.004.

Ghasemi, L., Nouri, L., Mohammadi Nafchi, A., Al-Hassan, A.A., 2022. The effects of encapsulated probiotic bacteria on the physicochemical properties, staling, and viability of probiotic bacteria in gluten-free bread. Journal of Food Processing and Preservation 46 (3). Available at https://doi.org/10.1111/jfpp.16359.

Gheorghita, R., Anchidin-Norocel, L., Filip, R., Dimian, M., Covasa, M., 2021. Applications of biopolymers for drugs and probiotics delivery. Polymers 13 (16), 2729. Available at https://doi.org/10.3390/polym13162729.

Goktas, H., Dikmen, H., Bekiroglu, H., Cebi, N., Dertli, E., Sagdic, O., 2022. Characteristics of functional ice cream produced with probiotic *Saccharomyces boulardii* in combination with *Lactobacillus rhamnosus* GG. LWT 153, 112489. Available at https://doi.org/10.1016/j.lwt.2021.112489.

Goodman, C., Keating, G., Georgousopoulou, E., Hespe, C., Levett, K., 2021. Probiotics for the prevention of antibiotic-associated diarrhoea: a systematic review and meta-analysis. BMJ Open 11 (8). Available at https://doi.org/10.1136/bmjopen-2020-043054, http://bmjopen.bmj.com/content/early/by/section.

Gueimonde, M., Sánchez, B., 2012. Enhancing probiotic stability in industrial processes. Microbial Ecology in Health and Disease 23 (0). Available at https://doi.org/10.3402/mehd.v23i0.18562.

Gueimonde, M., Sánchez, B., de los Reyes-Gavilán, C.G., Margolles, A., 2013. Antibiotic resistance in probiotic bacteria. Frontiers in Microbiology 4. Available at https://doi.org/10.3389/fmicb.2013.00202.

Hamdy, A.M., Ahmed, M.E., Mehta, D., Elfaruk, M.S., Hammam, A.R.A., El-Derwy, Y.M.A., 2021. Enhancement of low-fat Feta cheese characteristics using probiotic bacteria. Food Science and Nutrition 9 (1), 62–70. Available at https://doi.org/10.1002/fsn3.1889.

Happel, A.U., Barnabas, S.L., Froissart, R., Passmore, J.S., 2018. Weighing in on the risks and benefits of probiotic use in HIV-infected and immunocompromised populations. Beneficial Microbes 9 (2), 239–246. Available at https://doi.org/10.3920/bm2017.0106.

He, J., Zhou, T., Young, J.C., Boland, G.J., Scott, P.M., 2010. Chemical and biological transformations for detoxification of trichothecene mycotoxins in human and animal food chains: a review. Trends in Food Science and Technology 21 (2), 67–76. Available at https://doi.org/10.1016/j.tifs.2009.08.002.

Heller, K.J., Bockelmann, W., Schrezenmeir, J., Devrese, M., 2003. Cheese and its potential as a probiotic food. Handbook of Fermented Functional Foods. Taylor and Francis, Germany, pp. 203–226. Available at https://doi.org/10.1201/9780203009727-13, http://www.tandfebooks.com/doi/book/9780203009727.

Hempel, S., Newberry, S., Ruelaz, A., Wang, Z., Miles, J.N.V., Suttorp, M.J. et al. 2011. Safety of Probiotics to Reduce Risk and Prevent or Treat Disease.

Holzapfel, W.H., Haberer, P., Geisen, R., Björkroth, J., Schillinger, U., 2001. Taxonomy and important features of probiotic microorganisms in food and nutrition. The American Journal of Clinical Nutrition 73 (2), 365s. Available at https://doi.org/10.1093/ajcn/73.2.365s.

Homayoni Rad, A., Akbarzadeh, F., Mehrabany, E.V., 2012. Which are more important: prebiotics or probiotics? Nutrition (Burbank, Los Angeles County, Calif.) 28 (11-12), 1196–1197. Available at https://doi.org/10.1016/j.nut.2012.03.017.

Homayoni Rad, A., Vaghef Mehrabany, E., Alipoor, B., Vaghef Mehrabany, L., 2016. The comparison of food and supplement as probiotic delivery vehicles. Critical Reviews in Food

Science and Nutrition 56 (6), 896–909. Available at https://doi.org/10.1080/10408398.2012.733894, www.tandf.co.uk/journals/titles/10408398.asp.

Homayouni, A., Azizi, A., Javadi, M., Mahdipour, S., Ejtahed, H., 2012. Factors influencing probiotic survival in ice cream: a review. International Journal of Dairy Science 7 (1), 1–10. Available at https://doi.org/10.3923/ijds.2012.1.10, http://scialert.net/qredirect.php?doi=ijds.2012.1.10&linkid=pdf.

Homayouni, A., Javadi, M., Ansari, F., Pourjafar, H., Jafarzadeh, M., Barzegar, A., 2018. Advanced methods in ice cream analysis: a review. Food Analytical Methods 11 (11), 3224–3234. Available at https://doi.org/10.1007/s12161-018-1292-0.

Homayouni, A., Ansari, F., Azizi, A., Pourjafar, H., Madadi, M., 2020a. Cheese as a potential food carrier to deliver probiotic microorganisms into the human gut: a review. Current Nutrition and Food Science 16 (1), 15–28. Available at https://doi.org/10.2174/1573401314666180817101526.

Homayouni, A., Bagheri, N., Mohammad-Alizadeh-Charandabi, S., Kashani, N., Mobaraki-Asl, N., Mirghafurvand, M., et al., 2020b. Prevention of gestational diabetes mellitus (GDM) and probiotics: mechanism of action: a review. Current Diabetes Reviews 16 (6), 538–545. Available at https://doi.org/10.2174/1573399815666190712193828.

Homayouni Rad, A., Torab, R., Ghalibaf, M., Norouzi, S., Mehrabany, E.V., 2013. Might patients with immune-related diseases benefit from probiotics? Nutrition (Burbank, Los Angeles County, Calif.) 29 (3), 583–586. Available at https://doi.org/10.1016/j.nut.2012.10.008.

Homayouni-Rad, A., Soroush, A.R., Khalili, L., Norouzi-Panahi, L., Kasaie, Z., Ejtahed, H.S., 2016. Diabetes management by probiotics: current knowledge and future perspectives. International Journal for Vitamin and Nutrition Research 86 (5-6), 215–227. Available at https://doi.org/10.1024/0300-9831/a000273, https://econtent.hogrefe.com/doi/pdf/10.1024/0300-9831/a000273.

Homayouni Rad, a, Aghebati Maleki, L., Samadi Kafil, H., Abbasi, A., Khani, N., 2021. Postbiotics as a safe alternative to live probiotic bacteria in the food and pharmaceutical industries. Scientific Journal of Kurdistan University of Medical Sciences 26 (4), 132–157. Available at https://doi.org/10.52547/sjku.26.4.132.

Homayouni-Rad, A., Abedi Soleimani, R., Khani, N., 2023. Can postbiotics prevent or improve SARS-CoV-2? Current Nutrition and Food Science 19 (8), 756–757. Available at https://doi.org/10.2174/1573401318666221004112500.

Hood, S.K., Zoitola, E.A., 1988. Effect of low pH on the ability of *Lactobacillus acidophilus* to survive and adhere to human intestinal cells. Journal of Food Science 53 (5), 1514–1516. Available at https://doi.org/10.1111/j.1365-2621.1988.tb09312.x.

Karimi, R., Mortazavian, A.M., Da Cruz, A.G., 2011. Viability of probiotic microorganisms in cheese during production and storage: a review. Dairy Science and Technology 91 (3), 283–308. Available at https://doi.org/10.1007/s13594-011-0005-x.

Kebary, K.M.K., Kamaly, K.M., Mailam, M.A., Maamoon, A.G., 2021. Improving the health benefits and quality of Labneh using probiotic bacteria. Menoufia Journal of Food and Dairy Sciences 6 (1), 1–16. Available at https://doi.org/10.21608/mjfds.2021.149236.

Khalili, M., Rad, A.H., Khosroushahi, A.Y., Khosravi, H., Jafarzadeh, S., 2020. Application of probiotics in folate bio-fortification of yoghurt. Probiotics and Antimicrobial Proteins 12 (2), 756–763. Available at https://doi.org/10.1007/s12602-019-09560-7.

Khani, N., Abedi Soleimani, R., Noorkhajavi, G., Abedi Soleimani, A., Abbasi, A., Homayouni Rad, A., 2022. Postbiotics as potential promising tools for SARS-CoV-2 disease adjuvant therapy. Journal of Applied Microbiology 132 (6), 4097–4111. Available at https://doi.org/10.1111/jam.15457.

Kieliszek, M., Pobiega, K., Piwowarek, K., Kot, A.M., 2021. Characteristics of the proteolytic enzymes produced by lactic acid bacteria. Molecules (Basel, Switzerland) 26 (7). Available at https://doi.org/10.3390/molecules26071858, https://www.mdpi.com/1420-3049/26/7/1858/pdf.

Kim, S., Yoon, Y., Oh, S., 2020. Suggestion of a safety evaluation procedure to improve probiotic safety. Journal of Dairy Science and Biotechnology 38 (2), 99–111. Available at https://doi.org/10.22424/jdsb.2020.38.2.99.

Korčok, D., 2020. Microbiological aspects of cleaning validation during the production of dietary supplements. Tehnika 75 (2), 255–260. Available at https://doi.org/10.5937/tehnika2002255k.

Kowlgi, N.G., Chhabra, L., 2015. D-lactic acidosis: an underrecognized complication of short bowel syndrome. Gastroenterology Research and Practice 2015. Available at https://doi.org/10.1155/2015/476215, http://www.hindawi.com/journals/grp/.

Lebreton, F., Depardieu, F., Bourdon, N., Fines-Guyon, M., Berger, P., Camiade, S., et al., 2011. D-Ala-D-Ser VanN-type transferable vancomycin resistance in *Enterococcus faecium*. Antimicrobial Agents and Chemotherapy 55 (10), 4606–4612. Available at https://doi.org/10.1128/AAC.00714-11, http://aac.asm.org/cgi/reprint/55/10/4606.

Li, G., Walker, M.J., De Oliveira, D.M.P., 2023. Vancomycin resistance in *Enterococcus* and *Staphylococcus aureus*. Microorganisms 11 (1). Available at https://doi.org/10.3390/microorganisms11010024, www.mdpi.com/journal/microorganisms.

Liliana, S.C., Vladimir, V.C., 2013. Probiotic encapsulation. African Journal of Microbiology Research 7 (40), 4743–4753. Available at https://doi.org/10.5897/AJMR2013.5718.

Luckow, T., Delahunty, C., 2004. Which juice is 'healthier'? A consumer study of probiotic non-dairy juice drinks. Food Quality and Preference 15 (7-8), 751–759. Available at https://doi.org/10.1016/j.foodqual.2003.12.007.

Magra, T., Soultos, N., Dovas, C., Papavergou, E., Lazou, T., Apostolakos, I., et al., 2021. Dry fermented sausages with total replacement of fat by extra virgin olive oil emulsion and indigenous lactic acid bacteria. Food Technology and Biotechnology 59 (3), 267–281. Available at https://doi.org/10.17113/ftb.59.03.21.7114, http://www.ftb.com.hr/images/pdfarticles/2021/July-September/FTB-59-267.pdf.

Maleki, D., Homayouni, A., Khalili, L., Golkhalkhali, B., 2016. Probiotics in cancer prevention, updating the evidence. Probiotics, Prebiotics, and Synbiotics: Bioactive Foods in Health Promotion. Elsevier Inc, Iran, pp. 781–791. Available at https://doi.org/10.1016/B978-0-12-802189-7.00059-9, http://www.sciencedirect.com/science/book/9780128021897.

Marín, L., Estella, B., Cardona, M.A., Restrepo, M., 2020. Development of a Baby Food Pureed with Probiotic Microorganisms for an Infant from 6 to 12 Months.

Marteau, T.M., Lerman, C., 2001. Genetic risk and behavioural change. British Medical Journal 322 (7293), 1056–1059. Available at https://doi.org/10.1136/bmj.322.7293.1056, http://www.bmj.com/.

Mehanna, N., Salama, S., Arafa, M., 2021. Impact of some ingredients and the processing method on composition and quality of probiotic Labneh. Romanian Biotechnological Letters 26 (3), 2587–2593. Available at https://doi.org/10.25083/rbl/26.3/2587.2593.

Min, M., Bunt, C.R., Mason, S.L., Hussain, M.A., 2019. Non-dairy probiotic food products: an emerging group of functional foods. Critical Reviews in Food Science and Nutrition 59 (16), 2626–2641. Available at https://doi.org/10.1080/10408398.2018.1462760.

Mohammadi, N., Ahari, H., Fahimdanesh, M., Zanjani, M.A.K., Anvar, A., Shokri, E., 2012. Survival of alginate-prebiotic microencapsulated *Lactobacillus acidophilus* in mayonnaise sauce. Iranian Journal of Veterinary Medicine 6, 259–264.

Mojikon, F.D., Kasimin, M.E., Molujin, A.M., Gansau, J.A., Jawan, R., 2022. Probiotication of nutritious fruit and vegetable juices: an alternative to dairy-based probiotic functional products. Nutrients 14 (17), 3457. Available at https://doi.org/10.3390/nu14173457.

Monteiro, S.S., Beserra, Y.A.S., Oliveira, H.M.L., de Bittencourt Pasquali, M.A., 2020. Production of probiotic passion fruit (*Passiflora edulis* Sims f. Flavicarpa Deg.) drink using *Lactobacillus reuteri* and microencapsulation via spray drying. Foods 9 (3). Available at https://doi.org/10.3390/foods9030335, https://www.mdpi.com/2304-8158/9/3/335.

Nedovic, V., Kalusevic, A., Manojlovic, V., Levic, S., Bugarski, B., 2011. An overview of encapsulation technologies for food applications. Procedia Food Science 1, 1806–1815. Available at https://doi.org/10.1016/j.profoo.2011.09.265.

Nikolaeva, O., Andreeva, A., Altynbekov, O., Mishukovskaya, G., Ismagilova, E., 2020. Probiotic drugs impact on the innate immunity factors. Journal of Global Pharma Technology 12 (1), 38–45. Available at http://www.jgpt.co.in/.

Nyanzi, R., Jooste, P.J., Buys, E.M., 2021. Invited review: probiotic yogurt quality criteria, regulatory framework, clinical evidence, and analytical aspects. Journal of Dairy Science 104 (1), 1–19. Available at https://doi.org/10.3168/jds.2020-19116.

Oak, S.J., Jha, R., 2019. The effects of probiotics in lactose intolerance: a systematic review. Critical Reviews in Food Science and Nutrition 59 (11), 1675–1683. Available at https://doi.org/10.1080/10408398.2018.1425977.

Oggioni, M.R., Pozzi, G., Valensin, P.E., Galieni, P., Bigazzi, C., 1998. Recurrent septicemia in an immunocompromised patient due to probiotic strains of *Bacillus subtilis*. Journal of Clinical Microbiology 36 (1), 325–326. Available at https://doi.org/10.1128/jcm.36.1.325-326.1998.

Oh, Y.J., Nam, K., Kim, Y., Lee, S.Y., Kim, H.S., Kang, J.I., et al., 2021. Effect of a nutritionally balanced diet comprising whole grains and vegetables alone or in combination with probiotic supplementation on the gut microbiota. Preventive Nutrition and Food Science 26 (2), 121–131. Available at https://doi.org/10.3746/pnf.2021.26.2.121.

Oliveira, M., Carvalho, M., Teixeira, P., 2023. Characterization of the toxigenic potential of *Bacillus cereus* sensu lato Isolated from raw berries and their products. Foods 12 (21), 4021. Available at https://doi.org/10.3390/foods12214021.

Ono, M., Ono, A., 2015. Impacts of the FoSHU (Food for Specified Health Uses) system on food evaluations in Japan. Journal of Consumer Marketing 32 (7), 542–550. Available at https://doi.org/10.1108/jcm-09-2014-1144.

Patel, A.R., 2017. Probiotic fruit and vegetable juices- recent advances and future perspective. International Food Research Journal 24 (5), 1850–1857. Available at http://www.ifrj.upm.edu.my/24%20(05)%202017/(2).pdf.

Pereira, A.L.F., Maciel, T.C., Rodrigues, S., 2011. Probiotic beverage from cashew apple juice fermented with *Lactobacillus casei*. Food Research International 44 (5), 1276–1283. Available at https://doi.org/10.1016/j.foodres.2010.11.035.

Pimentel, T.C., de Oliveira, L.I.G., de Souza, R.C., Magnani, M., 2022. Probiotic ice cream: a literature overview of the technological and sensory aspects and health properties. International Journal of Dairy Technology 75 (1), 59–76. Available at https://doi.org/10.1111/1471-0307.12821.

Pineiro, M., Stanton, C., 2007. Probiotic bacteria: legislative framework—requirements to evidence basis. The Journal of Nutrition 137 (3), 850S. Available at https://doi.org/10.1093/jn/137.3.850S.

Rad, A.H., Javadi, M., Kafil, H.S., Pirouzian, H.R., Khaleghi, M., 2019. The safety perspective of probiotic and non-probiotic yoghurts: a review. Food Quality and Safety 3 (1), 9–14. Available at https://doi.org/10.1093/fqsafe/fyz006.

Rad, A.H., Abbasi, A., Kafil, H.S., Ganbarov, K., 2020a. Potential pharmaceutical and food applications of postbiotics: a review. Current Pharmaceutical Biotechnology 21 (15), 1576–1587. Available at https://doi.org/10.2174/1389201021666200516154833.

Rad, H., Kafil, H.S., Zavoshti, H.F., Shahbazi, N., Abbasi, A., 2020b. Therapeutically effects of functional postbiotic foods. Clinical Excellence 10, 33–52.

Rad, A.H., Aghebati-Maleki, L., Kafil, H.S., Abbasi, A., 2021a. Molecular mechanisms of postbiotics in colorectal cancer prevention and treatment. Critical Reviews in Food Science and Nutrition 61 (11), 1787–1803. Available at https://doi.org/10.1080/10408398.2020.1765310.

Rad, A.H., Maleki, L.A., Kafil, H.S., Zavoshti, H.F., Abbasi, A., 2021b. Postbiotics as promising tools for cancer adjuvant therapy. Advanced Pharmaceutical Bulletin 11 (1), 1–5. Available at https://doi.org/10.34172/apb.2021.007, https://apb.tbzmed.ac.ir/Article/apb-28481.

Rani, K., Kaur, G., Ali, S.A., 2023. Probiotic-prebiotic therapeutic potential: a new horizon of microbial biotherapy to reduce female reproductive complications. PharmaNutrition 24, 100342. Available at https://doi.org/10.1016/j.phanu.2023.100342.

Ranjbar, F., Akbarzadeh, F., Homayouni, A., 2016. Probiotics usage in heart disease and psychiatry. Probiotics, Prebiotics, and Synbiotics: Bioactive Foods in Health Promotion. Elsevier Inc., Iran, pp. 807–811. Available at https://doi.org/10.1016/B978-0-12-802189-7.00061-7, http://www.sciencedirect.com/science/book/9780128021897.

Rasika, D.M.D., Vidanarachchi, J.K., Luiz, S.F., Azeredo, D.R.P., Cruz, A.G., Ranadheera, C.S., 2021. Probiotic delivery through non-dairy plant-based food matrices. Agriculture 11 (7), 599. Available at https://doi.org/10.3390/agriculture11070599.

Rezende, E.S.V., Lima, G.C., Naves, M.M.V., 2021. Dietary fibers as beneficial microbiota modulators: a proposed classification by prebiotic categories. Nutrition (Burbank, Los Angeles County, Calif.) 89, 111217. Available at https://doi.org/10.1016/j.nut.2021.111217.

Ridlon, J.M., Kang, D.J., Hylemon, P.B., Bajaj, J.S., 2014. Bile acids and the gut microbiome. Current Opinion in Gastroenterology 30 (3), 332–338. Available at https://doi.org/10.1097/MOG.0000000000000057, www.lww.com/store/products?0267-1379.

Rivera-Espinoza, Y., Gallardo-Navarro, Y., 2010. Non-dairy probiotic products. Food Microbiology 27 (1), 1–11. Available at https://doi.org/10.1016/j.fm.2008.06.008.

Roberfroid, M., Gibson, G.R., Hoyles, L., McCartney, A.L., Rastall, R., Rowland, I., et al., 2010. Prebiotic effects: metabolic and health benefits. British Journal of Nutrition 104 (S2), S1. Available at https://doi.org/10.1017/s0007114510003363.

Rodrigues, F.J., Cedran, M.F., Bicas, J.L., Sato, H.H., 2020. Encapsulated probiotic cells: Relevant techniques, natural sources as encapsulating materials and food applications – a narrative review. Food Research International 137, 109682. Available at https://doi.org/10.1016/j.foodres.2020.109682.

Roe, A.L., Boyte, M.-E., Elkins, C.A., Goldman, V.S., Heimbach, J., Madden, E., et al., 2022. Considerations for determining safety of probiotics: a USP perspective. Regulatory Toxicology and Pharmacology 136, 105266. Available at https://doi.org/10.1016/j.yrtph.2022.105266.

Roy, D., 2005. Technological aspects related to the use of bifidobacteria in dairy products. Le Lait 85 (1-2), 39–56. Available at https://doi.org/10.1051/lait:2004026.

Sabahi, S., Homayouni Rad, A., Aghebati-Maleki, L., Sangtarash, N., Ozma, M.A., Karimi, A., et al., 2023. Postbiotics as the new frontier in food and pharmaceutical research. Critical Reviews in Food Science and Nutrition 63 (26), 8375–8402. Available at https://doi.org/10.1080/10408398.2022.2056727.

Said M.S., E. Tirthani, E. Lesho, Enterococcus infections. 2021.

Salminen, M.K., Tynkkynen, S., Rautelin, H., Saxelin, M., Vaara, M., Ruutu, P., et al., 2002. *Lactobacillus bacteremia* during a rapid increase in probiotic use of *Lactobacillus rhamnosus* GG in Finland. Clinical Infectious Diseases 35 (10), 1155–1160. Available at https://doi.org/10.1086/342912.

Salminen, M.K., Rautelin, H., Tynkkynen, S., Poussa, T., Saxelin, M., Valtonen, V., et al., 2004. *Lactobacillus bacteremia*, clinical significance, and patient outcome, with special focus on probiotic *L. rhamnosus* GG. Clinical Infectious Diseases 38 (1), 62–69. Available at https://doi.org/10.1086/380455.

Sameer, B., Ganguly, S., Khetra, Y., Sabikhi, L., 2020. Development and characterization of probiotic buffalo milk ricotta cheese. LWT 121, 108944. Available at https://doi.org/10.1016/j.lwt.2019.108944.

Sanders, M.E., 2011. Impact of probiotics on colonizing microbiota of the gut. Journal of Clinical Gastroenterology 45 (3), S115. Available at https://doi.org/10.1097/mcg.0b013e318227414a.

Sanders, M.E., Akkermans, L.M.A., Haller, D., Hammerman, C., Heimbach, J.T., Hörmannsperger, G., et al., 2014. Safety assessment of probiotics for human use. Gut Microbes 1 (3), 164–185. Available at https://doi.org/10.4161/gmic.1.3.12127.

Seifert, A., Kashi, Y., Livney, Y.D., 2019. Delivery to the gut microbiota: a rapidly proliferating research field. Advances in Colloid and Interface Science 274, 102038. Available at https://doi.org/10.1016/j.cis.2019.102038.

Shafie, M., Homayouni Rad, A., Mohammad-Alizadeh-Charandabi, S., Mirghafourvand, M., 2022. The effect of probiotics on mood and sleep quality in postmenopausal women: a triple-blind randomized controlled trial. Clinical Nutrition ESPEN 50, 15–23. Available at https://doi.org/10.1016/j.clnesp.2022.06.005.

Shah, N.P., Lankaputhra, W.E.V., Britz, M.L., Kyle, W.S.A., 1995. Survival of *Lactobacillus acidophilus* and Bifidobacterium bifidum in commercial yoghurt during refrigerated storage. International Dairy Journal 5 (5), 515–521. Available at https://doi.org/10.1016/0958-6946(95)00028-2.

Shahbandeh M., Global probiotics market value 2022-2027. 2023.

Shenderov, B.A., 2013. Metabiotics: novel idea or natural development of probiotic conception. Microbial Ecology in Health and Disease 24 (0). Available at https://doi.org/10.3402/mehd.v24i0.20399.

Shewale, R.N., Sawale, P.D., Khedkar, C.D., Singh, A., 2014. Selection criteria for probiotics: a review. International Journal of Probiotics and Prebiotics 9 (1-2), 17–22. Available at http://www.newcenturyhealthpublishers.com/probiotics_and_prebiotics/index.php.

Soltani, M., Hekmat, S., Ahmadi, L., 2018. Microbial and sensory evaluation of probiotic yoghurt supplemented with cereal/pseudo-cereal grains and legumes. International Journal of Dairy Technology 71 (S1), 141–148. Available at https://doi.org/10.1111/1471-0307.12389.

Sotoudegan, F., Daniali, M., Hassani, S., Nikfar, S., Abdollahi, M., 2019. Reappraisal of probiotics' safety in human. Food and Chemical Toxicology 129, 22–29. Available at https://doi.org/10.1016/j.fct.2019.04.032.

Steenson, L.R., Klaenhammer*, T.R., Swaisgood, H.E., 1987. Calcium alginate-immobilized cultures of lactic streptococci are protected from bacteriophages. Journal of Dairy Science 70 (6), 1121–1127. Available at https://doi.org/10.3168/jds.s0022-0302(87)80121-2.

Talon, R., Leroy, S., Lebert, I., Giammarinaro, P., Chacornac, J., Latorremoratalla, M., et al., 2008. Safety improvement and preservation of typical sensory qualities of traditional dry fermented sausages using autochthonous starter cultures. International Journal of Food Microbiology 126 (1-2), 227–234. Available at https://doi.org/10.1016/j.ijfoodmicro.2008.05.031.

Tamminen, 2013. Fermentation of carrot juice by probiotics: viability and preservation of adhesion. International Journal of Biotechnology for Wellness Industries. Available at https://doi.org/10.6000/1927-3037.2013.02.01.2.

Teame, T., Wang, A., Xie, M., Zhang, Z., Yang, Y., Ding, Q., et al., 2020. Paraprobiotics and postbiotics of probiotic lactobacilli, their positive effects on the host and action mechanisms: a review. Frontiers in Nutrition 7. Available at https://doi.org/10.3389/fnut.2020.570344.

Tewari, A., Abdullah, S., 2015. Bacillus cereus food poisoning: International and Indian perspective. Journal of Food Science and Technology 52 (5), 2500–2511. Available at https://doi.org/10.1007/s13197-014-1344-4, http://www.springerlink.com/content/121580/.

Thakur, N., Rokana, N., Panwar, H., 2016. Probiotics, selection criteria, safety and role in health and. Journal of Innovative Biology 3, 259–270.

Trinchieri, V., Laghi, L., Vitali, B., Parolin, C., Giusti, I., Capobianco, D., et al., 2017. Efficacy and safety of a multistrain probiotic formulation depends from manufacturing. Frontiers in Immunology 8. Available at https://doi.org/10.3389/fimmu.2017.01474.

Tripathi, M.K., Giri, S.K., 2014. Probiotic functional foods: survival of probiotics during processing and storage. Journal of Functional Foods 9 (1), 225–241. Available at https://doi.org/10.1016/j.jff.2014.04.030.

Tsilingiri, K., Rescigno, M., 2013. Postbiotics: what else? Beneficial Microbes 4 (1), 101–107. Available at https://doi.org/10.3920/bm2012.0046.

Tsilingiri, K., Barbosa, T., Penna, G., Caprioli, F., Sonzogni, A., Viale, G., et al., 2012. Probiotic and postbiotic activity in health and disease: comparison on a novel polarised ex-vivo organ culture model. Gut 61 (7), 1007–1015. Available at https://doi.org/10.1136/gutjnl-2011-300971.

Vaghef-Mehrabany, E., Alipour, B., Homayouni-Rad, A., Sharif, S.-K., Asghari-Jafarabadi, M., Zavvari, S., 2014. Probiotic supplementation improves inflammatory status in patients with rheumatoid arthritis. Nutrition (Burbank, Los Angeles County, Calif.) 30 (4), 430–435. Available at https://doi.org/10.1016/j.nut.2013.09.007.

Valero-Cases, E., Cerdá-Bernad, D., Pastor, J.-J., Frutos, M.-J., 2020. Non-dairy fermented beverages as potential carriers to ensure probiotics, prebiotics, and bioactive compounds arrival to the gut and their health benefits. Nutrients 12 (6), 1666. Available at https://doi.org/10.3390/nu12061666.

Van Reenen, C.A., Dicks, L.M.T., 2011. Horizontal gene transfer amongst probiotic lactic acid bacteria and other intestinal microbiota: what are the possibilities? A review. Archives of Microbiology 193 (3), 157–168. Available at https://doi.org/10.1007/s00203-010-0668-3.

Warnecke, F., Luginbühl, P., Ivanova, N., Ghassemian, M., Richardson, T.H., Stege, J.T., et al., 2007. Metagenomic and functional analysis of hindgut microbiota of a wood-feeding higher termite. Nature 450 (7169), 560–565. Available at https://doi.org/10.1038/nature06269.

Wash, P., Batool, A., Mulk, S., Nazir, S., Yasmin, H., Mumtaz, S., et al., 2022. Prevalence of antimicrobial resistance and respective genes among *Bacillus* spp., a versatile bio-fungicide. International Journal of Environmental Research and Public Health 19 (22). Available at https://doi.org/10.3390/ijerph192214997, http://www.mdpi.com/journal/ijerph.

Wegh, C.A.M., Geerlings, S.Y., Knol, J., Roeselers, G., Belzer, C., 2019. Postbiotics and their potential applications in early life nutrition and beyond. International Journal of Molecular Sciences 20 (19), 4673. Available at https://doi.org/10.3390/ijms20194673.

Yamamura, R., Inoue, K.Y., Nishino, K., Yamasaki, S., 2023. Intestinal and fecal pH in human health. Frontiers in Microbiomes 2. Available at https://doi.org/10.3389/frmbi.2023.1192316, https://www.frontiersin.org/journals/microbiomes.

Yang, Y., Zhou, B., Zhang, S., Si, L., Liu, X., Li, F., 2023. Prebiotics for depression: how does the gut microbiota play a role? Frontiers in Nutrition 10. Available at https://doi.org/10.3389/fnut.2023.1206468.

Yoon, K.Y., Woodams, E.E., Hang, Y.D., 2005. Fermentation of beet juice by beneficial lactic acid bacteria. LWT - Food Science and Technology 38 (1), 73–75. Available at https://doi.org/10.1016/j.lwt.2004.04.008.

Zarić, D.B., Bulatović, M.L., Rakin, M.B., Krunić, T.Ž., Lončarević, I.S., Pajin, B.S., 2016. Functional, rheological and sensory properties of probiotic milk chocolate produced in a ball mill. RSC Advances 6 (17), 13934–13941. Available at https://doi.org/10.1039/c5ra21363k.

Zavišić, G., Popović, M., Stojkov, S., Medić, D., Gusman, V., Jovanović Lješković, N., et al., 2023. Antibiotic resistance and probiotics: knowledge gaps, market overview and preliminary screening. Antibiotics 12 (8). Available at https://doi.org/10.3390/antibiotics12081281, http://www.mdpi.com/journal/antibiotics.

Zendeboodi, F., Khorshidian, N., Mortazavian, A.M., da Cruz, A.G., 2020. Probiotic: conceptualization from a new approach. Current Opinion in Food Science 32, 103–123. Available at https://doi.org/10.1016/j.cofs.2020.03.009.

Zheng, D.-W., Li, R.-Q., An, J.-X., Xie, T.-Q., Han, Z.-Y., Xu, R., et al., 2020. Prebiotics-encapsulated probiotic spores regulate gut microbiota and suppress colon cancer. Advanced Materials 32 (45). Available at https://doi.org/10.1002/adma.202004529.

Zielińska, D., Sionek, B., Kołożyn-Krajewska, D., 2018. Safety of probiotics. Diet, Microbiome and Health. Elsevier, Poland, pp. 131–161. Available at https://doi.org/10.1016/B978-0-12-811440-7.00006-5, http://www.sciencedirect.com/science/book/9780128114407.

Zommiti, M., Chikindas, M.L., Ferchichi, M., 2020. Probiotics—live biotherapeutics: a story of success, limitations, and future prospects—not only for humans. Probiotics and Antimicrobial Proteins 12 (3), 1266–1289. Available at https://doi.org/10.1007/s12602-019-09570-5.

Section II

New trends on oral probiotic

Chapter 8

Oral probiotics and prebiotics: general definitions and concepts

Sahar A. Khairy and Mervat A. Fouad
National Nutrition Institute, Cairo, Egypt

8.1 Introduction

Probiotics have been considered an economical and safe alternative for the treatment of a large number of chronic diseases and the improvement of human health (Yadav et al., 2022). They are known to modulate the host immunity and protect from several infectious and noninfectious diseases. Colonization, killing of pathogens, and induction of host cells are a few of the important probiotic attributes that affect several functions of the host. In addition, prebiotics and nondigestible food substances selectively promote the growth of probiotics and human health through nutrient enrichment and modulation of gut microbiota and the immune system. Pro- and prebiotics play a key role in neuroimmune processes. Their beneficial effects on health are linked to interactions of the gastrointestinal tract, immune system, and neurologic systems.

Probiotics and prebiotics have evolved as promising therapeutic techniques to treat several disease conditions associated with the gastrointestinal tract. The use of probiotics and prebiotics is currently useful in the mitigation of various diseases such as COVID-19, congenital heart disease, diarrhea, inflammatory bowel disease, hypertension, genitourinary tract infection, colon cancer, immune system defense, mineral absorption, allergic disorders, and atopic dermatitis (Manzoor et al., 2022).

8.2 Definition and classification of probiotic active and inactive

8.2.1 Probiotic

The field of probiotics is rapidly evolving, and new definitions, concepts, and strains are emerging all the time. The term "probiotic" was first introduced in 1965 by Lilly and Stillwell. Their initial definition described probiotics as

Evaluating Oral Probiotics and Prebiotics. DOI: https://doi.org/10.1016/B978-0-443-29835-6.00010-1

"substances that promote the growth of other microorganisms." This definition was later refined by Parker in 1974, who defined probiotics as "organisms and substances which contribute to intestinal microbial balance" (Schrezenmeir and de Vrese, 2001). Then, the term probiotic was defined in 2001 by an Expert Consultation of the Food and Agricultural Organization of the United Nations and the World Health Organization (FAO/WHO), as "Live microorganisms, which when administered in adequate amounts, exert a beneficial effect on the host's health." However, a slight modification in probiotics' definition is accomplished by the International Scientific Association for Probiotics and Prebiotics (ISAPP) as "Live microorganisms that, when administered in adequate amounts, confer a health benefit on the host." The definition emphasizes the need for probiotics to be alive, to be consumed in sufficient quantities, and to provide a demonstrable health benefit (O'Toole et al., 2017).

In recent years, findings show that inactivated, nonviable, and even ruptured cells of probiotics have beneficial effects on the host on one hand, and some concerns about safety problems of live microbial cells on the other hand have caused the incorporation of some new terms to probiotic terminology, such as paraprobiotics and postbiotics. These terms, focused on the healthful effects of inactive and nonviable probiotic cells in different forms, are confusing and ambiguous due to several reasons. A new terminology to cover all healthful impacts of live probiotic cells and their derivatives in all possible forms includes "true probiotic" (=TP: viable and active microbial cells), "pseudoprobiotic" (=PP: viable but inactive cell in the forms of vegetative or spore), and "ghost probiotic" (=GP: nonviable cell in the forms of intact or ruptured). Mentioned types or states of probiotics (comprising all probiotic areas) along with their health impacts and mechanisms of action were tersely discussed in recent reviews. According to the new definition, "probiotic" can be defined as "viable or nonviable microbial cell (vegetative or spore; intact or ruptured) that is potentially healthful to the host."

New probiotics are isolated mainly from the indigenous human intestinal microbiome. Focused on the treatment of specific and systemic diseases, regulation of inflammation, metabolic diseases, and obesity, helping in the production of vitamins of the host, and increasing the immune system, a single strain can help in different diseases. Development of specific drugs and production of secondary metabolites such as vitamins. They are divided into immunobiotics, psychobiotics, oncobiotics, and pharmabiotics. Functional components such as paraprobiotics, postbiotics, probioceuticals, or probiotaceuticals are in development (Kumari et al., 2021).

8.2.2 Psychobiotics

Psychobiotics are defined as a special class of probiotics that when ingested exert mental health benefits through an interaction with the gut microbiota, stimulating the production of short-chain fatty acids, enteroendocrine hormones, antiinflammatory cytokines, and neurotransmitters such as gamma-aminobutyric acid, serotonin, acetylcholine, catecholamines, and norepinephrine, which help regulate mood, cognitive, memory, and learning functions (Sharma and Shukla, 2016; Zendeboodi et al., 2020). An ability to modulate the expression of neurochemical receptors, such as endocannabinoids, is also attributed to psychobiotics.

8.2.3 Oncobiotics

Oncobiotics are probiotics that enhance the response to cancer immunotherapy, helping the immune system eliminate tumor cells through cytotoxicity stimulated by biological agents aimed at preventing apoptosis or lymphocyte inhibition (Zitvogel et al., 2016). Some studies suggest that oncobiotic probiotics play an important role in the immune response after chemotherapy and radiotherapy treatment, thus helping the host to avoid future relapses (Gopalakrishnan et al., 2018).

8.2.4 Pharmabiotics

Pharmabiotics are probiotics that can function as pharmacologically active ingredients with preventive and curative potential. Pharmabiotics are used for the adjunctive treatment of various diseases, such as acute diarrhea and irritable bowel syndrome. Studies have determined that specific pharmabiotics can improve many lower gastrointestinal tract symptoms in adults (Belkaid and Hand, 2014; Navarro, 2021).

8.2.5 Paraprobiotics or phantom probiotics or ghost probiotics

Taverniti and Guglielmetti (2011) described paraprobiotics as inactive, dead, and nonviable cells, or cell fractions, from intact or broken probiotics that provide health benefits to the recipient (Taverniti and Guglielmetti, 2011). Other authors have shown that cells from probiotics, even when dead or inactivated, continue to have a beneficial effect on the host, such as decreasing the inflammatory response, as well as preventing the adhesion of pathogenic bacteria (Zendeboodi et al., 2020). They are also attributed to the attenuation of colitis, stimulation of the intestinal immune system, and antiadhesion capacity against several pathogens in CaCo-2 experimental models (Nataraj et al., 2020). Some paraprobiotics produce compounds such as teichoic acid, peptidoglycan-derived muropeptides, exopolysaccharides, and biosurfactants bound to their cell wall (Nataraj et al., 2020).

8.2.6 Postbiotics

Postbiotics are defined as a complex mixture of molecules that are part of the products or by-products derived from probiotic metabolism (Aguilar-Toalá et al., 2018; Salminen et al., 2021). Postbiotics are also preparations of inanimate microorganisms and/or their components that confer health benefits to the host. This definition encompasses intact nonviable microbial cells, cell fragments (such as cell walls), and may include metabolic by-products produced during fermentation. However, the presence of metabolites is not a mandatory component of postbiotics (Vinderola et al., 2024).

The term "components" in this context refers to cellular structures such as cell walls, peptidoglycans, teichoic acids, and surface proteins. These structural elements can play significant roles in eliciting beneficial physiological responses in the host. Therefore postbiotics are not strictly required to be cell-free; they can include these inanimate cellular components (Vinderola et al., 2022).

Postbiotic molecules have a known chemical formula, prolonged storage stability, and an ability to trigger various treatment mechanisms for diseases such as inflammation, adhesion of pathogens to the gastrointestinal tract, obesity, coronary artery disease, cancer, and oxidative stress (Aguilar-Toalá et al., 2018). Some examples of postbiotic molecules are enzymes, peptides, proteins, exopolysaccharides, organic acids, serine protease inhibitors (serpin), short-chain fatty acids, and cell wall components such as teichoic acid, peptidoglycan, hydrogen peroxide, reuterin, diacetyl, bacteriocins, phenol, amino acids, benzoic acid, phenylactic acid, and vitamins (Jastrząb et al., 2021). There is scientific evidence attributing different antimicrobial, antioxidant, and immunomodulatory functions to postbiotics, properties that could positively affect the homeostasis of the gut microbiota, as well as signaling pathways and metabolic pathways of the host, thus affecting specific physiological, immunological, neuronal, hormonal, biological, regulatory, and metabolic reactions (Salminen et al., 2021).

8.2.7 Probioceuticals

Probioceuticals are biologically active substances derived from the metabolism of probiotics that are attributed with health benefits, as well as the treatment and prevention of diseases. An example of probioceutical compounds is exopolysaccharides, which are long-chain polymers of sugars with antioxidant and immunomodulatory properties (Kumar et al., 2020).

8.2.8 Pseudoprobiotics

Pseudoprobiotics are viable, but not active, probiotic cells that were inactivated by environmental factors such as high or low temperatures, extreme pH, lack of nutrients, and low water activity, so they have low or no growth rates. Based on these characteristics, they cannot be classified as traditional or true probiotics since they need to be viable and active cells. The advantages that these pseudoprobiotics have are that they can function in an inactive manner as paraprobiotics or be activated and used as traditional probiotics, with the advantages that both classifications give them (Zendeboodi et al., 2020).

8.3 Candidate microorganisms for new probiotics

Among the species that have been reported as new probiotics are *Akkermansia miciniphila* (Castañeda–Guillot, 2019), *Ruminococcus bromii* (Kumari et al., 2021), *Roseburia intestinalis* (Kasahara et al., 2018). *Anaerobutyricum hallii* (Kumari et al., 2021), *Faecalibacterium prausnitzii* (Castañeda–Guillot, 2019; Kumari et al., 2021), *Oscillospira* (Yang et al., 2021), *Christensenella minuta* (Castañeda–Guillot, 2019), *Heyndrickxia coagulans* (formerly known as *Bacillus coagulans*) (Narsing Rao et al., 2023) (EFSA Panel on Biological Hazards (BIOHAZ), 2008), *Lacticaseibacillus paracasei* (Orlando, 2012), *Pediococcus acidilactici* (Klaenhammer, 1993; Jha et al., 2020), and *Lacticaseibacillus rhamnosus GR-1* (Zheng et al., 2020; Petrova et al., 2018; Petrova et al., 2016).

Lactiplantibacillus plantarum (Lactobacillus plantarum, 2018), *Bifidobacterium bifidum* (Palmer et al., 2007).

The main characteristic of these microorganisms is that they are isolated from the human gastrointestinal tract. Therefore they must be resistant to a low pH in the intestinal tract, have no resistance to antibiotics, and exercise pathogen control, although they may be sensitive to oxygen.

Ghannam (2022) found that *L. Paracasei* supernatant edible film was used as an alternative for packaging, providing the main features that these biodegradable packaging should meet towards specific uses for the preservation and improvement of various food products.

8.4 Prebiotics

Prebiotics are nondigestible food substances that selectively promote the growth of probiotics and human health through nutrient enrichment and modulation of gut microbiota and the immune system. Recently, prebiotics have been reviewed and defined by (ISAPP) as *a substrate that is selectively utilized by host microorganisms conferring a health benefit*. This definition expands the concept of prebiotics to possibly include noncarbohydrate substances. These are generally short-chain carbohydrates that escape digestion but are used as substrates for the growth of probiotics in the upper gastrointestinal tract. The industrial development of prebiotics is a priority for improving food quality and human health. Several functional foods containing prebiotics have been used to develop biscuits, candies, tabletop sweeteners, frozen yoghurt, etc. Foods for specified health use (FOSHU) have used several prebiotics in specified food items, like lactulose, FOS, lacto-sucrose, and oligosaccharides.

8.5 Synbiotics

Synbiotics refer to a mixture of prebiotics and probiotics for the improvement of human health. In symbiotic food products, probiotic bacteria selectively utilize the prebiotics as a substrate for their growth. Synbiotics are of two types, complementary and synergistic. A complementary symbiotic consists of a probiotic and a prebiotic that confer one or more health benefits but do not require codependent functions. A synergistic synbiotic contains a substrate that is selectively utilized by coadministered microorganism(s). There are several reports available suggesting that synbiotic foods positively stimulate the host's health and nutrition. It was observed that synbiotics not only increased the number of probiotic bacteria, *Bifidobacterium* and *Lactobacillus,* in the fecal sample with a concomitant reduction in coliform but also increased the level of digestive enzymes such as lactase, lipase, and sucrase and isomaltose. Recently, it has been reported that the use of symbiotics has significantly decreased several cardiovascular risk factors, metabolic syndrome prevalence, and markers of insulin resistance in elderly patients. Other possible mixtures, which sit outside of the synbiotic definition, will likely be developed in the future, as food industry trends around gut health show no sign of slowing. Such mixtures may include combinations of live or killed microorganisms, coupled with potential microbiome-modulating compounds such as fermentable substrates, vitamins, and minerals, phytochemicals, and other plant-based materials. Fermented foods can be considered to be naturally occurring microbiome-targeted mixtures (Cunningham et al., 2021) delivering microbes and microbial substrates together, with a range of bioactive fermentation metabolites (Marco et al., 2017). The increasing popularity of such foods will likely drive everyday consumer recognition of prebiotics, probiotics, and symbiotics moving forward.

8.6 Mechanism of action of probiotics

Probiotics are based on the understanding that the human gut is home to a complex ecosystem of microorganisms, collectively known as the gut microbiome. This microbiome plays a vital role in various bodily functions, including digestion, immune function, and nutrient absorption. Today, probiotics' s definition is understood as encompassing a broader range of potential benefits beyond just gut health.

Mechanisms of action of probiotics and prebiotics are complex, diverse, heterogeneous, and often strain- and compound-specific. While many have been described, there remain calls for increased understanding, especially structure–function explanations of observed health effects and long-term influences. Probiotics interact with both the host and the microbiome via molecular effectors present on the cell structure or secreted as metabolic products. Probiotic metabolites can act on the microbiota through cross-feeding interactions, changes in the gastrointestinal microenvironment (e.g., pH lowering), competition for nutrients and binding sites, and inhibition of growth via the production of strain-specific antibacterial compounds, including bacteriocins (Lebeer et al., 2018; Kleerebezem et al., 2019; Plaza-Diaz et al., 2019; Monteagudo-Mera et al., 2019). Such microbiota-directed effects contribute to the ability of probiotics to mediate health benefits in pathogen overgrowth states such as vaginal and oral dysbiosis (Zendeboodi et al., 2020).

Concerning host cells, probiotic effector molecules can interact directly with receptors in intestinal epithelial, enteroendocrine, and immune cells, as well as vagal afferent fibers. These interactions produce local gut effects, such as enhancement of intestinal barrier integrity and inflammation (e.g., via Toll-like receptors), as well as systemic effects via host immune, endocrine, and nervous system mediators (Kumari et al., 2021; Sharma and Shukla, 2016; Zendeboodi et al., 2020).

Probiotics can also perform enzymatic metabolism of host compounds such as bile salts and ingested xenobiotics (Sharma and Shukla, 2016). Specific probiotic surface-associated effector molecules include pili, lipoteichoic acids, exopolysaccharides, and various surface-layer proteins, many of which are strain-specific and therefore mediate the delivery of strain-specific effects (Kumari et al., 2021; Zendeboodi et al., 2020).

Classical prebiotic effects are mediated through the consumption of the substrate by specific groups within the microbiota, promoting their growth and metabolic activity. Provision of substrate to select group/s of bacteria can also indirectly influence other bacterial groups within the microbiome—promoting growth through cross-feeding interactions as well as inhibitory effects via pathogen displacement.

Resulting changes in microbial composition and metabolite concentrations from prebiotic administration impact host epithelial, immune, nervous, and endocrine signaling and mediate health benefits such as improvements in bowel function, immune response, glucose and lipid metabolism, bone health, and regulation of appetite and satiety (Gibson et al., 2017). Chief by-products of bacterial prebiotic metabolism are the SCFAs acetate, butyrate, and propionate, which are well recognized to interact with these host systems and facilitate many prebiotic effects (Blaak et al., 2020). In addition to nutritive effects on microbes, prebiotic molecules are also recognized to interact directly with host receptors, modulating immune and gut epithelial cell signaling with local effects on inflammation and barrier function (Zendeboodi et al., 2020).

8.7 Sources and strains of probiotics

Still relevant after 19 years, the FAO/WHO definition of probiotics can be translated into four simple and pragmatic criteria allowing one to conclude if specific strains of microorganisms qualify as a probiotic for use in foods and dietary supplements. Probiotic strains must be (1) sufficiently characterized; (2) safe for the intended use; (3) supported by at least one positive human clinical trial conducted according to generally accepted scientific standards or as per recommendations and provisions of local/national authorities when applicable; and (4) alive in the product at an efficacious dose throughout the shelf life (Binda et al., 2020).

Traditionally, the probiotic organisms that feature these in products have been mainly sourced from the gut or traditional fermented foods, such as pickles, yoghurts, and kefir grains. Thus the majority of the probiotics sold and used, both in probiotic research and commercial probiotic development, are from a limited list of genera, which mainly include *Lactobacillus* spp., which are lactic acid bacteria (LAB), and *Bifidobacterium* spp. These strains were often found in fermented foods such as yogurt, kimchi, sauerkraut, and kombucha.

Developments in affordable complete genome sequencing and powerful cultivation methods have allowed the isolation and characterization of a new range of microorganisms from human microbiomes with potential health benefits and the opportunity to be developed as next-generation probiotics (NGPs). NGPs include non-LAB strains such as Bacillus and Saccharomyces. These strains have been shown to have various health benefits, including immune modulation and improved gut barrier function. The scientific evidence supporting the health benefits of probiotics is still evolving. While there is strong evidence for some strains and specific health conditions, more research is needed to understand the full potential of probiotics. Various bacteria, such as Roseburia intestinalis, Faecalibacterium prausnitzii, Eubacterium spp., Bacteroides spp., and Akkermansia muciniphila, have been isolated from the human gut with growing interest in their probiotic potential (O'Toole et al., 2017; Brodmann et al., 2017). Live A. muciniphila is already on the market in a multispecies synbiotic preparation, containing inulin, Bifidobacterium longum subsp. Infants and other anaerobic bacteria (Clostridium beijerinckii, Clostridium butyricum, and Anaerobutyricum hallii) were shown to improve glucose levels in type 2 diabetics (Perraudeau et al., 2020).

In conclusion, many different strains of bacteria can be considered probiotics. The most popular probiotic genus sources and their application as probiotics are extensive in fermented food products, nonfermented food items, as well as functional and nutraceutical dietary supplements, including

- *Lactococcus lactis*
- *Lactobacillus acidophilus*
- *Lactobacillus casei*

- *Lactobacillus plantarum*
- *Lactobacillus rhamnosus*
- *Lactobacillus reuteri*
- *Bifidobacterium bifidum*
- *Bifidobacterium lactis*
- *Bifidobacterium infantis*
- *Bifidobacterium longum*
- *Bifidobacterium breve*
- *Saccharomyces boulardii*
- *Streptococcus thermophilus*
- *Roseburia intestinalis*
- *Faecalibacterium prausnitzii*
- *Eubacterium* spp.
- *Bacteroides* spp.
- *Akkermansia muciniphila*
- *Bacillus coagulans*
- *Bacillus subtilis*
- *Bacillus clausii*
- *Enterococcus faecium*
- *Leuconostoc*
- *Propionibacterium freudenreichii*
- *Pediococcus acidilactici*

8.8 Sources of prebiotics

Currently, a narrow range of confirmed prebiotic substances exists, with Galatians and fructans (e.g., inulin) dominating the market. The desire to stimulate a wider group of commensal organisms has allowed the development of novel candidate prebiotic compounds. These will likely include carbohydrate-based substances derived from plants—the source of traditional prebiotics such as inulin—but may also include those that mimic animal-derived substrates (oligosaccharides present in milk; O-linked glycans present in mucins), yeast-based substances, and many noncarbohydrate substances, including polyphenolics, fatty acids, herbs, and other micronutrients. Over 8000 known polyphenols exist in plants, vegetables, and fruits, and many reach the colon intact, to be utilized by resident microorganisms (Cardona et al., 2013). Some polyphenols have been shown to have prebiotic potential, such as cranberry-rich extracts stimulating *A. muciniphila* (Anhê et al., 2015), or providing antimicrobial action against pathogens (Kumar Singh et al., 2019).

8.9 Uses and benefits of probiotics and prebiotics

Probiotics are now being seen as a powerful tool for personalized, preventive, and therapeutic medicine. Probiotics have a centuries-long history of safe use but have only been recognized as being of economic value during the 20th century and are dominated by food companies, nutritional supplement companies, and dedicated probiotic production companies.

The specific strain of probiotics chosen for a particular individual will depend on their individual health needs and goals, so it is important to consult with a healthcare professional before taking any probiotic supplements, as they may not be suitable for everyone. There are some points to consider:

Dosage: The recommended dosage of probiotics varies depending on the strain and the desired health benefit.
Safety: Probiotics are generally considered safe for healthy individuals, but side effects such as bloating and gas may occur in some people.
Sources: Probiotics can be found in fermented foods such as yogurt, kimchi, sauerkraut, and kombucha, as well as in probiotic supplements. Not all strains of bacteria are equally effective as probiotics. The specific strain or combination of strains used in a probiotic product will depend on the intended health benefit. It is important to choose probiotic supplements that are from reputable manufacturers and have undergone rigorous testing to ensure their safety and efficacy.

Probiotic functionality is mainly based on bacterial interaction with the host gut microbiota through several actions, including

- Enhancing the intestinal barrier function by regulating the proinflammatory cytokines and chemokines' profile.
- Improving the intestinal barrier selectivity through higher mucin, immunoglobulin A (IgA), and defensins production.
- Increasing the production of vitamins, minerals, short-chain fatty acids (SCFAs), and growth regulators to strengthen the intestinal epithelium layer.
- Promoting antiangiogenic factors, cytokines (IL-2 and IL12), and antioxidants.
- Reducing the intestinal pH.
- Increasing antiinflammatory molecule formation to enhance the immune system.
- Modifying intestinal microbiota.
- Regulating mechanisms of apoptosis and cell differentiation.
- Retarding tyrosine kinase and other deleterious pathways.

Recent preclinical and clinical trials highlighted the significance of probiotics intake in the prevention and treatment of gastrointestinal diseases (salmonellosis, traveler's diarrhea, antibiotic-associated diarrhea [AAD], acute diarrhea, Helicobacter pylori gastroenteritis, irritable bowel syndrome [IBS], inflammatory bowel diseases [IBDs], hyperlipidemia and hypercholesterolemia, cancer, lactose intolerance [LI], autoimmune diseases, as well as bone, neurodegenerative, and metabolic disorders).

Probiotics are responsible for mediating a wide variety of health problems, as they are involved in regulating intestinal homeostasis, suppressing procarcinogenic enzymatic activities, interfering with the ability of pathogenic bacteria to infect the intestinal mucosa, improving nutrient bioavailability, reducing symptoms of lactose intolerance, positively influencing the urogenital flora, reducing infections in the gastrointestinal tract, and improving the regulation of intestinal motility (Cunningham et al., 2021).

8.10 Next-generation probiotics

A new era in probiotic research is emerging, enabling scientists to develop personalized probiotic formulations tailored to the unique health needs, preferences, and clinical conditions of individual consumers.

These organisms are referred to as NGPs, also termed live biotherapeutic products (LBPs). NGPs also fit well within the US Food and Drug Administration (FDA) definition of LBP: "a biological product that: (1) contains live organisms, such as bacteria; (2) applies to the prevention, treatment, or cure of a disease or condition of human beings; and (3) is not a vaccine." Emerging research has identified several microorganisms as promising candidates for NGPs, offering potential health benefits beyond traditional strains. Notable examples, in Table 8.1 show the important NGPs and their health benefits.

- Many NGPs are anaerobic and oxygen-sensitive, making them harder to produce and stabilize in supplements.
- Most are in clinical or preclinical development—some are being tested as live biotherapeutic products (LBPs).
- Companies and researchers are working on encapsulation, freeze-drying, and other tech to deliver them effectively.
- The studies underscore the emerging potential of NGPs in addressing metabolic and mental health conditions. However, further research and clinical trials are necessary to fully establish their efficacy and safety profiles.

8.11 Microbiome and personalized medicine

Alteration and disturbance in microbial communities (often referred to as dysbiosis) are associated with diseases ranging from infectious (*Clostridium difficile* infection) to inflammatory (inflammatory bowel disease, rheumatoid arthritis) and metabolic (diabetes, obesity) diseases, suggesting an important role for them in the pathogenesis of multifactorial conditions. The human microbiome can be used as a primary diagnostic biomarker, and researchers today focus on its therapeutic role. The microbiota of each human body organ is unique, and its effects on inflammation and cancer are also distinct in each organ, as well as understanding the changes in the interpersonal microbiome

TABLE 8.1 Important next-generation probiotics and their health benefits.

Genus/Species	Health benefits
1. *Akkermani muciniphila*	Promising in obesity (reduce body weight), diabetes, and metabolic syndrome (improve insulin sensitivity, lower plasma cholesterol)
	Depommier et al. (2019), Jian et al. (2023)
2. *Faecalibacterium prausnitzii*	IBD and depression. Potential role in mental health interventions
	Averina et al. (2024), Hao et al. (2019)
3. *Christensenella minuta*	Obesity, metabolic disorders; modulates lipid metabolism and energy harvest
	Waters and Ley (2019), Ignatyeva (2024)
4. *Bacteroides fragilis*	Autism spectrum disorder (ASD); pediatric ASD
	Lin et al. (2024)
5. *Parabacteroides distasonis*	Obesity (reduces weight gain; metabolic disorders and Insulin resistance (improve glucose tolerance and reduce inflammation))
	Wang et al. (2019), Sun et al. (2023)
6. *Eubacterium hallii*	Supports insulin sensitivity and gut–brain satiety signaling
	Udayappan et al. (2016)
7. *Clostridium butyricum*	Enhances secretion of GLP-1, a gut hormone regulating hunger and glucose; insulin sensitivity; neuroprotective potential
	Wang et al. (2023), Jia et al. (2017)
8. *Bacteroide fragilis*	Potential in treating colitis and neurological disorders
	Chan et al. (2019), Sun et al. (2019)
9. *Heyndrickxia coagulans* formerly known as *Bacillus coagulans*	Increased immune response to viral challenge; treatment of Clostridium difficile-associated disease
	Liang et al. (2024), Baron (2009), Fitzpatrick (2013)

(*Continued*)

TABLE 8.1 (Continued)

Genus/Species	Health benefits
10. Roseburia spp.	Gut antiinflammatory effects, supports colonocyte health. Low abundance linked to gut dysbiosis, obesity, type 2 diabetes
	Jaime Cloyd (2025), Tamanai-Shacoori et al. (2017)
11. Clostridium clusters IV and XIVa (e.g., *Clostridium leptum* and *Clostridium butyricum*)	Promote regulatory T cells (Tregs); maintain intestinal immune tolerance. Useful in IBD, allergies, and autoimmune conditions.
	Hussein (2021), Guo et al. (2020)
12. Lactobacillus reuteri	On the edge of NGP classification: stimulates oxytocin production and social behaviors in mice. Shown to reduce stress-related behavior and promote resilience.
	Mu et al. (2018)
13. Lactic acid bacteria innovations • Novel strains of *Lactobacillus* and *Bifidobacterium*, such as – *Lactobacillus plantarum* WCFS1 – *Bifidobacterium longum subsp.* infantis EVC001 (used in infant nutrition)	Engineered strains for vitamin production or specific immune responses
	Sanders et al. (2019)
14. Next-gen yeast probiotics	Continues to evolve in formulations for gut health, antibiotic-associated diarrhea
Saccharomyces boulardii CNCM I-745	McFarland (2010), Kelesidis and Pothoulakis (2012)
15. Spore-forming *Bacillus* spp.	Heat-resistant, good for food and supplement delivery, gut health and immune benefits
E.g., Bacillus coagulans, Bacillus subtilis	Elshaghabee et al. (2017), Maathuis et al. (2010)

and the frequency of microbial populations in different positions in organs, leading to information potentially related to the development of diseases such as cancer

An important aspect of the microbiome is its resilience and plasticity, making it more mutable than human cells. It maintains a relatively stable steady state despite temporary insults such as travel, diet, and antibiotics, and it can be modified by dietary and other-directed therapies. Further, the interindividual variability in composition and metabolic capacity of the microbiome plays an important role in interactions with the environment, resulting in the development of disease, as well as response to treatment and development of adverse events. These make it a targetable factor, all pointing towards the importance of incorporating the microbiome in precision medicine.

New therapeutic approaches, called personalized medicine, have opened a new window in medical science, and the association between microbiome and personalized medicine seems to be one of the most interesting aspects of subsequent research and has a pivotal perspective on the treatment of diseases such as cancer (Kashyap et al., 2017).

The importance of incorporating the microbiome as a component of personalized or precision medicine is to improve diagnosis, reduce disease risk, and optimize early detection and treatment (Kashyap et al., 2017). Microbial fingerprints could serve as precise, noninvasive, accessible, and economic tools that could be used for personalized disease diagnosis, including phenotypes, severity, and prognosis. The role of the microbiome in the metabolism of many chemical compounds makes it a key player in determining drug availability, efficacy, and toxicity, making it indispensable for developing personalized drug therapies. Finally, the ability to manipulate the microbiome makes it appealing for developing personalized treatment approaches by using precision microbiome targeting approaches. The use of approaches targeting specific microbial pathways tailored to an individual's microbiota may enable the development of treatment for multifactorial disorders such as inflammatory bowel disease, obesity, and diabetes mellitus. The development of precision probiotics using genetic engineering approaches, next-generation prebiotics resulting from a better understanding of metabolic interactions among members of the microbial ecosystem, and personalized dietary therapies tailored to an individual's microbiota will form the new frontier in the field of personalized medicine (Table 8.2).

8.12 Disease prevention using probiotics

A study was conducted that showed the strongest evidence in favor of probiotics lies in the prevention or treatment of five disorders: necrotizing enterocolitis, acute infectious diarrhea, acute respiratory tract infections, antibiotic-associated diarrhea, and infant colic. Probiotic mechanisms of action include the inhibition of bacterial adhesion; enhanced mucosal barrier function; modulation of the innate and adaptive immune systems (including induction of tolerogenic dendritic cells and regulatory T cells); secretion of

TABLE 8.2 **Probiotic strains and proposed health benefits.**

Probiotic strain	Proposed health benefits	References
Akkermansia muciniphila	Gut barrier integrity, metabolic health, obesity	Cani et al. (2022), *Nature Reviews Gastroenterology*
Faecalibacterium prausnitzii	Antiinflammatory effects, IBD management	Martín et al. (2021), *Microorganisms*
Clostridium butyricum	Butyrate production, gut immune modulation	Guo et al. (2020), *Frontiers in Microbiology*
Bacillus coagulans	Digestive support, immune response	Majeed et al. (2021), *Beneficial Microbes*
Escherichia coli Nissle 1917	Diarrhea, IBD, microbiome balance	Sonnenborn and Schulze (2020), *Therap Adv Gastroenterol*
Bacteroides fragilis (nontoxigenic)	Immunomodulation, gut homeostasis	Round and Mazmanian (2020), *Cell Host & Microbe*
Lactococcus lactis	Antiinflammatory, skin and respiratory health	Lebeer et al. (2018), *Trends in Microbiology*
Roseburia hominis	SCFA production, gut health	Patterson et al. (2022), *Gut Microbes*
Next-gen synbiotics (engineered or personalized strains)	Targeted gut modulation, chronic disease support	Zmora et al. (2021), *Cell*

bioactive metabolites; and regulation of the enteric and central nervous systems (Liu et al., 2018) (Table 8.3).

8.13 Therapeutic and functional use of probiotics

Probiotics mostly belong to Gram-positive bacteria, having a natural habitat in the human gut. Probiotics may alleviate gastrointestinal dysbiosis, lower serum cholesterol, ameliorate cancer, and prevent allergic and autoimmune disorders. Nearly 2000 clinical research studies regarding various prophylactic and therapeutic health benefits related to probiotics have been published. In addition, an outstanding pace was observed in exploring various molecular mechanisms by which probiotics interact with the host and exert their impact

TABLE 8.3 Preventive effects of probiotics.

Health area	Key findings	Probiotic strains	Study type/population	Source
Colorectal cancer	Regular yogurt consumption (≥2 servings/week) linked to 20% reduced risk of *Bifidobacterium*-positive proximal colon cancer	*Bifidobacterium*, *Lactobacillus*	Cohort study; >151,000 participants	Food and Wine
Cognitive decline	*Lactobacillus rhamnosus* GG improved cognitive scores in individuals with mild cognitive impairment over 3 months	*Lactobacillus rhamnosus* GG	Clinical trial; older adults	News-Medical
Antibiotic-associated diarrhea (AAD)	Probiotics reduced AAD incidence by 38% without increasing adverse events	Various strains	Metaanalysis; 9312 adults	Journal of Clinical Gastroenterology
Early childhood caries	Probiotic supplementation lowered caries incidence from 17.5% to 8.5% in preschoolers	Various strains	Metaanalysis; 2363 children	PubMed
Oxidative stress	Probiotics improved antioxidant biomarkers (e.g., GSH and TAC) and reduced MDA levels, especially in individuals with glucose intolerance	Various strains	Umbrella metaanalysis	Frontiers in Nutrition
Immune function	Probiotics may shorten colds, enhance flu vaccine efficacy, and aid in managing intestinal diseases like C. diff infections	*Lactobacillus*, *Bifidobacterium*, *Saccharomyces*	Review article	Verywell Health
Cardiovascular health	Probiotics associated with improved cholesterol levels, blood pressure, and inflammation markers	Various strains	Review article	PubMed
Preterm infant health	Multistrain probiotics reduced mortality and morbidity in preterm infants more effectively than single strains	*Lactobacillus*, *Bifidobacterium* combinations	Systematic review and network metaanalysis	JAMA Pediatrics

on the target cells. However, there is still a gap in translating the several recognized probiotics' health benefits mechanisms into target therapies with specific probiotics.

Probiotics and prebiotics have emerged as safe, viable options for various therapeutic strategies. They have key functional properties that could help meet the majority of our nutritional and therapeutic supplementation needs. These bacteria have demonstrated promising results in clinical trials for a variety of diseases and illnesses, including diarrhea, IBS, and food allergies. Furthermore, the role of probiotics in the prevention and treatment of diabetes, obesity, cancer, and disorders caused by pathogenic bacteria is an intriguing and fast-developing study area. Dietary probiotic supplementation is most commonly associated with dairy products, although probiotics can also be found in fermented nondairy foods. Probiotics and prebiotics have unquestionable physiological, nutritional, and immunologic benefits, including their involvement in fighting viral illnesses such as seasonal viral illnesses. More clinical and laboratory studies into their significant contribution to COVID-19 infection are still needed (Manzoor et al., 2022). The health effects of probiotics are often strain-specific. Effective outcomes depend on appropriate dosages and duration of probiotic intake. Probiotics are generally safe for healthy individuals, but those with compromised immune systems should consult healthcare providers before use (Table 8.4).

8.14 Safety and possible adverse effects (gap/problem)

The more commonly exploited strains/species among the lactobacilli and bifidobacteria have been accepted as having Generally Regarded as Safe (GRAS) status in the United States or have been granted Qualified Presumption of Safety status by the European Food Safety Authority (EFSA). Other probiotics currently available in the marketplace, such as Saccharomyces and Bacillus spp. Some strains are more effective than others, and some may not be suitable for everyone.

It is important to note that not all probiotics are created equal. Some strains are more effective than others, and some may not be suitable for everyone. Probiotics are used as functional foods or supplements in the prevention or treatment of some diseases, but besides the beneficial effects, they may have negative effects if not appropriately used (Sotoudegan et al., 2019). Receiving probiotics in high-risk populations poses some health complications. Along with the genetic characteristics of intestinal microbiota in each human case, probiotics and their effects on adults are influenced by environmental factors, diet, and use of antibiotics. Hence, viewing nonidentical outcomes in different age groups of consumers would be predictable events, and therefore precaution is necessary to be taken just before receiving probiotics. Here are some key points to consider when evaluating prebiotics:

TABLE 8.4 Recent studies on the therapeutic and functional uses of probiotics across various health conditions.

Health condition	Probiotic strains	Key findings	Source
Cognitive function in Alzheimer's disease (AD)	*Lactobacillus* spp. (*L. acidophilus, L. casei, L. fermentum, L. plantarum, L. salivarius, L. lactis*); *Bifidobacterium* spp. (*B. bifidum, B. lactis, B. breve, B. longum*)	Improved cognitive function in AD patients through modulation of gut–brain axis and neurotransmitter production	The Egyptian Journal of Neurology, Psychiatry and Neurosurgery
Multiple sclerosis (MS)	Various probiotic strains (not specified)	Altered gut microbiome leading to improved immune responses and reduced inflammation in MS patients	Biomedicine and Pharmacotherapy
Autism spectrum disorders (ASD)	Various probiotic strains (not specified)	Probiotic supplementation improved core and associated behavioral symptoms in individuals with ASD	Child and Adolescent Psychiatry and Mental Health
Functional gastrointestinal disorders in toddlers	*Lactobacillus reuteri*	Prevented and managed conditions like functional constipation, infantile colic, regurgitation, and diarrhea in toddlers	International Journal of Gastroenterology
Parkinson's disease	*Lactobacillus plantarum* PS128; *Bifidobacterium bifidum* BGN4; *Bifidobacterium longum* BORI	Improved cognitive and memory functions, and alleviated constipation in Parkinson's patients	Journal of Translational Medicine

(*Continued*)

TABLE 8.4 (Continued)

Health condition	Probiotic strains	Key findings	Source
General immune function	*Lactobacillus rhamnosus* GG; various *Lactobacillus* and *Bifidobacterium* spp.	Enhanced immune responses, reduced inflammation, and improved gut barrier function	Frontiers in Microbiology
Mental health (anxiety and depression)	Probiotics from fermented foods like kimchi and miso	Reduced anxiety and depression symptoms by positively impacting gut bacteria and neurotransmitter production	EMBO Molecular Medicine
Age-related cognitive decline	*Lactobacillus rhamnosus* GG (LGG)	Improved cognitive function in middle-aged and older adults with mild cognitive impairment after 12 weeks of supplementation	Health.com
Neurological disorders	Various probiotic strains (not specified)	Probiotics may alleviate symptoms of neurological disorders through antiinflammatory properties and neurotransmitter production	Biomedicine and Pharmacotherapy
Immune system modulation in infants	Various probiotic strains (not specified)	Probiotics interact with intestinal immune cells to maintain immunologic equilibrium and promote health in infants	American Journal of Pediatrics

- Type of prebiotic: Different prebiotics have different effects on the gut microbiome. Common types include inulin, fructooligosaccharides (FOS), galactooligosaccharides (GOS), and resistant starch. Choose one based on your desired health goals, considering the specific benefits of each type.
- Dosage and source: Research suggests optimal dosages for specific prebiotics and their effectiveness depending on the source (e.g., food vs supplements). Start with lower doses and adjust gradually while monitoring your digestive response.
- Purity and potential contaminants: Ensure the prebiotic is free of impurities or additives that could negatively impact your gut health. Choose trusted brands with high-quality products.
- Synergy with probiotics: Certain prebiotics may work synergistically with specific probiotic strains, enhancing their effectiveness. Research compatible combinations for potential benefits.
- Here are some key points to consider when evaluating probiotics:
- Strain-specific benefits: Different probiotic strains offer unique benefits. Focus on strains with research supporting your specific health concerns, such as digestion, immunity, or mental health.
- Colony-forming units (CFUs): The number of live bacteria measured in CFUs determines the potential effectiveness of a probiotic supplement. Choose products with adequate CFUs based on your needs.
- Delivery format: Probiotics can be found in capsules, tablets, yogurt, fermented foods, etc. Consider factors like shelf life, convenience, and potential for degradation based on the format.
- Potential side effects: While generally safe, some probiotics may cause initial digestive discomfort for certain individuals. Start with low doses and monitor your tolerance.
- Interactions with medications: Certain medications may interact with probiotics, impacting their effectiveness or causing side effects. Consult your doctor if you take any medications before starting probiotics.

General evaluation points:

- Clinical evidence: Look for prebiotics and probiotics supported by robust clinical research demonstrating their effectiveness for your desired health outcomes.
- Long-term effects: While short-term benefits may be promising, consider studies investigating the long-term safety and efficacy of specific prebiotics and probiotics.
- Individual response: Your response to prebiotics and probiotics will be unique. Monitor your reactions and adjust dosages or types as needed.
- Consultation with healthcare professionals: Consult your doctor or a registered dietitian before taking any prebiotics or probiotics, especially if you have underlying health conditions.

References

Aguilar-Toalá, J.E., Garcia-Varela, R., Garcia, H.S., Mata-Haro, V., González-Córdova, A.F., Vallejo-Cordoba, B., et al., 2018. Postbiotics: an evolving term within the functional foods field. Trends in Food Science and Technology 75, 105–114. Available at https://doi.org/10.1016/j.tifs.2018.03.009.

Anhê, F.F., Roy, D., Pilon, G., Dudonné, S., Matamoros, S., Varin, T.V., et al., 2015. A polyphenol-rich cranberry extract protects from diet-induced obesity, insulin resistance and intestinal inflammation in association with increased *Akkermansia* spp. population in the gut microbiota of mice. Gut 64 (6), 872–883. Available at https://doi.org/10.1136/gutjnl-2014-307142, http://gut.bmj.com/content/64/6/872.full.pdf+html.

Averina, O.V., Poluektova, E.U., Zorkina, Y.A., Kovtun, A.S., Danilenko, V.N., 2024. Human gut microbiota for diagnosis and treatment of depression. Federation International Journal of Molecular Sciences. 25 (11). Available at https://doi.org/10.3390/ijms25115782, http://www.mdpi.com/journal/ijms.

Baron, M., 2009. Original research: a patented strain of *Bacillus coagulans* increased immune response to viral challenge. Postgraduate Medicine 121 (2), 114–118. Available at https://doi.org/10.3810/pgm.2009.03.1971, http://www.postgradmed.com/beta/images/pdf/pgm_03_2009/1971.pdf?view_pdf=I+Agree+to+these+terms.

Belkaid, Y., Hand, T.W., 2014. Role of the microbiota in immunity and inflammation. Cell 157 (1), 121–141. Available at https://doi.org/10.1016/j.cell.2014.03.011, https://www.sciencedirect.com/journal/cell.

Binda, S., Hill, C., Johansen, E., Obis, D., Pot, B., Sanders, M.E., et al., 2020. Criteria to qualify microorganisms as "probiotic" in foods and dietary supplements. Frontiers in Microbiology 11. Available at https://doi.org/10.3389/fmicb.2020.01662, https://www.frontiersin.org/journals/microbiology#.

Blaak, E.E., Canfora, E.E., Theis, S., Frost, G., Groen, A.K., Mithieux, G., et al., 2020. Short chain fatty acids in human gut and metabolic health. Beneficial Microbes 11 (5), 411–455. Available at https://doi.org/10.3920/bm2020.0057.

Brodmann, T., Endo, A., Gueimonde, M., Vinderola, G., Kneifel, W., de Vos, W.M., et al., 2017. Safety of novel microbes for human consumption: practical examples of assessment in the European Union. Frontiers in Microbiology 8. Available at https://doi.org/10.3389/fmicb.2017.01725, http://journal.frontiersin.org/article/10.3389/fmicb.2017.01725/full.

Cardona, F., Andrés-Lacueva, C., Tulipani, S., Tinahones, F.J., Queipo-Ortuño, M.I., 2013. Benefits of polyphenols on gut microbiota and implications in human health. Journal of Nutritional Biochemistry 24 (8), 1415–1422. Available at https://doi.org/10.1016/j.jnutbio.2013.05.001.

Castañeda–Guillot, C., 2019. Probióticos de nueva generación. Belize Journal of Medicine 8 (2), 26–33.

Chan, J.L., Wu, S., Geis, A.L., Chan, G.V., Gomes, T.A.M., Beck, S.E., et al., 2019. Non-toxigenic *Bacteroides fragilis* (NTBF) administration reduces bacteria-driven chronic colitis and tumor development independent of polysaccharide A. Mucosal Immunology 12 (1), 164–177. Available at https://doi.org/10.1038/s41385-018-0085-5, http://www.nature.com/mi/lib/index.html.

Cunningham, M., Azcarate-Peril, M.A., Barnard, A., Benoit, V., Grimaldi, R., Guyonnet, D., et al., 2021. Shaping the future of probiotics and prebiotics. Trends in Microbiology 29 (8), 667–685. Available at https://doi.org/10.1016/j.tim.2021.01.003, www.elsevier.com/locate/tim.

Depommier, C., Everard, A., Druart, C., Plovier, H., Van Hul, M., Vieira-Silva, S., et al., 2019. Supplementation with *Akkermansia muciniphila* in overweight and obese human volunteers: a proof-of-concept exploratory study. Nature Medicine 25 (7), 1096–1103. Available at https://doi.org/10.1038/s41591-019-0495-2, http://www.nature.com/nm/index.html.

EFSA Panel on Biological Hazards (BIOHAZ), 2008. The maintenance of the list of QPS microorganisms intentionally added to food or feed - Scientific Opinion of the Panel on Biological Hazards. EFSA Journal 6 (12), 923. Available at https://doi.org/10.2903/j.efsa.2008.923.

Elshaghabee, F.M.F., Rokana, N., Gulhane, R.D., Sharma, C., Panwar, H., 2017. *Bacillus* as potential probiotics: status, concerns, and future perspectives. Frontiers in Microbiology 8. Available at https://doi.org/10.3389/fmicb.2017.01490, https://fjfsdata01prod.blob.core.windows.net/articles/files/271541/pubmed-zip/.versions/1/.package-entries/fmicb-08-01490/fmicb-08-01490.pdf.

Fitzpatrick, L.R., 2013. Probiotics for the treatment of *Clostridium difficile* associated disease. World Journal of Gastrointestinal Pathophysiology 4 (3), 47. Available at https://doi.org/10.4291/wjgp.v4.i3.47.

Ghannam, R.B., 2022. Biosynthesis of Edible Films to Control Food-born Pathogens and Spoilage Microorganisms of *Solanaceous crops*.

Gibson, G.R., Hutkins, R., Sanders, M.E., Prescott, S.L., Reimer, R.A., Salminen, S.J., et al., 2017. Expert consensus document: the International Scientific Association for Probiotics and Prebiotics (ISAPP) consensus statement on the definition and scope of prebiotics. Nature Reviews Gastroenterology and Hepatology 14 (8), 491–502. Available at https://doi.org/10.1038/nrgastro.2017.75, http://www.nature.com/nrgastro/index.html.

Gopalakrishnan, V., Helmink, B.A., Spencer, C.N., Reuben, A., Wargo, J.A., 2018. The influence of the gut microbiome on cancer, immunity, and cancer immunotherapy. Cancer Cell 33 (4), 570–580. Available at https://doi.org/10.1016/j.ccell.2018.03.015, https://www.journals.elsevier.com/cancer-cell.

Guo, P., Zhang, K., Ma, X., He, P., 2020. Clostridium species as probiotics: potentials and challenges. Journal of Animal Science and Biotechnology 11 (1). Available at https://doi.org/10.1186/s40104-019-0402-1.

Hao, Z., Wang, W., Guo, R., Liu, H., 2019. Faecalibacterium prausnitzii (ATCC 27766) has preventive and therapeutic effects on chronic unpredictable mild stress-induced depression-like and anxiety-like behavior in rats. Psychoneuroendocrinology 104, 132–142. Available at https://doi.org/10.1016/j.psyneuen.2019.02.025.

Hussein, L.A., 2021. Novel prebiotics and next-generation probiotics: opportunities and challenges. Functional Foods and Nutraceuticals in Metabolic and Non-communicable Diseases. Elsevier, Egypt, pp. 431–457. Available at https://www.sciencedirect.com/book/9780128198155, 10.1016/B978-0-12-819815-5.00055-0.

Ignatyeva, O., 2024. New insight into the interplay between probiotics, prebiotics, postbiotics, and gut microbiota in glycolipid metabolism, inflammation, and associated metabolic disorders. Frontiers in Microbiology. 14 (2023).

Jaime Cloyd, N.D., 2025. Understanding the Gut Bacteria Roseburia: A Comprehensive Guide to Health Benefits and Testing Patient Levels. 2025.

Jastrząb, R., Graczyk, D., Siedlecki, P., 2021. Molecular and cellular mechanisms influenced by postbiotics. International Journal of Molecular Sciences 22 (24), 13475. Available at https://doi.org/10.3390/ijms222413475.

Jha, R., Das, R., Oak, S., Mishra, P., 2020. Probiotics (direct-fed microbials) in poultry nutrition and their effects on nutrient utilization, growth and laying performance, and gut health: a systematic review. Animals: An Open Access Journal from MDPI 10 (10), 1863. Available at https://doi.org/10.3390/ani10101863.

Jian, H., Liu, Y., Wang, X., Dong, X., Zou, X., 2023. Akkermansia muciniphila as a next-generation probiotic in modulating human metabolic homeostasis and disease progression: a role mediated by gut–liver–brain axes? International Journal of Molecular Sciences 24 (4), 3900. Available at https://doi.org/10.3390/ijms24043900.

Jia, L., Li, D., Feng, N., Shamoon, M., Sun, Z., Ding, L., et al., 2017. Anti-diabetic effects of *Clostridium butyricum* CGMCC0313.1 through promoting the growth of gut butyrate-producing bacteria in type 2 diabetic mice. Scientific Reports 7 (1). Available at https://doi.org/10.1038/s41598-017-07335-0, www.nature.com/srep/index.html.

Kasahara, K., Krautkramer, K.A., Org, E., Romano, K.A., Kerby, R.L., Vivas, E.I., et al., 2018. Interactions between *Roseburia intestinalis* and diet modulate atherogenesis in a murine model. Nature Microbiology 3 (12), 1461–1471. Available at https://doi.org/10.1038/s41564-018-0272-x, www.nature.com/nmicrobiol/.

Kashyap, P.C., Chia, N., Nelson, H., Segal, E., Elinav, E., 2017. Microbiome at the frontier of personalized medicine. Mayo Clinic Proceedings 92 (12), 1855–1864. Available at https://doi.org/10.1016/j.mayocp.2017.10.004, http://www.journals.elsevier.com/mayo-clinic-proceedings.

Kelesidis, T., Pothoulakis, C., 2012. Efficacy and safety of the probiotic *Saccharomyces boulardii* for the prevention and therapy of gastrointestinal disorders. Therapeutic Advances in Gastroenterology 5 (2), 111–125. Available at https://doi.org/10.1177/1756283x11428502.

Klaenhammer, T.R., 1993. Genetics of bacteriocins produced by lactic acid bacteria. FEMS Microbiology Reviews 12 (1-3), 39–85. Available at https://doi.org/10.1111/j.1574-6976.1993.tb00012.x.

Kleerebezem, M., Binda, S., Bron, P.A., Gross, G., Hill, C., van Hylckama Vlieg, J.E., et al., 2019. Understanding mode of action can drive the translational pipeline towards more reliable health benefits for probiotics. Current Opinion in Biotechnology 56, 55–60. Available at https://doi.org/10.1016/j.copbio.2018.09.007, http://www.elsevier.com/locate/copbio.

Kumar Singh, A., Cabral, C., Kumar, R., Ganguly, R., Kumar Rana, H., Gupta, A., et al., 2019. Beneficial effects of dietary polyphenols on gut microbiota and strategies to improve delivery efficiency. Nutrients 11 (9), 2216. Available at https://doi.org/10.3390/nu11092216.

Kumari, M., Singh, P., Nataraj, B.H., Kokkiligadda, A., Naithani, H., Azmal Ali, S., et al., 2021. Fostering next-generation probiotics in human gut by targeted dietary modulation: an emerging perspective. Food Research International 150 (Pt A), 110716. Available at https://doi.org/10.1016/j.foodres.2021.110716.

Kumar, R., Bansal, P., Singh, J., Dhanda, S., 2020. Purification, partial structural characterization and health benefits of exopolysaccharides from potential probiotic *Pediococcus acidilactici* NCDC 252. Process Biochemistry 99, 79–86. Available at https://doi.org/10.1016/j.procbio.2020.08.028.

Lactobacillus plantarum. Viticulture and Enology. 2018.

Lebeer, S., Bron, P.A., Marco, M.L., Van Pijkeren, J.P., O'Connell Motherway, M., Hill, C., et al., 2018. Identification of probiotic effector molecules: present state and future perspectives. Current Opinion in Biotechnology 49, 217–223. Available at https://doi.org/10.1016/j.copbio.2017.10.007, http://www.elsevier.com/locate/copbio.

Liang, J., Li, C., Chen, Z., Guo, F., Dou, J., Wang, T., et al., 2024. Progress of research and application of *Heyndrickxia coagulans* (*Bacillus coagulans*) as probiotic bacteria. Frontiers in Cellular and Infection Microbiology 14. Available at https://doi.org/10.3389/fcimb.2024.1415790, http://www.frontiersin.org/Cellular_and_Infection_Microbiology/archive.

Lin, C.-hui, Zeng, T., Lu, C.-wei, Li, D.-yang, Liu, Y.-ying, Li, B.-mei, et al., 2024. Efficacy and safety of *Bacteroides fragilis* BF839 for pediatric autism spectrum disorder: a randomized clinical trial. Frontiers in Nutrition 11. Available at https://doi.org/10.3389/fnut.2024.1447059.

Liu, Y., Tran, D.Q., Rhoads, J.M., 2018. Probiotics in disease prevention and treatment. Journal of Clinical Pharmacology 58. Available at https://doi.org/10.1002/jcph.1121. S164, Available from: http://onlinelibrary.wiley.com/journal/10.1002/(ISSN)1552-4604.

Maathuis, A., Keller, D., Farmer, S., 2010. Survival and metabolic activity of the GanedenBC 30 strain of *Bacillus coagulans* in a dynamic in vitro model of the stomach and small intestine. Beneficial Microbes 1 (1), 31–36. Available at https://doi.org/10.3920/bm2009.0009.

Manzoor, S., Wani, S.M., Ahmad Mir, S., Rizwan, D., 2022. Role of probiotics and prebiotics in mitigation of different diseases. Nutrition 96, 111602. Available at https://doi.org/10.1016/j.nut.2022.111602.

Marco, M.L., Heeney, D., Binda, S., Cifelli, C.J., Cotter, P.D., Foligné, B., et al., 2017. Health benefits of fermented foods: microbiota and beyond. Current Opinion in Biotechnology 44, 94–102. Available at https://doi.org/10.1016/j.copbio.2016.11.010, http://www.elsevier.com/locate/copbio.

McFarland, L.V., 2010. Systematic review and meta-analysis of *Saccharomyces boulardii* in adult patients. World Journal of Gastroenterology 16 (18), 2202. Available at https://doi.org/10.3748/wjg.v16.i18.2202.

Monteagudo-Mera, A., Rastall, R.A., Gibson, G.R., Charalampopoulos, D., Chatzifragkou, A., 2019. Adhesion mechanisms mediated by probiotics and prebiotics and their potential impact on human health. Applied Microbiology and Biotechnology 103 (16), 6463–6472. Available at https://doi.org/10.1007/s00253-019-09978-7, link.springer.de/link/service/journals/00253/index.htm.

Mu, Q., Tavella, V.J., Luo, X.M., 2018. Role of *Lactobacillus reuteri* in human health and diseases. Frontiers in Microbiology 9. Available at https://doi.org/10.3389/fmicb.2018.00757, https://www.frontiersin.org/articles/10.3389/fmicb.2018.00757/full.

Narsing Rao, M.P., Banerjee, A., Liu, G.-H., Thamchaipenet, A., 2023. Genome-based reclassification of *Bacillus acidicola*, *Bacillus pervagus* and the genera Heyndrickxia, Margalitia and Weizmannia. International Journal of Systematic and Evolutionary Microbiology 73 (7). Available at https://doi.org/10.1099/ijsem.0.005961.

Nataraj, B.H., Ali, S.A., Behare, P.V., Yadav, H., 2020. Postbiotics-parabiotics: the new horizons in microbial biotherapy and functional foods. Microbial Cell Factories 19 (1). Available at https://doi.org/10.1186/s12934-020-01426-w, http://www.microbialcellfactories.com/home/.

Navarro, D., 2021. Terapias complementarias en diarrea aguda. Archivos Venezolanos de Puericultura y Pediatría 84, 62–71.

O'Toole, P.W., Marchesi, J.R., Hill, C., 2017. Next-generation probiotics: the spectrum from probiotics to live biotherapeutics. Nature Microbiology 2. Available at https://doi.org/10.1038/nmicrobiol.2017.57, www.nature.com/nmicrobiol/.

Orlando, A., 2012. Nutrition and Cancer 64 (7), 1103–1111.

Palmer, C., Bik, E.M., DiGiulio, D.B., Relman, D.A., Brown, P.O., 2007. Development of the human infant intestinal microbiota. PLoS Biology 5 (7), 1556–1573. Available at https://doi.org/10.1371/journal.pbio.0050177, http://biology.plosjournals.org/archive/1545-7885/5/7/pdf/10.1371_journal.pbio.0050177-L.pdf.

Perraudeau, F., McMurdie, P., Bullard, J., Cheng, A., Cutcliffe, C., Deo, A., et al., 2020. Improvements to postprandial glucose control in subjects with type 2 diabetes: a multicenter, double blind, randomized placebo-controlled trial of a novel probiotic formulation. BMJ Open

Diabetes Research and Care 8 (1), e001319. Available at https://doi.org/10.1136/bmjdrc-2020-001319.

Petrova, M.I., Lievens, E., Verhoeven, T.L.A., MacKlaim, J.M., Gloor, G., Schols, D., et al., 2016. The lectin-like protein 1 in *Lactobacillus rhamnosus* GR-1 mediates tissue-specific adherence to vaginal epithelium and inhibits urogenital pathogens. Scientific Reports 6. Available at https://doi.org/10.1038/srep37437, www.nature.com/srep/index.html.

Petrova, M.I., Macklaim, J.M., Wuyts, S., Verhoeven, T., Vanderleyden, J., Gloor, G.B., et al., 2018. Comparative genomic and phenotypic analysis of the vaginal probiotic *Lactobacillus rhamnosus* GR-1. Frontiers in Microbiology 9. Available at https://doi.org/10.3389/fmicb.2018.01278, https://www.frontiersin.org/articles/10.3389/fmicb.2018.01278/full.

Plaza-Diaz, J., Ruiz-Ojeda, F.J., Gil-Campos, M., Gil, A., 2019. Mechanisms of action of probiotics. Advanced Nutrition 10 (suppl_1), S49–S66. Available at 10.1093/advances/nmy063.

Salminen, S., Collado, M.C., Endo, A., Hill, C., Lebeer, S., Quigley, E.M.M., et al., 2021. The International Scientific Association of Probiotics and Prebiotics (ISAPP) consensus statement on the definition and scope of postbiotics. Nature Reviews Gastroenterology and Hepatology 18 (9), 649–667. Available at https://doi.org/10.1038/s41575-021-00440-6, http://www.nature.com/nrgastro/index.html.

Sanders, M.E., Merenstein, D.J., Reid, G., Gibson, G.R., Rastall, R.A., 2019. Probiotics and prebiotics in intestinal health and disease: from biology to the clinic. Nature Reviews Gastroenterology and Hepatology 16 (10), 605–616. Available at https://doi.org/10.1038/s41575-019-0173-3, http://www.nature.com/nrgastro/index.html.

Schrezenmeir, J., de Vrese, M., 2001. Probiotics, prebiotics, and synbiotics—approaching a definition. The American Journal of Clinical Nutrition 73 (2), 361s. Available at https://doi.org/10.1093/ajcn/73.2.361s.

Sharma, M., Shukla, G., 2016. Metabiotics: one step ahead of probiotics; an insight into mechanisms involved in anticancerous effect in colorectal cancer. Frontiers in Microbiology 7. Available at https://doi.org/10.3389/fmicb.2016.01940.

Sotoudegan, F., Daniali, M., Hassani, S., Nikfar, S., Abdollahi, M., 2019. Reappraisal of probiotics' safety in human. Food and Chemical Toxicology 129, 22–29. Available at https://doi.org/10.1016/j.fct.2019.04.032.

Sun, Y., Nie, Q., Zhang, S., He, H., Zuo, S., Chen, C., et al., 2023. Parabacteroides distasonis ameliorates insulin resistance via activation of intestinal GPR109a. Nature Communications 14 (1). Available at https://doi.org/10.1038/s41467-023-43622-3.

Sun, F., Zhang, Q., Zhao, J., Zhang, H., Zhai, Q., Chen, W., 2019. A potential species of next-generation probiotics? The dark and light sides of Bacteroides fragilis in health. Food Research International 126, 108590. Available at https://doi.org/10.1016/j.foodres.2019.108590.

Tamanai-Shacoori, Z., Smida, I., Bousarghin, L., Loreal, O., Meuric, V., Fong, S.B., et al., 2017. Roseburia spp.: a marker of health? Future Microbiology 12 (2), 157–170. Available at https://doi.org/10.2217/fmb-2016-0130, http://www.futuremedicine.com/loi/fmb.

Taverniti, V., Guglielmetti, S., 2011. The immunomodulatory properties of probiotic microorganisms beyond their viability (ghost probiotics: proposal of paraprobiotic concept). Genes and Nutrition 6 (3), 261–274. Available at https://doi.org/10.1007/s12263-011-0218-x.

Udayappan, S., Manneras-Holm, L., Chaplin-Scott, A., Belzer, C., Herrema, H., Dallinga-Thie, G.M., et al., 2016. Oral treatment with Eubacterium hallii improves insulin sensitivity in db/db mice. NPJ Biofilms and Microbiomes 2. Available at https://doi.org/10.1038/npjbiofilms.2016.9, https://www.nature.com/npjbiofilms/.

Vinderola, G., Sanders, M.E., Cunningham, M., Hill, C., 2024. Frequently asked questions about the ISAPP postbiotic definition. Frontiers in Microbiology 14. Available at https://doi.org/10.3389/fmicb.2023.1324565.

Vinderola, G., Sanders, M.E., Salminen, S., 2022. The concept of postbiotics. Foods 11 (8), 1077. Available at https://doi.org/10.3390/foods11081077.

Wang, Xl, Chen, Wj, Jin, R., Xu, X., Wei, J., Huang, H., et al., 2023. Engineered probiotics *Clostridium butyricum*-pMTL007-GLP-1 improves blood pressure via producing GLP-1 and modulating gut microbiota in spontaneous hypertension rat models. Microbial Biotechnology 16 (4), 799–812. Available at https://doi.org/10.1111/1751-7915.14196, http://onlinelibrary.wiley.com/journal/10.1111/(ISSN)1751-7915.

Wang, K., Liao, M., Zhou, N., Bao, L., Ma, K., Zheng, Z., et al., 2019. Parabacteroides distasonis alleviates obesity and metabolic dysfunctions via production of succinate and secondary bile acids. Cell Reports 26 (1), 222. Available at https://doi.org/10.1016/j.celrep.2018.12.028, http://www.sciencedirect.com/science/journal/22111247.

Waters, J.L., Ley, R.E., 2019. The human gut bacteria Christensenellaceae are widespread, heritable, and associated with health. BMC Biology 17 (1). Available at https://doi.org/10.1186/s12915-019-0699-4, http://www.biomedcentral.com/bmcbiol/.

Yadav, M.K., Kumari, I., Singh, B., Sharma, K.K., Tiwari, S.K., 2022. Probiotics, prebiotics and synbiotics: safe options for next-generation therapeutics. Applied Microbiology and Biotechnology 106 (2), 505–521. Available at https://doi.org/10.1007/s00253-021-11646-8, link.springer.de/link/service/journals/00253/index.htm.

Yang, J., Li, Y., Wen, Z., Liu, W., Meng, L., Huang, H., 2021. Oscillospira - a candidate for the next-generation probiotics. Gut Microbes 13 (1). Available at https://doi.org/10.1080/19490976.2021.1987783.

Zendeboodi, F., Khorshidian, N., Mortazavian, A.M., da Cruz, A.G., 2020. Probiotic: conceptualization from a new approach. Current Opinion in Food Science 32, 103–123. Available at https://doi.org/10.1016/j.cofs.2020.03.009, http://www.journals.elsevier.com/current-opinion-in-food-science/.

Zheng, J., Wittouck, S., Salvetti, E., Franz, C.M.A.P., Harris, H.M.B., Mattarelli, P., et al., 2020. A taxonomic note on the genus Lactobacillus: description of 23 novel genera, emended description of the genus *Lactobacillus beijerinck* 1901, and union of Lactobacillaceae and Leuconostocaceae. International Journal of Systematic and Evolutionary Microbiology 70 (4), 2782–2858. Available at https://doi.org/10.1099/ijsem.0.004107, https://www.microbiologyresearch.org/docserver/fulltext/ijsem/70/4/2782_ijsem004107.pdf?expires=1588756143&id=id&accname=guest&checksum=EF2CECD3AEE31ABAFD0AFB3955240396.

Zitvogel, L., Ayyoub, M., Routy, B., Kroemer, G., 2016. Microbiome and anticancer immunosurveillance. Cell 165 (2), 276–287. Available at https://doi.org/10.1016/j.cell.2016.03.001.

Chapter 9

Nonviable probiotic ingestion strategy: research facts and usage in drugs, supplements, and food industries

Mahmoud Yolmeh[1] and Anderson S. Sant'Ana[2]
[1]*Department of Food Engineering, Faculty of Engineering, Inonu University, Malatya, Türkiye,*
[2]*Department of Food Science and Nutrition, Faculty of Food Engineering, University of Campinas, Campinas, São Paulo, Brazil*

9.1 Introduction

In recent years, the exploration of probiotics has expanded beyond live microorganisms, leading to a growing interest in their inactivated form, known as paraprobiotics. These nonviable probiotic cells offer a promising alternative to traditional probiotics, with potential applications in the food, supplements, and pharmaceutical industries as health-promoting agents. As science uncovers more about the benefits and mechanisms of paraprobiotics, these inactivated microbial cells are emerging as a compelling strategy for enhancing health and wellness without the challenges associated with live probiotics (Yolmeh et al., 2026a).

9.1.1 Definition of paraprobiotic

Paraprobiotics are inactivated probiotics that, despite being dead, can benefit the host. Unlike probiotics, which must be alive at the time of consumption (and after it inside the consumer) to exert their effects, paraprobiotics can be administered after being rendered nonviable through processes such as heat treatment, irradiation, and ultrasound. The term paraprobiotic dates to around 2011 when Taverniti and Guglielmetti (2011) introduced these inactivated cells as an immunomodulator agent to improve the immune system. This inactivation not only can extend the shelf life of products containing these inactivated cells but also eliminates concerns related to the stability and viability of live probiotic strains during processing and storage (Lee et al., 2023).

Evaluating Oral Probiotics and Prebiotics. DOI: https://doi.org/10.1016/B978-0-443-29835-6.00009-5

It should be noted that in addition to "paraprobiotic," there is another term in this field (probiotic derivatives) called postbiotic, which refers to the metabolites of probiotics and is defined as the metabolites of probiotics. Therefore in obtaining paraprobiotics, the possibility of producing postbiotics should also be considered. This possibility can be investigated using analytical methods such as flow cytometry (FC), and postbiotic production should be prevented as much as possible (Vallejo-Cordoba et al., 2020; Nataraj et al., 2020). Section 7 of this chapter (analytical techniques for the quality control of paraprobiotics) explains this differentiation.

9.2 The importance of paraprobiotics in health and industry

The health-promoting benefits (HPBs) of paraprobiotics are at the forefront of current studies on probiotic derivatives in various research fields. Fig. 9.1 shows a review of the existing scientific literature, using "paraprobiotics" as a search term for the topic, demonstrating a notable number of publications over the past decade across various research fields. According to this literature review, most research conducted is in the "Food Science and Technology" field, with 25% of the articles (out of 130; Source: Web of Sciences). This high number of studies confirms the great interest of researchers in this field in this subject, along with its high importance and innovation.

The studies attributed to the HPBs of paraprobiotics suggest that they can show immunomodulatory (Danladi et al., 2022; Neissi et al., 2024), antiinflammatory (Kostelac et al., 2022; Rezaie et al., 2024), antioxidant (Aydin et al., 2021; Hyun et al., 2024), anticancer (Elham et al., 2022; Akan et al., 2025), antimicrobial (Yolmeh et al., 2020; Kürşad İncili et al., 2023; Yolmeh et al., 2025), and gut microbiota-improving (Wu et al., 2020; Lee et al., 2024) activities (Table 9.1). This is crucial, particularly for immunocompromised

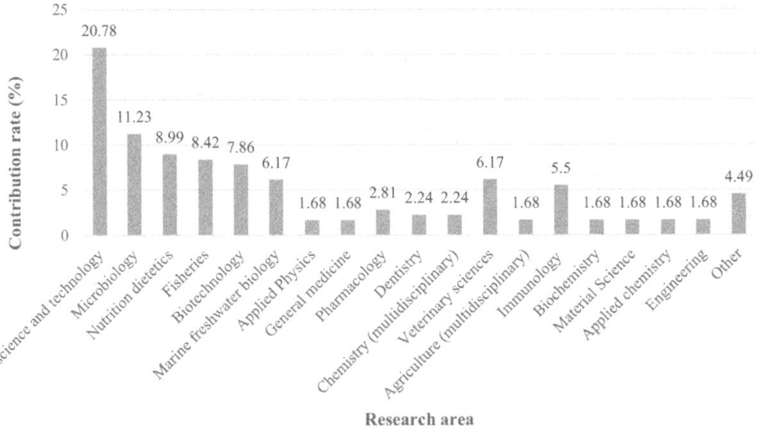

FIGURE 9.1 Current state of the scientific literature on "paraprobiotics": distribution of papers published in the past decade (2014–24) in the main research area.

TABLE 9.1 Health-promoting benefits (HPBs) of paraprobiotics and their main mechanisms.

Benefit	Microorganism	Inactivation method	Carrier food	Key mechanism	References
Immunomodulatory	*Lactiplantibacillus plantarum; Bacillus subtilis subsp. spizizenii; L. plantarum L-137*	Heat (60°C for 30 min); Gamma irradiation (19 KGy); Heat (–)	Fish feed; fish feed; carp feed	Production of antimicrobial peptides, tightening the epithelial junction, production of mucosal immunoglobulin, and modulating inflammatory reaction	Neissi et al. (2024), Dash et al. (2015), Yassine et al. (2021)
Antimicrobial activity	*Pediococcus acidilactici* B-LC-20; *L. plantarum* MIUG BL21 and *L. paraplantarum* MIUG BL74; *Bacillus coagulans* ATCC 7050	Heat (70°C for 40 min); Ohmic heating (at the power of 10 kW and voltage gradient of 20 V/cm for 15 min); Heat (60°C for 30 min)	Meatball; biopolymeric matrix; shrimp	Leaking of cellular contents through disrupting the integrity of microbial cell membrane; creating an unfavorable environment for pathogenic microbes by lowering the pH through organic acids; inhibiting the adhesion of pathogens to the intestinal surface	Kürşad İncili et al. (2023), Păcularu-Burada et al. (2024), Priya et al. (2025)
Gut microbiota improvement	*B.animalis*; *L. acidophilus*; *Bacillus siamensis* LF4	Gamma irradiation (2.5 KGy); Heat (121°C for 15 min); Heat (–)	Wheat-durum pasta; broiler chicken feed; soybean meal	Increasing the abundance of beneficial microbes such as *Bifidobacterium* and *Lactobacillus*, through inhibiting the colonization of pathogens; creating unsuitable environmental conditions for pathogens; competing with pathogens for nutrients and adhesion sites	Almada et al. (2021), Tukaram et al. (2022), Liu et al. (2024)

(*Continued*)

TABLE 9.1 (Continued)

Benefit	Microorganism	Inactivation method	Carrier food	Key mechanism	References
Enhancement of gut barrier function	*Limosilactobacillus fermentum* HF06; *L. delbrueckii CIDCA 133; Lacticaseibacillus rhamnosus CGMCC1.3724*	Heat (121°C for 15 min); Heat (121°C for 15 min); Heat (120°C for 15 min)	Drinking water; PBS solution (0.1 M); saline (NaCl 0.9%) solution	Physically and chemically strengthening the intestinal lining; modulating the activity of immune cells	Batista et al. (2022), Liu et al. (2024), Trindade et al. (2023)
Antioxidant activity	*L. plantarum; Bacillus subtilis subsp. spizizenii*	Heat (75°C for 30 min); Gamma irradiation (19 KGy)	Soy dairy-free beverage; fish feed	Neutralizing reactive oxygen species (ROS); chelating metal ions that catalyze the formation of ROS	Neissi et al. (2024), Zahidah et al. (2024)
Antiinflammatory activity	*L.rhamnosus* GG; Kefir paraprobiotics; *L.acidophilus* PIN7; *L. plantarum* K8	Ultraviolet (39-W germicidal UV lamp for 5 min at 10 cm); Heat (70°C for 30 min); Heat (-) and lysozyme treatment; Heat (90 °C for 30 min)	Drinking water; grape seed flour; drinking water; high-fat diet of mice	Decrease in pro-inflammatory mediators	Lopez et al. (2008), Seo et al. (2020), Kye et al. (2022), Lim et al. (2022)
Anticancer activity	*B. animalis* subsp. *lactis* BB-12; *L. casei* 39	Heat (121°C for 15 min); Heat (121°C for 15 min)	Yogurt; Kefir	Inducing apoptosis and inhibiting the proliferation of cancer cells through their bioactive molecules, such as peptidoglycans, lipoteichoic acids, and exopolysaccharides	Akan et al. (2025), Parvarei and Mortazavian (2022)

individuals with problems that may be exacerbated by administering live microorganisms (probiotics). Since paraprobiotics are nonviable, they offer a safer alternative, minimizing the risk of infections and other side effects of probiotics, for example, concerns regarding the potential for transfer of resistance genes from probiotics to pathogens (Mehta et al., 2023).

On the industrial scale, unlike probiotics, paraprobiotics do not require refrigeration and are less sensitive to environmental changes, making them easier to handle and incorporate into products that undergo heat processing. In addition to the HPBs, paraprobiotics may offer several practical advantages over probiotics in functional foods. For example, in a promising study, *Lactobacillus acidophilus* ATCC SD 5221 and *Bifidobacterium lactis* BB-12 paraprobiotics have been successfully used to produce functional yogurt (Parvarei et al., 2021b). Their results showed that the yogurt containing paraprobiotics had better physical properties, such as water-holding capacity (Parvarei et al., 2021b), and organoleptic properties, such as sensorial total score (Parvarei et al., 2021a), compared to the probiotic types. In another encouraging study, *Lactiplantibacillus plantarum* 299 v has been used to produce a functional soy dairy-free beverage, which has shown the highest taste and overall impression values compared to other samples, even more than the probiotic sample, in addition to significant antioxidant activity (Zahidah et al., 2024). With these multiple potential benefits, paraprobiotics open new possibilities for creating functional food, supplements, and pharmaceutical products that can deliver health and practical benefits, especially in industries where it is not possible to use probiotics in products, because of their limitations (Barros et al., 2020; Siciliano et al., 2021).

It is worth noting that obtaining paraprobiotics should be targeted. Indeed, since the method (and its condition) affects the benefits of paraprobiotics, their preparation should be optimized according to their specific characteristics (Yolmeh et al., 2025; Almada et al., 2021b).

9.3 The science behind paraprobiotics: mechanism of beneficial actions

The health benefits of paraprobiotics are primarily mediated through their interactions with the host's immune system and gut microbiota. Despite being nonviable, paraprobiotics retain structural components and bioactive molecules that can trigger specific responses in the host. The main mechanisms of action include immune system modulation, anticancer potential, enhancement of the gut barrier function, and antimicrobial activity. These effects contribute to the overall maintenance of health and the prevention of diseases (Teame et al., 2020). Understanding these mechanisms is important for developing effective paraprobiotic-based interventions in related industries, such as the pharmaceutical and food industries.

This section delves into the science behind the beneficial actions of paraprobiotics, exploring the pathways through which they exert their HPBs.

9.3.1 Immunomodulatory

Paraprobiotics can show an immunomodulatory role by interacting with their structural components, such as cell wall fragments, peptidoglycans, and surface proteins, with the host's immune cells. These components can stimulate both the innate and adaptive parts of the immune system (Mehta et al., 2023).

- **Activation of immune cells:** Paraprobiotic's components can interact with immune cells like macrophages, dendritic cells, and T cells. For example, the lipoteichoic acid and peptidoglycan in the cell walls can bind to pattern recognition receptors such as Toll-like receptors on immune cells. This interaction leads to the activation of signaling pathways that result in the production of cytokines and other immune mediators. The secretion of cytokines such as interleukin-10 (IL-10) and transforming growth factor-beta (TGF-β) promotes an antiinflammatory response, while interleukin-12 (IL-12) and interferon-gamma (IFN-γ) can enhance the body's ability to fight infections (Taverniti and Guglielmetti, 2011; Blanchet et al., 2021).
- **Promotion of regulatory T cells:** Some paraprobiotics can promote the induction of regulatory T cells that play a key role in preserving immune tolerance and inhibiting excessive inflammatory responses. This immunomodulatory mechanism of paraprobiotics is important, particularly in conditions like autoimmune diseases and allergies, where immune regulation is compromised (Taverniti and Guglielmetti, 2011).
- **Enhancement of phagocytic activity:** Paraprobiotics can enhance the phagocytic activity of immune cells like macrophages and neutrophils. This process involves engulfing and destroying pathogens and other foreign particles, aiding the body's defense mechanisms. The presence of paraprobiotics can prime immune cells to be more effective in identifying and eliminating harmful microorganisms (Dash et al., 2015).

9.3.2 Enhancement of gut barrier function

The gut barrier is an important section of the immune system, serving as the first line of defense against pathogens or other harmful substances (Ohland and MacNaughton, 2010). Paraprobiotics can contribute to the maintenance and enhancement of this barrier through various mechanisms:

- **Tight junction integrity:** The intestinal epithelium is composed of a layer of cells connected by tight junctions, which regulate the permeability of the intestinal barrier. Paraprobiotics have been shown to influence the expression and assembly of tight junction proteins, such as claudin, occludin, and zonula occludens. These inactivated cells help to reduce intestinal permeability (known as "leaky gut") by strengthening these tight junctions, thereby inhibiting the translocation of toxins and pathogens into the bloodstream (Batista et al., 2022; Liu et al., 2024).

- **Mucin production:** Mucins form a protective mucus layer on the surface of the intestinal epithelium. This mucus layer acts as a physical barrier, trapping pathogens and preventing their adherence to the epithelial cells. Paraprobiotics can stimulate the production of mucins, enhance the protective mucus layer, and contribute to gut health. This activity is important, particularly in preventing the colonization of pathogens and promoting the growth of beneficial gut microbiota (Batista et al., 2020; Trindade et al., 2023).

9.3.3 Antimicrobial activity and microbiota modulation

Paraprobiotics can act with direct antimicrobial effects against pathogenic microorganisms, as well as affect the activity and composition of the gut microbiota (Yolmeh et al., 2025; Almada et al., 2021b). This dual action helps maintain a healthy microbial balance in the gut.

- **Competitive exclusion of pathogens:** Although paraprobiotics are nonviable and do not colonize in the gut, they can still inhibit the adhesion of pathogens to the surface of the intestine. This occurs through a process known as competitive exclusion, where paraprobiotics compete with pathogens for binding sites on the epithelial cells. By preventing pathogens from adhering to and invading the gut lining, paraprobiotics help to reduce the risk of infections and gastrointestinal disorders (Yolmeh et al., 2026a).
- **Presence of bioactive compounds:** Paraprobiotics retain the ability to release bioactive compounds, such as short-chain fatty acids (SCFAs), even after inactivation. SCFAs, including acetate, propionate, and butyrate, are important metabolites that have various health benefits. They serve as an energy source for intestinal cells, enhance gut barrier function, and have antiinflammatory and antimicrobial properties. By promoting the release of SCFAs, paraprobiotics contribute to gut health and the inhibition of pathogenic bacteria (Teame et al., 2020).
- **Modulation of gut microbiota:** The interaction between paraprobiotics and the gut microbiota can lead to improvement in the composition and activity of the microbial community. Paraprobiotics can promote the growth of beneficial bacteria, such as *Bifidobacterium* and *Lactobacillus* species, while inhibiting the growth of harmful microorganisms. This modulation of the gut microbiota can result in a more balanced microbial environment, which is attributed to improved digestion, reduced risk of gastrointestinal diseases, and enhanced immune function (Yolmeh et al., 2026b).

9.3.4 Antiinflammatory effects

Paraprobiotics have shown antiinflammatory effects, which are particularly beneficial in managing conditions characterized by chronic inflammation, such as inflammatory bowel disease (IBD) and allergies (Kostelac et al., 2022). These inactivated cells can show this activity by these mechanisms:

- **Inhibition of pro-inflammatory cytokines:** Paraprobiotics can reduce the production of pro-inflammatory cytokines like tumor necrosis factor-alpha (TNF-α), interleukin-1 beta (IL-1β), and interleukin-6 (IL-6). These cytokines play a key role in the inflammatory response and are often elevated in conditions like IBD and rheumatoid arthritis. Paraprobiotics can help to alleviate inflammation by downregulating the gene expression of these cytokines (Kumar et al., 2023).
- **Promotion of antiinflammatory mediators:** In addition to inhibiting pro-inflammatory cytokines, paraprobiotics can enhance the production of antiinflammatory mediators, such as IL-10 and TGF-β. These mediators help to regulate the immune response and maintain a state of immune tolerance, reducing the likelihood of excessive or chronic inflammation. This antiinflammatory action is useful, particularly in conditions like allergies and autoimmune diseases, in which the immune system is dysregulated (Kostelac et al., 2022).

9.3.5 Potential role in metabolic health

Emerging studies suggest that paraprobiotics may also have a role in metabolic health, particularly in obesity, type 2 diabetes, and metabolic syndrome.

- **Improvement of insulin sensitivity:** Some studies have indicated that paraprobiotics can improve glucose metabolism and insulin sensitivity. This effect may be mediated through the reduction of systemic inflammation and the modulation of gut hormones, such as glucagon-like peptide-1. Paraprobiotics may help in the management of metabolic disorders, such as type 2 diabetes, by enhancing insulin sensitivity (Arellano-García et al., 2024).
- **Regulation of lipid metabolism:** Paraprobiotics have been shown to influence lipid metabolism, potentially reducing serum cholesterol and triglyceride levels. This effect may be related to SCFAs, which can inhibit cholesterol synthesis in the liver and promote the excretion of bile acids. Paraprobiotics may contribute to cardiovascular health and the prevention of atherosclerosis by modulating lipid metabolism (López-Almada et al., 2024).

9.4 Paraprobiotic obtaining methods

The type of paraprobiotic obtaining method is crucial to ensuring their efficacy, safety, and stability for use in various health applications. These methods are designed to inactivate live probiotic cells while preserving as much of their bioactive components and structural integrity as possible, which are responsible for their HPBs (Yolmeh et al., 2024). This section will explore different methods used to obtain paraprobiotics, ranging from conventional to advanced methods, and discuss the importance of choosing the right technique to maximize HPB. Conventional methods, such as heat treatment and UV irradiation, are more straightforward and widely used, while advanced methods like sonication offer more precise control of the inactivation process. Each method has its advantages and limitations, and the choice of method can significantly influence the characteristics and efficacy of the resulting paraprobiotic (Almada et al., 2021).

9.4.1 Conventional methods

9.4.1.1 Heat treatment

Heat treatment is the most common method used to inactivate probiotics. This method involves exposing probiotic cultures to high temperatures for a specific period, effectively killing the cells while preserving their structural components. The temperature and duration of the thermal process must be carefully controlled to prevent the degradation of essential bioactive molecules (Barros et al., 2021). Based on the type of strain and desired outcome (beneficial properties), temperatures ranging from 60°C to 121°C are typically used.

- **Advantages**: This method is cost-effective, easy to perform, and can be applied on a large scale. It also ensures that the paraprobiotics retain cell wall structures, peptidoglycans, and surface proteins, which are important for their HPBs.
- **Limitations**: High temperatures can denature bioactive compounds, such as enzymes and proteins, which may reduce the health benefits of the paraprobiotic. Therefore it is crucial to optimize the temperature and exposure time to minimize damage to these components.

9.4.1.2 Chemical inactivation (lysozyme)

Chemical inactivation involves using formaldehyde, alcohol, or organic acids to kill probiotic cells. This method can be highly effective, but it requires precise control to avoid altering the cell structures or leaving behind harmful residues. Chemical agents can inactivate probiotics by disrupting the cell membrane, denaturing proteins, or interfering with cellular metabolism (Almada et al., 2021; Lee et al., 2022).

- **Advantages**: Chemical inactivation does not require high temperatures, which helps in preserving heat-sensitive bioactive components.
- **Limitations**: The use of chemicals raises safety concerns, particularly for food and supplement applications. It is essential to ensure that no harmful

residues remain in the final product. Additionally, some chemicals may alter the structural integrity of cells, potentially reducing their HPBs.

9.4.2 Advanced methods

9.4.2.1 Spray drying

Spray drying is an advanced technique that involves atomizing a liquid probiotic suspension into a hot air cylinder, where the droplets rapidly dry and form a powder containing inactivated cells. This method combines heat treatment with dehydration, effectively inactivating the probiotics while encapsulating them in a protective matrix (Lee et al., 2023).

- **Advantages**: Spray drying can be useful for producing paraprobiotics with a long shelf life and easily scaled up for industrial production.
- **Limitations**: High temperatures during this process can damage sensitive bioactive molecules. It is essential to optimize drying conditions to balance effective inactivation and the preservation of components attributed to HPB.

9.4.2.2 Gamma irradiation

This process is an innovative method for obtaining paraprobiotics, which uses high-energy gamma rays to inactivate probiotics while maintaining their structural integrity and bioactive components. Gamma irradiation has gained attention for its precision, scalability, and ability to produce safe, high-quality paraprobiotics (Almada et al., 2021; Porfiri et al., 2022).

Advantages: Gamma irradiation inactivates bacteria without destroying their essential bioactive components, such as cell wall polysaccharides, proteins, and DNA fragments. Unlike extreme heat processes, gamma irradiation minimizes thermal degradation, ensuring that the HPBs of the probiotics remain as much as possible.

Limitations: Setting up facilities for gamma irradiation requires a high initial investment and high costs. In addition, while gamma irradiation preserves most bioactive compounds, high doses may cause minor structural alterations in certain components. These changes could affect the overall bioactivity of the paraprobiotics, necessitating careful optimization of irradiation levels.

9.4.2.3 High-pressure processing

High-pressure processing is a nonthermal technique that uses high hydrostatic pressure to inactivate microorganisms. During HPP, probiotic cells are subjected to pressures ranging from 100 to 600 MPa, which disrupts their cellular structures and metabolic processes, leading to cell death. This method preserves the nutritional and bioactive components of the cells, as it does not involve high temperatures (Georget et al., 2015; Sehrawat et al., 2021).

Advantages: HPP preserves the nutritional and bioactive components of probiotic cells, as it does not involve heat. This method is effective in maintaining the functional integrity of the cells and can be used for heat-sensitive strains.

Limitations: The equipment required for HPP is expensive, and the process needs to be carefully controlled to ensure uniform inactivation. Additionally, high-pressure can alter the physical properties of the cells, which may affect their interaction with the host's immune system.

9.4.2.4 Sonication

Sonication uses ultrasonic waves to create cavitation bubbles in a liquid medium, which collapse and generate intense shear forces that disrupt cell membranes and inactivate probiotics. This method can effectively kill cells without using heat or chemicals, preserving the bioactive components of the cells (Brandão et al., 2021; Gholian et al., 2024).

- **Advantages**: Sonication is a nonthermal method that can effectively kill cells without using chemicals, preserving the structural and bioactive components of the cells.
- **Limitations**: The effectiveness of sonication depends on various factors, including frequency, intensity, and duration. Overexposure can damage sensitive components, reducing the beneficial potential of paraprobiotics.

9.5 Optimization of inactivation conditions

Recent studies have disclosed that the type and conditions of paraprobiotic obtaining methods are crucial for their HPBs (Yolmeh et al., 2025; Almada et al., 2021b; Yolmeh et al., 2024). These studies revealed that these processes need to be optimized to prepare paraprobiotics with the maximum targeted HPB. Factors such as temperature, pressure, exposure time, and the presence of protective agents, especially in relatively extreme methods, can influence the outcome of the inactivation process (Yolmeh et al., 2025; Yolmeh et al., 2024). However, this issue has been ignored by some researchers who studied obtaining paraprobiotics and their HPBs. For this purpose, some statistical optimization methods, such as response surface methodology (RSM), can be useful and efficient. RSM is a powerful mathematical tool widely used to optimize processes in the food industry and biotechnology. This statistical method evaluates the interactions between multiple variables and their combined effect on selected responses, enabling efficient process optimization with fewer experimental trials, compared to other statistical methods like full factorial, which can reduce process cost and time while increasing process efficiency (Yolmeh, Ferreira et al., 2026).

It is worth noting that in the optimization process, it is important to select the membrane integrity as one of the responses, which can be measured using the FC method, because measuring this parameter is useful in distinguishing between paraprobiotics and postbiotics. Section 9.7 discusses this method and how to differentiate it.

The temperature must be carefully controlled in intense heat processes to avoid denaturing proteins and other bioactive compounds. A balance must be struck between achieving inactivation and preserving the functional integrity of the cells. For instance, adding protective agents like sugars or proteins during heat treatment can help to maintain the structural integrity and bioactivity of paraprobiotics, enhancing their efficacy. These agents can form a protective barrier around the cells, reducing damage during processes like heat treatment or freeze–thaw cycling (Archacka et al., 2019). A comparative study on the protective ability of some saccharides (skimmed milk, inulin, maltodextrin, and sucrose) on *L. plantarum* cells over freeze-drying and storage has shown that skimmed milk provides the highest survival, up to 91%; however, despite good cryoprotectant performance, inulin shows the least protection over storage, with <1% cell survival (Oluwatosin et al., 2022). Additionally, specific characteristics of the probiotic strain, such as its thermal tolerance, cell wall composition, and sensitivity to environmental stressors, must be considered when selecting and optimizing the inactivation method. Strain-specific optimization ensures that paraprobiotics retain their unique health benefits and functional properties (Mathipa and Thantsha, 2017).

9.6 Safety considerations

Safety is paramount when choosing an inactivation method, particularly for paraprobiotics used directly or indirectly by humans or animals. Methods involving chemicals or extreme physical conditions must be thoroughly evaluated to ensure that no harmful residues or by-products remain in the final product. The inactivation process should render the cells nonviable without compromising their safety for human consumption. For example, UV irradiation is a nonchemical method that can effectively inactivate probiotics without introducing residues. However, it is essential to ensure that the process does not generate UV-induced mutagens or other harmful compounds (Song, Mohseni and Taghipour, 2019).

9.7 Analytical techniques for the quality control of paraprobiotics

Quality control is critical in obtaining paraprobiotics, ensuring that they maintain their intended HPBs, safety, and stability. Along with traditional methods, such as colony-forming unit (CFU) count to evaluate the loss of bacterial cultivability, advanced analytical techniques to study the structural integrity and inactivation efficacy. One of the most valuable tools in this context is FC, which has proven to be highly effective in evaluating the quality of paraprobiotics throughout the production process.

9.7.1 Flow cytometry in obtaining paraprobiotics

FC is a powerful analytical technique that can rapidly analyze the physical and chemical characteristics of individual cells in a heterogeneous mixture. This technique is particularly useful for quality control in obtaining paraprobiotics, as it allows for the assessment of cell viability, membrane integrity, and the presence of specific cell surface markers. By utilizing FC, researchers can ensure that paraprobiotic obtaining methods have the desired properties of safety and efficacy.

9.7.1.1 Principle of flow cytometry in obtaining paraprobiotics

FC works by passing cells in a single file through a laser beam within a fluid stream. As each cell passes through the laser, it scatters light and may emit fluorescence if stained with specific stains. Detectors measure the scattered light and fluorescence, providing information on various cellular properties such as size, granularity, and the presence of specific markers (Adan et al., 2017).

One of the primary uses of FC in paraprobiotic quality control is to determine the proportion of viable, nonviable, and damaged cells within a sample. By using a couple of fluorescent stains that selectively mark alive or dead cells, such as propidium iodide (PI) and Syto 9, FC can distinguish between viable cells with intact membranes and nonviable cells that have been effectively inactivated (Yolmeh et al., 2024; Barros et al., 2021).

- **Propidium iodide (PI)**: PI is a nucleic acid stain that can only penetrate cells with compromised membranes, indicating dead or inactivated cells. Cells that are PI-positive are considered nonviable, making this stain useful for confirming successful inactivation in paraprobiotic production.
- **Syto 9**: This stain marks all cells (alive and dead), allowing for a total cell count. When combined with PI, Syto 9 can help differentiate between alive (Syto 9-positive/PI-negative) and dead (Syto 9-positive/PI-positive) cells.

In addition to alive and dead proportions, it is possible to measure the proportions of damaged (Syto 9-positive/PI-positive) and debris (Syto 9-negative/PI-negative) using the FC method. This fact is important to differentiate paraprobiotic and postbiotic concepts. As defined, postbiotics are mainly known as probiotic metabolites and can be classified as debris (PI-negative) (Yolmeh et al., 2024).

9.8 Conclusions

A review of the literature on the use of paraprobiotics disclosed an increasing number of studies, particularly in the field of "Food Science and Technology," including some promising research suggesting that the HPBs of paraprobiotics may surpass those of their original probiotics. Since the biological and physical activities of paraprobiotics are significantly influenced by the type and conditions of the obtaining method, optimizing these processes is critical. FC can be useful for this purpose, as it enables real-time monitoring of the cell membrane integrity and inactivation of probiotics.

References

Adan, A., Alizada, G., Kiraz, Y., Baran, Y., Nalbant, A., 2017. Flow cytometry: basic principles and applications. Critical Reviews in Biotechnology 37 (2), 163–176. Available at https://doi.org/10.3109/07388551.2015.1128876.

Akan, E., Yavaş, A., Dikme, M., 2025. Bioactive potentials of paraprobiotic kefir: enhanced protein hydrolysis and anticancer efficacy. International Dairy Journal 163, 106169. Available at https://doi.org/10.1016/j.idairyj.2024.106169.

Almada, C.N., Almada-Érix, C.N., Bonatto, M.S., Pradella, F., dos Santos, P., Abud, Y.K.D., et al., 2021. Obtaining paraprobiotics from *Lactobacilus acidophilus*, *Lacticaseibacillus casei* and *Bifidobacterium animalis* using six inactivation methods: impacts on the cultivability, integrity, physiology, and morphology. Journal of Functional Foods 87. Available at https://doi.org/10.1016/j.jff.2021.104826, http://www.elsevier.com/wps/find/journaldescription.cws_home/717426/description#description.

Almada, C.N., Almada-Érix, C.N., Costa, W.K.A., Graça, J.S., Cabral, L., Noronha, M.F., et al., 2021a. Wheat-durum pasta added of inactivated *Bifidobacterium animalis* decreases glucose and total cholesterol levels and modulates gut microbiota in healthy rats. International Journal of Food Sciences and Nutrition 72 (6), 781–793. Available at https://doi.org/10.1080/09637486.2021.1877261, http://www.tandfonline.com/loi/iijf20.

Almada, C.N., Almada-Érix, C.N., Roquetto, A.R., Santos-Junior, V.A., Cabral, L., Noronha, M.F., et al., 2021b. Paraprobiotics obtained by six different inactivation processes: impacts on the biochemical parameters and intestinal microbiota of Wistar male rats. International Journal of Food Sciences and Nutrition 72 (8), 1057–1070. Available at https://doi.org/10.1080/09637486.2021.1906211, http://www.tandfonline.com/loi/iijf20.

Archacka, M., Białas, W., Dembczyński, R., Olejnik, A., Sip, A., Szymanowska, D., et al., 2019. Method of preservation and type of protective agent strongly influence probiotic properties of *Lactococcus lactis*: a complete process of probiotic preparation manufacture and use. Food Chemistry 274, 733–742. Available at https://doi.org/10.1016/j.foodchem.2018.09.033.

Arellano-García, L.I., Portillo, M.P., Martínez, J.A., Courtois, A., Milton-Laskibar, I., 2024. Postbiotics for the management of obesity, insulin resistance/type 2 diabetes and NAFLD. Beyond microbial viability. Critical Reviews in Food Science and Nutrition 1–24. Available at https://doi.org/10.1080/10408398.2024.2437143.

Aydin, B., Çiydem, T., Kaya, E., Açik, L., 2021. Evaluation of the antioxidant effects of postbiotics and paraprobiotics in lactic acid bacteria isolated from traditional fermented sausages. European Journal of Science and Technology. Available at https://doi.org/10.31590/ejosat.1011409.

Barros, C.P., Guimarães, J.T., Esmerino, E.A., Duarte, M.C.K., Silva, M.C., Silva, R., et al., 2020. Paraprobiotics and postbiotics: concepts and potential applications in dairy products. Current Opinion in Food Science 32, 1–8. Available at https://doi.org/10.1016/j.cofs.2019.12.003, http://www.journals.elsevier.com/current-opinion-in-food-science/.

Barros, C.P., Pires, R.P.S., Guimarães, J.T., Abud, Y.K.D., Almada, C.N., Pimentel, T.C., et al., 2021. Ohmic heating as a method of obtaining paraprobiotics: impacts on cell structure and viability by flow cytometry. Food Research International 140. Available at https://doi.org/10.1016/j.foodres.2020.110061, www.elsevier.com/inca/publications/store/4/2/2/9/7/0.

Batista, V.L., da Silva, T.F., de Jesus, L.C.L., Coelho-Rocha, N.D., Barroso, F.A.L., Tavares, L.M., et al., 2020. Probiotics, prebiotics, synbiotics, and paraprobiotics as a therapeutic alternative for intestinal mucositis. Frontiers in Microbiology 11. Available at https://doi.org/10.3389/fmicb.2020.544490, https://www.frontiersin.org/journals/microbiology#.

Batista, V.L., De Jesus, L.C.L., Tavares, L.M., Barroso, F.L.A., Fernandes, L.Jd.S., Freitas, Ad.S., et al., 2022. Paraprobiotics and postbiotics of *Lactobacillus delbrueckii* CIDCA 133 mitigate 5-FU-induced intestinal inflammation. Microorganisms 10 (7). Available at https://doi.org/10.3390/microorganisms10071418, www.mdpi.com/journal/microorganisms.

Blanchet, F., Rault, L., Peton, V., Le Loir, Y., Blondeau, C., Lenoir, L., et al., 2021. Heat inactivation partially preserved barrier and immunomodulatory effects of *Lactobacillus gasseri* LA806 in an in vitro model of bovine mastitis. Beneficial Microbes 12 (1), 95. Available at https://doi.org/10.3920/bm2020.0146.

Brandão, L.R., De Brito Alves, J.L., Da Costa, W.K.A., Ferreira, G.D.A.H., De Oliveira, M.P., Gomes Da Cruz, A., et al., 2021. Live and ultrasound-inactivated: *Lacticaseibacillus casei* modulate the intestinal microbiota and improve biochemical and cardiovascular parameters in male rats fed a high-fat diet. Food and Function 12 (12), 5287–5300. Available at https://doi.org/10.1039/d1fo01064f, http://pubs.rsc.org/en/journals/journal/fo.

Danladi, Y., Loh, T.C., Foo, H.L., Akit, H., Md Tamrin, N.A., Naeem Azizi, M., 2022. Effects of postbiotics and paraprobiotics as replacements for antibiotics on growth performance, carcass characteristics, small intestine histomorphology, immune status and hepatic growth gene expression in broiler chickens. Animals: An Open Access Journal from MDPI 12 (7), 917. Available at https://doi.org/10.3390/ani12070917.

Dash, G., Raman, R.P., Pani Prasad, K., Makesh, M., Pradeep, M.A., Sen, S., 2015. Evaluation of paraprobiotic applicability of *Lactobacillus plantarum* in improving the immune response and disease protection in giant freshwater prawn, *Macrobrachium rosenbergii* (de Man, 1879). Fish and Shellfish Immunology 43 (1), 167–174. Available at https://doi.org/10.1016/j.fsi.2014.12.007, http://www.elsevier.com/inca/publications/store/6/2/2/8/3/2/index.htt.

Elham, N., Naheed, M., Elahe, M., Hossein, M.M., Majid, T., 2022. Selective cytotoxic effect of probiotic, paraprobiotic and postbiotics of *L. casei* strains against colorectal cancer cells: invitro studies. Brazilian Journal of Pharmaceutical Sciences 58. Available at https://doi.org/10.1590/s2175-97902022e19400, https://www.scielo.br/j/bjps/a/4MTX7b3DDQycwHNmyQSBnZH/?lang=en&format=pdf.

Georget, E., Sevenich, R., Reineke, K., Mathys, A., Heinz, V., Callanan, M., et al., 2015. Inactivation of microorganisms by high isostatic pressure processing in complex matrices: a review. Innovative Food Science and Emerging Technologies 27, 1–14. Available at https://doi.org/10.1016/j.ifset.2014.10.015.

Gholian, M.M., Babaei, A., Zendeboodi, F., Mortazavian, A.M., Koushki, V., 2024. Effect of different inactivation condition on *Lactobacillus gasseri* and *Lactobacillus plantarum*: culturability, cell integrity and morphology. LWT 197, 115915. Available at https://doi.org/10.1016/j.lwt.2024.115915.

Hyun, J.H., Woo, I.K., Kim, K.T., Park, Y.S., Kang, D.K., Lee, N.K., et al., 2024. Heat-treated paraprobiotic *Latilactobacillus sakei* KU15041 and *Latilactobacillus curvatus* KU15003 show an antioxidant and immunostimulatory effect. Journal of Microbiology and Biotechnology 34 (2), 358–366. Available at https://doi.org/10.4014/jmb.2309.09007, https://www.jmb.or.kr/journal/list.html?pn=current_issue.

Kostelac, D., Gerić, M., Gajski, G., Frece, J., 2022. Probiotic and paraprobiotic derivates exhibit anti-inflammatory and genoprotective effects during induced stress. Journal of Applied Microbiology 133 (2), 819–829. Available at https://doi.org/10.1111/jam.15595.

Kumar, H., Dhalaria, R., Guleria, S., Cimler, R., Choudhary, R., Singh Dhanjal, D., et al., 2023. To exploring the role of probiotics, plant-based fermented products, and paraprobiotics as anti-inflammatory agents in promoting human health. Journal of Agriculture and Food Research 14, 100896. Available at https://doi.org/10.1016/j.jafr.2023.100896.

Kye, Y.J., Lee, S.Y., Kim, H.R., Lee, B.H., Park, J., Park, M.S., et al., 2022. Lactobacillus acidophilus PIN7 paraprobiotic supplementation ameliorates DSS-induced colitis through anti-inflammatory and immune regulatory effects. Journal of Applied Microbiology 132 (4), 3189–3200. Available at https://doi.org/10.1111/jam.15406, https://academic.oup.com/jambio.

Kürşad İncili, G., Akgöl, M., Karatepe, P., Kanmaz, H., Kaya, B., Tekin, A., et al., 2023. Inhibitory effect of bioactive compounds derived from freeze-dried paraprobiotic of *Pediococcus acidilactici* against food-borne pathogens: in-vitro and food model studies. Food Research International 170, 113045. Available at https://doi.org/10.1016/j.foodres.2023.113045.

Lee, S.Y., Lee, B.H., Park, J.H., Park, M.S., Ji, G.E., Sung, M.K., 2022. Bifidobacterium bifidum BGN4 paraprobiotic supplementation alleviates experimental colitis by maintaining gut barrier and suppressing nuclear factor kappa b activation signaling molecules. Journal of Medicinal Food 25 (2), 146–157. Available at https://doi.org/10.1089/jmf.2021.K.0150, www.liebertonline.com/jmf.

Lee, N., Park, Y.S., Kang, D.K., Paik, H.D., 2023. Paraprobiotics: definition, manufacturing methods, and functionality. Food Science and Biotechnology 32 (14), 1981–1991. Available at https://doi.org/10.1007/s10068-023-01378-y, https://www.springer.com/journal/10068.

Lee, H.J., Tran, M.T.H., Le, M.H., Justine, E.E., Kim, Y.J., 2024. Paraprobiotic derived from *Bacillus velezensis* GV1 improves immune response and gut microbiota composition in cyclophosphamide-treated immunosuppressed mice. Frontiers in Immunology 15. Available at https://doi.org/10.3389/fimmu.2024.1285063, https://www.frontiersin.org/journals/immunology#.

Lim, J.-J., Jung, A.-H., Joo Suh, H., Choi, H.-S., Kim, H., 2022. *Lactiplantibacillus plantarum* K8-based paraprobiotics prevents obesity and obesity-induced inflammatory responses in high fat diet-fed mice. Food Research International 155, 111066. Available at https://doi.org/10.1016/j.foodres.2022.111066.

Liu, C., Qi, X., Li, D., Zhao, L., Li, Q., Mao, K., et al., 2024. *Limosilactobacillus fermentum* HF06-derived paraprobiotic and postbiotic alleviate intestinal barrier damage and gut microbiota disruption in mice with ulcerative colitis. Journal of the Science of Food and Agriculture 104 (3), 1702–1712. Available at https://doi.org/10.1002/jsfa.13057, http://onlinelibrary.wiley.com/journal/10.1002/(ISSN)1097-0010.

Liu, Z.Y., Yang, H.L., Li, S., Cai, G.H., Ye, J.D., Zhang, C.X., et al., 2024. Paraprobiotic and postbiotic forms of Bacillus siamensis improved growth, immunity, liver and intestinal health in *Lateolabrax maculatus* fed soybean meal diet. Fish and Shellfish Immunology 145.

Available at https://doi.org/10.1016/j.fsi.2024.109370, https://www.sciencedirect.com/science/journal/10504648.

Lopez, M., Li, N., Kataria, J., Russell, M., Neu, J., 2008. Live and ultraviolet-inactivated *Lactobacillus Rhamnosus* GG decrease flagellin-induced interleukin-8 production in Caco-2 cells. The Journal of Nutrition 138 (11), 2264–2268. Available at https://doi.org/10.3945/jn.108.093658.

López-Almada, G., Mejía-León, M.E., Salazar-López, N.J., 2024. Probiotic, postbiotic, and paraprobiotic effects of *Lactobacillus rhamnosus* as a modulator of obesity-associated factors. Foods 13 (22), 3529. Available at https://doi.org/10.3390/foods13223529.

Mathipa, M.G., Thantsha, M.S., 2017. Probiotic engineering: towards development of robust probiotic strains with enhanced functional properties and for targeted control of enteric pathogens. Gut Pathogens 9 (1). Available at https://doi.org/10.1186/s13099-017-0178-9, http://www.gutpathogens.com/.

Mehta, J.P., Ayakar, S., Singhal, R.S., 2023. The potential of paraprobiotics and postbiotics to modulate the immune system: a review. Microbiological Research 275, 127449. Available at https://doi.org/10.1016/j.micres.2023.127449.

Nataraj, B.H., Ali, S.A., Behare, P.V., Yadav, H., 2020. Postbiotics-parabiotics: the new horizons in microbial biotherapy and functional foods. Microbial Cell Factories 19 (1). Available at https://doi.org/10.1186/s12934-020-01426-w, http://www.microbialcellfactories.com/home/.

Neissi, A., Heidarieh, M., Shahhosseini, G., Zahed, H.M., Kalantari-Hesari, A., 2024. Improvement of the immunological and antioxidant capacity of rainbow trout (*Oncorhynchus mykiss*) by using the paraprobiotic *Bacillus subtilis* subsp. spizizenii. Aquaculture Reports 38, 102304. Available at https://doi.org/10.1016/j.aqrep.2024.102304.

Ohland, C.L., MacNaughton, W.K., 2010. Probiotic bacteria and intestinal epithelial barrier function. American Journal of Physiology - Gastrointestinal and Liver Physiology 298 (6), G807. Available at https://doi.org/10.1152/ajpgi.00243.2009Canada, http://ajpgi.physiology.org/cgi/reprint/298/6/G807.

Oluwatosin, S.O., Tai, S.L., Fagan-Endres, M.A., 2022. Sucrose, maltodextrin and inulin efficacy as cryoprotectant, preservative and prebiotic – towards a freeze dried *Lactobacillus plantarum* topical probiotic. Biotechnology Reports 33, e00696. Available at https://doi.org/10.1016/j.btre.2021.e00696.

Parvarei, M.M., Fazeli, M.R., Mortazavian, A.M., Sarem Nezhad, S., Mortazavi, S.A., 2021a. Comparative effect of probiotic and paraprobiotic addition on rheological and sensory properties of yoghurt. International Journal of Dairy Technology 74 (1), 95–106. Available at https://doi.org/10.1111/1471-0307.12727, http://onlinelibrary.wiley.com/journal/10.1111/(ISSN)1471-0307.

Parvarei, M.M., Fazeli, M.R., Mortazavian, A.M., Sarem Nezhad, S., Mortazavi, S.A., Golabchifar, A.A., et al., 2021b. Comparative effects of probiotic and paraprobiotic addition on microbiological, biochemical and physical properties of yogurt. Food Research International 140. Available at https://doi.org/10.1016/j.foodres.2020.110030, www.elsevier.com/inca/publications/store/4/2/2/9/7/0.

Parvarei, M.M., Mortazavian, A.M., 2022. Cytotoxic effect of paraprobiotic yogurts on cancer cell lines. Indian Journal of Microbiology Research 9 (2), 106–113. Available at https://doi.org/10.18231/j.ijmr.2022.019, https://www.ijmronline.org/journal-article-file/16912.

Porfiri, L., Burtscher, J., Kangethe, R.T., Verhovsek, D., Cattoli, G., Domig, K.J., et al., 2022. Irradiated non-replicative lactic acid bacteria preserve metabolic activity while exhibiting diverse immune modulation. Frontiers in Veterinary Science 9. Available at https://doi.org/10.3389/fvets.2022.859124, journal.frontiersin.org/journal/veterinary-science.

Priya, P.S., Pavithra, V., Vaishnavi, S., Almutairi, B.O., Arokiyaraj, S., et al., 2025. Application of *Bacillus Coagulans* as paraprobiotic against acute hepatopancreatic necrosis disease in shrimp. Probiotics and Antimicrobial Proteins 17 (3), 1536–1550. Available at https://doi.org/10.1007/s12602-024-10230-6.

Păcularu-Burada, B., Grigore-Gurgu, L., Cotârleț, M., Vasile, A.M., Nistor, O.V., Cîrciumaru, A., et al., 2024. Freeze-dried biotics based on *Lactiplantibacillus plantarum* and *Lactiplantibacillus paraplantarum* with enhanced selected biological activity. LWT 203, 116339. Available at https://doi.org/10.1016/j.lwt.2024.116339.

Rezaie, N., Aghamohammad, S., Gholizadeh Khiavi, E.H.A., Pourshafie, M.R., Talebi, M., Rohani, M., 2024. Comparison of novel native probiotics and paraprobiotics in modulating oxidative stress and inflammation in DSS-induced colitis: implications for enhanced therapeutic strategies in high fat diet. BMC Immunology 25 (1). Available at https://doi.org/10.1186/s12865-024-00678-0, https://bmcimmunol.biomedcentral.com/.

Sehrawat, R., Kaur, B.P., Nema, P.K., Tewari, S., Kumar, L., 2021. Microbial inactivation by high pressure processing: principle, mechanism and factors responsible. Food Science and Biotechnology 30 (1), 19–35. Available at https://doi.org/10.1007/s10068-020-00831-6, http://www.springerlink.com/content/n3031612t262/.

Seo, K.-H., Jeong, J., Kim, H., 2020. Synergistic effects of heat-killed kefir paraprobiotics and flavonoid-rich prebiotics on western diet-induced obesity. Nutrients 12 (8), 2465. Available at https://doi.org/10.3390/nu12082465.

Siciliano, R.A., Reale, A., Mazzeo, M.F., Morandi, S., Silvetti, T., Brasca, M., 2021. Paraprobiotics: a new perspective for functional foods and nutraceuticals. Nutrients 13 (4), 1225. Available at https://doi.org/10.3390/nu13041225.

Song, K., Mohseni, M., Taghipour, F., 2019. Mechanisms investigation on bacterial inactivation through combinations of UV wavelengths. Water Research 163, 114875. Available at https://doi.org/10.1016/j.watres.2019.114875.

Taverniti, V., Guglielmetti, S., 2011. The immunomodulatory properties of probiotic microorganisms beyond their viability (ghost probiotics: proposal of paraprobiotic concept). Genes and Nutrition 6 (3), 261–274. Available at https://doi.org/10.1007/s12263-011-0218-x.

Teame, T., Wang, A., Xie, M., Zhang, Z., Yang, Y., Ding, Q., et al., 2020. Paraprobiotics and postbiotics of probiotic Lactobacilli, their positive effects on the host and action mechanisms: a review. Frontiers in Nutrition 7. Available at https://doi.org/10.3389/fnut.2020.570344.

Trindade, L.M., Torres, L., Matos, I.D., Miranda, V.C., de Jesus, L.C.L., Cavalcante, G., et al., 2023. Paraprobiotic *Lacticaseibacillus rhamnosus* protects intestinal damage in an experimental murine model of mucositis. Probiotics and Antimicrobial Proteins 15 (2), 338–350. Available at https://doi.org/10.1007/s12602-021-09842-z, https://www.springer.com/journal/12602.

Tukaram, N.M., Biswas, A., Deo, C., Laxman, A.J., Monika, M., Tiwari, A.K., 2022. Effects of paraprobiotic as replacements for antibiotic on performance, immunity, gut health and carcass characteristics in broiler chickens. Scientific Reports 12 (1). Available at https://doi.org/10.1038/s41598-022-27181-z, https://www.nature.com/srep/.

Vallejo-Cordoba, B., Castro-López, C., García, H.S., González-Córdova, A.F., Hernández-Mendoza, A., 2020. Postbiotics and paraprobiotics: a review of current evidence and emerging trends. Advances in Food and Nutrition Research 94, 1–34. Available at https://doi.org/10.1016/bs.afnr.2020.06.001, http://www.elsevier.com/wps/find/bookdescription.cws_home/704624/description#description.

Wu, X., Teame, T., Hao, Q., Ding, Q., Liu, H., Ran, C., et al., 2020. Use of a paraprobiotic and postbiotic feed supplement (HWF™) improves the growth performance, composition and function of gut microbiota in hybrid sturgeon (*Acipenser baerii* x *Acipenser schrenckii*. Fish and Shellfish Immunology 104, 36–45. Available at https://doi.org/10.1016/j.fsi.2020.05.054.

Yassine, T., Khalafalla, M.M., Mamdouh, M., Elbialy, Z.I., Salah, A.S., Ahmedou, A., et al., 2021. The enhancement of the growth rate, intestinal health, expression of immune-related genes, and resistance against suboptimal water temperature in common carp (*Cyprinus carpio*) by dietary paraprobiotics. Aquaculture Reports 20, 100729. Available at https://doi.org/10.1016/j.aqrep.2021.100729.

Yolmeh, M., Alves, P.C.M., Vilela, M.M.D.S., Sant'Ana, A.S., 2024. Impacts of ohmic-heating on the surface characteristics, membrane integrity, morphology, and chemical compounds of *Bifidobacterium animalis* paraprobiotic. Food Bioscience 59. Available at https://doi.org/10.1016/j.fbio.2024.104141, https://www.sciencedirect.com/science/journal/22124292.

Yolmeh, M., Junior, W.J. F.L., Amorim-Neto, D.P., Reis, V.H.D.O.T., dos Santos, A., Alves, P.C.M., Cazarin, C.B.B., Vilela, M.M.S., Sant'Ana, A.S., 2026a. Paraprobiotics retain their ability to counteract diet-induced dysbiosis during shelf-life: an efficacy study correlating fecal metabolomic and microbiome changes. Food Bioscience 80. Available at https://doi.org/10.1016/j.fbio.2026.108947.

Yolmeh, M., Ferreira, B.M., Amaral, Y.M.S., Alves, P.C.M., de Castro, R.J.S., dos Santos Vilela, M.M., Sant'Ana, A.S., 2026b. Lactiplantibacillus plantarum Paraprobiotics Produced by Thermosonication: A Flow Cytometry and Antidiabetic Activity Assessment. 19(3), 154. Food and Bioprocess Technology 19 (3). Available at https://doi.org/10.1007/s11947-016-1855-2, https://doi.org/10.1007/s11947-026-04253-w.

Yolmeh, M., Sforça, M.L., Sant'Ana, A.S., 2025. Antimicrobial properties of *Bifidobacterium animalis* subsp. lactis Bb-12 paraprobiotic obtained using ohmic heating against *Salmonella enterica*: a metabolomic approach. International Journal of Food Microbiology 429, 111016. Available at https://doi.org/10.1016/j.ijfoodmicro.2024.111016.

Yolmeh, Mahmoud, Khomeiri, Morteza, Ghaemi, Ezzatollah, 2020. High-efficiency anti-enterotoxigenic activity of Lactobacillus on staphylococcal enterotoxins biosynthesis. LWT 125, 109266. Available at https://doi.org/10.1016/j.lwt.2020.109266.

Zahidah, I., Bölek, S., Terzioğlu, Ö.T., Adıgüzel, S., 2024. Determination of the effects of novel paraprobiotic supplement of *Lactobacillus plantarum* on soy dairy-free beverage by physicochemical, antioxidant, sensory analyses, and Raman spectroscopy technique. John Wiley and Sons Inc, Turkey. Journal of Food Science 89 (11), 7189–7202. Available at https://doi.org/10.1111/1750-3841.17365, http://onlinelibrary.wiley.com/journal/10.1111/(ISSN)1750-3841.

Chapter 10

A critical approach to probiotic effectiveness and technological challenges when incorporated as viable or nonviable (postbiotic)

Sourav Misra[1], Sanchita Biswas Murmu[1], Abhisek Pradhan[2], P. Sivamma[3], Shubham Mandliya[4], and Basavaprabhu Haranahalli Nataraj[5]

[1]*ICAR – National Institute of Natural Fibres Engineering and Technology, Kolkata, West Bengal, India,* [2]*Agricultural and Food Engineering Department, Indian Institute of Technology Kharagpur, Kharagpur, West Bengal, India,* [3]*ICAR - National Institute of Seed Science & Technology, Kushmaur, Mau, Uttar Pradesh, India,* [4]*Department of Food and Nutritional Sciences, University of Reading, Reading, England, United Kingdom,* [5]*Southern Regional Station, ICAR-National Dairy Research Institute, Adugodi, Bengaluru, Karnataka, India*

10.1 Introduction

The FAO/WHO (2020) defines probiotics as living organisms that, when provided in adequate quantities, enhance the health of the host. They are delivered to the consumer in the form of dietary supplements (in capsule, tablet, or powder form) or food products (nonfermentable/nonfermentable). The use of probiotics is becoming more prevalent as it is cheap and provides the necessary solution to consumers with widespread knowledge. The human body consists of many kinds of microbes; harmful ones lead to illnesses and diseases due to the presence of toxic compounds, which accumulate and thus deplete the immune system. To address this issue, utilizing higher doses of probiotics presents an alternative approach to antibiotic treatment (Wang et al., 2022). The predominant probiotic bacteria are *Lactobacillus* and Bifidobacteria strains, which are beneficial to health if used in the range of 10^{6-7} CFU/mL or g in the final product during consumption (Tsai et al., 2019). Probiotic food comes under the category of functional foods with a compound annual growth rate (CAGR) of 8.3% by 2026, reaching $91.1 billion from $61.1 billion in 2023 (Liang et al., 2023).

With increasing awareness, various bioactives have also been incorporated into these foods, such as natural extracts, prebiotics, minerals, vitamins, flavors, and essential oils (Misra et al., 2023). The addition of probiotic cells

Evaluating Oral Probiotics and Prebiotics. DOI: https://doi.org/10.1016/B978-0-443-29835-6.00003-4

and these ingredients could influence their preservation capabilities and sensory characteristics of food products, while maintaining the metabolic activity of the chemicals and microorganisms throughout processing and storage. In this context, encapsulation serves as a viable strategy to address the challenges at hand. To date, the trapping of individual species, whether bioactive compounds or probiotic bacteria, has proven to be an effective preservation technique. This method helps prevent deterioration while facilitating a precise release in the gastrointestinal tract in suitable proportions (Misra et al., 2024). Orally ingested probiotics must endure challenging conditions while traversing the gastrointestinal (GI) passage and exhibit the potential to affect gut microbiota. The human body's several physiological functions were influenced by the gut microbiota. The probiotic intake containing natural bioactive components presents an eco-friendly alternative to medications for regulating blood pressure, enhancing mental well-being, strengthening immunity, and addressing sleep disorders, especially during the COVID-19 epidemic (Misra et al., 2022a, Misra et al., 2022b).

Nonetheless, certain studies indicate that the coencapsulation of multiple bioactive compounds has a synergistic effect, enhancing bioactivity and functionality relative to a single component (Vu et al., 2023). This concept has gained considerable interest in the pharmaceutical sector and shows potential as a micro-delivery vehicle for bioactive components and probiotic strains, specifically targeting the colon via food products. The coencapsulation of prebiotics like polydextrose and inulin, along with probiotics, has shown efficacy in retention of their viability throughout the GIT. Furthermore, the incorporation of antioxidants like tocopherol and ascorbic acid into capsules has safeguarded the probiotics against a harmful environment, including low pH and oxidative stress, at the time of processing as well as storage (Atia et al., 2017).

Bioactive components such as apple skin polyphenols, anthocyanin pigments, and black bean extract have recently been coencapsulated along with probiotics using freeze and spray drying, and coextrusion processes for incorporation into the food products. This was done to improve the functional qualities of milk drinks, yogurt, cheese, and strawberry nectar to shield the probiotics from unfavorable processing conditions and to ensure that they are able to pass through acidic gastrointestinal conditions (Morsy et al., 2022).

10.2 Functional foods

A food is deemed functional if it not only provides sufficient nutritional effects but also has a positive impact on one or more specific body processes, contributing to a lower risk of disease or an enhanced state of health and well-being (Birch and Bonwick, 2019). Food products that include one or more bioactive ingredients that have physiological effects and improve consumers' health are termed "functional foods" (Zeisel, 1999). There is a difference between fortified foods and functional foods; fortified food contains certain ingredients rich in minerals and vitamins and can be incorporated as per the recommendations and guidelines by the authorizing body, while functional foods have a superior quality over common foods, providing health benefits (Reque and Brandelli, 2021). The market for natural products is dominated by functional foods, followed by functional drinks and food supplements. They can be further divided into natural (vegetables and fruits), altered (whole grains and fiber), enriched (probiotics), fortified (mineral and vitamin additions), or enhanced (salmon with more beneficial fatty acids or eggs with more omega-3 fatty acids) products (Kaur and Singh, 2017; Santini et al., 2018; Misra et al., 2023) claim that the clinical data supporting the safety and effectiveness of nutraceuticals sets them apart from food supplements (Fig. 10.1).

10.3 Probiotics and their mechanism of action

As per FAO/WHO (2020), probiotics are live bacteria that, when given in adequate quantity, promote the host's health. Numerous probiotic species have been utilized over time to improve the function of the gut barrier, produce antimicrobial components, and immunoprotective responses, thereby preventing and treating conditions like gastric cancer, diabetes, and inflammatory bowel disease (IBD). These responses lead to the removal of pathogens like Salmonella, Helicobacter pylori, and rotaviruses and modulate the immunity of the host (Razavi et al., 2021). The probiotication in food matrix has increased significantly with consumer awareness, as it improves the digestive system (Nithya et al., 2023; Misra et al., 2025a).

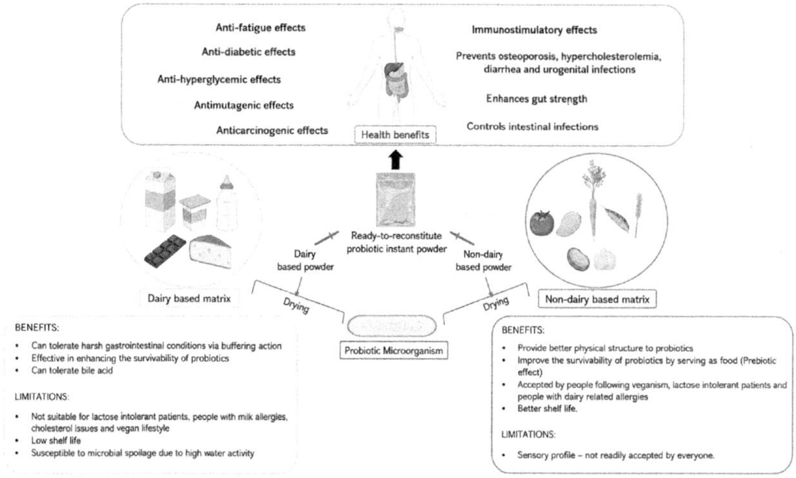

FIGURE 10.1 Health benefits exerted by ready-to-reconstitute synbiotic powder in human well-being. Source*: Adapted from Chaturvedi et al. (2018).*

Strains of *Lactobacillus* and *Bifidobacterium* are widely used probiotics; nevertheless, other species, including a nonpathogenic strain of *E. coli* Nissle 1917 and *B. cereus*, have also been used as probiotics to achieve goals like preventing ulcerative colitis relapse, enhancing the immunity of body's immunity, and producing favorable impacts on inflammatory or allergy sicknesses (Solanki et al., 2013). Probiotics must be metabolically active throughout the stomach-to-intestine transit in sufficient amounts to achieve these positive health benefits (Sanz, 2007) and should not be less than $10^{6\text{-}7}$ CFU/ml or g during consumption for their functionality (Adhikari et al., 2000), which can be an alternative to medicines for the prevention of several illnesses (Misra et al., 2021).

The word "Probiotics" encompasses other related concepts, including prebiotics, postbiotics, and synbiotics. The term "Postbiotics" is characterized by by-products of metabolism derived from probiotics or nonviable bacterial products, including hydrogen peroxide, bacteriocins, organic acids, diacetyl, ethanol, and acetaldehydes (Vallianou et al., 2020). According to Rastall and Gibson (2015), prebiotics are nutrients that have the ability to alter the gut microbiota and promote the development of advantageous microbes. Natural sources such as vegetables, fruits, and grains may be used to extract prebiotics. Inulin, lactulose, fructooligosaccharides, arabinoxylan, galactooligosaccharides, and resistant starch-1, -2, -3, and -4 are a few examples of common prebiotics (Hutkins et al., 2016). Because of the activity of the gut bacteria, these sugars ferment and provide energy to the colon's epithelial cells. Prebiotics and probiotics combine to produce synbiotics. The synergistic effect occurs when prebiotics and probiotics collaborate within a living system, yielding multiple advantages, including the enhancement of probiotic survival, disease prevention, and improvements in nutritional status. Due to their ability to improve gut health, commercial functional foods, including symbiotic foods, are gaining popularity. A study Makino et al. (2010) confirmed that the consumption of fermented food containing probiotic bacteria prevents infection in the upper respiratory tract. The improvement of immune function through probiotics can be explained by their mechanisms of action within the host's gastrointestinal tract. Before selecting probiotic strains, it is essential to examine their advantageous characteristics, including tolerance to acidic and bile salt environments and antibacterial efficacy against pathogenic bacteria.

Probiotics have a strain-dependent mechanism with three different modes of action. The interaction of host immune cells with probiotics results in immunological regulation (Burgain et al., 2011). Second, probiotics have the potential to compete with infections, preventing them from adhering to the gut; third, probiotics may change dietary components, toxins, and bile salts (Frakolaki et al., 2021). The metabolites produced by probiotics control the human immune system, including peptides, amino acids, and vitamins, potentially reducing the risk of infections caused by the severe acute respiratory syndrome coronavirus 2 (SARS-CoV-2) (Koirala and Anal, 2021).

10.4 Novel strategies to enhance the viability of probiotics in food products

10.4.1 Microencapsulation and edible coating technology

The general health benefits of probiotics may not be demonstrated by every strain to its full potential because of the frequent survival loss during storage of foods and gastrointestinal (GI) transit caused by exposure to multiple stresses. Survival loss is controlled using microencapsulation of bacteria in a coating solution ranging in size from 0.2 to 5000 μm. The polymeric matrix used for encapsulation should offer good physical protection against various gastrointestinal (GI) stresses, such as low pH (1–2.5) and high bile stress (0.8%–1%). The polymer should not be toxic or immunogenic to either the probiotic bacteria or the host. It should not demonstrate antagonistic activity against the encapsulated bacteria. The encapsulated matrices interact through ionically or covalently cross-linked polymer networks or are not cross-linked, such as spray-dried materials. Several food-grade polysaccharides and protein-based materials/matrices have been used for the microencapsulation of probiotics, including alginates, chitosan, xanthan, gum arabic, maltodextrin, caseins, and whey proteins.

Microencapsulation techniques are categorized as emulsion, extrusion, and spray drying techniques. However, extrusion and emulsion technologies have been the most extensively studied and researched methods. The controlled release of probiotic cells encapsulated within the matrices at the target site (intestine) may help probiotic bacterial colonization in the gut (Cook et al., 2012). The coating material acts as a barrier between the bacterium and its environment, providing a protective core that shields it from harsh external conditions. This coating is also known as a wall, shell, membrane, or carrier material. The presence of the wall polymers above the probiotic bacterial cell plays a key role in the functionality of the capsule materials. The encapsulation may be arranged in various forms, entrapping the active compounds. If the coating material is edible, it should be interactive concerning the active substances, effectively encapsulate along with the retention of core materials within the capsule, and protect against adverse environmental conditions while remaining sustainable and cost-effective. Additionally, edible coatings should offer resistance to low pH, moisture content, differential pressure, exchange of gases (levels of O_2/CO_2), and stress due to thermal exposure.

Biopolymer-based edible coating materials are commonly used for encapsulating probiotic cells. These include polysaccharides (starch, cellulose derivatives, chitosan, and alginate), lipids (waxes and fats), proteins (gelatin, soy protein, zein, collagen, egg white, whey protein concentrate and isolate, and casein) (Pech-Canul et al., 2020). In one study, pullulan was employed as a carrier matrix in microencapsulation of *Lactiplantibacillus plantarum* CRD7 through spray drying, with the prebiotic of isomalto-oligosaccharides. The results demonstrated that the microencapsulated *L. plantarum* maintained viability above 7.5 log CFU/g under harsh gastrointestinal conditions, including acid and bile salt exposure. Moreover, the encapsulated *L. plantarum* successfully fermented milk into curd, reaching a final pH of 4.62 (Misra et al., 2025b).

10.4.2 Electrospinning technology

Electrospraying and electrospinning are two processes known for their simplicity, versatility, and mild processing conditions. These methods have gained significant attention for the immobilization of probiotics, enhancing survival in adverse surroundings, which ensures targeted delivery within the GI tract with minimal loss. In these techniques, a feed solution containing both probiotics and polymers is extruded by a needle with the help of a syringe pump. After the emergence of the droplet, the formation of a Taylor cone at the tip of the needle occurs with the balancing of electrical repulsion and surface tension. Exceeding the repulsive forces over the surface tension and viscoelastic forces, jets break, leading to the formation of micro or nanostructures that are directed toward a collector (Seethu et al., 2023).

Electrohydrodynamic techniques can be categorized based on the nozzle type used: uniaxial, coaxial, and triaxial. Among these, electrospinning has been extensively studied, with various classifications such as uniaxial electrospinning, coaxial electrospinning, and emulsion electrospinning. This method primarily produces micro and nanoscale fibers with desirable structural and functional properties, including porosity, tunable diameters, and a high surface-to-volume ratio, with excellent entrapment efficiency for bioactives. Electrospinning has demonstrated promising potential for encapsulation of probiotics, with uniaxial electrospinning being the most commonly employed approach. While synthetic polymers (e.g., PVP, PVA, and PEO) are often used as carrier materials, biopolymers (e.g., proteins, polysaccharides, and pullulan) and additives (e.g., lactose, glycerol, skim milk, mannitol, and prebiotics) have also been explored as safer alternatives. Recent advancements in electrospinning have introduced modifications such as binary coaxial electrospinning, further enhancing probiotic stability and viability (Feng et al., 2023).

Encapsulated probiotics produced through electrohydrodynamic techniques have been confirmed to be nontoxic and safe. A study evaluating the safety of electrohydrodynamically encapsulated *Lacticaseibacillus rhamnosus* CRD11 and *L. plantarum* CRD7 followed guidelines of ICMR/DBT, assessing parameters such as hemolysis, mucin degradation, virulence factors, antimicrobial susceptibility, and biogenic amines. The outcomes indicated that these probiotic strains were free from toxic determinants (Varada et al., 2023). Moreover, even after encapsulation via electrospinning, these probiotics retained their in vivo functional efficacy, including immunomodulatory effects, gut microbiome modulation, and antioxidant activity (Varada et al., 2024).

10.4.3 3D-printing technology

Extrusion-based 3D-printing is an innovative technology increasingly used to create semisolid food products, for example, pastes, gels, and dough, with customized designs. This process utilizes computer-aided design (CAD) to layer printing inks, integrating specialized software for precise fabrication. Several 3D food printing technologies include selective sintering, extrusion-based printing (EBP), binder jetting, and inkjet printing. The 3D-printing method enables the customization of probiotic dosage and strain selection, facilitating the development of personalized probiotic food products tailored to individual health needs. The integration of live probiotic cells with biomaterials in 3D-printing is a promising advancement in regenerative medicine and nutraceutical delivery. This technique enhances probiotic protection against environmental stressors, improves their functionality, and ensures targeted delivery with controlled release (Sadeghi et al., 2024). Key advantages of 3D food printing include hypercustomization, waste minimization, and the ability to implement production with minimal facility modifications, making it a suitable approach for manufacturing customized functional probiotic foods (Kuo et al., 2022).

Recent research has explored the use of exopolysaccharides derived from microbial actions and biofilms as innovative bioinks for probiotic 3D-printing. These bioinks hold diversified applications for the formulation of next-generation functional foods, eco-friendly bioremediation solutions, and advanced biomaterials. However, most 3D-printed food materials need postprocessing steps like frying, baking, freezing, or drying, which can alter their shape, texture, and size. These factors must be carefully considered during product formulation. In a study investigating 3D-printed probiotic foods, a "honeycomb" structure maintained viable probiotic counts exceeding 10^6 CFU/g when baked at 145°C with a targeted 6% moisture reduction. Another work examined the probiotics' survivability in mashed potato (MP) formulations printed using several nozzle diameters (0.6–1.4 mm) at temperatures varying from 25°C to 55°C. Results showed that extrusion through a small nozzle of diameter 0.6 mm resulted in a slight reduction in viability of the probiotic from 9.93 to 9.74 log CFU/g. However, when MP was held in a heated nozzle at 55°C for 45 minutes, a significant decrease in viability was observed (from 10.07 to 7.99 log CFU/g). Notably, probiotic viability remained stable over a storage of 12 days (Li et al., 2020). These outcomes showed the suitability of 3D-printing as a novel tool to develop functional foods with probiotics, offering new possibilities for personalized nutrition and enhanced probiotic delivery.

10.5 Utilization of novel strategies for food products probiotication

With the emergence of functional foods aiming to deliver health advantages to humanity, a rapid rise in probiotic food products was observed. The durability of probiotic strains in food items is influenced by a variety of parameters, including the kind of strain, culture preparation method, cell inoculum state, storage temperature, and oxygen level (Ying et al., 2013). In comparison to wet gelled formulations, microencapsulated probiotic powder formulations have been shown to be more convenient. Nevertheless, the preservation of probiotics and the retention of their functions in various food matrices are contingent upon the selection of a suitable microencapsulation technique and coating materials (Fig. 10.2).

10.5.1 Beverages

Pourjafar et al. (2020) placed 1 g of *L. acidophilus* in 10 mL of Iranian Doogh beverage after encapsulating it using the extrusion process with sodium alginate and chitosan. Probiotic viability was found to be higher in the microencapsulated beverage (7.4×10^7 CFU/g) than in free cells (5.8×10^4 CFU/g) over a period of storage (for up to 42 days at 5°C) without a discernible change in the acidity, pH, or sensory qualities of beverages.

Olivares et al. (2019) investigated the use of probiotics *L. casei* encapsulated in orange juice using extrusion technology and sodium alginate. Despite the juice's lower pH (3.45), the authors found that probiotics had a higher viability (10^6 CFU/mL) at 4°C for 28 days. The inclusion of ascorbic acid, an antioxidant, antibacterial, and essential component for probiotics, allowed for this increased concentration of survival. According to Ying et al. (2013), the addition of spray-dried microencapsulated *L. rhamnosus* GG (1% w/v) to probiotic apple juice improved survival at higher temperatures of 25°C

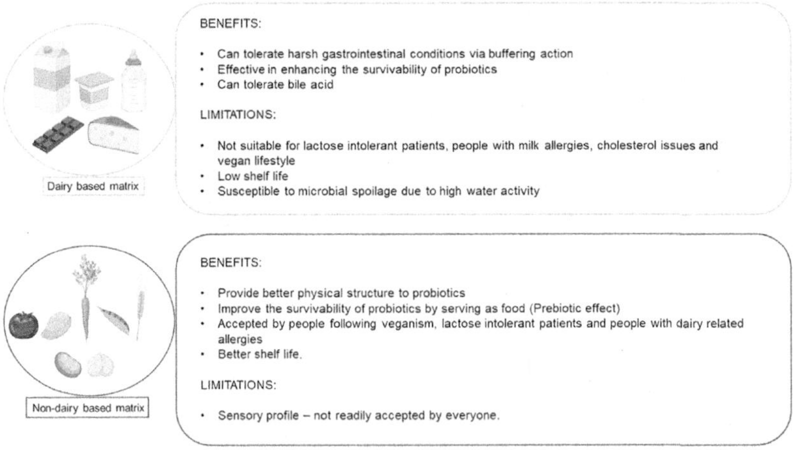

FIGURE 10.2 A summarized presentation illustrating the benefits and drawbacks of dairy and nondairy matrices for delivery of probiotic food powders.
Adapted from Chaturvedi et al. (2018).

and lower pH than the lowering of temperature. Dias et al. (2018) found an increase in inulin concentration that enhanced the survival of spray-dried *Bifidobacteria animalis* (exceeding 6.5 log CFU/g) in juice stored for 30 days at temperatures of 4°C and 25°C, while maltodextrin resulted in a nonviability of bacteria in passion fruit juice stored for the same period.

10.5.2 Cheese

The use of bacterial strains that have been microencapsulated in various cheeses has been the subject of several investigations. Cheese's comparatively high fat content and strong ability to buffer against harmful GI conditions make it an ideal choice for probiotic microbes (Frakolaki et al., 2021). The higher bacterial viability (>6 log CFU/mL) was found in spreadable goat ricotta cheese, and properties, including reduced gumminess, adhesiveness, and proteolysis, were enhanced by *L. acidophilus* encapsulation through freeze-drying with chitosan and alginate (Lopes et al., 2021). Vasile et al. (2020) used freeze-drying for encapsulation of *L. casei*, biopolymers like inulin, chitosan, and whey protein isolate, and black bean aqueous extract. This resulted in a two-log enhancement in the survival of bacterial cells, along with increased phytochemical stability and biological variables, including higher α-amylase inhibitory activity and α-glucosidase in soft cheese held for 21 days. A study by Özer et al. (2008) found no difference in extrusion and emulsion technology for microencapsulation for Kesar cheese in terms of bacterial count, proteolysis, and organoleptic properties. *Bifidobacterium bifidum* BB-12 and *L. acidophilus* LA-5 microencapsulated in cheese exhibit viability greater than 10^7 CFU/g, but nonencapsulated probiotics exhibited a continual decline owing to scalding during the 90-day storage period.

10.5.3 Yogurt

Probiotic yogurts have been formulated by incorporating live strains to enhance their medicinal efficacy (Chen and Chen, 2007). It is an inadequate vehicle for probiotics owing to its lower pH (4.2–4.6). The issue has been addressed by many researchers by utilizing microencapsulation of probiotic strains. During 28 days of storage, the encapsulated probiotic strain *L. acidophilus* had a higher viability (7 log CFU/g) than the free cells for yoghurt due to the coating of xanthan gum and whey protein. However, the use of microencapsulated probiotics enhanced the yoghurt samples' firmness, viscosity, and adhesiveness, while reducing syneresis (Khorshidi et al., 2021). Using extrusion technology, De Prisco et al. (2017) encapsulated *L. delbrueckii* in chitosan-alginate capsules for yogurt starter culture. Probiotics survived the 28-day storage period (7.70 log CFU/g) and digestion (2 hours in gastrointestinal conditions) with 100% viability, whereas free cells died. The syneresis and texture profile of yogurt were adversely affected by the supplementation of encapsulated *Lactobacillus* with lactitol and galactooligosaccharides by the process of extrusion during storage, as observed by Li et al. (2020). However, the viability of the yogurt samples was improved by 9.95 log CFU/g in comparison to the control sample. Adhikari et al. (2003) observed that the sensory quality was delayed when encapsulated bifidobacteria were added into agitated yogurt, resulting in a granular texture with particle sizes spanning from 22 to 50 μm. In simulated gastrointestinal conditions, the viability was improved by the inclusion of prebiotic constituents, including resistant starch and glycerol, as encapsulating material (Sultana et al., 2000). However, the probiotic cells were not properly safeguarded. This enhances the food functionalities by stimulating the probiotic growth through a synergistic effect of prebiotics and probiotics.

10.5.4 Ice cream

Probiotic proliferation in ice cream is influenced by pH because bacteria prefer to thrive better in environments that are neutral. The probiotic bacteria's metabolic activity is impacted by fermented ice cream, but the loss of strain viability in frozen dairy desserts is mostly instigated by elevated osmotic pressure, oxygen toxicity, and frozen damage (Frakolaki et al., 2021). Alginate, inulin, resistant starch, and chitosan as prebiotics, along with *L. plantarum*, were incorporated into ice cream via electrosprayed microcapsules by Li et al. (2020) and observed higher vitality (7.82 log CFU/g) up to 90 days. By speeding up their development and proliferation, coencapsulating probiotics and prebiotics increases their survivability under the challenging circumstances of the food matrix and digestion (Misra et al., 2026). The encapsulated *L. casei* hydrogel developed from whey protein concentrate and calcium alginate showed a superior survival rate (>8 log CFU/mL) than the free cells after 80 days. The form, size, and kind of coating materials utilized for encapsulation had a significant influence on the viscosity of ice cream.

10.5.5 Other food products

"Synbiotic" mousse was developed by dos Santos et al. (2019) using spray drying by encapsulating inulin and *L. acidophilus* and exposing it to the in vitro gastrointestinal condition. After 6 hours of in vitro research, a decline in bacterial cell count was observed with a highest of 7.4 log cycle in free cells, followed by 3 log cycle in mousse containing free cells, 2 log cycle in microencapsulated cells, and 1.3 log cycle in mousse containing microencapsulated cells.

Hossain et al. (2021) prepared chocolate with *L. casei* (freeze-dried) encapsulated with fructooligosaccharides, alginate, and cocoa powder. The higher survival was found in the gastrointestinal digestion process (8 log CFU/g) and during storage (>7 log CFU/g) after 90 days at 25°C. Ozturk et al. (2021) stored dark and milk chocolates at 4°C for 180 days and found the viability of 6.90 log CFU/g in dark and 7.12 log CFU/g in milk chocolates without compromising the chocolates' moisture or sensory quality when encapsulated with *S. thermophiles* with gum Arabic, cellobiose, carboxymethylcellulose, and pectin. During storage, the improved survivability of probiotics may be attributed to the increased content of carbohydrate and protein in milk chocolate.

Muzzafar and Sharma (2018) developed the creamy biscuit where microcapsulated powder was incorporated into the cream. This powder was developed using freeze-drying of wall materials (maltodextrin, xanthan gum, and sunflower oil) and probiotics (*B. bifidum, L. rhamnosus, and L. acidophilus*). The viability of probiotic biscuits (exceeding 8 log CFU/g) was sustained for a period of 8 weeks when stored at 25°C. Additionally, probiotic biscuits exhibited better taste, with higher fat and fiber content, providing good overall acceptability. González-Forte et al. (2014) also prepared functional biscuits by encapsulating *L. plantarum* and *inulin* in alginate solution using the emulsification method. The survivability of the probiotic was significantly increased from 8.7 log CFU/mL to 7.8 log CFU/mL in the uncoated sample during storage for up to one month. Furthermore, during the simulated GI tract, a viability of 5.5 log CFU/g was observed, which was significantly higher than 3.5 log CFU/g viability in the control sample. Altamirano-Fortoul et al. (2012) evaluated the physicochemical parameters of starch-coated bread baked at 180°C for 16 minutes after incorporating *L. acidophilus* to the surface. The thermal process did not influence the probiotic count, which demonstrated a survival rate of 63.2%. The bread that was developed demonstrated comparable qualities to the standard bread, including improved acceptability and a reduced failure force.

An Italian salami was developed by adding a spray-dried powder (de Oliveira Gomes et al., 2021). The powder was prepared using xanthan gum, β-cyclodextrin, and alginate with *B. animalis*. Higher survivability (>8 log CFU/g) was achieved during storage (up to 45 days at 25°C) when 0.02% curing salt was added with encapsulated probiotic. This was accomplished without compromising the salami's textural qualities, nutritional profile, fatty acid profile, or lipid oxidation, and the probiotication enhanced its organoleptic qualities. Camargo et al. (2021) used an emulsification technique using nutmeg, black pepper, garlic, and curing salt to encapsulate *B. bifidum* in coppa. After 30 days of ripening, viability ranged from 10.6 to 7.3 log CFU/g. Additionally, probiotic-added coppa showed reduced oxidation of lipid and increased weight loss during storage, in contrast to control samples.

A 3D-printed synbiotic-composite flour construct was developed by Yoha et al. (2021) utilizing ajwain seeds, fried gram, green gram, and barnyard millet. *L. plantarum* was incorporated into the flour construct and encapsulated using various technologies. The 3D-printed foods were subjected to microwave, hot air, and freeze-drying techniques. The spray-freeze-dried product demonstrated the maximum probiotic viability, exhibiting survival of 79% and 96% during simulated digestion and storage up to 35 days, respectively.

Rajam et al. (2015) prepared noodles by incorporating the freeze-dried powder. This powder was developed using denatured whey protein isolate and fructooligosaccharide as carrier material for encapsulating *L. plantarum*. The survivability of probiotics in hot-dried noodles decreased from 93.63% to 62.42%. The overall acceptability of the probiotic noodles was higher, despite a minor brown coloration.

10.6 Functional food development via coencapsulation

The therapeutic product development with encapsulation of bioactives with probiotics has received very little attention. Ice cream formulated with the incorporation of coencapsulated *B. bifidum* 231 and *L. helveticus* 194 alongside fructooligosaccharides achieved the higher viability after 90 days of storage, without compromising appearance, texture, color, or overall acceptability (Sahitya et al., 2013). Anthocyanin and polyphenol content increased by 12% and 26%, respectively, when comicroencapsulated black currant extract and *L. casei* were added at (5%, w/w) in yoghurt (Enache et al., 2020). However, flavonoid content decreased by 12%, and probiotic bacterial viability increased during storage for 21 days. Comicrocapsules demonstrated better preservation of bacterial cells than free cells and microencapsulated cells. The coencapsulation of prebiotic Hi-maize and *L. acidophilus* coated with chitosan only caused a reduction of probiotic count by 1.0-log cycle in synbiotic yoghurt (Iyer and Kailasapathy, 2005). Holkem et al. (2019) added *B. animalis* subsp. *Lactis* (probiotic) and cinnamon extract (proanthocyanidin-rich) coencapsulated in sugarcane juice, resulting in a pH drop, an increment in soluble solids and thereby density, and a rise in phenolic content due to the addition. A reduction in proanthocyanidin concentration occurred because of the interaction of polyphenols with probiotics in the food matrix, accompanied by an increase in viscosity and enhanced bacterial viability, all while retaining sensory attributes throughout a 28-day storage period.

10.7 Probiotic food products packaging

The choice of packaging material has a substantial effect on the stability of the nutraceutical foods fortified with probiotics (Fig. 10.2). When handling the probiotic products, plastic materials with a good oxygen permeability should be avoided because certain anaerobes, especially bifidobacteria, are more sensitive to oxygen. For the preservation of probiotics in yogurt, optimal conditions were prepared by utilizing an oxygen-scavenging agent in packaging materials to improve the oxygen barrier properties. Thicker glass containers demonstrated enhanced survivability of *L. acidophilus* and bifidobacteria in yogurt samples when compared to plastic containers (Shah, 2000). The use of vacuum packaging, active packaging, and oxygen scavengers with improved moisture and air barrier qualities could offset the higher cost associated with glass containers (Tripathi and Giri, 2014). It has been demonstrated that applying prebiotics like oligosaccharides and inulin to edible coatings or films increases the bacterial viability in foods (Asaithambi et al., 2021). Nitrogen, an inert gas, kept the stored product from oxidizing, ensuring the probiotics' survival in the corn flakes. Relative humidity, temperature, and the crystallinity of the packaging material may significantly affect the permeability of the material, which may influence the survival rate of the probiotic strains that are encapsulated (Priyadarsini et al., 2022; Jose et al., 2024).

10.8 Technological challenges in probiotic food development

Using quick beverage powder in conjunction with the probiotic and prebiotic effects is definitely a consumer-friendly strategy with several advantages. But some processing difficulties are also related to such products that need to be comprehended before commercial use (Muzzafar and Sharma, 2018). One of the main justifications for probiotic microencapsulation is that the cells are shielded by microbes from harsh intestinal circumstances, including reduced pH, exposure to bile salts and acids, etc. But numerous difficulties occurred during the probiotic microencapsulation development process. These consist of the ability of the capsule material to endure in novel matrices, such as goods made from wheat and legumes, and the efficiency of the microcapsule in enhancing probiotic survival in the face of various stressors, including lowering pH, drying temperature, and storage conditions. Furthermore, the quick breakdown of the microcapsules in digestive fluids is a major problem, subjecting them to bile salts and stomach acid. Another difficulty is microencapsulating probiotics in powders like yoghurt or curd powder and fermented foods like sauces, juices, and beverages. This is mostly due to the items' low pH and the organic acids and antibacterial substances they contain (Zhang et al., 2020). Several investigations into the stability of probiotics in microencapsulation under stressful circumstances have concluded that to overcome this difficulty, novel strategies must be developed, like using matrices that can shield the cells against acid, using different bacterial strains that can withstand acidic conditions better, or even using spores that can withstand heat, etc. Also, the final powdered mixture faces numerous difficulties throughout the drying process of both the food product and the probiotic bacteria. The probiotic's viability is reduced by stressful conditions such as high drying temperatures (inlet and exit 200–205 and 85°C–90°C, respectively) and low moisture content (less than 4%) of the produced product. Furthermore, because juice powders include a lot of sugar and organic acids, the dried powders also have a tendency to adhere to the chamber walls, which lowers yield and causes operational issues. The glass transition temperature is linked to a number of issues with reconstituted powders of dairy origin, including particle aggregation, caking, and rehydration challenges. The food industry uses a variety of bulking agents, including proteins, gum Arabic, modified starches, and maltodextrins, to address the stickiness problem. Stickiness can also be minimized by optimizing consolidated particle size and temperature of stickiness using a sticky curve. These aids also serve as encapsulating agents, allowing various spray-dried items to maintain qualities, including antioxidant activity, nutritional value, and attractiveness. It is interesting to note that these compounds, which are also prebiotics, promote the probiotics during the storage and drying processes. Moreover, food matrices' composition is also impacted by drying. Compared to nondairy powders, dairy-based food powders lose more water activity, which lowers their quality and shortens their shelf life. The impact of using the biophysical and sensory effects of using soy milk rather than skim milk on the characteristics of sesame ice cream was examined (Ahanian et al., 2014). The findings demonstrated that soy milk-added ice cream as the primary component exhibited higher proximate values and several health advantages.

10.9 Evolution of the market of probiotic foods

Worldwide probiotic business remains in a developmental phase and is anticipated to experience significant growth since individuals become attentive to health and opt for preventative medical treatment. Increasing demand for functional foods as well as nutritional supplements drives the market for probiotic ingredients in developing Asian countries due to their growing populations and disposable incomes. Major industries have been making significant investments in research and development for a new product portfolio for several years. Notable industries include Yakult Honsha Co., Ltd. from Japan, as well as Chr. Hansen A/S from Denmark formulated various probiotics and received a patent, which claims several health advantages. The probiotic ingredient market is divided into three segments based on their use: animal feed, dietary supplements, and food and beverage. In 2015 the food and beverage segment was claimed to be the biggest in the probiotic market, holding over 85% of the total income. This segment includes both dairy and nondairy products, baked goods, cereals, fermented meat items as well as dried food probiotics. Among all these goods, dairy products made up a significant portion during 2015. The market for fermented meat items is anticipated to grow at the fastest rate. During 2015, the market demand was found to be value of $33.19 billion. With the fast expansion of the food and beverage sector in the developing nations, including India, China, and Japan, the Asia-Pacific probiotic market is forecasted to expand at the greatest CAGR (compound annual growth rate). According to Markets and Markets (2016), the Middle East, Africa, South America, and Central America are anticipated to exhibit modest development over the projected timeframe. Increased clinical evidence about probiotics' effectiveness in treating diseases and maintaining health will accelerate the market's expansion in the healthcare industry. Among the main characteristics of probiotics are the ability to treat urogenital infections, intestinal inflammation, and act as an antibiotic in treating diarrhea by countering the harmful bacteria in the gut. In light of treatments related to lactose intolerance, oxalate metabolism, lipid metabolism, ulcerative colitis, irritable bowel syndrome (IBS), Helicobacter pylori, IBDs, necrotizing enterocolitis, allergic rhinitis, eczema, as well as infectious diarrhea, the probiotic industry anticipates substantial expansion in the food segment.

10.9.1 Global products

According to a market survey on available probiotic products worldwide, nondairy probiotic goods are gradually becoming part of humans' everyday diet (Table 10.1). The fruit and vegetable industry, in particular, is providing consumers with a variety of products. Modern technology has made it possible to produce several shelf-stable nondairy products using nonconventional/modern matrices such as chewing gum, kombucha tea, sugar, as well as baking mix, in addition to cereals, fruit, and vegetables. Interestingly, because of its thermophilic nature, *Bacillus coagulans* and its spores are also being used more frequently in commercial products along with *Lactobacillus* (Table 10.2).

10.9.2 Global regulatory environment

Probiotics' regulatory environment varies globally. Japan has the longest established probiotic market due to its food for specified health use. However, the Asia-Pacific is noted as having great permissive legislation. Both Australia and New Zealand hold a permissive regulatory environment that allows health claims to be made on an individual basis. Toward the west, both the United States and Canada likewise possess fairly lenient regulatory requirements. According to Degnan (2008), the Canadian government has accepted the "probiotics" list and permits health claims such as "provides microorganisms that contribute to a healthy gut flora." Whereas Latin American Chile permits labels such as "may help stimulate the immune system." The EU has stricter regulations. Other than "aiding lactose digestion," probiotic microbes have no accepted health claims. Foods, Nestle, Danisco A/S, Danone, Chr. Hansen, Arla, Probi, and Lallemand Inc. are significant players in the market (Aparimita et al., 2025).

TABLE 10.1 Brief glance at the effect of incorporation of probiotics in physicochemical properties of synbiotic food products.

Probiotic strains	Products	Functionalities	References
Bifidobacterium lactis Bb12 Lactiplantibacillus *plantarum* a7	Bread	• Increase in the specific volume and moisture content	Zhang et al. (2014)
L. plantarum	Oats milk drink	• Significant increase in the antioxidant activity and good viability of probiotics	Ravindran and S (2021)
Lactococcus lactis SKL 13	Composite dough	• Better dough properties in terms of improved textural (decrease in adhesiveness, cohesiveness, chewiness, and increase in springiness) and pasting properties (decrease in setback viscosity) and exhibited higher viscoelasticity	Misra et al. (2024)
L. plantarum TISTR 2075	Cereal extract powder	• Improved functional properties of powder after spray drying	Savedboworn and Wanchaitanawong (2015)
Lacticaseibacillus *rhamnosus* GG	Bread	• Enhanced textural and sensory properties	Qandashtant et al. (2020)
L. plantarum ATCC 8014	Almond milk powder	• Desired antioxidant activity and phenolic content	(Lipan et al. 2020)
Lactococcus lactis SKL 13	Multigrain biscuits	• Improved physicochemical and sensory properties during the storage period	Misra et al. (2025)
L. plantarum	Sohiong fruit powder	• Acceptable changes in flow properties and organoleptic properties	Vivek et al. (2020)
Bacillus coagulans	Spray-dried raisin powder	• Acceptable in terms of bioactivity with good viability of probiotics	Nazari Kermanshahi et al. (2021)
L. acidophilus and *Bifidobacterium lactis*	Moringa leaf juice powder	• Increased powder yield and bioactivity	(Looi et al. 2019)

TABLE 10.2 Global commercial probiotic food products.

Types	Food matrix	Brand/trade name	Place of origin	Active probiotic culture
Fruit and vegetable matrices	Fruit soda	Obi, Probiotic Soda	USA	Bacillus *coagulans*
	Cold pressed fruit and vegetables	Welo Probiotic cold pressed drink	Canada	*B. coagulans*
	Ginger and turmeric root	Welo Probiotic ferments	Canada	*B. coagulans*
	Dry plums	Mariani Premium Probiotic Prunes	California, USA	*B. coagulans*
	Fermented organic sugarcane molasses incorporated with special herbal tea	Vita Biosa 10+	Canada	Bifidobacterium *lactis*, *Bifidobacterium animalis*, *Bifidobacterium longum*, Lactococcus *casei*, *Lactococcus acidophilus*, Lacticaseibacillus *rhamnosus*, *Lactococcus lactis* subsp. *lactis* biovar. *diacetylactis*, *L. lactis* subsp. *lactis*, *S. thermophilus*, *L. pseudomesenteroides*
Cereal matrices	Oatmeal bar	Pop Culture Probiotic	California, USA	*B. coagulans*
	Probiotic + prebiotic bar	Truth Bars	USA	*B. coagulans*
	Cereal bar	Macrolife Macrogreens Superfood Bars	California, USA	*L. acidophilus*, *L. bulgaricus*, *L. rhamnosus*, *B. breve*, *B. longus*,

(Continued)

TABLE 10.2 (Continued)

Types	Food matrix	Brand/trade name	Place of origin	Active probiotic culture
Other matrices	Kombucha	KeVita Master Brew Kombucha	Oxnard, USA	*B. coagulans*
	Cultured coconut milk yogurt	SO DELICIOUS	Europe and North America	*L. bulgaricus, L. rhamnosus, L. plantarum, L. paracasei, S. thermophilus* and *B. lactis*

10.10 Conclusions

Probiotics have the ability to improve gut health by consuming the probiotic-enriched synbiotic food products. Food matrices act as a safeguard to protect the probiotics from the adverse conditions during processing, storage, and passage through the GI tract. To retain the viability of probiotics, various probiotication processes have been widely used, such as encapsulation, electrospinning, and 3D-printing techniques, along with the incorporation of prebiotics in the encapsulation matrix. Worldwide probiotic business remains in a developmental phase and is anticipated to experience significant growth since individuals become attentive to health and opt for preventative medical treatment. Increasing demand for functional foods as well as nutritional supplements drives the market for probiotic ingredients in developing Asian countries due to their growing populations and disposable incomes.

References

Adhikari, K., Mustapha, A., Grün, I.U., Fernando, L., 2000. Viability of microencapsulated Bifidobacteria in set yogurt during refrigerated storage. Journal of Dairy Science 83 (9), 1946–1951. Available at https://doi.org/10.3168/jds.s0022-0302(00)75070-3.

Adhikari, K., Mustapha, A., Grün, I.U., 2003. Survival and metabolic activity of microencapsulated *Bifidobacterium longum* in stirred yogurt. Journal of Food Science 68 (1), 275–280.

Ahanian, B., Pourahmad, R., Mirahmadi, F., 2014. Effect of substituting soy milk instead of skim milk on physico-chemical and sensory properties of sesame ice cream. Indian Journal of Science Research 7 (1), 1134–1143.

Altamirano-Fortoul, R., Moreno-Terrazas, R., Quezada-Gallo, A., Rosell, C.M., 2012. Viability of some probiotic coatings in bread and its effect on the crust mechanical properties. Food Hydrocolloids 29 (1), 166–174. Available at https://doi.org/10.1016/j.foodhyd.2012.02.015.

Aparimita, S., Misra, S., Panigrahi, C., Mandliya, S., Mishra, H.N., 2025. High intensity ultrasonication of fermentative Jun tea: Evaluation of physicochemical, microbiological and rheological properties. Food Physics 2, 100051.

Asaithambi, N., Singh, S.K., Singha, P., 2021. Current status of non-thermal processing of probiotic foods: a review. Journal of Food Engineering 110567.

Atia, A., Gomma, A.I., Fliss, I., Beyssac, E., Garrait, G., Subirade, M., 2017. Molecular and biopharmaceutical investigation of alginate–inulin synbiotic coencapsulation of probiotic to target the colon. Journal of Microencapsulation 34 (2), 171–184. Available at https://doi.org/10.1080/02652048.2017.1313330.

Birch, C.S., Bonwick, G.A., 2019. Ensuring the future of functional foods. International Journal of Food Science and Technology 54 (5), 1467–1485. Available at https://doi.org/10.1111/ijfs.14060.

Burgain, J., Gaiani, C., Linder, M., Scher, J., 2011. Encapsulation of probiotic living cells: from laboratory scale to industrial applications. Journal of Food Engineering 104 (4), 467–483. Available at https://doi.org/10.1016/j.jfoodeng.2010.12.031.

Camargo, V.P., Catanio, N., de Marins, A.R., R.d.C. Bergamasco, R.G., Gomes, A.C., Feihrmann, 2021. Physicochemical and sensory characteristics of Coppa with *Bifidobacterium animalis* ssp. Lactis (BB12) probiotic. Acta Scientiarum - Technology 43. Available at https://doi.org/10.4025/ACTASCITECHNOL.V43I1.55119, http://periodicos.uem.br/ojs/index.php/ActaSciTechnol.

Chaturvedi, S., Khartad, A., Chakraborty, S., 2021. The potential of non-dairy synbiotic instant beverage powder: Review on a new generation of healthy ready-to-reconstitute drinks. Food Bioscience 42, 101195.

Chen, M.J., Chen, K.N., 2007. Applications of probiotic encapsulation in dairy products. Encapsulation and Controlled Release Technologies in Food Systems. Wiley Blackwell, Taiwan, pp. 83–112. Available at http://doi.org/10.1002/9780470277881.ch4, http://onlinelibrary.wiley.com/book/10.1002/9780470277881.

Cook, M.T., Tzortzis, G., Charalampopoulos, D., Khutoryanskiy, V.V., 2012. Microencapsulation of probiotics for gastrointestinal delivery. Journal of Controlled Release 162 (1), 56–67. Available at https://doi.org/10.1016/j.jconrel.2012.06.003.

De Prisco, A., van Valenberg, H.J.F., Fogliano, V., Mauriello, G., 2017. Microencapsulated starter culture during yoghurt manufacturing, effect on technological features. Food and Bioprocess Technology 10 (10), 1767–1777. Available at https://doi.org/10.1007/s11947-017-1946-8, http://www.springer.com/life+sci/food+science/journal/11947.

de Oliveira Gomes, B., de Mesquita Oliveira, C., de Marins, A.R., Gomes, R.G., Feihrmann, A.C., 2021. Application of microencapsulated probiotic *Bifidobacterium animalis* ssp. lactis BB-12 in Italian salami. Journal of Food Processing and Preservation 45 (10). Available at https://doi.org/10.1111/jfpp.15841, http://onlinelibrary.wiley.com/journal/10.1111/(ISSN)1745-4549.

Dias, C.O., dos Santos Opuski de Almeida, J., Pinto, S.S., de Oliveira Santana, F.C., Verruck, S., Müller, C.M.O., et al., 2018. Development and physico-chemical characterization of microencapsulated bifidobacteria in passion fruit juice: a functional non-dairy product for probiotic delivery. Food Bioscience 24, 26–36. Available at https://doi.org/10.1016/j.fbio.2018.05.006, http://www.journals.elsevier.com/food-bioscience/.

dos Santos, D.X., Casazza, A.A., Aliakbarian, B., Bedani, R., Saad, S.M.I., Perego, P., 2019. Improved probiotic survival to in vitro gastrointestinal stress in a mousse containing *Lactobacillus acidophilus* La-5 microencapsulated with inulin by spray drying. LWt 99, 404–410.

Enache, I.M., Vasile, A.M., Enachi, E., Barbu, V., Stănciuc, N., Vizireanu, C., 2020. Co-microencapsulation of anthocyanins from black currant extract and lactic acid bacteria in biopolymeric matrices. Molecules (Basel, Switzerland) 25 (7). Available at https://doi.org/10.3390/molecules25071700, https://www.mdpi.com/1420-3049/25/7/1700/pdf.

Feng, K., Huangfu, L., Liu, C., Bonfili, L., Xiang, Q., Wu, H., et al., 2023. Electrospinning and electrospraying: emerging techniques for probiotic stabilization and application. Polymers 15 (10), 2402. Available at https://doi.org/10.3390/polym15102402.

Frakolaki, G., Giannou, V., Kekos, D., Tzia, C., 2021. A review of the microencapsulation techniques for the incorporation of probiotic bacteria in functional foods. Critical Reviews in Food Science and Nutrition 61 (9), 1515–1536.

González-Forte, L., Bruno, E., Martino, M., 2014. Application of coating on dog biscuits for extended survival of probiotic bacteria. Animal Feed Science and Technology 195, 76–84.

Holkem, A.T., José, E., Neto, S., Nakayama, M., Souza, C.J. F., Thomazini, M., et al. 2019. Sugarcane juice with Co-encapsulated *Bifidobacterium animalis* subsp. lactis BLC1 and Proanthocyanidin-rich cinnamon extract.

Hossain, M.N., Ranadheera, C.S., Fang, Z., Ajlouni, S., 2021. Impact of encapsulating probiotics with cocoa powder on the viability of probiotics during chocolate processing, storage, and in vitro gastrointestinal digestion. Journal of Food Science 86 (5), 1629–1641. Available at https://doi.org/10.1111/1750-3841.15695, http://onlinelibrary.wiley.com/journal/10.1111/(ISSN)1750-3841.

Hutkins, R.W., Krumbeck, J.A., Bindels, L.B., Cani, P.D., Fahey, G., Goh, Y.J., et al., 2016. Prebiotics: why definitions matter. Current Opinion in Biotechnology 37, 1–7. Available at https://doi.org/10.1016/j.copbio.2015.09.001, http://www.elsevier.com/locate/copbio.

Jose, N., Ray, D.P., Misra, S., Nayak, L., Ammayappan, L., 2024. Microencapsulation and nanoencapsulation of fungicidal and insecticidal agents for grain packaging and storage. Journal of Stored Products Research 109, 102468.

Iyer, C., Kailasapathy, K., 2005. Effect of co-encapsulation of probiotics with prebiotics on increasing the viability of encapsulated bacteria under in vitro acidic and bile salt conditions and in yogurt. Journal of Food Science 70 (1) M18. Available at https://doi.org/10.1111/j.1365-2621.2005.tb09041.x, http://onlinelibrary.wiley.com/journal/10.1111/(ISSN)1750-3841.

Kaur, N., Singh, D.P., 2017. Deciphering the consumer behaviour facets of functional foods: a literature review. Appetite 112, 167–187. Available at https://doi.org/10.1016/j.appet.2017.01.033, http://www.elsevier.com/inca/publications/store/6/2/2/7/8/5/index.htt.

Khorshidi, M., Heshmati, A., Taheri, M., Karami, M., & Mahjub, R. 2021. Effect of whey protein- and xanthan-based coating on the viability of microencapsulated Lactobacillus acidophilus and physiochemical, textural, and sensorial properties of yogurt. Food Science & Nutrition.

Koirala, S., Anal, A.K., 2021. Probiotics-based foods and beverages as future foods and their overall safety and regulatory claims. Future Foods 3, 100013. Available at https://doi.org/10.1016/j.fufo.2021.100013.

Kuo, C.-C., Clark, S., Qin, H., Shi, X., 2022. Development of a shelf-stable, gel-based delivery system for probiotics by encapsulation, 3D printing, and freeze-drying. LWT 157, 113075. Available at https://doi.org/10.1016/j.lwt.2022.113075.

Liang, D., Wu, F., Zhou, D., Tan, B., Chen, T., 2023. Commercial probiotic products in public health: current status and potential limitations. Critical Reviews in Food Science and Nutrition 1–22.

Lipan, L., Rusu, B., Sendra, E., Hernández, F., Vázquez-Araújo, L., Vodnar, D.C., et al., 2020. Spray drying and storage of probiotic-enriched almond milk: probiotic survival and physicochemical properties. Journal of the Science of Food and Agriculture 100 (9), 3697–3708. Available at https://doi.org/10.1002/jsfa.10409, http://onlinelibrary.wiley.com/journal/10.1002/(ISSN)1097-0010.

Li, H., Zhang, T., Li, C., Zheng, S., Li, H., Yu, J., 2020. Development of a microencapsulated synbiotic product and its application in yoghurt. LWT 122, 109033. Available at https://doi.org/10.1016/j.lwt.2020.109033.

Looi, Y.F., Ong, S.P., Julkifle, A., Alias, M.S., 2019. Effects of pretreatment and spray drying on the physicochemical properties and probiotics viability of Moringa (*Moringa oleifera* Lam) leaf juice powder. Journal of Food Processing and Preservation 43 (4). Available at https://doi.org/10.1111/jfpp.13915, http://onlinelibrary.wiley.com/journal/10.1111/(ISSN)1745-4549.

Lopes, L.A.A., Pimentel, T.C., Carvalho, Rd.S.F., Madruga, M.S., Galvão, Md.S., Bezerra, T.K.A., Barão, C.E., et al., 2021. Spreadable goat Ricotta cheese added with *Lactobacillus acidophilus* La-05: Can microencapsulation improve the probiotic survival and the quality parameters? Food Chemistry 346. Available at https://doi.org/10.1016/j.foodchem.2020.128769, www.elsevier.com/locate/foodchem.

Makino, S., Ikegami, S., Kume, A., Horiuchi, H., Sasaki, H., Orii, N., 2010. Reducing the risk of infection in the elderly by dietary intake of yoghurt fermented with *Lactobacillus delbrueckii* ssp. bulgaricus OLL1073R-1. British Journal of Nutrition 104 (7), 998–1006. Available at https://doi.org/10.1017/s000711451000173x.

Misra, S., Pandey, P., Mishra, H.N., 2021. Novel approaches for co-encapsulation of probiotic bacteria with bioactive compounds, their health benefits and functional food product development: a review. Trends in Food Science & Technology 109, 340–351.

Misra, S., Rayaguru, K., Dash, S.K., Mohanty, S., Panigrahi, C., 2022a. Efficacy of microwave irradiation in enhancing the shelf life of groundnut (*Arachis hypogaea* L.). Journal of Stored Products Research 97, 101957.

Misra, S., Pandey, P., Dalbhagat, C.G., Mishra, H.N., 2022b. Emerging technologies and coating materials for improved probiotication in food products: a review. Food and Bioprocess Technology 15 (5), 998–1039. Available at https://doi.org/10.1007/s11947-021-02753-5, http://www.springer.com/life+sci/food+science/journal/11947.

Misra, S., Pandey, P., Panigrahi, C., Mishra, H.N., 2023. A comparative approach on the spray and freeze drying of probiotic and Gamma-aminobutyric acid as a single entity: characterization and evaluation of stability in simulated gastrointestinal conditions. Food Chemistry Advances 3. Available at https://doi.org/10.1016/j.focha.2023.100385, https://www.journals.elsevier.com/food-chemistry-advances.

Misra, S., Pandey, P., Panigrahi, C., Mishra, H.N., 2023. Evaluation of potentiality of erythritol on improving the physicochemical, functional, and pasting properties, along with the storability of multigrain flour using chemometric approach. Journal of Stored Products Research 101, 102088.

Misra, S., Mandliya, S., Pandey, P., Panigrahi, C., Dalbhagat, C.G., Mishra, H.N., 2024. Effect of spray-and freeze-dried microcapsules containing probiotics and γ-aminobutyric acid on nutritional, physicochemical, textural, pasting, rheological, and microstructural characteristics of composite dough. Food and Bioprocess Technology 17 (2), 464–478.

Misra, S., Pandey, P., Mishra, H.N., 2025. Probiotication in multigrain dough and biscuits with the incorporation of erythritol: evaluation of techno-functional properties using chemometric approach. Food Science and Technology International 31 (2), 142–154.

Misra, S., Kumar, S., Mishra, H.N., 2025a. Optimisation of Process Parameters for Development of Probiotic-GABA Enriched Nutri Bar by Response Surface Methodology (RSM) and Modelling Using Artificial Neural Networks (ANNs): Characterization and Sensory Evaluation by Fuzzy Logic Analysis. Probiotics and antimicrobial proteins 17 (4), 2611–2628.

Misra, S., Mandliya, S., Mishra, H.N., 2025b. Evaluation of Stability of Coencapsulated Probiotics, γ-Aminobutyric Acid (GABA), and Physicochemical Properties of Functional Bar Matrix during Storage and Digestion Using a Chemometric Approach. ACS Food Science & Technology 5 (6), 2371–2380.

Misra, S., Kukde, R.B., Rout, R.K., Sivamma, P., Kumar, S., 2026. Microencapsulation: A Customized Practice for Food Waste Minimiza-tion," Food Waste Valorization and the Circular Economy. CRC Press.

Morsy, M.K., Morsy, O.M., Abdelmonem, M.A., Elsabagh, R., 2022. Anthocyanin-colored microencapsulation effects on survival rate of *Lactobacillus rhamnosus* GG, color stability, and sensory parameters in strawberry nectar model. Food and Bioprocess Technology 15 (2), 352–367. Available at https://doi.org/10.1007/s11947-022-02758-8, http://www.springer.com/life+sci/food+science/journal/11947.

Muzzafar, A., Sharma, V., 2018. Microencapsulation of probiotics for incorporation in cream biscuits. Journal of Food Measurement and Characterization 12 (3), 2193–2201.

Nazari Kermanshahi, S., Sharifan, A., Yousefi, S., 2021. Physicochemical, microstructural, bioactivity and viability characteristics of probiotic spray-dried raisin powder. Journal of Food Measurement and Characterization 15, 633–642.

Nithya, A., Misra, S., Panigrahi, C., Dalbhagat, C.G., Mishra, H.N., 2023. Probiotic potential of fermented foods and their role in non-communicable diseases management: An understanding through recent clinical evidences. Food Chemistry Advances 3, 100381.

Olivares, A., Soto, C., Caballero, E., Altamirano, C., 2019. Survival of microencapsulated *Lactobacillus casei* (prepared by vibration technology) in fruit juice during cold storage. Electronic Journal of Biotechnology 42, 42–48. Available at https://doi.org/10.1016/j.ejbt.2019.10.002.

Ozturk, B., Elvan, M., Ozer, M., Tellioglu Harsa, S., 2021. Effect of different microencapsulating materials on the viability of *S. thermophilus* CCM4757 incorporated into dark and milk chocolates. Food Bioscience 44, 101413. Available at https://doi.org/10.1016/j.fbio.2021.101413.

Özer, B., Uzun, Y.S., Kirmaci, H.A., 2008. Effect of microencapsulation on viability of *Lactobacillus acidophilus* LA-5 and *Bifidobacterium bifidum* BB-12 during Kasar cheese ripening. International Journal of Dairy Technology 61 (3), 237–244.

Pech-Canul, A.de la C., Ortega, D., García-Triana, A., González-Silva, N., Solis-Oviedo, R.L., 2020. A brief review of edible coating materials for the microencapsulation of probiotics. Coatings 10 (3), 197. Available at https://doi.org/10.3390/coatings10030197.

Pourjafar, H., Noori, N., Gandomi, H., Basti, A.A., Ansari, F., 2020. Viability of microencapsulated and non-microencapsulated Lactobacilli in a commercial beverage. Biotechnology Reports 25. Available at https://doi.org/10.1016/j.btre.2020.e00432, http://www.journals.elsevier.com/biotechnology-reports/.

Priyadarsini, D., Rayaguru, K., Misra, S., Dash, S.K., 2022. Effect of drying techniques on physicochemical properties of oyster mushroom (Pleurotus sajor-caju). Journal of Food Processing and Preservation 46 (7), e16598.

Qandashtant, R.A., Salehi, E.A., Sani, A.M., Sangatash, M.M., Safari, O., 2020. Investigation on quality properties of traditional bulk bread covered with probiotics and soybean oil edible coating. Acta Alimentaria 49 (2), 144–153.

Rajam, R., Kumar, S.B., Prabhasankar, P., Anandharamakrishnan, C., 2015. Microencapsulation of *Lactobacillus plantarum* MTCC 5422 in fructooligosaccharide and whey protein wall systems and its impact on noodle quality. Journal of Food Science and Technology 52 (7), 4029–4041. Available at https://doi.org/10.1007/s13197-014-1506-4, http://www.springerlink.com/content/121580/.

Rastall, R.A., Gibson, G.R., 2015. Recent developments in prebiotics to selectively impact beneficial microbes and promote intestinal health. Current Opinion in Biotechnology 32, 42–46. Available at https://doi.org/10.1016/j.copbio.2014.11.002, http://www.elsevier.com/locate/copbio.

Ravindran, S., S, R., 2021. Probiotic oats milk drink with microencapsulated *Lactobacillus plantarum*–an alternative to dairy products. Nutrition & Food Science 51 (3), 471–482.

Razavi, S., Janfaza, S., Tasnim, N., Gibson, D.L., Hoorfar, M., 2021. Microencapsulating polymers for probiotics delivery systems: preparation, characterization, and applications. Food Hydrocolloids 120. Available at https://doi.org/10.1016/j.foodhyd.2021.106882, https://www.sciencedirect.com/science/journal/0268005X.

Reque, P.M., Brandelli, A., 2021. Encapsulation of probiotics and nutraceuticals: applications in functional food industry. Trends in Food Science and Technology 114, 1–10. Available at https://doi.org/10.1016/j.tifs.2021.05.022, http://www.elsevier.com/wps/find/journaldescription.cws_home/601278/description#description.

Sadeghi, A., Can Karaca, A., Ebrahimi, M., Assadpour, E., Jafari, S.M., 2024. The 3D printed probiotic products; an emerging category of the functional foods for the next-generations. Trends in Food Science and Technology 148. Available at https://doi.org/10.1016/j.tifs.2024.104526, https://www.sciencedirect.com/science/journal/09242244.

Sahitya, R.M., Reddy, K.K., Mallikarjuna Reddy, P.V., Madhava Rao, T., 2013. Evaluation of viability of co-encapsulated pre-and certain probiotics in ice cream during frozen storage. International Journal of Food, Agriculture and Veterinary Sciences 3 (1), 141–147.

Santini, A., Cammarata, S.M., Capone, G., Ianaro, A., Tenore, G.C., Pani, L., et al., 2018. Nutraceuticals: opening the debate for a regulatory framework. British Journal of Clinical Pharmacology 84 (4), 659–672. Available at https://doi.org/10.1111/bcp.13496, http://www.bjcp-journal.com/.

Sanz, Y., 2007. Ecological and functional implications of the acid-adaptation ability of Bifidobacterium: a way of selecting improved probiotic strains. International Dairy Journal 17 (11), 1284–1289. Available at https://doi.org/10.1016/j.idairyj.2007.01.016.

Savedboworn, W., Wanchaitanawong, P., 2015. Viability and probiotic properties of *Lactobacillus plantarum* TISTR 2075 in spray-dried fermented cereal extracts. Maejo International Journal of Science and Technology 9 (3), 382.

Seethu, B.G., Aditya, S.P., Devikrishna, P., Kulkarni, H., Emerald, M.E., Grover, C., et al., 2023. Electrospinning technologies for encapsulation of probiotics. Materials Science and Engineering in Food Product Development. wiley, India, pp. 279–301. Available at http://doi.org/10.1002/9781119860594.ch14, https://www.wiley.com/en-us/Materials+Science+and+Engineering+in+Food+Product+Development-p-9781119860358.

Shah, N.P., 2000. Probiotic bacteria: selective enumeration and survival in dairy foods. Journal of Dairy Science 83 (4), 894–907. Available at https://doi.org/10.3168/jds.s0022-0302(00)74953-8.

Solanki, H.K., Pawar, D.D., Shah, D.A., Prajapati, V.D., Jani, G.K., Mulla, A.M., et al., 2013. Development of microencapsulation delivery system for long-term preservation of probiotics as biotherapeutics agent. BioMed Research International 2013. Available at https://doi.org/10.1155/2013/620719.

Sultana, K., Godward, G., Reynolds, N., Arumugaswamy, R., Peiris, P., Kailasapathy, K., 2000. Encapsulation of probiotic bacteria with alginate–starch and evaluation of survival n simulated gastrointestinal conditions and in yoghurt. International Journal of Food Microbiology 62 (1-2), 47–55. Available at https://doi.org/10.1016/s0168-1605(00)00380-9.

Tripathi, M.K., Giri, S.K., 2014. Probiotic functional foods: survival of probiotics during processing and storage. Journal of Functional Foods 9 (1), 225–241. Available at https://doi.org/10.1016/j.jff.2014.04.030.

Tsai, Y.L., Lin, T.L., Chang, C.J., Wu, T.R., Lai, W.F., Lu, C.C., et al., 2019. Probiotics, prebiotics and amelioration of diseases. Journal of Biomedical Science 26 (1). Available at https://doi.org/10.1186/s12929-018-0493-6, http://www.jbiomedsci.com.

Vallianou, N., Stratigou, T., Christodoulatos, G.S., Tsigalou, C., Dalamaga, M., 2020. Probiotics, prebiotics, synbiotics, postbiotics, and obesity: current evidence, controversies, and perspectives. Current Obesity Reports 9 (3), 179–192. Available at https://doi.org/10.1007/s13679-020-00379-w.

Varada, V.V., Panneerselvam, D., Pushpadass, H.A., Mallapa, R.H., Ram, C., Kumar, S., 2023. In vitro safety assessment of electrohydrodynamically encapsulated *Lactiplantibacillus plantarum* CRD7 and Lacticaseibacillus rhamnosus CRD11 for probiotics use. Current Research in Food Science 6. Available at https://doi.org/10.1016/j.crfs.2023.100507, https://www.journals.elsevier.com/current-research-in-food-science.

Varada, V.V., Kumar, S., Balaga, S., Thanippilly, A.J., Pushpadass, H.A., Rashmi, M.H., et al., 2024. Oral delivery of electrohydrodynamically encapsulated *Lactiplantibacillus plantarum* CRD7 modulates gut health, antioxidant activity, and cytokines-related inflammation and immunity in mice. Food and Function 15 (21), 10761–10781. Available at https://doi.org/10.1039/d4fo02732a, http://pubs.rsc.org/en/journals/journal/fo.

Vasile, M.A., Milea, A., Enachi, E., Barbu, V., Cîrciumaru, A., Bahrim, G.E., et al., 2020. Functional enhancement of bioactives from black beans and lactic acid bacteria into an innovative food ingredient by comicroencapsulation. Food and Bioprocess Technology 13 (6), 978–987. Available at https://doi.org/10.1007/s11947-020-02451-8, http://www.springer.com/life+sci/food+science/journal/11947.

Vivek, K., Mishra, S., Pradhan, R.C., 2020. Characterization of spray dried probiotic Sohiong fruit powder with *Lactobacillus plantarum*. Lwt 117, 108699.

Vu, A.T., Kha, T.C., Phan, H.T., 2023. Encapsulation of bioactive compounds from germinated mung bean by freeze-drying, release kinetics, and storage stability. Foods 13 (1), 100.

Wang, S., Ren, H., Zhong, H., Zhao, X., Li, C., Ma, J., et al., 2022. Combined berberine and probiotic treatment as an effective regimen for improving postprandial hyperlipidemia in type 2 diabetes patients: a double blinded placebo controlled randomized study. Gut Microbes 14 (1). Available at https://doi.org/10.1080/19490976.2021.2003176.

World Health Organization. 2020. Hearts: technical package for cardiovascular disease management in primary health care.

Ying, D., Schwander, S., Weerakkody, R., Sanguansri, L., Gantenbein-Demarchi, C., Augustin, M.A., 2013. Microencapsulated *Lactobacillus rhamnosus* GG in whey protein and resistant starch matrices: Probiotic survival in fruit juice. Journal of Functional Foods 5 (1), 98–105.

Yoha, K.S., Anukiruthika, T., Anila, W., Moses, J.A., Anandharamakrishnan, C., 2021. 3D printing of encapsulated probiotics: effect of different post-processing methods on the stability of *Lactiplantibacillus plantarum* (NCIM 2083) under static in vitro digestion conditions and during storage. LWT 146. Available at https://doi.org/10.1016/j.lwt.2021.111461.

Zeisel, S.H., 1999. Regulation of 'nutraceuticals. Science (New York, N.Y.) 285 (5435), 1853–1855. Available at https://doi.org/10.1126/science.285.5435.1853.

Zhang, L., Huang, S., Ananingsih, V.K., Zhou, W., Chen, X.D., 2014. A study on *Bifidobacterium lactis* Bb12 viability in bread during baking. Journal of Food Engineering 122 (1), 33–37. Available at https://doi.org/10.1016/j.jfoodeng.2013.08.029.

Zhang, C., Quek, S.Y., Fu, N., Su, Y., Kilmartin, P.A., Chen, X.D., 2020. Storage stability and in vitro digestion of microencapsulated powder containing fermented noni juice and probiotics. Food Bioscience 37. Available at https://doi.org/10.1016/j.fbio.2020.100740.

Chapter 11

The quality factors for the assessment of nonviable probiotic (postbiotic)

Arantxa Almada-Corral[1], Lourdes Santiago-López[1], Belinda Vallejo-Cordoba[1], Aarón F. González-Córdova[1], Hugo S. Garcia[2], and Adrian Hernández-Mendoza[1]

[1]*Laboratorio de Química y Biotecnología de Productos Lácteos, Centro de Investigación en Alimentación y Desarrollo, Hermosillo, Sonora, Mexico,* [2]*Unidad de Investigación y Desarrollo de Alimentos, Instituto Tecnológico de Veracruz, Veracruz, México*

11.1 Introduction

In recent years, not only have probiotics been thoroughly studied, but also other newer concepts, such as paraprobiotics and postbiotics associated with probiotics, have been explored. For probiotics, the accepted definition is "live microorganisms that, when administered in adequate amounts, confer health benefits on the host" (Hill et al., 2014), while paraprobiotics refers to "non-viable microbial cells (inactivated and/or dead) which, when administered in adequate amounts, confer a benefit on the human consumer" (Taverniti and Guglielmetti, 2011). However, the definition of postbiotics remains controversial. In 2021, the International Scientific Association for Probiotics and Prebiotics (ISAPP) defined postbiotics as "preparations of inanimate microorganisms and/or their components that confer a health benefit on the host." However, the word "inanimate" was controversial because it was considered to be "an inert material." Therefore, other definitions have been proposed that exclude inactivated or inanimate microorganisms, instead defining them as products or metabolic byproducts (well-defined single molecules or mixtures) secreted/released after bacterial lysis, with health-promoting properties (Aguilar-Toalá et al., 2021).

Some molecules recognized as postbiotics are enzymes, peptides, vitamins, teichoic acids, peptidoglycan-derived muropeptides, secreted and extracellular polysaccharides, cell lysates, s-layer proteins, fatty acids, and short-chain fatty acids (Vinderola et al., 2022; Nataraj et al., 2020). The concentration and type of postbiotics are strain-dependent, as well as the culture medium and the

Evaluating Oral Probiotics and Prebiotics. DOI: https://doi.org/10.1016/B978-0-443-29835-6.00021-6

handling of the bacteria after propagation (Liang and Xing, 2023). These components can be obtained by enzymatic methods, technological processes, nonthermal treatments, emerging treatments, or a mix of these (Żółkiewicz et al., 2020).

The effect of postbiotics has been demonstrated in several studies, as well as their efficacy, stability, and safety (Isaac-Bamgboye et al., 2024; Moradi et al., 2021). Currently, postbiotics have been shown to have positive effects, including antiinflammatory, immunomodulatory, antiproliferative, antioxidant, hypocholesterolemic, and antiobesogenic (Islam et al., 2024). In this chapter, we provide a comprehensive summary of quality factors and safety technical criteria for evaluating the safety of nonviable probiotics and their impacts on human health.

11.2 Obtention of postbiotics

Production of postbiotics requires no viable cells. The steps described for their preparation are (1) selection of microorganisms, (2) growth conditions, (3) harvesting, and (4) analysis and characterization (Asif et al., 2023). Postbiotics have a clear chemical structure and several advantages related to their absorption, metabolism, distribution, and excretion, which promote communication with various tissues and support local and systemic biological responses (Aggarwal et al., 2022). However, to release postbiotics, various laboratory techniques have been used to disrupt microbial cell walls (Rad et al., 2021). The techniques for their obtention can be enzymatic treatments, technological (e.g., pasteurization and sterilization), nonthermal (e.g., ionizing radiation, ultraviolet light, high hydrostatic pressure, ultrasound, and dehydration), emerging treatments such as pulsed electric fields, ohmic heating, supercritical fluids-CO_2, or combinations thereof (Fig. 11.1) (Dunand et al., 2019).

Heat treatment has been used to release postbiotics. Postbiotics were prepared from heat-treated *Lactobacillus casei* ATCC 393 and *Bifidobacterium bifidum* BGN4, or *Bifidobacterium* sp. CN2 at 100°C, 40 minutes, or 95°C, 30 minutes, respectively. In addition to thermal treatment, extraction processes have been complemented with sonication (Lee et al., 2002; Tiptiri-Kourpeti et al., 2016). Lipoteichoic acids were isolated from *Lactobacillus plantarum* K8 by *n*-butanol extraction; glycolipids from *Lactiplantibacillus plantarum* B-01 by applying supercritical fluid technology using CO_2, or using lysozyme/mutanolysin for postbiotics derived from *Lacticaseibacillus casei* CRL 431 (Rakhuba et al., 2009; Kim et al., 2011; Aguilar-Toalá et al., 2021).

11.2.1 Characterization of postbiotics

Qualitative and quantitative analyses of postbiotics are necessary; specialized equipment and various concentrations and purification steps are usually

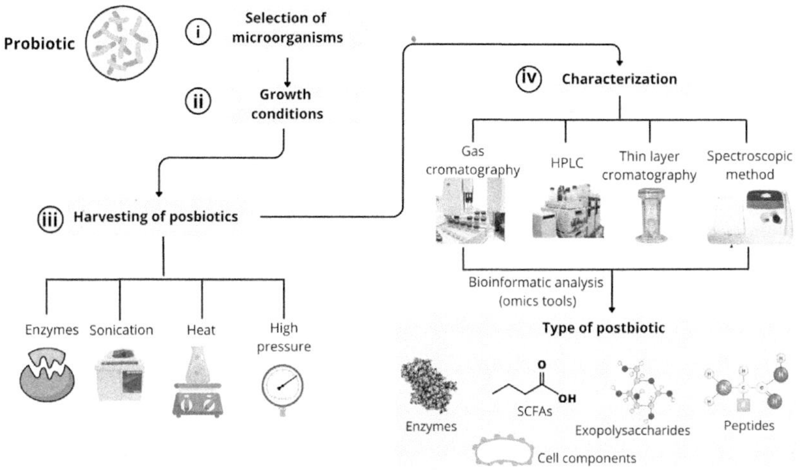

FIGURE 11.1 Pathways for the obtention of postbiotics.

required. For this, analytical methods have been used, among them, gas chromatography (GC), high-performance liquid chromatography (HPLC), thin-layer chromatography (TLC), and spectroscopic methods (Barros et al., 2020; Thorakkattu et al., 2022). Using these analytical methods, many compounds have been characterized, including the volatile profile of *Lacticaseibacillus casei* containing 62 analytes was identified by headspace solid-phase microextraction GC–MS, and SCFAs were analyzed by GC (Ricci et al., 2018). Other methods, including nuclear magnetic resonance spectroscopy, were used for antifungal and antioxidant compounds (Arasu et al., 2015).

Even though these techniques can be used to identify and quantify postbiotics, specific protocols must be followed to enable the discovery and characterization of new postbiotics. However, further research is required to optimize culture conditions and analytical methods (Aguilar-Toalá et al., 2018).

11.2.2 Classification of postbiotics

Postbiotics can be categorized according to their chemical nature (proteins, carbohydrates, lipids), their bioactivities (e.g., immunomodulatory, antioxidant, antiinflammatory, and antimicrobial), and their origin and production process (Asif et al., 2023). Another way of classification is by microbial source: compounds derived from *Lactobacillus, Bifidobacterium*, and yeasts (Duarte et al., 2022) (Table 11.1).

TABLE 11.1 Type of postbiotics and their physiological effect.

Metabolites	Physiological effect	Response	References
Cell wall componentsPeptidoglycan-derived plasmalogenLipoteichoic acidsTeichoics acidsCell surface proteins	ImmunomodulationAntiinflammatory	Enhance the production of IL-8 in Caco-2 cells.Decrease IL-12 but enhance IL-10 in a TNBS-induced colitis mouse model.	Lebeer et al. (2012) and Sokol et al. (2008)
CarbohydratesExopolysaccharides	AntioxidantAnticholesterolemicAntitumor	Increase antioxidant enzymes (catalase, glutathione peroxidase, superoxide dismutase), low lipid peroxidation.Prevent the absorption of cholesterol.Inhibit or slow down the proliferation of HT-29 tumor cells.	Qi and Tang (2020), Khalil et al. (2018), Wang et al. (2014), Riaz Rajoka et al. (2019)
Proteins/peptides (p40, p75 molecule, lactocepin, glutathione, bacteriocins)	AntioxidantAntimicrobian	Enhance the antioxidant activity by the DPPH method.Reduction of the cell concentration of *S. aureus*.	Jalali et al. (2024)
Lipids (acetate, propionate, butyrate, lactate, dimethyl acetyl-derived plasmologen)	AntiinflammatoryAntimicrobianAntiobesogenicAntitumor	Regulate the expression of nuclear factor-kappa B, decrease proinflammatory cytokines.Inhibit the growth of *Klebsiella oxytoca* by	Canfora et al. (2015), Rad et al. (2021)

		lysing the cell wall.Regulation of lipid metabolism, glucose homeostasis, and insulin sensitivity.Activate apoptosis pathway in colon cancer cell lines and reduce tumor cell growth.	
Vitamins (biotin, riboflavin, ascorbic acid)	Immunomodulation	Regulate the production of T-cells and produce IL-10 cytokine	Haileselassie et al. (2016)
Organic acids (3-phenyllactiv acid and propionic)	AntimicrobialAntiinflammatory	Inhibit the growth of *E. coli* strains.Regulate the expression of PGE-2 and IL-8 in human colon epithelial HT-29 cells and modulate the expression of IL-1β, IL-6, TNF-α, and IL-10.	De Marco et al. (2018)

DPPH, 2,2-Diphenyl-1-picrylhydrazyl; *EPS*, Exopolysaccharides; *SCFA*, Short- chain fatty acids.

11.3 Mechanism of action of postbiotics

The presence of EPS, LTA, and S-layer protein exhibits antioxidant properties. These components inhibited reactive oxygen species (ROS) generation, scavenged hydroxyl radicals, and improved antioxidant enzyme gene expression and activity (Zhao et al., 2024). For instance, the *L. plantarum* RG14 postbiotics reduced malondialdehyde (MDA) production and increased serum and ruminal GSH peroxidase (GPX) enzyme activity, while superoxide dismutase (SOD) levels remained unaffected (Izuddin et al., 2020). *Lacticaseibacillus paracasei* ET-22 and *Bifidobacterium lactis* BL-99 postbiotics obtained by thermal treatment showed antiinflammatory activity by suppressing the expression of TNF-α. Some metabolites identified by HMDB and Metlin were amino acids, organic acids, amines, nucleotides, lipids, and peptides (Sungur et al., 2017). *Lactobacillus reuteri* DSM17938 reduced UV-R-induced inflammatory cytokines such as IL-6 and IL-8; meanwhile, *Akkermansia muciniphila* postbiotics were aimed at preventing obesity and improving glucose metabolism disorders (Wu et al., 2023; Khmaladze et al., 2019). Furthermore, the postbiotics can (1) modulate resident microbiota, (2) enhance the epithelial barrier function, (3) modulate the local and systemic immune response, (4) modulate the systemic metabolic response, and (5) improve the systemic signaling via the nervous system (Fig. 11.2) (Salminen et al., 2021).

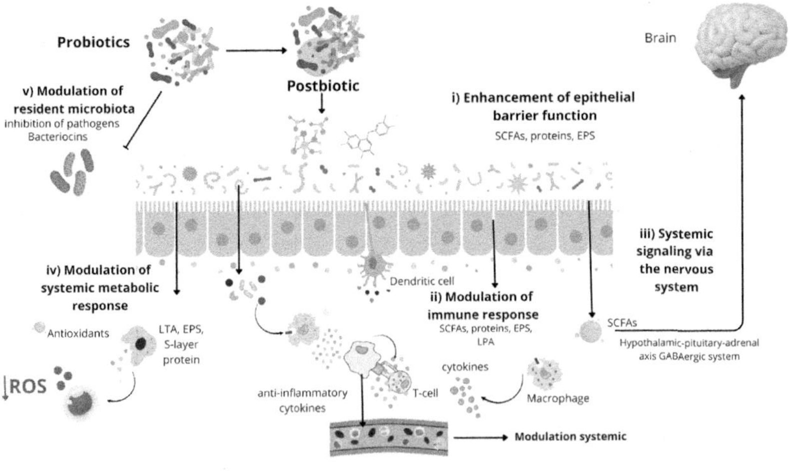

FIGURE 11.2 Mechanisms of action of postbiotics. *EPS*, Exopolysaccharides; *LPA*, lipoteichoic acids; *ROS*, reactive oxygen species; *SCFA*, short-chain fatty acids.

11.3.1 Enhancement of epithelial barrier function

Postbiotics, such as lactic acid or bacteriocins, can exhibite antimicrobial activity or produce molecules that affect quorum sensing and quorum quenching; they also produce SCFAs via lactic acid metabolism or provide adhesion sites against pathogenic microorganisms (Lukic et al., 2017). The epithelial barrier function can be mediated by secreted proteins (Msp1/p75, MsP1/p40, HM0539), or the presence of EPS; this promotes barrier function and an antiinflammatory effect by signaling the MAPK and AKT pathways, or the presence of SCFAs can stimulate the formation of tight junctions, or via lipoxygenases (Schiavi et al., 2016; Gao et al., 2019; Feng et al., 2018; Engevik et al., 2019).

11.3.2 Immunomodulation, local, and systemic

This mechanism can be mediated by interactions with receptors such as Toll-like receptors (TLRs), nucleotide-binding oligomerization domain (NOD) receptors, or C-type lectins (Wicherska-Pawłowska et al., 2021). Some metabolites, such as SCFAs and histamine, can influence the suppression of NF-kB, a transcription factor in the inflammatory process. Particularly, metabolites such as lipoteichoic acids (LPA), fimbriae or pili, and β-glucans (*Saccharomyces cerevisiae*) bind to TLR2, lipopolysaccharide (*E. coli* Nissle) to TLR4 or TLR2, and sometimes to TLR6 to LPA (Mehta et al., 2023). Other receptors, such as GPR31, which bind to lactic acid, are also involved. For the stimulation of CD4+T and CD4+ CD8+ intraepithelial lymphocytes, indole-derivatized tryptophan from *Limosilactobacillus reuteri* has been demonstrated (Thomas et al., 2012). Additionally, the NF-kB pathway should be suppressed by other metabolites, such as histamine and SCFAs (Ozma et al., 2022). In particular, SCFAs bound to G protein-coupled receptors (GPCRs) inhibit TNF-α, IL-1β, IL-6, and nitric oxide (NO) and enhance IL-10 in macrophages (Nakkarach et al., 2021), as well as components of peptidoglycan (Salva et al., 2021); meanwhile, for EPS-derived of *Lactobacillus* spp have been used in cervical cancer HeLa cells (Sungur et al., 2017).

11.3.3 Improved systemic signaling via brain–gut–microbiota

Based on the above, bifacial communication involves the brain, the enteric nervous system, the gut-associated immune system, and the enteric neuroendocrine system, which are responsible for human physiological functions, such as sleep, emotions, and mood, and intestinal health (Binda et al., 2024). Furthermore, the hypothalamic–pituitary–adrenal axis (HPA) can be modulated by probiotics and postbiotics. For instance, some neurotransmitters such as gamma-aminobutyric acid (GABA), secreted by *Lacticaseibacillus rhamnosus* YS9 and *Lacticaseibacillus casei* or *Levilactobacillus brevis* KU15152, can modulate the neural route GABAergic system and HPA axis through the expression of GABAA and GABAB receptors (Yunes et al., 2016; Bock et al., 2023; Lin, 2013). Other effects, such as the modulation of the 5-HT system, have been attributed to the presence of serotonin and norepinephrine secreted by *Lactobacillus helveticus* 100ash (Oleskin et al., 2014), and histamine secreted by *Lactobacillus reuteri* ATCC PTA6475, which decreases the expression of proinflammatory cytokines, then the brain-derived neurotrophic factor (BDNF) (Hemarajata et al., 2013). SCFA could stimulate enterochromaffin cells to produce serotonin. For instance, butyrate maintains intestinal barrier integrity and not only modulates the expression of BDNF and the 5-HT system but also decreases the expression of cytokines and inflammation, thereby contributing to the prevention of anxiety (Flik et al., 2015).

11.3.4 Modulation of systemic metabolic response

An imbalance between energy intake and consumption causes obesity. In this case, the SCFAs activate GPCR43 in adipose tissue and inhibit excessive energy intake and insulin secretion. For instance, acetic acid stimulates peptide YY (PYY) production; meanwhile, propionic acid stimulates the production of PYY and glucagon-like peptide-1 in human colon cells (van der Beek et al., 2016; Pingitore et al., 2017). These two hormones signal the brain, indicating satiety.

In contrast, diabetes mellitus is a chronic disease characterized by elevated blood sugar levels. In this case, the gluconeogenesis and sympathetic nerve activity may be attributed to propionic and butyric acids (Bouter et al., 2018). Moreover, EPSs derived from *Bifidobacterium animalis* IPLA R1 decrease insulin in the serum of rats fed a high-fat diet, and EPSs derived from *Lactobacillus paracasei* M7 decrease blood cholesterol levels (Bhat and Bajaj, 2019). These effects suggest the potential of postbiotics for usein metabolic diseases.

11.3.5 Modulation of resident microbiota

The microbiota of each person is unique; however, when subjected to external changes, the balance of the microbiota community can be affected, leading to dysregulation of bodily functions and diseases (Hou et al., 2022). Evidence to date has demonstrated that the microbiota is associated with the development of certain diseases, including cancer, diabetes, respiratory diseases, brain disorders, chronic kidney disease, liver disease, inflammatory bowel disease, and heart disease (Lau et al., 2017; Hou et al., 2022). Concerning the interaction between postbiotics and microbiota, the intake of fermented milk with *Lactobacillus gasseri* CP2305 increased the concentration of SCFAs and *Clostridium cluster* IV. This study was conducted in patients with constipation or frequent bowel movements (Sawada et al., 2016).

Metabolites, such as organic acids and bacteriocins, can inhibit pathogens. Furthermore, lactic acid can be metabolized, resulting in the production of SCFAs by resident microbiota (Laverde Gomez et al., 2019). For instance, the presence of pili and the protein p40/p75 released by *Lactobacillus* exhibited antagonistic effects (Teame et al., 2020). EPS can be used as a fermentable substrate. This effect has been attributed to the release of EPS by *Bifidobacterium* strains, which causes changes in microbiota diversity. Moreover, EPS released by *L. plantarum* strains has demonstrated a bifidogenic effect (Das et al., 2014).

However, the effects attributed to postbiotics have been demonstrated in either in vitro or in vivo studies. At this point, clinical studies demonstrating their effectiveness are limited. Therefore, the study of postbiotics remains a topic of interest to various research groups. In addition to their effects, it is important to consider the factors that influence the effectiveness of postbiotics.

11.4 Evaluation of quality factors of postbiotic products

The capacity of postbiotics can be evaluated not only by their efficacy, stability, and functional capacity but also by the quantity and quality of the scientific evidence supporting their health benefits (Isaac-Bamgboye et al., 2024). Additionally, the source of bacterial postbiotics, as well as their production, standardization, and purification process, can strongly affect their quality (Rafique et al., 2023).

Unlike probiotics, a fixed dose for postbiotics has not been established; additionally, for economic reasons, it offers a longer shelf life, easy storage and transportation, and may not require refrigeration. In addition, different processes have been increasingly standardized to maximize the amount of postbiotics obtained (Żółkiewicz et al., 2020). A safety criterion is that postbiotics eliminate the possibility of acquiring antibiotic-resistance genes and virulence factors (Imperial and Ibana, 2016). On the other hand, these compounds are considered safe, attributed to the proven reduced risk of microbial translocation and infection, as well as to immunodeficient, infant, and vulnerable subjects (Deshpande et al., 2018).

11.4.1 Challenges in postbiotic production

Yield optimization. For optimal optimization of postbiotic production, it is necessary to study the different metabolic routes and the strain type (concentration, lysis method). In some studies, it has been suggested that combining chemical and physical methods can yield better postbiotics (Thorakkattu et al., 2022). In this step, it is important to establish all these criteria when scaling up from laboratory to industrial production, since at this stage, economic costs increase while compound stability must be maintained (Wegh et al., 2019). Stability refers to the postbiotic's ability to maintain its integrity, efficacy, and quality during storage and subsequent use. This quality can be lost or affected by changes in the food matrix and the storage conditions to which it is subjected. If this were to happen, some of the postbiotics ' multiple bioactivities could be reduced (Moradi et al., 2021).

Due to the nature of the components in intracellular contents (e.g., fatty acids, peptides, enzymes, and amino acids), these components are sensitive to various factors such as temperature, humidity, and light. That is why it is recommended that, once these components are obtained, they be stored under refrigeration (2°C–8°C), frozen (−20°C or −80°C), or freeze-dried. These conditions will help to preserve the stability of the postbiotics (Vinderola et al., 2022). Additionally, oxygen and moisture can affect postbiotic stability, so packaging is important. For this purpose, it is essential to use vacuum-sealed, dark containers to prevent light from entering and packaging resistant to UV-B radiation. During storage, it is advisable to monitor the stability of the compound, potential microbial contamination or other contaminants, and pH values (De Almeida et al., 2023; Prajapati et al., 2023).

Safety considerations. Given the nature of postbiotics, they pose a low risk due to the absence of viable bacteria, thus helping to prevent bacteremia or fungemia (Yelin et al., 2019; Prajapati et al., 2023).

Contaminants are unwanted substances that may be present in the postbiotic, whether they are microorganisms, chemicals, or physical particles. If they are present in the product, they can interfere with mechanisms of action, but postbiotics have been reported to contribute to the biodegradation of harmful chemical contaminants, such as mycotoxins and biogenic amines. They can counteract some of these effects (Isaac-Bamgboye et al., 2024).

The postbiotics of lactic acid bacteria have been evaluated for their ability to reduce biogenic amine production by pathogens (i.e., *Salmonella paratyphi* A and *Escherichia coli*). In this study, the postbiotic concentration from 25% to 50% reduced the presence of cadaverine and putrescine (Yilmaz et al., 2022).

11.4.2 Biosecurity

In recent years, the limitations of some probiotics have spurred the development of postbiotics as a novel functional ingredient (Mishra et al., 2024). Therefore, several scientific experiments have been conducted to evaluate the absorption, metabolism, and safety of postbiotics in humans (Tomasik and Tomasik, 2020). An example was the study carried out by Vandenplas et al. (2020), who conducted a double-blind, randomized, controlled, multicountry, two-arm parallel trial (NCT03476889), assigning 215 infants <14 days of age to two groups (a control group with milk formula and a test group with milk formula added with postbiotics). In this study, no significant differences were detected in side effects or gastrointestinal tolerance between the groups studied, concluding that supplementation with postbiotics maintained satisfactory and safe child growth.

A systematic review by Malagón-Rojas et al. (2020) reported that in three of seven randomized controlled trials involving infants under the age of 5 years, ingestion of inactivated *L. acidophilus* was associated with increased rates of stomach distention, severe dehydration, and vomiting. However, no side effects were reported in other clinical trials.

Zagato et al. (2014) demonstrated that infant formulas prepared with fermented milk and metabolic products of *L. paracasei* CBA L74 had strong antiinflammatory activity in vitro and protected against colitis or enteric pathogens in vivo. Thus, the results have been quite favorable, indicating that the application of postbiotics can be a reliable alternative to mitigate risk factors associated with live probiotic bacteria and offers an appropriate strategy for the treatment of intestinal diseases (Moghaddam et al., 2019). Furthermore, studies show that postbiotics do not interact with the food matrix, so there is no apparent risk of acquiring antibiotic-resistance genes and no undesirable changes in flavor and aroma (Nataraj et al., 2020).

All of this may be due to the fact that postbiotics possess a known chemical composition, low toxicity, stability, and a longer shelf life (Chen et al., 2024). Therefore, they can be safe substitutes for live probiotic bacteria commonly used in the food and pharmaceutical industries. The physiological characteristics and durability of postbiotic metabolites have thus been used in biopreservation, packaging, biofilm removal, and degradation of toxic contaminants (Aggarwal et al., 2022). As there are very few studies reporting side effects derived from postbiotics (Sabahi et al., 2023), it is advisable to continue discussing and conducting in vitro or in vivo studies and, eventually, clinical trials to provide valuable information on this topic.

11.5 Applications

In recent years, there has been growing interest among the population in functional foods, including those containing prebiotics, probiotics, and postbiotics.

Evidence from several studies has demonstrated the effects of postbiotics, including antibacterial activity against pathogenic bacteria, disease prevention (metabolic disorders, inflammatory bowel diseases, mental health, modulation of immune response), and help in preventing food spoilage (Rad et al., 2021). Postbiotics can even be useful as food supplements, as prophylactic drugs, or as complementary treatments. Therefore, the main interested parties are companies involved in the chemical, food, pharmaceutical, and cosmetic industries (Liang and Xing, 2023).

11.5.1 Food industry

One of the main interested parties in the advancement and study related to postbiotics is the food industry, all this due to the fact that it is easy to use postbiotics in foods and raw materials before any thermal process without compromising their functions, because they are stable at different temperatures and pH ranges (Thorakkattu et al., 2022). Fermentation is the most common process used in the main applications of postbiotics. *Lactobacillus and Bifidobacterium* strains are the ones commonly used as producer strains (Chaluvadi et al., 2015). Some metabolites benefit manufacturers both technically and economically; for example, exopolysaccharides play an important role in dairy products, as they can improve the physicochemical and sensory qualities of food products, such as by reducing their moisture content. Food products, such as yogurt, fermented drinks, or snacks, are carriers for postbiotics, which have been analyzed as a pathway for the consumption of short-chain fatty acids derived from *L. plantarum* strains or *B. breve*, which help to improve the integrity of the intestinal barrier or lower glucose in the intestinal tract (Thorakkattu et al., 2022).

Bacteriocins can inhibit a wide range of pathogenic and spoilage microorganisms (Perez et al., 2014). Some bacterial peptides have also proven to be effective preservatives in various foods, such as infant formulas and dairy products (Chen and Hoover, 2003). Therefore, the use of postbiotics is a natural alternative to chemical preservatives; in fact, this practice is known as "biopreservation" (Isaac-Bamgboye et al., 2024).

11.5.2 Cosmetic industry

In recent years, the cosmetic industry has also shown strong interest in the development, study, and incorporation of postbiotics into various products. Currently, some companies have attempted to incorporate postbiotics into the matrix of conventional pharmaceutical products due to their stable pharmacodynamic profiles and beneficial effects in clinical applications (Aguilar-Toalá et al., 2018). Furthermore, compared to probiotic bacteria, postbiotics have a longer shelf life, greater safety, and do not require viability in topical formulation, making them an innovative approach within the cosmetic ingredients market (Duarte et al., 2022).

Human skin is the first line of defense against external agents; it is considered both a physical and an immunological barrier. There are bacteria in the microbiota, such as *S. epidermidis* and *S. aureus*, that vary from person to person (Lee et al., 2019). Diseases or conditions that affect the skin are generally due to dysbiosis in the skin microbiota. Scientific evidence indicates that topical postbiotics reduce the abundance of bacteria that cause inflammation and acne (Aggarwal et al., 2022). It has also been observed that postbiotics support the skin's barrier function, help maintain hydration, protect against environmental aggressors, and promote cell regeneration and skin repair; all the above, in addition to antimicrobial effects, help to control the growth of pathogenic microorganisms on the skin (Aslan and Çelebi, 2023). For instance, postbiotic topical application decreases the presence of *Staphylococcus epidermidis*, *S. aureus*, *Streptococcus pyogenes*, and *Cutibacterium acnes* in skin (Aggarwal et al., 2022).

11.5.3 Agricultural and animal health nutrition

Postbiotics can be applied in agriculture, as they have improved nutrient availability, control diseases, and enhance plant growth. They also contribute to plant resistance to factors such as drought and a balance in soil microbial communities (Fiodor et al., 2021; Vassileva et al., 2020; Aggarwal et al., 2022). Then, the use of postbiotics supports sustainable farming practices.

In animal nutrition, postbiotics improve gastrointestinal health and promote better digestion and nutrient absorption (Zhong et al., 2022). Therefore, the animals that ingest postbiotics have better growth rates, feed efficiency, and illness resistance. As an additional advantage, postbiotics could serve as antibiotic replacements in livestock farming and, consequently, in meat products with no antibiotics (Rahman et al., 2022).

11.6 Conclusions

Future investigations are necessary to assess the use of postbiotics for human health and their applications in related areas. For instance, improve production techniques or a combination of techniques that improve yield. Furthermore, it is important to consider the bacterial type, the purity of the compounds obtained, the chemical structure, and potential functionality, which have been demonstrated both in vitro and in vivo; most importantly, clinical trials are necessary to demonstrate their potential effect. The inclusion of postbiotics in food or other areas, such as agriculture or animal health, offers significant advantages over probiotics, due to greater stability and improved storage conditions. An important stage in the development of functional foods is compliance with regulatory requirements; in this case, organizations such as the Food and Drug Administration and the European Food Safety Authority have not issued any specific statements regarding postbiotics.

Acknowledgments

Almada-Corral thanks the Secretary of Science, Humanities, Technology and Innovation (SecihtiTI) of Mexico for the graduate scholarship provided and research grant (No. 319922).

References

Aggarwal, S., Sabharwal, V., Kaushik, P., Joshi, A., Aayushi, A., Suri, M., 2022. Postbiotics: from emerging concept to application. Frontiers in Sustainable Food Systems 6. Available at https://doi.org/10.3389/fsufs.2022.887642, www.frontiersin.org/journals/sustainable-food-systems#.

Aguilar-Toalá, J.E., Garcia-Varela, R., Garcia, H.S., Mata-Haro, V., González-Córdova, A.F., Vallejo-Cordoba, B., et al., 2018. Postbiotics: an evolving term within the functional foods field. Trends in Food Science and Technology 75, 105–114. Available at https://doi.org/10.1016/j.tifs.2018.03.009.

Aguilar-Toalá, J.E., Arioli, S., Behare, P., Belzer, C., Berni Canani, R., Chatel, J.M., et al., 2021. Postbiotics—when simplification fails to clarify. Nature Reviews Gastroenterology and Hepatology 18 (11), 825–826. Available at https://doi.org/10.1038/s41575-021-00521-6, http://www.nature.com/nrgastro/index.html.

Arasu, M.V., Al-Dhabi, N.A., Rejiniemon, T.S., Lee, K.D., Huxley, V.A.J., Kim, D.H., et al., 2015. Identification and characterization of *Lactobacillus brevis* P68 with antifungal, antioxidant and probiotic functional properties. Indian Journal of Microbiology 55 (1), 19–28. Available at https://doi.org/10.1007/s12088-014-0495-3, https://www.springer.com/journal/12088.

Asif, A., Afzaal, M., Shahid, H., Saeed, F., Ahmed, A., Shah, Y.A., et al., 2023. Probing the functional and therapeutic properties of postbiotics in relation to their industrial application. Food Science and Nutrition 11 (8), 4472–4484. Available at https://doi.org/10.1002/fsn3.3465.

Aslan, İ., Çelebi, L.T., 2023. Postbiotics cosmetic formulation: in vitro efficacy studies on a microbiome friendly antiperspirant. Journal of Research in Pharmacy 27 (5), 2095–2105. Available at https://doi.org/10.29228/jrp.488.

Barros, C.P., Guimarães, J.T., Esmerino, E.A., Duarte, M.C.K., Silva, M.C., Silva, R., et al., 2020. Paraprobiotics and postbiotics: concepts and potential applications in dairy products. Current Opinion in Food Science 32, 1–8. Available at https://doi.org/10.1016/j.cofs.2019.12.003, http://www.journals.elsevier.com/current-opinion-in-food-science/.

Bhat, B., Bajaj, B.K., 2019. Hypocholesterolemic potential and bioactivity spectrum of an exopolysaccharide from a probiotic isolate *Lactobacillus paracasei* M7. Bioactive Carbohydrates and Dietary Fibre 19, 100191. Available at https://doi.org/10.1016/j.bcdf.2019.100191.

Binda, S., Tremblay, A., Iqbal, U.H., Kassem, O., Le Barz, M., Thomas, V., et al., 2024. Psychobiotics and the microbiota–gut–brain axis: where do we go from here? Microorganisms 12 (4). Available at https://doi.org/10.3390/microorganisms12040634, www.mdpi.com/journal/microorganisms.

Bock, H.J., Lee, N.K., Paik, H.D., 2023. Neuroprotective effects of heat-killed *Levilactobacillus brevis* KU15152 on H_2O_2 induced oxidative stress. Journal of Microbiology and Biotechnology 33 (9), 1189–1196. Available at https://doi.org/10.4014/jmb.2304.04045, https://www.jmb.or.kr/journal/list.html?page=2&sort=&scale=10&key=&keyword=&s_v=33&s_n=9&pn=vol&TG=vol&Type=&year=2023&sm=&jname=.

Bouter, K.E.C., Bakker, G.J., Levin, E., Hartstra, A.V., Kootte, R.S., Udayappan, S.D., et al., 2018. Differential metabolic effects of oral butyrate treatment in lean versus metabolic syndrome subjects. Clinical and Translational Gastroenterology 9 (5), e155. Available at https://doi.org/10.1038/s41424-018-0025-4.

Canfora, E.E., Jocken, J.W., Blaak, E.E., 2015. Short-chain fatty acids in control of body weight and insulin sensitivity. Nature Reviews Endocrinology 11 (10), 577–591. Available at https://doi.org/10.1038/nrendo.2015.128, http://www.nature.com/nrendo/index.html.

Chaluvadi, S., Hotchkiss, A.T., Yam, K.L., 2015. Gut Microbiota: Impact of Probiotics, Prebiotics, Synbiotics, Pharmabiotics, and Postbiotics on Human Health Probiotics, Prebiotics, and Synbiotics: Bioactive Foods in Health Promotion. Elsevier Inc., United States, pp. 515–523.

Available at http://www.sciencedirect.com/science/book/9780128021897, 10.1016/B978-0-12-802189-7.00036-8.

Chen, H., Hoover, D.G., 2003. Bacteriocins and their food applications. Comprehensive Reviews in Food Science and Food Safety 2 (3), 82–100. Available at https://doi.org/10.1111/j.1541-4337.2003.tb00016.x.

Chen, Q., Li, X.J., Xie, W., Su, Z.A., Qin, G.M., Yu, C.H., 2024. Postbiotics: emerging therapeutic approach in diabetic retinopathy. Frontiers in Microbiology 15. Available at https://doi.org/10.3389/fmicb.2024.1359949, https://www.frontiersin.org/journals/microbiology#.

Das, D., Baruah, R., Goyal, A., 2014. A food additive with prebiotic properties of an α-d-glucan from *Lactobacillus plantarum* DM5. International Journal of Biological Macromolecules 69, 20–26. Available at https://doi.org/10.1016/j.ijbiomac.2014.05.029.

De Almeida, C.V., Antiga, E., Lulli, M., 2023. Oral and topical probiotics and postbiotics in skincare and dermatological therapy: a concise review. Microorganisms 11 (6), 1420. Available at https://doi.org/10.3390/microorganisms11061420.

De Marco, S., Sichetti, M., Muradyan, D., Piccioni, M., Traina, G., Pagiotti, R., et al., 2018. Probiotic cell-free supernatants exhibited anti-inflammatory and antioxidant activity on human gut epithelial cells and macrophages stimulated with LPS. Evidence-Based Complementary and Alternative Medicine 2018 (1). Available at https://doi.org/10.1155/2018/1756308.

Deshpande, G., Athalye-Jape, G., Patole, S., 2018. Para-probiotics for preterm neonates—the next frontier. Nutrients 10 (7), 871. Available at https://doi.org/10.3390/nu10070871.

Duarte, M., Oliveira, A.L., Oliveira, C., Pintado, M., Amaro, A., Madureira, A.R., 2022. Current postbiotics in the cosmetic market—an update and development opportunities. Applied Microbiology and Biotechnology 106, 5879–5891. Available at https://doi.org/10.1007/s00253-022-12116-5, https://www.springer.com/journal/253.

Dunand, E., Burns, P., Binetti, A., Bergamini, C., Peralta, G.H., Forzani, L., et al., 2019. Postbiotics produced at laboratory and industrial level as potential functional food ingredients with the capacity to protect mice against Salmonella infection. Journal of Applied Microbiology 127 (1), 219–229. Available at https://doi.org/10.1111/jam.14276.

Engevik, M.A., Luk, B., Chang-Graham, A.L., Hall, A., Herrmann, B., Ruan, W., et al., 2019. Bifidobacterium dentium fortifies the intestinal mucus layer via autophagy and calcium signaling pathways. mBio 10 (3). Available at https://doi.org/10.1128/mBio.01087-19, https://mbio.asm.org/content/10/3/e01087-19.full.pdf.

Feng, Y., Wang, Y., Wang, P., Huang, Y., Wang, F., 2018. Short-chain fatty acids manifest stimulative and protective effects on intestinal barrier function through the inhibition of NLRP3 inflammasome and autophagy. Cellular Physiology and Biochemistry 49 (1), 190–205. Available at https://doi.org/10.1159/000492853.

Fiodor, A., Singh, S., Pranaw, K., 2021. The contrivance of plant growth promoting microbes to mitigate climate change impact in agriculture. Microorganisms 9 (9), 1841. Available at https://doi.org/10.3390/microorganisms9091841.

Flik, G., Folgering, J.H.A., Cremers, T.I.H.F., Westerink, B.H.C., Dremencov, E., 2015. Interaction between brain histamine and serotonin, norepinephrine, and dopamine systems: in vivo microdialysis and electrophysiology study. Journal of Molecular Neuroscience 56 (2), 320–328. Available at https://doi.org/10.1007/s12031-015-0536-3.

Gao, J., Li, Y., Wan, Y., Hu, T., Liu, L., Yang, S., et al., 2019. A novel postbiotic from *Lactobacillus rhamnosus* GG with a beneficial effect on intestinal barrier function. Frontiers in Microbiology 10. Available at https://doi.org/10.3389/fmicb.2019.00477.

Haileselassie, Y., Navis, M., Vu, N., Qazi, K.R., Rethi, B., Sverremark-Ekström, E., 2016. Postbiotic modulation of retinoic acid imprinted mucosal-like dendritic cells by probiotic

Lactobacillus reuteri 17938 in vitro. Frontiers in Immunology 7. Available at https://doi.org/10.3389/fimmu.2016.00096, http://journal.frontiersin.org/article/10.3389/fimmu.2016.00096/full.

Hemarajata, P., Gao, C., Pflughoeft, K.J., Thomas, C.M., Saulnier, D.M., Spinler, J.K., et al., 2013. Lactobacillus reuteri-specific immunoregulatory gene rsiR modulates histamine production and immunomodulation by Lactobacillus reuteri. Journal of Bacteriology 195 (24), 5567–5576. Available at https://doi.org/10.1128/jb.00261-13.

Hill, C., Guarner, F., Reid, G., Gibson, G.R., Merenstein, D.J., Pot, B., et al., 2014. The International Scientific Association for Probiotics and Prebiotics consensus statement on the scope and appropriate use of the term probiotic. Nature Reviews Gastroenterology and Hepatology 11 (8), 506–514. Available at https://doi.org/10.1038/nrgastro.2014.66.

Hou, K., Wu, Z.-X., Chen, X.-Y., Wang, J.-Q., Zhang, D., Xiao, C., Zhu, D., Koya, J.B., Wei, L., Li, J., Chen, Z.-S., 2022. Microbiota in health and diseases. Signal Transduction and Targeted Therapy 7, 135. Available at https://doi.org/10.1038/s41392-022-00974-4.

Imperial, I.C.V.J., Ibana, J.A., 2016. Addressing the antibiotic resistance problem with probiotics: reducing the risk of its double-edged sword effect. Frontiers in Microbiology 7. Available at https://doi.org/10.3389/fmicb.2016.01983, http://journal.frontiersin.org/article/10.3389/fmicb.2016.01983/full.

Isaac-Bamgboye, F.J., Mgbechidinma, C.L., Onyeaka, H., Isaac-Bamgboye, I.T., Chukwugozie, D.C., 2024. Exploring the potential of postbiotics for food safety and human health improvement. Journal of Nutrition and Metabolism 2024. Available at https://doi.org/10.1155/2024/1868161, www.hindawi.com/journals/jnume/.

Islam, F., Azmat, F., Imran, A., Zippi, M., Hong, W., Tariq, F., et al., 2024. Role of postbiotics in food and health: a comprehensive review. CYTA - Journal of Food 22 (1). Available at https://doi.org/10.1080/19476337.2024.2386412, http://www.tandfonline.com/toc/tcyt20/current.

Izuddin, W.I., Humam, A.M., Loh, T.C., Foo, H.L., Samsudin, A.A., 2020. Dietary postbiotic *Lactobacillus plantarum* improves serum and ruminal antioxidant activity and upregulates hepatic antioxidant enzymes and ruminal barrier function in post-weaning lambs. Antioxidants 9 (3). Available at https://doi.org/10.3390/antiox9030250, https://www.mdpi.com/2076-3921/9/3/250/pdf.

Jalali, S., Mojgani, N., Haghighat, S., Sanjabi, M.R., Sarem-Nezhad, S., 2024. Investigation of antimicrobial and antioxidant properties of postbiotics produced by *Lactobacillus rhamnosus* and *Limosilactobacillus* reuteri and their potential application in surface decontamination of red meat. LWT 209, 116758. Available at https://doi.org/10.1016/j.lwt.2024.116758.

Khalil, E., Abd Manap, M., Mustafa, S., Alhelli, A., Shokryazdan, P., 2018. Probiotic properties of exopolysaccharide-producing *Lactobacillus* strains isolated from tempoyak. Molecules 23, 398. Available at 10.3390/molecules23020398.

Khmaladze, I., Butler, É., Fabre, S., Gillbro, J.M., 2019. Lactobacillus reuteri DSM 17938—A comparative study on the effect of probiotics and lysates on human skin. Experimental Dermatology 28 (7), 822–828. Available at https://doi.org/10.1111/exd.13950, http://onlinelibrary.wiley.com/journal/10.1111/(ISSN)1600-0625.

Kim, H.G., Lee, S.Y., Kim, N.R., Lee, H.Y., Ko, M.Y., Jung, B.J., et al., 2011. *Lactobacillus plantarum* lipoteichoic acid down-regulated *Shigella flexner*i peptidoglycan-induced inflammation. Molecular Immunology 48 (4), 382–391. Available at https://doi.org/10.1016/j.molimm.2010.07.011.

Lau, K., Srivatsav, V., Rizwan, A., Nashed, A., Liu, R., Shen, R., et al., 2017. Bridging the gap between gut microbial dysbiosis and cardiovascular diseases. Nutrients 9 (8), 859. Available at https://doi.org/10.3390/nu9080859.

Laverde Gomez, J.A., Mukhopadhya, I., Duncan, S.H., Louis, P., Shaw, S., Collie-Duguid, E., et al., 2019. Formate cross-feeding and cooperative metabolic interactions revealed by transcriptomics in co-cultures of acetogenic and amylolytic human colonic bacteria. Environmental Microbiology 21 (1), 259–271. Available at https://doi.org/10.1111/1462-2920.14454, http://onlinelibrary.wiley.com/journal/10.1111/(ISSN)1462-2920.

Lee, M.J., Zang, Z., Choi, E.Y., Shin, H.K., Ji, G.E., 2002. Cytoskeleton reorganization and cytokine production of macrophages by bifidobacterial cells and cell-free extracts. Journal of Microbiology and Biotechnology 12 (3), 398–405.

Lee, Y.B., Byun, E.J., Kim, H.S., 2019. Potential role of the microbiome in acne: a comprehensive review. Journal of Clinical Medicine 8 (7). Available at https://doi.org/10.3390/jcm8070987, https://www.mdpi.com/2077-0383/8/7/987/pdf.

Lebeer, S., Claes, I., Tytgat, H.L.P., Verhoeven, T.L.A., Marien, E., von Ossowski, I., Reunanen, J., Palva, A., de Vos, W.M., De Keersmaecker, S.C.J., Vanderleyden, J., 2012. Functional analysis of lactobacillus rhamnosus GG pili in relation to adhesion and immunomodulatory interactions with intestinal epithelial cells. Applied and Environmental Microbiology 78, 185–193. Available at https://doi.org/10.1128/aem.06192-11.

Liang, B., Xing, D., 2023. The current and future perspectives of postbiotics. Probiotics and Antimicrobial Proteins 15 (6), 1626–1643. Available at https://doi.org/10.1007/s12602-023-10045-x.

Lin, Q., 2013. Submerged fermentation of *Lactobacillus rhamnosus* YS9 for γ-aminobutyric acid (GABA) production. Brazilian Journal of Microbiology 44 (1), 183–187. Available at https://doi.org/10.1590/S1517-83822013000100028.

Lukic, J., Chen, V., Strahinic, I., Begovic, J., Lev-Tov, H., Davis, S.C., et al., 2017. Probiotics or pro-healers: the role of beneficial bacteria in tissue repair. Wound Repair and Regeneration 25 (6), 912–922. Available at https://doi.org/10.1111/wrr.12607, http://onlinelibrary.wiley.com/journal/10.1111/(ISSN)1524-475X.

Malagón-Rojas, J.N., Mantziari, A., Salminen, S., Szajewska, H., 2020. Postbiotics for preventing and treating common infectious diseases in children: a systematic review. Nutrients 12 (2), 389. Available at https://doi.org/10.3390/nu12020389.

Mehta, J.P., Ayakar, S., Singhal, R.S., 2023. The potential of paraprobiotics and postbiotics to modulate the immune system: a review. Microbiological Research 275. Available at https://doi.org/10.1016/j.micres.2023.127449, www.urbanfischer.de/journals/microbiolres/microbio.htm.

Mishra, B., Mishra, A.K., Mohanta, Y.K., Yadavalli, R., Agrawal, D.C., Reddy, H.P., et al., 2024. Postbiotics: the new horizons of microbial functional bioactive compounds in food preservation and security. Food Production, Processing and Nutrition 6 (1). Available at https://doi.org/10.1186/s43014-023-00200-w, https://www.springer.com/journal/43014.

Moghaddam, A.F., Sarlak, Z., Hosseini, H., 2019. Application of probiotics in aflatoxins risk reduction in foods: a review. International Journal of Environmental Sciences and Natural Resources 22 (3), 88–96. Available at https://doi.org/10.19080/IJESNR.2019.22.556087.

Moradi, M., Molaei, R., Guimarães, J.T., 2021. A review on preparation and chemical analysis of postbiotics from lactic acid bacteria. Enzyme and Microbial Technology 143, 109722. Available at https://doi.org/10.1016/j.enzmictec.2020.109722.

Nakkarach, A., Foo, H.L., Song, A.A.L., Mutalib, N.E.A., Nitisinprasert, S., Withayagiat, U., 2021. Anti-cancer and anti-inflammatory effects elicited by short chain fatty acids produced by *Escherichia coli* isolated from healthy human gut microbiota. Microbial Cell Factories 20 (1). Available at https://doi.org/10.1186/s12934-020-01477-z, http://www.microbialcellfactories.com/home/.

Nataraj, B.H., Ali, S.A., Behare, P.V., Yadav, H., 2020. Postbiotics-parabiotics: the new horizons in microbial biotherapy and functional foods. Microbial Cell Factories 19 (1). Available at https://doi.org/10.1186/s12934-020-01426-w, http://www.microbialcellfactories.com/home/.

Oleskin, A.V., Zhilenkova, O.G., Shenderov, B.A., Amerhanova, A.M., Kudrin, V.S., Klodt, P.M., 2014. Lactic-acid bacteria supplement fermented dairy products with human behavior-modifying neuroactive compounds. Journal of Pharmacy and Nutrition Sciences 4 (3), 199–206. Available at https://doi.org/10.6000/1927-5951.2014.04.03.5, http://www.lifescienceglobal.com/pms/index.php/jpans/article/download/2223/pdf_11.

Ozma, M.A., Abbasi, A., Akrami, S., Lahouty, M., Shahbazi, N., Ganbarov, K., et al., 2022. Postbiotics as the key mediators of the gut microbiota-host interactions. Infezioni in Medicina 30 (2), 180–193. Available at https://doi.org/10.53854/liim-3002-3, http://www.infezmed.it.

Perez, R.H., Zendo, T., Sonomoto, K., 2014. Novel bacteriocins from lactic acid bacteria (LAB): various structures and applications. Microbial Cell Factories 13. Available at https://doi.org/10.1186/1475-2859-13-S1-S3, http://www.microbialcellfactories.com/content/13/S1/S3.

Pingitore, A., Chambers, E.S., Hill, T., Maldonado, I.R., Liu, B., Bewick, G., et al., 2017. The diet-derived short chain fatty acid propionate improves beta-cell function in humans and stimulates insulin secretion from human islets in vitro. Diabetes, Obesity and Metabolism 19 (2), 257–265. Available at https://doi.org/10.1111/dom.12811, http://onlinelibrary.wiley.com/journal/10.1111/(ISSN)1463-1326.

Prajapati, N., Patel, J., Singh, S., Yadav, V.K., Joshi, C., Patani, A., et al., 2023. Postbiotic production: harnessing the power of microbial metabolites for health applications. Frontiers in Microbiology 14. Available at https://doi.org/10.3389/fmicb.2023.1306192, https://www.frontiersin.org/journals/microbiology#.

Qi, P., Tang, Z., 2020). The Nrf2 molecule trigger antioxidant defense against acute benzo(a) pyrene exposure in the thick shell mussel Mytilus coruscus. Aquatic Toxicology 226, 105554. Available at 10.1016/j.aquatox.2020.105554.

Rad, A.H., Aghebati-Maleki, L., Kafil, H.S., Abbasi, A., 2021. Molecular mechanisms of postbiotics in colorectal cancer prevention and treatment. Critical Reviews in Food Science and Nutrition 61 (11), 1787–1803. Available at https://doi.org/10.1080/10408398.2020.1765310, https://www.tandfonline.com/loi/bfsn20.

Rahman, M.R.T., Fliss, I., Biron, E., 2022. Insights in the development and uses of alternatives to antibiotic growth promoters in poultry and swine production. Antibiotics 11 (6). Available at https://doi.org/10.3390/antibiotics11060766, https://www.mdpi.com/2079-6382/11/6/766/pdf?version=1654169283.

Rakhuba, D., Novik, G., Dey, E.S., 2009. Application of supercritical carbon dioxide (scCO_2) for the extraction of glycolipids from *Lactobacillus plantarum* B-01. Journal of Supercritical Fluids 49 (1), 45–51. Available at https://doi.org/10.1016/j.supflu.2008.11.016.

Rafique, N., Jan, S.Y., Dar, A.H., Dash, K.K., Sarkar, A., Shams, R., Pandey, V.K., Khan, S.A., Amin, Q.A., Hussain, S.Z., 2023. Promising bioactivities of postbiotics: A comprehensive review. Journal of Agriculture and Food Research 14, 100708. Available at 10.1016/j.jafr.2023.100708.

Riaz Rajoka, M.S., Mehwish, H.M., Fang, H., Padhiar, A.A., Zeng, X., Khurshid, M., et al., 2019. Characterization and anti-tumor activity of exopolysaccharide produced by *Lactobacillus kefiri* isolated from Chinese kefir grains. Journal of Functional Foods 63. Available at https://

doi.org/10.1016/j.jff.2019.103588, http://www.elsevier.com/wps/find/journaldescription.cws_home/717426/description#description.

Ricci, A., Levante, A., Cirlini, M., Calani, L., Bernini, V., Del Rio, D., et al., 2018. The influence of viable cells and cell-free extracts of *Lactobacillus casei* on volatile compounds and polyphenolic profile of elderberry juice. Frontiers in Microbiology 9. Available at https://doi.org/10.3389/fmicb.2018.02784.

Sabahi, S., Homayouni Rad, A., Aghebati-Maleki, L., Sangtarash, N., Ozma, M.A., Karimi, A., et al., 2023. Postbiotics as the new frontier in food and pharmaceutical research. Critical Reviews in Food Science and Nutrition 63 (26), 8375–8402. Available at https://doi.org/10.1080/10408398.2022.2056727, https://www.tandfonline.com/loi/bfsn20.

Salminen, S., Collado, M.C., Endo, A., Hill, C., Lebeer, S., Quigley, E.M.M., et al., 2021. The International Scientific Association of Probiotics and Prebiotics (ISAPP) consensus statement on the definition and scope of postbiotics. Nature Reviews Gastroenterology and Hepatology 18 (9), 649–667. Available at https://doi.org/10.1038/s41575-021-00440-6, http://www.nature.com/nrgastro/index.html.

Sawada, D., Sugawara, T., Ishida, Y., Aihara, K., Aoki, Y., Takehara, I., et al., 2016. Effect of continuous ingestion of a beverage prepared with *Lactobacillus gasseri* CP2305 inactivated by heat treatment on the regulation of intestinal function. Food Research International 79, 33–39. Available at https://doi.org/10.1016/j.foodres.2015.11.032.

Salva, S., Tiscornia, I., Gutiérrez, F., Alvarez, S., Bollati-Fogolín, M., 2021. Lactobacillus rhamnosus postbiotic-induced immunomodulation as safer alternative to the use of live bacteria. Cytokine 146, 155631. Available at https://doi.org/10.1016/j.cyto.2021.155631.

Schiavi, E., Gleinser, M., Molloy, E., Groeger, D., Frei, R., Ferstl, R., et al., 2016. The surface-associated exopolysaccharide of *Bifidobacterium longum* 35624 plays an essential role in dampening host proinflammatory responses and repressing local TH17 responses. Applied and Environmental Microbiology 82 (24), 7185–7196. Available at https://doi.org/10.1128/aem.02238-16.

Sokol, H., Pigneur, B., Watterlot, L., Lakhdari, O., Bermúdez-Humarán, L.G., Gratadoux, J.-J., et al., 2008. *Faecalibacterium prausnitzii* is an anti-inflammatory commensal bacterium identified by gut microbiota analysis of Crohn disease patients. Proceedings of the National Academy of Sciences 105 (43), 16731–16736. Available at https://doi.org/10.1073/pnas.0804812105.

Sungur, T., Aslim, B., Karaaslan, C., Aktas, B., 2017. Impact of exopolysaccharides (EPSs) of *Lactobacillus gasseri* strains isolated from human vagina on cervical tumor cells (HeLa). Anaerobe 47, 137–144. Available at https://doi.org/10.1016/j.anaerobe.2017.05.013.

Taverniti, V., Guglielmetti, S., 2011. The immunomodulatory properties of probiotic microorganisms beyond their viability (ghost probiotics: proposal of paraprobiotic concept). Genes & Nutrition 6 (3), 261–274. Available at https://doi.org/10.1007/s12263-011-0218-x.

Teame, T., Wang, A., Xie, M., Zhang, Z., Yang, Y., Ding, Q., et al., 2020. Paraprobiotics and postbiotics of probiotic Lactobacilli, their positive effects on the host and action mechanisms: a review. Frontiers in Nutrition 7. Available at https://doi.org/10.3389/fnut.2020.570344.

Thomas, C.M., Hong, T., van Pijkeren, J.P., Hemarajata, P., Trinh, D.V., Hu, W., et al., 2012. Histamine derived from probiotic Lactobacillus reuteri suppresses TNF via modulation of PKA and ERK signaling. PLoS One 7 (2). Available at https://doi.org/10.1371/journal.pone.0031951, http://www.plosone.org/article/fetchObjectAttachment.action?uri=info%3Adoi%2F10.1371%2Fjournal.pone.0031951&representation=PDF.

Thorakkattu, P., Khanashyam, A.C., Shah, K., Babu, K.S., Mundanat, A.S., Deliephan, A., et al., 2022. Postbiotics: current trends in food and pharmaceutical industry. Foods 11 (19). Available at https://doi.org/10.3390/foods11193094, www.mdpi.com/journal/foods.

Tiptiri-Kourpeti, A., Spyridopoulou, K., Santarmaki, V., Aindelis, G., Tompoulidou, E., Lamprianidou, E.E., et al., 2016. *Lactobacillus casei* exerts anti-proliferative effects accompanied by apoptotic cell death and up-regulation of TRAIL in colon carcinoma cells. PLoS One 11 (2), e0147960. Available at https://doi.org/10.1371/journal.pone.0147960.

Tomasik, P., Tomasik, P., 2020. Probiotics, non-dairy prebiotics and postbiotics in nutrition. Applied Sciences 10 (4), 1470. Available at https://doi.org/10.3390/app10041470.

van der Beek, C.M., Canfora, E.E., Lenaerts, K., Troost, F.J., Damink, S.W.M.O., Holst, J.J., et al., 2016. Distal, not proximal, colonic acetate infusions promote fat oxidation and improve metabolic markers in overweight/obese men. Clinical Science 130 (22), 2073–2082. Available at https://doi.org/10.1042/CS20160263, http://www.clinsci.org/content/ppclinsci/130/22/2073.full.pdf.

Vandenplas, Y., de Halleux, V., Arciszewska, M., Lach, P., Pokhylko, V., Klymenko, V., et al., 2020. A partly fermented infant formula with postbiotics including 3′-GL, specific oligosaccharides, 2′-FL, and milk fat supports adequate growth, is safe and well-tolerated in healthy term infants: a double-blind, randomised, controlled, multi-country trial. Nutrients 12 (11), 3560. Available at https://doi.org/10.3390/nu12113560.

Vassileva, M., Flor-Peregrin, E., Malusá, E., Vassilev, N., 2020. Towards better understanding of the interactions and efficient application of plant beneficial prebiotics, probiotics, postbiotics and synbiotics. Frontiers in Plant Science 11. Available at https://doi.org/10.3389/fpls.2020.01068.

Vinderola, G., Sanders, M.E., Salminen, S., Szajewska, H., 2022. Postbiotics: the concept and their use in healthy populations. Frontiers in Nutrition 9. Available at https://doi.org/10.3389/fnut.2022.1002213, journal.frontiersin.org/journal/nutrition.

Wang, K., Li, W., Rui, X., Chen, X., Jiang, M., Dong, M., 2014. Characterization of a novel exopolysaccharide with antitumor activity from *Lactobacillus plantarum* 70810. International Journal of Biological Macromolecules 63, 133–139. Available at https://doi.org/10.1016/j.ijbiomac.2013.10.036.

Wegh, C.A.M., Geerlings, S.Y., Knol, J., Roeselers, G., Belzer, C., 2019. Postbiotics and their potential applications in early life nutrition and beyond. International Journal of Molecular Sciences 20 (19). Available at https://doi.org/10.3390/ijms20194673, https://www.mdpi.com/1422-0067/20/19/4673/pdf.

Wicherska-Pawłowska, K., Wróbel, T., Rybka, J., 2021. Toll-like receptors (TLRs), NOD-like receptors (NLRs), and RIG-I-like receptors (RLRs) in innate immunity. TLRs, NLRs, and RLRs ligands as immunotherapeutic agents for hematopoietic diseases. International Journal of Molecular Sciences 22 (24), 13397. Available at https://doi.org/10.3390/ijms222413397.

Wu, Z.Q., Chen, X.M., Ma, H.Q., Li, K., Wang, Y.L., Li, Z.J., 2023. *Akkermansia muciniphila* cell-free supernatant improves glucose and lipid metabolisms in *Caenorhabditis elegans*. Nutrients 15 (7). Available at https://doi.org/10.3390/nu15071725, http://www.mdpi.com/journal/nutrients/.

Yelin, I., Flett, K.B., Merakou, C., Mehrotra, P., Stam, J., Snesrud, E., et al., 2019. Genomic and epidemiological evidence of bacterial transmission from probiotic capsule to blood in ICU patients. Nature Medicine 25 (11), 1728–1732. Available at https://doi.org/10.1038/s41591-019-0626-9, http://www.nature.com/nm/index.html.

Yilmaz, N., Özogul, F., Moradi, M., Fadiloglu, E.E., Šimat, V., Rocha, J.M., 2022. Reduction of biogenic amines formation by foodborne pathogens using postbiotics in lysine-decarboxylase broth. Journal of Biotechnology 358, 118–127. Available at https://doi.org/10.1016/j.jbiotec.2022.09.003, www.elsevier.com/locate/jbiotec.

Yunes, R.A., Poluektova, E.U., Dyachkova, M.S., Klimina, K.M., Kovtun, A.S., Averina, O.V., et al., 2016. GABA production and structure of gadB / gadC genes in *Lactobacillus* and *Bifidobacterium strains* from human microbiota. Anaerobe 42, 197–204. Available at https://doi.org/10.1016/j.anaerobe.2016.10.011.

Zagato, E., Mileti, E., Massimiliano, L., Fasano, F., Budelli, A., Penna, G., et al., 2014. *Lactobacillus paracasei* CBA L74 metabolic products and fermented milk for infant formula have anti-inflammatory activity on dendritic cells in vitro and protective effects against colitis and an enteric pathogen in vivo. PLoS One 9 (2), e87615. Available at https://doi.org/10.1371/journal.pone.0087615.

Zhao, X., Liu, S., Li, S., Jiang, W., Wang, J., Xiao, J., et al., 2024. Unlocking the power of postbiotics: a revolutionary approach to nutrition for humans and animals. Cell Metabolism 36 (4), 725–744. Available at https://doi.org/10.1016/j.cmet.2024.03.004.

Zhong, Y., Wang, S., Di, H., Deng, Z., Liu, J., Wang, H., 2022. Gut health benefit and application of postbiotics in animal production. Journal of Animal Science and Biotechnology 13 (1). Available at https://doi.org/10.1186/s40104-022-00688-1.

Żółkiewicz, J., Marzec, A., Ruszczyński, M., Feleszko, W., 2020. Postbiotics—a step beyond pre- and probiotics. Nutrients 12 (8), 2189. Available at https://doi.org/10.3390/nu12082189.

Chapter 12

Aspects of failure in nonviable probiotic ingestion via drugs, supplements, and foods

Chiraz Amrouche
Department of Outpatient Clinic and Functional Explorations, National Institute of Nutrition and Food Technology of Tunis, Tunis, Tunisia

12.1 Introduction, definition, and basic concepts

Improvement of human health by modification of microbial interactions has emerged as a growing priority for health experts, food manufacturers, and consumers alike. Nonviable microbiological expressions like "paraprobiotics" and "postbiotics" are commonly used to characterize the healthy impacts of microbial intervention. Before the International Scientific Association of Probiotics and Prebiotics (ISAPP) panel (Salminen et al., 2021), many different nomenclatures had been applied to nonviable microorganisms that have their positive effects following ingestion in effective dosages. These were heat-killed probiotics, heat-treated probiotics, nonviable probiotics, paraprobiotics, or postbiotics. ISAPP's panel recommended that the term "postbiotic" be used by scientists and commercial enterprises to describe them. The term combines "biotic," meaning "relating to or produced by living things," with "post," "after," suggesting "after life" or nonliving substances. Together, these terms suggest "after life," that is, nonliving organisms. They classified a postbiotic as a "preparation of inanimate microorganisms and/or their constituents which benefit the host's health." In this chapter, we will deal with nonviable probiotics, which are previously referred to as paraprobiotics (Taverniti and Guglielmetti, 2011).

Nonviable probiotics have more positive features over live probiotics: slighter risk of infection, especially among vulnerable groups like immunocompromised subjects, zero probability of antibiotic resistance gene transfer, and higher stability in processing and storage (Salminen et al., 2021; Marco et al., 2021; Lee et al., 2016; Bernbom et al., 2019; Yelin et al., 2019). Taking into account these advantages, nonviable probiotics are being used in raising

Evaluating Oral Probiotics and Prebiotics. DOI: https://doi.org/10.1016/B978-0-443-29835-6.00001-0

diverse applications like foods, drinks, nutritional supplements, and drugs (Kamalledin et al., 2024; Lahtinen, 2012; Siciliano et al., 2021). However, there are valid scientific and regulatory problems that need to be overcome, notwithstanding their greater popularity.

This chapter will try to address the limitations of using nonviable probiotics in medications, food supplements, and food and propose ways of overcoming such limitations.

12.2 Limitations and challenges

Use of nonviable probiotics has several limitations and challenges, which are primarily scientific understanding, regulation, and practical application.

12.2.1 Impact of inactivation techniques on biological activity

The inactivation process is significant to render probiotics nonviable yet preserve their health-giving features. It is essential to understand how the different inactivation processes affect cell structure and biological activity to develop efficient nonviable probiotics.

12.2.1.1 Heat inactivation (thermal treatment)

Heat is an extremely old and universal method for inactivating probiotics. Exposing them to 60°C–100°C temperature relaxes proteins and breaks cell membranes. While this is an effective way to kill probiotics, it still has the disadvantage of inducing enzymatic failure and cellular disruptions that reduce their therapeutic effectiveness (Shokryazdan et al., 2017; Gänzle et al., 2019).

12.2.1.2 Freeze-drying (lyophilization)

This method allows for long-term storage of probiotics at room temperature. The probiotics are frozen, and the water is removed by sublimation without affecting their shape. This may cause a loss of biological activity during the process (Møller et al., 2021; Karam et al., 2020).

12.2.1.3 High-pressure processing

HPP is a nonthermal process where probiotics go through pressures as high as 600 MPa, rupturing the cell membrane and inactivating them. If the exposure is prolonged, the final product loses its health benefits (Carrias et al., 2020; Hendriksen et al., 2019).

12.2.1.4 Gamma irradiation

This technique utilizes high-energy radiation to kill probiotic DNA, making them useless. Although effective in sterilization, this technique may at the same time damage the components responsible for the beneficial effect of the nonviable probiotic (Demir et al., 2022; Tichaczek-Goska et al., 2020).

12.2.1.5 Chemical inactivation

Chemicals such as formaldehyde or alcohol disrupt the cell membrane and inhibit metabolic processes. The safety threats occur when chemical residues are left on the product (Pinho et al., 2021; Morteza et al., 2020).

12.2.1.6 Ultraviolet radiation

UV light irradiation makes probiotics nonviable by destroying DNA. Fast and efficient, especially on liquids or thin films, it might be prolonged in time, which affects the degree of DNA damage, so that reducing beneficial effects (Araujo et al., 2021; Goto et al., 2022).

12.2.1.7 Sonication (ultrasonic inactivation)

Ultrasonic waves generate collapsing bubbles, which produce forces to rupture probiotic cells. This mechanism can generate free radicals that induce further damage to cells. Therefore an appropriate balance is necessary to avoid excessive damage (Dols et al., 2020; Jun et al., 2021).

Choosing the right method to inactivate probiotics involves a trade-off. Chemicals and heat alter probiotics and potentially weaken their health benefits. Heat-free methods such as freeze drying, HPP, UV, and sonication offer more suitable alternatives but have more serious drawbacks. Finding the most suitable method for each application is essential to ensure efficacy, health benefits, and safety.

12.2.2 Less well-characterized mechanisms of action

As they are increasingly known to have health benefits, however, the mechanism by which nonviable probiotics work at the molecular level is unknown. Nonviable probiotics lack the capacity to colonize the gut or crowd out pathogens, unlike live probiotics. Research suggests that their benefits might be due to bioactive fractions like cell wall fragments, surface proteins, exopolysaccharides, and metabolites that influence the immune system and gut microbiota. Yet the detailed mechanisms remain to be investigated. Better expertise in these mechanisms is needed to enable their safe and well-regulated use in the food and pharmaceutical sectors (Adams, 2021; Moradi et al., 2022).

12.2.3 Challenges in detection and quantification

Nonviable probiotics cannot be counted with standard microbiology methods relying on cell proliferation and colony growth (Siciliano et al., 2021). Culture-independent methods, such as polymerase chain reaction (PCR), can detect live and dead cells but do not specifically quantify nonviable cells (Soejima et al., 2020). Certain recent studies reported that neither traditional nor molecular methods proved effective in excluding nonviable cells from probiotic products ().

12.2.4 Lack of standardized dosage for nonviable probiotics

Determining the appropriate dosage of nonviable probiotics is crucial to their therapeutic potential. Since their biological activity depends on specific components and metabolites, dosage levels must be optimized to achieve the required health benefits. However, the optimal dosage is far from established, and hence it is challenging to formulate standard therapeutic guidelines (Sewell et al., 2022b). Dosage ranges need to be standardized to provide consistent product formulations and make certain recommendations to healthcare practitioners and consumers (Cifuentes et al., 2023; Müller et al., 2021a).

12.2.5 Safety issues

12.2.5.1 Microbial contamination

Even though nonviable probiotics are inactive, microbial contamination during manufacturing is an issue, especially in immunocompromised patients (Müller et al., 2022). Microbial analyses and contamination controls must be rigorous throughout the manufacturing chain (Williams et al., 2023).

12.2.5.2 Purity, potency, and composition

Ensuring the strength, purity, and composition of nonviable probiotics is essential to ensure their safety and efficacy (Pérez et al., 2021). The product must be free of unwanted microorganisms, and the amount of nonviable microbes must be confirmed to produce therapeutic effects (Zhang et al., 2020).

12.2.5.3 Allergens and contaminants

Allergen testing is necessary to prevent allergic reactions, while contaminant testing for heavy metals or trace chemicals ensures the product's safety (Sewell et al., 2022a).

12.2.6 Lack of standardization and quality control

With the growing popularity of nonviable probiotics, it is a major challenge to prove their safety, efficacy, and reliability. Currently, there are no standard guidelines for their manufacture, quality control, or quantification in different formulations (Kok et al., 2021a). Differences in manufacturing and storage conditions can lead to differences in product quality and clinical response. Strict quality control measures need to be followed at every step, from manufacture to distribution, to ensure uniformity, safety, and efficacy (Sewell et al., 2022c).

Manufacturing nonviable probiotics is not a simple process. Everything from the source of raw materials to how they are inactivated and formulated can impact their final properties. Without strict protocols, product batches will be heterogeneous, not all offering the same benefits, making it difficult for healthcare professionals and consumers to trust their efficacy (Perez et al., 2021).

The stability of nonviable probiotics is also affected by storage conditions, namely, temperature and humidity (Liu et al., 2020). If noncompliant, these conditions might decline their potentialities and lead to possible contamination.

To ensure product safety and efficacy, rigorous quality control measures must be adopted from production to distribution. Developing standardized regulations will help ensure reliable products and thus gain the trust of consumers and physicians in the growing nonviable probiotics sector.

12.2.7 Health claims and regulation

It is difficult to control nonviable probiotics, especially when it is a question of determining their health benefits. Most countries have strict controls over products that purport to prevent or treat disease (Schwiertz et al., 2020; Müller et al., 2021b). Nonviable probiotics do not quite fill either of these categories, and the approvals for health claims are at a standstill. There must be strict guidelines to substantiate health claims through sound evidence for the sake of safe and effective products (Kok et al., 2021b; Hefni et al., 2022). Worldwide standardization of definitions would also avoid research chaos, labeling, and claims (Pérez et al., 2023).

12.2.8 Research gaps

Despite their promising potential, there is little long-term research to assess the safety and efficacy of nonviable probiotics, particularly in vulnerable populations. Although short-term studies have shown some benefits, such as immune modulation and regulation of gut health, the long-term impact remains understudied (Zhang et al., 2023; Pérez et al., 2022; Müller et al., 2021c). More studies are needed to ascertain their effect on vulnerable populations like the elderly, immunocompromised individuals, and infants (Ceci et al., 2020).

12.3 Recommendations

Recommendations flow easily from challenges. First, scientists should carry out more long-term clinical studies to identify with precision the effects of nonviable probiotics on gut microbiota, immune function, and overall health, especially in vulnerable groups. Then, authorities have to establish and maintain high-quality control to ensure that nonviable probiotics are safe and also effective. Equal dosages, purity, and freedom from contamination have to be guaranteed. Moreover, further research studies need to be conducted to determine the best inactivation processes that retain beneficial properties and ensure safety and stability. In addition, legislators should impose clear and consistent guidelines regarding the approval, labeling, and health claims of nonviable probiotics. These laws are put in place to preserve consumer safety and market clarity. Finally, it is wise to conduct more clinical trials to test the efficacy and safety of nonviable probiotics in high-risk groups such as infants, the elderly, and pregnant women.

12.4 Conclusion

Nonviable probiotics appear to hold great promise for health. However, it is imperative to understand the limitations of their inactivation procedures, to institute strict quality controls, and to continue long-term studies to ensure their safety and efficacy when administered in drugs, dietary supplements, and foods.

Conflicts of interest

The author declares no conflict of interest.

References

Adams, C.A., 2021. The non-viable probiotics: a review of mechanisms, bioactivity, and applications. Trends in Food Science & Technology 112, 491–502.

Araujo, A.M., et al., 2021. Ultraviolet radiation as an inactivation method for probiotic strains: efficacy and limitations. International Journal of Food Microbiology 338, 108961.

Bernbom, N., et al., 2019. Paraprobiotics: non-viable probiotics and their health benefits. International Journal of Food Science & Technology 54 (7), 2231–2243.

Carrias, A., et al., 2020. High pressure processing: a method for probiotic inactivation. Journal of Food Science 85 (3), 567–576.

Ceci, A., et al., 2020. Safety and efficacy of non-viable probiotics in elderly and immunocompromised individuals: a systematic review. Clinical Nutrition 39 (5), 1681–1692.

Cifuentes, M., et al., 2023. Non-viable probiotics: dosage standardization and clinical outcomes. Gut Microbes 14 (1), 142–152.

Demir, N., et al., 2022. Gamma irradiation as an inactivation method for probiotic microorganisms: a review. Food Control 132, 108473.

Dols, J., et al., 2020. The use of sonication for inactivation of probiotics: effect on viability and biological activity. Food Control 110, 107–113.

Gänzle, M.G., et al., 2019. Effect of thermal processing on probiotic bacteria in fermented dairy products. International Dairy Journal 94, 75–85.

Goto, M., et al., 2022. UV inactivation of probiotic bacteria: a review of applications in food and beverage industries. Food Bioprocess Technology 15 (8), 1259–1268.

Hefni, M.E., et al., 2022. Non-viable probiotics and health claims: regulatory pathways and evidence. European Journal of Clinical Nutrition 76 (6), 868–876.

Hendriksen, N.B., et al., 2019. Effect of high-pressure processing on the viability of probiotic microorganisms in dairy products. International Journal of Food Science & Technology 54 (4), 1121–1130.

Jun, K., et al., 2021. Ultrasonic inactivation of probiotic microorganisms: a review on mechanisms and applications. Journal of Food Science 86 (5), 2295–2307.

Kamalledin, M.S., Farhoodi, M., Mofid, V., Homayouni-Rad, A., Milani, et al., 2024. Effects of adding non-viable *Lacticaseibacillus casei* and *Lactobacillus acidophilus* on physicochemical, microbial, chemical and sensory attributes of probiotic doogh. Applied Food Biotechnology 11 (1), e14. Available at https://doi.org/10.22037/afb.v11i1.44105.

Karam, R.S., et al., 2020. Lyophilization of probiotics: a review of the impact on stability and bioactivity. Critical Reviews in Food Science and Nutrition 60 (17), 3000–3015.

Kok, R.M., et al., 2021a. Standardizing the production and quality control of non-viable probiotics: challenges and strategies. Journal of Food Science and Technology 58 (5), 1895–1905.

Kok, R.M., et al., 2021b. Regulatory challenges of non-viable probiotics: bridging gaps for health claims approval. International Journal of Food Science & Technology 56 (7), 2854–2863.

Lahtinen, S.J., 2012. Probiotic viability – does it matter? Microbial Ecology in Health and Disease 23, 18567. Available at https://doi.org/10.3402/mehd.v23i0.18567.

Lee, Y., et al., 2016. Health benefits of non-viable probiotics. International Journal of Food Sciences and Nutrition 67 (5), 626–633.

Liu, M., et al., 2020. Temperature and humidity effects on the stability of non-viable probiotics: a systematic review. Food Research International 134, 109–115.

Marco, M.L., et al., 2021. Postbiotics and gut health: what is the evidence? Microorganisms 9 (10), 2229.

Møller, P.L., et al., 2021. The impact of freeze-drying on the viability and biological activity of probiotics. Food Research International 141, 110–118.

Moradi, M., et al., 2022. Postbiotic components and their proposed mechanisms of action: a critical review. Trends in Food Science & Technology 120, 21–34.

Morteza, N., et al., 2020. Formaldehyde and alcohol in probiotic inactivation: safety and efficiency considerations. Food and Bioprocess Technology 13 (9), 1575–1583.

Müller, T., et al., 2021a. Standardizing the dosage of non-viable probiotics for food and therapeutic applications. Frontiers in Nutrition 8, 674–681.

Müller, T., et al., 2021c. The mechanisms of action of non-viable probiotics: insights into mucosal immunity and gut permeability. Journal of Probiotics and Health 10 (2), 245–258.

Müller, T., et al., 2021b. Substantiating health claims for non-viable probiotics: current framework. Journal of Probiotics and Health 9 (4), 154–163.

Müller, T., et al., 2022. Microbial contamination and quality control in non-viable probiotics. Journal of Food Protection 85 (5), 794–801.

Perez, A., et al., 2021. Regulatory considerations for the production and approval of non-viable probiotics in food supplements. Food Regulatory Research 11 (3), 75–82.

Pérez, A., et al., 2021. Purity and potency in non-viable probiotics: regulatory challenges and solutions. Frontiers in Microbiology 12, 717611.

Pérez, A., et al., 2022. Non-viable probiotics and their impact on gut microbiota composition: a long-term clinical trial approach. Microorganisms 10 (8), 1651.

Pérez, A., et al., 2023. Standardizing definitions for probiotics: defining the future of paraprobiotics. Journal of Food and Drug Administration 23 (1), 1–8.

Pinho, C., et al., 2021. Chemical inactivation of probiotics: potential risks and benefits. Journal of Applied Microbiology 131 (4), 1395–1406.

Salminen, S., Collado, M.C., Endo, A., et al., 2021. The International Scientific Association of Probiotics and Prebiotics (ISAPP) consensus statement on the definition and scope of postbiotics. Nature Reviews Gastroenterology & Hepatology 18, 649–667. Available at https://doi.org/10.1038/s41575-021-00440-6.

Schwiertz, A., et al., 2020. Global regulatory frameworks for probiotics: Focus on non-viable probiotics. Food Control 112, 107123.

Sewell, H.F., et al., 2022b. The therapeutic use of non-viable probiotics: determining effective dosages for clinical applications. Clinical Microbiology and Infection 28 (5), 647–655.

Sewell, H.F., et al., 2022a. Ensuring the safety of probiotics: the impact of allergens and contaminants. International Journal of Food Microbiology 348, 109–118.

Sewell, H.F., et al., 2022c. Quality control and stability of non-viable probiotics: ensuring safety and efficacy through standardized protocols. International Journal of Food Microbiology 348, 109–118.

Shokryazdan, P., et al., 2017. Heat treatment and probiotics: potential benefits and risks. Journal of Microbiology and Biotechnology 27 (5), 815–823.

Siciliano, R.A., Reale, A., Mazzeo, M.F., Morandi, S., Silvetti, T., Brasca, M., 2021. Paraprobiotics: a new perspective for functional foods and nutraceuticals. Nutrients 13, 1225. Available at https://doi.org/10.3390/nu13041225.

Soejima, T., et al., 2020. Exclusive use of digital PCR allows an absolute assay of heat-killed Lactobacilli in foods targeting multiple copies of 16S rDNA. Scientific Reports 10, 12691.

Taverniti, V., Guglielmetti, S., 2011. The immunomodulatory properties of probiotic microorganisms beyond their viability (ghost probiotics: proposal of paraprobiotic concept). Genes Nutrition 6 (3), 261–274. Available at https://doi.org/10.1007/s12263-011-0218-x.

Tichaczek-Goska, D., et al., 2020. Impact of gamma radiation on probiotic bacteria. Food Research International 127, 108627.

Williams, J., et al., 2023. Assessing microbial contamination in non-viable probiotics: pathogen control strategies. Microorganisms 11 (7), 1622.

Yelin, I., et al., 2019. Genomic and epidemiological evidence of bacterial transmission from probiotic capsule to blood in ICU patients. Nature Medicine 25, 1728–1732.

Zhang, X., et al., 2020. Quality control of probiotics: a look at allergenicity, contamination, and potency testing. Food Control 116, 107319.

Zhang, X., et al., 2023. Long-term effects of non-viable probiotics on immune modulation and gut health in at-risk populations. Frontiers in Immunology 14, 846231.

Section III

Trends on prebiotic and synbiotic ingestion instead of probiotic

Chapter 13

Incorporation of prebiotics in medicine, supplement, and food product: health effectiveness, technological challenges, consumer perspectives, and future trends

Ali Jafari[1], Elaheh Karimzadeh-Soureshjani[2], Daryoush Kiani-Shamsabadi[2], Hooriyeh Mohammadpour[3], and Helia Mardani[4]

[1]*Student Research Committee, Department of Community Nutrition, Faculty of Nutrition Sciences and Food Technology, National Nutrition and Food Technology Research Institute, Shahid Beheshti University of Medical Sciences, Tehran, Iran,* [2]*Student Research Committee, Abadan University of Medical Sciences, Abadan, Khuzestan, Iran,* [3]*Department of Food, Environmental and Nutritional Sciences, Universita Degli Studi di Milano, Milan, Italy,* [4]*Students' Scientific Research Center (SSRC), Tehran University of Medical Sciences, Tehran, Iran*

13.1 Introduction

Prebiotics, as defined by the International Scientific Association for Probiotics and Prebiotics (ISAPP), are selectively metabolized by host microorganisms, resulting in measurable health benefits (Gibson et al., 2017). The substances are commonly indigestible to humans but fermentable by gut microbes, such as fructo-oligosaccharides (FOS), galacto-oligosaccharides (GOS), inulin, resistant starch, and beta-glucans. They selectively stimulate the growth of beneficial intestinal bacteria (Miqdady et al., 2020).

On the other hand, probiotics are microorganisms, and when consumed in sufficient and clinically validated doses, they induce beneficial effects on the host through mechanisms such as colonization, metabolic interaction, or immune system modulation. A requirement for classifying a microorganism as a probiotic is its viability at the time of ingestion, as only live cells can apply the intended physiological effects within the host (Sanders et al., 2019).

Evaluating Oral Probiotics and Prebiotics. DOI: https://doi.org/10.1016/B978-0-443-29835-6.00004-6

Synbiotics are products that combine probiotics and prebiotics to produce a synergistic effect that is greater than the sum of their individual components. They can be classified as either complementary, where the probiotic and prebiotic components are administered separately, or synergistic, where the components are mixed to ensure the optimal survival of the living microorganisms and the effective utilization of the prebiotics (Swanson et al., 2020).

Postbiotics are defined as preparations of inactivated microorganisms and/or their components that confer health benefits to the host, even in the absence of viable cells (Salminen et al., 2021). These products may include nonviable cells, cell wall fragments, metabolites, or secreted bioactive compounds, provided that their health-promoting efficacy has been scientifically demonstrated (Vinderola et al., 2023).

In recent years, there has been a growing interest in the use of prebiotics within the pharmaceutical and food industries. This surge is driven by their diverse benefits, including improvements in host health, enhanced production stability, and added value to final products. Unlike probiotics, which require the survival of live microorganisms during processing, storage, and digestion, prebiotics offer the advantage of chemical stability, making them a practical and reliable choice for industrial applications (Markowiak and Śliżewska, 2017).

From a biotechnological standpoint, prebiotics are process-stable, allowing for large-scale production under varying conditions of temperature, pH, and mechanical stress. This stability enables manufacturers to incorporate prebiotics into products such as beverages, dairy foods, breakfast cereals, and supplements without compromising their functional benefits. Additionally, prebiotics are more cost-effective than probiotics, as they do not require specialized storage conditions like refrigeration or modified atmosphere packaging (Roberfroid, 2007).

The rising industrial demand for prebiotics is largely attributable to their capacity for host-selective modulation of the gut microbiota. Unlike some probiotics, whose effects can be influenced by host-specific factors, prebiotics can selectively increase the number of resident beneficial microorganisms, such as *Bifidobacterium* and *Lactobacillus*. Due to this, they are more stable and less variable in their effects on the host's metabolic and immune processes. They have also been discovered in a few preliminary studies to have the potential to sustain mental health (Gibson et al., 2017; Slavin, 2013). The development of newer prebiotic foods like oligosaccharides of novel plant origin, digestion-resistant polysaccharides, and multipurpose soluble fibers has brought forth a new possibility for food product development. These innovations, combined with a rapidly expanding body of scientific research highlighting the role of prebiotics in improving metabolic, inflammatory, and neuropsychiatric conditions, have positioned them as premier functional ingredients in the global health and wellness market (Bindels et al., 2015; Rastall and Gibson, 2015; Carlson et al., 2018).

This chapter provides a clear framework for the role and applications of prebiotics in the pharmaceutical and food industries. It begins by examining key definitions and scientific distinctions between prebiotics and related concepts such as probiotics, synbiotics, and postbiotics, thereby establishing a clear conceptual foundation for subsequent studies. Subsequently, the factors driving the growing industrial demand for prebiotics are examined, with emphasis on their health-enhancing properties, innovations in manufacturing technologies, and evolving market dynamics. The chapter then discusses the role of prebiotics within the human microbiota, offering a thorough explanation of how they influence gut flora balance and the stimulation of a number of distinct biological processes. The chapter also discusses scientific evidence on the health-improving efficacy of prebiotics in humans, together with a comparative analysis of the same in relation to other microbiota-altering agents. Technical and practical problems concerning prebiotic production, formulation, and stability maintenance in industrial products are dealt with at length. Consumer behavior, market conditions, and economic factors influencing the manufacture and acceptance of prebiotic products are analyzed.

In the last section, future trends, research opportunities, and development directions are debated to provide a clear vision in this field. The chapter concludes with an outline of key points and an emphasis on the importance of applied applications of prebiotics.

The chapter is structured to systematically guide the reader through the theoretical framework, followed by an examination of the challenges, opportunities, and industrial applications of prebiotics. It concludes with an overview of the current state of research and potential future directions in the field (Fig. 13.1).

13.2 Prebiotics in the microbiome context

13.2.1 Definition and criteria

Prebiotic ideas have changed significantly during the past few decades, resulting in more detailed definitions and criteria shaped by advances in microbiome science, gastrointestinal physiology, and clinical nutrition. In general terms, a prebiotic is now "a substrate selectively utilized by host microorganisms and provides a range of health benefits." This definition, introduced in 2017 by the International Scientific Association for Probiotics and Prebiotics (ISAPP) (Gibson et al., 2017), is based on a modern, function-oriented approach with emphasis on selective modulation of the microbiota and quantifiable effects on host health (Sanders et al., 2019).

Prior to the new ISAPP definition, the Food and Agriculture Organization of the United Nations (FAO) had suggested an operational definition that emphasized compounds' lack of digestibility in the upper gastrointestinal tract and fermentability by beneficial gut microbes of the host (Pineiro et al., 2008).

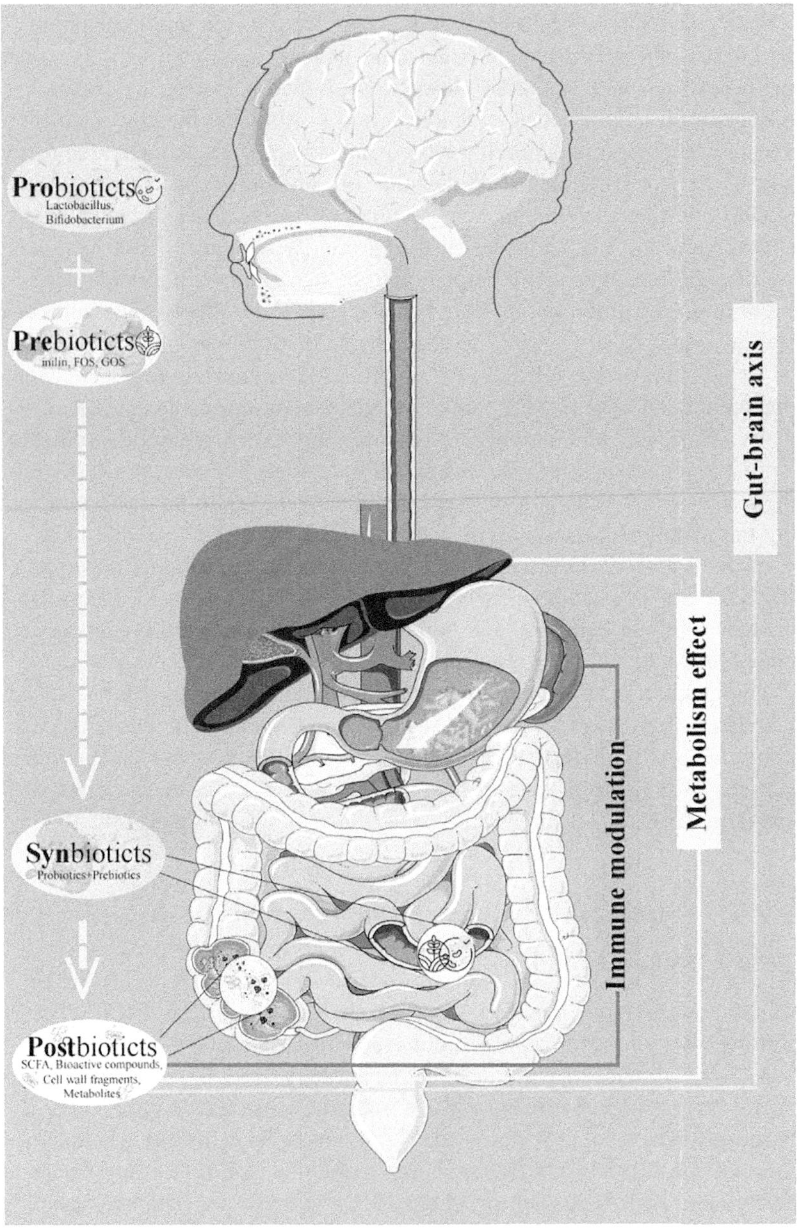

FIGURE 13.1 Conceptual framework showing the interaction of probiotics, prebiotics, synbiotics, and postbiotics in human health.

Although the FAO definition primarily focused on fiber-like compounds, it lacked sufficient specificity about the benefits to the health of the host and selective utilization by microbes. Relative to the ISAPP standards, the ISAPP standards specifically demand that a compound be: (1) resistant to digestion in the upper part of the gut; (2) fermentable by the gut microbiota; and (3) induce a specific health effect in the host by selectively altering the composition or function of the microbiota (Gibson et al., 2017).

As microbiome science progresses, the criteria for including compounds as prebiotics are becoming increasingly strict. It is essential to demonstrate that the compound can resist gastric acidity, hydrolysis by mammalian enzymes, and absorption in the small intestine. Furthermore, the compound must be fermentable by colon microbiota and should demonstrate the ability to selectively stimulate health-promoting strains, including *Bifidobacterium*, *Lactobacillus*, and newer beneficial taxa such as *Faecalibacterium prausnitzii* (Bindels et al., 2015; Scott et al., 2013). Additionally, changes in microbial composition or activity should correlate with measurable improvements in host health, including enhancements in immune function, better intestinal mucosal barrier integrity, or reduced inflammation (Slavin, 2013).

The second key distinction in current definitions is the inclusion of noncarbohydrate molecules. Polyphenols and certain fatty acids have been proposed as prebiotic candidates based on their ability to selectively modulate microbial composition and systemic effects. This extended paradigm broadens the concept of prebiotics beyond the realm of traditional fermentable fibers to include bioactive molecules with defined microbiome interactions (Rastmanesh, 2011; Cardona et al., 2013).

It is important to note that not all fermentable substances are prebiotics. It is the property of selective microbial stimulation that separates true prebiotics from substrates fermented in a nonspecific, wide manner by a large range of potentially pathogenic microorganisms. Therefore the term prebiotic should be applied only to substrates with clear proof of selective activities upon the microbiome that are consistent with proven health benefits (Davani-Davari et al., 2019).

13.2.2 Prebiotic categories

The most extensively studied group of prebiotics is the fructans, specifically inulin and FOS. These compounds are naturally present in foods such as chicory root, garlic, and leek, consisting of linear polymers of fructose, which promote the growth of Bifidobacteria and Lactobacilli species (Roberfroid, 2007). Their fermentation by gut microbiota results in the production of short-chain fatty acids (SCFAs), particularly acetate and butyrate, which play vital roles in colon health and have systemic metabolic effects.

GOS are among the most significant prebiotics incorporated in infant formulas, added most notably for their bifidogenic nature. GOS triggers the growth of Bifidobacteria within the gut microbiota exclusively, conditioning the microbial pattern of the infant closer to what is experienced by breastfed infants. They have a role in regulating the immune system, decreasing the prevalence of gastrointestinal infections, and improving intestinal epithelial barrier function, particularly in infancy and old age. In addition, the evidence further indicates the modulatory effect of GOS on relieving the risk of allergic diseases and promoting mucosal immunity maturation, with specific clinical value as functional ingredients for foodstuffs (Vandenplas et al., 2015).

Xylo-oligosaccharides (XOSs) are prebiotic oligosaccharides that consist of 2–10 xylose monomers bonded by β-(1→4) linkages. As a result of increased resistance to enzymatic hydrolysis in the upper gut, XOSs remain intact and reach the large intestine. In the colon, XOSs are selectively fermented by beneficial gut microbiota such as bifidobacteria and lactobacilli, which enhance their growth. This activity helps maintain gut microbial balance and promotes the production of SCFAs, which are crucial for intestinal metabolism and physiology. Additionally, XOSs have potential roles in immune and metabolic regulation, including the management of blood glucose levels and lipid metabolism. One of the most significant advantages of XOS is their environmentally friendly and cost-effective production from lignocellulosic biomass and agricultural waste. This process not only reduces waste in nature but also broadens their applications in the industrial and pharmaceutical sectors, allowing for further development as functional foods and dietary supplements (Valladares-Diestra et al., 2023).

Resistant starches constitute another important category of prebiotics, and they are the fractions of starch that are resistant to digestion in the small intestine but are fermented in the colon. These not only increase butyrate production but also were recently shown to increase insulin sensitivity and reduce inflammatory markers (Martínez et al., 2010). Resistant starch, a complex carbohydrate that is resistant to hydrolysis by the enzymes of the small intestine, belongs to five wide categories of unique structural and functional characteristics. RS1 is found primarily in physically inaccessible food matrices such as whole grains and legumes. It is resistant because of the compact structural organization that hinders enzymatic access. RS2 consists of naturally occurring granular starches that are present in raw, unripe food items like green bananas and raw potatoes. Its crystalline composition renders it resistant to enzymatic digestion. RS3 is produced by heating and subsequent refrigeration of starchy foods like rice or potatoes, in which retrogradation processes reorganize starch molecules into more digestion-resistant forms. RS4 consists of chemically modified starches structurally altered, often by esterification or cross-linking, to enhance resistance to digestive enzymes. RS5 is a result of complexation between fatty acids and starch, leading to a highly resistant structure toward enzymatic hydrolysis. All types of resistant starch, according to chemical structure, possess a characteristic pattern of fermentation in the colon and contribute to the production of SCFAs, such as butyrate, at varying rates and capacities. These SCFAs are important in gut health and in the modulation of immune function (Nugent, 2005).

One of the more recently accepted categories is polyphenols. Even though they are traditionally known for their antioxidant effect, polyphenols possess prebiotic-like activity. Certain polyphenols, such as those in green tea, berries, and cocoa, are not digested until they undergo microbial biotransformation in the colon. This metabolic process leads to the generation of metabolites that control the composition and functional properties of gut microbiota (Ozdal et al., 2016). Interestingly, these chemicals are associated with improved cognitive function, metabolic regulation, and a reduction of inflammation, with mounting evidence for their prebiotic effect. In addition, by stimulating bioactive metabolite formation, such as SCFAs and other antiinflammatory compounds, polyphenols can preserve gut integrity and elicit systemic protective effects (Dueñas et al., 2015).

These diverse groups of prebiotics differ in chemical makeup, fermentation rates, microbial substrates, as well as their impact on the host's physiological factors. Their combined or synergistic effect, taken collectively, traditionally referred to as prebiotic blends, is being more and more considered both clinically and industrially (Bindels et al., 2015). Ongoing research into novel prebiotics, particularly those of underutilized natural origin or produced through the biotechnological process, is poised to drive innovative leaps in the manufacture of functional foods and dietary supplements.

13.2.3 Natural sources versus industrially produced prebiotics

Prebiotics are derived from huge quantities of natural sources that vary from various vegetables, fruits, and whole grains to legumes. These natural sources, as intricate matrices constituted by various fibers and bioactive molecules, may exert significant synergistic effects on the gut microbiota and host well-being. For instance, inulin and FOS are naturally present in chicory root, garlic, onion, and asparagus; likewise, GOS are naturally found in human and bovine milk (Gibson et al., 2017; Swennen et al., 2006). It is important to note that the concentration and composition of prebiotics in these sources can vary greatly depending on the plant species, environmental conditions, and stage of ripeness (Holscher, 2017).

Industrial production of prebiotics focuses on generating compounds with defined purity and dosage under controlled conditions. Techniques such as enzymatic hydrolysis, microbial fermentation, and chemical synthesis facilitate the large-scale manufacturing of key prebiotics, including inulin, FOS, and GOS. These industrial processes ensure consistent quality, precise dosing, and seamless integration into food and pharmaceutical formulations. Standardized manufacturing not only supports clinical application but also enables rigorous scientific evaluation. By specifying concentration and composition with precision, these processes allow for controlled clinical trials and comprehensive assessments of safety and efficacy (Davani-Davari et al., 2019).

However, the absence of a food matrix that is intricate and other synergistic molecules in industrial foods may impact their overall efficacy. This has led to intensive studies on the combination of the consumption of natural prebiotic sources with industrial products in a bid to achieve both the virtue of dietary complexity as well as exact dosing simultaneously (Bindels et al., 2015).

The growing consumer preference for natural ingredients and clean-label products has driven manufacturers to innovate low-impact extraction technologies that more effectively preserve the bioactive compounds found in raw materials. The technologies aim to minimize chemical and physical alteration of prebiotic compounds while maximizing their biological activity and offering products that are more natural and of higher quality, meeting the increasing market demand and regulatory requirements for consumer health. On the other hand, industrial processes with large-scale production capabilities, guaranteed quality, and precision-controlled dosage, particularly in the forms of functional foods and supplements, continue to be at the forefront of the pervasive supply of prebiotics. These qualities enable strict scientific evaluation of clinical effects, safety control, and higher efficacy in population-based as well as specific nutritional programs. Thus the integration of the advantages of natural compound preservation with industrial production is a primary area of activity in current R&D in the prebiotic area to provide final products not only with consistent clinical performance but also fulfilling different consumer needs (Carlson et al., 2018).

13.2.4 How prebiotics differ functionally from probiotics

Prebiotics and probiotics are distinct yet complementary approaches to regulating the gut microbiome and enhancing host health. Although both contribute to the maintenance of intestinal homeostasis, they differ fundamentally in their mechanisms of action, origins, and functional roles. Probiotics consist of live microorganisms that, when administered in adequate amounts, confer health benefits by directly influencing microbial composition and activity. In contrast, prebiotics are nondigestible compounds that selectively stimulate the growth and activity of beneficial gut bacteria, thereby indirectly supporting microbiome balance and overall physiological well-being.

Probiotics are live microbial preparations introduced in adequate amounts, which confer a beneficial physiological effect on the host (Marco et al., 2021). These health-promoting microorganisms, usually members of the genera *Lactobacillus* and *Bifidobacterium*, colonize the gut directly or travel through the gut transiently. They exert their effects by competing with harmful microorganisms, promoting the mucosal defense mechanism, and modulating local and systemic immune responses (Sanders et al., 2019). The efficacy of probiotics depends very much on being strain-specific, their survival at the site of action, and host factors, all of which are important determinants in successful colonization and health outcomes (Plaza-Diaz et al., 2019).

In contrast, prebiotics are selectively utilized by favorable microbial communities in the gut to be fermented in a manner that is beneficial to the host, indirectly affecting microbiota composition and function (Gibson et al., 2017). Prebiotics are different from probiotics in that they do not contain live cells; they are usually indigestible carbohydrates or bioactive molecules that cannot be digested by enzymes in the upper gastrointestinal tract and instead reach the colon intact (Bindels et al., 2015). Prebiotics improve the growth and metabolic activity of beneficial bacteria such as *Bifidobacterium* and *F. prausnitzii* selectively by inducing the production of SCFAs and other metabolites that enhance intestinal barrier function, regulate inflammation, and contribute to systemic metabolic functions (Koh et al., 2016).

Probiotics introduce beneficial external microbes to the gut microbiome and can potentially restore microbial equilibrium following dysbiosis or infection by pathogens. Instead, prebiotics are dietary substrates that stimulate the growth and activity of the endogenous beneficial bacteria, thereby nutritionally powering an advantageous microbial community. This disparity leads to divergent applications: probiotics are primarily used for short-term use and reseeding of microbiota, whereas prebiotics are engaged more in long-term regulation and preservation of an optimal microbiome (Markowiak and Śliżewska, 2017).

Synbiotic foods, as synergistic blends of prebiotics and probiotics, are designed to enhance the viability of probiotics in the gut and improve their biological activity. The formulations, by providing specific and selective nutrient substrates for probiotic growth, provide optimal probiotic growth and activity of the beneficial microbes, thereby imparting double health effects on the host. Synbiotics, in addition to improving probiotic survivability, also modulate the gut microbial environment and the generation of health-benefiting metabolites, that is, SCFAs. It is of extreme significance to maintain microbial balance and immune and metabolic function. Synbiotics are a novel integrated approach for maintaining gut health, combining the synergistic effects of prebiotics and probiotics. Backed by scientific evidence, synbiotics are widely utilized in clinical nutrition and can enhance patients' quality of life (Swanson et al., 2020) (Table 13.1).

13.3 Health effectiveness of prebiotics and comparative insights

13.3.1 Biological mechanisms of prebiotics

13.3.1.1 Fermentation and short-chain fatty acid production

Prebiotics have their physiologic effect primarily through fermentation by gut microbiota. During fermentation, they yield enormous amounts of metabolites,

TABLE 13.1 Classification and sources of common prebiotics, their target microorganisms, and physiological roles.

Prebiotic	Source	Natural/ industrial	Molecular structure	Physiological mechanism	Target microbes	Health benefits	References
HMOs	Human breast milk	Natural	Lactose-based oligosaccharides with GlcNAc branches and fucose or sialic acid substitutions	Selective prebiotic nourishment with immunomodulatory glycan presentation, epithelial barrier maturation, and pathogen competitive inhibition	*Bifidobacterium* spp.	Immune system ontogenesis, neurodevelopmental support, infection prophylaxis, and microbiome establishment programming	Corona et al. (2021)
Lactulose	Synthesized from lactose	Industrial	Disaccharide composed of galactose and fructose linked by β-1,4 glycosidic bond	Colonic fermentation to SCFAs with pH acidification, competitive pathogen exclusion, and osmotic laxative properties	*Bifidobacterium* spp., *Lactobacillus* spp.	Constipation management, enhanced calcium bioavailability, metabolic homeostasis support, and digestive tract optimization	Karakan et al. (2021)
Polyphenols	Fruits, vegetables, cereals, beans, tea, coffee, honey, red wine	Both	Aromatic compounds with multiple hydroxyl groups and complex phenolic structures	Limited small intestinal absorption with extensive colonic biotransformation, bidirectional microbiome interaction, SCFA stimulation, antiinflammatory cascades, and antimicrobial activity	*Bifidobacterium* spp., *Akkermansia muciniphila*, *Faecalibacterium prausnitzii*	Cardiometabolic risk reduction, immunocompetence enhancement, systemic inflammation attenuation, metabolic flux optimization, and broad-spectrum antimicrobial effects	Wang et al. (2022)

XOS	Corn cobs, bamboo shoots	Both	Xylose oligomeric chains (DP 3–7) with β-(1→4) glycosidic linkages	Selective prebiotic fermentation with colonic pH modulation, enhanced microbial ecological diversity, and targeted beneficial bacterial proliferation	*Bifidobacterium* spp.	Lipid metabolism enhancement, GI inflammatory response attenuation, and metabolic syndrome risk reduction	Zhao et al. (2024)
Inulin	*Cichorium intybus*, garlic, Jerusalem artichoke	Both	β-(2→1) linked polyfructan polymer with variable DP (2–60)	Bifidogenic fermentation substrate with robust SCFA production, glucose homeostasis regulation, lipid profile optimization, and antiinflammatory mediation	*Bifidobacterium* spp.	Glycemic control enhancement, dyslipidemia correction, systemic inflammation reduction, and insulin sensitivity improvement	Alonso-Allende et al. (2024)
RS	Legumes, green bananas, cooled potatoes	Both	Glucose polymer with α-(1→4) and α-(1→6) glycosidic linkages resistant to enzymatic hydrolysis	Colonic SCFA biosynthesis with intestinal barrier fortification, metabolic regulation, and gut microbiome structural modulation	*Bifidobacterium* spp., *Ruminococcus* spp., *Faecalibacterium* spp.	Colorectal cancer risk mitigation, insulin sensitivity enhancement, weight management support, and metabolic health optimization	Chen et al. (2024)

(*Continued*)

TABLE 13.1 (Continued)

Prebiotic	Source	Natural/ industrial	Molecular structure	Physiological mechanism	Target microbes	Health benefits	References
FOS	Chicory roots, Jerusalem artichoke tubers, yacon, onion	Both	Short-chain fructose oligomers with β-(2→1) linkages (DP 3–5)	Selective beneficial bacterial nourishment, enhanced mineral solubilization and absorption, metabolic pathway modulation, and immune system calibration	*Bifidobacterium* spp., *Lactobacillus* spp.	Lipid profile improvement (cholesterol and triglyceride reduction), immunocompetence augmentation, mineral bioavailability enhancement (calcium, magnesium), weight management, and comprehensive metabolic regulation	Fernandes (2025)
GOS	Human and bovine milk	Both	Galactose oligomeric chains (DP 2–8) with terminal glucose unit (β-GOS configuration)	Targeted *Bifidobacterium* and *Lactobacillus* stimulation, gut–brain axis modulation, cortisol regulation, and neuroendocrine system optimization	*Bifidobacterium* spp., *Lactobacillus* spp.	Immune response optimization, psychobiotic effects (anxiety and depression reduction), infant gut microbiome maturation, and neuropsychological well-being enhancement	Johnstone and Cohen Kadosh (2025)

DP, Degree of polymerization; *FOS*, fructo-oligosaccharides; *GI*, gastrointestinal; *GlcNAc*, N-acetylglucosamine; *GOS*, galacto-oligosaccharides; *HMOs*, human milk oligosaccharides; *RS*, resistant starch; *SCFAs*, short-chain fatty acids; *XOS*, xylo-oligosaccharides.

particularly SCFAs such as acetate, propionate, and butyrate. SCFAs not only contribute a primary energy substrate to intestinal epithelial cells, such as colonocytes, but also exert widespread local and systemic actions through mechanisms involving the activation of G-protein–coupled receptors (GPCRs), inhibition of histone deacetylases, and gene regulation. SCFAs potentially have an effect on cognitive function and mood through immune, hormonal, neural, and endocrine mechanisms. They are able to affect brain activity by activation of neurotransmitter release, modulation of the HPA axis, and systemic immune responses. Evidence is also mounting for SCFAs being involved in the modulation of the neuroimmune system, particularly during states of inflammation (Koh et al., 2016; Dalile et al., 2019; Silva et al., 2020).

Fermentation begins when nondigestible carbohydrates such as inulin, FOS, and GOS are passed through the enzymatic breakdown of the upper gastrointestinal tract and reach the colon, where they are fermented by symbiotic anaerobic bacteria (Slavin, 2013). Bacterial populations such as *Bifidobacterium*, *Lactobacillus*, *Faecalibacterium*, and *Roseburia* are particularly efficient at fermenting these and producing SCFAs (Louis and Flint, 2017).

Among the SCFAs, butyrate is of particular relevance in that it is the colonocyte's preferred fuel and assists with maintaining intestinal barrier integrity by enhancing the expression of tight junction proteins and stimulating mucus secretion (Peng et al., 2009; Kelly et al., 2015). Butyrate is also responsible for epigenetic regulation through HDAC enzyme inhibition. Butyrate inhibition of HDAC enzymes leads to increased histone acetylation, which relaxes the chromatin and increases the accessibility of transcription factors to DNA. The outcome is the alteration of gene expression patterns with the activation of anti-inflammatory genes and immune response modulators in the intestinal epithelial cells and immune cells. Through this epigenetic mechanism, butyrate not only maintains the integrity of the mucosal layer but also suppresses inappropriate inflammatory responses that are responsible for inflammatory bowel disease as well as other systemic disorders (Davie, 2003).

Propionate, which is one of the most important SCFAs, is formed by some bacterial genera such as *Bacteroides* and *Veillonella* mainly through the fermentation of undigested carbohydrates. After being formed in the colon, this metabolite enters the portal circulation and is transported to the liver. Within this organ, propionate is tasked with regulating important metabolic processes, primarily by inhibiting the use of lactate and alanine in gluconeogenesis, thereby assisting blood glucose control. Moreover, there is evidence that propionate is capable of reducing hepatic cholesterol biosynthesis, a process mediated by the inhibition of β-Hydroxy β-methylglutaryl-CoA (HMG-CoA) reductase activity, the rate-controlling enzyme in cholesterol biosynthesis. Additionally, propionate acts as a signaling molecule by stimulating G protein-coupled receptors GPR41 and GPR43 in immune and epithelial cells, thereby modulating antiinflammatory pathways and systemic metabolic regulation. Propionate hence occupies a pivotal position in host metabolism, gut microbiota, and dietary interaction, placing it as a determinant of energy homeostasis and lipid metabolism (den Besten et al., 2013). Recent research has shown that this fatty acid significantly modulates immune responses by regulating dendritic cell maturation and enhancing the differentiation of regulatory T cells (Tregs) (Arpaia et al., 2013).

The most prevalent SCFA produced by intestinal microbiota is acetate, which is produced by a wide range of gastrointestinal bacteria. The metabolite is rapidly absorbed and transported to the systemic circulation. It has been proven in studies to pass through the blood–brain barrier and directly act on the appetite centers of the hypothalamus. Acetate triggers some of the signaling pathways in neurons in the region, enhancing hunger-suppressing signals and reducing food consumption. Furthermore, acetate plays a crucial role in regulating the production of neurotransmitters and hormones that are part of hunger and satiety. Such a mechanism is implicated in the complex relationship between the gut microbiota and the brain, which is crucial for maintaining energy balance and metabolic homeostasis (Frost et al., 2014).

The SCFA profile produced through fermentation is different depending on the type of prebiotic consumed, gut microbial content, and factors related to the host, such as diet and disease status. Increased production of SCFAs by gut microbiota occurs when the diet is composed of high-fermentable fiber. These SCFAs, especially acetate, propionate, and butyrate, play significant roles in generating an antiinflammatory intestinal environment that is responsible for maintaining gut mucosal health and neuronal protection. Such an environment strengthens the essential gut–brain axis by reducing chronic systemic inflammation and regulating immune responses. Furthermore, SCFAs also regulate different metabolic processes by stimulating specific receptors on host cells and may influence body weight regulation, enhance insulin sensitivity, and maximize glucose metabolism. These functions are crucial for preventing and treating metabolic diseases, including obesity and type 2 diabetes. SCFAs have even been shown to cross the blood–brain barrier and can directly influence brain function by modulating hormones and neurotransmitters that regulate appetite and energy metabolism (Macfarlane and Macfarlane, 2011).

There is now emerging evidence that SCFAs are crucial in controlling the activity of microglia, the immune cells of the central nervous system. Through their effects on the maturation, function, and inflammatory responses of microglia, SCFAs assist in the maintenance of brain homeostasis. SCFAs are also able to modulate the HPA axis, thus regulating the host's stress response, which is a critical factor in controlling physiological responses to environmental stressors and maintaining mental health. Furthermore, SCFAs, particularly butyrate, enhance the expression of brain-derived neurotrophic factor (BDNF) in key regions of the brain, such as the hippocampus, which is implicated in neural plasticity and enhancement of cognitive and emotional functions. Such a constellation of molecular and systemic effects renders SCFAs fascinating molecules with therapeutic potential in the management of symptoms of anxiety, depression, and other inflammation- and stress-related psychiatric illnesses (Erny et al., 2015; van de Wouw et al., 2018; Stilling et al., 2016).

13.3.1.2 Immune modulation

Among the most significant functions exerted by prebiotics in favoring host health is their ability to modulate and control the immune system. These effects are largely produced through alterations in the composition and functionality of the gut microbiota, with subsequent influences on both innate and adaptive immunity (Rooks and Garrett, 2016; Kundu et al., 2017). The prebiotic fermentation and subsequent production of metabolites such as SCFAs are critical in this context because these metabolites have the capacity to initiate cellular signaling pathways that lead to inflammation and immunity (Trompette et al., 2014).

Specifically, gut microbiota-produced metabolites, particularly SCFAs such as butyrate and propionate, are tasked with taking the lead role in the regulation of mucosal immunity. SCFAs play a role in interacting with G protein-coupled receptor family surface receptors, such as GPR43 (found on immune cells such as neutrophils and macrophages) and GPR41, thereby eliciting intracellular signaling pathways that activate antiinflammatory processes. Additionally, the metabolites suppress HDACs, the epigenetic controllers of gene expression, elevating the biosynthesis of the antiinflammatory cytokine IL-10. However, butyrate has been found to be a key controller in the colon for immune homeostasis and the regulation of chronic inflammation (Maslowski and Mackay, 2011; Smith et al., 2013). This leads to an increase in the number of Tregs, which are crucial for immune balance and the prevention of autoimmunity (Arpaia et al., 2013).

Prebiotics can help reduce the translocation of pro-inflammatory molecules and pathogens by enhancing intestinal barrier function, which contributes to lowering systemic inflammation (Cani and Everard, 2016). The effect is noteworthy in chronic inflammatory diseases such as inflammatory bowel disease (IBD) and metabolic syndrome, where prebiotics have been reported to be adjunct therapies that reduce the inflammatory burden (Macia et al., 2012). Multiple studies indicate that prebiotic consumption is associated with reduced concentrations of inflammatory markers, such as TNF-α and IL-6, which can improve immune function in both patients and healthy individuals (Bindels et al., 2015).

The precise mechanisms of such immune regulation are not fully understood yet; however, emerging evidence indicates a complex interplay between microbiota, fermentation metabolites, and host immune cells, in which prebiotics play a key role (Belkaid and Hand, 2014).

13.3.1.3 Gut barrier function improvement

The intestinal barrier is the main defense against the invasion of toxic environmental compounds, microorganisms, and food antigens, and its integrity is essential for the host's overall health. Prebiotics play an important role in strengthening the structure and function of this barrier by beneficially modulating the activity and composition of the gut microbiota and inducing the production of fermentation metabolites such as SCFAs (Vinelli et al., 2022).

Prebiotics significantly improve intestinal barrier function by increasing the expression of tight junction proteins such as claudins, occludin, and zonulin. This elevation makes the barrier more robust by reducing permeability across the intestine and excluding undesirable substances injurious to the body from entering the circulation (Suzuki, 2013; Mankertz and Schulzke, 2007). Butyrate, the most important prebiotic fermentation end product, governs the expression of such proteins in a positive manner by regulating the cell's signaling pathways. It also serves as a principal energy source for epithelial cells, to assist in lining and repairing the intestinal lining. In addition, studies have shown that the consumption of prebiotics such as FOS leads to the upregulation of MUC2 mucin by goblet cells. The mucin forms a tight layer on the surface of the intestines that prevents direct contact between microorganisms and toxic compounds with the epithelial cells. This protective function is crucial to prevent mucosal inflammation and maintain intestinal immune homeostasis (Paone et al., 2022; Tonetti et al., 2024).

Multiple studies have revealed that the consumption of prebiotics leads to the relief of intestinal barrier dysfunction symptoms, including IBD (Looijer-van Langen and Dieleman, 2009). Furthermore, improvements in intestinal barrier function act against systemic inflammatory responses and improve glucose and lipid metabolism, which are key in the prevention and management of metabolic diseases (Cani et al., 2007).

13.3.1.4 Systemic effects

Prebiotics, in addition to their localized effect in the gut, exert a significant effect on the regulation of systemic metabolic processes, including lipid and glucose metabolism. These effects are largely via alterations in microbiota composition and microbial metabolite production, such as SCFAs (Canfora et al., 2015).

SCFAs, especially propionate and butyrate, play a vital role in regulating key energy metabolism signaling pathways. They achieve this by activating the metabolic receptors GPR41 and GPR43 in various tissues, including the liver, muscle, and fat. These processes lead to increased insulin sensitivity, decreased lipogenesis, and enhanced fatty acid oxidation, ultimately resulting in improved blood glucose levels and reduced fat storage (den Besten et al., 2013).

Furthermore, prebiotics can improve metabolic health by reducing systemic inflammation, which plays a key role in the development of insulin resistance and metabolic syndrome. Research has shown that prebiotic diets can lower plasma levels of lipopolysaccharides (LPS), which are endotoxins that lead to a condition known as "metabolic endotoxemia."

Metabolic endotoxemia is associated with heightened intestinal permeability and the activation of inflammatory pathways such as TLR4, the prime instigators of obesity and insulin resistance. Prebiotic supplementation increases a more ideal gut microbiota composition by increasing the growth of such beneficial bacteria as Bifidobacteria, thereby decreasing the production of LPS. This reduction in LPS subsequently attenuates systemic inflammatory responses and enhances insulin sensitivity (Cani et al., 2007). Other clinical trials have also revealed that prebiotic intake results in reduced body weight, improved lipid profiles, and reduced insulin resistance in patients with type 2 diabetes and metabolic syndrome, both in children and adults (Dewulf et al., 2013; Zhang et al., 2015). The precise mechanisms also involve alterations in the secretion of gut hormones that modulate appetite, such as peptide YY (PYY) and glucagon-like peptide-1 (GLP-1). These hormones increase satiety signals and also improve glucose regulation (Psichas et al., 2015). By their actions on the central nervous system and pancreatic beta cells, they play a crucial role in the regulation of energy balance and glucose metabolism (Cornu et al., 2009; De Silva and Bloom, 2012).

As prebiotics directly impact the structure and function of the microbial community, their specific addition to the diet is not only viewed as a safe and natural approach but also as an evidence-based therapy in preventing disorders that are associated with microbial loss of diversity, including metabolic diseases. Therefore the use of prebiotics can result in the promotion of general well-being, particularly through the restoration of crucial functions that are associated with microbial metabolites such as SCFAs (Deehan and Walter, 2016).

13.3.2 Clinical effects of prebiotics, probiotics, and their combinations

13.3.2.1 Evidence from randomized controlled trials and metaanalyzes

There has been a substantial rise in high-quality studies, including randomized controlled trials (RCTs) and metaanalyzes, that investigate the therapeutic effectiveness of prebiotics, probiotics, and synbiotics in promoting human health. These therapies have been researched for a wide range of conditions, particularly gastrointestinal diseases, metabolic disorders, allergic disorders, and psychiatric diseases.

In irritable bowel syndrome (IBS), multiple RCTs and metaanalyzes have found specific probiotic strains to be therapeutically effective in mitigating the symptoms. In a metaanalysis of 45 RCTs with over 5000 IBS patients, Goodoory et al. (2023) found that supplementation with probiotics significantly reduced the clinical severity of IBS symptoms. In a twist, symptoms such as abdominal pain, bloating, and bowel dysfunction improved significantly. In addition, there was a significant improvement in the quality of life of the patients, pointing to the importance of such an effect in the long-term control of the condition. The probiotics employed in the study were generally prepared with more than a single strain of bacteria, and from the study, it was noted that multistrain preparations, rather than single-strain products, were more effective at symptom reduction. Side effects were generally mild, highlighting the excellent safety profile of these supplements. Prebiotics like FOS and GOS have also demonstrated improvement in symptoms of IBS in several trials (Yoo et al., 2024).

In IBD, including ulcerative colitis and Crohn's disease, the use of synbiotics as an adjunct therapeutic approach is increasingly recognized. A metaanalysis conducted by Zhang et al. (2021), which included 17 RCTs, demonstrated that synbiotic supplementation was associated with significant improvements in disease activity indices. These improvements encompassed reduced intestinal inflammation and accelerated healing of the damaged mucosa. Furthermore, synbiotic formulations containing probiotic strains from the genus *Bifidobacterium*, particularly *Bifidobacterium breve*, combined with prebiotics such as inulin and GOS, showed the greatest efficacy in improving clinical indices and gut microbiota function. These findings highlight the beneficial impact and safety of synbiotics as complementary agents in the management of inflammatory bowel diseases (Zhang et al., 2021).

Beyond their well-established benefits for gastrointestinal health, new evidence from randomized controlled clinical trials demonstrates the efficacy of probiotics and synbiotics in eliciting aspects of metabolic syndrome. In one double-blind, placebo-controlled, randomized trial by Basafa-Roodi et al. (2024), schizophrenic patients receiving supplementation with synbiotics experienced significant decreases in fasting glucose, triglycerides, and waist circumference. Additionally, levels of HDL cholesterol, which serve as a protective factor against metabolic syndrome, increased significantly. The findings are consistent with the claims of synbiotics as a safe and effective intervention for enhancing metabolic markers in specific populations, including psychiatric patients (Basafa-Roodi et al., 2024). Dietary prebiotic fibers, especially inulin-type fructans, have been shown to reduce glucose and lipid levels in type 2 diabetic patients (Pourghassem Gargari et al., 2013).

In a 2025 systematic review and metaanalysis by Asad et al. of 54 RCTs involving a total of 3,923 patients with clinical depression or anxiety, probiotic and prebiotic treatment was demonstrated to substantially reduce symptoms of depression and anxiety. The average effect size for depressive symptoms was reported as an SMD of –0.29 (95% CI: –0.39 to –0.19), and for anxiety symptoms, an SMD of –0.26 (95% CI: –0.38 to –0.14), reflecting the mild-to-moderate efficacy of such substances. Subgroup analysis revealed that prebiotics were outperformed by probiotics, while interventions for longer periods than eight weeks were more beneficial. Mechanistically, such benefits are delivered by the gut–brain axis through a range of mechanisms that include modulating the gut microbiota, reducing systemic inflammation, and creating neuroactive metabolites such as SCFAs (Asad et al., 2025). A 2020 systematic review and metaanalysis of 29 randomized trials with 3,423 children by Jiang et al. showed that probiotic supplementation profoundly reduced the risk of atopic dermatitis in children. The effect was particularly potent for instances where intervention was at prenatal or early postnatal ages. Mechanisms suggested to provide this advantage involve modulation of the mucosal immune response, Th1/Th2 balance reconstitution, and induction of immune tolerance through gut microbiota interaction. Excessive variability across studies and incongruent results were also reported according to intervention duration and strains of probiotics utilized (Jiang et al., 2020).

Evidence from well-designed randomized controlled trials and metaanalyzes supports the therapeutic potential of prebiotics, probiotics, and synbiotics across various clinical conditions. However, variability in strain selection, dosage, and duration of therapy presents a significant challenge. Overall, the risk–benefit ratio for these interventions is generally favorable. These findings emphasize the necessity for more rigorous, personalized research to enhance their clinical effectiveness.

13.3.2.2 Synbiotic formulations: rationale and outcomes

Synbiotics, or synergistic combinations of prebiotics and probiotics, are constructed specifically to enhance the viability and activity of the host's beneficial microbes. This is achieved by the combined administration of live microbial strains and selectively fermentable substrates that support their proliferation. The rationale for the use of synbiotics stems from the fact that the colonization and action of probiotics, particularly in dysbiotic or inflammatory intestinal disease, are typically compromised in the absence of specific nutritional supplementation (Swanson et al., 2020).

Clinical responses to synbiotic treatment have been well investigated in many fields of health. In a multicenter, double-blind, randomized, placebo-controlled clinical trial, the efficacy of a nine-strain synbiotic product was compared in IBS patients in a primary care clinic setting. Two hundred eighty-nine IBS patients were randomly assigned to receive either the synbiotic product or placebo for eight weeks. Participants in the intervention group took one capsule a day that contained a balanced mix of nine probiotic strains and prebiotic fibers. Probiotic strains found in the synbiotic were *Lactobacillus acidophilus*, *Lactiplantibacillus plantarum*, *Lacticaseibacillus casei*, *Lacticaseibacillus rhamnosus*, *Bifidobacterium breve*, *Bifidobacterium longum*, *Bifidobacterium bifidum*, *Streptococcus thermophilus*, and

Lactococcus lactis. The prebiotic contained FOS fibers. This synbiotic mixture was selected in particular not only to enhance microbial diversity and probiotic viability within the gut but also to enhance their functional performance by providing an ideal nutritional substrate. Trial results indicated that patients receiving the synbiotic had decreased composite severity of IBS symptoms of pain, bloating, and bowel frequency compared with the placebo group. These improvements continued, at least in part, even after the supplementation was stopped. Health-related quality-of-life scores also significantly improved in the synbiotic group (Sommermeyer et al., 2024). In IBD, synbiotics have been associated with reduced mucosal inflammation and improved remission maintenance. A significant study found that a synbiotic combination of *Bifidobacterium longum* and psyllium significantly reduced colonic inflammation and improved histological scores in patients with ulcerative colitis (Fujimori et al., 2009).

In addition to causing gastrointestinal illness, synbiotics have shown metabolic effects in patients with obesity and metabolic syndrome. The daily intake of synbiotic *B. animalis* subsp. *lactis* MN Gup with the supplementation of GOS and XOS over 12 weeks in obesity patients resulted in changes in mucosal inflammation and metabolic indicators like body fat percentage, LDL-C, and waist circumference, which decreased. Additionally, the levels of appetite-suppressing hormones (PYY, CCK, OXM) increased, the gut-favorable microbiome species such as *Bifidobacterium* and *Romboutsia* were enhanced, and the bile acid chenodeoxycholic acid (CDCA) saw an increase (Niu et al., 2024).

Synbiotics have also been encouraging in the treatment of allergic disease. Oral ingestion of a synbiotic product containing the probiotic strains *Lactobacillus acidophilus*, *Bifidobacterium bifidum*, and *Bifidobacterium lactis* and the prebiotic FOS in a double-blind randomized study involving 64 adult patients with allergic rhinitis caused improvement in symptoms, immune modulation, and quality-of-life improvement (Faridzadeh et al., 2023). Furthermore, a randomized double-blind placebo-controlled clinical trial was conducted among 75 patients undergoing chronic hemodialysis for 12 weeks. The test synbiotic product consisted of a prebiotic fraction, FOS, and probiotic microbes *Lactobacillus acidophilus*, *Bifidobacterium bifidum*, *Bifidobacterium longum*, and *Lactobacillus rhamnosus*. The results showed that the synbiotic group had lower scores in depression, anxiety, and stress compared to the placebo group (Haghighat et al., 2021) (Table 13.2).

13.3.3 Dosage, safety, and tolerability

13.3.3.1 Dose–response relationships

A crucial consideration in determining the therapeutic impact of prebiotics and probiotics is the dose–response relationship. The term is used to refer to the

TABLE 13.2 Summary of clinical trials comparing prebiotics, probiotics and synbiotics in health applications.

Intervention type	Study design	Country	Disease/ Condition	Population	Mean age	Participants (*n*)	Mechanism of action	Clinical outcome	References
Prebiotics									
GOS	RCT	United Kingdom	Travelers' diarrhea	Adolescents and adults	≥16 years	523	Selective microbiota modulation through competitive exclusion, enhanced epithelial barrier integrity via anti-adhesion mechanisms, and augmented mucosal immunocompetence	Significant reduction in diarrhea episodes ≥1 day duration with prophylactic efficacy demonstrated	Hasle et al. (2017)
Oligofructose-enriched inulin	RCT	Belgium	UC	Adults	18–65 years	25	ITF-mediated functional microbiome restructuring with preserved taxonomic diversity, enhanced SCFA biosynthesis, and restored epithelial homeostasis	Clinically significant amelioration of UC symptomatology with improved disease activity indices	Valcheva et al. (2019)

(Continued)

TABLE 13.2 (Continued)

Intervention type	Study design	Country	Disease/ Condition	Population	Mean age	Participants (*n*)	Mechanism of action	Clinical outcome	References
Oligofructose and inulin	RCT	Norway	T2DM	Adults and elderly	41–71 years	25	Modulated fecal microbiota architecture with enhanced saccharolytic fermentation, increased total SCFA synthesis (acetate and propionate predominant), and improved metabolic flux	Optimized gut microbiome composition with enhanced microbial fermentation capacity in T2DM patients	Birkeland et al. (2020)
Roasted *Plantago major* seed	RCT	Iran	UC	Adolescents and adults	15–70 years	61	Polyphenolic compounds (tannins, coumarins, flavonoids) mediate anti-inflammatory cascades, visceral hypersensitivity modulation, and mucosal healing acceleration	Comprehensive symptom amelioration, including reduced rectal bleeding, pain intensity, and GI distress with adjunctive therapeutic efficacy	Baghizadeh et al. (2021)
GOS	Clinic-al trial	United Kingdom	UC	Adolescents and adults	16–65 years	17	Targeted bifidogenic effect with increased *Bifidobacterium* and *Christensenellaceae* abundance, enhanced colonic epithelial integrity	Improved stool consistency parameters, reduced loose stool frequency and urgency with enhanced bowel movement quality	Wilson et al. (2021)

Oat fiber, inulin, FOS, marshmallow extract	RCT	Spain	Constipation	Adults	18–65 years	74	Synergistic microbiota modulation with enhanced SCFA production, improved colonic motility, and optimized fecal water content regulation	Significant functional improvement in chronic constipation with sustained therapeutic benefit after 8-week intervention	Terrén Lora et al. (2024)
Probiotics									
L. rhamnosus GG; *S. boulardii*; *B. clausii*; bacterial mixture; *E. faecium* SF68	RCT	Italy	Acute diarrhea	Children	3–36 months	571	Competitive pathogen exclusion, enhanced epithelial barrier function, localized immune system modulation, and restoration of eubiotic microenvironment	Strain-specific efficacy: *L. rhamnosus* GG and bacterial mixture demonstrated superior clinical outcomes (78.5 and 70 vs 115 hours duration reduction)	Canani et al. (2007)
Lactobacilli and *bifidobacteria*	RCT	United Kingdom	AAD	Elderly	≥65 years	2941	Antibiotic-induced dysbiosis restoration through competitive colonization and metabolic niche occupation	Multistrain formulation demonstrated no significant preventive efficacy against AAD in elderly populations	Allen et al. (2013)

(Continued)

TABLE 13.2 (Continued)

Intervention type	Study design	Country	Disease/ Condition	Population	Mean age	Participants (n)	Mechanism of action	Clinical outcome	References
L. reuteri	Clinical Trial	Australia	Infant colic	Infants	<3 months	167	Gut–brain axis modulation through neurotransmitter regulation and visceral hypersensitivity reduction	No therapeutic benefit observed in community-based breastfed and formula-fed infants with colic	Sung et al. (2014)
L. reuteri	RCT	Mexico	Diarrhea	Children	6–36 months	336	Enhanced mucosal immunity, pathogen competitive exclusion, and systemic immune system optimization	Significant reduction in diarrheal episode frequency and duration with concurrent respiratory infection prevention	Gutierrez-Castrellon et al. (2014)
L. reuteri	RCT	Poland	AAD	Children	<18 years	250	Antibiotic-associated microbiome restoration and competitive pathogen inhibition	No significant difference in AAD incidence or diarrheal episode occurrence compared to placebo	Kołodziej and Szajewska (2017)

L. plantarum	RCT	Poland	AAGS	Children	1–11 years	438	Epithelial barrier strengthening and localized anti-inflammatory response modulation	LP299V demonstrated no superior efficacy compared to placebo for loose stools or abdominal symptom incidence	Olek et al. (2017)
L. rhamnosus and *L. reuteri*	Clinic-al Trial	Ukraine	Infant colic	Infants	4–12 weeks	172	Microbiome normalization, gastroesophageal reflux reduction, and enhanced gastric motility regulation	Significant reduction in crying duration and fussiness with improved feeding tolerance in exclusively breastfed infants	Gerasimov et al. (2018)
Multistrain LAB and bifidobacteria	RCT	Italy	CD with IBS	Adults	≥18 years	109	Targeted microbiota restructuring with enhanced *Bifidobacterium* colonization and immune homeostasis restoration	Six-week intervention demonstrated significant IBS symptom severity reduction in CD patients on GFD	Francavilla et al. (2019)
B. clausii UBBC-07	RCT	India	Acute diarrhea	Children	6 month-s-5 years	119	Bacteriocin production, antipathogenic enzyme secretion, competitive epithelial adhesion, and mucosal immune enhancement	Significant reduction in diarrheal duration, frequency, and improved stool consistency with enhanced recovery	Sudha et al. (2019)

(*Continued*)

TABLE 13.2 (Continued)

Intervention type	Study design	Country	Disease/ Condition	Population	Mean age	Participants (*n*)	Mechanism of action	Clinical outcome	References
Lactobacillus and *Bifidobacterium* spp.	RCT	Jordan	UC	Adults	18–65 years	30	Immunomodulatory effects with improved hemoglobin, hematocrit, reduced CRP, enhanced IgA, and IL-10 upregulation	Significant UC remission induction with favorable hematological and inflammatory marker profiles	Agraib et al. (2022)
Synbiotics									
L. rhamnosus, *B. lactis* Bb12, inulin	RCT	Iran	Cardiom-etabolic risk factors	Children	6–18 years	70	Synergistic lipid metabolism optimization through enhanced SCFA production and improved insulin sensitivity pathways	Significant triglyceride and LDL-cholesterol reduction with comprehensive cardiometabolic risk factor improvement	Safavi et al. (2013)
L. rhamnosus, inulin	RCT	Canada	Obesity	Adults	18–55 years	153	Leptin regulation modulation with enhanced *Lachnospiraceae* abundance and improved metabolic homeostasis	Clinically significant weight reduction with sustained metabolic benefits and improved body composition	Sanchez et al. (2014)

BMDO, *B. animalis* ssp. *lactis*	RCT	Germany	Diarrhea and febrile infections	Infants	0–14 years	413	Enhanced mineral bioavailability, selective beneficial bacterial proliferation, pH homeostasis, elevated fecal sIgA, and immune system maturation	Promoted optimal infant growth trajectory with improved gut health biomarkers and enhanced tolerance profile	Radke et al. (2017)
L. plantarum, FOS	RCT	India	Sepsis	Infants	3 days	4556	Targeted *L. plantarum* colonization enhancement with competitive pathogen exclusion and immune system priming	Remarkable reduction in neonatal sepsis-related mortality with significant clinical impact	Panigrahi et al. (2017)
Multistrain probiotics, FOS	RCT	Türkiye	UC	Adults	≥18 years	40	Anti-inflammatory cascade activation with reduced CRP levels, decreased ESR, and comprehensive clinical and endoscopic improvement	Significant UC clinical activity amelioration with measurable inflammatory marker reduction	Kamarlı Altun et al. (2019)

(*Continued*)

TABLE 13.2 (Continued)

Intervention type	Study design	Country	Disease/ Condition	Population	Mean age	Participants (*n*)	Mechanism of action	Clinical outcome	References
Multistrain probiotics, FOS	RCT	Iran	UC	Adolescents and adults	18–60 years	60	Comprehensive microbiota modulation, mucosal barrier fortification, immune homeostasis restoration, reduced intestinal permeability, and maintained gut homeostasis	Substantial UC symptom reduction with restored intestinal barrier function and improved disease management	Amiriani et al. (2020)
L. plantarum, *L. acidophilus*, *L. reuteri* PBS072, inulin, FOS	RCT	Italy	MetS and systemic inflam-mation	Elderly	65–80 years	60	Enhanced lactobacilli viability with systemic anti-inflammatory effects, improved insulin sensitivity, and cardiovascular risk reduction	Significant MetS prevalence reduction with improved cardiovascular risk profile and insulin resistance markers	Cicero et al. (2021)

AAD, Antibiotic-associated diarrhea; *AAGS*, antibiotic-associated gastrointestinal symptoms; *BMDO*, bovine milk-derived oligosaccharides; *CD*, celiac disease; *CRP*, C-reactive protein; *ESR*, erythrocyte sedimentation rate; *FOS*, fructo-oligosaccharides; *GFD*, gluten-free diet; *GI*, gastrointestinal; *GOS*, galacto-oligosaccharides; *IBS*, irritable bowel syndrome; *IgA*, immunoglobulin A; *IL-10*, interleukin-10; *ITF*, inulin-type fructans; *LAB*, lactic acid bacteria; *LDL*, low-density lipoprotein; *MetS*, metabolic syndrome; *RCT*, randomized controlled trial; *SCFA*, short-chain fatty acids; *sIgA*, secretory immunoglobulin A; *T2DM*, type 2 diabetes mellitus; *UC*, ulcerative colitis.

alteration in therapeutic or biological response according to the quantity of supplements taken. It has been firmly established through numerous studies that responses can vary exceedingly between doses; therefore the determination of the appropriate dose is even more critical for each various preparation and therapeutic goal (Ouwehand, 2017).

For example, in a randomized double-blind placebo-controlled prospective clinical trial conducted by Tandon et al. (2019), the dose–response effects of FOS on human gut microflora were investigated. The study was directed toward assessing the effect of various doses of FOS on gut microbiota composition. Subjects were assigned to four groups: placebo, low dose (2.5 g/day), medium dose (5 g/day), and high dose (10 g/day), and received the intervention. Results indicated that there was a significant increase in the numbers of populations of beneficial bacteria such as *Bifidobacterium* spp. and *Lactobacillus* spp. in response to FOS intake, which had a clear dose-related trend. Increasing doses led to greater efficacy and more pronounced changes in gut microbial populations. Specifically, the dose of 10 g/day produced the highest amount of increase in beneficial bacteria that did not lead to appreciable gastrointestinal side effects. Such findings emphasize the importance of determining the ideal dose of prebiotics to achieve maximum therapeutic benefits and gut health (Tandon et al., 2019).

There are various clinical trials that indicate that the typical dosing interval of probiotics is usually between 10^8 and 10^{11} colony-forming units (CFU) per day. The dose–response relationship of probiotics is nonlinear and multidimensional, meaning that increasing the dose does not always lead to a proportionate increase in clinical benefits. In fact, doses beyond a certain level may not yield additional advantages for gut health. The effectiveness of probiotics can vary based on several factors, including the type of strain, specific clinical conditions, and individual patient characteristics.

Generally, consuming probiotics in typical quantities is safe and well-tolerated. However, in some instances, higher doses may cause mild gastrointestinal side effects such as bloating, gas, or discomfort. Thus rather than simply increasing the dosage, it is more effective to identify an optimal dose that offers the greatest therapeutic benefit while ensuring safety. This approach considers the specific probiotic strain, intended therapeutic outcome, and the patient's profile, thereby maximizing clinical benefits while minimizing the risk of side effects (Sanders et al., 2019).

The response to the dose and form of microbiome-based treatment is influenced by numerous factors specific to each individual. Basal gut microbiota composition, preexisting clinical conditions, age, diet, and drugs all contribute to an individual's sensitivity and response to various doses of probiotics or other microbiome therapies. In patients with type 2 diabetes and metabolic syndrome, such reactions are highly variable, and this highlights the need to accept personalized approaches when selecting the dose and mode of treatment. These personalized approaches, taking into account the unique features of each individual, can have the potential to enhance treatment efficacy and yield superior clinical outcomes (Shapiro et al., 2017).

Along with the determination of dose, the administration interval and duration of treatment are also important factors in determining the dose–response relationship. It has been substantiated that regular and sustained ingestion of prebiotics and probiotics taken repeatedly and continuously is necessary to stabilize beneficial changes in the gut microbiota and to achieve long-term clinical impacts (Smolinska et al., 2025). Moreover, discontinuation of treatment may lead to a rapid reversion of the microbial composition to its original state. Therefore therapeutic protocols should be customized based on the specific disease and individual responses to achieve optimal outcomes (Chandrasekaran et al., 2024).

13.3.3.2 Side effects and their mitigation

Probiotic therapy is generally regarded as safe, although some individuals have reported mild gastrointestinal side effects (Doron and Snydman, 2015). Similarly, prebiotics are also considered safe overall, but they may cause mild, temporary gastrointestinal issues such as gas and bloating, particularly when taken in larger doses or by more sensitive individuals (Davani-Davari et al., 2019). These effects may be caused by microbial fermentation activity and gas production within the gut (Slavin, 2013).

It can be noted that the severity of such side effects depends on the quantity and the nature of the probiotic or prebiotic used. For instance, excessive levels of some prebiotics, such as FOS, can lead to overstimulation of fermentation, producing bloating and gas, whereas low doses can reduce these adverse effects without compromising the curative effects (Marteau and Seksik, 2004).

One of the most effective ways of minimizing side effects is to begin with small doses and gradually build up. This way, the gut is allowed time to adjust to the changes in microbial populations and heightened fermentation, which will eventually minimize adverse side effects (Hill et al., 2014a). Recommending the intake of probiotics with food can help minimize gastrointestinal symptoms. This is because food in the stomach and intestines reduces the negative effects of gastric acid and bile on probiotics. Additionally, food acts as a buffer, helping to regulate sudden changes in pH and stomach motility (Tompkins et al., 2011).

Research indicates that responses to prebiotics and probiotics vary significantly among patients and depend on their clinical condition. Those with a history of specific gastrointestinal illnesses such as IBS, IBD, or advanced gut microbiota dysbiosis may have discordant and even magnified reactions to such supplements. Genetic predisposition and food intolerance also impact probiotic effectiveness and tolerance. This response heterogeneity highlights the requirement for personalized approaches to prebiotic and probiotic prescription. Careful evaluation of patient status before treatment, including consideration of a meticulous assessment of medical history, intestinal flora, diet, and environment, is imperative. Based on this information, to optimize therapeutic outcomes and reduce potential side effects, the formulation and dosage of the supplement should be tailored to the subject's condition. Moreover, close observation during therapy to determine clinical response and microbiota changes enables effective adjustment of the therapeutic regimen to achieve the best outcome. This tailored and personalized method guarantees that the use of probiotics and prebiotics is not only safe but also effectively promotes sustainable health improvements in patients' gastrointestinal systems (Brüssow, 2019).

Clinical trials have further demonstrated that the consumption of some strains of probiotics can even diminish gastrointestinal as well as inflammatory symptoms substantially. One of them is *Bifidobacterium infantis* 35624, which was specifically researched in RCTs. This strain can potentially influence host inflammatory processes beyond the gastrointestinal tract and can modulate the immune response in diseases involving chronic intestinal inflammation and gastrointestinal disorders. The outcome has been that treatment with *B. infantis* 35624 resulted in a considerable decrease in clinical symptoms such as abdominal pain, bloating, and other gastrointestinal disturbances. In addition, this strain has been associated with exceedingly rare and well-tolerated side effects, which reveal that it is an effective and safe intervention for improving the quality of life in patients with gastrointestinal disorders (Groeger et al., 2013).

13.3.3.3 Individual variability

It has been proposed with new evidence that the effectiveness of colonization by probiotics and their biological activity is not uniform across individuals but is actually strongly influenced by host- and microbiome-associated factors. In a landmark paper by Zmora et al. (2018), it was shown that empirical delivery of probiotics resulted in extremely variable mucosa colonization of the gut among individuals. This resistance to colonization was strongly associated with the indigenous characteristics of the gut microbiota and host immune profile, thus forming the need for a personalized approach to probiotic supplementation (Zmora et al., 2018). Johnson et al. (2019), using longitudinal and daily sampling, demonstrated that regular dietary use defines individual microbiota profiles and that diet-microbiome interaction varies greatly from person to person. Cumulatively, these findings suggest that probiotic efficacy should not be assumed for everyone but rather interpreted in light of each individual's unique microbial ecology and dietary choices, paving the way for microbiome-guided precision therapies (Johnson et al., 2019).

Age has also been identified as a major force behind the host response to prebiotics and probiotics. Alterations in microbial content and a reduction in microbiota diversity with increasing age could influence the potency of these supplements. Older persons, for instance, possess reduced levels of protective bacteria such as *Bifidobacterium*; therefore varying dosing or supplements might be necessary to provide the therapeutic effects (Badal et al., 2020; O'Toole and Jeffery, 2015).

Diet plays a key role in defining an individual's microbiota and, consequently, their sensitivity to prebiotics and probiotics. Dietary fiber, polyphenols, and other nutrient intake may create conditions that are favorable for the development and activity of beneficial microorganisms and augment supplementation activity. Recent studies have demonstrated that combining dietary consumption with bioactive supplementation, particularly with fiber-based diets, is related to increased production of SCFAs and improved immune functions (Holscher, 2017; Wan et al., 2021).

Furthermore, the health status of a person and the concomitant use of medicines could also influence the extent and quality of the reaction to these agents. For example, the use of antibiotics could drastically alter the flora of bacteria, making it necessary to change the dosage and type of probiotic given (Becattini et al., 2016).

Because of the vast and complex nature of individual variability, current research is increasingly directed toward the development of personalized approaches that dictate the type and level of supplements based on each person's microflora, age, and lifestyle. These precision strategies can be anticipated to enhance therapeutic outcomes while minimizing side effects (Mejía-Caballero et al., 2021).

13.3.3.4 Regulatory safety profiles

Regulatory safety assessment of probiotics and prebiotics constitutes a vital component in the development and public acceptance of these products, particularly given their growing introduction as health-promoting agents. In the United States, the Food and Drug Administration (FDA) assigns the "Generally Recognized as Safe" (GRAS) status to specific probiotic strains and prebiotic compounds following a comprehensive evaluation of published scientific data and expert consensus. Several widely used probiotic strains, including *Lactobacillus acidophilus* and *Bifidobacterium longum*, along with prebiotics such as GOS, have received this status, confirming their safety for consumption at defined dosages (FDA USFaDA, 2025; Kumar et al., 2015; GRAS Notice No. 671: Galacto-oligosaccharides GOS, 2016).

The Presumption of Safety (QPS) system is used by the European Food Safety Authority (EFSA) within the European Union and is similar to GRAS but only for the safety evaluation of microorganisms. QPS status is conferred on a strain-by-strain basis and relies on the ability to be used safely without any pathogenicity attributes, antibiotic resistance determinants, or genomic instability. EFSA has brought a number of synbiotic ingredients, such as

Lactobacillus rhamnosus and *Bifidobacterium animalis* subsp. *lactis* strains, under the QPS status (Hazards EPanel oB et al., 2023).

At the international level, the Food and Agriculture Organization (FAO) and the World Health Organization (WHO) of the United Nations have issued guidelines requesting an overall safety assessment. These include precise strain identification, in vitro and in vivo toxicity, and examination for potential translocation or systemic effects (FAO/WHO, 2002). The expert consensus also emphasizes the need for postmarket surveillance and product stability testing for the continued safety and efficacy of probiotic products, especially for susceptible groups such as infants, the elderly, and immunocompromised individuals (Hill et al., 2014b).

However, the absence of a single international regulatory framework persists because there are significant divergences in the categorization of synbiotics, specifically, whether they should be categorized as dietary supplements, functional food, or medicine. This type of categorization determines the level of scrutiny and the quality of evidence needed for approval. Thus there is a strong need for a globally coordinated strategy to enable consistent safety assessment and regulation across markets.

13.4 Technological challenges in industrial use

13.4.1 Formulation and stability issues

13.4.1.1 Sensitivity of prebiotics to heat, pH, and moisture

Prebiotics' technological application in functional foods and pharmaceuticals is limited by their physicochemical susceptibility to ambient conditions such as heat, pH, and humidity. The abovementioned parameters can significantly affect the structural stability, fermentability, and overall activity of prebiotic materials during production, storage, and transit through the gastrointestinal tract. For instance, GOS are very stable at mild temperatures; however, prolonged exposure to high temperatures, for example, those encountered in baking or extrusion processing, could lead to depolymerization and loss of their functional attributes, that is, their ability to selectively stimulate microbial growth (Mei et al., 2022).

pH stability is equally a key issue to be considered, especially in acidic pH products such as fruit drinks or gastric juice. Although inulin-type fructans are resistant to acid to some extent, they undergo hydrolysis under very low pH with a loss of molecular weight and decreased prebiotic activity (Kolida et al., 2002). This is one of the main issues in colon-targeted drug delivery system development, so that premature degradation will prevent the delivery of the compound to the colon in a nontampered state. Moisture sensitivity is also one major issue in encapsulation and storage operations. Hygroscopic prebiotics such as resistant starch and polydextrose can absorb atmospheric water and cause caking, microbial contamination, and reduced powder flowability, leading to decreased product uniformity and shelf life (Patel and Goyal, 2011; Saad et al., 2013).

To mitigate these threats, different technological innovations have been proposed, such as encapsulation, controlled atmosphere spray drying, and packaging with desiccants (Figueroa-González et al., 2011). Whatever technological innovation is employed, however, a basic understanding of the thermal degradation kinetics and acid–base hydrolysis of the individual prebiotic molecules is essential to optimize formulation parameters and preserve functionality across different industrial platforms.

13.4.1.2 Microencapsulation technologies and delivery systems

Microencapsulation has emerged as an important method to improve the delivery and stability of probiotics, particularly in adverse gastrointestinal conditions and industrial processing. Microencapsulation facilitates controlled release by the administration of probiotic drugs, masking bitter tastes, and shielding sensitive structures from environmental aggression in the modes of heat, pH fluctuation, and enzymatic activity (Anal and Singh, 2007; Sbehat et al., 2022).

Among various techniques, spray drying is widely employed as the process lends itself to easy scaling up, is cost-effective, and is compatible with heat-stable excipients such as maltodextrin, gum arabic, and whey protein isolates. Using this technique, microparticles can be created that are stable and preserve the prebiotic physicochemical stability of inulin under heat treatment and storage (Premi and Sharma, 2017; Bhagwat et al., 2020; Misra et al., 2023).

On the other hand, extrusion and coacervation processes are finding growing relevance in synbiotic product development, whereby prebiotics are encapsulated along with probiotics to ensure controlled and synergistic delivery into the colon. Such systems are made using biopolymer coatings such as alginate, chitosan, or hydroxypropyl methylcellulose (HPMC) that exhibit pH-dependent release properties, thereby ensuring prebiotics are released predominantly in the distal gut where fermentation by the commensal microbiota is primarily located (Guo et al., 2024; Vijayaram et al., 2024).

One of the primary factors in encapsulation design is the fulfillment of the coating-to-core ratio, matrix porosity, and optimization of release kinetics, parameters that collectively impact the dynamics of colonization of the gut microbiota. Models of the gastrointestinal tract have demonstrated that encapsulated prebiotics are more fermentable, along with the capability to stimulate bifidogenic and butyrogenic microbial communities, selectively better than their free-form counterparts (Calderón-Oliver and Ponce-Alquicira, 2022).

13.4.1.3 Coformulation with probiotics

Prebiotic–probiotic coformulation, referred to as synbiotics, offers broad synergy for gut health by virtue of co-expression of selective access to biological substrates and exposure to live, active beneficial microorganisms. Such a combination does prove to be a monumental challenge in the area of formulation, where the stability and viability of probiotic cultures need to be maintained during manufacturing and storage processes, without compromising the structural integrity and fermentability of prebiotic ingredients.

One of the major challenges in developing synbiotics is the difference in the physicochemical requirements of probiotics and prebiotics. For instance, probiotic bacteria such as *Lactobacillus acidophilus* and *Bifidobacterium longum* require low water activity and an oxygen barrier to remain viable. On the other hand, the majority of prebiotics, such as FOS, are hygroscopic in nature and tend to absorb water from the environment. This absorption of water can alter the microenvironment and, in the process, increase the metabolic rate and oxygen sensitivity of the probiotic strains, ultimately compromising their stability (Tripathi and Giri, 2014; Burgain et al., 2011).

Studies also show that prefreezing conditions and cryoprotectant application play a crucial role in maintaining the viability of this probiotic. Freezing may cause injury to the cell membrane through the formation of ice crystals intra- and extracellularly, whereas freeze-drying can cause overdehydration and damage to vital cellular structures. The extent of cellular damage during these processes is highly pretreatment dependent, that is, salt concentration, environmental pH, and the presence or absence of cryoprotective agents. Some cryoprotectants, including maltodextrin, sorbitol, and milk proteins, have been found to experimentally protect the cell membrane quite effectively and significantly enhance the viability of *Lactobacillus rhamnosus* GG (LGG). These findings emphasize that the functional stability of probiotics in processed products is achieved by an optimal degree of prefreezing conditions, freezing temperature, and the type of cryoprotective compound. These data can be very useful in the formulation of industrial synbiotic products, particularly those subjected to thermal processing or intended for long-term storage (Wang et al., 2025). Environmental factors such as pH, water activity (aw), oxygen availability, salt concentration, and food matrix composition can significantly affect the viability and functional activity of lactic acid bacteria (LAB) in foods. pH reduction and the presence of organic acids can lead to damage in bacterial cell membranes, resulting in a loss of the viable cell population. Low water activity, particularly in dry or semidry foods, generates osmotic stress that is lethal to cells. Temperature fluctuations during processing and storage, that is, freezing or drying, can also cause structural and functional cell damage to LAB. Natural antimicrobial compounds as well as certain food additives can also suppress the growth of LAB. Thus selecting the appropriate food matrix, employing protection technologies like microencapsulation, and closely controlling physicochemical parameters during food formulation are essential for maintaining the viability and activity of probiotic strains (Sionek et al., 2024).

13.4.2 Interactions with the food/medicine matrix

13.4.2.1 Binding with minerals, fats, or proteins

Prebiotics, particularly oligosaccharides and indigestible fibers such as inulin, FOS, and GOS, are capable of engaging in nonselective interactions with dietary and endogenous molecules, including minerals, lipids, and proteins, which can influence their bioavailability, structural integrity, and physiological efficacy (Roberfroid et al., 2010; Guo et al., 2018).

One of the main objections to the application of prebiotics is their ability to complex with critical minerals such as calcium, magnesium, iron, and zinc. Certain prebiotics, particularly those that are of high molecular weight and have higher levels of polymerization, as established in laboratory, animal, and human trials, have a tendency to bind with divalent cations and form complexes that hinder bioavailability. These interactions can inhibit the absorption of minerals in the upper small intestine, particularly when there is a high intake of fiber or when the diet includes antinutritional factors like phytates. In contrast, the impact of prebiotics on mineral absorption is also very much associated with the prebiotic species used, the dosage consumed, and the primary fermentation site within the gut. For instance, certain oligosaccharides such as inulin and FOS can activate calcium and magnesium absorption from the colon by lowering colonic pH, increasing the generation of SCFAs, and facilitating the solubilization of mineral salts. These findings suggest that the initial negative interactions of prebiotics with ions may be antagonized later in the gut by compensatory absorption mechanisms. Yet the formation of poorly soluble iron and zinc complexes in the upper intestine remains recognized as creating a barrier to mineral bioavailability in prebiotic-supplemented diets (Scholz-Ahrens et al., 2007; Rossander et al., 1992).

Studies have shown that certain soluble fibers and prebiotics can suppress the reabsorption of bile acids in the small intestine, thereby leading to reduced absorption of dietary fat and cholesterol and enhanced fecal loss. This effect is one of the ways through which lipid metabolism is positively regulated (Slavin, 2013).

Protein–protein interactions are more complex. One study investigated the interaction of whey protein and inulin as a prebiotic in a model system. The results revealed that noncovalent interactions, primarily hydrogen bonding and van der Waals forces, existed between whey protein and inulin, leading to secondary structure changes of the protein and increased system stability. FTIR scans showed a significant reduction in the bands of the protein's amide groups, which indicated structural change through binding to inulin. Particle size in the mixed solutions was also lower than in the pure protein solutions, which assisted in the enhancement of colloidal stability. The surface charge measurements also confirmed that inulin is capable of changing the surface charge of the protein and thus preventing particle aggregation. For viscosity, the addition of inulin enhanced solution viscosity, something that can be a significant contributor to improvements in sensory and functional food product quality. Furthermore, these interactions also improved whey protein's emulsifying and foamability, traits that are highly sought after in industrial applications. At the molecular level, the interactions between whey protein and inulin are governed by physical forces such as hydrogen bonding and van der Waals forces that collectively impart some positive effects on the structural stability and functional attributes of food systems (Guo et al., 2018).

13.4.2.2 Impact on taste, texture, and viscosity

Their application in food products has the potential to alter organoleptic properties such as flavor, texture, and viscosity and indirectly influence consumer acceptance. Inulin and FOS are two common prebiotics that add sweetness and can be substituted for fats in dairy products. By doing so, they enhance creaminess without providing caloric value (Franck, 2002; Meyer et al., 2011). It has also been observed that the addition of inulin significantly increases the viscosity and consistency of desserts (Tárrega and Costell, 2006).

Flavor interactions are also noteworthy. Although GOS and XOS usually have a neutral taste, at high concentrations and depending on the food matrix, they may impart a mild bitterness or metallic aftertaste (Mussatto and Mancilha, 2007).

13.4.2.3 Gastrointestinal release kinetics

Prebiotic release in the gut, their rate of release, and kinetics are determinative factors dictating the effectiveness of the compounds in influencing the gut microbiota and in achieving optimal therapeutic outcomes. In an experiment, nonfat yogurt samples were made by adding various prebiotics such as inulin, FOS, and GOS. The growth rate and pattern of probiotic bacteria throughout fermentation were regulated by monitoring the pH shift and microbial populations. The results confirmed that different prebiotics can influence fermentation velocity and enhance microbial activity, such that some prebiotics induce an increased rate of fermentation, making it possible to attain optimal pH earlier (Oliveira et al., 2009).

The nature and extent of prebiotic fermentation by the gut microbiota are determined by a variety of factors, including the chemical structure of the prebiotic, the microbial population structure of the community, and environmental parameters like temperature. Short-chain molecules are more rapidly fermented, often leading to the rapid production of SCFAs and byproduct gases that can cause gastrointestinal discomfort. In contrast, long-chain prebiotics like inulin are fermented slowly and continuously to generate sustained SCFAs and enhance microbial stability and balance. Reducing the rate of fermentation and regulating it is inextricably linked to beneficial physiological effects like the maintenance of intestinal mucosal integrity, regulation of immunity, and improvement in energy metabolism (Gibson and Rastall, 2006).

Innovative formulations such as microencapsulation and the use of biopolymeric matrices like alginate and chitosan can provide controlled release kinetics of prebiotics. These strategies ensure the reduction of side effects caused by sudden release or ineffective targeting of the site (Shanuke et al., 2025) (Fig. 13.2).

13.4.3 Cold chain and shelf life considerations

13.4.3.1 Comparison with probiotic cold-chain dependence

One of the major challenges during the production and distribution of probiotics is their extreme sensitivity to the cold chain. In a study on probiotic survival in yogurts with cold chain breaks, the temperature of storage was

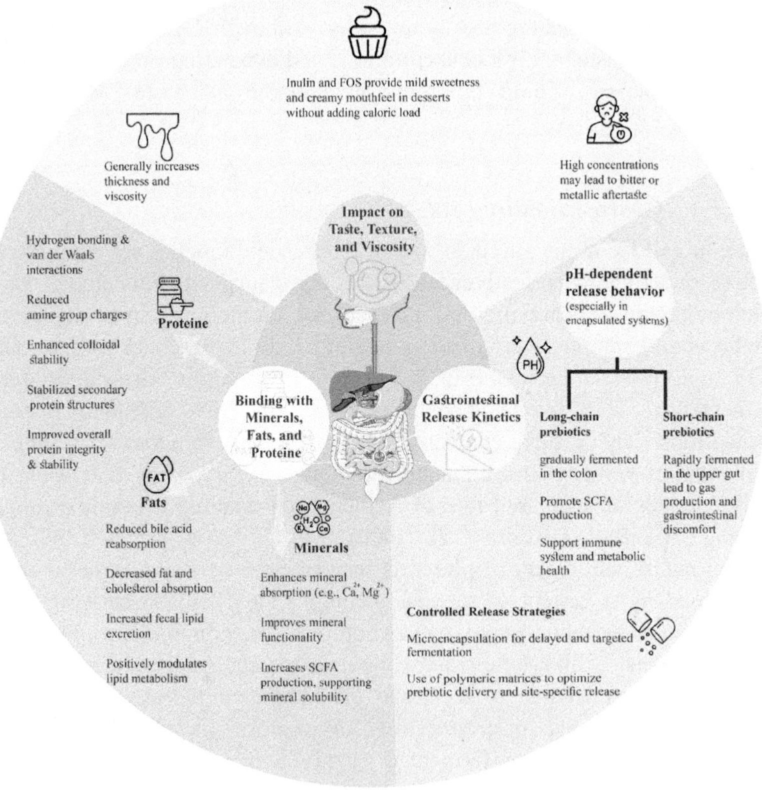

FIGURE 13.2 Technological flowchart of incorporating prebiotics in foods and supplements.

identified as a major determinant of probiotic viability. Storage at room temperature caused a significant reduction in viability compared to refrigerated storage. Different probiotic strains showed differing levels of heat-resistance. The lowest amount was found in *Bifidobacterium lactis* Bb12, while the highest was assigned to *Lactobacillus rhamnosus* HN001. This study emphasized the need for an uninterrupted cold chain to guarantee the integrity and effectiveness of probiotic yogurts (Ferdousi et al., 2013). In another research study, three probiotic yogurts were stored at 4°C or kept at room temperature (RT) for 6 or 24 hours. The cells were cultured and incubated at 37°C for 48 hours. The results showed that the yogurt inoculated with *Lactobacillus johnsonii* had a significant decrease in CFUs after 6 hours at room temperature, which increased further after 24 hours. The CFU for the other yogurt containing *Lactobacillus* GG also reduced after 6 hours and dropped even further after 24 hours at room temperature. In the third yogurt, with *Lactobacillus acidophilus*, only 53.8% of CFUs remained viable after 6 hours at room temperature; after 24 hours, only about one-quarter of the CFUs were present. These findings emphasize the significant effect of temperature on the survival of probiotics (Scharl et al., 2011).

Recent studies have shown that the majority of probiotic foods and nutraceuticals require cold storage, typically refrigeration, to preserve bacterial viability. This reliance on temperature-controlled logistics poses a significant challenge for global manufacturing and distribution. Interruptions in the cold chain can lead to substantial reductions in live bacterial counts, ultimately compromising the product's clinical efficacy and functional performance. Moreover, this temperature sensitivity contributes to increased costs across transportation, warehousing, and distribution, particularly in regions with limited access to refrigeration infrastructure.

To address these challenges, research has increasingly focused on technological innovations that reduce or eliminate the need for continuous cold-chain logistics. Among these, stabilization techniques such as freeze-drying have demonstrated effectiveness in extending shelf life under ambient conditions. Additionally, advancements in heat- and humidity-resistant formulations, along with protective packaging solutions, including multilayer capsules, resilient coating films, and desiccants, have proven successful in maintaining probiotic viability without constant refrigeration (Fenster et al., 2019).

13.4.3.2 Long-term storage and labeling challenges

One of the key concerns in the commercial application of prebiotics is ensuring their physicochemical stability during prolonged storage and under varying environmental conditions. During storage, most cells are unable to multiply. Furthermore, weak incubation culture conditions can also adversely affect their preservation (Visciglia et al., 2022). Prebiotics such as inulin and GOS have the natural ability to neutralize reactive oxygen species (ROS) and bile acids.

These compounds could potentially neutralize ROS directly or prevent damage to probiotics through the inhibition of bile acid uptake (You et al., 2022). Experimental evidence has proven that techniques such as lyophilization, microencapsulation, and controlled drying effectively preserve probiotic viability. Additionally, storing them in milder conditions promotes their stability and longevity (Moumita et al., 2018). Synbiotics research has shown that the incorporation of FOS at the 4% (wt) level had the most significant positive effect on the preservation of probiotic viability. The survival of the probiotic upon freeze-drying after having been stored in the FOS-supplemented formula was approximately 83.4%, which represented about 28% higher than FOS-free capsules (Thinkohkaew et al., 2025).

Regulatory-wise, prebiotic labeling is a significant challenge. Several organizations, such as the ISAPP, have proposed frameworks for defining prebiotics in terms of certain health outcomes and modulation of the microbiota (Hutkins et al., 2016). Regulatory authorities' adoption and use of these frameworks vary significantly by region. For example, in the European Union, the European Food Safety Authority (EFSA) is strict in approving prebiotic-related health claims. In accordance with Article 13 of Regulation No 1924/2006, EFSA evaluates health claims and approves these only upon the submission of significant human evidence (Authority EFS, 2011). EFSA requires well-designed, high-quality controlled human studies. For example, research on inulin derived from chicory indicates that consuming at least 12 grams per day increases bowel movement frequency, establishing a causal relationship with the health effect (EFSA Panel on Dietetic Products N, Allergies, 2011). Additionally, evaluations of dossiers such as those for FOS have shown that, without rigorous human studies, claims are not approved (EFSA Panel on Dietetic Products N, Allergies, 2015). Evidence from animal studies or in vitro experiments alone is insufficient; such data can only be used to explain mechanisms, not to prove efficacy. Overall, no official health claim using the term "prebiotic" has been authorized in Europe. However, some countries, like Italy, permit limited use of the term "prebiotic" under specific conditions, even without full EFSA approval (Tuohy et al., 2024). In the United States, the FDA has not yet established an independent definition for prebiotics, leading to a variety of marketing claims and confusion among consumers. This issue has been thoroughly addressed in the consensus statement issued by the ISAPP (Swanson et al., 2020).

13.5 Consumer perspectives and market dynamics

13.5.1 Consumer understanding and perceptions of prebiotics

Despite growing scientific evidence of prebiotics' role in modulating host–microbiome interactions and their related health benefits, public interest and understanding of the term "prebiotic" remain low. In a 2022 study, Precup et al. examined Romanian consumers' awareness, understanding, and interest in prebiotics using two complementary data sources: Google Trends search data from 2018 to 2021 and a web-based survey of 303 individuals across two age groups (15–24 and 25–64 years). The survey revealed that while 74% of respondents had encountered the term "prebiotic," only about 40% rated their knowledge between 6 and 10 on a 10-point scale, indicating a moderate to high perceived familiarity. In contrast, nearly 22% rated their understanding between 1 and 3, suggesting minimal or unclear comprehension. Crucially, despite widespread recognition of the term, only one participant could accurately define it. Most respondents confused prebiotics with related concepts such as dietary fiber or probiotics, underscoring a significant gap between perceived and actual knowledge. Age, education, and health background were key differentiators in knowledge levels. The 25–39 age group demonstrated the highest understanding, while the youngest cohort (Rastmanesh, 2011; Cardona et al., 2013; Davani-Davari et al., 2019; Vandenplas et al., 2015; Valladares-Diestra et al., 2023; Martínez et al., 2010; Nugent, 2005; Ozdal et al., 2016; Dueñas et al., 2015; Swennen et al., 2006) showed the lowest. Participants with university or postgraduate degrees exhibited significantly greater awareness than those with lower educational attainment. Moreover, individuals with academic backgrounds in health or medical sciences outperformed their peers from other disciplines. Notably, respondents with a history of chronic illness displayed higher levels of knowledge, suggesting that personal health experiences may drive greater engagement with nutritional information and proactive health-seeking behavior (Precup et al., 2022). Another study found that the acceptability of nondairy products containing probiotics and prebiotics largely depends on individuals' awareness of the health benefits associated with these ingredients. People who are health-conscious and more thoughtful about their food choices are more likely to consume such products. The study indicates a positive and significant correlation between health consciousness and the acceptance of these biotics (Cosme et al., 2022).

A cross-sectional online survey conducted among 603 participants from diverse regions of Turkey revealed key demographic patterns, with a predominance of women, university graduates, and urban residents. The study aimed to assess public awareness and consumption behaviors related to probiotics and prebiotics. Results indicated a high level of familiarity with probiotics, with 72.1% of respondents recognizing the term. In contrast, awareness of prebiotics was markedly lower, with only 32.3% demonstrating familiarity. Conceptual confusion was widespread, as many participants misidentified prebiotics as either probiotics or dietary fiber. This knowledge gap was reflected in consumption habits: while 61.5% reported having consumed probiotic products, only 20.4% had knowingly consumed prebiotic products. The most influential sources of information were healthcare professionals, particularly physicians and pharmacists, and media outlets. Based on these findings, the authors emphasized the urgent need to enhance public understanding of prebiotics through targeted nutritional education. They advocated for collaborative efforts between health authorities and the food industry to develop specialized educational initiatives aimed at improving nutritional literacy and correcting prevalent misconceptions (İnce Palamutoğlu et al., 2023).

A study found that packaging serves as a crucial link between producers and consumers. It must not only ensure the safety and shelf life of the product but also foster consumer confidence by providing essential information. This information includes health benefits, storage instructions, active ingredients, and usage recommendations. Graphic design, color, symbols, and functional labels significantly influence consumer understanding and purchasing decisions. However, packaging also has its limitations. These include the lack of clear standards for health-related labeling, potential misinterpretation due to complex scientific terminology, and the misuse of terms like "natural," "organic," or "healthy" without supporting scientific evidence. To improve packaging effectiveness, it is recommended to use simple language, easy-to-read graphic symbols, and official seals (Siddiqui et al., 2023).

13.5.2 Demand trends across regions

Global demand for prebiotics is extremely geographically varied based on factors such as cultural dietary habits, regulatory climatic conditions, consumer awareness, and the innovation of functional foods. The global prebiotics market was valued at approximately USD 6.05 billion in 2021 and is projected to reach around USD 21.2 billion by 2030, reflecting a robust compound annual growth rate (CAGR) of 14.9% over the forecast period from 2022 to 2030 (Research, 2022). Another report indicates that the market for prebiotic fiber was approximately USD 8.97 billion in 2024 and will grow to USD 28.62 billion by 2034, with a CAGR of approximately 12.5% (Research, 2023).

In 2023 the European market represented approximately USD 3.499 billion in revenue and had approximately 41.1% of the overall market share for the year. It is anticipated to grow to approximately USD 8.487 billion by 2030, with a growth rate of approximately 13.5% from 2024 to 2030 (Research GV, 2023a). It was worth approximately USD 1.44 billion in 2023 and is estimated to be approximately USD 3.645 billion in 2030, with a CAGR of approximately 14.2% between 2024 and 2030. Additionally, North America holds the largest market share and is still one of the key markets globally (Research GV, 2023b). The Asian market was valued at approximately USD 2.20 billion in 2024 and is projected to grow at a compound annual growth rate (CAGR) of approximately 12.4% between 2025 and 2030 (Research, 2025).

13.5.3 Common misconceptions about prebiotics

Despite the growing confusion, consumer awareness of prebiotics is limited and is often accompanied by a set of misconceptions. One of the most common is that the concept has been reduced to the equation "fiber = prebiotic." This is partly because many prebiotics are indeed types of dietary fiber but in no way are all fibers prebiotic. Simply being fiber does not necessarily lead to prebiotic effects; instead, proof of selective utilization by host microbiota and a health benefit must be demonstrated. ISAPP has described a prebiotic as "a substrate that is selectively utilized by host microorganisms conferring a health benefit." The definition emphasizes a specific functional requirement that is not true for all dietary fibers (Gibson et al., 2017).

Many fibers are not fermentable, such as cellulose, lignin, and certain insoluble fibers; they add bulk to stool and promote intestinal motility. All prebiotics, on the other hand, are fermentable and are degraded to SCFAs in the colon. Additionally, fibers are fermented by a broad spectrum of bacterial species, resulting in diffuse and generalized effects on the gut microbiota. Prebiotics, such as FOS and GOS, are fermented in a selective manner by some genera like *Bifidobacterium* and *Lactobacillus*, and their selectivity is among their most characteristic functional traits. The second significant distinction is in their metabolic action. Not only do prebiotics produce health-benefiting metabolites (e.g., SCFAs), but they have also been shown to stimulate immunity, increase calcium absorption, and suppress inflammation. Fibers, on the other hand, exert more nonspecific effects at the physiological level, such as relief of constipation, lowering of cholesterol, and helping to manage blood glucose levels and control weight (Holscher, 2017).

Another misconception is the incorrect belief that all nondigestible carbohydrates are inherently prebiotics. This view overlooks the importance of microbial selectivity and the clinical evidence supporting health benefits. A prime example of this is resistant starch: While it is resistant to enzymatic digestion, it does not qualify as a prebiotic unless its effectiveness in promoting health and selective microbial utilization is established (Bindels et al., 2015).

Moreover, most consumers assume that all prebiotics act in the same manner, regardless of their chemical composition or source. Actually, the prebiotic action strongly relies on substrate type. Variations in chain length, degree of branching, glycosidic linkages, and solubility can significantly impact microbial targets and metabolic outcomes. For instance, long-chain inulin can reach more distal regions of the colon, affecting a different microbial community compared to short-chain oligofructose, which is fermented in the proximal region right away (Kolida et al., 2002).

As previously mentioned, there is limited reliable information and a clear definition of prebiotics among the public. Many people often confuse prebiotics with dietary fiber or use the terms interchangeably. This misunderstanding is further complicated by regional differences in regulations. For instance, the EFSA has been more stringent regarding health claims associated with prebiotics and requires credible, specific human evidence to support these claims (Authority EFS, 2011).

13.5.4 Labeling, health claims, and transparency

The regulatory environment for prebiotic labeling and health claims varies significantly across different regions, resulting in product inconsistency and consumer distrust. In the United States, the FDA has not yet established a legal definition for prebiotics. This lack of regulation has led to various claims on product labels, further confusing consumers (Deehan et al., 2024).

The European Union follows a stringent regulatory framework through the EFSA, which assesses health claims based on established guidelines. However, EFSA has not authorized any specific health claims related to prebiotics. As a result, most manufacturers in the EU refrain from making direct health claims and instead use structure/function statements, such as indicating that their products support or maintain normal digestive function (Tuohy et al., 2024).

The terminology used for labels can be confusing. Many foods contain ingredients like dietary fiber, yet they often overlook their prebiotic activity. However, there are significant differences in the fermentability and microbial selectivity of these compounds (Davani-Davari et al., 2019). ISAPP and other scientific groups have thus recommended that labeling should be functional, that is, that the ingredient has been demonstrated to have prebiotic effects under human clinical conditions (Gibson et al., 2017). The other concern is the lack of measurable data on product labeling. Most consumer products do not provide information about the actual amount of prebiotic fibers they contain, even though their effectiveness usually depends on the dose. Without this information, consumers cannot determine whether they are consuming an adequate amount (Holscher, 2017).

Transparency should involve clearly disclosing synergistic ingredients such as synbiotics. In the United States, there is no official legal definition of "synbiotics" according to the FDA. As a result, many products containing synbiotics are categorized as dietary supplements, leading to limited and vague regulations regarding their health claims. This regulatory gap has resulted in inconsistent and sometimes unsubstantiated claims on product labels, which can confuse consumers and diminish trust in these products.

Similarly to the situation with probiotics, the EFSA requires robust and credible clinical evidence to support synbiotic claims. However, many synbiotic products lack sufficient information about their ingredient compositions. This deficiency compromises the scientific validity of their claims and makes it difficult for consumers to make informed decisions. It is essential to specifically detail the ingredients, the types and amounts of probiotic strains and prebiotics, and their ratios to maintain the scientific integrity of the claims and to enhance consumer awareness (Hussian et al., 2024) (Table 13.3).

TABLE 13.3 Comparison of market-approved health claims for prebiotics across major jurisdictions.

Regulatory authority	Legal framework	Approved prebiotics	Authorized health claims	Target demographics	Evidence requirements	Approval status	Implementation year	Regulatory conditions
EFSA (EFSA Panel on Dietetic Products N, Allergies, 2015; Commission E, 2021)	Article 13.5 Proprietary Health Claims Regulation	Native chicory inulin	"Chicory inulin contributes to the maintenance of normal defecation by increasing stool frequency"	General population	Six RCTs (*n*=86 subjects) demonstrating consistent stool frequency increase with ≥12 g/day dosage	Approved	2015	Minimum effective dose: 12 g/day; BENEO Orafti granted 5-year market exclusivity
FDA (Administration USFaD, 2024a,b)	Dietary Fiber Definition Framework	Inulin-type fructans, GOS, dietary fiber	No FDA-authorized prebiotic health claims	General population	Structure/function claims permitted under enforcement discretion; 30-day FDA notification required for manufacturers	Permitted under enforcement discretion	2024	Manufacturers may petition for dietary fiber classification; enforcement discretion pending final rulemaking
Health Canada - Conventional Foods (Association GP, 2025)	Food & Drugs Act / Dietary Fiber Labeling Policy (CFIA/HC)	Inulin, FOS, GOS, resistant starch, acacia gum, IMO	Digestive health, bowel regularity, beneficial gut bacteria support	General population (unless otherwise specified)	Companies must maintain substantiating evidence on file; subject to CFIA/HC audit	Permitted without premarket approval	Ongoing	Fiber content thresholds: 2 g/4 g/6 g per serving; "prebiotic" term must not imply universal fiber prebiotic properties

(*Continued*)

TABLE 13.3 (Continued)

Regulatory authority	Legal framework	Approved prebiotics	Authorized health claims	Target demographics	Evidence requirements	Approval status	Implementation year	Regulatory conditions
Health Canada - NHP (Canada, 2024)	NHP Regulations; NHPID/ LNHPD Licensing	Inulin, GOS, FOS (as medicinal ingredients)	Monograph-based claims: "Prebiotic," "Supports healthy digestive system," "Relieves constipation"	Adults ≥18 years (2–15 g/ day per inulin monograph)	Formal evidence submission per NNHPD guidance; risk-stratified evidence requirements	Premarket authorization required (NPN)	Inulin monograph: 2024; Bimuno GOS: 2023	Fermentable carbohydrates targeting microbiota classified as medicinal ingredients; dose-specific authorizations
Japan MHLW (Ministry of Health LaW, n.d.)	FOSHU System	Oligosaccharides	"Helps improve intestinal microflora and maintain good gastrointestinal condition"	General population	Human physiological efficacy studies plus comprehensive safety and quality assessments	Approved	1991	MHLW evaluation mandatory for safety, efficacy, and quality; FOSHU certification seal required; human physiological data essential

CFIA, Canadian Food Inspection Agency; *EFSA*, European Food Safety Authority; *FDA*, U.S. Food and Drug Administration; *FOS*, fructo-oligosaccharides; *FOSHU*, Foods for Specified Health Uses; *GOS*, galacto-oligosaccharides; *HC*, Health Canada; *IMO*, isomalto-oligosaccharides; *LNHPD*, Licensed Natural Health Products Database; *MHLW*, Ministry of Health, Labour and Welfare; *NHP*, Natural Health Products; *NHPID*, Natural Health Products Ingredients Database; *NNHPD*, Natural and Non-prescription Health Products Directorate; *NPN*, Natural Product Number; *RCTs*, randomized controlled trials.

13.6 Future trends and research directions

13.6.1 Next-generation prebiotics

In the past couple of years, scientific attention has shifted away from traditional prebiotics and toward new multifunctional substances that have newly appeared on the horizon. Leading the way here is one of the groups known as polyphenols, those plant metabolites with heteroaromatic complexity, as famous for their antioxidant action as for their functional properties in interaction with the gut microbiota, which is a target point for current research. In contrast to classical fibers, which are mostly fermented in the proximal colon, polyphenols, due to their comparative resistance to enzymatic breakdown and absorption in the higher GI tract, reach the distal colon in high concentrations. There, they undergo microbial fermentation and metabolism. The ultimate products of microbial metabolism are phenolic acids and minor metabolites that can exhibit a range of biological effects, including modulation of microbiota composition, inhibition of pathogenic bacteria, and activation of beneficial species. Polyphenols have also been shown to improve the growth and relative proportion of beneficial bacteria like *F. prausnitzii*, *Akkermansia muciniphila*, and *Bifidobacterium* spp. and could decrease harmful bacteria like *Clostridium perfringens*. These microbial alterations are not only associated with elevated inflammatory markers and metabolic health but also with reduced gut permeability, modulation of mucosal immunity, and even alleviation of conditions secondary to metabolic syndrome. Furthermore, the polyphenol-gut microbiota interaction is mutualistic: Polyphenols may modulate microbial composition, but they themselves require microbial metabolism to be bioactive and demonstrate their prebiotic-like effects. This feature positions polyphenols as next-generation active prebiotics, or "noncarbohydrate prebiotics," that can be utilized strategically in intelligent diet interventions to promote gut health (Cardona et al., 2013; Wan et al., 2021).

Research has demonstrated that the interaction between polyphenols and dietary fiber can enhance their bioaccessibility and increase their bioavailability in the colon. This interaction promotes fermentation by gut microorganisms, leading to higher levels of beneficial metabolites. As a result, this can produce antiinflammatory, antioxidant, and even antipathological effects. Polyphenols help maintain microbial balance by encouraging the growth of beneficial bacteria such as *Bifidobacterium* spp. and *Lactobacillus* spp., while inhibiting pathogenic bacteria. These effects are mediated through several mechanisms, including the modulation of intestinal pH, the inhibition of specific microbial enzymes, and the induction of antimicrobial metabolite production, such as SCFAs. Consequently, consuming polyphenol-rich foods like pomegranates, cranberries, blueberries, green tea, and cocoa on a daily basis can promote the gut microbiota's shift toward an antiinflammatory pathway, ultimately contributing to overall well-being (Nemzer et al., 2025).

Additionally, bioactive peptides are a recent generation of prebiotics. These are typically derived from foods rich in proteins (e.g., whey protein or glycomacropeptides) through enzymatic hydrolysis. Certain specifically selected peptides have also been shown to promote selective growth in organisms like *Bifidobacterium* and are associated with increased markers of intestinal epithelial barrier function and inflammation (Ashaolu and Ashaolu, 2021; Rackerby et al., 2024). Studies have also found that they act as prebiotics through the selective growth of gut-friendly microorganisms like *Lactobacillus*, leading to increased microbial diversity and improved microbial balance in the gut. Bioactive peptides also exhibit direct biological activities, such as strong antioxidant activities that counteract oxidative stress in intestinal epithelial cells. The other characteristic feature of these peptides is their bioactive property to lower blood pressure by inhibiting angiotensin-converting enzyme (ACE) activity. Antihypertensive action, due to the inhibition of the renin-angiotensin-aldosterone system and reduction of oxidative and inflammatory responses, is also potentially cardioprotective (Wijesekara et al., 2024).

Synthetic oligosaccharides represent a second category of new prebiotics. They are produced in large quantities through biotechnology or enzymatic processes. The advancement of various technologies has garnered particular interest, including the use of recombinant enzyme technology, metabolic engineering of production microbes, advanced bioreactor systems, and enzymatic catalysis in nonaqueous environments. These technologies not only enhance manufacturing yield but also facilitate the intelligent design of oligosaccharide structures with optimal biological attributes. Synthetic mimics may be created that induce microbial diversity, enhance intestinal barrier integrity, reduce systemic inflammation, and exhibit antipathogenic activity (Ahmed and Goyal, 2024).

It has been shown through research that GOS can selectively enhance the proliferation of beneficial gut bacteria such as *Lactobacillus* spp. and *Bifidobacterium* spp. and elevate the levels of SCFAs; these play significant roles in the sustenance of gastrointestinal health and the strengthening of the immune system. Antiinflammatory action, microbial balance restoration, and the prevention of metabolic and gastrointestinal diseases associated with the intake of GOS have also been well documented. As structurally and functionally engineerable prebiotics, GOS are more versatile than traditional prebiotics, and they have broad application potential in foods, dietary supplements, and pharmaceuticals targeting gut health. Challenges still remain, however, such as optimizing yields in production, rendering purification cost-effective, and the need for more clinical trials to confirm their health-promoting effects (Zeng et al., 2023).

Under a research study using a new chemical method referred to as FITDOG (Fenton's Initiation Toward Defined Oligosaccharide Groups), researchers promoted controlled and selective degradation of intricate plant cell walls, such as sugar beet pulp (SBP), into an extensive variety of oligosaccharides. The resulting product was referred to as Sugar Beet Oligosaccharides (SBO) and is a mixture with varying oligosaccharide structures derived from arabinose and galactose. The experimental results showed that SBO had a greater ability than SBP to induce the growth of beneficial bacteria such as *Bacteroides* and *Bifidobacterium*. Genomic and glycomic analyzes have shown that commonly used enzymes, such as extracellular arabinofuranosidases, which are typically necessary for breaking down arabinose molecules in other prebiotics, are not required for the metabolic process of SBO. This indicates that the oligosaccharides produced through FITDOG possess new and uniquely characterized glycan structures, which initiate distinct metabolic processes in the gut microbiota. When SBO is cocultivated with the probiotic Bifidobacterium longum in a fermentation culture, it results in enhanced probiotic stability. This suggests that SBO has the potential to serve as a synergistic ingredient in synbiotic foods (Masarweh et al., 2025).

13.6.2 Personalized nutrition and microbiome-guided approaches

13.6.2.1 Microbiome profiling and artificial intelligence in prebiotic recommendation

Studies emphasizing the microbiome's central role in human health and disease highlight that much of the existing research relies more on statistical association rather than causality. This focus on association, coupled with the paucity of methodological rigor, inadequate study designs, and the absence of robust analytical frameworks, has undermined the scientific validity of findings and the translational potential of research. AI can be a very powerful tool for multiomic data integration and hypothesis generation at scale. However, in the absence of transparency in model architecture, training data bias, or independent empirical validation, AI-generated conclusions can be misleading, if not harmful. Therefore authors stress the need to complement AI-generated analysis with validated biological data, controlled preclinical models, and human interventional trials (Caminero et al., 2025).

AI, in the form of machine learning and deep learning, has demonstrated immense potential to probe large-scale, multidimensional data for the identification of obscured structures of microbial communities, the identification of rare microbial species, and the prediction of functional interactions between bacteria and host phenotypes. Random Forest and Support Vector Machines (SVM) algorithms have been used for classification, whereas unsupervised methods like clustering have been employed to investigate microbial community structures. Graph Neural Networks (GNNs) have also been utilized in representing intricate microbial interactions. These have proven highly beneficial in the prediction of disease states and responses to treatment. AI is increasingly being used to optimize and personalize probiotic therapy. Probiotic formulas with multiple strains have shown efficacy in alleviating symptoms of diseases such as ulcerative colitis, type 2 diabetes, rheumatoid arthritis, and Alzheimer's disease. AI is revolutionizing the selection of optimal probiotic strains for individual patients. By integrating genomic, metabolomic, and clinical datasets, AI algorithms can accurately forecast the effectiveness of specific probiotic combinations, identify potential adverse reactions, and design fully personalized therapeutic strategies tailored to each person's unique biological profile (D'Urso and Broccolo, 2024).

Prebiotic science is often hindered by the prevalence of the vast complexity of prebiotic chemical and biological systems, the vast diversity of the human gut microbiota, and the difficulty in making sense of multilayered big data that is generated through genomic, metabolomic, and proteomic research. Deep learning and machine learning models have the ability to abstract complex patterns, predict microbial behavior, and recognize metabolic interactions beyond the realm of traditional research techniques. In particular, the employment of advanced models in biomarker identification involving prebiotic effects on the microbiome and host health allows for formulation optimization and dosage determination. Prebiotic architecture simulation and modeling through artificial intelligence technologies support the development of target molecules in a swift manner that can influence the microbiome and enhance immune function. These processes involve converting and looping experimental data into understandable models to utilize for clinical trials. Additionally, ethical considerations and technological developments are also evolving while creating AI for prebiotics. Granting algorithmic transparency, ensuring privacy, and minimizing data bias are described as the key principles for the ethical use of artificial intelligence in this area (Al-Akayleh et al., 2025).

Despite such promising capabilities, some challenges do exist. One such challenge is the nonstandardization of data collection and processing, which leads to inconsistencies in model outputs. On the other hand, most deep learning models have noninterpretable decision systems, which undermine their clinical credibility. Effective integration of multiomic datasets with good-quality clinical data remains a monumental hurdle. A separate sidebar on unintended consequences. Ethical concerns such as the privacy of data, bias in algorithms, and accountability in medical decision-making are other significant considerations that must be addressed in the application of AI to microbiome research (D'Urso and Broccolo, 2024).

Over the last few years, much progress has been made in discovering and investigating the human microbiome, which has helped significantly in the construction of personalized dietary recommendations, for instance, the use of prebiotics. Three broad categories of tools for investigating the relationship between the microbiome and responses to nutritional interventions were identified by a study. First, in silico modeling programs, including statistical models, machine learning models, and network-based models of the metabolic network, are used to discover and predict correlations between microbial content, metabolites, and physiological responses. Second, in vitro approaches include human fecal sample-derived simulated cultures within bioreactor systems, and more advanced technology, for instance, organ-on-a-chip, simulating the gut epithelium microenvironment and communication with the microbiota, enabling high-throughput functional and physiological screenings. Third, in vivo studies include animal models, observational human studies, and controlled dietary trials that aim to reveal cause-and-effect relationships and to examine hypotheses. The authors also note the rapid growth of personalized nutrition platforms that offer diet plans, probiotics, or prebiotic supplementation according to the analysis of a person's microbiome. However, they caution that the scientific evidence supporting such recommendations is limited, and the majority of the systems require more stringent clinical and biological validation. The findings emphasize the importance of integrating various kinds of data and technology. It is through the synergy of computer modeling, in vitro testing, and human data that an integrated model can be developed to predict an individual's response to nutritional intervention. Such an approach not only buttresses public health and disease prevention but also lays the foundation for nutritional, pharmaceutical, or supplemental interventions, depending on individuals' biological, microbial, and behavioral characteristics (Gibbons et al., 2022).

As technology advances further and clinical research progresses, the coming together of microbiome profiling and artificial intelligence promises to bring about a revolution in personalized prebiotic recommendations.

13.6.2.2 Integration with omics technologies

The integration of microbiome data with high-throughput omics technologies, such as genomics, transcriptomics, proteomics, and metabolomics, has unparalleled enhanced the understanding of prebiotic function and its impact on human health. The integration of big omics datasets with machine learning software could enable the design of tailored prebiotics for individual patients, thereby cementing the emphasis on "organ-targeted precision prebiotics." Herein, the use of predictive modeling to screen effective compounds and evaluate responses from individual patients is an emerging direction in research as well as in therapeutic strategies (Al-Adham et al., 2025).

Although metagenomics has been a productive approach to finding out the microbial community structure, it is not possible to obtain detailed functional data regarding the microorganisms and how they interact with the host through it alone. Therefore the integration of omics technologies has come into focus as an adjunct approach to studying metabolic functions, signaling processes, and complex host–microbe interactions. Multiomics platforms are helpful in discovering biological processes under specific disease states, biomarker identification, and the development of more targeted therapies. In particular, the integration of transcriptomics with metabolomics allows for enhanced elucidation of active biological pathways and the dynamic microbiome response to diet or medicine. However, due to the enormity and complexity of multiomics data, machine learning and artificial intelligence algorithms are required for data integration as well as to uncover hidden patterns. There are still challenges that remain, including concerns about standardization, very high costs, and analytical complexity. Yet, the authors suggest that the future of microbiome research is largely dependent on the broader deployment of these multitechnology platforms (Yang et al., 2025).

13.6.3 Sustainability and industrial innovation

13.6.3.1 Prebiotics from agri-food waste and by-products

Global estimates indicate that nearly 40% of fruits and vegetables are lost before reaching the point of sale, with an additional 50% wasted postsale. These residues, often left behind on farms, in processing facilities, or throughout distribution chains, represent a significant drain on resources, contribute to environmental degradation, and incur substantial economic costs.

Repurposing these by-products as sources of functional bioactive compounds offers a powerful solution. This strategy not only mitigates food waste but also enhances bio-efficiency, reduces dependency on finite natural resources, and supports the development of healthier, value-added food products. Such an approach is fully aligned with the principles of the Circular Bioeconomy, promoting sustainability, resource regeneration, and innovation across the agri-food and nutraceutical sectors (Pop et al., 2021). Agricultural and food industry wastes in the form of pomace, peels, and seed residues are low-cost, easily available, and rich sources of raw materials for prebiotic molecule production. These prebiotic substances are derived and applied on a large scale to enrich food products such as beverages, snacks, and dairy products, as well as in the production of dietary supplements and even active and intelligent packaging. Additionally, their application in a circular bioeconomy concept is emphasized, in which wastes are not only exploited for the production of prebiotic substances but also for energy production and secondary materials for maximum total added value (Vazquez-Olivo et al., 2019; How and Nyam, 2024).

Studies confirmed that the use of agro-industrial waste as renewable feedstocks for the production of prebiotics is an economical and environmentally sustainable method to reduce environmental pollution and enhance the efficiency of bioresource utilization. The majority of biomass wastes, like fruit peels, cereal brans, pomace, and other polysaccharide-bound by-products, are employed as substrates to yield prebiotics via enzymatic processes. Enzymes such as hydrolases are key to converting raw materials into bioactive oligosaccharides. Enzymatic systems are advantageous over conventional chemical systems in having a lower reaction temperature, improved selectivity, fewer toxic by-products, and higher environmental compatibility (Gonçalves et al., 2023).

The content and functionality of fiber differ according to the plant species, developmental stage, processing parameters, and plant part (e.g., peel, pomace, and seed). Vegetable wastes tend to have more insoluble fibers such as cellulose and lignin, while fruits provide a wider range of both soluble and insoluble fibers. For instance, apple pomace can contain between 4.4 and 47.3 g of fiber per 100 g of dry matter, often much higher than in the respective fruit. These fibers not only have prebiotic potential but also contribute to texture improvement, stability, moisture retention, and glycemic index in food products. A few other wastes that have been studied are orange residues, carrot residues, cabbage waste, artichokes, and even cactus pears, all of which contain a unique fiber profile that makes them available and inexpensive for food fortification. Prebiotic fibers from these wastes have been shown to be able to induce the growth of beneficial gut bacteria such as *Bifidobacterium* and *Lactobacillus*. By generating SCFAs, they can positively influence glucose and lipid metabolism, strengthen gut barrier function, reduce systemic inflammation, and enhance immune function. Additional benefits include anticancer properties, weight management, cholesterol modulation, and enhanced absorption of phenolic compounds. Certain fruit and vegetable wastes may also contain antinutritional compounds such as phytic acid, tannins, oxalates, or cyanogenic glycosides. Raw or excessive consumption of these can inhibit mineral absorption or cause gastrointestinal issues. Apart from this, crop borders are more likely to contain chemical residues like pesticides, mycotoxins, and heavy metals. Hence, extraction processes must include risk analysis and detoxification procedures.

Mechanical pressing, nanofiltration, hot water extraction, enzymatic hydrolysis, and green processes with no chemical solvent utilization are some of the extraction methods to obtain prebiotic fibers and substances from residues. These methods preserve the structure of bioactive compounds, minimize chemical contamination, and allow direct application in food industries with minimal processing. In the majority of cases, they are added to bread, biscuits, yogurt, cheese, meat products, and beverages, with benefits that encompass high fiber content, enhanced microbial stability, increased satiety, and reduced glycemic index (Pop et al., 2021).

13.6.3.2 Clean-label and plant-based formulation trends

The growth of the prebiotic business is evident as product launches containing prebiotics increased by approximately 190% between 2019 and 2023 (Ltd, 2024). Among these, there has been a growing trend toward utilizing new plant sources, which are also categorized as clean-label products (Ltd, 2024).

Therefore 73% of the people who use plant-based food have stated that clean-label labels are tremendously important to them, and they indeed search for ingredient transparency (Magazine, 2025). Analysis revealed that clean-label products prefer natural functional ingredients instead of artificial coloring, preservatives, or flavor. It also highlighted the inclusion of these ingredients in developing biodegradable and active packaging. The discovery showed that the type of packaging is not just biodegradable but also exhibits antibacterial or antioxidant properties, thereby enhancing shelf life and food safety (Alexandri et al., 2022).

It has also been confirmed through research that consumers' positive attitudes toward clean-label products have a strong and significant influence on their purchase intentions for them; that is, the more positive the consumers' beliefs about the healthiness, quality, and safety of the products, the more likely they will be to buy them. Additionally, perceived behavioral control, consisting of people's belief in their ability to purchase these goods and their availability, has a significant influence on purchase intentions. Moreover, subjective norms, or social pressures from society, family, and friends, strengthen the intention to buy clean-label products. It has been demonstrated that to develop the clean-label products market, it is necessary to make efforts to raise consumer awareness and knowledge, actively resolving health concerns in trustworthy marketing and informational programs. Also, considering the impact of subjective norms and social pressures can help in creating effective advertising programs and market development campaigns (Chang and Chen, 2022).

13.6.3.3 Smart packaging for microbiome-targeted products

In the present era, with greater emphasis on microbiome health and personalized nutrition, smart packaging has been one of the central developments. Smart packaging is a packaging system capable of sensing, monitoring, and responding to changes in the ambient environment and the condition of the product. These kinds of packages generally fall into three broad categories: active packaging, intelligent packaging with alerting, and traceable packaging. A number of technologies have been developed, including the use of absorbers, catalysts, and active coatings, with the aim of prolonging shelf life and maintaining product quality. Smart packaging includes sensors, indicators, and communication systems that are capable of detecting changes in quality, spoilage, or environmental parameters such as temperature, humidity, and the presence of specific gases and conveying the information to consumers or manufacturers. The traceability in packaging utilizes modern technologies such as two-dimensional barcodes, radio frequency identification (RFID), and blockchain, which offer supply chain tracing, product authenticity, and quality assurance. The technologies play a significant role in providing transparency and consumer confidence (Drago et al., 2020).

Smart packaging is able to sense, control, and respond to changes in the environment, while bio-based packaging consists of renewable, biodegradable, and sustainable materials. Additionally, new technologies such as nanoparticles, nanocomposites, and nanostructures have been introduced to enhance the mechanical, thermal, and antimicrobial properties of packaging. Silver nanoparticles, zinc oxide, and silicates extend food shelf life by inhibiting microbial growth. These innovations have proven benefits like longer freshness, reduced food wastage, and improved consumer safety (Chaudhary et al., 2022).

Biosensors have also been suggested to be among the most crucial components of intelligent packaging. Biosensors employ biological molecules such as enzymes, antibodies, or DNA in a way that makes them recognize physical, chemical, and microbiological changes. Biosensors are capable of detecting the presence of contaminants, bacteria, or gas composition changes and relaying the information obtained in a readable format to the end-user or manufacturer. Various biosensor uses have also been indicated, including colorimetric sensors to detect meat spoilage, optical strips to detect ammonia gas, and electrochemical sensors to detect microbial contamination.

The Internet of Things (IoT) connecting packaging to digital networks enables monitoring, recording data in real time, and communication along the supply chain. Sensors in a packaging system enable the measurement of conditions like temperature, humidity, impact, or pressure change and transmit this data via RFID or NFC to a smartphone or central server. It provides improved inventory management, waste reduction, and visibility along the supply chain (Bhatlawande et al., 2024).

Technical, economic, and regulatory issues are also being taken into consideration in the case of smart packaging development. Some of the significant barriers include high production costs, technical complexities, the need for standardization, food contact material safety, and consumer acceptance (Drago et al., 2020).

13.7 Conclusion

This comprehensive chapter illuminates the profound role of prebiotics in advancing modern food systems, medical therapeutics, and nutraceuticals, positioning them as pivotal agents for enhancing gut health and overall well-being through selective fermentation that nurtures beneficial microbiota and generates SCFAs like acetate, propionate, and butyrate. These SCFAs bolster intestinal integrity, modulate immune responses, regulate metabolism, and even influence cognitive and emotional health via the gut–brain axis, making prebiotics versatile for preventing and treating conditions such as IBS, IBD, metabolic syndrome, and mental health disorders, particularly in synbiotic formulations. Industrial prebiotics—including inulin, FOS, GOS, XOS, and resistant starches—offer superior stability against heat, acidity, and processing, surpassing probiotics in ease of incorporation into foods and supplements, though challenges like gastrointestinal side effects at high doses, variable host responses, and inconsistent global regulations highlight the need for standardized guidelines to ensure safety and efficacy. In comparison, while probiotics deliver live microbes for direct, short-term gut restoration—often requiring cold storage and facing viability hurdles—prebiotics provide enduring, indirect stimulation of native microbes with greater resilience and formulation flexibility, rendering synbiotics an optimal hybrid for comprehensive microbial support. Looking ahead, the evolution of prebiotics hinges on innovative integrations, such as novel sources like polyphenols and algal polysaccharides, advanced delivery technologies like microencapsulation and targeted systems, and consumer-centric designs that prioritize clean labels, sensory appeal, and transparent health claims in everyday products like dairy, cereals, and beverages. Ultimately, by fostering multidisciplinary collaboration in science, technology, and market strategies, prebiotics stand poised to transform microbiome-targeted solutions into accessible, effective tools for personalized nutrition and sustained health empowerment.

References

Administration USFaD. Questions and Answers on Dietary Fiber. USA; 2024a.

Administration USFaD. Structure/function claims — Nutrition, labeling & critical food information. USA; 2024b.

Agraib, L.M., Yamani, M.I., Tayyem, R., Abu-Sneineh, A.T., Rayyan, Y.M., 2022. Probiotic supplementation induces remission and changes in the immunoglobulins and inflammatory response in active ulcerative colitis patients: a pilot, randomized, double-blind, placebo-controlled study. Clinical Nutrition ESPEN 51, 83–91.

Ahmed, J., Goyal, A., 2024. Emerging trends in novel technological advancements in oligosaccharide production and their potential applications as prebiotics with beneficial effects. In: Verma, P. (Ed.), Industrial Microbiology and Biotechnology: An Insight into Current Trends. Springer Nature Singapore, Singapore, pp. 153–174.

Al-Adham, I.S.I., Agha, A., Al-Akayleh, F., Al-Remawi, M., Jaber, N., Al Manasur, M., et al., 2025. Prebiotics beyond the gut: omics insights, artificial intelligence, and clinical trials in organ-specific applications. Probiotics Antimicrob Proteins.

Al-Akayleh F., Al-Remawi M., Abdel-Rahem R.A., Al-Adham I.S.I., Aburub F., Agha A.S.A.A., editors. AI-driven strategies in prebiotic research: addressing challenges and advancing human health. 2025 1st International Conference on Computational Intelligence Approaches and Applications (ICCIAA); 2025 28-30 April 2025.

Alexandri, M., Kachrimanidou, V., Papapostolou, H., Papadaki, A., Kopsahelis, N., 2022. Sustainable food systems: the case of functional compounds towards the development of clean label food products. Foods 11 (18).

Allen, S.J., Wareham, K., Wang, D., Bradley, C., Hutchings, H., Harris, W., et al., 2013. Lactobacilli and bifidobacteria in the prevention of antibiotic-associated diarrhoea and Clostridium difficile diarrhoea in older inpatients (PLACIDE): a randomised, double-blind, placebo-controlled, multicentre trial. Lancet 382 (9900), 1249–1257.

Alonso-Allende, J., Milagro, F.I., Aranaz, P., 2024. Health effects and mechanisms of inulin action in human metabolism. Nutrients 16 (17).

Amiriani, T., Rajabli, N., Faghani, M., Besharat, S., Roshandel, G., Akhavan Tabib, A., et al., 2020. Effect of lactocare® synbiotic on disease severity in ulcerative colitis: a randomized placebo-controlled double-blind clinical trial. Middle East Journal of Digestive Diseases 12 (1), 27–33.

Anal, A.K., Singh, H., 2007. Recent advances in microencapsulation of probiotics for industrial applications and targeted delivery. Trends in Food Science and Technology 18 (5), 240–251.

Arpaia, N., Campbell, C., Fan, X., Dikiy, S., van der Veeken, J., deRoos, P., et al., 2013. Metabolites produced by commensal bacteria promote peripheral regulatory T-cell generation. Nature 504 (7480), 451–455.

Asad, A., Kirk, M., Zhu, S., Dong, X., Gao, M., 2025. Effects of prebiotics and probiotics on symptoms of depression and anxiety in clinically diagnosed samples: systematic review and meta-analysis of randomized controlled trials. Nutrition Reviews 83 (7), e1504–e1520.

Ashaolu, T.J., Ashaolu, J.O., 2021. Prebiotic peptides, their formation, fermentation in the gut, and health implications. Biotechnology Progress 37 (3), e3142.

Association GP, 2025. Canada Regulatory Status (Prebiotic Regulatory Guide — Canada). Global Prebiotic Association.

Authority EFS. Consolidated list of Article 13 health claims 2011.

Badal, V.D., Vaccariello, E.D., Murray, E.R., Yu, K.E., Knight, R., Jeste, D.V., et al., 2020. The gut microbiome, aging, and longevity: a systematic review. Nutrients 12 (12).

Baghizadeh, A., Davati, A., Heidarloo, A.J., Emadi, F., Aliasl, J., 2021. Efficacy of Plantago major seed in management of ulcerative colitis symptoms: a randomized, placebo controlled, clinical trial. Complementary Therapies in Clinical Practice 44, 101444.

Basafa-Roodi, P., Jazayeri, S., Hadi, F., Paghaleh, S.J., Khosravi-Darani, K., Malakouti, S.K., 2024. Effects of synbiotic supplementation on the components of metabolic syndrome in patients with schizophrenia: a randomized, double-blind, placebo-controlled trial. BMC Psychiatry 24 (1), 669.

Becattini, S., Taur, Y., Pamer, E.G., 2016. Antibiotic-induced changes in the intestinal microbiota and disease. Trends in Molecular Medicine 22 (6), 458–478.

Belkaid, Y., Hand, T.W., 2014. Role of the microbiota in immunity and inflammation. Cell 157 (1), 121–141.

Bhagwat, A., Bhushette, P., Annapure, U.S., 2020. Spray drying studies of probiotic Enterococcus strains encapsulated with whey protein and maltodextrin. Beni-Suef University Journal of Basic and Applied Sciences 9 (1), 33.

Bhatlawande, A.R., Ghatge, P.U., Shinde, G.U., Anushree, R.K., Patil, S.D., 2024. Unlocking the future of smart food packaging: biosensors, IoT, and nano materials. Food Science and Biotechnology 33 (5), 1075–1091.

Bindels, L.B., Delzenne, N.M., Cani, P.D., Walter, J., 2015. Towards a more comprehensive concept for prebiotics. Nature Reviews Gastroenterology and Hepatology 12 (5), 303–310.

Birkeland, E., Gharagozlian, S., Birkeland, K.I., Valeur, J., Måge, I., Rud, I., et al., 2020. Prebiotic effect of inulin-type fructans on faecal microbiota and short-chain fatty acids in type 2 diabetes: a randomised controlled trial. European Journal of Nutrition 59 (7), 3325–3338.

Brüssow, H., 2019. Probiotics and prebiotics in clinical tests: an update. F1000Research 8.

Burgain, J., Gaiani, C., Linder, M., Scher, J., 2011. Encapsulation of probiotic living cells: From laboratory scale to industrial applications. Journal of Food Engineering 104 (4), 467–483.

Calderón-Oliver, M., Ponce-Alquicira, E., 2022. The role of microencapsulation in food application. Molecules (Basel, Switzerland) 27 (5).

Caminero, A., Tropini, C., Valles-Colomer, M., Shung, D.L., Gibbons, S.M., Surette, M.G., et al., 2025. Credible inferences in microbiome research: ensuring rigour, reproducibility and relevance in the era of AI. Nature Reviews Gastroenterology and Hepatology.

Canada, H., 2024. Natural Health Product – Inulin (Monograph). Health Canada – Natural and Non-prescription Health Products Directorate.

Canani, R.B., Cirillo, P., Terrin, G., Cesarano, L., Spagnuolo, M.I., De Vincenzo, A., et al., 2007. Probiotics for treatment of acute diarrhoea in children: randomised clinical trial of five different preparations. BMJ (Clinical Research) 335 (7615), 340.

Canfora, E.E., Jocken, J.W., Blaak, E.E., 2015. Short-chain fatty acids in control of body weight and insulin sensitivity. Nature Reviews Endocrinology 11 (10), 577–591.

Cani, P.D., Everard, A., 2016. Talking microbes: when gut bacteria interact with diet and host organs. Molecular Nutrition and Food Research 60 (1), 58–66.

Cani, P.D., Amar, J., Iglesias, M.A., Poggi, M., Knauf, C., Bastelica, D., et al., 2007. Metabolic endotoxemia initiates obesity and insulin resistance. Diabetes 56 (7), 1761–1772.

Cardona, F., Andrés-Lacueva, C., Tulipani, S., Tinahones, F.J., Queipo-Ortuño, M.I., 2013. Benefits of polyphenols on gut microbiota and implications in human health. The Journal of Nutritional Biochemistry 24 (8), 1415–1422.

Carlson, J.L., Erickson, J.M., Lloyd, B.B., Slavin, J.L., 2018. Health effects and sources of prebiotic dietary fiber. Current Developments in Nutrition 2 (3), nzy005.

Chandrasekaran, P., Weiskirchen, S., Weiskirchen, R., 2024. Effects of probiotics on gut microbiota: an overview. International Journal of Molecular Sciences 25 (11).

Chang, M.Y., Chen, H.S., 2022. Understanding consumers' intentions to purchase clean label products: evidence from Taiwan. Nutrients 14 (18).

Chaudhary, V., Punia Bangar, S., Thakur, N., Trif, M., 2022. Recent advancements in smart biogenic packaging: reshaping the future of the food packaging industry. Polymers (Basel) 14 (4).

Chen, Z., Liang, N., Zhang, H., Li, H., Guo, J., Zhang, Y., et al., 2024. Resistant starch and the gut microbiome: exploring beneficial interactions and dietary impacts. Food Chemistry 21, 101118.

Cicero, A.F.G., Fogacci, F., Bove, M., Giovannini, M., Borghi, C., 2021. Impact of a short-term synbiotic supplementation on metabolic syndrome and systemic inflammation in elderly patients: a randomized placebo-controlled clinical trial. European Journal of Nutrition 60 (2), 655–663.

Commission E, 2021. Commission Implementing Regulation (EU) 2021/686 of 23 April 2021 authorising a health claim made on foods, other than those referring to the reduction of disease risk and to children's development and health, and amending Regulation (EU) No 432/2012. Official Journal of the European Union. pp. 8–10 L 143(27.4.2021).

Cornu, M., Yang, J.Y., Jaccard, E., Poussin, C., Widmann, C., Thorens, B., 2009. Glucagon-like peptide-1 protects beta-cells against apoptosis by increasing the activity of an IGF-2/IGF-1 receptor autocrine loop. Diabetes 58 (8), 1816–1825.

Corona, L., Lussu, A., Bosco, A., Pintus, R., Cesare Marincola, F., Fanos, V., et al., 2021. Human milk oligosaccharides: a comprehensive review towards metabolomics. Children (Basel) 8 (9).

Cosme, F., Inês, A., Vilela, A., 2022. Consumer's acceptability and health consciousness of probiotic and prebiotic of non-dairy products. Food Research International 151, 110842.

Dalile, B., Van Oudenhove, L., Vervliet, B., Verbeke, K., 2019. The role of short-chain fatty acids in microbiota-gut-brain communication. Nature Reviews Gastroenterology and Hepatology 16 (8), 461–478.

Davani-Davari, D., Negahdaripour, M., Karimzadeh, I., Seifan, M., Mohkam, M., Masoumi, S.J., et al., 2019. Prebiotics: definition, types, sources, mechanisms, and clinical applications. Foods 8 (3).

Davie, J.R., 2003. Inhibition of histone deacetylase activity by butyrate. The Journal of Nutrition 133 (7 Suppl), 2485s–2493ss.

Deehan, E.C., Walter, J., 2016. The fiber gap and the disappearing gut microbiome: implications for human nutrition. Trends in Endocrinology and Metabolism: TEM 27 (5), 239–242.

Deehan, E.C., Al Antwan, S., Witwer, R.S., Guerra, P., John, T., Monheit, L., 2024. Revisiting the concepts of prebiotic and prebiotic effect in light of scientific and regulatory progress-a consensus paper from the global prebiotic association. Advances in Nutrition 15 (12), 100329.

den Besten, G., van Eunen, K., Groen, A.K., Venema, K., Reijngoud, D.J., Bakker, B.M., 2013. The role of short-chain fatty acids in the interplay between diet, gut microbiota, and host energy metabolism. Journal of Lipid Research 54 (9), 2325–2340.

De Silva, A., Bloom, S.R., 2012. Gut hormones and appetite control: a focus on PYY and GLP-1 as therapeutic targets in obesity. Gut Liver 6 (1), 10–20.

Dewulf, E.M., Cani, P.D., Claus, S.P., Fuentes, S., Puylaert, P.G., Neyrinck, A.M., et al., 2013. Insight into the prebiotic concept: lessons from an exploratory, double blind intervention study with inulin-type fructans in obese women. Gut 62 (8), 1112–1121.

Doron, S., Snydman, D.R., 2015. Risk and safety of probiotics. Clinical Infectious Diseases: an Official Publication of the Infectious Diseases Society of America 60 (Suppl 2), S129–S134.

Drago, E., Campardelli, R., Pettinato, M., Perego, P., 2020. Innovations in smart packaging concepts for food: an extensive review. Foods 9 (11).

Dueñas, M., Cueva, C., Muñoz-González, I., Jiménez-Girón, A., Sánchez-Patán, F., Santos-Buelga, C., et al., 2015. Studies on modulation of gut microbiota by wine polyphenols: from isolated cultures to omic approaches. Antioxidants (Basel) 4 (1), 1–21.

D'Urso, F., Broccolo, F., 2024. Applications of artificial intelligence in microbiome analysis and probiotic interventions—an overview and perspective based on the current state of the art. Applied Sciences 14 (19).

EFSA Panel on Dietetic Products N, Allergies, 2011. Scientific Opinion on the substantiation of health claims related to fructooligosaccharides (FOS) from sucrose and decreasing potentially pathogenic gastro intestinal microorganisms (ID 774), changes in short chain fatty acid (SCFA) production and pH in the gastro-intestinal tract (ID 775), changes in bowel function (ID 775, 778), reduction of gastro intestinal discomfort (ID 775, 778), increase in calcium and/or magnesium absorption leading to an increase in magnesium and/or calcium retention (ID 776, 777), maintenance of normal blood LDL cholesterol concentrations (ID 805) and maintenance of normal (fasting) blood concentrations of triglycerides (ID 805) pursuant to Article 13(1) of Regulation (EC) No 1924/2006. EFSA Journal 9 (4), 2023.

EFSA Panel on Dietetic Products N, Allergies, 2015. Scientific opinion on the substantiation of a health claim related to "native chicory inulin" and maintenance of normal defecation by increasing stool frequency pursuant to Article 13.5 of Regulation (EC) No 1924/2006. EFSA Journal 13 (1), 3951.

Erny, D., Hrabě de Angelis, A.L., Jaitin, D., Wieghofer, P., Staszewski, O., David, E., et al., 2015. Host microbiota constantly control maturation and function of microglia in the CNS. Nature Neuroscience 18 (7), 965–977.

FAO/WHO, 2002. Guidelines for the Evaluation of Probiotics in Food. Food and Agriculture Organization of the United Nations (FAO) and World Health Organization (WHO), London, Ontario, Canada.

Faridzadeh, A., Yadegari, Y., Bakhshaee, M., Kabiri, M., Mohammadi, M., Khoshkhui, M., et al., 2023. The role of synbiotics as adjunctive agents in the treatment of allergic rhinitis: a randomized controlled trial. Health Science Reports 6 (10), e1571.

(FDA) USFaDA. Recently Published GRAS Notices and FDA Letters 2025 Available from: https://www.fda.gov/food/gras-notice-inventory/recently-published-gras-notices-and-fda-letters.

Fenster, K., Freeburg, B., Hollard, C., Wong, C., Rønhave Laursen, R., Ouwehand, A.C., 2019. The production and delivery of probiotics: a review of a practical approach. Microorganisms 7 (3).

Ferdousi, R., Rouhi, M., Mohammadi, R., Mortazavian, A.M., Khosravi-Darani, K., Homayouni Rad, A., 2013. Evaluation of probiotic survivability in yogurt exposed to cold chain interruption. Iran Journal of Pharmaceutical Research 12 (Suppl), 139–144.

Fernandes, P., 2025. Fructooligosaccharides (FOSs): a condensed overview. Compounds 5 (2).

Figueroa-González, I., Quijano, G., Ramírez, G., Cruz-Guerrero, A., 2011. Probiotics and prebiotics–perspectives and challenges. Journal of the Science of Food and Agriculture 91 (8), 1341–1348.

Francavilla, R., Piccolo, M., Francavilla, A., Polimeno, L., Semeraro, F., Cristofori, F., et al., 2019. Clinical and microbiological effect of a multispecies probiotic supplementation in celiac patients with persistent IBS-type symptoms: a randomized, double-blind, placebo-controlled, multicenter trial. Journal of Clinical Gastroenterology 53 (3), e117–e125.

Franck, A., 2002. Technological functionality of inulin and oligofructose. The British Journal of Nutrition 87 (Suppl 2), S287–S291.

Frost, G., Sleeth, M.L., Sahuri-Arisoylu, M., Lizarbe, B., Cerdan, S., Brody, L., et al., 2014. The short-chain fatty acid acetate reduces appetite via a central homeostatic mechanism. Nature Communications 5, 3611.

Fujimori, S., Gudis, K., Mitsui, K., Seo, T., Yonezawa, M., Tanaka, S., et al., 2009. A randomized controlled trial on the efficacy of synbiotic versus probiotic or prebiotic treatment to improve the quality of life in patients with ulcerative colitis. Nutrition (Burbank, Los Angeles County, Calif.) 25 (5), 520–525.

Gerasimov, S., Gantzel, J., Dementieva, N., Schevchenko, O., Tsitsura, O., Guta, N., et al., 2018. Role of *Lactobacillus rhamnosus* (FloraActive™) 19070-2 and *Lactobacillus reuteri* (FloraActive™) 12246 in infant colic: a randomized dietary study. Nutrients 10 (12).

Gibson, G., Rastall, R., 2006. Prebiotics: Development & Application 2006.

Gibson, G.R., Hutkins, R., Sanders, M.E., Prescott, S.L., Reimer, R.A., Salminen, S.J., et al., 2017. Expert consensus document: The International Scientific Association for Probiotics and Prebiotics (ISAPP) consensus statement on the definition and scope of prebiotics. Nature Reviews Gastroenterology and Hepatology 14 (8), 491–502.

Gibbons, S.M., Gurry, T., Lampe, J.W., Chakrabarti, A., Dam, V., Everard, A., et al., 2022. Perspective: leveraging the gut microbiota to predict personalized responses to dietary, prebiotic, and probiotic interventions. Advances in Nutrition 13 (5), 1450–1461.

Goodoory, V.C., Khasawneh, M., Black, C.J., Quigley, E.M.M., Moayyedi, P., Ford, A.C., 2023. Efficacy of probiotics in irritable bowel syndrome: systematic review and meta-analysis. Gastroenterology. 165 (5), 1206–1218.

Gonçalves, D.A., González, A., Roupar, D., Teixeira, J.A., Nobre, C., 2023. How prebiotics have been produced from agro-industrial waste: an overview of the enzymatic technologies applied and the models used to validate their health claims. Trends in Food Science and Technology 135, 74–92.

GRAS Notice No. 671: Galacto-oligosaccharides (GOS). 2016.

Groeger, D., O'Mahony, L., Murphy, E.F., Bourke, J.F., Dinan, T.G., Kiely, B., et al., 2013. *Bifidobacterium infantis* 35624 modulates host inflammatory processes beyond the gut. Gut Microbes 4 (4), 325–339.

Guo, M., Wang, H., Wang, C., 2018. Interactions between whey protein and inulin in a model system. Journal of Food Science and Technology 55 (10), 4051–4058.

Guo, Y., Liu, Y., Chen, K., Cai, L., Huang, S., Zhang, Y., 2024. Super gastro-resistant microcapsules based on $CaCO_3$ nanocrystal buffered alginate/pectin composites for colon-targeted probiotic delivery: in vitro and in vivo evaluation. Advanced Composites and Hybrid Materials 7 (6), 207.

Gutierrez-Castrellon, P., Lopez-Velazquez, G., Diaz-Garcia, L., Jimenez-Gutierrez, C., Mancilla-Ramirez, J., Estevez-Jimenez, J., et al., 2014. Diarrhea in preschool children and *Lactobacillus reuteri*: a randomized controlled trial. Pediatrics 133 (4), e904–e909.

Haghighat, N., Mohammadshahi, M., Shayanpour, S., Haghighizadeh, M.H., Rahmdel, S., Rajaei, M., 2021. The effect of synbiotic and probiotic supplementation on mental health parameters in patients undergoing hemodialysis: a double-blind, randomized, placebo-controlled trial. Indian Journal of Nephrology 31 (2), 149–156.

Hasle, G., Raastad, R., Bjune, G., Jenum, P.A., Heier, L., 2017. Can a galacto-oligosaccharide reduce the risk of traveller's diarrhoea? A placebo-controlled, randomized, double-blind study. Journal of Travel Medicine: Official Publication of the International Society of Travel Medicine and the Asia Pacific Travel Health Association 24 (5).

Hazards EPanel oB, Koutsoumanis, K., Allende, A., Álvarez-Ordóñez, A., Bolton, D., Bover-Cid, S., et al., 2023. Update of the list of qualified presumption of safety (QPS) recommended microorganisms intentionally added to food or feed as notified to EFSA. EFSA Journal 21 (1), e07747.

Hill, C., Guarner, F., Reid, G., Gibson, G.R., Merenstein, D.J., Pot, B., et al., 2014a. The International Scientific Association for Probiotics and Prebiotics consensus statement on the scope and appropriate use of the term probiotic. Nature Reviews Gastroenterology and Hepatology 11 (8), 506–514.

Hill, C., Guarner, F., Reid, G., Gibson, G.R., Merenstein, D.J., Pot, B., et al., 2014b. Expert consensus document. The International Scientific Association for Probiotics and Prebiotics consensus statement on the scope and appropriate use of the term probiotic. Nature Reviews Gastroenterology and Hepatology 11 (8), 506–514.

Holscher, H.D., 2017. Dietary fiber and prebiotics and the gastrointestinal microbiota. Gut Microbes 8 (2), 172–184.

How, Y.H., Nyam, K.L., 2024. Reutilization of fruit waste as potential prebiotic for probiotic or food-grade microorganisms in food applications: a review. Probiotics and Antimicrobial Proteins.

Hussian, T., Lokesh, Saini, A., Tripathy, S., Dureja, H., 2024. Regulatory aspects of synbiotics-based delivery system. In: Dua, K. (Ed.), Synbiotics in Human Health: Biology to Drug Delivery. Springer Nature Singapore, Singapore, pp. 495–506.

Hutkins, R.W., Krumbeck, J.A., Bindels, L.B., Cani, P.D., Fahey Jr., G., Goh, Y.J., et al., 2016. Prebiotics: why definitions matter. Current Opinion in Biotechnology 37, 1–7.

İnce Palamutoğlu, M., Bilgi, E., Horzum, M., Kılıç, Z., Karaca, E., 2023. Determination of consumers' knowledge levels and consumption status on probiotic and prebiotic products.

Jiang, W., Ni, B., Liu, Z., Liu, X., Xie, W., Wu, I.X.Y., et al., 2020. The role of probiotics in the prevention and treatment of atopic dermatitis in children: an updated systematic review and meta-analysis of randomized controlled trials. Paediatric Drugs 22 (5), 535–549.

Johnson, A.J., Vangay, P., Al-Ghalith, G.A., Hillmann, B.M., Ward, T.L., Shields-Cutler, R.R., et al., 2019. Daily sampling reveals personalized diet-microbiome associations in humans. Cell Host and Microbe 25 (6), 789–802.e5.

Johnstone, N., Cohen Kadosh, K., 2025. A randomised controlled trial of the effects of galacto-oligosaccharides on the gut brain-axis of young females. Brain, Behavior, and Immunity 129, 573–584.

Kamarlı Altun, H., Akal Yıldız, E., Akın, M., 2019. Effects of synbiotic therapy in mild-to-moderately active ulcerative colitis: a randomized placebo-controlled study. The Turkish Journal of Gastroenterology: the Official Journal of Turkish Society of Gastroenterology 30 (4), 313–320.

Karakan, T., Tuohy, K.M., Janssen-van Solingen, G., 2021. Low-dose lactulose as a prebiotic for improved gut health and enhanced mineral absorption. Frontiers in Nutrition 8, 672925.

Kelly, C.J., Zheng, L., Campbell, E.L., Saeedi, B., Scholz, C.C., Bayless, A.J., et al., 2015. Crosstalk between microbiota-derived short-chain fatty acids and intestinal epithelial hif augments tissue barrier function. Cell Host and Microbe 17 (5), 662–671.

Koh, A., De Vadder, F., Kovatcheva-Datchary, P., Bäckhed, F., 2016. From dietary fiber to host physiology: short-chain fatty acids as key bacterial metabolites. Cell 165 (6), 1332–1345.

Kolida, S., Tuohy, K., Gibson, G.R., 2002. Prebiotic effects of inulin and oligofructose. The British Journal of Nutrition 87 (Suppl 2), S193–S197.

Kołodziej, M., Szajewska, H., 2017. *Lactobacillus reuteri* DSM 17938 in the prevention of antibiotic-associated diarrhoea in children: protocol of a randomised controlled trial. BMJ Open 7 (1), e013928.

Kumar, H.S.S., Vergagen, H., Rowland, I., Heimbach, J., Banares, S., Young, T., et al. 2015. Novel probiotics and prebiotics: road to the market. Available from: https://isappscience.org/publications/novel-probiotics-and-prebiotics-road-to-the-market/.

Kundu, P., Blacher, E., Elinav, E., Pettersson, S., 2017. Our gut microbiome: the evolving inner self. Cell 171 (7), 1481–1493.

Looijer-van Langen, M.A., Dieleman, L.A., 2009. Prebiotics in chronic intestinal inflammation. Inflammatory Bowel Diseases 15 (3), 454–462.

Louis, P., Flint, H.J., 2017. Formation of propionate and butyrate by the human colonic microbiota. Environmental Microbiology 19 (1), 29–41.

Ltd, P., 2024. Prebiotic Trend, Opportunities, and Challenges. PharmaLinea Ltd. Available at https://pharmalinea.com/intelhub/prebiotic-trend-opportunities-challenges/.

Macfarlane, G.T., Macfarlane, S., 2011. Fermentation in the human large intestine: its physiologic consequences and the potential contribution of prebiotics. Journal of Clinical Gastroenterology 45 (Suppl), S120–S127.

Macia, L., Thorburn, A.N., Binge, L.C., Marino, E., Rogers, K.E., Maslowski, K.M., et al., 2012. Microbial influences on epithelial integrity and immune function as a basis for inflammatory diseases. Immunological Reviews 245 (1), 164–176.

Magazine, F.B., 2025. Clean-Label Ingredients and Tech Are Redefining Plant-Based USA: Food & Beverage Magazine. Available from: https://www.fb101.com/clean-label-ingredients-and-tech-are-redefining-plant-based/.

Mankertz, J., Schulzke, J.D., 2007. Altered permeability in inflammatory bowel disease: pathophysiology and clinical implications. Current Opinion in Gastroenterology 23 (4), 379–383.

Marco, M.L., Sanders, M.E., Gänzle, M., Arrieta, M.C., Cotter, P.D., De Vuyst, L., et al., 2021. The International Scientific Association for Probiotics and Prebiotics (ISAPP) consensus statement on fermented foods. Nature Reviews Gastroenterology and Hepatology 18 (3), 196–208.

Markowiak, P., Śliżewska, K., 2017. Effects of probiotics, prebiotics, and synbiotics on human health. Nutrients 9 (9).

Marteau, P., Seksik, P., 2004. Tolerance of probiotics and prebiotics. Journal of Clinical Gastroenterology 38 (6 Suppl), S67–S69.

Martínez, I., Kim, J., Duffy, P.R., Schlegel, V.L., Walter, J., 2010. Resistant starches types 2 and 4 have differential effects on the composition of the fecal microbiota in human subjects. PLoS One 5 (11), e15046.

Masarweh, C., Maldonado-Gomez, M., Paviani, B., Bhattacharya, M., Weng, C.Y., Suarez, C., et al., 2025. Generation of novel prebiotic oligosaccharide pools from fiber drives biological insight in bacterial glycan metabolism. Applied and Environmental Microbiology 91 (3), e0207724.

Maslowski, K.M., Mackay, C.R., 2011. Diet, gut microbiota and immune responses. Nature Immunology 12 (1), 5–9.

Mei, Z., Yuan, J., Li, D., 2022. Biological activity of galacto-oligosaccharides: a review. Frontiers in Microbiology 13, 993052.

Mejía-Caballero, A., Salas-Villagrán, V.A., Jiménez-Serna, A., Farrés, A., 2021. Challenges in the production and use of probiotics as therapeuticals in cancer treatment or prevention. Journal of Industrial Microbiology and Biotechnology 48 (9-10).

Meyer, D., Bayarri, S., Tárrega, A., Costell, E., 2011. Inulin as texture modifier in dairy products. Food Hydrocolloids 25 (8), 1881–1890.

Ministry of Health LaW. n.d. Food for Specified Health Uses (FOSHU) — Food with Health Claims, Food for Special Dietary Uses, and Nutrition Labeling. Japan: Ministry of Health, Labour and Welfare.

Miqdady, M., Al Mistarihi, J., Azaz, A., Rawat, D., 2020. Prebiotics in the infant microbiome: the past, present, and future. Pediatric Gastroenterology, Hepatology and Nutrition 23 (1), 1–14.

Misra, S., Pandey, P., Panigrahi, C., Mishra, H.N., 2023. A comparative approach on the spray and freeze drying of probiotic and gamma-aminobutyric acid as a single entity: characterization and evaluation of stability in simulated gastrointestinal conditions. Food Chemistry Advances 3, 100385.

Moumita, S., Das, B., Hasan, U., Jayabalan, R., 2018. Effect of long-term storage on viability and acceptability of lyophilized and spray-dried synbiotic microcapsules in dry functional food formulations. LWT 96, 127–132.

Mussatto, S.I., Mancilha, I.M., 2007. Non-digestible oligosaccharides: a review. Carbohydrate Polymers 68 (3), 587–597.

Nemzer, B.V., Al-Taher, F., Kalita, D., Yashin, A.Y., Yashin, Y.I., 2025. Health-improving effects of polyphenols on the human intestinal microbiota: a review. International Journal of Molecular Sciences 26 (3).

Niu, X., Zhang, Q., Liu, J., Zhao, Y., Shang, N., Li, S., et al., 2024. Effect of synbiotic supplementation on obesity and gut microbiota in obese adults: a double-blind randomized controlled trial. Frontiers in Nutrition 11, 1510318.

Nugent, A.P., 2005. Health properties of resistant starch. Nutrition Bulletin 30 (1), 27–54.

O'Toole, P.W., Jeffery, I.B., 2015. Gut microbiota and aging. Science (New York, N.Y.) 350 (6265), 1214–1215.

Olek, A., Woynarowski, M., Ahrén, I.L., Kierkuś, J., Socha, P., Larsson, N., et al., 2017. Efficacy and safety of *Lactobacillus plantarum* DSM 9843 (LP299V) in the prevention of antibiotic-associated gastrointestinal symptoms in children-randomized, double-blind, placebo-controlled study. The Journal of Pediatrics 186, 82–86.

Oliveira, R.P.S., Florence, A.C.R., Silva, R.C., Perego, P., Converti, A., Gioielli, L.A., et al., 2009. Effect of different prebiotics on the fermentation kinetics, probiotic survival and fatty acids profiles in nonfat symbiotic fermented milk. International Journal of Food Microbiology 128 (3), 467–472.

Ouwehand, A.C., 2017. A review of dose-responses of probiotics in human studies. Beneficial Microbes 8 (2), 143–151.

Ozdal, T., Sela, D.A., Xiao, J., Boyacioglu, D., Chen, F., Capanoglu, E., 2016. The reciprocal interactions between polyphenols and gut microbiota and effects on bioaccessibility. Nutrients 8 (2), 78.

Panigrahi, P., Parida, S., Nanda, N.C., Satpathy, R., Pradhan, L., Chandel, D.S., et al., 2017. A randomized synbiotic trial to prevent sepsis among infants in rural India. Nature 548 (7668), 407–412.

Paone, P., Suriano, F., Jian, C., Korpela, K., Delzenne, N.M., Van Hul, M., et al., 2022. Prebiotic oligofructose protects against high-fat diet-induced obesity by changing the gut microbiota, intestinal mucus production, glycosylation and secretion. Gut Microbes 14 (1), 2152307.

Patel, S., Goyal, A., 2011. Functional oligosaccharides: production, properties and applications. World Journal of Microbiology and Biotechnology 27 (5), 1119–1128.

Peng, L., Li, Z.R., Green, R.S., Holzman, I.R., Lin, J., 2009. Butyrate enhances the intestinal barrier by facilitating tight junction assembly via activation of AMP-activated protein kinase in Caco-2 cell monolayers. The Journal of Nutrition 139 (9), 1619–1625.

Pineiro, M., Asp, N.G., Reid, G., Macfarlane, S., Morelli, L., Brunser, O., et al., 2008. FAO Technical meeting on prebiotics. Journal of Clinical Gastroenterology 42 (Suppl 3 Pt 2), S156–S159.

Plaza-Diaz, J., Ruiz-Ojeda, F.J., Gil-Campos, M., Gil, A., 2019. Mechanisms of action of probiotics. Advances in Nutrition 10 (suppl_1), S49–s66.

Pop, C., Suharoschi, R., Pop, O.L., 2021. Dietary fiber and prebiotic compounds in fruits and vegetables food waste. Sustainability 13 (13).

Pourghassem Gargari, B., Dehghan, P., Aliasgharzadeh, A., Asghari Jafar-Abadi, M., 2013. Effects of high performance inulin supplementation on glycemic control and antioxidant status in women with type 2 diabetes. Diabetes and Metabolism Journal 37 (2), 140–148.

Precup, G., Pocol, C.B., Teleky, B.E., Vodnar, D.C., 2022. Awareness, knowledge, and interest about prebiotics-a study among romanian consumers. International Journal of Environmental Research and Public Health 19 (3).

Premi, M., Sharma, H.K., 2017. Effect of different combinations of maltodextrin, gum arabic and whey protein concentrate on the encapsulation behavior and oxidative stability of spray dried drumstick (*Moringa oleifera*) oil. International Journal of Biological Macromolecules 105 (Pt 1), 1232–1240.

Psichas, A., Sleeth, M.L., Murphy, K.G., Brooks, L., Bewick, G.A., Hanyaloglu, A.C., et al., 2015. The short chain fatty acid propionate stimulates GLP-1 and PYY secretion via free fatty acid receptor 2 in rodents. International Journal of Obesity 39 (3), 424–429.

Rackerby, B., Le, H.N.M., Haymowicz, A., Dallas, D.C., Park, S.H., 2024. Potential prebiotic properties of whey protein and glycomacropeptide in gut microbiome. Food Science of Animal Resources 44 (2), 299–308.

Radke, M., Picaud, J.C., Loui, A., Cambonie, G., Faas, D., Lafeber, H.N., et al., 2017. Starter formula enriched in prebiotics and probiotics ensures normal growth of infants and promotes gut health: a randomized clinical trial. Pediatric Research 81 (4), 622–631.

Rastall, R.A., Gibson, G.R., 2015. Recent developments in prebiotics to selectively impact beneficial microbes and promote intestinal health. Current Opinion in Biotechnology 32, 42–46.

Rastmanesh, R., 2011. High polyphenol, low probiotic diet for weight loss because of intestinal microbiota interaction. Chemico-Biological Interactions 189 (1-2), 1–8.

Research P. 2023. Prebiotic Ingredients Market Size, Share, and Trends 2024 to 2034. Ottawa, Canada.

Research GV. 2022. Prebiotics Market Size, Share & Trends Analysis Report By Ingredients (FOS, Inulin, GOS, MOS), By Application (Food & Beverages, Dietary Supplements, Animal Feed), By Region, And Segment Forecasts, 2022–2030.

Research GV. 2023a. Europe Prebiotics Market Size & Outlook, 2023–2030. USA.

Research GV. 2023b. North America Prebiotics Market Size & Outlook, 2023–2030. USA.

Research GV. 2025. Asia Pacific Prebiotics Market Size, Share & Trends Analysis Report By Product, By Source (Roots, Fruits & Vegetables, Cereal & Grains, Others), By Form, By Functionality, By Application, By Country, And Segment Forecasts, 2025–2030. USA.

Roberfroid, M., 2007. Prebiotics: the concept revisited. The Journal of Nutrition 137 (3 Suppl 2), 830s–837ss.

Roberfroid, M., Gibson, G.R., Hoyles, L., McCartney, A.L., Rastall, R., Rowland, I., et al., 2010. Prebiotic effects: metabolic and health benefits. The British Journal of Nutrition 104 (Suppl 2), S1–S63.

Rooks, M.G., Garrett, W.S., 2016. Gut microbiota, metabolites and host immunity. Nature Reviews Immunology 16 (6), 341–352.

Rossander, L., Sandberg, A.S., Sandström, B., 1992. The influence of dietary fibre on mineral absorption and utilisation. In: Schweizer, T.F., Edwards, C.A. (Eds.), Dietary Fibre — A Component of Food: Nutritional Function in Health and Disease. Springer London, London, pp. 197–216.

Saad, N., Delattre, C., Urdaci, M., Schmitter, J.M., Bressollier, P., 2013. An overview of the last advances in probiotic and prebiotic field. LWT - Food Science and Technology 50 (1), 1–16.

Safavi, M., Farajian, S., Kelishadi, R., Mirlohi, M., Hashemipour, M., 2013. The effects of synbiotic supplementation on some cardio-metabolic risk factors in overweight and obese children: a randomized triple-masked controlled trial. International Journal of Food Sciences and Nutrition 64 (6), 687–693.

Salminen, S., Collado, M.C., Endo, A., Hill, C., Lebeer, S., Quigley, E.M.M., et al., 2021. The International Scientific Association of Probiotics and Prebiotics (ISAPP) consensus statement on the definition and scope of postbiotics. Nature Reviews Gastroenterology and Hepatology 18 (9), 649–667.

Sanchez, M., Darimont, C., Drapeau, V., Emady-Azar, S., Lepage, M., Rezzonico, E., et al., 2014. Effect of *Lactobacillus rhamnosus* CGMCC1.3724 supplementation on weight loss and maintenance in obese men and women. The British Journal of Nutrition 111 (8), 1507–1519.

Sanders, M.E., Merenstein, D.J., Reid, G., Gibson, G.R., Rastall, R.A., 2019. Probiotics and prebiotics in intestinal health and disease: from biology to the clinic. Nature Reviews Gastroenterology and Hepatology 16 (10), 605–616.

Sbehat, M., Mauriello, G., Altamimi, M., 2022. Microencapsulation of probiotics for food functionalization: an update on literature reviews. Microorganisms 10 (10).

Scharl, M., Geisel, S., Vavricka, S.R., Rogler, G., 2011. Dying in yoghurt: the number of living bacteria in probiotic yoghurt decreases under exposure to room temperature. Digestion 83 (1-2), 13–17.

Scholz-Ahrens, K.E., Ade, P., Marten, B., Weber, P., Timm, W., Açil, Y., et al., 2007. Prebiotics, probiotics, and synbiotics affect mineral absorption, bone mineral content, and bone structure. The Journal of Nutrition 137 (3 Suppl 2), 838s–846ss.

Scott, K.P., Gratz, S.W., Sheridan, P.O., Flint, H.J., Duncan, S.H., 2013. The influence of diet on the gut microbiota. Pharmacological Research: the Official Journal of the Italian Pharmacological Society 69 (1), 52–60.

Shanuke, D.S., Ranasinghage, N., Illippangama, A.U., Kulathunga, J., Bandara, M.D., 2025. Co-encapsulation of probiotics and prebiotics: techniques and applications in food fortification. Food Science and Nutrition 13 (6), e70426.

Shapiro, H., Suez, J., Elinav, E., 2017. Personalized microbiome-based approaches to metabolic syndrome management and prevention. Journal of Diabetes 9 (3), 226–236.

Siddiqui, S.A., Khan, S., Mehdizadeh, M., Nirmal, N.P., Khanashyam, A.C., Fernando, I., et al., 2023. Consumer studies focus on prebiotics, probiotics, and synbiotics in food packaging: a review. Current Food Science and Technology Reports 1 (1), 13–29.

Silva, Y.P., Bernardi, A., Frozza, R.L., 2020. The role of short-chain fatty acids from gut microbiota in gut-brain communication. Frontiers in Endocrinology 11, 25.

Sionek, B., Szydłowska, A., Trząskowska, M., Kołożyn-Krajewska, D., 2024. The impact of physicochemical conditions on lactic acid bacteria survival in food products. Fermentation 10 (6), 298.

Slavin, J., 2013. Fiber and prebiotics: mechanisms and health benefits. Nutrients 5 (4), 1417–1435.

Smith, P.M., Howitt, M.R., Panikov, N., Michaud, M., Gallini, C.A., Bohlooly, Y.M., et al., 2013. The microbial metabolites, short-chain fatty acids, regulate colonic Treg cell homeostasis. Science (New York, N.Y.) 341 (6145), 569–573.

Smolinska, S., Popescu, F.D., Zemelka-Wiacek, M., 2025. A review of the influence of prebiotics, probiotics, synbiotics, and postbiotics on the human gut microbiome and intestinal integrity. Journal of Clinical Medicine 14 (11).

Sommermeyer, H., Chmielowiec, K., Bernatek, M., Olszewski, P., Kopczynski, J., Piątek, J., 2024. Effectiveness of a balanced nine-strain synbiotic in primary-care irritable bowel syndrome patients-a multi-center, randomized, double-blind, placebo-controlled trial. Nutrients 16 (10).

Stilling, R.M., van de Wouw, M., Clarke, G., Stanton, C., Dinan, T.G., Cryan, J.F., 2016. The neuropharmacology of butyrate: the bread and butter of the microbiota-gut-brain axis? Neurochemistry International 99, 110–132.

Sudha, M.R., Jayanthi, N., Pandey, D.C., Verma, A.K., 2019. Bacillus clausii UBBC-07 reduces severity of diarrhoea in children under 5 years of age: a double blind placebo controlled study. Beneficial Microbes 10 (2), 149–154.

Sung, V., Hiscock, H., Tang, M.L., Mensah, F.K., Nation, M.L., Satzke, C., et al., 2014. Treating infant colic with the probiotic Lactobacillus reuteri: double blind, placebo controlled randomised trial. BMJ (Clinical Research) 348, g2107.

Suzuki, T., 2013. Regulation of intestinal epithelial permeability by tight junctions. Cellular and Molecular Life Sciences: CMLS 70 (4), 631–659.

Swanson, K.S., Gibson, G.R., Hutkins, R., Reimer, R.A., Reid, G., Verbeke, K., et al., 2020. The International Scientific Association for Probiotics and Prebiotics (ISAPP) consensus statement on the definition and scope of synbiotics. Nature Reviews Gastroenterology and Hepatology 17 (11), 687–701.

Swennen, K., Courtin, C.M., Delcour, J.A., 2006. Non-digestible oligosaccharides with prebiotic properties. Critical Reviews in Food Science and Nutrition 46 (6), 459–471.

Tandon, D., Haque, M.M., Gote, M., Jain, M., Bhaduri, A., Dubey, A.K., et al., 2019. A prospective randomized, double-blind, placebo-controlled, dose-response relationship study to investigate efficacy of fructo-oligosaccharides (FOS) on human gut microflora. Scientific Report 9 (1), 5473.

Tárrega, A., Costell, E., 2006. Effect of inulin addition on rheological and sensory properties of fat-free starch-based dairy desserts. International Dairy Journal 16 (9), 1104–1112.

Terrén Lora, A., Penadés, B.F., López Oliva, S., Arponen, S., Okutan, G., Sánchez Niño, G.M., et al., 2024. Supplementation with probiotics, prebiotics, and synbiotics in patients with chronic functional constipation: a randomized, double-blind, placebo-controlled pilot clinical trial. Gastroenterology Report 12, goae101.

Thinkohkaew, K., Jonjaroen, V., Niamsiri, N., McClements, D.J., Panya, A., Suppavorasatit, I., et al., 2025. Fabrication of synbiotic carbohydrate polymer-based microcapsules: Effect of prebiotics on probiotic viability during freeze-drying, gastrointestinal transit and storage. Carbohydrate Polymers 359, 123582.

Tompkins, T.A., Mainville, I., Arcand, Y., 2011. The impact of meals on a probiotic during transit through a model of the human upper gastrointestinal tract. Beneficial Microbes 2 (4), 295–303.

Tonetti, F.R., Eguileor, A., Llorente, C., 2024. Goblet cells: guardians of gut immunity and their role in gastrointestinal diseases. eGastroenterology 2 (3).

Tripathi, M.K., Giri, S.K., 2014. Probiotic functional foods: survival of probiotics during processing and storage. Journal of Functional Foods 9, 225–241.

Trompette, A., Gollwitzer, E.S., Yadava, K., Sichelstiel, A.K., Sprenger, N., Ngom-Bru, C., et al., 2014. Gut microbiota metabolism of dietary fiber influences allergic airway disease and hematopoiesis. Nature Medicine 20 (2), 159–166.

Tuohy, K., Vaughan, E.E., Harthoorn, L.F., Blaak, E.E., Burnet, P.W.J., Busetti, A., et al., 2024. Prebiotics in food and dietary supplements: a roadmap to EU health claims. Gut Microbes 16 (1), 2428848.

Valcheva, R., Koleva, P., Martínez, I., Walter, J., Gänzle, M.G., Dieleman, L.A., 2019. Inulin-type fructans improve active ulcerative colitis associated with microbiota changes and increased short-chain fatty acids levels. Gut Microbes 10 (3), 334–357.

Valladares-Diestra, K.K., de Souza Vandenberghe, L.P., Vieira, S., Goyzueta-Mamani, L.D., de Mattos, P.B.G., Manzoki, M.C., et al., 2023. The potential of xylooligosaccharides as prebiotics and their sustainable production from agro-industrial by-products. Foods 12 (14).

van de Wouw, M., Boehme, M., Lyte, J.M., Wiley, N., Strain, C., O'Sullivan, O., et al., 2018. Short-chain fatty acids: microbial metabolites that alleviate stress-induced brain-gut axis alterations. The Journal of Physiology 596 (20), 4923–4944.

Vandenplas, Y., Zakharova, I., Dmitrieva, Y., 2015. Oligosaccharides in infant formula: more evidence to validate the role of prebiotics. The British Journal of Nutrition 113 (9), 1339–1344.

Vazquez-Olivo, G., Gutiérrez-Grijalva, E.P., Heredia, J.B., 2019. Prebiotic compounds from agro-industrial by-products. Journal of Food Biochemistry 43 (6), e12711.

Vijayaram, S., Sinha, R., Faggio, C., Ringø, E., Chou, C.C., 2024. Biopolymer encapsulation for improved probiotic delivery: advancements and challenges. AIMS Microbiology 10 (4), 986–1023.

Vinderola, G., Sanders, M.E., Cunningham, M., Hill, C., 2023. Frequently asked questions about the ISAPP postbiotic definition. Frontiers in Microbiology 14, 1324565.

Vinelli, V., Biscotti, P., Martini, D., Del, B.C., Marino, M., Meroño, T., et al., 2022. Effects of dietary fibers on short-chain fatty acids and gut microbiota composition in healthy adults: a systematic review. Nutrients 14 (13).

Visciglia, A., Allesina, S., Amoruso, A., De Prisco, A., Dhir, R., Bron, P.A., et al., 2022. Assessment of shelf-life and metabolic viability of a multi-strain synbiotic using standard and innovative enumeration technologies. Frontiers in Microbiology 13, 989563.

Wan, M.L.Y., Co, V.A., El-Nezami, H., 2021. Dietary polyphenol impact on gut health and microbiota. Critical Reviews in Food Science and Nutrition 61 (4), 690–711.

Wang, X., Qi, Y., Zheng, H., 2022. Dietary polyphenol, gut microbiota, and health benefits. Antioxidants (Basel) 11 (6).

Wang, J., Wu, P., Dhital, S., Yu, A., Chen, X.D., 2025. Impact of freezing and freeze drying on *Lactobacillus rhamnosus* GG survival: mechanisms of cell damage and the role of pre-freezing conditions and cryoprotectants. Foods 14 (10).

Wijesekara, T., Abeyrathne, E., Ahn, D.U., 2024. Effect of bioactive peptides on gut microbiota and their relations to human health. Foods 13 (12).

Wilson, B., Eyice, Ö., Koumoutsos, I., Lomer, M.C., Irving, P.M., Lindsay, J.O., et al., 2021. Prebiotic galactooligosaccharide supplementation in adults with ulcerative colitis: exploring the impact on peripheral blood gene expression, gut microbiota, and clinical symptoms. Nutrients 13 (10).

Yang, S.Y., Han, S.M., Lee, J.Y., Kim, K.S., Lee, J.E., Lee, D.W., 2025. Advancing gut microbiome research: the shift from metagenomics to multi-omics and future perspectives. Journal of Microbiology and Biotechnology 35, e2412001.

Yoo, S., Jung, S.C., Kwak, K., Kim, J.S., 2024. The role of prebiotics in modulating gut microbiota: implications for human health. International Journal of Molecular Sciences 25 (9).

You, S., Ma, Y., Yan, B., Pei, W., Wu, Q., Ding, C., et al., 2022. The promotion mechanism of prebiotics for probiotics: a review. Frontiers in Nutrition 9, 1000517.

Zeng, M., van Pijkeren, J.P., Pan, X., 2023. Gluco-oligosaccharides as potential prebiotics: Synthesis, purification, structural characterization, and evaluation of prebiotic effect. Comprehensive Reviews in Food Science and Food Safety 22 (4), 2611–2651.

Zhang, C., Yin, A., Li, H., Wang, R., Wu, G., Shen, J., et al., 2015. Dietary modulation of gut microbiota contributes to alleviation of both genetic and simple obesity in children. EBioMedicine 2 (8), 968–984.

Zhang, X.F., Guan, X.X., Tang, Y.J., Sun, J.F., Wang, X.K., Wang, W.D., et al., 2021. Clinical effects and gut microbiota changes of using probiotics, prebiotics or synbiotics in inflammatory bowel disease: a systematic review and meta-analysis. European Journal of Nutrition 60 (5), 2855–2875.

Zhao, X., Wang, Q., Wang, T., Su, Y., Huang, C., Lai, C., et al., 2024. Evaluation of prebiotic ability of xylo-oligosaccharide fractions with different polymerization degrees from bamboo shoot shells. Food and Bioproducts Processing 143, 202–211.

Zmora, N., Zilberman-Schapira, G., Suez, J., Mor, U., Dori-Bachash, M., Bashiardes, S., et al., 2018. Personalized gut mucosal colonization resistance to empiric probiotics is associated with unique host and microbiome features. Cell 174 (6), 1388–1405.e21.

Chapter 14

The economy of prebiotics in drugs, supplements, and foods, and the future trend

Erhan Keyvan[1], Erdi Şen[1], Nilay Keyvan[2], and Muammer Goncuoglu[3]
[1]*Department of Food Hygiene and Technology, Faculty of Veterinary Medicine, Burdur Mehmet Akif Ersoy University, Burdur, Türkiye,* [2]*Burdur Representative Office, Turkish Standards Institute, Burdur, Türkiye,* [3]*Department of Food Hygiene and Technology, Faculty of Veterinary Medicine, Ankara University, Ankara, Türkiye*

14.1 Introduction

In recent years, people have adopted new and diverse habits, shifting away from routine practices in various aspects of life. Advances in the food industry and changes in lifestyles have altered expectations, particularly in food consumption. As interest grows in the connection between dietary habits, health, and longevity, numerous studies have emerged in the fields of food science and nutrition. Increasing awareness has made it clear to consumers that daily nutrition, as part of one's overall diet, plays a critical role in health. Consequently, there is a growing trend toward functional foods—those that offer additional benefits beyond basic nutrition, such as energy, mineral salts, trace elements, and vitamins.

Today, many consumers seek functional foods that help prevent disruptions in the gut microbiota, which is linked to various health issues. Recent genomic studies have significantly expanded current knowledge regarding the gut microbiota. The relationship between gut health and the risk of disease, along with the impact of dietary components on microbiota modulation, has become a prominent area of discussion. As a result, functional foods and additives that positively affect the composition or activity of the microbiota have become major topics of research in recent years (Kasote et al., 2021; Martinez et al., 2015; Topolska et al., 2021).

The development of functional foods containing prebiotic ingredients has become a key focus in the food industry, emerging as a promising market. This shift is driven not only by economic potential but also by scientific evidence

Evaluating Oral Probiotics and Prebiotics. DOI: https://doi.org/10.1016/B978-0-443-29835-6.00018-6

supporting the benefits of such foods. Today's consumers are increasingly aware of the relationship between balanced nutrition and health, which has led to changing food consumption priorities. They are now seeking foods that provide both nutritional value and health benefits. In response, prebiotics like inulin and fructooligosaccharides (FOS) are being incorporated into a variety of foods and beverages, including dairy products, bread, cereals, and dietary supplements. These additions allow consumers to choose foods that can offer positive health effects.

Prebiotics have been shown to provide numerous health benefits, making them attractive to the food processing and pharmaceutical industries. Their benefits include supporting the development of starter cultures, promoting intestinal health, and aiding in the prevention of cancer, obesity, and constipation. Prebiotics may also help modulate the immune system positively. Recent trends have focused on innovative sources of prebiotics, such as food waste, incorporating prebiotics into the diet, and exploring their biotechnological synthesis. Researchers worldwide are working to improve access to prebiotics, making them more widely recognized and consumed by the general public (Burgain et al., 2011; Coman et al., 2012; Rolim, 2015; Choque Delgado et al., 2011).

The human gastrointestinal tract hosts a complex and dynamic population of commensal bacteria that have coevolved with the host in a symbiotic relationship. This intestinal microbiota, a dynamic community of microorganisms, plays critical roles in metabolic, immunological, and physiological processes essential for health. In healthy individuals, the microbiota is predominantly composed of bacteria from the Bacteroidetes and Firmicutes phyla. The human body's systems, functions, and cycles are closely linked to the microbiota, though its composition can vary from person to person. Any disruption to the microbiota can significantly impact health and disease processes. Longterm dietary habits are crucial for shaping and maintaining a healthy gut microbiota, with most microorganisms residing in the large intestine—home to an estimated 10^{14} bacteria. Factors such as diet, age, lifestyle, ethnicity, environmental toxins, and overall health influence the composition of the microbiota (De Filippo et al., 2010; Bäckhed, 2011; Kaewarsar et al., 2023).

Once ingested, prebiotics pass through the upper digestive tract intact and reach the colon, where they are fermented by the gut microbiota, producing health promoting metabolites known as short chain fatty acids (SCFAs). Prebiotics, in this context, are nondigestible food components that improve host health by selectively stimulating the growth and/or activity of beneficial microorganisms in the colon. The main prebiotic components such as inulin, FOS, galactooligosaccharides (GOS), and lactulose achieve their effects through the SCFA produced during fermentation. Although prebiotics are manufactured on an industrial scale, consuming them directly through food can be equally effective. Common dietary sources of prebiotics include grains like wheat, barley, oats, and corn; fruits and root vegetables such as onions, garlic, bananas, endive, asparagus, and leeks; as well as legumes, nuts, breast milk, and yogurt (Slavin, 2013; Amorim et al., 2020; Ayaz, 2021; Ricke, 2021; Singh et al., 2012).

14.2 Definition and classification of prebiotics

The concept of prebiotics has continuously evolved since it was first defined in 1995. Gibson and Roberfroid (1995) initially described prebiotics as "a nondigestible food ingredient that beneficially affects the host by selectively stimulating the growth and/or activity of one or a limited number of bacteria already resident in the colon." Later, in 2004, this definition was expanded to "selectively fermented ingredients that allow specific changes in both the composition and/or activity of the digestive tract microbiota that provide benefits to the well being and health of the host" (Gibson et al., 2004). According to this updated definition, prebiotics must meet three basic criteria: They must confer health benefits by resisting digestion, be fermented by intestinal microorganisms, and selectively stimulate the growth and/or activity of beneficial bacteria. Although the prebiotic effect provides strong evidence for health benefits, in vitro analyzes show that more research is needed to fully understand the effects of prebiotics alone (Cardoso et al., 2021).

In 2008 the Food and Agriculture Organization (FAO) of the United Nations redefined prebiotics as "a nonviable food component that confers health benefits to the host in conjunction with the modulation of the microbiota" (Pineiro et al., 2008). This definition emphasized the application of prebiotics beyond the gastrointestinal tract, but it did not fully encapsulate the concept, as it removed the selective fermentation criterion and failed to exclude antibiotics. To address these gaps, Gibson et al. (2010) redefined dietary prebiotics as "a selectively fermented ingredient that induces specific changes in the composition and/or activity of the digestive tract microbiota, thereby providing host health benefit(s)." Later research, such as that by Dewulf et al. (2013) and Bindels et al. (2015), provided increasing evidence that prebiotics stimulate more than one microbial taxon rather than a single species. Consequently, the definition was refined to describe prebiotics as nondigestible compounds metabolized by intestinal microorganisms, leading to changes in microbiota composition and/or activity, thereby benefiting the host's health.

While this definition highlighted the functional effects of prebiotics on the microbiota, it restricted prebiotics to interactions with the intestinal microbiota and removed the need for selective fermentation. In response to ongoing scientific developments, the International Scientific Association for Probiotics and Prebiotics (ISAPP) convened an expert panel to further revise the definition of prebiotics. This updated definition, established by Gibson et al. (2017), describes prebiotics as "a substrate selectively utilized by host microorganisms, providing a health benefit." This broader definition remains consistent with earlier concepts of prebiotics while expanding the scope to include potential prebiotic substances beyond carbohydrates, such as plant polyphenols. It also opens the door to applications outside the gastrointestinal tract. Advances in molecular based tools, especially in microbiome research, have further supported this expanded definition (Cardoso et al., 2021).

Several fundamental criteria must be met for a food ingredient to be classified as a prebiotic:

- It should resist digestion and absorption in the upper gastrointestinal tract, reaching the colon intact. Ideally, prebiotics should persist throughout the large intestine, offering benefits that extend distally.
- It must be selectively metabolized by probiotic microorganisms. A key criterion for classification as a prebiotic is selective fermentation by potentially beneficial bacteria in the colon.
- It should support the growth and activity of intestinal microflora. Prebiotic activity refers to a substrate's ability to foster the growth of beneficial organisms over other organisms, compared to nonprebiotic substrates like glucose. Carbohydrates demonstrate positive prebiotic activity if they are selectively metabolized by probiotic strains alongside glucose but not by other intestinal bacteria.
- Additionally, prebiotics must offer added health benefits to humans when consumed and remain stable during food production processes (Gibson et al., 2004; Wang, 2009; Oliveira et al., 2011).

Perhaps most importantly, prebiotics must provide health benefits in addition to their selective effects on microbial groups, particularly probiotics. SCFAs, including acetate, propionate, and lactate (the precursor to butyrate), are crucial metabolites produced by prebiotics. These compounds play a key role in modulating colonocyte function and are linked to various health outcomes, such as maintaining intestinal homeostasis, energy acquisition, immune function, blood lipid levels, appetite control, and metabolic activity. The importance of these impacts will be explored in greater detail in the subsequent sections (Roberfroid, 2002; Koh et al., 2016; Boets et al., 2017; Gibson et al., 2017; Cardoso et al., 2021).

Earlier studies on prebiotic types suggested that prebiotics were mainly oligosaccharide carbohydrates, including xylooligosaccharides (XOS), galactooligosaccharides (GOS), lactulose, inulin, and its derivatives such as fructooligosaccharides (FOS). However, recent research indicates that prebiotics can also include noncarbohydrate substances that meet the prebiotic criteria, such as polyphenols isolated from fruits, particularly berries (Ashwini et al., 2019; Gu et al., 2019; Jiao et al., 2019; Khangwal and Shukla, 2019; Zhao et al., 2021; Yin et al., 2022; You et al., 2022). Besides indigestible carbohydrates like oligosaccharides, some peptides, proteins, and lipids also possess prebiotic properties. Many prebiotics, including fructooligosaccharides (FOS), galactooligosaccharides (GOS), isomaltoligosaccharides (IMO), and

TABLE 14.1 Types and sources of prebiotics.

Type	Source
Inulin	Chicory root, Jerusalem artichoke, garlic
Fructooligosaccharides (FOS)	Bananas, onions, asparagus, leeks
Galactooligosaccharides (GOS)	Chickpeas, lentils, infant formulas
Lactulose Resistant Starch	Synthetic (derived from lactose)
	Green bananas, raw potatoes, legumes
Arabinoxylan	Wheat, rye, barley
Pectin	Apples, citrus fruits, berries
Beta Glucans	Oats, barley, mushrooms
Polyphenols	Berries, tea, red wine, dark chocolate
Human Milk Oligosaccharides (HMOs)	Human breast milk

xylooligosaccharides (XOS), are either extracted from or synthesized via depolymerization of plant and algal polysaccharides (Table 14.1). Commercial preparations of oligosaccharides, such as soybean oligosaccharides (SOS), GOS, and XOS, are widely available (Courtin et al., 2009; Wang, 2009; Ozyurt and Otles, 2014). As prebiotic preparation processes improve, new types of prebiotics are constantly being developed, particularly polysaccharides, polyphenols, and polypeptide polymers, which hold significant research potential. Prebiotics can be classified into different groups based on their chemical structure and source.

14.2.1 Inulin type fructans

14.2.1.1 Inulin

Inulin is found in chicory root, Jerusalem artichoke, dandelion greens, and garlic.

- Source: Inulin is a type of soluble fiber found in various plants, with chicory root being one of the richest sources. Other sources include Jerusalem artichoke, dandelion greens, garlic, and leeks.
- Structure: Inulin is a linear chain of fructose molecules with a terminal glucose molecule. The length of the chain can vary, affecting its solubility and fermentability.
- Function: Inulin serves as a prebiotic by resisting digestion in the small intestine and reaching the colon intact. There, it is fermented by beneficial bacteria, promoting their growth and activity.
- Health benefits: In addition to supporting gut health, inulin has been associated with various health benefits, including improved mineral absorption, blood sugar regulation, and potential weight management.

14.2.1.2 *Fructooligosaccharides*

Present in various fruits and vegetables like bananas, onions, leeks, and asparagus.

- Source: FOS are naturally occurring compounds found in a variety of fruits and vegetables, such as bananas, onions, leeks, asparagus, and garlic.
- Structure: FOS are short chain polymers of fructose molecules with a terminal glucose unit.
- Function: Like inulin, FOS act as prebiotics by resisting digestion in the small intestine and reaching the colon, where they are fermented by beneficial bacteria.
- Health benefits: FOS have been studied for their potential to promote gut health, enhance mineral absorption, and modulate immune function. They may also contribute to the production of SCFAs in the colon.

In summary, inulin type fructans, including inulin and fructooligosaccharides (FOS), are prebiotics found in certain plants. They contribute to gut health by serving as a food source for beneficial bacteria in the colon, promoting their growth and metabolic activity. The fermentation of inulin type fructans in the colon is associated with various health benefits beyond gut health.

14.2.2 Galactooligosaccharides

GOSs are found naturally in some legumes, such as chickpeas and lentils. They are often added to certain foods and used as an ingredient in prebiotic supplements.

- Source: GOSs are naturally found in certain legumes, including chickpeas and lentils. They can also be produced synthetically and are often added to food products and infant formulas.
- Structure: GOSs are oligosaccharides composed of galactose units, typically with a terminal glucose unit.
- Function: Similar to other prebiotics, GOSs resist digestion in the small intestine and reach the colon intact. There, they serve as a substrate for the growth of beneficial bacteria, contributing to a balanced gut microbiota.
- Health benefits: GOSs have been studied for their potential to support gut health, enhance the growth of Bifidobacteria, and modulate the composition of the gut microbiota. They may also have positive effects on immune function.

14.2.3 Lactulose

Lactulose is a synthetic disaccharide derived from lactose. It is resistant to digestion in the small intestine, reaching the colon where it serves as a substrate for beneficial bacteria.

- Source: Lactulose is a synthetic disaccharide derived from lactose, which is the sugar found in milk.

- Structure: It is composed of two sugar molecules, galactose and fructose, linked together.
- Function: Lactulose is not fully absorbed in the small intestine. Instead, it passes into the colon, where it can be fermented by beneficial bacteria. This fermentation process produces SCFAs and promotes the growth of beneficial microorganisms.
- Health benefits: Lactulose is often used as a laxative due to its ability to draw water into the colon, softening stools. In addition to its laxative effects, the fermentation of lactulose contributes to a healthier gut environment.

14.2.4 Resistant starch

RS is a type of starch that resists digestion in the small intestine. It is found in foods like green bananas, raw potatoes, and certain legumes.

- Source: Resistant starch is a type of starch that escapes digestion in the small intestine and reaches the colon intact. It is found in certain foods, including green bananas, raw potatoes, legumes, and some grains
- Structure: Resistant starch has a structure that allows it to resist enzymatic digestion in the small intestine, leading to its fermentation in the colon.
- Function: As a prebiotic, resistant starch serves as a substrate for beneficial bacteria in the colon. The fermentation process produces SCFAs, which contribute to gut health.
- Health benefits: Resistant starch has been associated with various health benefits, including improved insulin sensitivity, enhanced satiety, and the production of SCFAs that support colon health.

14.2.5 Arabinoxylan

AX is a hemicellulose found in the cell walls of grains, including wheat and rye. It acts as a prebiotic to support the growth of beneficial bacteria in the gut.

- Source: Arabinoxylan is a type of hemicellulose found in the cell walls of many grains, including wheat, rye, barley, and oats.
- Structure: It is a complex polysaccharide consisting of a backbone of xylose molecules with arabinose side chains.
- Function: Arabinoxylan acts as a prebiotic by reaching the colon intact and serving as a substrate for the fermentation by beneficial bacteria. The breakdown of arabinoxylan by these bacteria produces SCFAs.
- Health benefits: In addition to supporting gut health, arabinoxylan has been studied for its potential to have antiinflammatory effects and to modulate the immune system. It may contribute to the production of SCFAs, which play a role in maintaining the health of the colon.

14.2.6 Pectin

Pectin is a soluble fiber found in fruits like apples, citrus fruits, and berries. It can be fermented by gut bacteria, promoting their growth.

- Source: Pectin is a soluble fiber found in various fruits, particularly in the peels and cores. Good sources include apples, citrus fruits, berries, and apricots.
- Structure: Pectin is a complex polysaccharide composed of repeating units of galacturonic acid. Its structure varies, and it can form gels under certain conditions.
- Function: Pectin acts as a prebiotic by resisting digestion in the small intestine and reaching the colon intact. There, it serves as a substrate for the fermentation by beneficial bacteria.
- Health benefits: Beyond its prebiotic effects, pectin has been associated with several health benefits, including improved cholesterol levels, blood sugar regulation, and potential antiinflammatory effects. It may also contribute to a feeling of fullness.

14.2.7 Beta glucans

BGs are found in oats, barley, and certain mushrooms. They are known for their immune modulating properties and potential prebiotic effects.

- Source: Beta glucans are polysaccharides found in the cell walls of certain grains, such as oats and barley, as well as in certain mushrooms, including shiitake and maitake.
- Structure: Beta glucans are linear or branched glucose polymers linked by beta glycosidic bonds.
- Function: Beta glucans act as prebiotics by reaching the colon intact and serving as a substrate for the fermentation by beneficial bacteria. They contribute to the production of SCFAs.
- Health benefits: Beta glucans are wellknown for their immune modulating properties. Additionally, their prebiotic effects may contribute to gut health by promoting the growth of beneficial bacteria. They have also been studied for their potential role in managing cholesterol levels and supporting cardiovascular health.

14.2.8 Polyphenols

PPs are plant compounds found in foods like berries, tea, red wine, and dark chocolate. While primarily known for their antioxidant properties, some polyphenols may also have prebiotic effects.

- Source: Polyphenols are a diverse group of plant compounds found in a variety of foods, including fruits, vegetables, tea, coffee, red wine, dark chocolate, and certain herbs and spices.

- Structure: Polyphenols have complex chemical structures, and they are classified into different subclasses, such as flavonoids, phenolic acids, and lignans.
- Function: While primarily known for their antioxidant properties, some polyphenols have prebiotic effects. They can reach the colon intact, where they may influence the composition and activity of the gut microbiota.
- Health benefits: Polyphenols have been associated with various health benefits, including antiinflammatory and antioxidant effects. Their interactions with the gut microbiota may contribute to overall gut health, and they may also have implications for immune function.

14.2.9 Human milk oligosaccharides

HMOs are unique to human breast milk. They serve as prebiotics for the development of a healthy gut microbiota in infants.

- Source: HMOs are a group of complex oligosaccharides found exclusively in human breast milk.
- Structure: HMOs are diverse in structure, consisting of various sugar units. They are unique to human milk and serve as a source of nutrition for the developing infant.
- Function: HMOs act as prebiotics by promoting the growth of specific beneficial bacteria, such as *Bifidobacteria*, in the infant's gut. They contribute to the establishment of a healthy gut microbiota in early life.
- Health benefits: HMOs are thought to play a crucial role in supporting the infant's immune system, protecting against infections, and influencing the development of a balanced gut microbiome. They are not digested by the infant but serve as food for beneficial bacteria.

These classifications are based on the chemical nature of prebiotics and their sources. It's important to note that the field of prebiotics is dynamic, and ongoing research may reveal new compounds with prebiotic properties. Additionally, the effects of prebiotics can vary among individuals, and the overall impact on health is a subject of ongoing investigation.

While the categories mentioned earlier encompass naturally occurring prebiotics found in various foods, the food industry has also developed commercial prebiotics for use in food products and supplements. These commercial prebiotics often aim to provide specific health benefits and enhance the nutritional profile of the products. Here are some examples of commercially used prebiotics:

- Fructooligosaccharides (FOS) and inulin
  These are commonly used as prebiotic ingredients in a variety of food products, including dairy, cereals, bars, and beverages.
- Galactooligosaccharides (GOSs)
  GOSs are often used as a prebiotic in infant formulas to mimic the prebiotic content found in human breast milk.

- Lactulose
  Lactulose is not only used as a prebiotic but is also employed as a laxative in the food and pharmaceutical industries.
- Resistant starch
  Modified forms of resistant starch may be used as prebiotic ingredients in certain food products, including baked goods and snacks.
- Arabinoxylan
  Extracts of arabinoxylan may be utilized in the food industry to enhance the prebiotic content of certain products.
- Pectin
  Modified pectin is sometimes used as a prebiotic ingredient in the food industry, particularly in the production of certain jams, jellies, and baked goods.
- Beta glucans
  Beta glucans from oats, barley, or mushrooms may be incorporated into food products to provide prebiotic effects.
- Polyphenols

Certain polyphenol rich extracts or compounds may be used as prebiotics in functional foods, beverages, or supplements (Gibson et al., 2010, 2017; Cardoso et al., 2021; Muzaffar et al., 2021; You et al., 2022).

14.2.10 New and candidate prebiotics

Exploring new prebiotics is crucial for understanding their potential to support intestinal health and maintain a balanced microbiota. The field of prebiotic research is dynamic, with a growing number of candidates under investigation. Below are some promising prebiotics.

14.2.10.1 Xylo oligosaccharides

XOSs are derived from xylan in plant fibers. They are being studied for their ability to selectively stimulate beneficial bacteria in the gut.

14.2.10.2 Mannan oligosaccharides

MOSs are extracted from yeast cell walls. They show potential to promote gut health by encouraging the growth of beneficial bacteria and supporting the immune system.

14.2.10.3 Glucomannan

GM is a soluble fiber found in the roots of the konjac plant and is being explored for its prebiotic effects and its role in fostering beneficial bacteria.

14.2.10.4 Lactosaccharose

LS is formed through enzymatic synthesis of lactose and sucrose and has been investigated for its prebiotic properties and potential applications in food products.

14.2.10.5 Alginate oligosaccharides

AO is obtained from seaweed and may influence gut microbiota composition, thus supporting intestinal health.

14.2.10.6 Isomalto oligosaccharides

IMOs consist of a mixture of short chain carbohydrates used in some food products. They are currently under investigation for their impact on gut microbiota and health.

14.2.10.7 Polydextrose

PD is a synthetic glucose polymer that resists digestion in the small intestine and reaches the colon, where it serves as a substrate for beneficial bacteria.

14.2.10.8 Acacia gum

AG is also known as gum arabic and is a soluble fiber from acacia tree sap. Studies have explored its potential prebiotic properties and effects on the gut microbiota (Cardoso et al., 2021; Cunningham et al., 2021; You et al., 2022; Biswal et al., 2017).

While the prebiotic potential of these substances is being investigated, it is crucial to acknowledge that further research is needed to fully understand their impact on gut microbiota and overall health. With continuous advances in the field, new candidates may emerge, providing additional benefits. The identification of novel prebiotics often involves examining various compounds and their influence on the gut microbiota. Below are key areas of interest in the search for potential prebiotics:

14.2.10.9 Plant based fibers

Ongoing research is exploring a wide range of plant based fibers, such as those from fruits, vegetables, and whole grains, for their prebiotic properties. These may include lesser-known fibers with unique structures and potential health benefits.

14.2.10.10 Microbial derived compounds

Certain compounds produced by microorganisms, particularly bacterial fermentation products, are being investigated for their prebiotic effects. These metabolites may positively influence the intestinal environment.

14.2.10.11 Synthetic compounds

Researchers are also developing synthetic prebiotics tailored for specific functions. These compounds are designed to resist digestion in the upper gastrointestinal tract and selectively promote the growth of beneficial bacteria in the colon.

- Nanostructures
  Nanotechnology is an emerging field that could revolutionize prebiotics. Researchers are investigating nanostructures that can act as carriers for prebiotics, protecting them during digestion and delivering them to the colon for fermentation.
- Protein derived prebiotics

Some peptides and proteins are being studied for their prebiotic potential. These could include proteins from unconventional sources or modified protein structures that enhance resistance to digestion (Cardoso et al., 2021; Cunningham et al., 2021; You et al., 2022; Biswal et al., 2017).

It is important to note that the use of commercial prebiotics in food products is highly regulated, and their introduction into the market is generally based on scientific evidence supporting their safety and efficacy. Additionally, the specific prebiotic chosen for a product may depend on the desired properties and targeted health benefits. Consumers should always check product labels for information regarding added prebiotics and consult healthcare professionals for specific advice on incorporating them into their diet.

14.3 Mode of action of prebiotics

Prebiotics exert their effects through various mechanisms that contribute to the maintenance of a healthy gut microbiota. Key mechanisms of action are discussed in the following sections.

14.3.1 Selective fermentation

Prebiotics are selectively fermented in the colon by beneficial bacteria, such as bifidobacteria and lactobacilli. These bacteria possess enzymes that breakdown complex carbohydrates into smaller molecules, producing beneficial metabolites in the process.

14.3.2 Production of short chain fatty acids

The fermentation of prebiotics by gut bacteria leads to the production of SCFAs, such as acetate, propionate, and butyrate. SCFAs play a crucial role in maintaining the integrity of the intestinal lining, modulating immune function, and serving as an energy source for colonocytes (cells lining the colon).

14.3.3 pH modulation

SCFA production lowers the pH of the colonic environment, creating an acidic milieu that inhibits the growth of potentially harmful bacteria and favors the proliferation of beneficial bacteria.

14.3.4 Development of beneficial bacteria

Prebiotics support the growth and activity of beneficial bacteria, particularly bifidobacteria and lactobacilli. These bacteria help establish a balanced and diverse gut microbiota, which is associated with various health benefits.

14.3.5 Stimulation of mucus production

Certain prebiotics, especially specific fibers, can enhance mucus production in the colon. Mucus forms a protective barrier for the intestinal lining and facilitates the interaction between the gut microbiota and the host.

14.3.6 Modulation of the gut–brain axis

The gut–brain axis refers to the bidirectional communication between the gut and the central nervous system. Prebiotics can influence this communication, potentially affecting mood, cognitive function, and stress responses.

14.3.7 Immune modulation

By influencing the composition of the gut microbiota, prebiotics can modulate the immune system. A balanced and diverse gut microbiota has been linked to improved immune function and sensitivity.

14.3.8 Antiinflammatory effects

SCFAs, particularly butyrate, have been shown to exert antiinflammatory effects in the gut. Prebiotics contribute to the production of these antiinflammatory metabolites (Guarino et al., 2020; Megur et al., 2022; You et al., 2022; Ansari et al., 2023).

It is important to note that the effects of prebiotics can vary depending on factors such as the specific type of prebiotic used, individual differences in gut microbiota composition, and an individual's diet and lifestyle. Ongoing research continues to uncover new insights into the mechanisms of action of prebiotics and their potential health benefits (Fig. 14.1).

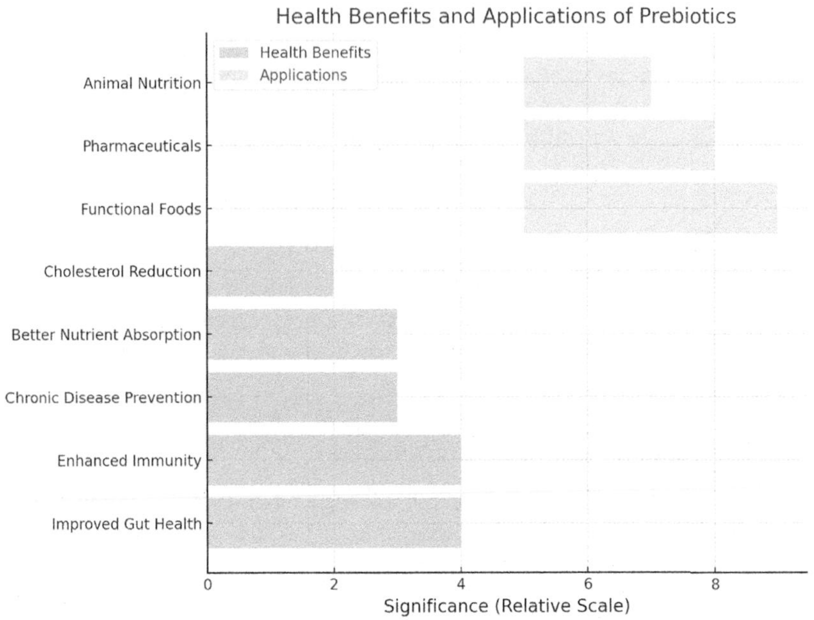

FIGURE 14.1 Health benefits and applications of prebiotics.

14.4 Effects of prebiotics on gastrointestinal health

The gut microbiota refers to the vast and diverse population of microorganisms residing in the human gastrointestinal tract. This ecosystem plays a critical role in various aspects of health, including digestion, nutrient absorption, immune response, and metabolism. Maintaining a balanced gut microbiota is essential for overall health, and prebiotics have emerged as key regulators with significant benefits (Quigley, 2019). The gut microbiome contains the largest bacterial population in the gastrointestinal tract and employs multiple mechanisms that enhance immune function (Vitetta et al., 2014). Prebiotics, in particular, promote the growth of beneficial microorganisms, such as bifidobacteria and lactobacilli, contributing to a more diverse and healthier gut environment (Gibson and Collins, 1999).

Prebiotics are classified as dietary fibers (DF), though not all dietary fibers exhibit prebiotic properties (Guarino et al., 2020). DF is a type of carbohydrate that humans cannot digest due to the lack of necessary digestive enzymes. Naturally occurring sources of DF include fruits, vegetables, and grains (He et al., 2022). Studies have demonstrated the positive impact of DF on managing serious health conditions such as cardiovascular diseases, obesity, diabetes mellitus, and cancer (Wang et al., 2015; Luo et al., 2017; Cheng et al., 2017; Soliman, 2019).

Irritable bowel syndrome (IBS) is a gastrointestinal disorder characterized by chronic abdominal symptoms linked to gut microbiota imbalance, affecting the normal functioning of the digestive system. While the exact causes of IBS are not fully understood, it is associated with sociological, economic, and psychological factors that contribute to a decreased quality of life and the development of anxiety and depression (Simon et al., 2021).

14.5 Prebiotics in pharmaceuticals

Prebiotics have shown immense potential in the pharmaceutical industry due to their health promoting properties. They are being investigated for their roles in maintaining gut health, preventing colitis, inhibiting cancer, enhancing immune function, and reducing cholesterol levels (Patel and Goyal, 2012). The pharmaceutical applications of prebiotics are diverse, ranging from their use in starter culture formulations to their incorporation in various therapeutic interventions. Prebiotics play a crucial role in modulating the intestinal microbiota, which is essential for maintaining gut health. They increase the production of SCFAs, which have been linked to a reduced risk of developing diseases such as colon cancer, irritable bowel syndrome, type 2 diabetes mellitus, and obesity (Farias et al., 2019). The ability of prebiotics to enhance gut health makes them valuable in the prevention and management of gastrointestinal diseases. Prebiotics have been shown to improve immune function by stimulating the growth of beneficial bacteria that produce metabolites with immunomodulatory effects. These metabolites can enhance the body's defense mechanisms, making prebiotics a promising component in cancer prevention strategies (Patel and Goyal, 2012). The antiinflammatory properties of prebiotics further contribute to their potential in reducing the risk of chronic diseases. The consumption of prebiotics has been associated with a reduction in cholesterol levels, which is beneficial for cardiovascular health. By promoting the growth of bacteria that metabolize cholesterol, prebiotics can help lower the risk of cardiovascular diseases. This makes them an attractive ingredient in the development of pharmaceutical products aimed at improving heart health (Patel and Goyal, 2012).

14.6 Prebiotics as food additives

The food industry has recognized the potential of prebiotics as functional ingredients that can enhance the nutritional, physical–chemical, and sensory characteristics of food products. The incorporation of prebiotics into food additives is a growing trend, driven by consumer demand for healthier food options. Prebiotics can improve the nutritional profile of food products by increasing the content of beneficial compounds such as SCFAs, vitamins, and minerals. The use of microorganisms and enzymes in the synthesis of nondigestible carbohydrates can further enhance the variety and availability of prebiotic ingredients in the market (Farias et al., 2019). The addition of prebiotics to food products can improve their sensory attributes, such as taste and texture, making them more appealing to consumers. Innovative technologies, such as high hydrostatic pressure, ultrasound, and microwave processing, are being explored to incorporate prebiotics into food products without compromising their stability and functionality (Hurtado-Romero et al., 2020). In animal studies, prebiotics have been effective in enhancing growth performance, improving carcass traits, and regulating intestinal microbiota. For instance, a phytobiotic-prebiotic feed additive containing carob pulp, chicory, and fenugreek significantly improved growth indicators and carcass quality in fattening pigs while promoting beneficial bacteria and inhibiting pathogenic bacteria (Juhász et al., 2023). Similarly, prebiotics have shown potential in limiting the establishment of foodborne pathogens such as *Campylobacter* in poultry, thereby contributing to safer food production (Kim et al., 2019; Gaggìa et al., 2010; Roto et al., 2015). Prebiotics are not only beneficial for health but also offer several technological advantages when used as food additives. They improve the sensory characteristics of food products, such as taste and texture, and enhance the stability of foams, emulsions, and mouthfeel in various food applications like dairy products and bread (Al-Sheraji et al., 2013). For example, an α-d-glucan from *Lactobacillus plantarum* DM5 demonstrated excellent prebiotic activities and technological properties, making it a potential additive for food products (Das et al., 2014). Despite the growing interest and promising results, there are still several research gaps and challenges in the field of prebiotics. The mechanisms of their beneficial impacts on human health are complex and not fully understood. Future research should focus on elucidating these mechanisms and exploring the synergistic effects of combining prebiotics with probiotics (synbiotics) to enhance their health benefits (Hasan et al., 2019).

14.7 Prebiotics in foods

The application of prebiotics in food products is a rapidly expanding field, driven by the increasing consumer focus on gut health and the demand for ingredients that promote well being. This growth is closely linked to the rise of functional foods, which provide health benefits beyond basic nutrition. Prebiotics play a crucial role in this market by enhancing gut health, boosting immune function, and potentially lowering the risk of chronic diseases (Hurtado-Romero et al., 2020).

Emerging technologies are pivotal in improving the extraction and incorporation of prebiotics into food products. These advancements enhance prebiotic stability, extraction efficiency, and concentration of bioactive compounds, enabling more innovative applications in the functional foods market (Balthazar et al., 2022). This technological progress is expected to drive the development of a broader range of prebiotic enriched foods.

In addition to their health benefits, prebiotics offer significant technological advantages in food production. They improve sensory attributes such as taste and texture and enhance the stability of food systems, including foams, emulsions, and mouthfeel. This makes prebiotics valuable in various food pplications like dairy products, bread, and other processed foods (Charalampopoulos and Rastall, 2012; Al-Sheraji et al., 2013; Florowska et al., 2016).

Research in this area is continually evolving, with ongoing efforts aimed at identifying new prebiotic compounds and understanding their specific health benefits. As this research progresses, it will likely further expand the range of prebiotic containing food products available to consumers, solidifying the role of prebiotics in the future of functional foods.

14.8 Economy of prebiotics

Nowadays, consumers are increasingly turning to products containing probiotic bacteria or prebiotics, which are part of the growing functional food market (Ballini et al., 2023). According to the research, the global average per capita expenditure on functional foods is expected to be € 26. However, spending varies depending on the location. Asia is the world's largest market for functional foods, with Japanese customers paying an average of € 150 per year. The second biggest worldwide market is in the United States and Canada, which account for 25% of the global market. In Europe, functional foods are popular in numerous countries, including Germany (14%), France (13%), and Italy (11%) (Sgroi et al., 2024). It is reported that the global functional food market was worth USD 175 billion in 2020 and is estimated to exceed USD 309 billion by 2027. On a global scale, Japan is the largest functional food market, with 1700 functional foods launched. Although the US and European markets appear relatively less developed, these three markets account for 90% of overall sales. Merging of various categories in the market, rapid increase in the number of functional foods and beverages, cobranding collaborations between component suppliers and manufacturers, increased focus on scientific verification and substantiation of claims, increasing demand for sustainable and environmentally friendly foods, expansion of active nutrition products as consumers become more health conscious, the emergence and diffusion of innovative dosage and administration forms, a new paradigm in health and wellness with the rise of millennials are the key trends driving activity and growth in the functional food and beverage market (Pérez-Marroquín et al., 2023). Concerns about growing medical and healthcare expenses, the desire to improve quality of life in old age as life expectancy continues to rise, and scientific proof that functional foods are good for health are all driving the expansion of the functional food sector. Due to the high costs of healthcare in the fight against chronic diseases and lower costs compared to the development of pharmaceutical products, the pharmaceutical industry favors the development of functional food products. These factors are among the key elements shaping the development and growth in the production of functional foods in the global market. Optimizing the functional food development process would boost the food industry's inventive potential and competitiveness by encouraging manufacturers to create high quality, healthier, and more sustainable foods that fulfill consumer and market demands. Furthermore, using industrial waste as a functional component will allow you to give a value added product that is always available, low cost, and high in bioactive chemicals while lowering waste management expenses (Alongi and Anese, 2021; Tadesse and Emire, 2020; Domínguez Díaz et al., 2020; Wu et al., 2022).

The market potential of functional food products depends on personal factors such as age, gender, health awareness, familiarity, income, and education, as well as cultural and social factors. Additionally, price, taste, brand, packaging, and labeling are also criteria for functional food products. Among these factors, labeling, health motivation, and customers' opinions about functional food items have the most effect on their purchase decisions. In fact, the more positive information provided on the label of a functional food product, the greater the purchase interest in that product (Domínguez Díaz et al., 2020).

The rise in chronic health issues, the adoption of new consumption patterns, and the growing demand for natural products have all contributed to the acceleration of the prebiotic trend. As the advantages of prebiotics have become more widely known, interest in prebiotic containing foods has grown dramatically. In parallel with the increasing demand, steps are being taken to increase prebiotic production to meet consumer needs. Companies in the food industry have responded to this need by incorporating prebiotics into their goods and bringing prebiotic preparations to market. This increases the diversity and accessibility of prebiotics in the market and increases their economic value every year (Correa et al., 2024; Farias et al., 2019). The global market for prebiotic compounds is growing. In 2021 the global prebiotics market was valued at USD 6.05 billion and is expected to reach USD 20.71 billion by 2030. The global prebiotics market is projected to grow at a CAGR of 14.9% between 2022 and 2030 (Correa et al., 2024; Kumar et al., 2024).

Prebiotic and postbiotic patents generally focused on food, feed additives, the prevention and treatment of gastrointestinal disorders, targeted medicine delivery, and enhancing the effects of drugs. A search of a total of 2215 global prebiotic and postbiotic patents between 2001 and 2020 clearly shows the development trend of these substances. The increase in the number of patents by 1.45% in the period 2011–20 compared to the period 2001–10 is the most important indicator of the interest in prebiotics and postbiotics. The widespread use of prebiotics and postbiotics in areas such as medicine, food, healthcare, and animal husbandry presents significant growth prospects for the increase in patent applications. The development of new technologies and methods in this field creates promising opportunities for prebiotics. Therefore prebiotic and postbiotic patents can stimulate the development of new technologies. The United States has become a center of innovation for prebiotic and postbiotic patent applications, representing 43.1% of total global applications. Patent applications are filed both in technologically advanced countries and in countries with weaker industries, which enables these products to become widespread in the global retail market (Zang et al., 2024).

The availability of raw materials for prebiotic manufacture, the manufacturing procedures, and the storage of functional foods are the primary problems confronting industrial scale prebiotic production and are significant variables influencing the worldwide market value of prebiotics (Ahire et al., 2022). Prebiotics are naturally present in some plant and animal foods. However, due to the significant interest in prebiotics, the concentration achieved by natural extraction is low, restricting their commercial applicability. As a result, new sources of raw materials are being investigated and examined. XOS production from Lignocellulosic Biomass can be given as an example. Lignocellulosic biomass is of great interest to the industry due to its huge production volume, global distribution, and cheap market price (Valladares-Diestra et al., 2023).

Different techniques are used for prebiotic production. Because of the ongoing growth in demand for prebiotics, innovations in manufacturing processes are being implemented to meet the industrial demand. Therefore instead of costly and low efficient methods, different methods that offer higher efficiency and lower cost are preferred. The most often used technique is enzymatic hydrolysis, which is very efficient and repeatable. Enzymatic hydrolysis of xylans yields XOS, whereas β-galactosidase acts as a biocatalyst to make GOS from lactose (Hurtado-Romero et al., 2020; Ibrahim, 2021).

The prebiotic market predominantly comprises fructans (e.g., inulin) and galactans (Cunningham et al., 2021). Market expectations for inulin, FOS, and XOS are USD 1.4 billion, USD 2.37 billion, and USD 99 million, respectively, with annual growth rates of 6.4%, 10%, and 4.4% in 2019–20, respectively. The FOS market is expected to total USD 3.9 billion by 2027. In the global inulin market, approximately 246.5 thousand metric tons of products were processed in 2013, and the sector is expected to reach a market value of approximately USD 3.3 billion by 2026 with an annual growth rate of 8.29%. FOS and inulin are being incorporated into many foods because they are readily available from many natural sources, have several proven and significant health benefits, have been extensively researched, have versatile physicochemical and technological properties, are used as fat and sugar substitutes in the food industry, improve texture, and enhance the sensory properties of the product, resulting in an increase in global demand. The GOS market has shown significant growth since the launch of the first commercial GOS product by Yakult Pharmaceuticals Industry Co. Ltd. in Japan in 1989. The global GOS market is estimated to be worth USD 570 million in 2021 and is expected to grow at a compound annual growth rate of approximately 6% from 2021 to 2026. In 2021 GOS was estimated to be the second most popular prebiotic among consumers in the US prebiotic market. In fact, infant nutrition constitutes the largest market for GOS. It is also found as a prebiotic component in various functional foods. In addition, it is also used as a raw material in cosmetics, pharmaceuticals, and nutritional supplements. The XOS market is predicted to attain $130 million by 2025. Globally, XOS is most commonly used in feed (49.6%), the medical and health sectors (25.4%), the food and beverage industry (23.2%), and other sectors (1.8%). XOS powder (95%), XOS powder (70%), XOS powder (35%), XOS powder (20%), and XOS sirup (70%) are the characteristics of XOS products sold in the market. The market price of XOS ranges between $25 and $50 per kilogram in the United States, depending on the product's purity level. XOS consumption is highest in the United States, followed by Germany and Japan, with market shares of 34%, 13%, and 1.5%, respectively (Correa et al., 2024; Ferreira et al., 2023; Ibrahim, 2021; Illippangama et al., 2022; Maráz et al., 2022; Ruiz-Ramírez and Jiménez-Flores, 2023; Yoo et al., 2024; Verma et al., 2024). XOSs are a prebiotic that is demanded in the food market as they favorably support health. Considering that this market will grow, different raw material sources and production techniques are being investigated for the production of XOS (Valladares-Diestra et al., 2023).

IMO is one of the preferred prebiotics, with an annual market size of 69,000 metric tons (Bajury et al., 2018). Due to high market demand, IMOs are commercially produced through the enzymatic modification of starch, as providing sufficient quantities of naturally occurring IMOs for commercial use is uneconomical (Rini et al., 2024). Among all oligosaccharides, IMO accounts for 60% of the food and beverage market share. IMO's global market generated sales revenues of USD 850 million in 2023 and is expected to reach USD 1.69 billion by the end of 2033 (Kumar et al., 2023). Market growth reveals a significant increase in demand for IMOs, leading to a greater need for alternative sources and more advanced production strategies (Palaniappan and Emmambux, 2023).

The worldwide market size in 2018 was USD 1.28 billion, owing to increased interest in the food, pharmaceutical, and healthcare industries, as well as the acceleration of polyphenol research in recent years. It is also projected to reach USD 2.08 billion by 2025. In 2018 polyphenols from grape seeds dominated the global market, followed by polyphenols from green tea, apples, cocoa, and coffee beans. Functional beverages, foods, and dietary supplements are recognized as the main areas where polyphenols are used. In particular, China and Japan are recognized as the largest regions of the polyphenol market (Cardoso et al., 2021).

The absence of technology in large scale HMO production and rigorous government regulations have hindered the expansion of the HMO sector. The global market size of HMOs reached USD 19.3 million in 2019 and is expected to continue growing at a CAGR of 23%. The main use of HMOs is in infant formulas, followed by functional food beverages and food supplements. The HMO market is largely dominated by the North American region (Cardoso et al., 2021; Zhao, 2021).

Nowadays, treating illnesses requires significant financial outlays. The utilization of prebiotics in illness prevention is a prevalent practice. As a result, the health and economic impacts of prebiotics in illness therapy are being examined through model studies for different disease prevention. A study found that using infant formula with a particular prebiotic mix is a very cost effective strategy to avoid atopic dermatitis in the Netherlands. Prebiotic baby formula was found to cost 51 euros more than nonprebiotic formula (Lenoir-Wijnkoop et al., 2012).

Prebiotics are also used in the nutrition and welfare improvement of farm animals (pigs, poultry, calves, and fish) and domestic animals (cats and dogs) (Van Loo, 2012). By adding prebiotics to poultry feed, treatment costs decrease, disease and mortality rates decline, and production increases. Furthermore, the utilization of prebiotics reduces the development of antibiotic resistance by decreasing antibiotic consumption (Nugusa and Kasa, 2024). A study revealed that including prebiotics into the diet for broilers reduced production and feed costs (Ibrahim et al., 2021).

14.9 Future perspective

The most comprehensive and advanced research on prebiotics has focused on inulin type fructans and galactooligosaccharides (GOS) (Roberfroid et al., 2010). Typically, this research has involved a combination of in vitro, animal, and human feeding studies, with bifidobacteria being selectively induced. According to Rastall and Gibson (2015), while in vitro and animal studies are important for understanding mechanisms, human trials are crucial for validating these effects. As a result, fructooligosaccharides (FOS) and GOS are currently among the leading prebiotics. However, they emphasized that further validation from human studies using advanced methodologies is needed to fully confirm the prebiotic effects. These promising avenues of research could have significant implications for long term weight management and overall consumer health if successful. Additionally, they noted that prebiotic effects observed in supplements like GOS or fructans must be confirmed in their final delivery systems.

As molecular mechanisms continue to evolve, future research should focus on innovative prebiotic probiotic interactions and unravel the genetic and bacterial profiles of hosts that may influence treatment efficacy. Although all prebiotics are fibers, not all fibers are prebiotics due to variations in their sources, structures, and functional properties. Therefore further studies are necessary to better characterize fibers and their effects on gut microbiota composition and function (Biswal et al., 2017).

Cardoso et al. (2021) highlighted that both industry and academia are working on more integrated, cost effective, and environmentally sustainable strategies for prebiotic production. They suggested that using abundant and inexpensive agricultural residues could offer viable alternatives to traditional substrates for enzyme and prebiotic production. Spacova et al. (2020) further noted that emerging tools enabling realtime human studies and monitoring how prebiotics integrate with the existing microbiota will be pivotal in advancing the field. These tools could also elucidate how prebiotics interact with various levels of host biology, including the immune system and metabolism. Ultimately, an integrative approach would facilitate personalized medicine by establishing dose–response relationships and aligning the processing of prebiotics with desired health outcomes.

Scott et al. (2019) pointed out that new prebiotic candidates and alternative sources, such as agricultural waste from non Western regions, may offer unique health benefits and foster positive social, economic, and environmental impacts. They also identified polyphenols and other compounds beyond current prebiotic carbohydrates as promising next generation prebiotics. However, the main challenge remains associating microbial composition, activity, and metabolic outputs with measurable health benefits.

These insights underline the dynamic and multifaceted nature of prebiotic research. Ongoing advancements in technology and a deeper understanding of the gut microbiome's role in health will continue to shape the field. Future perspectives highlight the exciting potential for prebiotics to influence nutrition, health, and disease prevention. Continued collaboration among scientists, healthcare professionals, and the food industry will be critical in unlocking the full potential of prebiotics to enhance overall well being.

From an economic standpoint, the future of prebiotics in pharmaceuticals, food additives, and food products is promising. The health benefits of prebiotics, combined with advances in production technologies, are driving their integration into a wide range of products. As consumer awareness of gut health grows, the demand for prebiotics is expected to increase, fostering further innovation and expansion in both the food and pharmaceutical industries.

References

Ahire, E.D., Sharma, N., Gupta, P.C., Khairnar, S., Surana, K., Ahire, B., et al., 2022. Developing formulations of prebiotics and probiotics. In: Kesharwani, R.K., Rao, T.J.M., Keservani, R.K. (Eds.), Prebiotics and Probiotics in Disease Regulation and Management. Wiley, pp. 271–290. Available at https://doi.org/10.1002/9781394167227.ch10.

Alongi, M., Anese, M., 2021. Re-thinking functional food development through a holistic approach. Journal of Functional Foods 81, 104466. Available at https://doi.org/10.1016/j.jff.2021.104466.

Al-Sheraji, S.H., Ismail, A., Manap, M.Y., Mustafa, S., Yusof, R.M., Hassan, F.A., 2013. Prebiotics as functional foods: a review. Journal of Functional Foods 5 (4), 1542–1553. Available at https://doi.org/10.1016/j.jff.2013.08.009.

Amorim, C., Silvério, S.C., Cardoso, B.B., Alves, J.I., Pereira, M.A., Rodrigues, L.R., 2020. In vitro fermentation of raffinose to unravel its potential as prebiotic ingredient. LWT 126, 109322. Available at https://doi.org/10.1016/j.lwt.2020.109322.

Ansari, F., Neshat, M., Pourjafar, H., Jafari, S.M., Samakkhah, S.A., Mirzakhani, E., 2023. The role of probiotics and prebiotics in modulating of the gut-brain axis. Frontiers in Nutrition 10. Available at https://doi.org/10.3389/fnut.2023.1173660.

Ashwini, A., Ramya, H.N., Ramkumar, C., Reddy, K.R., Kulkarni, R.V., Abinaya, V., et al., 2019. Reactive mechanism and the applications of bioactive prebiotics for human health: review. Journal of Microbiological Methods 159, 128–137. Available at https://doi.org/10.1016/j.mimet.2019.02.019.

Ayaz, Z., 2021. Prebiotics and their health benefits. The Journal of Turkish Family Physician 12 (4), 201–206. Available at https://doi.org/10.15511/tjtfp.21.00493.

Bäckhed, F., 2011. Programming of host metabolism by the gut microbiota. Annals of Nutrition and Metabolism 58 (Suppl. 2), 44–52. Available at https://doi.org/10.1159/000328042.

Bajury, D.M., Nashri, S.M., King Jie Hung, P., Sarbini, S.R., 2018. Evaluation of potential prebiotics: a review. Food Reviews International 34 (7), 639–664. Available at https://doi.org/10.1080/87559129.2017.1373287.

Ballini, A., Charitos, I.A., Cantore, S., Topi, S., Bottalico, L., Santacroce, L., 2023. About functional foods: the probiotics and prebiotics state of art. Antibiotics 12 (4), 635. Available at https://doi.org/10.3390/antibiotics12040635.

Balthazar, C.F., Guimarães, J.F., Coutinho, N.M., Pimentel, T.C., Ranadheera, C.S., Santillo, A., et al., 2022. The future of functional food: emerging technologies application on prebiotics, probiotics and postbiotics. Comprehensive Reviews in Food Science and Food Safety 21 (3), 2560–2586. Available at https://doi.org/10.1111/1541-4337.12962.

Bindels, L.B., Delzenne, N.M., Cani, P.D., Walter, J., 2015. Towards a more comprehensive concept for prebiotics. Nature Reviews Gastroenterology and Hepatology 12 (5), 303–310. Available at https://doi.org/10.1038/nrgastro.2015.47.

Biswal, P., Pal, A., Das, A.P., 2017. Chapter 3 - Current Trends and Future Prospective of Prebiotics as Therapeutic Food. In: Holban, A.M., Grumezescu, A.M. (Eds.), Microbial Production of Food Ingredients and Additives. Academic Press, pp. 57–88. Available at https://doi.org/10.1016/B978-0-12-811520-6.00003-9.

Boets, E., Gomand, S.V., Deroover, L., Preston, T., Vermeulen, K., De Preter, V., et al., 2017. Systemic availability and metabolism of colonic-derived short-chain fatty acids in healthy subjects: a stable isotope study. Journal of Physiology 595 (2), 541–555. Available at https://doi.org/10.1113/JP272613.

Burgain, J., Gaiani, C., Linder, M., Scher, J., 2011. Encapsulation of probiotic living cells: from laboratory scale to industrial applications. Journal of Food Engineering 104 (4), 467–483. Available at https://doi.org/10.1016/j.jfoodeng.2010.12.031.

Cardoso, B.B., Amorim, C., Silvério, S.C., Rodrigues, L.R., 2021. Novel and emerging prebiotics: advances and opportunities. Advances in Food and Nutrition Research 95, 41–95. Available at https://doi.org/10.1016/bs.afnr.2020.08.001.

Charalampopoulos, D., Rastall, R.A., 2012. Prebiotics in foods. Current Opinion in Biotechnology 23 (2), 187–191. Available at https://doi.org/10.1016/j.copbio.2011.12.028.

Cheng, L., Zhang, X., Hong, Y., Li, Z., Li, C., Gu, Z., 2017. Characterisation of physicochemical and functional properties of soluble dietary fibre from potato pulp obtained by enzyme-assisted extraction. International Journal of Biological Macromolecules 101, 1004–1011. Available at https://doi.org/10.1016/j.ijbiomac.2017.03.156.

Choque Delgado, G.T., Tamashiro, W.M.D.S.C., Junior, M.R.M., Moreno, Y.M.F., Pastore, G.M., 2011. The putative effects of prebiotics as immunomodulatory agents. Food Research International 44 (10), 3167–3173. Available at https://doi.org/10.1016/j.foodres.2011.07.032.

Coman, M.M., Cecchini, C., Verdenelli, M.C., Silvi, S., Orpianesi, C., Cresci, A., 2012. Functional foods as carriers for SYNBIO®, a probiotic bacteria combination. International Journal of Food Microbiology 157 (3), 346–352. Available at https://doi.org/10.1016/j.ijfoodmicro.2012.06.003.

Correa, A.D.C., Lopes, M.S., Perna, R.F., Silva, E.K., 2024. Fructan-type prebiotic dietary fibers: clinical studies reporting health impacts and recent advances in their technological application in bakery, dairy, meat products and beverages. Carbohydrate Polymers 323. Available at https://doi.org/10.1016/j.carbpol.2023.121396.

Courtin, C.M., Swennen, K., Verjans, P., Delcour, J.A., 2009. Heat and pH stability of prebiotic arabinoxylooligosaccharides, xylooligosaccharides and fructooligosaccharides. Food Chemistry 112 (4), 831–837. Available at https://doi.org/10.1016/j.foodchem.2008.06.039.

Cunningham, M., Azcarate-Peril, M.A., Barnard, A., Benoit, V., Grimaldi, R., Guyonnet, D., et al., 2021. Shaping the future of probiotics and prebiotics. Trends in Microbiology 29 (8), 667–685. Available at https://doi.org/10.1016/j.tim.2021.01.003.

Das, D., Baruah, R., Goyal, A., 2014. A food additive with prebiotic properties of an α-d-glucan from *Lactobacillus plantarum* DM5. International Journal of Biological Macromolecules 69, 20–26. Available at https://doi.org/10.1016/j.ijbiomac.2014.05.029.

De Filippo, C., Cavalieri, D., Di Paola, M., Ramazzotti, M., Poullet, J.B., Massart, S., et al., 2010. Impact of diet in shaping gut microbiota revealed by a comparative study in children from Europe and rural Africa. Proceedings of the National Academy of Sciences 107 (33), 14691–14696. Available at https://doi.org/10.1073/pnas.1005963107.

Dewulf, E.M., Cani, P.D., Claus, S.P., Fuentes, S., Puylaert, P.G.B., Neyrinck, A.M., et al., 2013. Insight into the prebiotic concept: Lessons from an exploratory, double blind intervention study with inulin-type fructans in obese women. Gut 62 (8), 1112–1121. Available at https://doi.org/10.1136/gutjnl-2012-303304.

Domínguez Díaz, L., Fernández-Ruiz, V., Cámara, M., 2020. An international regulatory review of food health-related claims in functional food products labeling. Journal of Functional Foods 68, 103896. Available at https://doi.org/10.1016/j.jff.2020.103896.

Farias, D.D.P., de Araújo, F.F., Neri-Numa, I.A., Pastore, G.M., 2019. Prebiotics: trends in food, health and technological applications. Trends in Food Science and Technology 93, 23–35. Available at https://doi.org/10.1016/j.tifs.2019.09.004.

Ferreira, V.C., Barroso, T.L.C.T., Castro, L.E.N., da Rosa, R.G., de Siqueira Oliveira, L., 2023. An overview of prebiotics and their applications in the food industry. European Food Research and Technology 249 (11), 2957–2976. Available at https://doi.org/10.1007/s00217-023-04341-7.

Florowska, A., Krygier, K., Florowski, T., Dłużewska, E., 2016. Prebiotics as functional food ingredients preventing diet-related diseases. Food and Function 7 (5), 2147–2155. Available at https://doi.org/10.1039/c5fo01459j.

Gaggìa, F., Mattarelli, P., Biavati, B., 2010. Probiotics and prebiotics in animal feeding for safe food production. International Journal of Food Microbiology 141, S15. Available at https://doi.org/10.1016/j.ijfoodmicro.2010.02.031.

Gibson, G.R., Roberfroid, M.B., 1995. Dietary modulation of the human colonic microbiota: introducing the concept of prebiotics. Journal of Nutrition 125 (6), 1401–1412. Available at https://doi.org/10.1093/jn/125.6.1401.

Gibson, G.R., Collins, M.D., 1999. Concept of balanced colonic microbiota, prebiotics and synbiotics. Probiotics, Other Nutritional Factors and Intestinal Microflora 42, 139–153.

Gibson, G.R., Probert, H.M., Loo, J.V., Rastall, R.A., Roberfroid, M.B., 2004. Dietary modulation of the human colonic microbiota: updating the concept of prebiotics. Nutrition Research Reviews 17 (2), 259–275. Available at https://doi.org/10.1079/NRR200479.

Gibson, G.R., Scott, K.P., Rastall, R.A., Tuohy, K.M., Hotchkiss, A., Dubert-Ferrandon, A., et al., 2010. Dietary prebiotics: current status and new definition. Food Science and Technology Bulletin: Functional Foods 7 (1), 1–19. Available at https://doi.org/10.1616/1476-2137.15880.

Gibson, G.R., Hutkins, R., Sanders, M.E., Prescott, S.L., Reimer, R.A., Salminen, S.J., et al., 2017. Expert consensus document: The International Scientific Association for Probiotics and Prebiotics (ISAPP) consensus statement on the definition and scope of prebiotics. Nature Reviews Gastroenterology and Hepatology 14 (8), 491–502. Available at https://doi.org/10.1038/nrgastro.2017.75.

Gu, J., Thomas-Ahner, J.M., Riedl, K.M., Bailey, M.T., Vodovotz, Y., Schwartz, S.J., et al., 2019. Dietary black raspberries impact the colonic microbiome and phytochemical metabolites in mice. Molecular Nutrition and Food Research 63 (8). Available at https://doi.org/10.1002/mnfr.201800636.

Guarino, M.P.L., Altomare, A., Emerenziani, S., Di Rosa, C., Ribolsi, M., Balestrieri, P., et al., 2020. Mechanisms of action of prebiotics and their effects on gastro-intestinal disorders in adults. Nutrients 12 (4), 1037. Available at https://doi.org/10.3390/nu12041037.

Hasan, M.T., Je Jang, W., Lee, J.M., Lee, B.J., Hur, S.W., Gu Lim, S., et al., 2019. Effects of immunostimulants, prebiotics, probiotics, synbiotics, and potentially immunoreactive feed additives on olive flounder (*Paralichthys olivaceus*): a review. Reviews in Fisheries Science and Aquaculture 27 (4), 417–437. Available at https://doi.org/10.1080/23308249.2019.1622510.

He, Y., Wang, B., Wen, L., Wang, F., Yu, H., Chen, D., et al., 2022. Effects of dietary fiber on human health. Food Science and Human Wellness 11 (1), 1–10. Available at https://doi.org/10.1016/j.fshw.2021.07.001.

Hurtado-Romero, A., Del Toro-Barbosa, M., Garcia-Amezquita, L.E., García-Cayuela, T., 2020. Innovative technologies for the production of food ingredients with prebiotic potential: modifications, applications, and validation methods. Trends in Food Science and Technology 104, 117–131. Available at https://doi.org/10.1016/j.tifs.2020.08.007.

Ibrahim, N.S.K., Ahmed, A.-B.N.S., Abdel-raheem, G.S.E., 2021. Impact of dietary supplementation of prebiotics on the growth performance and immunity in broilers fed low protein diets. Assiut Veterinary Medical Journal 0 (0), 103–119. Available at https://doi.org/10.21608/avmj.2021.100505.1027.

Ibrahim, O.O., 2021. Technological aspects of fructo-oligosaccharides (Fos), production processes, physiological properties, applications and health benefits. Journal of Food Chemistry and Nanotechnology 7 (2), 41–46. Available at https://doi.org/10.17756/jfcn.2021-111.

Illippangama, A.U., Jayasena, D.D., Jo, C., Mudannayake, D.C., 2022. Inulin as a functional ingredient and their applications in meat products. Carbohydrate Polymers 275. Available at https://doi.org/10.1016/j.carbpol.2021.118706.

Jiao, X., Wang, Y., Lin, Y., Lang, Y., Li, E., Zhang, X., et al., 2019. Blueberry polyphenols extract as a potential prebiotic with anti-obesity effects on C57BL/6 J mice by modulating the gut microbiota. The Journal of Nutritional Biochemistry 64, 88–100. Available at https://doi.org/10.1016/j.jnutbio.2018.07.008.

Juhász, Á., Molnár-Nagy, V., Bata, Z., Tso, K.-H., Posta, K., 2023. Phytobiotic-prebiotic feed additive containing a combination of carob pulp, chicory, and fenugreek improve growth performance, carcass traits, and fecal microbiota of fattening pigs. Animals 13 (23), 3621. Available at https://doi.org/10.3390/ani13233621.

Kaewarsar, E., Chaiyasut, C., Lailerd, N., Makhamrueang, N., Peerajan, S., Sirilun, S., 2023. Optimization of mixed inulin, fructooligosaccharides, and galactooligosaccharides as prebiotics for stimulation of probiotics growth and function. Foods 12 (8), 1591. Available at https://doi.org/10.3390/foods12081591.

Kasote, D., Tiozon, R.N., Sartagoda, K.J.D., Itagi, H., Roy, P., Kohli, A., et al., 2021. Food processing technologies to develop functional foods with enriched bioactive phenolic compounds in cereals. Frontiers in Plant Science 12. Available at https://doi.org/10.3389/fpls.2021.771276.

Khangwal, I., Shukla, P., 2019. Potential prebiotics and their transmission mechanisms: recent approaches. Journal of Food and Drug Analysis 27 (3), 649–656. Available at https://doi.org/10.1016/j.jfda.2019.02.003.

Kim, S.A., Jang, M.J., Kim, S.Y., Yang, Y., Pavlidis, H.O., Ricke, S.C., 2019. Potential for prebiotics as feed additives to limit foodborne campylobacter establishment in the poultry gastrointestinal tract. Frontiers in Microbiology 10. Available at https://doi.org/10.3389/fmicb.2019.00091.

Koh, A., De Vadder, F., Kovatcheva-Datchary, P., Bäckhed, F., 2016. From dietary fiber to host physiology: short-chain fatty acids as key bacterial metabolites. Cell 165 (6), 1332–1345. Available at https://doi.org/10.1016/j.cell.2016.05.041.

Kumar, R., Næss, G., Sørensen, M., 2024. Xylooligosaccharides from lignocellulosic biomass and their applications as nutraceuticals: a review on their production, purification, and characterization. Journal of the Science of Food and Agriculture 104 (13), 7765–7775. Available at https://doi.org/10.1002/jsfa.13523.

Kumar, S., Tissopi, T., Mutturi, S., 2023. The successful synthesis of industrial isomaltooligosaccharides lies in the use of transglycosylating α-glucosidases: a review. Carbohydrate Polymer Technologies and Applications 5, 100325. Available at https://doi.org/10.1016/j.carpta.2023.100325.

Lenoir-Wijnkoop, I., van Aalderen, W.M.C., Boehm, G., Klaassen, D., Sprikkelman, A.B., Nuijten, M.J.C., 2012. Cost-effectiveness model for a specific mixture of prebiotics in The Netherlands. The European Journal of Health Economics 13 (1), 101–110. Available at https://doi.org/10.1007/s10198-010-0289-4.

Luo, X., Wang, Q., Zheng, B., Lin, L., Chen, B., Zheng, Y., et al., 2017. Hydration properties and binding capacities of dietary fibers from bamboo shoot shell and its hypolipidemic effects in mice. Food and Chemical Toxicology 109 (Pt 2), 1003–1009. Available at https://doi.org/10.1016/j.fct.2017.02.029.

Maráz, A., Kovács, Z., Benjamins, E., Pázmándi, M., 2022. Recent developments in microbial production of high-purity galacto-oligosaccharides. World Journal of Microbiology and Biotechnology 38 (6). Available at https://doi.org/10.1007/s11274-022-03279-4.

Martinez, R.C.R., Bedani, R., Saad, S.M.I., 2015. Scientific evidence for health effects attributed to the consumption of probiotics and prebiotics: an update for current perspectives and future challenges. British Journal of Nutrition 114 (12), 1993–2015. Available at https://doi.org/10.1017/S0007114515003864.

Megur, A., Daliri, E.B.M., Baltriukienė, D., Burokas, A., 2022. Prebiotics as a tool for the prevention and treatment of obesity and diabetes: classification and ability to modulate the gut microbiota. International Journal of Molecular Sciences 23 (11), 6097. Available at https://doi.org/10.3390/ijms23116097.

Muzaffar, K., Jan, R., Ahmad Bhat, N., Gani, A., Ahmed Shagoo, M., 2021. Chapter 25 - Commercially available probiotics and prebiotics used in human and animal nutrition. In: Dhanasekaran, D., Sankaranarayanan, A. (Eds.), Advances in Probiotics: Microorganisms in Food and Health. Academic Press, pp. 417–435. Available at https://doi.org/10.1016/B978-0-12-822909-5.00025-3.

Nugusa, G.A., Kasa, R.W., 2024. Review on probiotics and prebiotics' in healthcare poultry's prevention of intestinal bacterial infection. Acta Entomology and Zoology 5 (1), 68–76. Available at https://doi.org/10.33545/27080013.2024.v5.i1a.128.

Oliveira, R.P.D.S., Perego, P., Oliveira, M.N.D., Converti, A., 2011. Effect of inulin as prebiotic and synbiotic interactions between probiotics to improve fermented milk firmness.

Journal of Food Engineering 107 (1), 36–40. Available at https://doi.org/10.1016/j.jfoodeng.2011.06.005.

Özyurt, V.H., Ötleş, S., 2014. Prebiyotikler: metabolizma için önemli bir gıda bileseni. Akademik Gıda/Academic Food Journal 12 (1), 115–123. Available at https://dergipark.org.tr/tr/download/article-file/1186526.

Palaniappan, A., Emmambux, M.N., 2023. The challenges in production technology, health-associated functions, physico-chemical properties and food applications of isomaltooligosaccharides. Critical Reviews in Food Science and Nutrition 63 (19), 3821–3837. Available at https://doi.org/10.1080/10408398.2021.1994522.

Patel, S., Goyal, A., 2012. The current trends and future perspectives of prebiotics research: a review. 3 Biotech 2 (2), 115–125. Available at https://doi.org/10.1007/s13205-012-0044-x.

Pérez-Marroquín, X.A., Estrada-Fernández, A.G., García-Ceja, A., Aguirre-Álvarez, G., León-López, A., 2023. Agro-food waste as an ingredient in functional beverage processing: sources, functionality, market and regulation. Foods 12 (8), 1583. Available at https://doi.org/10.3390/foods12081583.

Pineiro, M., Asp, N.-G., Reid, G., Macfarlane, S., Morelli, L., Brunser, O., et al., 2008. FAO technical meeting on prebiotics. Journal of Clinical Gastroenterology 42 (Supplement 3), S156. Available at https://doi.org/10.1097/mcg.0b013e31817f184e.

Quigley, E.M.M., 2019. Prebiotics and probiotics in digestive health. Clinical Gastroenterology and Hepatology 17 (2), 333–344. Available at https://doi.org/10.1016/j.cgh.2018.09.028.

Rastall, R.A., Gibson, G.R., 2015. Recent developments in prebiotics to selectively impact beneficial microbes and promote intestinal health. Current Opinion in Biotechnology 32, 42–46. Available at https://doi.org/10.1016/j.copbio.2014.11.002.

Ricke, S.C., 2021. Prebiotics and alternative poultry production. Poultry Science 100 (7), 101174. Available at https://doi.org/10.1016/j.psj.2021.101174.

Rini, D.M., Xu, W., Suzuki, T., 2024. Current research on the role of isomaltooligosaccharides in gastrointestinal health and metabolic diseases. Preventive Nutrition and Food Science 29 (2), 93–105. Available at https://doi.org/10.3746/pnf.2024.29.2.93.

Roberfroid, M., 2002. Functional food concept and its application to prebiotics. Digestive and Liver Disease 34 (2), S105. Available at https://doi.org/10.1016/S1590-8658(02)80176-1.

Roberfroid, M., Gibson, G.R., Hoyles, L., McCartney, A.L., Rastall, R., Rowland, I., et al., 2010. Prebiotic effects: metabolic and health benefits. British Journal of Nutrition 104 (2), S1. Available at https://doi.org/10.1017/S0007114510003363.

Rolim, P.M., 2015. Development of prebiotic food products and health benefits. Food Science and Technology (Campinas) 35 (1), 3–10. Available at https://doi.org/10.1590/1678-457X.6546.

Roto, S.M., Rubinelli, P.M., Ricke, S.C., 2015. An introduction to the avian gut microbiota and the effects of yeast-based prebiotic-type compounds as potential feed additives. Frontiers in Veterinary Science 2. Available at https://doi.org/10.3389/fvets.2015.00028.

Ruiz-Ramírez, S., Jiménez-Flores, R., 2023. Invited review: properties of β-galactosidases derived from *Lactobacillaceae* species and their capacity for galacto-oligosaccharide production. Journal of Dairy Science 106 (12), 8193–8206. Available at https://doi.org/10.3168/jds.2023-23392.

Scott, K.P., Grimaldi, R., Cunningham, M., Sarbini, S.R., Wijeyesekera, A., Tang, M.L.K., et al., 2019. Developments in understanding and applying prebiotics in research and practice—an ISAPP conference paper. Journal of Applied Microbiology 128 (4), 934–949. Available at https://doi.org/10.1111/jam.14424.

Sgroi, F., Sciortino, C., Baviera-Puig, A., Modica, F., 2024. Analyzing consumer trends in functional foods: a cluster analysis approach. Journal of Agriculture and Food Research 15, 101041. Available at https://doi.org/10.1016/j.jafr.2024.101041.

Simon, E., Călinoiu, L.F., Mitrea, L., Vodnar, D.C., 2021. Probiotics, prebiotics, and synbiotics: Implications and beneficial effects against irritable bowel syndrome. Nutrients 13 (6), 2112. Available at https://doi.org/10.3390/nu13062112.

Singh, R., Sharma, P.K., Malviya, R., 2012. Prebiotics: future trends in health care. Mediterranean Journal of Nutrition and Metabolism 5 (2), 81–90. Available at https://doi.org/10.1007/s12349-011-0065-8.

Slavin, J., 2013. Fiber and prebiotics: mechanisms and health benefits. Nutrients 5 (4), 1417–1435. Available at https://doi.org/10.3390/nu5041417.

Soliman, G.A., 2019. Dietary fiber, atherosclerosis, and cardiovascular disease. Nutrients 11 (5), 1155. Available at https://doi.org/10.3390/nu11051155.

Spacova, I., Dodiya, H.B., Happel, A.U., Strain, C., Vandenheuvel, D., Wang, X., et al., 2020. Future of probiotics and prebiotics and the implications for early career researchers. Frontiers in Microbiology 11. Available at https://doi.org/10.3389/fmicb.2020.01400.

Tadesse, S.A., Emire, S.A., 2020. Production and processing of antioxidant bioactive peptides: a driving force for the functional food market. Heliyon 6 (8). Available at https://doi.org/10.1016/j.heliyon.2020.e04765.

Topolska, K., Florkiewicz, A., Filipiak-Florkiewicz, A., 2021. Functional food-consumer motivations and expectations. International Journal of Environmental Research and Public Health 18 (10), 5327. Available at 10.3390/ijerph18105327.

Valladares-Diestra, K.K., de Souza Vandenberghe, L.P., Vieira, S., Goyzueta-Mamani, L.D., de Mattos, P.B.G., Manzoki, M.C., et al., 2023. The potential of xylooligosaccharides as prebiotics and their sustainable production from agro-industrial by-products. Foods 12 (14). Available at https://doi.org/10.3390/foods12142681.

Van Loo, J., 2012. Inulin-Type Fructans as Prebiotics. In: Gibson, G.R., Rastall, R.A. (Eds.), Prebiotics: Development and Application. John Wiley and Sons, Belgium, pp. 57–100. Available at https://doi.org/10.1002/9780470023150.ch3.

Verma, P., Kaushik, R., Sirohi, R., 2024. Insights into recent advances in agro-waste derived xylooligosaccharides: production, purification, market, and health benefits. Biomass Conversion and Biorefinery. Available at https://doi.org/10.1007/s13399-024-05993-5.

Vitetta, L., Briskey, D., Alford, H., Hall, S., Coulson, S., 2014. Probiotics, prebiotics and the gastrointestinal tract in health and disease. Inflammopharmacology 22 (3), 135–154. Available at https://doi.org/10.1007/s10787-014-0201-4.

Wang, C.H., Qiao, C., Wang, R.C., Zhou, W.P., 2015. Dietary fiber intake and pancreatic cancer risk: a meta-analysis of epidemiologic studies. Scientific Reports 5. Available at https://doi.org/10.1038/srep10834.

Wang, Y., 2009. Prebiotics: present and future in food science and technology. Food Research International 42 (1), 8–12. Available at https://doi.org/10.1016/j.foodres.2008.09.001.

Wu, I.H., Liang, C., Ip, C.Y., 2022. Involvement theory with market segmentation: effect of false functional food advertising on purchase intention. Foods 11 (7). Available at https://doi.org/10.3390/foods11070978.

Yin, X., Cai, T., Liu, C., Ma, Y., Hu, J., Jiang, J., et al., 2022. A novel solvothermal biorefinery for production of lignocellulosic xylooligosaccharides, fermentable sugars and lignin nano-particles in biphasic system. Carbohydrate Polymers 295, 119901. Available at https://doi.org/10.1016/j.carbpol.2022.119901.

Yoo, S., Jung, S.C., Kwak, K., Kim, J.S., 2024. The role of prebiotics in modulating gut microbiota: implications for human health. International Journal of Molecular Sciences 25 (9). Available at https://doi.org/10.3390/ijms25094834.

You, S., Ma, Y., Yan, B., Pei, W., Wu, Q., Ding, C., et al., 2022. The promotion mechanism of prebiotics for probiotics: a review. Frontiers in Nutrition 9. Available at https://doi.org/10.3389/fnut.2022.1000517.

Zang, T., Han, L., Lu, Z., Tan, L., Liang, D., Shen, X., et al., 2024. The history and prediction of prebiotics and postbiotics: a patent analysis. Nutrients 16 (3), 380. Available at https://doi.org/10.3390/nu16030380.

Zhao, C., 2021. Oligosaccharides in food. In: Xiao, J., Sarker, S.D., Asakawa, Y. (Eds.), Handbook of Dietary Phytochemicals. Springer Nature, Singapore, pp. 1465–1499. Available at https://doi.org/10.1007/978-981-15-4148-3_32.

Zhao, J., Zhang, X., Zhou, X., Xu, Y., 2021. Selective production of xylooligosaccharides by xylan hydrolysis using a novel recyclable and separable furoic acid. Frontiers in Bioengineering and Biotechnology 9. Available at https://doi.org/10.3389/fbioe.2021.660266.

Chapter 15

Critical approach on effectiveness of prebiotic against probiotic in enhancing and promoting the native probiotic microflora in human body

Rami M. Althnaibat[1], Muhammad Alu'datt[2], and Diana Alhaddad[1]
[1]*Nutrition and Food Technology Department, Faculty of Agriculture, Mutah University, Karak, Jordan,* [2]*Department of Nutrition and Food Technology, Faculty of Agriculture, Jordan University of Science and Technology, Irbid, Jordan*

15.1 Introduction

Microbiota and epithelium cells in the animal body's gut are the main tools responsible for the digestion and absorption of foods and nutrients. (Rajilić-Stojanović and de Vos, 2014). Microflora play an important role in the gut's permeability, thus regulating some important physiological processes like digestion, absorption, metabolism, and immune system (Derrien et al., 2019; Razali et al., 2023). In the last decades, the microflora in the gastrointestinal tract has gained a lot of attention from researchers and scientists. Because of their major role in the human body's health through helping in nutrients (food) digestion and regulating some of the physiological activities (Magwira and Taylor, 2018; Marrs et al., 2021). However, the qualitative and quantitative composition of microbiota will reflect on the health of animal bodies (Liu et al., 2019). Practically, most of the sites in the human body are colonized by microorganisms leading to many interactions and effects on the human body's organs. The approximate number of microflora in the adult intestinal tract is about 4×10^{13}, which means more than the human body cells (Ribera et al., 2024; Charbonneau et al., 2016; Pushpanathan et al., 2019).

The imbalance between bad and good gut bacteria (intestinal dysbiosis) mainly depends on the consumed food, gender, age, bad lifestyle, and drugs (Xia et al., 2023; Pan et al., 2024). Intestinal dysbiosis is related to diverse

Evaluating Oral Probiotics and Prebiotics. DOI: https://doi.org/10.1016/B978-0-443-29835-6.00006-X

pathogenic species and microbial diseases; consequently, a solid understanding of the gastrointestinal microbial balance system is very important for keeping good health in the animal bodies (Pan et al., 2024). However, the healthy microorganisms help in producing short-chain fatty acids (SCFAs) like butyrate working as a reducing agent against intestinal inflammation (Taylor et al., 2020; Jimenez et al., 2022; Tamura et al., 2017).

Excellent examples of good and beneficial foods are foods containing prebiotics and probiotics. Even though prebiotics and probiotics have different roles in the human body, both are very important for keeping healthful native microflora in the gastrointestinal tract (Tran and Li, 2022).

The native probiotic microflora in the human body is stimulated by several means, including taking an appropriate dose of probiotics and/or prebiotics. To achieve the desired effects, which are providing health benefits, the quantity must be sufficient because it works in a way with different effectiveness depending on the amount of the dose, the type of strain, and the components that were used to form a product supported by a specific probiotic (Anadón et al., 2021). In this chapter, the prebiotics, probiotics, and synbiotics with more focus on the means, health benefits, major sources, and their beneficial effects on native microflora are studied.

15.2 Prebiotics

Prebiotics are defined as chemical compounds or natural foods with a high content of fibers that cannot be digested in the gastrointestinal tracts, which makes it a good food for the probiotics to survive and do their job effectively (Anadón et al., 2021). Prebiotics are nondigestible food compounds like fiber, such as cellulose, hemicellulose, and pectin, generated from plants' cell walls and complex carbohydrates that contribute to the growth and activation of several good bacteria in the large intestine (Taghizadeh Ghassab et al., 2024). The most popular prebiotic foods that exert many health benefits include whole grains, vegetables, fruits, breast milk, yogurt, cereals, some juicy drinks, infants' foods, inulin, soybean, oat, and unrefined wheat or barley. In addition to some types of bread, cakes, biscuits, pasta, sauces, cereal bars, and some types of candies and table sweeteners have been studied (Anadón et al., 2021; Kazemi et al., 2019; Yadav et al., 2022; You et al., 2022).

As shown in Table 15.1, the main examples of prebiotics as nondigestible oligosaccharides are xylo-oligosaccharides (XOS) found in dairy products, legumes, and soybean, which exhibit several functions like antioxidant, antitumor, and antiinflammatory prospects (Odriozola et al., 2024). Galactooligosaccharides (GOS), which reduce the incidence of diarrhea and constipation, have a role in the absorption of magnesium, calcium, and iron, in addition to their role in the reduction of blood cholesterol levels (Odriozola et al., 2024; Derrien et al., 2019; Wopereis et al., 2018). Inulin is a type of oligosaccharide found in garlic, banana, and wheat that limits the growth of pathogenic bacteria

TABLE 15.1 The main known prebiotics (complex carbohydrates and polyphenols).[a]

Complex carbohydrates	Polyphenols
Xylo-oligosaccharides (XOS)	Flavonoids
Pectin	Lignans
Inulin	Polyphenolic amides
Fructooligosaccharides (FOS)	Phenolic acids
Galactooligosaccarides (GOS)	Stilbenes
Resistant starch	Dyferuloylmethane (curcumin)
Arabinoxylan-oligosaccharides	Tannins
β-glucans	
Arabinoxylan	
Raffinose	

[a]*(Ribera et al., 2024; Odriozola et al., 2024; Derrien et al., 2019; Wopereis et al., 2018; Anadón et al., 2021; Kazemi et al., 2019; Yadav et al., 2022; You et al., 2022).*

via the production of lactic acid therefore reducing the pH of the large intestine and also exhibits anticarcinogenic and antiinflammatory agents (Wopereis et al., 2018).

Fructooligosaccharides (FOS) are found in barley, wheat, and onion. It promotes the growth of some favorite bacteria in the colon such as *Bifidobacterium*, and FOS can reduce the blood cholesterol level (Ribera et al., 2024). Pectin is a type of nondigestible fiber found in plant cell walls such as apples and broccoli; it is beneficial in the reduction of some carcinogenic diseases (McCabe et al., 2015). Polyphenol is derived from plant metabolism with a prebiotic effect. Polyphenol compounds can be divided according to their source of origin and their functions into phenolic acids and polyphenolic amides. Phenolic acid is found in nuts, berries, and pomegranates; meanwhile, polyphenolic amides are found in chili peppers (Hussein, 2022; Kumar et al., 2024). Flavonoids are found in blueberries, plums, grapes, and black soybeans. Finally, the tannins are found in raspberry, blackberry, tea, coffee, and ginger (Hussein, 2022).

The gut microbiota should digest and absorb these compounds (as prebiotics) to increase their bioavailability in human bodies and exert several beneficial biological functions like as prevention of certain types of cancer and cardiovascular diseases, reducing blood pressure, antioxidant activity, exerting antiinflammatory, decreasing cholesterol level, working as antioxidative stress, and improving the immune system (Hussein, 2022; Borase et al., 2022; Lelia and Suharoschi, 2022).

Prebiotics are a strong competitor that resists both acids and bile salts present in the digestive system, which makes it more able to survive. The activity rate of native microflora can be determined by the rate of fermentation activity; with more fermentation activity, the native probiotic microflora grows faster. Also, the prebiotics affect the formation of useful products in large quantities such as SCFAs (Jan et al., 2024). Prebiotics can be used for long periods as their effectiveness will not stop, unlike probiotics. Prebiotics are not harmful to the microorganisms inside the body if they are consumed in an appropriate quantity, but if they are not, an inappropriate quantity may cause side effects such as diarrhea (De Vos et al., 2022). The mechanism of prebiotics that healthily enhance beneficial bacteria is through antagonistic mechanisms and enhances native probiotics by supplying them with energy through the products resulting from the fermentation process such as acetic acid, butyric acid, and propionic acid. While probiotics, if consumed for long periods, will become ineffective to native microflora (De Vos et al., 2022). Nondigestible fiber acts as food for the gut microflora. Prebiotics exhibit different documented functions such as relieving symptoms associated with gastrointestinal disorders, like diarrhea, preventing colon cancer, and decreasing the level of cholesterol reported in several studies after consumption of fermented milk products combined with probiotic strains like *Lactobacillus reuteri* and *Lactobacillus bulgaricus* (Borase et al., 2022; Lelia and Suharoschi, 2022). However, the optimum prebiotic should be resistant to hydrolyzing enzymes such as bile acids and stomach acidity and other gut enzymes, moisture, oxygen, and pH, easily fermented by the microflora, and absorbed in the lower part of the gastrointestinal tract.

15.3 Probiotics

For 2000 years ago, experimentally rather than theoretically, beneficial microorganisms, especially bacteria, have been utilized in fermented foods as beneficial food products like fermented dairy products and pickles (Xi et al., 2017; Baliyan et al., 2021). During the last 2000 years, hundreds of scientists have studied microorganisms and probiotics and defined and documented them over different periods with different names and definitions. Scientists have discovered different types of microbes and their benefits, leading to advances in defining beneficial microbes; thus the term probiotic was born (Mokoena et al., 2016; Baliyan et al., 2021).

The formal term of probiotic was initiated in 1965. Probiotics are supplements containing a live beneficial bacterium that can offer many health benefits to the flora in the gut. Probiotics are microorganisms that contribute to the gut microflora with therapeutic and beneficial effects on our health (Gänzle, 2019; Jan et al., 2024; Lee and Hase, 2014). Recently, in 2001, the Food and Agricultural Organization (FAO) and World Health Organization (WHO) worked together in a committee of expert scientists to redefine probiotics as "live microorganisms which, when supplied in adequate amounts, promote health benefits in the host organism" (Pyle, 2022; Borase et al., 2022; Sampaio et al., 2022).

Probiotics are nonpathogenic favorable microorganisms that stimulate and exhibit a positive effect on the microflora in different circumstances in the gut (Jan et al., 2024). Probiotics have many vital functions with beneficial effectiveness such as host immunity and reducing the level of cholesterol (Sampaio et al., 2022). The balance of the probiotics in the gut also plays a functional role in human health affecting nutrition, metabolism, and native microflora activity, and decreasing the objecting of the human body to different types of illnesses like type 2 diabetes, cardiovascular diseases, and limiting the obesity (You et al., 2022). Whereas the disturbance of the balance of the gut microflora has a relationship with some gastrointestinal disorders such as inflammatory bowel disease (IBD), irritable bowel syndrome (IBS), and other metabolic irritable syndromes (You et al., 2022; Yadav et al., 2022).

Several factors such as a high-carbohydrate diet, the percentage of pesticides in food, and antibiotic usage change the kinds of gut microbiota and make them unresponsive to medicines, affecting the persistence and the ratio of the gut microflora (probiotic). However, this can lead to colonize the unwanted or bad bacteria in the gut (Gibson et al., 2015; Berding et al., 2018). The two main starter bacteria used in dairy fermentation products such as milk and cheese are *Streptococcus thermophiles* and *Lactobacillus bulgaricus*. These two bacteria give texture, aroma, and flavor to dairy products due to the large amounts of lactic acid production during processing. *Streptococcus thermophiles* and *Lactobacillus bulgaricus* are considered good bacteria; therefore they enhance the microflora in the human gastrointestinal tract (Berding et al., 2018; Kashtanova et al., 2016). Several conditions are encountered by the bacteria from oral consumption until the digestion process reaches the target tissue such as biochemical factors like pH, digestive enzymes, the usage of antimicrobial medicines, and bile acids (Kashtanova et al., 2016; Gibson et al., 2015; Doré and Blottière, 2015).

The main examples of probiotic food are fermented food such as sauerkraut, yogurt, pickles, sourdough bread, and buttermilk. However, the most common probiotic microbiota is illustrated in Table 15.2 such as lactic acid bacteria (LAB), mainly *Bifidobacterium (longum, adolescentis, bifidum, animalis, and breve)* and *Lactobacillus* (*casei, acidophilus, fermentum, reuteri, johnsonii, gasseri, plantarum, paracasei, salivarius*, and *rhamnosus*) as well as *Bacillus coagulans* and *Entercoccus faecium* (Pushpanathan et al., 2019; Rajilić-Stojanović and de Vos, 2014; Turnbaugh et al., 2009; Derrien et al., 2019; Cooke et al., 2024). These microorganisms are the most dominant, added either in combination (more than one bacterial strain) or singly to food such as fermented milk products like yogurt. The generated probiotic product exerts a variety of benefits depending on the type and number of microorganisms used in the probiotic products (Razali et al., 2023). Once these strains, or one of them, are enrolled in the food with enough amounts ($>10^9$ CFU/mL),

TABLE 15.2 The main known traditional and new-generation probiotics, and the main probiotics' foods.[a]

Traditional probiotics	New-generation probiotics	Probiotic foods
Bacillus *Lactobacillus* *Bifidobacterium* *Streptococcus* *Thermophiles* *Enterococcus* *Lactococcus lactis*	*Eubacterium hallii/ Anaerobutyricum hallii* *Faecalibacterium prausnitzii* *Clostridium butyricum* *Akkermansia muciniphila*	*Chucrut (Sauerkraut), Kefir, Miso, Doenjang, Natto Tempeh (soya), Yoghurt, Kimchi, Sourdough bread, and Kombucha*

[a]*(Flint et al., 2012; Walker and Lawley, 2013; Odriozola et al., 2024; Pushpanathan et al., 2019; Rajilić-Stojanović and de Vos, 2014; Turnbaugh et al., 2009; Derrien et al., 2019; Cooke et al., 2024; Doré and Blottière, 2015; Mooser et al., 2018).*

then this food will be considered a healthy food (Borase et al., 2022; Patra et al., 2016). Several studies showed the beneficial functions of probiotics in the prevention of cancer, urinary and respiratory tract infections, obesity, type 2 diabetes, the prevention of diarrhea, and cholesterol levels. Furthermore, the addition of unique flavors is food due to the production of some organic acids (De Vos et al., 2022; Patra et al., 2016). The mentioned functions are associated with different mechanisms, like shutting down the adhesion sites of unfavorable microorganisms, competition for nutrients with the pathogenic bacteria, releasing inhibitory substances for pathogenic bacteria such as organic acids, H_2O_2, and blocking the toxin's receptors (De Vos et al., 2022; Patra et al., 2016).

Food or supplements may contain probiotics like lactic acid bacteria; the main objective is to improve the intestinal microbial balance in a beneficial way to reduce digestive disorders. This effect is related to feeding the microorganisms in the digestive system with more strains and forming them by providing the nutrients (Walker and Lawley, 2013). People can consume the probiotics from raw vegetables or in the form of fermented dairy products, like yogurt (Xia et al., 2023). The effect of some probiotic intake is short, and its effectiveness on microflora in the human body does not continue for a long time and disappears hours after stopping taking them, due to the inability of some of them to remain in the intestines. This is attributed to their sensitivity to the stresses that occur during the digestion process, and some of them may become ineffective and have no effect in the long-term (Xia et al., 2023; Pan et al., 2024).

The main advantage of prebiotics is their effectiveness in staying longer than probiotics in the digestive system. In addition, each of them (pre and probiotics) affects the native microflora in different places in the gastrointestinal tract; for example, probiotic enhances the probiotic microflora present in the small intestine, while prebiotic enhances the one present in the large intestine and colon (Jan et al., 2024). Because of the many differences in activity and enhancement of prebiotics and probiotics, mixing them creates a very effective mixture (Jan et al., 2024; De Vos et al., 2022). As it has a very large and integrated effect, it improves the viability and function of native gut bacteria by creating a favorable environment for their growth.

15.4 Synbiotics

The term integration of probiotics and prebiotics is known as symbiotic, which improves the growth and the balance of the beneficial microflora and influences microbiota diversity. Synbiotics increase and balance the level of microbiota, relieve some gastrointestinal disorders like irritable bowel syndrome and antibiotic-associated diarrhea, reduce intestinal inflammation and infections, enhance liver functions, and reduce the incidence of infections therefore improving the immune system. Some studies showed that the existence of synbiotics increases the ability of the intestine to absorb vitamins and minerals affecting the energy balance and leading to weight reduction (Ribera et al., 2024; Yadav et al., 2022). Synbiotics such as yogurt fortified with dietary supplements contain synbiotics such as the presence of inulin or fructooligosaccharides and a *Bifidobacterium* strain as a probiotic.

Maintaining the stability of products containing synbiotics during storage and transport to keep their effectiveness and availability, individual variability factors like lifestyle and genetics, and careful selection of materials and beneficial microorganism strains are the main challenges facing synbiotic preparation (Ribera et al., 2024; Yadav et al., 2022). Synbiotics create an environment in which native species can grow as a single unit, which helps in the formation of an environment that grows with high microbial diversity and flexibility and in greater numbers. It is also related to gut health and the production of an important food, which is SCFAs, in a way that includes stimulating microbes. For example, butyrate increases the rate of fat production, thus providing the digestive system with strength, and this makes it a strong competitor against any pathogenic microbe (Kashtanova et al., 2016; Turnbaugh et al., 2009). The reason behind that is the ability of synbiotics to activate the original probiotic group in the body, which reduces the risk of infection. Some studies have confirmed that synbiotics enhance the survival of probiotics and enhance natural microflora. For example, adding fructooligosaccharide as a prebiotic to *Lactobacillus acidophilus* improved its survival and digestive stability (Yadav et al., 2022).

Functional foods based on synbiotics have rapidly increased commercially, despite several types of research need to address the benefits of synbiotics as well as recently created, also to indicate different individual responses for consuming synbiotic due to the differences in their gut microflora environments or microflora diversity moreover, to show the optimum dose of synbiotic should be taken to meet individual need depending on diverse population, lifestyles, diseases, age, and microbiota profile (Razali et al., 2023). Generally, the consumed diet is the main factor in modifying the native microbiota. The prebiotics and probiotics are the main foods involved in modulating and improving the microflora in positive aspects leading to boost the human body health (Rosenbaum et al., 2015; Canfora et al., 2015).

As mentioned above, prebiotics and probiotics work and have advantages related to native microflora. Consuming nondigestible food compounds like fiber may lead to improvement in the microflora in the body, whereas consuming food rich in sugar and digestible carbohydrates will negatively affect the microflora in the body (Kashtanova et al., 2016; Turnbaugh et al., 2009). On the other hand, eating known probiotics food like fermented food products, including beneficial bacteria such as lactic acid bacteria (mainly *Bifidobacterium* and *Lactobacillus),* will increase the number of good bacteria and enhance the gut microflora, meanwhile consuming the wrong bacteria regularly may lead to enhance the growth and colonize of bad bacteria higher than the good bacteria playing an undesirable role in the imbalance between bad and good gut bacteria (Doré and Blottière, 2015; Mooser et al., 2018). Thus focusing on the consumed food and mixing the prebiotics and the probiotics in the meals may help and lead to offset the undesirable growth of bad bacteria against the good bacteria, in addition to mixing prebiotics and probiotics, which will improve the gut microflora and enhance the body's health (Walker and Lawley, 2013).

The incorporation of prebiotics and probiotics is the best approach for animal and human health because they work in tandem. The prebiotics will offer the optimum conditions and fuel for the good gut microflora leading to encouraging their growth. Meanwhile, consuming the probiotics with the availability of prebiotics as a good food source required to grow and survive will enhance the healthy native microflora against the undesirable bad bacteria (Walker and Lawley, 2013; Odriozola et al., 2024). The combination of probiotics and prebiotics is a critical field of research and discussion within the gut microflora health area. Understanding and investigating the effectiveness of prebiotics against probiotics in enhancing and promoting the native probiotic microflora in the human body is a nuanced approach and needs acute exploration of their mechanisms of action separately and synergistic effects (Odriozola et al., 2024).

Table 15.3 demonstrates that coingestion of prebiotics, probiotics, and synbiotics gives different health-promoting properties with variation between them depending on their properties and characteristics. The tolerance level for storage conditions varies between prebiotics and probiotics. The prebiotics as nondigestible fibers are more resistant to storage and transport conditions and keep their activity and efficacy (Charbonneau et al., 2016); however, the

TABLE 15.3 The main differences between prebiotics, probiotics, and synbiotics.

Properties	Prebiotics	Probiotics	Synbiotics
Tenuousness to storage conditions	Less tenuousness	Tenuousness	Less tenuousness
Sensitivity to stomach acids	Less sensitive	Sensitive	Less sensitive
Effect duration	Long-term	Short-term	Both
Microbial diversity	Broader influence	Strain-specific	Broader influence
Action	Sustainable benefits	Specific conditions	Sustainable benefits
Flexibility	Flexible	Not flexible	Flexible

(Ribera et al., 2024; Odriozola et al., 2024; Derrien, Alvarez and de Vos, 2019; Wopereis et al., 2018; Anadón et al., 2021; Kazemi et al., 2019; Yadav et al., 2022; You et al., 2022; Flint et al., 2012; Walker and Lawley, 2013; Odriozola et al., 2024; Razali et al., 2023).

probiotics as live microorganisms are tenuous and have less resistance to storage and transport conditions because they need optimum conditions to survive like food, optimum temperature, and optimum acidity, and any changes in storage or transport conditions of probiotics may affect their availability. On the other hand, the synbiotics contain both properties of pre and probiotics, leading to a boost in their resistance to different storage and transport conditions (Razali et al., 2023).

Regarding sensitivity to gastrointestinal tract conditions, mainly stomach acids, lacking specific enzymes to hydrolyze the prebiotic's polymers in the human gastrointestinal tract makes them resistant to gastric and pancreatic peptidases and escape the digestion and reach the large intestine. Meanwhile, the live organisms in probiotics are sensitive and affected by the acidity of the stomach (Razali et al., 2023; Ribera et al., 2024). Prebiotics take a long time to show their effects and benefits to nourish and improve the growth of the present gut microbiota depending on the age, health situation, and individual dietary habits. Probiotics take a short range of time and have a faster effect on restoring the microbial balance by boosting the beneficial bacteria (Charbonneau et al., 2016). Probiotic efficacy depends on the sustainability of microorganisms to stay active against gut conditions like stomach acidity and the usage of antibiotics restoring gut balance, which significantly affects the percentage of beneficial bacteria in the gut, as well as maintaining proper storage conditions. The synbiotics will have a combination of properties and will work in the short and long-term on beneficial intestine bacteria (Ribera et al., 2024; Yadav et al., 2022).

Probiotics may be strain-specific because their support for good bacteria will be through surviving microorganisms; that is mean depending on the individual's existing bacteria, the probiotics incorporation will vary significantly. But the prebiotics, in general, have a wider impact by encouraging and supporting the growth and multiplication of diverse desirable microflora, since they are fermentable fiber and produce many useful SCFAs (Yadav et al., 2022).

The targeted action or approach of prebiotics is different from the probiotics approach; the prebiotics are supported by sustainable resources that help in promoting the microflora growth, and most likely this bacterium is available in the large intestine already. On the other side, probiotics are helpful for specific conditions and interventions. For instance, In cases where it is necessary to restore the microbial balance, the gastrointestinal tract can be supported by introducing beneficial external bacteria. With all the above differences and variations, the prebiotics will be more flexible than probiotics, and synbiotics can be the best option to improve the microflora in the human body (Razali et al., 2023; Ribera et al., 2024). However, there are thousands of in vitro research studes and hundreds of in vivo research studes, but clinical research is limited, and just a handful of research studies were published. Thus more research (mainly clinical research) is needed to confirm the best approach for using and consuming the prebiotics, probiotics, and synbiotics to enhance and improve the original microflora in human and animal guts. In summary, hundreds of in vitro and in vivo research studies studied and confirmed the efficacy of prebiotics and probiotics in improving and elevating the native microflora, whereas randomized clinical trials are very limited. The foods and integrated food products rich in synbiotics (a mixture of prebiotics and probiotics) are comprehensive and sufficient by combining prebiotics and probiotics to effectively influence and enhance the original bacteria present inside the human body, making them a powerful tool for gut and digestive health for the longest possible period.

References

Anadón, A., Ares, I., Martínez-Larrañaga, M.R., Martínez, M.A., 2021. Prebiotics: safety and toxicity considerations. Nutraceuticals: Efficacy, Safety and Toxicity. Elsevier, Spain, pp. 1061–1080. Available at http://doi.org/10.1016/B978-0-12-821038-3.00064-1, https://www.sciencedirect.com/book/9780128210383.

Baliyan, N., Mukhia, S., Kumari, P., Dindhoria, K., Mukhia, S., Kumar, S., et al., 2021. Probiotics in fermented products and supplements. Current Developments in Biotechnology and Bioengineering: Technologies for Production of Nutraceuticals and Functional Food Products. Elsevier, India, pp. 73–107. Available at http://doi.org/10.1016/B978-0-12-823506-5.00014-X, https://www.sciencedirect.com/book/9780128235065.

Berding, K., Holscher, H.D., Arthur, A.E., Donovan, S.M., 2018. Fecal microbiome composition and stability in 4- to 8-year old children is associated with dietary patterns and nutrient intake. Journal of Nutritional Biochemistry 56, 165–174. Available at https://doi.org/10.1016/j.jnutbio.2018.01.002.

Borase, H., Dwivedi, M.K., Krishnamurthy, R., Patil, S., 2022. Probiotics: health safety considerations. Probiotics in the Prevention and Management of Human Diseases: A Scientific Perspective. Elsevier, India, pp. 449–463. Available at http://doi.org/10.1016/B978-0-12-823733-5.00020-9, https://www.sciencedirect.com/book/9780128237335.

Canfora, E.E., Jocken, J.W., Blaak, E.E., 2015. Short-chain fatty acids in control of body weight and insulin sensitivity. Nature Reviews Endocrinology 11 (10), 577–591. Available at https://doi.org/10.1038/nrendo.2015.128, http://www.nature.com/nrendo/index.html.

Charbonneau, M.R., O'Donnell, D., Blanton, L.V., Totten, S.M., Davis, J.C.C., Barratt, M.J., et al., 2016. Sialylated milk oligosaccharides promote microbiota-dependent growth in models of infant undernutrition. Cell 164 (5), 859–871. Available at https://doi.org/10.1016/j.cell.2016.01.024, https://www.sciencedirect.com/journal/cell.

Cooke, C.G., Gibb, Z., Grupen, C.G., Schemann, K., Deshpande, N., Harnett, J.E., 2024. Effect of probiotics and prebiotics on the composition of the equine fecal and seminal microbiomes and sperm quality: a pilot study. Journal of Equine Veterinary Science 135. Available at https://doi.org/10.1016/j.jevs.2024.105032, https://www.sciencedirect.com/science/journal/07370806.

De Vos, W.M., Tilg, H., Van Hul, M., Cani, P.D., 2022. Gut microbiome and health: mechanistic insights. Gut 71 (5), 1020–1032. Available at https://doi.org/10.1136/gutjnl-2021-326789, http://gut.bmj.com/content/by/year.

Derrien, M., Alvarez, A.S., de Vos, W.M., 2019. The gut microbiota in the first decade of life. Trends Microbiology 27 (12), 997–1010. Available at https://doi.org/10.1016/j.tim.2019.08.001, www.elsevier.com/locate/tim.

Doré, J., Blottière, H., 2015. The influence of diet on the gut microbiota and its consequences for health. Current Opinion in Biotechnology 32, 195–199. Available at https://doi.org/10.1016/j.copbio.2015.01.002, http://www.elsevier.com/locate/copbio.

Flint, H.J., Scott, K.P., Duncan, S.H., Louis, P., Forano, E., 2012. Microbial degradation of complex carbohydrates in the gut. Gut Microbes 3 (4), 289–306. Available at http://doi.org/10.4161/gmic.19897UnitedKingdom, http://www.landesbioscience.com/journals/gutmicrobes/FlintGUTMIC3-4.pdf?nocache=342263602.

Gänzle, M., 2019. Fermented foods. Food Microbiology: Fundamentals and Frontiers. wiley, Canada, pp. 855–900. Available at http://doi.org/10.1128/9781555819972.ch33, http://onlinelibrary.wiley.com/book/10.1002/9781683670476.

Gibson, M.K., Crofts, T.S., Dantas, G., 2015. Antibiotics and the developing infant gut microbiota and resistome. Current Opinion in Microbiology 27, 51–56. Available at https://doi.org/10.1016/j.mib.2015.07.007, http://www.elsevier.com/locate/mib.

Hussein, L.A., 2022. Novel Prebiotics and Next-Generation Probiotics: Opportunities and Challenges. Elsevier BV, pp. 431–457. Available at http://doi.org/10.1016/b978-0-12-819815-5.00055-0.

Jan, T., Negi, R., Sharma, B., Kumar, S., Singh, S., Kumar Rai, A., et al., 2024. Next generation probiotics for human health: an emerging perspective. Heliyon 10 (16), e35980. Available at https://doi.org/10.1016/j.heliyon.2024.e35980.

Jimenez, M.E., O'Donovan, C.M., Ullivarri, M.F.D., Cotter, P.D., 2022. Microorganisms present in artisanal fermented food from South America. Frontiers in Microbiology 13.

Available at https://doi.org/10.3389/fmicb.2022.941866, https://www.frontiersin.org/journals/microbiology#.

Kashtanova, D.A., Popenko, A.S., Tkacheva, O.N., Tyakht, A.B., Alexeev, D.G., Boytsov, S.A., 2016. Association between the gut microbiota and diet: fetal life, early childhood, and further life. Nutrition 32 (6), 620–627. Available at https://doi.org/10.1016/j.nut.2015.12.037, https://www.elsevier.com/locate/nut.

Kazemi, A., Noorbala, A.A., Azam, K., Eskandari, M.H., Djafarian, K., 2019. Effect of probiotic and prebiotic vs placebo on psychological outcomes in patients with major depressive disorder: a randomized clinical trial. Clinical Nutrition 38 (2), 522–528. Available at https://doi.org/10.1016/j.clnu.2018.04.010, http://www.elsevier-international.com/journals/clnu/.

Kumar, A., Kumar, A., Gehlot, R., Singh, D., Chaudhary, T., 2024. Probiotics and prebiotics from fruit waste. Adding Value to Fruit Wastes: Extraction, Properties, and Industrial Applications. Elsevier, India, pp. 261–290. Available at http://doi.org/10.1016/B978-0-443-13842-3.00010-1, https://www.sciencedirect.com/book/9780443138423.

Lee, W.J., Hase, K., 2014. Gut microbiota-generated metabolites in animal health and disease. Nature Chemical Biology 10 (6), 416–424. Available at https://doi.org/10.1038/nchembio.1535.

Lelia, P.O., Suharoschi, R., 2022. Emerging Food Processing Technologies: Probiotics and Prebiotics. Elsevier BV, pp. 509–536. Available at http://doi.org/10.1016/b978-0-323-85052-0.00008-8.

Liu, R.T., Walsh, R.F.L., Sheehan, A.E., 2019. Running head: prebiotics, probiotic. Depression, Anxiety 02915.

Magwira, C.A., Taylor, M.B., 2018. Composition of gut microbiota and its influence on the immunogenicity of oral rotavirus vaccines. Vaccine 36 (24), 3427–3433. Available at https://doi.org/10.1016/j.vaccine.2018.04.091, www.elsevier.com/locate/vaccine.

Marrs, T., Jo, J.-H., Perkin, M.R., Rivett, D.W., Witney, A.A., et al., 2021. Gut microbiota development during infancy: impact of introducing allergenic foods. Journal of Allergy and Clinical Immunology 147 (2), 613. Available at https://doi.org/10.1016/j.jaci.2020.09.042.

McCabe, L., Britton, R.A., Parameswaran, N., 2015. Prebiotic and probiotic regulation of bone health: role of the intestine and its microbiome. Current Osteoporosis Reports 13 (6), 363–371. Available at https://doi.org/10.1007/s11914-015-0292-x, http://www.springerlink.com/content/1544-1873/.

Mokoena, M.P., Mutanda, T., Olaniran, A.O., 2016. Perspectives on the probiotic potential of lactic acid bacteria from African traditional fermented foods and beverages. Food and Nutrition Research 60 (1), 29630. Available at https://doi.org/10.3402/fnr.v60.29630.

Mooser, C., Gomez de Agüero, M., Ganal-Vonarburg, S.C., 2018. Standardization in host–microbiota interaction studies: challenges, gnotobiology as a tool, and perspective. Current Opinion in Microbiology 44, 50–60. Available at https://doi.org/10.1016/j.mib.2018.07.007, http://www.elsevier.com/locate/mib.

Odriozola, A., González, A., Odriozola, I., Álvarez-Herms, J., Corbi, F., 2024. Microbiome-based precision nutrition: prebiotics, probiotics and postbiotics. Advanced Genetics 111, 237–310. Available at https://doi.org/10.1016/bs.adgen.2024.04.001, https://www.sciencedirect.com/science/bookseries/00652660.

Pan, R., Chen, Y., Lin, G., Wang, H., Xu, X., Wang, L., et al., 2024. Imbalances in the gut microbiota and metabolome are associated with functional constipation in pregnancy. Food Bioscience 59, 103852. Available at https://doi.org/10.1016/j.fbio.2024.103852.

Patra, J.K., Das, G., Paramithiotis, S., Shin, H.S., 2016. Kimchi and other widely consumed traditional fermented foods of Korea: a review. Frontiers in Microbiology 7. Available at https://doi.org/10.3389/fmicb.2016.01493, http://journal.frontiersin.org/article/10.3389/fmicb.2016.01493/full.

Pushpanathan, P., Mathew, G.S., Selvarajan, S., Seshadri, K.G., Srikanth, P., 2019. Gut microbiota and its mysteries. Indian Journal of Medical Microbiology 37 (2), 268–277. Available at https://doi.org/10.4103/ijmm.ijmm_19_373.

Pyle, S., 2022. 2.23 - Human gut microbiota and the influence of probiotics, prebiotics, and micronutrients. Comprehensive Gut Microbiota 2, 271–288. Available at https://doi.org/10.1016/B978-0-12-819265-8.00076-0, https://www.sciencedirect.com/book/9780128220368.

Rajilić-Stojanović, M., de Vos, W.M., 2014. The first 1000 cultured species of the human gastrointestinal microbiota. FEMS Microbiology Reviews. 38 (5), 996–1047. Available at https://doi.org/10.1111/1574-6976.12075, http://femsre.oxfordjournals.org/.

Razali, F.N., Teoh, W.Y., Ramli, M.Z., Loo, C.Y., Gnanaraj, C., 2023. Role of prebiotics, probiotics, and synbiotics in the management of colonic disorders. Advanced Drug Delivery Systems for Colonic Disorders. Elsevier, Malaysia, pp. 243–270. Available at http://doi.org/10.1016/B978-0-443-14044-0.00002-8, https://www.sciencedirect.com/book/9780443140440.

Ribera, C., Sánchez-Ortí, J.V., Clarke, G., Marx, W., Mörkl, S., Balanzá-Martínez, V., 2024. Probiotic, prebiotic, synbiotic and fermented food supplementation in psychiatric disorders: a systematic review of clinical trials. Neuroscience and BioBehavioral Reviews 158, 105561. Available at https://doi.org/10.1016/j.neubiorev.2024.105561, https://www.sciencedirect.com/science/journal/01497634.

Rosenbaum, M., Knight, R., Leibel, R.L., 2015. The gut microbiota in human energy homeostasis and obesity. Trends in Endocrinology and Metabolism 26 (9), 493–501. Available at https://doi.org/10.1016/j.tem.2015.07.002, https://www.elsevier.com/locate/tem.

Sampaio, K.B., Fusco, V., Alves, J.Ld.B., de Souza, E.L., 2022. Probiotics: concepts, evolution, and applications. Probiotics for Human Nutrition in Health and Disease. Elsevier, Brazil, pp. 3–24. Available at http://doi.org/10.1016/B978-0-323-89908-6.00019-4, https://www.sciencedirect.com/book/9780323899086.

Taghizadeh Ghassab, F., Shamlou Mahmoudi, F., Taheri Tinjani, R., Emami Meibodi, A., Zali, M.R., Yadegar, A., 2024. Probiotics and the microbiota-gut-brain axis in neurodegeneration: beneficial effects and mechanistic insights. Life Sciences 350, 122748. Available at https://doi.org/10.1016/j.lfs.2024.122748.

Tamura, K., Hemsworth, G.R., Déjean, G., Rogers, T.E., Pudlo, N.A., Urs, K., et al., 2017. Molecular mechanism by which prominent human gut bacteroidetes utilize mixed-linkage beta-glucans, major health-promoting cereal polysaccharides. Cell Reports 21 (2), 417–430. Available at https://doi.org/10.1016/j.celrep.2017.09.049.

Taylor, B.C., Lejzerowicz, F., Poirel, M., Shaffer, J.P., Jiang, L., Aksenov, A., et al., 2020. Consumption of fermented foods is associated with systematic differences in the gut microbiome and metabolome. mSystems 5 (2). Available at https://doi.org/10.1128/mSystems.00901-19, https://msystems.asm.org/content/msys/5/2/e00901-19.full.pdf.

Tran, N.T., Li, S., 2022. Potential role of prebiotics and probiotics in conferring health benefits in economically important crabs. Fish and Shellfish Immunology Reports 3, 100041. Available at https://doi.org/10.1016/j.fsirep.2021.100041.

Turnbaugh, P.J., Hamady, M., Yatsunenko, T., Cantarel, B.L., Duncan, A., Ley, R.E., et al., 2009. A core gut microbiome in obese and lean twins. Nature 457 (7228), 480–484. Available at https://doi.org/10.1038/nature07540.

Walker, A.W., Lawley, T.D., 2013. Therapeutic modulation of intestinal dysbiosis. Pharmacological Research 69 (1), 75–86. Available at https://doi.org/10.1016/j.phrs.2012.09.008.

Wopereis, H., Sim, K., Shaw, A., Warner, J.O., Knol, J., Kroll, J.S., 2018. Intestinal microbiota in infants at high risk for allergy: effects of prebiotics and role in eczema development. Journal of Allergy and Clinical Immunology 141 (4), 1334. Available at https://doi.org/10.1016/j.jaci.2017.05.054, http://www.elsevier.com/inca/publications/store/6/2/3/3/6/8/index.htt.

Xi, X., Huang, W., Hou, Q., Kwok, L.Y., Sun, Z., Sun, T., 2017. Assessment of microbial community structure in human colostrum and mature milk based on geographical location and delivery mode. Scientific Bulletin 62 (11), 745–747. Available at https://doi.org/10.1016/j.scib.2017.03.016.

Xia, B., Liu, X., Li, Z., Ren, J., Liu, X., 2023. The effects of microbiota-targeted approaches in inflammatory bowel disease: probiotics, probiotic foods, and prebiotics. Current Opinion in Food Science 49, 100956. Available at https://doi.org/10.1016/j.cofs.2022.100956.

Yadav, M.K., Kumari, I., Singh, B., Sharma, K.K., Tiwari, S.K., 2022. Probiotics, prebiotics and synbiotics: safe options for next-generation therapeutics. Applied Microbiology and Biotechnology 106 (2), 505–521. Available at https://doi.org/10.1007/s00253-021-11646-8, https://link.springer.de/link/service/journals/00253/index.htm.

You, S., Ma, Y., Yan, B., Pei, W., Wu, Q., Ding, C., et al., 2022. The promotion mechanism of prebiotics for probiotics: a review. Frontiers in Nutrition 9. Available at https://doi.org/10.3389/fnut.2022.1000517.

Chapter 16

An overview of the new generations of probiotics and their future trend in industrial products (drugs, supplements, and foods)

Zahra Hoseini Tavassol[1], Fatemeh-Sadat Hosseini[1], and Hanieh-Sadat Ejtahed[2]
[1]*Endocrinology and Metabolism Research Center, Endocrinology and Metabolism Clinical Sciences Institute, Tehran University of Medical Sciences, Tehran, Iran,* [2]*Obesity and Eating Habits Research Center, Endocrinology and Metabolism Clinical Sciences Institute, Tehran University of Medical Sciences, Tehran, Iran*

16.1 Introduction

The gastrointestinal tract plays a pivotal role in maintaining host health and homeostasis (Valdes et al., 2018; Sánchez et al., 2017). Dysbiosis, or the imbalance in the composition and function of the gut microbiota, has been linked to various diseases, including metabolic disorders, inflammatory conditions, and cancer (Sánchez et al., 2017; Carding et al., 2015; Zhang et al., 2022). Among the mechanisms by which the microbiota influences host physiology, the production of short-chain fatty acids (SCFAs) through the fermentation of dietary fibers stands out for its beneficial effects on glucose metabolism and immune modulation (Portincasa et al., 2022; Slavin, 2013; Bermudez-Brito et al., 2012). Probiotics, defined as live microorganisms that confer health benefits when administered in adequate amounts, have garnered increasing attention for their potential to restore microbial balance and improve disease outcomes (Hill et al., 2014; Zommiti et al., 2020). Traditional probiotics primarily include strains from the genera *Lactobacillus* and *Bifidobacterium*; however, emerging next-generation probiotics (NGPs) derived from commensal bacteria such as *Akkermansia muciniphila* and *Faecalibacterium prausnitzii* offer promising therapeutic avenues with

Evaluating Oral Probiotics and Prebiotics. DOI: https://doi.org/10.1016/B978-0-443-29835-6.00007-1

enhanced efficacy and targeted functions (Zhang et al., 2022; Jian et al., 2023; Singh and Natraj, 2021). Advances in genetic engineering have further expanded the probiotic repertoire, enabling the design of strains with specific functions such as antiinflammatory properties, pathogen inhibition, and metabolic regulation (Steidler, 2003; Hwang et al., 2017; Shen et al., 2022). Despite growing evidence supporting the benefits of probiotics, challenges remain regarding strain viability, safety, and personalized application, underscoring the need for continued research and optimized delivery systems (Wendel, 2022; Merenstein et al., 2023; Saarela, 2019).

This chapter provides a comprehensive overview of NGPs, highlighting their unique characteristics and therapeutic potential. It begins by exploring the critical role of the gut microbiota in health and diseases, followed by a comparison between traditional probiotics and novel commensal-derived strains. The chapter further delves into recent advances in genetic engineering that have enabled the development in designing probiotics with targeted functionalities. Finally, it addresses current challenges and future perspectives in the field, offering insights into how NGPs may revolutionize microbiome-based therapies.

16.2 A brief overview of the human microbiota, its role, and dysbiosis

Microbiota refers to the diverse community of microorganisms (bacteria, fungi, viruses, and archaea) that live in various parts of the human body, including the gut, skin, mouth, and respiratory tract. The gut microbiota plays a crucial role in digestion and metabolism by producing SCFAs like butyrate, which are important for gut health. It also interacts with the immune system to develop immune responses, compete with harmful microbes, and maintain tolerance to nonharmful substances and the integrity of the gut barrier. Dysbiosis refers to an imbalance or disruption in the composition of the microbiota, when harmful microorganisms outnumber beneficial ones, which can lead to various health issues (Carding et al., 2015; Portincasa et al., 2022; Gholami et al., 2023).

16.3 Introducing probiotics as one of the modulating approaches of microbiota

Modulating the microbiota aims to restore balance and improve health. This can be done through dietary changes such as consuming a diet rich in fiber, prebiotics, and fermented foods (like yogurt, kefir, and sauerkraut) to promote healthy microbiota. Prebiotics are nondigestible food ingredients, like inulin or fructooligosaccharides (FOS), that stimulate the growth of beneficial bacteria (Valdes et al., 2018; Slavin, 2013; Gibson et al., 2017). Probiotics are live microorganisms that, when consumed in adequate amounts, can confer health benefits by restoring the balance of the microbiota. Synbiotics refer to a combination of probiotics and prebiotics that work together to enhance the growth and activity of beneficial microorganisms (Pandey et al., 2015).

Traditional probiotics include species like *Lactobacillus* and *Bifidobacterium*. The benefits of probiotics include restoring gut balance, helping to correct dysbiosis, and improving digestive health, especially after antibiotic use. They also enhance immune function by supporting the immune system and protecting against infections, improving digestive health, and alleviating symptoms of IBS, diarrhea, and lactose intolerance. Additionally, probiotics may influence the gut-brain axis, potentially providing mental health benefits by helping with mood regulation and mental health (Pandey et al., 2015; Mayer et al., 2015; Didari et al., 2015).

16.4 Probiotic classification and characterization

The World Health Organization defines probiotics as "live microorganisms which when administered in adequate amounts confer a health benefit on the host" (Hill et al., 2014; Zommiti et al., 2020). These microorganisms, primarily bacteria and some yeasts, are naturally found in fermented foods, can be added to other food items, and are also accessible as dietary supplements. Probiotics are named using a specific alphanumeric strain designation that includes the genus, species, subspecies (if applicable), and a unique alphanumeric code (Guarner et al., 2017). The most common probiotics often used in probiotic products include *Lactobacillus* spp., *Bifidobacterium* spp., *Streptococcus* spp., *Enterococcus* spp., *Bacillus* spp., *Escherichia coli* spp., and *Saccharomyces boulardii* (Fijan, 2014; Roobab et al., 2020). Probiotics can be found in various sources, including fermented dairy products such as yogurt, kefir, and some cheeses, fermented vegetables such as sauerkraut, kimchi, and pickles, and fermented soy products such as miso and tempeh (Kaur et al., 2021). Probiotics can also be found in dietary supplements, which are available in various forms of capsules, tablets, powders (in sachets), or liquids (Kaur et al., 2021). The effectiveness of probiotics is determined by their viability, stability, and functionality. These are the main properties of probiotics. Probiotic viability refers to the ability of probiotic cells to remain alive and maintain their beneficial effects throughout their shelf life and during passage through the gastrointestinal tract. Stability refers to the ability of the product to maintain the viability of the probiotics (Wendel, 2022). Factors such as packaging, storage temperature, oxygen, water activity, chemical factors, pH, and light can affect the stability of probiotics during storage. Therefore assessing and enhancing the viability of probiotics is essential to ensure their effectiveness in providing the claimed health benefits (Wendel, 2022; Kaur et al., 2002; Gueimonde and Sánchez, 2012). To enhance the stability of probiotics, various strategies have been employed, including technological approaches for proper product formulation, controlled manufacturing, processing atmosphere, and microbiological parameters (Gueimonde and Sánchez, 2012; Terpou et al., 2019). The functionality of probiotics refers to their ability to confer a health benefit on the host, such as improving gut health, enhancing immune function, and reducing the risk of certain diseases (Sánchez et al., 2017).

The quality control and implementation of standards for introducing a new probiotic are important to ensure the safety and efficacy of the product. The microbial characterization of probiotics is a challenging task that requires a multidisciplinary approach. The taxonomic identity, biological safety, and product quality of probiotics must be characterized to ensure their effectiveness. Traditional probiotics are typically composed of lactic acid bacteria (LAB) and yeasts that have been used for centuries in fermented foods (Soemarie et al., 2021). These probiotics have been extensively studied and have demonstrated health benefits. However, traditional probiotics have limitations, including short-term effects, optimum quantity, survival and colonization, individual strain differences, safety concerns, and limited strain diversity (Hill et al., 2014; Merenstein et al., 2023; Shenderov et al., 2020). NGPs are being developed to overcome the limitations of traditional probiotics. These probiotics, which include genetically modified strains, spore-forming bacteria, and non-LAB probiotics, show promising potential for addressing various health concerns. Genetically modified strains can be engineered to enhance their functionality in the gastrointestinal tract. Spore-forming bacteria such as *Bacillus coagulans* demonstrate enhanced viability in the gastrointestinal tract compared to LAB. Furthermore, non-LAB probiotics such as *Escherichia coli* Nissle 1917 have been shown to have distinctive health benefits such as reducing the risk of inflammatory bowel disease (Zhang et al., 2022; Steidler, 2003; Bader et al., 2012; Zhao et al., 2022).

16.5 The next-generation probiotics

NGPs are an emerging class of beneficial microbes identified primarily from the human gut microbiome using advanced sequencing and techniques. Unlike traditional probiotics, NGPs are being investigated for targeted therapeutic roles in managing chronic and complex diseases. Novel Genetically Programmed bacteria, as another type of NGP, are genetically engineered microorganisms designed to carry out specific functions for health improvement and have applications in medicine, biotechnology, and environmental science. They can be categorized as synthetic biology-engineered bacteria and naturally occurring strain modifications.

16.5.1 Types of novel genetically programmed bacteria

16.5.1.1 Probiotic engineered bacteria

These are genetically modified to survive in the gut, interact with the immune system, and produce bioactive compounds for therapeutic applications (Duan et al., 2015).

16.5.1.2 Pathogen-targeting bacteria

Some strains are designed to seek out and kill pathogens within the body, such as modified *E. coli* engineered to detect and kill harmful bacteria (Hwang et al., 2017).

16.5.1.3 Tumor-targeting bacteria

Bacteria like *Salmonella* or *Clostridium* species are engineered to selectively target tumors, either by colonizing the tumor environment or producing anticancer compounds (Zhou et al., 2018).

16.5.1.4 Bacteria for metabolic disorders

Bacteria can be developed to combat metabolic diseases, such as strains of *Bacteroides* or *E. coli* engineered to produce insulin or other metabolic regulators in response to glucose levels (Fig. 16.1) (Shen et al., 2022).

16.5.2 Nnext-generation probiotic interaction with the host in different physiological states

In a healthy state, engineered probiotics can help maintain balance in the body by regulating the immune response (Zhang et al., 2023), promoting healthy digestion (Zhang et al., 2024), and preventing pathogen overgrowth. In infected hosts, these bacteria can combat infections, while in autoimmune diseases or chronic inflammatory states, they can modulate immune responses.

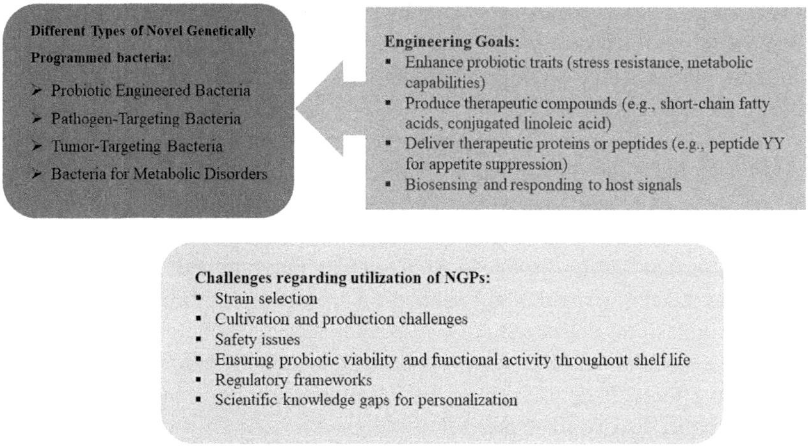

FIGURE 16.1 Overview of next-generation probiotics (NGPs): the major categories of novel genetically programmed bacteria, their engineering goals for enhanced therapeutic potential, and the key challenges faced in their development and application. *NGPs*, Next-generation probiotics.

In conditions such as diabetes or obesity, engineered bacteria can help regulate metabolic processes (Bai and Ouyang, 2006). Moreover, engineered bacteria can be used to target tumors, treat gastrointestinal disorders, combat infectious diseases, and potentially address neurological conditions. For example, they can produce antitumor agents within tumors and interact with the gut's immune system to reduce inflammation in gastrointestinal disorders. In infectious diseases, they can neutralize harmful organisms, and in neurological diseases, they may offer treatments through the gut-brain axis.

16.5.3 Strain selection

The efficiency of probiotics depends on the selection of the right strains. Traditional probiotic strains, including *Lactobacillus* and *Bifidobacterium* species, have been researched and utilized extensively. However, NGPs aim to expand the range of probiotic strains with unique properties (Bermudez-Brito et al., 2012). For example, recent studies have explored the potential of *Akkermansia muciniphila*, a mucin-degrading bacterium, as a NGP for the treatment of metabolic disorders (Jian et al., 2023). *Faecalibacterium prausnitzii*, *Bacteroides fragilis*, *Eubacterium hallii*, and *Roseburia* spp. are other taxa whose potential is being studied as NGPs (Zhang et al., 2022).

Some of the most promising NGPs include (Zhang et al., 2022):

Akkermansia muciniphila: Known for its role in enhancing gut barrier integrity, reducing inflammation, and improving metabolic health, including obesity and type 2 diabetes management. It is considered a key candidate for metabolic and inflammatory disorders.

Faecalibacterium prausnitzii: One of the most abundant and beneficial gut bacteria, it exhibits strong antiinflammatory properties, supports gut barrier function, and is linked to improved outcomes in inflammatory bowel disease and metabolic conditions.

Bacteroides fragilis: Recognized for its immune-modulating effects, particularly through polysaccharide A production, which helps reduce inflammation and may benefit central nervous system demyelinating diseases and gut health.

Eubacterium hallii: Produces SCFAs like butyrate, contributing to gut health and metabolic regulation, making it a promising candidate for metabolic and gastrointestinal disorders.

Christensenella minuta: Associated with lean body mass and beneficial metabolic effects, it helps increase SCFA production and may aid in obesity and metabolic disease management.

Prevotella copri: Postulated as a candidate for modulating metabolic and inflammatory pathways, though its role is still being explored.

Bacteroides thetaiotaomicron, Parabacteroides goldsteinii, Bacteroides uniformis, Bacteroides acidifaciens: These species are being studied for their roles in gut health, immune modulation, and metabolic benefits.

Clostridium butyricum: Known for butyrate production, it has potential in improving metabolic disorders and gut inflammation.

16.5.4 Therapeutic applications

In addition to their traditional applications, NGPs are being explored for a wide range of therapeutic applications. Probiotic-based interventions for mental health, oral health, cardiovascular health, and skin disorders are emerging areas of research (Park et al., 2018; Chugh et al., 2020; Kianmehr et al., 2022; Gao et al., 2023; Zhou et al., 2020; Gao et al., 2023). Also, the research on the effectiveness of new-generation probiotics on other health situations, including metabolic and inflammatory disorders, is still ongoing (Table 16.1) (Inchingolo et al., 2023; McFarland, 2021).

16.5.5 Next-generation probiotic future trends and challenges

NGPs are a promising advance in the field of probiotics, representing potentially improving health outcomes and resolving the challenges faced by traditional probiotics (Zhang et al., 2022). Designing these NGPs aims to improve the quality of life in patients with various health conditions. Strain selection, cultivation, and production challenges, safety issues, ensuring probiotic viability and functional activity throughout shelf life, regulatory frameworks, and scientific knowledge gaps for personalization are significant challenges regarding the utilization of NGPs. The current understanding of probiotics research and industry, as well as finding the future trends and challenges associated with NGPs, is important areas of focus (Fig. 16.1) (Zhang et al., 2022).

16.5.6 Next-generation probiotic, vulnerable populations, and individual capabilities

As the understanding of NGPs and their potential health benefits grows, there is a need to consider the safety and effectiveness of these products in vulnerable target populations, such as children, older adults, and individuals with compromised immune systems (Stadlbauer, 2015; Catania et al., 2022; Liu et al., 2023; Shi et al., 2023; Xue et al., 2023; Loy et al., 2023; He et al., 2023; Kou et al., 2023; Handajani et al., 2023; Shokri-Mashhadi et al., 2023). So further research and development are needed to confirm the safe and effective use of NGPs in different populations (Merenstein et al., 2023).

For children who are undergoing growth, exposure to novel compounds found in some NGPs could have unexpected long-term consequences on organ function and neurological development.

On the other hand, older adults often present with multiple comorbidities and may be on various medications, increasing the risk of unwanted

TABLE 16.1 Next-generation probiotics, properties, and effects.

Next-generation probiotics	Main properties	Applications	References
Akkermansia muciniphila	Metabolic health improvement, antiinflammatory effects	Diabetes management, obesity, metabolic diseases	Jian et al. (2023)
Faecalibacterium prausnitzii	Antiinflammatory effects, gut health promotion	Inflammatory bowel disease (IBD) treatment	Zhang et al. (2022)
Bacteroides spp.	Immune system regulation, lipid metabolism	Prevention and treatment of metabolic diseases	Singh and Natraj (2021)
Clostridium butyricum	Butyrate production, antiinflammatory effects	Gut health improvement, diarrhea and IBD treatment	Bader et al. (2012)
Engineered Probiotics (*E. coli* Nissle 1917)	Disease targeting, bio-drug production	Treatment of inflammatory diseases, cancer, diabetes	Hwang et al. (2017), Shen et al. (2022)

interactions with NGPs. Their aging physiological systems, such as compromised cardiovascular function, altered metabolism, and chronic conditions, remain largely unexplored in this field. Individuals with compromised immune systems—whether resulting from congenital conditions, illnesses, or medications—face increased risks. Therefore investigating the direct impact of NGPs in this group is crucial (Stadlbauer, 2015; Catania et al., 2022; Liu et al., 2023; Shi et al., 2023; Xue et al., 2023; Loy et al., 2023; He et al., 2023; Kou et al., 2023; Handajani et al., 2023; Shokri-Mashhadi et al., 2023).

So, research on the safety and effectiveness of the use of NGPs across vulnerable populations, by examining pharmacokinetics, long-term effects, specific health outcomes, dose–response relationships, and adhering to ethical standards, is essential.

16.5.7 Combining next-generation probiotics with prebiotics

It is suggested to use NGPs combined with prebiotics, which are nondigestible food ingredients that promote the growth of beneficial gut bacteria. This combination, known as synbiotics, offers synergistic effects and enhanced health benefits. Prebiotics optimize the survival, growth, and metabolic activity of probiotics, and their combined effects positively impact health by enhancing metabolic output such as SCFAs, improving gut barrier integrity, modulating immune responses, and providing targeted health benefits (Yadav et al., 2022). The synbiotics may be more effective than probiotics or prebiotics alone.

16.5.8 Synergy

A promising strategy for developing NGPs is to modulate the gut microbiota by using synergy between different strains for optimal growth and effectiveness, and adapting to specific environmental conditions, particularly oxygen levels (Khan et al., 2023; Kumari et al., 2021). Some of the most impactful NGPs are obligate or strict anaerobes (Zhang et al., 2022) and may require synergy with other microorganisms for their activity and effectiveness throughout their industrial production, formulation, and delivery to the anoxic environment of the human gut (Khan et al., 2023). This synergistic cultivation not only maintains their viability but also enhances their beneficial metabolic outputs and overall effectiveness in microbial therapies.

16.5.9 Regulation

One of the major challenges in developing NGPs is the fastidious growth of some strains, which require specific conditions for growth and survival. The optimization of growth conditions is necessary to obtain sufficient biomass yields, and novel strategies are required to preserve the viability of the final product. For example, some NGPs are strictly anaerobic and may require synergy with other bacteria for optimal growth (Khan et al., 2023). For example, *Akkermansia muciniphila*, *Faecalibacterium prausnitzii*, *Bacteroides fragilis*, and *Clostridium butyricum*, the highly prevalent and abundant human gut bacterium associated with health, have not yet been developed into probiotic formulations (Saarela, 2019; Khan et al., 2023; Kumari et al., 2021).

16.5.10 Safety

As probiotics research progresses, ensuring their safety and appropriate regulation becomes more important. NGPs may involve the use of genetically modified probiotics or novel bacterial species. Therefore it is crucial to assess the safety profiles of these new strains and establish an appropriate regulatory approach to ensure their safety for the consumer (Merenstein et al., 2023). For example, evaluating NGP potential for bacterial toxicity, developing antibiotic resistance, and acute and long-term risks should be considered for vulnerable target populations (Saarela, 2019).

16.5.11 Personalized probiotics

NGPs as a potential therapeutic agent could be among the leading strategies for personalized medicine. NGPs aim to find specific strains or combinations of strains that are most effective for personalized health management based on an individual's unique gut microbiota composition, genetics, dietary habits, and health condition (Singh and Natraj, 2021). For example, patients with a specific genetic predisposition to an inflammatory bowel disease might receive a personalized NGP supplement capable of modulating the specific inflammatory pathways implicated in their condition, based on their unique gut microbial profile. This personalized approach holds the promise of higher therapeutic efficacy with lower potential side effects (Singh and Natraj, 2021).

16.5.12 Delivery systems

Survival and colonization in the gastrointestinal tract of humans is one of the challenges associated with NGPs. NGPs focus on developing innovative delivery systems to enhance their viability and targeted delivery to specific sites within the gut. Different encapsulation techniques can help NGPs (NGPs) better survive manufacturing processes, such as heat and shear forces encountered during drying or formulation, while also maintaining viability against challenges in the gastrointestinal tract, like stomach acidity, bile salts, and digestive enzymes. Furthermore, specialized encapsulation techniques can enable targeted delivery to specific regions of the gut, such as the colon, optimizing their therapeutic effect for particular conditions (Agriopoulou et al., 2023).

16.6 Conclusion

Despite the potential benefits, several challenges need to be addressed for the successful development, industrial use, and commercialization of NGPs, and there is a lack of large-scale research to address them in various aspects, including those relevant to scientists, clinicians, and the regulatory system. These challenges include strain selection, therapeutic applications, vulnerable populations, synbiotics, synergy, regulation, safety, personalized probiotics, and delivery systems.

References

Agriopoulou, S., Tarapoulouzi, M., Varzakas, T., Jafari, S.M., 2023. Application of encapsulation strategies for probiotics: from individual loading to co-encapsulation. Microorganisms 11 (12), 2896. Available at https://doi.org/10.3390/microorganisms11122896.

Bader, J., Albin, A., Stahl, U., 2012. Spore-forming bacteria and their utilisation as probiotics. Beneficial Microbes 3 (1), 67–75. Available at https://doi.org/10.3920/bm2011.0039.

Bai, A.-P., Ouyang, Q., 2006. Probiotics and inflammatory bowel diseases. Postgraduate Medical Journal 82 (968), 376–382. Available at https://doi.org/10.1136/pgmj.2005.040899.

Bermudez-Brito, M., Plaza-Díaz, J., Muñoz-Quezada, S., Gómez-Llorente, C., Gil, A., 2012. Probiotic mechanisms of action. Annals of Nutrition and Metabolism 61 (2), 160–174. Available at https://doi.org/10.1159/000342079.

Carding, S., Verbeke, K., Vipond, D.T., Corfe, B.M., Owen, L.J., 2015. Dysbiosis of the gut microbiota in disease. Microbial Ecology in Health and Disease 26 (0). Available at https://doi.org/10.3402/mehd.v26.26191.

Catania, J., Pandit, N.G., Ehrlich, J.M., Zaman, M., Stone, E., Franceschi, C., et al., 2022. Probiotic supplementation for promotion of growth in children: a systematic review and meta-analysis. Nutrients 14 (1). Available at https://doi.org/10.3390/nu14010083, https://www.mdpi.com/2072-6643/14/1/83/pdf.

Chugh, P., Dutt, R., Sharma, A., Bhagat, N., Dhar, M.S., 2020. A critical appraisal of the effects of probiotics on oral health. Journal of Functional Foods 70, 103985. Available at https://doi.org/10.1016/j.jff.2020.103985.

Didari, T., Mozaffari, S., Nikfar, S., Abdollahi, M., 2015. Effectiveness of probiotics in irritable bowel syndrome: updated systematic review with meta-analysis. World Journal of Gastroenterology 21 (10), 3072–3084. Available at https://doi.org/10.3748/wjg.v21.i10.3072, http://www.wjgnet.com/1007-9327/pdf/v21/i10/3072.pdf.

Duan, F.F., Liu, J.H., March, J.C., 2015. Engineered commensal bacteria reprogram intestinal cells into glucose-responsive insulin-secreting cells for the treatment of diabetes. Diabetes 64 (5), 1794–1803. Available at https://doi.org/10.2337/db14-0635, http://diabetes.diabetesjournals.org/content/diabetes/64/5/1794.full.pdf.

Fijan, S., 2014. Microorganisms with claimed probiotic properties: an overview of recent literature. International Journal of Environmental Research and Public Health 11 (5), 4745–4767. Available at https://doi.org/10.3390/ijerph110504745.

Gao, T., Wang, X., Li, Y., Ren, F., 2023. The role of probiotics in skin health and related gut–skin axis: a review. Nutrients 15 (14), 3123. Available at https://doi.org/10.3390/nu15143123.

Gao, J., Zhao, L., Cheng, Y., Lei, W., Wang, Y., Liu, X., et al., 2023. Probiotics for the treatment of depression and its comorbidities: a systemic review. Frontiers in Cellular and Infection Microbiology 13. Available at https://doi.org/10.3389/fcimb.2023.1167116.

Gholami, H., Chmiel, J.A., Burton, J.P., Maleki Vareki, S., 2023. The role of microbiota-derived vitamins in immune homeostasis and enhancing cancer immunotherapy. Cancers 15 (4). Available at https://doi.org/10.3390/cancers15041300, http://www.mdpi.com/journal/cancers/.

Gibson, G.R., Hutkins, R., Sanders, M.E., Prescott, S.L., Reimer, R.A., Salminen, S.J., et al., 2017. Expert consensus document: The International Scientific Association for Probiotics and Prebiotics (ISAPP) consensus statement on the definition and scope of prebiotics. Nature Reviews Gastroenterology and Hepatology 14 (8), 491–502. Available at https://doi.org/10.1038/nrgastro.2017.75, http://www.nature.com/nrgastro/index.html.

Guarner, F., Sanders, M., Eliakim, R., Fedorak, R., Gangl, A., Garisch, J., 2017. Probiotics and Prebiotics. Milwaukee, WI: World Gastroenterology Organisation.

Gueimonde, M., Sánchez, B., 2012. Enhancing probiotic stability in industrial processes. Microbial Ecology in Health and Disease. 23 (0). Available at https://doi.org/10.3402/mehd.v23i0.18562.

Handajani, Y.S., Hengky, A., Schröder-Butterfill, E., Hogervorst, E., Turana, Y., 2023. Probiotic supplementation improved cognitive function in cognitively impaired and healthy older adults: a systematic review of recent trials. Neurological ScienceS 44 (4), 1163–1169. Available at https://doi.org/10.1007/s10072-022-06540-8.

He, X., Liu, W., Tang, F., Chen, X., Song, G., 2023. Effects of probiotics on autism spectrum disorder in children: a systematic review and meta-analysis of clinical trials. Nutrients 15 (6), 1415. Available at https://doi.org/10.3390/nu15061415.

Hill, C., Guarner, F., Reid, G., Gibson, G.R., Merenstein, D.J., Pot, B., et al., 2014. Expert consensus document: the international scientific association for probiotics and prebiotics consensus statement on the scope and appropriate use of the term probiotic. Nature Reviews Gastroenterology and Hepatology 11 (8), 506–514. Available at https://doi.org/10.1038/nrgastro.2014.66, http://www.nature.com/nrgastro/index.html.

Hwang, I.Y., Koh, E., Wong, A., March, J.C., Bentley, W.E., Lee, Y.S., et al., 2017. Engineered probiotic *Escherichia coli* can eliminate and prevent *Pseudomonas aeruginosa* gut infection in animal models. Nature Communications 8. Available at https://doi.org/10.1038/ncomms15028, http://www.nature.com/ncomms/index.html.

Inchingolo, F., Inchingolo, A.M., Malcangi, G., Leonardis, N.D., Sardano, R., Pezzolla, C., et al., 2023. The benefits of probiotics on oral health: systematic review of the literature. Pharmaceuticals 16 (9), 1313. Available at https://doi.org/10.3390/ph16091313.

Jian, H., Liu, Y., Wang, X., Dong, X., Zou, X., 2023. Akkermansia muciniphila as a next-generation probiotic in modulating human metabolic homeostasis and disease progression: a role mediated by gut–liver–brain axes? International Journal of Molecular Sciences 24 (4), 3900. Available at https://doi.org/10.3390/ijms24043900.

Kaur, I.P., Chopra, K., Saini, A., 2002. Probiotics: potential pharmaceutical applications. European Journal of Pharmaceutical Sciences 15 (1), 1–9. Available at https://doi.org/10.1016/s0928-0987(01)00209-3.

Kaur, S., Kaur, R., Rani, N., Sharma, S., Joshi, M., 2021. Sources and selection criteria of probiotics. Microorganisms for Sustainability 21, 27–43. Available from: Available at https://doi.org/10.1007/978-981-15-6795-7_2, https://www.springer.com/series/14379.

Khan, M.T., Dwibedi, C., Sundh, D., Pradhan, M., Kraft, J.D., Caesar, R., et al., 2023. Synergy and oxygen adaptation for development of next-generation probiotics. Nature 620 (7973), 381–385. Available at https://doi.org/10.1038/s41586-023-06378-w, https://www.nature.com/nature/.

Kianmehr, S., Jahani, M., Moazzen, N., Ahanchian, H., Khameneh, B., 2022. The potential of probiotics for treating skin disorders: a concise review. Current Pharmaceutical Biotechnology 23 (15), 1851–1863. Available at https://doi.org/10.2174/1389201023666220411090301.

Kou, J., Kang, H., Hu, L., Wang, D., Wang, S., Wang, Q., et al., 2023. Evaluation of improvement of cognitive impairment in older adults with probiotic supplementation: a systematic review and meta-analysis. Geriatric Nursing 54, 155–162. Available at https://doi.org/10.1016/j.gerinurse.2023.09.009.

Kumari, M., Singh, P., Nataraj, B.H., Kokkiligadda, A., Naithani, H., Azmal Ali, S., et al., 2021. Fostering next-generation probiotics in human gut by targeted dietary modulation: an emerging perspective. Food Research International 150 (Pt A), 110716. Available at https://doi.org/10.1016/j.foodres.2021.110716.

Liu, L., Wang, A., Shi, H., Tao, H., Nahata, M.C., 2023. Efficacy and safety of probiotics and synbiotics for functional constipation in children: a systematic review and meta-analysis of randomized clinical trials. Clinical Nutrition 42 (10), 1817–1826. Available at https://doi.org/10.1016/j.clnu.2023.08.015, http://www.elsevier-international.com/journals/clnu/.

Loy, M.H., Usseglio, J., Lasalandra, D., Gold, M.A., 2023. Probiotic use in children and adolescents with overweight or obesity: a scoping review. Childhood Obesity 19 (3), 145–159. Available at https://doi.org/10.1089/chi.2022.0059, www.liebertonline.com/obe.

Mayer, E.A., Tillisch, K., Gupta, A., 2015. Gut/brain axis and the microbiota. Journal of Clinical Investigation 125 (3), 926–938. Available at https://doi.org/10.1172/jci76304.

McFarland, L.V., 2021. Efficacy of single-strain probiotics versus multi-strain mixtures: systematic review of strain and disease specificity. Digestive Diseases and Sciences 66 (3), 694–704. Available at https://doi.org/10.1007/s10620-020-06244-z, www.wkap.nl/journalhome.htm/0163-2116.

Merenstein, D., Pot, B., Leyer, G., Ouwehand, A.C., Preidis, G.A., Elkins, C.A., et al., 2023. Emerging issues in probiotic safety: 2023 perspectives. Gut Microbes 15 (1). Available at https://doi.org/10.1080/19490976.2023.2185034, http://www.tandfonline.com/toc/kgmi20/current.

Pandey, K.R., Naik, S.R., Vakil, B.V., 2015. Probiotics, prebiotics and synbiotics- a review. Journal of Food Science and Technology 52 (12), 7577–7587. Available from: Available at https://doi.org/10.1007/s13197-015-1921-1, http://www.springerlink.com/content/121580/.

Park, C., Brietzke, E., Rosenblat, J.D., Musial, N., Zuckerman, H., Ragguett, R.M., et al., 2018. Probiotics for the treatment of depressive symptoms: an anti-inflammatory mechanism? Brain, Behavior, and Immunity 73, 115–124. Available at https://doi.org/10.1016/j.bbi.2018.07.006, http://www.elsevier.com/inca/publications/store/6/2/2/8/0/0/index.htt.

Portincasa, P., Bonfrate, L., Vacca, M., De Angelis, M., Farella, I., Lanza, E., et al., 2022. Gut microbiota and short chain fatty acids: implications in glucose homeostasis. International Journal of Molecular Sciences 23 (3), 1105. Available at https://doi.org/10.3390/ijms23031105.

Roobab, U., Batool, Z., Manzoor, M.F., Shabbir, M.A., Khan, M.R., Aadil, R.M., 2020. Sources, formulations, advanced delivery and health benefits of probiotics. Current Opinion in Food Science 32, 17–28. Available at https://doi.org/10.1016/j.cofs.2020.01.003, http://www.journals.elsevier.com/current-opinion-in-food-science/.

Saarela, M.H., 2019. Safety aspects of next generation probiotics. Current Opinion in Food Science 30, 8–13. Available at https://doi.org/10.1016/j.cofs.2018.09.001, http://www.journals.elsevier.com/current-opinion-in-food-science/.

Shenderov, B.A., Sinitsa, A.V., Zakharchenko, M.M., Lang, C., 2020. Drawbacks and negative consequences of traditional probiotics based on live microorganisms. Science and Business Media LLC 43–48. Available at https://doi.org/10.1007/978-3-030-34167-1_9.

Shen, H., Zhao, Z., Zhao, Z., Chen, Y., Zhang, L., 2022. Native and engineered probiotics: promising agents against related systemic and intestinal diseases. International Journal of Molecular Sciences 23 (2), 594. Available at https://doi.org/10.3390/ijms23020594.

Shi, J., Wang, Q., Ruan, G., Chen, Y., Zhao, M., Shi, D., et al., 2023. Efficacy of probiotics against dental caries in children: a systematic review and meta-analysis. Critical Reviews in Food Science and Nutrition 63 (29), 9977–9994. Available at https://doi.org/10.1080/10408398.2022.2077693.

Shokri-Mashhadi, N., Navab, F., Ansari, S., Rouhani, M.H., Hajhashemy, Z., Saraf-Bank, S., 2023. A meta-analysis of the effect of probiotic administration on age-related sarcopenia. Food Science and Nutrition 11 (9), 4975–4987. Available at https://doi.org/10.1002/fsn3.3515, onlinelibrary.wiley.com/journal/10.1002/%28ISSN%292048-7177.

Singh, T.P., Natraj, B.H., 2021. Next-generation probiotics: a promising approach towards designing personalized medicine. Critical Reviews in Microbiology 47 (4), 479–498. Available at https://doi.org/10.1080/1040841X.2021.1902940, https://www.tandfonline.com/loi/imby20.

Slavin, J., 2013. Fiber and prebiotics: mechanisms and health benefits. Nutrients 5 (4), 1417–1435. Available at https://doi.org/10.3390/nu5041417.

Soemarie, Y.B., Milanda, T., Barliana, M.I., 2021. Fermented foods as probiotics: a review. Journal of Advanced Pharmaceutical Technology and Research 12 (4), 335–339. Available from: Available at https://doi.org/10.4103/japtr.japtr_116_21, http://www.japtr.org.

Stadlbauer, V., 2015. Immunosuppression and probiotics: are they effective and safe. Beneficial Microbes 6 (6), 823–828. Available at https://doi.org/10.3920/bm2015.0065.

Steidler, L., 2003. Genetically engineered probiotics. Best Practice and Research Clinical Gastroenterology 17 (5), 861–876. Available at https://doi.org/10.1016/s1521-6918(03)00072-6.

Sánchez, B., Delgado, S., Blanco-Míguez, A., Lourenço, A., Gueimonde, M., Margolles, A., 2017. Probiotics, gut microbiota, and their influence on host health and disease. Molecular Nutrition and Food Research 61 (1). Available at https://doi.org/10.1002/mnfr.201600240.

Terpou, A., Papadaki, A., Lappa, I., Kachrimanidou, V., Bosnea, L., Kopsahelis, N., 2019. Probiotics in food systems: significance and emerging strategies towards improved viability and delivery of enhanced beneficial value. Nutrients 11 (7), 1591. Available at https://doi.org/10.3390/nu11071591.

Valdes, A.M., Walter, J., Segal, E., Spector, T.D., 2018. Role of the gut microbiota in nutrition and health. British Medical Journal 361, k2179. Available at https://doi.org/10.1136/bmj.k2179.

Wendel, U., 2022. Assessing viability and stress tolerance of probiotics—a review. Frontiers in Microbiology 12. Available at https://doi.org/10.3389/fmicb.2021.818468.

Xue, X., Yang, X., Shi, X., Deng, Z., 2023. Efficacy of probiotics in pediatric atopic dermatitis: a systematic review and meta-analysis. Clinical and Translational Allergy 13 (7). Available at https://doi.org/10.1002/clt2.12283.

Yadav, M.K., Kumari, I., Singh, B., Sharma, K.K., Tiwari, S.K., 2022. Probiotics, prebiotics and synbiotics: safe options for next-generation therapeutics. Applied Microbiology and Biotechnology. 106 (2), 505–521. Available at https://doi.org/10.1007/s00253-021-11646-8, link.springer.de/link/service/journals/00253/index.htm.

Zhang, H., Duan, Y., Cai, F., Cao, D., Wang, L., Qiao, Z., et al., 2022. Next-generation probiotics: microflora intervention to human diseases. BioMed Research International 2022. Available at https://doi.org/10.1155/2022/5633403, http://www.hindawi.com/journals/biomed/.

Zhang, S., Ma, J., Ma, Y., Yi, J., Wang, B., Wang, H., et al., 2024. Engineering probiotics for diabetes management: advances, challenges, and future directions in translational microbiology. International Journal of Nanomedicine 19, 10917–10940. Available from: Available at https://doi.org/10.2147/IJN.S492651, https://www.dovepress.com/article/download/96747.

Zhang, T., Zhang, J., Duan, L., 2023. The role of genetically engineered probiotics for treatment of inflammatory bowel disease: a systematic review. Nutrients 15 (7), 1566. Available at https://doi.org/10.3390/nu15071566.

Zhao, Z., Xu, S., Zhang, W., Wu, D., Yang, G., 2022. Probiotic *Escherichia coli* NISSLE 1917 for inflammatory bowel disease applications. Food and Function 13 (11), 5914–5924. Available at https://doi.org/10.1039/d2fo00226d.

Zhou, Z., Chen, X., Sheng, H., Shen, X., Sun, X., Yan, Y., et al., 2020. Engineering probiotics as living diagnostics and therapeutics for improving human health. Microbial Cell Factories 19 (1). Available at https://doi.org/10.1186/s12934-020-01318-z, http://www.microbialcellfactories.com/home/.

Zhou, S., Gravekamp, C., Bermudes, D., Liu, K., 2018. Tumour-targeting bacteria engineered to fight cancer. Nature Reviews Cancer 18 (12), 727–743. Available at https://doi.org/10.1038/s41568-018-0070-z.

Zommiti, M., Feuilloley, M.G.J., Connil, N., 2020. Update of probiotics in human world: a nonstop source of benefactions till the end of time. Microorganisms 8 (12), 1907. Available at https://doi.org/10.3390/microorganisms8121907.

Chapter 17

Regulations associated with the industrial prebiotic products: drugs, supplements, and food products

S. Mithul Aravind[1] and S. Chakkaravarthi[2]

[1]*Faculty of Agro-Industry, Prince of Songkla University, Hat Yai, Thailand,* [2]*Interdisciplinary Sciences, National Institute of Food Technology Entrepreneurship and Management-Kundli, Sonipat, Haryana, India*

17.1 Introduction

Human health, a top priority in modern society, is always dynamic and susceptible to profound changes, disorders, and diseases. Trending research links chronic disease with acute symptoms and the gut microbiota profile, broadening the horizons (Mithul Aravind et al., 2021). Inclining studies highlight the importance of gut flora homeostasis for overall well-being, including cardiovascular health, immune strength, neural and behavioral adaptations, and digestion. The potential gut microbial balance reduces disease risk and improves the overall host wellness. Concerns about gut health have been sparked by the rising prevalence of gastrointestinal tract (GIT) conditions, including constipation, inflammatory bowel disease (IBD), gastroesophageal reflux disease, and irritable bowel syndrome. It is well known that the use of prebiotic compounds soothes gut health (Mithul Aravind et al., 2021). Hence, clinical research uplifts the concept of prebiotics and reveals much about the triplex effect they exert, that is, the gut microbiota-prebiotics (metabolites)-host health. This, in turn, infers health benefits to the host, which are related, as shown in Fig. 17.1. Therefore, customers' demand and desire for prebiotic supplements is a direct reflection of these concerning health issues. The prebiotic demand is anticipated to sustain market expansion over the predicted time frame. The industrialization of commodities utilized in the food, nutrition, and pharmaceutical sectors has begun to leverage the now-possible technique of stimulating indigenous microbes in the body to achieve the

Evaluating Oral Probiotics and Prebiotics. DOI: https://doi.org/10.1016/B978-0-443-29835-6.00020-4

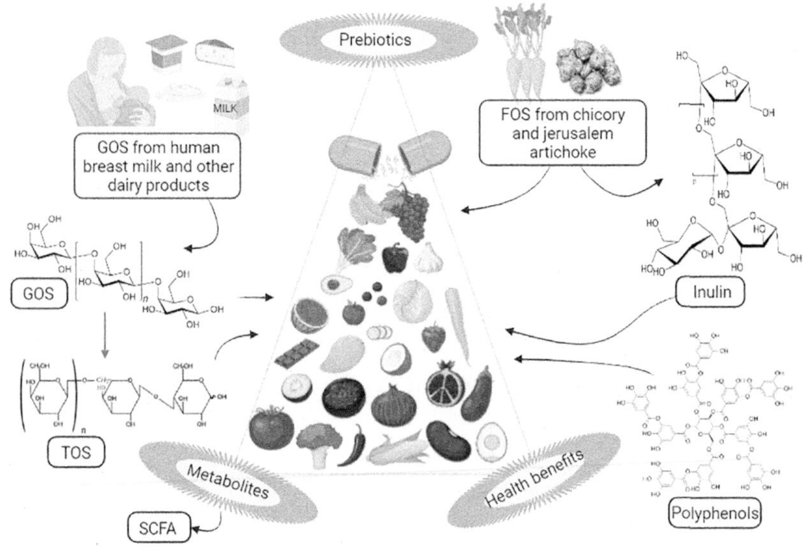

FIGURE 17.1 The triplex effect between prebiotics, metabolites, and health benefits: The triplex effect.

desired result. Prebiotics were estimated to be worth USD 6.05 billion globally in 2021, and between 2022 and 2030, the industry is projected to expand at a 14.9% compound annual growth rate (CAGR). Over the next seven years, development is expected to be aided by strong demand for supplements and the shifting consumer understanding of the health benefits of fiber and other bioactive prebiotic compounds.

Regulating costs and improving the quality of health and healthcare products are goals of research and applications in the field of functional foods, which have helped shape the present era. Research on functional foods is typically conducted through commodity-specific applications. Products evaluated and proven to have health benefits are identified as functional foods. Consumers are informed about the health-related benefits of these foods to prevent and treat diseases for which they are at higher risk, compared to regular foods. Hence, the primary approach to improving GIT health involves healthy feeding strategies that use prebiotics. This comprises the feed for the host and their associated gut commensals. Briefly, gut microbes utilize undigested residues (prebiotics) and selectively support their growth. The latter produce their corresponding metabolites and short-chain fatty acids (SCFAs), referred to as postbiotics, ultimately favoring the host's health (Davani-Davari et al., 2019). Industrial experts now target the vast array of scientifically proven, gut-mediated effects on host health. An abundant range of prebiotic products is produced in the market for commercial and societal welfare. The global prebiotic product accomplishment broadly comprises all the activities, commencing from composition fabrication to the effectiveness of the developed formulation post-consumption. Moreover, clinical trials strongly support the claim proposed by the manufacturer. Although the success is in-stream, customers or manufacturers themselves deviate from the concept of prebiotic regulations and confuse between dietary supplements, nutraceuticals, and functional foods. Hence, this chapter explains the regulatory concept of these terminologies and the industry-associated oral supplementation in relation to global regulations.

17.2 What are prebiotics?

The concept and consumption of prebiotics have been practiced for ages by the human race. Specifically, prebiotic intake begins in the infantile stage. That is, human breast milk is the first prebiotic food to be consumed as it comprises human milk oligosaccharides (HMO). HMO includes complicated galactosyllactose, lacto-N-tetraose, lacto-N-neotetraose, lactose derivatives with fucosylated and sialylated forms (Selvamani et al., 2023). Although the importance and characteristics are unclear to ancient humankind, the consumption of prebiotics remained ubiquitous. Various definitions have been updated and modified over the years as research has flourished to better understand the concept of prebiotics. For instance, the very first definition of prebiotics was drafted in 1995 as "a nondigestible food ingredient that beneficially affects the host by selectively stimulating the growth and activity of one or a limited number of bacteria in the colon, and thus improves host health" (Gibson and Roberfroid, 1995). However, according to the latest ISAPP, prebiotics are "a substrate that is selectively utilized by host microorganisms conferring a health benefit." However, they are not completely digested and absorbed in the small intestine (Ayyash and Liu, 2023). Thus, when understanding and innovation are fostered, the definition and perspective on the concept are broadened through deep knowledge. Most common prebiotic and prebiotic-like compounds used in the commercial run include the polyphenols, lactulose, dietary functional fibers from various plant sources comprising arabinoxylans, β-glucan, galacto-oligosaccharides (GOS), fructo-oligosaccharides (FOS), inulin, and isomalto-oligosaccharides (IMO), polydextrose, and trans-galacto-oligosaccharides (TOS), xylooligosaccharides (XOS) (Kaur et al., 2021; Davani-Davari et al., 2019). Alternatively, animal sources such as chitosan oligosaccharides (COS) and glucan complex (Lopez-Santamarina et al., 2020), and marine sources such as eicosapentaenoic acid, docosahexaenoic acid, and fucoidan are also widely used in the market today (Cardoso et al., 2021).

The ingested prebiotic compounds can offer health benefits to the host, directly or indirectly. Direct effects include regulation of bowel movements, stool quality, and consistency; toxin removal by adhering to the inner lining (mucosa) of the intestines (enterocytes); and lowering cholesterol, especially LDL. The indirect effect includes the host's beneficial action through gut flora modulation. The antimicrobial properties of certain prebiotic or prebiotic-like compounds, such as polyphenols, inhibit or suppress pathogenic growth and stimulate or favor beneficial bacteria. Moreover, these "beneficial" bacteria produce bioactive metabolites (Mithul Aravind et al., 2021; Vilas-Boas et al., 2021). Hence, the consumption patterns of dietary supplements, functional foods, nutraceuticals, and drugs determine the mediating role of consumption and the consequential effects.

Prebiotics are inert and resist upper-gut digestion. Manufacturers use clean processing, strategic ingredients, or compounds to design prebiotic products to avoid upper-gut digestion. This could be in powder, capsule, tablet, chew, or gummy form. Based on the dosage and nature of the product, it is categorized as a food, dietary supplement, or nutraceutical. If the designed product is categorized as a pharmaceutical drug, it is strictly reviewed for adequacy and effectiveness in clinical trials and in accordance with regulatory guidelines. Hence, most manufacturers favor producing dietary supplements or bioactive foods over nutraceuticals for prebiotics. For a compound to be prebiotic in nature, the dietary fiber shall possess a degree of polymerization value of ≥(3; it shall be resistant to acidic pH in the stomach, not be hydrolyzed by the endogenous enzymes in the upper GIT, not be absorbed or metabolized in the GIT, shall be metabolized by gut microbes, and selectively stimulates the growth/activity of one or limited number of species) (Davani-Davari et al., 2019; Brownawell et al., 2012).

17.3 Know the difference

17.3.1 Dietary supplements

Dietary supplements are food's additional companion to deliver selective (bioactive) compounds consumed by the host to achieve a desired health effect. These products specifically add to or supplement certain dietary nutrients (ingredients) or bioactive compounds to conventional food. They are not directly intended to treat or cure diseases. Instead, they support, improve, or maintain the host's health and prevent diseases or defects. Strictly, they cannot claim to cure or treat disease. They are self-regulated post-market by manufacturers. These are taken in the form of powders, liquids, syrups, gels, gel capsules, capsules, pills, or tablets (Dietary Supplement Health and Education Act, 1994), for example, vitamins, minerals, botanical components such as herbs, bioactive compounds such as biopeptides, collagen, and polyphenols (Cardoso et al., 2021). The US Dietary Supplement Health and Education Act (DSHEA) of 1994 stated the following regarding the sale and marketing of dietary supplements.

The term "dietary supplement" signifies according to multiple standards: (1) a product (with tobacco as an exception) meant to enhance the diet and containing one or more of the following dietary components: vitamins, minerals, botanicals, amino acids, or other related dietary substance to enhance the diet by raising the overall daily intake, or concentrate, metabolite, a component, extract, or mixtures of these substances; (2) a pill, tablet, capsule, sirup or gel-form product; (3) a product not meant to be consumed as a meal or diet item; (4) items labeled to be a "dietary supplement"; (5) The US Secretary of Health and Human Services authorized goods, such as recently developed drugs, approved antibiotics, or licensed biologics that were sold as food or dietary supplements before authorization or accreditation (Cencic and Chingwaru, 2010).

The main characteristic of a dietary supplement is to provide certain nutrients or bioactive compounds that are required by the host but are usually limited in their regular diet. Their objective is to maintain or promote the health of consumers and support the normal functioning of the organism in its natural state. In general, fitness freaks, athletes, and even ordinary individuals, including the elderly, are advised to take dietary supplements for a healthy and better lifestyle. Prebiotic compounds are widely used as dietary supplements, as they ease and soothe bowel movements, thereby suppressing constipation due to their bulkiness. This is mainly due to their indigestible, high-molecular-weight compounds, such as dietary fiber from oats, orange peel, or psyllium husk. For these reasons, prebiotic supplements are established for the long-term prevention of altered health status rather than for curative or substitutive action on physiological activity.

17.3.2 Nutraceuticals

Nutraceuticals, in other words, are dietary supplements sourced from natural sources. The term "Nutraceuticals" is a clipped form of "nutrition" and "pharmaceuticals." There are no proper, well-defined regulations for nutraceuticals, technically defined as such. This is highly synchronized with dietary supplements. However, nutraceuticals are mainly distinguished by the function that shall be tolerated in the diet, in addition to helping with disease prevention and medication (Cencic and Chingwaru, 2010). Nutraceuticals should enhance the diet and help mitigate and treat disease and dysfunction.

17.3.3 Functional foods

These are whole food items, usually considered part of a regular diet. They may profess specific positive effects on host health. For example, probiotic curd, other dairy products, and whole fruits are treated as functional foods. The regular and conventional foods, such as gravy, contain various natural bioactive ingredients, such as turmeric (curcumin), pepper (piperine), clove (eugenol), chili (capsaicin), garlic (allicin), and onions (quercetin and other flavonoids, sulfocysteine). Hence, they are called functional foods.

17.3.4 Drugs

They are chemicals, ideally formulated and intended to diagnose, treat, cure, or prevent diseases and disorders. They undergo rigorous testing and clinical trials before market release (Sanders et al., 2011). Drug products must endorse all aspects of compositional testing and chronic toxicity studies in clinical models. This ensures the effective dosage, intervention mode, and other dietary considerations.

It can be questioned if a nutrient used to treat a specific condition qualifies as a medicine, whereas a nutrient used to improve health (lower the risk of disease) qualifies as a functional food or dietary supplement. This intersection demonstrates the connections between functional foods and medications.

17.4 Manufacturing of prebiotics

Industrial operations include concentration, isolation, isomerization, and purification from the targeted sources to extract the specific compound matrix or ingredient. This could be sold as an ingredient or further formulated into a product through hydrolysis, chemical, and enzymatic production (Al-Sheraji et al., 2013). Industrial operations mainly rely on natural, undigestible carbohydrates and phenolic sources, such as rhizomes, roots, fruits, and vegetables. The list of prebiotics manufactured in different countries is presented in Table 17.1. Prebiotic production and manufacturing include different concepts and dynamic links broadly integrating the following, but not limited to, the chemistry of molecules and their rheology, biochemistry of hosts, including the health concerns, microbiology of the gut, industrial production mechanism, national and international standards and regulations, trade acceptance, and global consumer preferences. Manufacturers utilize various natural resources to produce prebiotic supplements. The use of prebiotics at the lab scale across different product formulations was described by Corbo et al. (2014). Alongside the production of prebiotics, their assessment is mandatory for manufacturers to obtain authorized claims (Bamigbade et al., 2022). The globally used commercial prebiotic compounds and their concurrent analysis are depicted in Fig. 17.2. For instance, a prebiotic must be associated with a selective health benefit by linking it to a specific change in gut bacterial profile (Brooks and Kalmokoff, 2012).

Analogically, the concept of prebiotic formulation is closely linked to healthcare professionals, industrial operators, consumers, and product formulators, including scientists. Although the concepts are converging across different domains, it is formidable to control and scrutinize the health-promoting claims of prebiotic products across diverse populations and

TABLE 17.1 Global prebiotic supplements product description.

Country	Company name	Product name	Prebiotic used	Consumption form
Australia	Nutra Life	ProBiotica	Inulin	Capsules
	Xemenry	Xemenry	Tapioca sirup, black elderberry anthocyanins	Capsules
	Nourishme Organics	Nourishme Organics	Hydrolysable guar gum, acacia fiber, GOS, FOS inulin, and pomegranate extract	Powder and capsules
	Natural evolution	Gut Health +	Resistant starch from green banana, inulin from artichoke, starch and fiber from sweet potato	Powder
Belgium	Cosucra	Cosucra	Chicory root fiber, inulin, polyfructose, oligofructose, HMO, GOS	Powder, tablet chews, gummies
Brazil	Blessy	Blessy	Fibers from apple, beetroot, flaxseed, and mint	Powder
	MadreLabs	HealthyBiom	Rice hulls, modified cellulose, IMO	Capsules
Canada	Left Coast Performance	MCT oil powder collagen peptides	Acacia fiber, collagen peptides	Powder
	Organika	Probiotic + Prebiotic	FOS	Powder
	Minaturals	Minaturals	Inulin	Powder
	RestoraLAX	RestoraFIBRE	Inulin	Gummies

(*Continued*)

TABLE 17.1 (Continued)

Country	Company name	Product name	Prebiotic used	Consumption form
	MetaMucil	MetaMucil Fiber	FOS and psyllium husk	Gummies
China	Cosun	Cosun	Inulin fiber	Powder and extract
Egypt	Generic	Cranby	Proanthocyanidins from cranberry and lingonberry	Capsules
England	Vita Bright	Vita Bright	Inositol, inulin, maca root, ginger root	Capsules
	Deeply	Deeply	Acacia gum, baobab fruits, chicory root, spinach fiber, Jerusalem artichoke and fruits and vegetable fibers from kiwi and carrot, rhizomes (ginger)	Juice
Germany	Beneo	Beneo	Inulin, oligofructose, scFOS, β-glucan, IMO, palatinose (isomaltulose)	Powder
	Suedzucker AG	Suedzucker AG	Chicory fiber, FOS, wheat, and broad bean proteins	Powder
India	The Good Bug	The Good Bug	Chicory root fiber, FOS, green pea fiber, GOS, IMO, polydextrose, a-cyclodextrin, glucomannan, polyphenols	Bars
	TATA Chemicals (NQ)	FOSSENCE	Short-chain FOS	Powder
	Sova	Sova	Inulin	Powder
Indonesia	Orafti	Orafti	Inulin fiber, and oligofructose	Powder
Israel	Galam	Galam	FOS, inulin	Powder, liquid and capsules

Japan	Yakult Pharmaceutical Industries Co. Ltd.	Yakult Pharmaceutical Industries Co. Ltd.	GOS	Sirup and Powder
	Meiji Holding Co. Ltd.	Meiji	Dietary fibers, collagen	Gummy
Scotland	The Prebiotic Company Ltd.	iófibrewater	Fruits and vegetables, dietary fibers	Juice
South Africa	Needleless	SmartPro	Psyllium Husk, Artemisia powder, Xanthan, marshmallow roots, and rhizoma dandelion roots	Powder
	Sabinsa	Sabinsa	Fenugreek fibers and turmeric starch	Powders and other farms
Switzerland	Biosynth	Biosynth	Oligosaccharides (agar, carrageenan, chitin, alginate, cellulose, pectin-derived, xylose, pentose, thiooligosaccharides, malto, manno and lactooligosaccharides, HMO), anthocyanidins, flavonoids, lignans, and catechins	Powders
Taiwan	Synbio Tech	Synbio-Oligo	FOS	Capsule or powder
Thailand	IMMOR	IMMOR probiotics	Inulin, FOS, and GOS	Capsules
	Nuven Naturals	Prebiotic Fiber Boost	FOS	Powder
The Netherlands	FrieslandCampina Ingredients	FrieslandCampina Ingredients	GOS, HMO, whey protein isolate, concentrates, and hydrolysate, lactoferrin, and soy hydrolysates	Powder, crystals and sirup.

(Continued)

TABLE 17.1 (Continued)

Country	Company name	Product name	Prebiotic used	Consumption form
Ukraine	Lactalis	Lactonia	Lactulose	Dairy-based beverage
USA	Codeage	Liposomal ovarian + Inositol	Inositol, wild blueberry complex, whole pomegranate extract, whole monk fruit extract	Powder
	Ora	Easy being green	Tapioca oligosaccharides	
		Trust your gut	Inulin from Jerusalem artichoke, pullulan, tapioca oligosaccharides	Capsules
	InnovixLabs	Broad Spectrum Prebiotic Fiber	Tapioca IMO, acacia gum, resistant starch from potato and tapioca, inulin from chicory, short-chain FOS from cane sugar, larch arabinogalactan, green banana powder, corn XOS, chitin-glycan from mushrooms and glucomannans from Konjac A	powder
	iHerb	Hyperbiotics	Fibers from acacia and Jerusalem artichoke, green banana flour	Powder

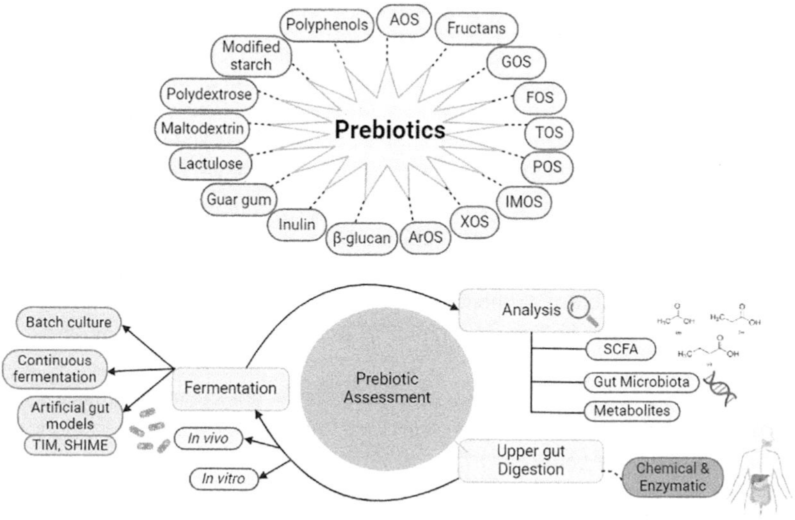

FIGURE 17.2 Commercially available prebiotic compounds and their assessment criteria. Commercial prebiotics and their assessment.

industrial production systems. Moreover, it is worth noting that healthcare professionals and other nutritional clinician associates differentiate among dietary supplements, drugs, and functional foods (Sanders et al., 2011). This is based on the dosage or concentration, formulations, and ultimate health effect. Based on their conceptualization, food technologists formulate the compositional ingredients and their associated qualitative testing (including regulatory compliance) for branding and marketing. The trade concept and marketing competence enable industrialists to claim higher product primacy. These edges are appraised by scientists and promoted by the companies. Every nation has peculiar standards and regulations for all products, including prebiotics.

17.5 Prebiotic regulatory guidelines

Regulations of a specific food or a claim that ought to be applied in conjunction with a food necessitate a description of the food as well as explicit standards for the execution of the claim. Food regulations examine and raise questions raised by manufacturers and preserve public health, consumer safety, and interests while not hindering creativity in food production and commerce. These laws and regulations must be unambiguous and grounded in clear empirical assessment, with a balance struck between lateral methods (where possible) for analysis and precise, product-category-specific standards and requirements.

When testimonials suggest that current regulations and legislation do not adequately address prebiotics and probiotics, and thus do not meet the demands of the public, manufacturers, and governing bodies, a need for specific laws or rules on prebiotics may be presumed. The customer must be properly and adequately notified through acceptable food labeling, safeguarded from counterfeit goods, and protected against deceptive assertions on food items that do not meet these standards. It is inappropriate to expect a comprehensive review of current prebiotic laws and regulations, as nutritional regulations vary significantly across nations. As a result, the following debate will center on the Codex Alimentarius principles and standards. Based on this, national food safety bodies and manufacturers set their own standards for producing prebiotics.

To date, the regulatory status of prebiotics is not clearly defined in the market based on their formulations and packaging, but they are marketable. From a comparison between different legislation, it is easily observable that while in some countries, the presence of a special prebiotic regulation is required as a necessary condition for marketing this class of food supplements, in other countries or inside the same legislative field, the same substances may not natively be subject to any specific recognition and then be marketed without classifying them in accordance with the prebiotic status. This creates a major perplexity in international trade.

17.6 Safety of prebiotic consumption

Since prebiotics are considered safe and have been used widely by humans for a long time, they are presumed to have negligible fatal effects. If the prebiotic compounds are inert (for both enzymatic and other chemical reactions) in the upper GIT, they react with the colonic microbiota and are fermented (Ayyash and Liu, 2023). Hence, the adverse complexity and convolution may result from osmotic activity. Various factors, including chain length, degree of polymerization, dosage level, and oral supplementation form, determine their effects on the host. In rare cases, negative conflicts comprising bloating, abdominal cramps and discomfort, flatulence, and osmotic diarrhea may reflect postconsumption. Immunocompromised individuals, geriatric patients, infants, sensitive and allergic consumers, individuals with severe gut dysbiosis or gut infection, pregnant women, etc., are advised to seek a proper, prescribed consultation before prebiotic consumption. Generally, for an active individual, a prebiotic consumption in the range of 2.5–10 g/day of dosage is reported to be effective (Davani-Davari et al., 2019). However, at least 3 g of a prebiotic must be administered to achieve practical health benefits in healthy adults (Swanson et al., 2020). GRAS limits usually rely on the safety of the ingredients or products rather than on the efficacy of the end use. Hence, only certain prebiotic ingredients, such as inulin, FOS, and GOS, are subject to GRAS limits. The functional activity of prebiotic compounds is strategically influenced by the matrix to which they are added in a dietary supplement, nutraceutical, or drug (Scott et al., 2020). Hence, deciding all the driving factors to formulate the prebiotic regulation is mandatory. Moreover, it would be more harmonized to establish proper regulation over the usage of selective prebiotic combinations after thorough clinical examinations.

The best available scientific findings ought to influence recommendations about public health initiatives and personal patient care (Sanders et al., 2011). Systematic reviews and meta-analyses provide a high degree of evidence and an objective summary of the body of knowledge on a particular subject, based on well-conducted randomized controlled trials. These findings can then be used to inform the development of clinical practice guidelines. Prebiotics are becoming increasingly popular, but there are not enough reliable, controlled research studies to assess their effects on any particular health issue (Rau et al., 2024). As a result, no globally recommended clinical practices for preventing or ameliorating a particular disease have included recommendations to use prebiotics (World Gastroenterology Organization, 2023). The findings of in vitro analysis may not translate well to clinical investigations because of the multitude of competing bacteria and substrates, and the interindividual heterogeneity of these parameters with diet and microbiome, accordingly. Although prebiotic benefits are observed in human studies, they may apply only to specific participant groups in which the right conditions are created by diet, the microbiota, or other personal traits. It is challenging to consistently select the right prebiotics for healthcare endpoints due to these interpretive concerns. When the prebiotic delivery matrix is altered, as in functional foods, additional variability in anticipated outcomes may be introduced, leading to occasional, unanticipated shifts in activity (Ballini et al., 2023).

17.7 Regulatory bodies and governing authorities of prebiotics

Food products are commercialized according to the country that registers them. All the regulations developed to address these products are focused on guaranteeing the quality of the products and the safety of the consumers to whom it is marketed (Yadav et al., 2022). Sanders and Levy (2011) stated that advocacy or counseling for prebiotic consumption among the American population was deemed negligible by the Dietary Guidelines Advisory Committee (DGAC). This may lead to the release of discrete products by manufacturers with misleading health claims. Hence, the focus of the regulations stems from the main objective of promoting and maintaining host health. Furthermore, the legislation imposes requirements to ensure that the products consumed will not harm the public.

To evaluate the health claims made for dietary supplements, the FDA was founded in the United States in 1990. The FDA has not yet approved prebiotics, and the supposed health benefits of changes in the microbiota have not yet been recognized as confirmed claims. Similarly, prebiotics and other functional foods are not subject to official laws in certain emerging Asian nations (Bamigbade et al., 2022). Although nutraceuticals are governed under the DSHEA, 1994, and the Food and Drug Administration Modernization Act (FDAMA), 1997, they are not entirely under the FDA's purview as dietary supplements. As sales of nutraceuticals and dietary supplements surged, the FDA established the Office of Dietary Supplement Programs (ODSP) in 1996 to step up regulation (Lordan et al., 2021). Nutraceuticals and novel products are obliged to notify the ODSP of any new dietary ingredients for authorization. Since these laws place the burden on producers to ensure product safety before production or marketing, serious concerns exist that they do not sufficiently protect consumers. Nutraceuticals and food supplements do not require registration or FDA clearance before manufacturers can produce or sell them. For the application of novel food products, such as dietary supplements, proposing a selective health claim shall meet the regulatory requirements and is primarily subject to health claim applications (Bech-Larsen and Scholderer, 2007). Additionally, permits must be obtained to market foods labeled with Nutrient Function Claims (NFC). Without justifiable cases aimed at protecting aspects such as the assurance of safe, high-quality, nutritious, nonmisleading food products and the freedom of movement of goods in the community, it is unsafe to label products as food supplements or to market them without prior approval. The unauthorized products, which claim to have any health effects, are considered illegal within the scope of drugs, as they are consumed as food products that are not classified as food.

There are variations in the regulatory landscape of prebiotics worldwide. With the passage of the Foods for Specified Health Use (FOSHU) Act in Japan in the 1980s, regulations on various functional foods were first introduced. The Japanese Ministry of Health, Labor, and Welfare certifies all food items recognized as having health advantages when the claims are supported by scientific evidence. After that, Japan permitted the commercial marketing of certain foods. However, in Europe, the Netherlands, the United Kingdom, and Sweden were the first countries to regulate functional foods before they joined the Joint Health Initiative (JHCI). The European Food Safety Authority (EFSA) was founded in 2002 to control the direct and indirect effects of food safety. The sole prebiotic that the EFSA has certified since its founding is chicory inulin, which carries an EU clinical assertion that it increases the frequency of stools produced, helping to maintain normal defecation. The CODEX Committee on Nutrition and Foods for Special Dietary Uses issued the definition in 2008, as did the WHO, and it is the same as those of EFSA, Food Standards Australia New Zealand (FSANZ), and Health Canada.

17.8 Regulatory guidelines for drugs and prebiotic supplements

The concept of a "health claim" is that a specific compound prevents or reduces the risk of certain diseases or death. For **drugs**, the basic concept is defined in the "standard of identity." It adheres to the Good Manufacturing

Practices regulatory framework and requires premarket review, chronic toxicity testing, and other clinical-level tests to be submitted to the FDA. The USFDA is responsible for regulating drugs and ensuring their safety and effectiveness. This includes a demonstration that they work as intended, purposive of treating a disease or controlling a physiological response. The FDA also oversees prescription and over-the-counter (OTC) drugs to ensure consumer safety. Historically, according to Section 201(g) of the Federal Food, Drug, and Cosmetic Act (FFDCA) has defined the term "drug" as "(1) articles recognized in the official US Pharmacopoeia, official Homeopathic Pharmacopoeia of the United States, or official National Formulary, or any supplement to any of them; and (2) articles intended for use in the diagnosis, cure, mitigation, treatment, or prevention of disease in man or other animals; and (3) articles (other than food) intended to affect the structure or any function of the body of man or other animals; and (4) articles intended for use as a component of any article specified in clauses (1), (2), or (3) of this subsection."

The regulated "drug" products are addressed by the statements relating to diagnosing, treating, or preventing disease and the structure/function of a phytochemical, a botanical, or a synthetic drug. Prebiotic drug products with nonmonosaccharide active ingredients are regulated as either drugs or biologic products, depending on the nature and intended use of the product. This classification is consistent with the FDA's proposal for classifying products intended to treat IBDs and other diseases by addressing the imbalance in the gut microbiota associated with these conditions. The intersection of drugs and dietary supplements is responsible for the treatment in the case of IBD. The prebiotic drug product is subject to the regulatory requirements for a new drug as defined in Section 201(p) of the FFDCA, as proposed to be used for the treatment of an intestinal disease (proposed §3311.5(d)(2)(ii)(A)). Marketing these products without approval from the FDA violates the Federal Food, Drug, and Cosmetic Act (FFDCA) (21 U.S.C. 331).

Reckitt Benckiser, UK, produces *Fybogel* (Hi-Fiber) sachets. This product contains psyllium husk and claims to "Gently relieve constipation." Here, the psyllium husk contains prebiotic dietary fiber and is used as a drug to relieve constipation. Similarly, *Agiolax* manufactures senna granules as a laxative to relieve constipation and support bowel health.

Food and dietary supplement products typically contain prebiotics to stimulate beneficial effects on the gastrointestinal flora of consumers. For such prebiotic **supplements**, the authorized regulations that guarantee their production and sale are covered by "dietary supplements" in the USA** and under the registered denomination; the European legislation directive 2002/46/CE, the "Compendium of Categorization" in Canada, and the legislation DMPS 218/01 in Brazil. The USFDA is authorized to ensure the safety, efficacy, and accurate labeling of various commodities that treat or prevent various health conditions for US consumers. In the dietary supplement sector, the FDA mandates the structural and functional requirements that govern the convergence of advertising and labeling worldwide, as well as side effects. The US Department of Agriculture (USDA) has established food quality and security rules parallel to those of the FDA for the effective functioning and screening of dietary supplements, including prebiotic products.

Using foods and dietary supplements to address the same issues is also allowed. However, these products are subjected to the corresponding premarket approval processes, specifically food additive or dietary ingredient notification approval for ingredients and health claim petition approval for their use. Besides, third-party verification, such as the United States Pharmacopeia (USP), states that the standard dietary supplement regulations are consistent with the USFDA based on scientific evidence. The USP system is global, and draft standards vary by geography and federal regulations. For instance, the USP in Hyderabad, India, is ISO 17025-accredited and sets and formulates standards for food ingredients and dietary standards (USP-India).

17.9 International perspectives on prebiotics regulatory status

The global norms known as the Codex Alimentarius, which govern food processing, manufacturing, trade, and nutritional supplements, are increasingly adopted by nations worldwide (Halabi, 2015). The highly demanding Codex standards and requirements for dietary supplements, particularly those adopted by several European nations, render it impossible for users to buy dietary supplements in quantities that are curative or meaningfully preventative (Cencic and Chingwaru, 2010).

17.9.1 The United States Food and Drug Administration (US FDA)

The United States has established different regulatory authorities that oversee and organize prebiotic products at the national and state levels. Prebiotic products developed for therapeutic purposes, such as the control of the intestinal environment or gut symptoms, are considered drugs and regulated by the US FDA (Cordaillat-Simmons et al., 2020). The FDA and DSHEA state that the industrial producer is strictly responsible for the safety of supplement products, including labeling and compositional regulatory compliance, before they are marketed to customers or consumers (Cencic and Chingwaru, 2010). However, Abdel-rahman et al. (2011) stated that DSHEA is responsible for ensuring pilot production, not restricting prebiotics on the manufacturer's end, and for warning about premarket risk and safety assurance.

The FDA does not have the authority to set safety standards and regulations for dietary supplements. The manufacturers must ensure this before dispatching it to the market (US FDA, 2023). The prebiotic supplements must possess a label as "Supplement Facts label *along with supplement identification tagline,*" indicating the kind of prebiotics used, dosage, serving size, serving count (per container), etc. (Centre for Food Safety and Applied Nutrition (CFSAN), USFDA). Furthermore, botanically sourced food ingredients (e.g., in prebiotic supplements) are inspected and authorized by the CFSAN (Abdel-rahman et al., 2011). Slavin (2013) mentioned the Dietary Guidelines for Americans Committee (DGAC), 2010, to examine the use of prebiotic products in America, which lacks evidence of the scientific relationship to the prebiotic formulation. However, recent advances have led to the classification of dietary fiber as a prebiotic, with research on this topic advancing rapidly.

The US FDA clearance is mandatory for supplement makers to make claims linked to disease and cure, which is the sole notable restriction on supplement marketing. Moreover, the FDA has authorized only a small number of supplement claims related to diseases (US FDA, 2023).

17.9.2 Europe

The Netherlands, the United Kingdom, and Sweden were among the countries that joined the functional food regulatory project, which eventually gave rise to JHCI. The EFSA was officially established in February 2002. All issues that impact the safety of food and feed, both directly and indirectly, are covered by EFSA's activities. Furthermore, EFSA oversees the confirmation of the scientific justification for the authorization of health claims within the EU. EFSA has just issued one EU health claim for chicory inulin, which states that it "aids in the persistence of usual defecation through elevated stool consistency." "Nondigestible carbohydrates trigger a decline in the postprandial glycemic state" is the second, broader health claim related to prebiotics (EFSA NDA Council, 2015). While prebiotic compounds developed after 1997 are considered unique and therefore require safety clearance in the EU under the unique Food Regulation, established prebiotics, such as fructans and GOS, are regarded as safe food ingredients. Besides, the comprehensive guidelines concerning the implementation of Federal EU Regulation 1234/2007 concerning authorizations for the incorporation of casein and caseinates in dairy products, such as cheese and other protein hydrolysates manufacturing, were established by Council Legislation (European Commission, EC) 760/2008 on July 31, 2008 (Phelan et al., 2009; Schaafsma, 2009).

17.9.3 Health Canada

The phrase "prebiotic" is only allowed for products that satisfy the requirement for an approved health claim. The phrase "prebiotic" on labels is regulated as an implied health claim (Canada Food Inspection Agency). However, the regulations are subject to restrictions that must be detailed, clear, and aligned with global standards.

17.9.4 Colombia and Costa Rica

According to the Central American Technical Regulation (RTCA) of Costa Rica and Columbia, a prebiotic substance must be a preferred substance by ≥1 species of beneficial bacteria in the large intestine or colon, be resistant to gastric acids, be fermentable by the intestinal microflora, be resistant to endogenous enzymatic hydrolysis, can produce changes in the lumen of the large intestine or the host organism showing health benefits, selectively stimulate the growth and activity of those bacteria that are associated with health and wellness.

17.9.5 The south-east countries

Inulin and FOS, two fibers found in chicory roots, have been granted a prebiotic claim by the Indonesian National Agency of Drug and Food Control and the Chinese Nutrition Society, rendering them the very first and only approved prebiotics in the nation.

In countries like Malaysia, Singapore, and Thailand, Inulin and FOS are widely accepted as prebiotic supplements by the respective national authorities (Brownawell et al., 2012). However, the specifications and regulations for product manufacturing are roughly outlined. Further, the Food Standards Australia New Zealand (FSANZ) broadly adopts the regulatory aspects from the FDA and the WHO.

17.9.6 India

The e-Platform from the Food Safety and Standards Association of India (FSSAI) distinctly categorized the regulations accordingly as follows: health supplements, nutraceuticals, food for special dietary use, food for special medical purposes, and prebiotic and probiotic foods. FSSAI strictly warns against the use of *Not for medicinal use* by dietary supplement manufacturers. A drug, as defined in articles (a), (b), and (h) of section 3 of the Drugs and Cosmetics Act, 1940 (23 of 1940), and guidelines established pursuant thereto, shall not be included in the items coming under these regulations. The products covered by these regulations cannot include any of the following: items described in Schedules E and E-1 of the Drugs and Cosmetics Rules, 1945; hormones, steroids; narcotic drugs; or psychotropic substances outlined in the Schedule of the Narcotic Drugs and Psychotropic Substances Act, 1985 (61 of 1985) and regulations made thereunder. Foods designed to have these unique qualities for babies under two years old must abide by the 2020 FSS (Food for Infant Nutrition) Regulations. Moreover, the Food Safety and Standards Regulations, 2011, thoroughly describe the differences between the definitions of food applications (FSSAI e-platform Comments, 2022).

17.9.7 Japan

The FOSHU Act, introduced in Japan in the 1980s, paved the way for the legal regulation of functional foods, notably prebiotic products, making Japan among the first nations worldwide to do so. Moreover, the term "Prebiotic" is not widely used; rather, "foods to modify gastrointestinal conditions" is preferred in the country. Under the Japanese Ministry of Health, Labour, and Welfare, foods with wellness claims may be sold if credible research supports them. Products that pass must then bear the FOSHU certification and label to be sold in Japan. Prebiotics can be categorized under "Food Related to Gastrointestinal Conditions," which includes dietary fiber, guar gum, ingestible dextrin, lactose, oligosaccharides, polydextrose, and psyllium seed coat as main ingredients, even though they are not stated expressly. The guidelines are followed distinctly and promptly for prebiotic production and foods for special uses.

17.10 Prebiotics—labeling regulations

Labeling has a marketing benefit. Inert materials are tested for their effects on specific intestinal bacteria rather than for technical or nutritional purposes. The resilience of gastrointestinal transit, influence on growth conditions, nontoxinogenicity, stability, and identity, and on prebiotic metabolic substrates (e.g., the ability to digest, formulation, dose, and degree of metabolism). The combination of technical and practical demands for food assessment raises critical problems in identifying and governing the research required to demostrate the labeled functionality of the food.

The FDA is still investigating the term "prebiotic," and any alteration to the microbiota might not be permitted as a regulated health claim. Official rules on prebiotics, aside from functional foods and health claims generally, are largely lacking in several developing Asian nations. This has enabled unlicensed food items and supplements to enter the market and make unsubstantiated claims to improve wellness or appearance (Starr, 2015). "Complementary and alternative medicine products" are subject to FDA regulation. Prebiotics are included in the "biologically based" group of foods under the Center for Complementary and Integrative Health (a subset of NIH), using the 1995 definition. Manufacturers can also self-affirm GRAS status for products labeled as prebiotics.

A food manufacturer can incorporate prebiotic ingredients into a food product to target the same clinically validated health areas, but would need to file a health claim petition to include the beneficial effects of prebiotics in the labeling and marketing of their new or reformulated food product. Various packaging and labeling laws directly impact dietary supplements. For example, DSHEA ensures the manufacturing of such products under GMP, proper labeling, highlighting the product as a "Dietary Supplement," ensuring the absence of any drug substance, and that it is intended for oral ingestion only. Similarly, nutrient support tables or fact sheets by Nutrition Labeling and Education Act (NLEA), false and misleading contents and claims by NLEA

FIGURE 17.3 Regulatory guidelines for the prebiotic supplements. A model for a standard global regulatory guideline for the prebiotic supplements.

and FFDCA, accuracy and precision in disclosing the supplementary content are foreseen by Fair Packaging and Labeling Act, facility registration by manufacturers with FDA is inspected through the Bioterrorism Act, and disclosing the possible allergen information or usage of certain allergen-specific ingredients by Food Allergen Labeling Act, and clarity on contact details for consumer complaints and feedback by the Dietary Supplement and Nonprescription Drug Consumer Protection Act.

Prebiotic ingredients can be supplied as food ingredients, dietary supplement ingredients, or drug products, depending on their intended use. The USFDA has established regulations for the labeling and marketing of dietary supplements, foods for special dietary uses by Japan, foods for particular nutritional uses, and drugs. FDA regulations focus on safety and efficacy, and the type of product, specifically nutraceuticals and dietary supplements, currently dictates the appropriate ingredient approval process. The intended use of the product drives the choice of the FDA approval process, which, in turn, determines the ingredient and labeling requirements.

Food scientists, microbiologists, dietitians, and pharmacists establish unambiguous criteria for developing prebiotic foods. In addition to the quality of the developed products, these specialists must ensure the products meet the legal standards of the country. An inherent difficulty is handling the many perspectives on health-related declarations on prebiotic products held by customers, medical professionals, legislators, lawmakers, and researchers (Sanders et al., 2011). The updated and developed criteria should serve as the foundation for guidelines on correctly applying the terms "prebiotic" and "probiotic" on product labels. Well-defined use of these phrases, based on controlled criteria, will boost customer confidence in the items and enable authorities to manage them. The tested prebiotic products can bear claims about their role in restoring the consumer's colonic flora equilibrium and improving health as part of a balanced diet.

The assertions that explicitly link probiotic or prebiotic ingestion to illness are prohibited. Assessments of the validity of illness risk mitigation claims are now conducted nationwide. Petitions for authorization of disease-related statements will need clear experimental evidence to convert the product into food for particular dietary purposes. A model for a standard global regulatory guideline for the prebiotic supplements could be drafted as shown in Fig. 17.3

17.11 Conclusion

It has been established that consumer acceptance of prebiotics supports gut bacteria and that they can be included in diets. Despite prebiotics being a growing segment of the food sector, there are significant differences in national and international regulations. While early research has identified an extensive range of potential candidate prebiotics from around the world, only a few have been explicitly shown to confer health benefits and have a proven mechanism of action, which are necessary to meet the prebiotic criterion. To prevent investigations into possible positive health effects on individuals, as well as organisms, from being stifled, regulations must be updated and improved to support the identification and marketing of novel prebiotics with distinct health benefits. The rising pandemic and alarming outbreaks may lead manufacturers to take advantage of and override their products' claims. Moreover, numerous discrete regulatory bodies with broad scopes create disorientation and perplexity among manufacturers and consumers. The subfederals and the distinct federal bodies, in disharmony with the international borders, favor manufacturers in counterfeiting the regulations, cluelessly. Hence, this chapter signals the strong demand for harmonized global regulatory guidelines for prebiotic formulations.

References

Abdel-rahman, A., Anyangwe, N., Carlacci, L., Casper, S., Danam, R.P., Enongene, E., et al., 2011. The safety and regulation of natural products used as foods and food ingredients. Toxicological Sciences 123 (2), 333–348. Available at https://doi.org/10.1093/toxsci/kfr198.

Al-Sheraji, S.H., Ismail, A., Manap, M.Y., Mustafa, S., Yusof, R.M., Hassan, F.A., 2013. Prebiotics as functional foods: a review. Journal of Functional Foods 5 (4), 1542–1553. Available at https://doi.org/10.1016/j.jff.2013.08.009.

Ayyash, M., Liu, S.-Q., 2023. Special Issue "Probiotics, prebiotics and functional foods: health benefits and biosafety" Microorganisms 11, 1218. Available at 10.3390/microorganisms11051218.

Ballini, A., Charitos, I.A., Cantore, S., Topi, S., Bottalico, L., Santacroce, L., 2023. About functional foods: the probiotics and prebiotics state of art. Antibiotics 12 (4) Available from. Available at https://doi.org/10.3390/antibiotics12040635, http://www.mdpi.com/journal/antibiotics.

Bamigbade, G.B., Subhash, A.J., Kamal-Eldin, A., Nyström, L., Ayyash, M., 2022. An updated review on prebiotics: insights on potentials of food seeds waste as source of potential prebiotics. Molecules (Basel, Switzerland) 27 (18) Available from. Available at https://doi.org/10.3390/molecules27185947, http://www.mdpi.com/journal/molecules.

Bech-Larsen, T., Scholderer, J., 2007. Functional foods in Europe: consumer research, market experiences and regulatory aspects. Trends in Food Science & Technology 18, 231–234. Available at 10.1016/j.tifs.2006.12.006.

Brooks, S.P.J., Kalmokoff, M.L., 2012. Prebiotics and probiotics: some thoughts on demonstration of efficacy within the regulatory sphere. Journal of AOAC International 95 (1), 2–4. Available from. Available at https://doi.org/10.5740/jaoacint.SGE_Brooks, http://docserver.ingentaconnect.com/deliver/connect/aoac/10603271/v95n1/s2.pdf?expires=1347066573&id=70363826&titleid=41000050&accname=Elsevier+BV&checksum=A16F9213DF308A4EEDA8F9444C193E7B.

Brownawell, A.M., Caers, W., Gibson, G.R., Kendall, C.W.C., Lewis, K.D., Ringel, Y., et al., 2012. Prebiotics and the health benefits of fiber: current regulatory status, future research, and goals. Journal of Nutrition 142 (5), 962–974. Available from. Available at https://doi.org/10.3945/jn.112.158147, http://jn.nutrition.org/content/142/5/962.full.pdf.

Cardoso, B.B., Amorim, C., Silvério, S.C., Rodrigues, L.R., 2021. Novel and emerging prebiotics: advances and opportunities. Advances in Food and Nutrition Research 95, 41–95. Available from. Available at https://doi.org/10.1016/bs.afnr.2020.08.001, http://www.elsevier.com/wps/find/bookdescription.cws_home/704624/description#description.

Cencic, A., Chingwaru, W., 2010. The role of functional foods, nutraceuticals, and food supplements in intestinal health. Nutrients 2 (6), 611–625. Available from. Available at https://doi.org/10.3390/nu2060611, http://www.mdpi.com/2072-6643/2/6/611/pdf.

Corbo, M.R., Bevilacqua, A., Petruzzi, L., Casanova, F.P., Sinigaglia, M., 2014. Functional beverages: the emerging side of functional foods. Comprehensive Reviews in Food Science and Food Safety 13, 1192–1206. Available at 10.1111/1541-4337.12109.

Cordaillat-Simmons, M., Rouanet, A., Pot, B., 2020. Live biotherapeutic products: the importance of a defined regulatory framework. Experimental & Molecular Medicine 52, 1397–1406. Available at 10.1038/s12276-020-0437-6.

Davani-Davari, D., Negahdaripour, M., Karimzadeh, I., Seifan, M., Mohkam, M., Masoumi, S.J., et al., 2019. Prebiotics: definition, types, sources, mechanisms, and clinical applications. Foods 8 (3) Available from. Available at https://doi.org/10.3390/foods8030092, https://www.mdpi.com/2304-8158/8/3.

Dietary Supplement Health and Education Act of, 1994. Dietary Supplement Health and Education Act of 1994. Public Law 103-417 103rd Congress.

EFSA NDA Council, 2015. Outcome of a public consultation on the Draft Scientific and Technical Guidance of the EFSA Panel on Dietetic Products, Nutrition and Allergies (NDA) on foods for special medical purposes in the context of Article 3 of Regulation (EU) No 609/2013. EFSA Supporting Publications 12. Available at 10.2903/sp.efsa.2015.en-904.

FSSAI e-platform Comments, 2022. Draft Notification on Food Safety and Standards (Health Supplements, Nutraceuticals, Food for Special Dietary Use, Food for Special Medical Purpose, and Prebiotic and Probiotic Food) Regulations. https://comments.fssai.gov.in/Bestviewwl.aspx?NOTIFICATION_ID=4121.

Gibson, G.R., Roberfroid, M.B., 1995. Dietary modulation of the human colonic microbiota: introducing the concept of prebiotics. The Journal of Nutrition 125, 1401–1412. Available at 10.1093/jn/125.6.1401.

Halabi, S.F., 2015. The codex alimentarius commission, corporate influence, and international trade: a perspective on FDA's global role. American Journal of Law & Medicine 41, 406–421. Available at 10.1177/0098858815591525.

Kaur, A.P., Bhardwaj, S., Dhanjal, D.S., Nepovimova, E., Cruz-martins, N., Kuča, K., et al., 2021. Plant prebiotics and their role in the amelioration of diseases. Biomolecules 11 (3), 1–28. Available from. Available at https://doi.org/10.3390/biom11030440, https://www.mdpi.com/2218-273X/11/3/440/pdf.

Lopez-Santamarina, A., Mondragon, A.D.C., Lamas, A., Miranda, J.M., Franco, C.M., Cepeda, A., 2020. Animal-origin prebiotics based on chitin: an alternative for the future? A critical review. Foods 9, 782. Available at 10.3390/foods9060782.

Lordan, R., Rando, H.M., Greene, C.S., 2021. Dietary supplements and nutraceuticals under investigation for COVID-19 prevention and treatment. mSystems 6. Available at 10.1128/msystems.00122-21.

Mithul Aravind, S., Wichienchot, S., Tsao, R., Ramakrishnan, S., Chakkaravarthi, S., 2021. Role of dietary polyphenols on gut microbiota, their metabolites and health benefits. Food Research International 142, 110189. Available at 10.1016/j.foodres.2021.110189.

Phelan, M., Aherne, A., FitzGerald, R.J., O'Brien, N.M., 2009. Casein-derived bioactive peptides: biological effects, industrial uses, safety aspects and regulatory status. International Dairy Journal 19 (11), 643–654. Available at https://doi.org/10.1016/j.idairyj.2009.06.001.

Rau, S., Gregg, A., Yaceczko, S., Limketkai, B., 2024. Prebiotics and probiotics for gastrointestinal disorders. Nutrients 16 (6), 778. Available at https://doi.org/10.3390/nu16060778.

Sanders, M.E., Levy, D.D., 2011. The science and regulations of probiotic food and supplement product labeling. Annals of the New York Academy of Sciences 1219 (1), E1. Available from. Available at https://doi.org/10.1111/j.1749-6632.2010.05956.x, http://www.blackwellpublishing.com/0077-8923.

Sanders, M.E., Heimbach, J.T., Pot, B., Tancredi, D., Lenoir-Wijnkoop, I., Lähteenmäki-Uutela, A., et al., 2011. Health claims substantiation for probiotic and prebiotic products. Gut Microbes 2 (3), 127–133. Available from. Available at https://doi.org/10.4161/gmic.2.3.16174, http://www.landesbioscience.com/journals/gutmicrobes/Sanders-GMIC2-3.pdf.

Schaafsma, G., 2009. Safety of protein hydrolysates, fractions thereof and bioactive peptides in human nutrition. European Journal of Clinical Nutrition 63 (10), 1161–1168. Available at https://doi.org/10.1038/ejcn.2009.56.

Scott, K.P., Grimaldi, R., Cunningham, M., Sarbini, S.R., Wijeyesekera, A., Tang, M.L.K., et al., 2020. Developments in understanding and applying prebiotics in research and practice—an ISAPP conference paper. Journal of Applied Microbiology 128 (4), 934–949. Available from. Available at https://doi.org/10.1111/jam.14424, https://academic.oup.com/jambio.

Selvamani, S., Kapoor, N., Ajmera, A., El Enshasy, H.A., Dailin, D.J., Sukmawati, D., et al., 2023. Prebiotics in new-born and children's health. Microorganisms 11 (10) Available from. Available at https://doi.org/10.3390/microorganisms11102453, www.mdpi.com/journal/microorganisms.

Slavin, J., 2013. Fiber and prebiotics: mechanisms and health benefits. Nutrients 5 (4), 1417–1435. Available at https://doi.org/10.3390/nu5041417.

Starr, R.R., 2015. Too little, too late: ineffective regulation of dietary supplements in the United States. American Journal of Public Health 105, 478–485. Available at 10.2105/ajph.2014.302348.

USP - India, USP–India. https://www.usp.org/usp-india. https://web.archive.org/web/20181128043111/https://health.gov/dietsuppisapp.htm.

Swanson, K.S., Gibson, G.R., Hutkins, R., Reimer, R.A., Reid, G., Verbeke, K., Scott, K.P., Holscher, H.D., Azad, M.B., Delzenne, N.M., Sanders, M.E., 2020. The International Scientific Association for Probiotics and Prebiotics (ISAPP) consensus statement on the definition and scope of synbiotics. Nature Reviews Gastroenterology & Hepatology 17, 687–701. Available at 10.1038/s41575-020-0344-2.

US FDA, 2023. Notifications for Structure/Function and Related Claims in Dietary Supplement Labeling. https://www.fda.gov/food/information-industry-dietary-supplements/notifications-structurefunction-and-related-claims-dietary-supplement-labeling.

Vilas-Boas, A.A., Pintado, M., Oliveira, A.L.S., 2021. Natural bioactive compounds from food waste: toxicity and safety concerns. Foods 10, 1564. Available at 10.3390/foods10071564.

World Gastroenterology Organization, February 2023. World Gastroenterology Organisation Global Guidelines - Probiotics and Prebiotics. https://www.worldgastroenterology.org/guidelines/probiotics-and-prebiotics/probiotics-and-prebiotics-english.

Yadav, M.K., Kumari, I., Singh, B., Sharma, K.K., Tiwari, S.K., 2022. Probiotics, prebiotics and synbiotics: Safe options for next-generation therapeutics. Applied Microbiology and Biotechnology 106, 505–521. Available at 10.1007/s00253-021-11646-8.

Chapter 18

The health effectiveness, technological challenges, and market of synbiotic solutions over self probiotic or self prebiotic approaches

Muhammad Afzaal[1], Farhan Saeed[2], Rimsha Gulzar[3], Ali Raza[4], Hafiz Abdullah[4], Muhammad Usman Butt[4], Sania Saleem[4], Abid Ali[4], Samreen Ahsan[5], and Huda Ateeq[6]

[1]*Department of Food Sciences, Government College University Faisalabad, Faisalabad, Pakistan,* [2]*Food Sciences, Government College University Faisalabad Pakistan, Faisalabad, Pakistan,* [3]*Department of Food Science, Government College University Faisalabad, Faisalabad, Pakistan,* [4]*Department of Food Science, Government College University, Food Safety & Biotechnology Lab, Faisalabad, Pakistan,* [5]*Institute of Food Science and Technology, Khwaja Fareed University of Engineering and Information Technology, Rahim Yar Khan, Pakistan,* [6]*Institute of Food Science and Technology, The University of Lahore, Lahore, Pakistan*

18.1 Introduction

The advancement in molecular techniques has shown and continues to reveal the significance of microbial symbionts in the evolution and function of their hosts. Still, there is disagreement over the idea of the multigenomic organism. Margulis coined the word "holobiont" to refer to the close relationships between microorganisms and hosts. In the last few years, the phrase has become increasingly often used to refer to a host and its microbial communities, which include all synbiotic viruses and microbes in a variety of diverse, context-dependent associations ranging from beneficial to parasitic (Sapp, 2016). The first account of synbiotic connections between microbes and plants dates back to 1697, when Malpighi documented the development of galls on roots. However, at the time, this was more of a curious issue than a significant scientific finding. Two centuries later, it was shown that these galls are nodules made of plant and bacterial (Rhizobiaceae) cells, and that they fix atmospheric N_2, supplying leguminous plants with the vital nutrient ammonia (Harman and

Evaluating Oral Probiotics and Prebiotics. DOI: https://doi.org/10.1016/B978-0-443-29835-6.00002-2

Uphoff, 2019). This word was first coined by Margulis to refer to various bacteria that unite asexually to form new, integrated hereditary symbionts; however, the idea has now been extended to more broadly characterize host organisms along with all of their microbial symbionts. As scientific research reveals the advantages of the human microbiome, it is believed that the majority of plants and animals have these types of microbial symbionts (O'Malley, 2017).

The composition of the microbiota changes dramatically at two times in early life: from birth to weaning and from weaning to maturity, driven by additional diversification of nutrition. Certain plant holobionts can function better and grow faster if the microbial communities surrounding, on, and inside their roots are deliberately changed. This can be achieved by introducing certain microbial strains or by employing various techniques that change the microflora of the roots (Thakur, 2020).

Dietary treatment has increased the use of functional foods, especially synbiotics, with the goal of enhancing the health of patients suffering from specific cancers. Variations in the human gut flora's makeup are frequently seen, and these variations might be caused by dietary and environmental factors that promote perioperative infections (Gillor et al., 2008). The natural and synbiotic relationship between the host and its microbiota in wildlife populations has been the subject of very few investigations. Few writers have looked into the possibility of using some archetypal symbionts, like lactic acid bacteria (LAB), as a sustainable and eco-friendly substitute to stop the spread of disease in wildlife (McKenzie et al., 2018). Knowledge of the gut microbiota, which is the community of microorganisms that colonize a site and create the microbiome, including bacteria, fungi, viruses, protozoa, and archaea, is necessary to comprehend the impact of prebiotics and probiotics in gastrointestinal illnesses.

18.2 Health benefits of synbiotic solutions

Symbiotic is a broader term for mutual relationships in which both living creatures get benefits, whereas synbiotic solutions refer only to the association of probiotics and prebiotics in gut microflora (Swanson et al., 2020). The synbiotic solution of prebiotics and probiotics offers enhanced health benefits and maintains homeostasis. Probiotics are considered health-beneficial bacteria; however, prebiotics found in fruits, vegetables, milk, and industrially produced foods include galacto oligosaccharides (GOs), xylooligosaccharides (XOs), and fructooligosaccharides (FOs). These prebiotics are derived from the processing of hemicellulose based on lignocellulosic material. The mix of probiotics and prebiotics prevents different diseases and substitutes as antibiotics against some infections (de Figueiredo and de Oliva-Neto, 2022). Scarpellini et al. (2022) reported that prebiotics (gut microbiota affecting food components) and probiotics (functional organisms) synbiotics relation boost the immune system to the next level against several diseases like diarrhea, constipation, cancer, inflammatory bowel disease (IBD), nonalcoholic fatty liver disease (NAFLD), irritable Bowel syndrome (IBS), obesity, atopic conditions, diabetes, neurological disorders, autoimmune disease, and psychiatric disorders.

18.2.1 Enhanced digestive health

Integration of probiotics and prebiotics, the synbiotic solution to maintain homeostasis, affects several cognitive functions to control gastrointestinal efficiency. Gut microbiota is adversely affected by the supplementation of synbiotics. The probiotics maintain gut health by facilitating the growth and effectiveness of prebiotics. Additionally, synbiotic association benefits can also provide further health benefits. The gut and microbiome interaction helps to sustain gastrointestinal stability and overall health (Fekete et al., 2024).

The synbiotic association of probiotics and probiotics affects the gastrointestinal (GI) tract and overall health. The disruption of the gut microbiota composition from the normal point known as "dysbiosis" can lead to various health issues, including digestive disorders. The synbiotic relationship with mutual benefits can provide more stable and cohabitating benefits. Gut health is influenced by utilizing the optimal amount of a probiotic and prebiotic mix (Leach, 2024). The probiotic cultures and prebiotic substrates provide a mixture of crucial importance in health-conscious foods. Traditionally, fermented foods are consumed due to their nutritional potential. The intake of synbiotics, probiotics, and prebiotics can support the overall health, particularly gut health. However, probiotics and prebiotics in meals support nutrition and health of oral health. When the gut-friendly bacteria and gut nutrients are assimilated, they produce metabolites that act as protective shields for the gut. The association of *Lactobacillus acidophilus* and fibers from pineapple has an excellent impact on the human gastrointestinal (GI) tract. A sufficient improvement in gut health can also improve the other body systems (Dahiya and Nigam, 2022). The gastrointestinal tract (GIT) hosts microorganisms having dual effects that can be dysbiotic and synbiotic. The effect of probiotics and probiotic consumption can establish and develop a well-regulated gut flora. Synergistically, to optimize the gut microflora, probiotics and prebiotics work. The self-probiotic approach to the functional bacteria in the gut and the self-prebiotic approach improve nutrient support. The synbiotic solution ensures that beneficial bacteria flourish in the gut, prolong their lives, colonize themselves, perform well, and, as a result, have a better oral cavity. This synbiotic solution offers a more efficient and stable way to sustain and enhance gut health (Olvera-Rosales et al., 2021). The consumption of a probiotic and prebiotic mix has numerous benefits related to the oral system, such as stability and integrity, inhibition of pathogen colonization, increased biosynthesis of SCFAs and vitamins, modification of bile salts, control of gut motility, boost enterocyte regeneration through stem cell stimulation, enzymatic breakdown of fibers, and detoxification of cancer-causing agents (carcinogens) or xenobiotics. All these functions promote the integrity of the gut and reduce the chances of gastrointestinal diseases. The optimum balance of human gut immunity can be achieved by the administration of probiotics and prebiotics. The therapeutic effect of the solution is emerging and safe in clinical settings to make sure the gut is healthy (Tsai et al., 2019). According to the investigation of Hassanien et al. (2024), the synbiotic doses show improvements in gut health by increasing the number of microflora in the intestinal tract.

This synbiotic solution treats inflammatory bowel disease by providing dietary factors and a specific environment. According to the study, it increases SCFA production and decreases inflammation significantly. The combination of fructooligosaccharides (prebiotic) and different beneficial bacteria includes *Lactobacillus casei, Lactobacillus acidophilus, Lactobacillus bulgaricus, Lactobacillus rhamnosus, Streptococcus thermophiles, Bifidobacterium longum, and Bifidobacterium breve*. The synbiotic capsules are taken two times a day for eight weeks to cure this disease. Patients who experience GI discomfort may be eligible for lower doses. The formulated synbiotic combinations safeguard the colon tissue, control mucosal production, and promote gut health (Roy and Dhaneshwar, 2023).

18.2.2 Immune system health

The association between microbiota and immunity plays a crucial role in the development and maintenance of the body from a health perspective. The role of probiotics in the gastrointestinal tract to enhance immune system health is very significant. The gut flora is crucial to stimulating, developing, and prolonging the human body's immune system. Disease-causing microorganisms (pathogens) weaken the immune system by disrupting gastrointestinal (GI) health (Dey et al., 2023). The sole effect is less than the enhanced effect of the synbiotic effect of probiotics and prebiotics on the immune system. This synbiotic association is able to tackle lactose intolerance, enhance immunity, stabilize gut flora, and reduce GI disorders. The combination of beneficial microorganisms and substrate(s) offers great health benefits to the host (human body) (Bisht et al., 2024).

Development of the immune system starts from birth, and then it develops to the mature level. Immunity of infants ultimately depends on the gut microbiota as it develops from *Enterobacteriaceae* to *Clostridium*, *Bifidobacterium*, and *Bacteroides* due to the unique composition of breast milk. According to this study, Bovine-milk-derived oligosaccharides BMOs (prebiotics) and *Bifidobacterium animalis sp. Lactis* Bb-12 (probiotics) are introduced in healthy infants to check changes compared to the breastfed. According to the study, the two different dosage levels are checked as probiotics at 10^7 CFU/g with BMOs at 8 g/L from birth to six months, and the same probiotics with Glucooligosaccharides (GOSs)/Fructooligosaccharides (FOSs) at 4 g/L from birth to 12 months. The results have related rates of diarrhea and febrile infection in breastfed infants. Likewise, the complexes of other probiotics with different prebiotics make significant solutions and are then introduced into the body of infants in a specific amount for a specific period, giving different clinical effects. The introduction of synbiotic solutions shows a better immune response against infections and treats several diseases. This retards pathogenic and toxin activity and introduces much healthier microbiota. However, these combinations of prebiotics and probiotics are not recommended for use in formulation (Lemoine et al., 2023).

Synergistic solutions of pre and probiotics play a fundamental role in increasing the strength of the immune response against infectious diseases. Production of mucus, antiinflammatory cytokines, and antimicrobial metabolites increases sufficiently (Sadiq et al., 2022). Supplementation of probiotics and prebiotics can be a viable option to treat obesity. The compositional and metabolic activity changes in the gut flora of obese hosts (human body) with the administration of the synbiotic solution of probiotics and prebiotics. The concentration of SCFAs increased to such a level by increasing the probiotics (*Bifidobacterium, Eubacterium rectale*, and *C. coccides)* with prebiotics (oligofructose and insulin). The composition of gut microbiota and SCFA production is linked with autoimmune and inflammatory functionality of the body, like colitis-associated carcinogenesis, diabetes type-1, intestinal inflammation, multiple sclerosis, and rheumatoid arthritis. Furthermore, SCFA's increased production level can modulate the activity of T-lymphocytes (da Silva et al., 2021). According to the study by Yeşilyurt et al. (2021), the involvement of a synbiotic solution increases the health of microbiota, especially the immune strength.

18.2.3 Cardiovascular health

Heart-related abnormalities are diagnosed and treated with the association of pro- and prebiotics. The most common cardiovascular diseases (CVD) are myocardial infarction, stroke, hypertension, cardiomyopathy, hypertensive disease, venous thrombosis, thromboembolic disease, and arrhythmia. Pathogen makes the immune response. The synbiotic solution makes the body stronger in hypercholesterolemic properties. According to clinical research, synbiotic supplements for 12 weeks ensure the body combats cardio-related diseases in hemodialysis patients. This suppresses gut microbial dysbiosis and is a positive approach to cardiovascular functionality (Olas, 2020).

Cardiovascular diseases (CVD) cause an increase in the death rate, unsurprisingly. The synbiotic (prebiotics and probiotics association) has improved outcomes in treating CVDs by strengthening microbiota. This reduces blood endotoxin, p-cresyl sulfate, and indoxyl-sulfate to lower levels (March et al., 2020). The most popular prebiotics, including galacto-oligosaccharides, xylooligosaccharides, raffinose oligosaccharides, isomaltooligosaccharides, arabinoxylan oligosaccharides, inulin, and lactulose, are oligosaccharides. Some other substances, such as polysaccharides, including pectin, starch, and dextrin, behave as prebiotics in the human microbiota. Self-probiotics include both bacteria and yeast. The synbiotic relation of pro- and prebiotics is responsible for the prevention of the expansion of pathogens. Moreover, it protects from reactive oxygen species (ROS) and is involved in boosting health by reducing blood cholesterol. Optimal dosages of supplementation for personal pro- and prebiotic therapy influence the well-being of cardiovascular health (Oniszczuk et al., 2021). Another study suggested the synbiotic supplementation of *Bacillus Coagulans*, *Lactobacillus acidophilus*, and *Lactobacillus rhamnosus* (probiotics) and Fructooligosaccharides (prebiotics) for 12 weeks to improve heart health (Velayati et al., 2023). According to Shen et al. (2023), the gut microbiota is a major recognized reason for CVD. The factors responsible for enhancing cardiovascular health include blood pressure, BMI, systemic inflammation, LDL, insulin resistance, and heart lipid contents, which are modified by pro- and prebiotic association.

The synbiotic environment in gut microbiota plays a fundamental role in the health of organisms. The healthy gut microbiota, due to the synbiotic association, prevents and diagnoses diseases. The gut flora and specific systemic organ (heart) relationship can influence cardiovascular health (Vasquez et al., 2019). Clinical studies prove that, including many beneficial functions, the pro- and prebiotic combination depresses the cardiometabolic risks. This positive approach lowers the cholesterol level, reduces ROS, and balances the composition of gut microbiota to increase immunity against cardiovascular diseases, especially. It also enlists the strategy of using synbiotics (probiotics and prebiotics) for CVD treatments (Pavlidou et al., 2022). McFarlane et al. (2021) reported that the probiotics and prebiotics formulation reflects cardiovascular health, much enhanced.

18.2.4 Neurological health

The health of gut microbiota is greatly influenced by pro- and prebiotics linkage, as gut microflora relates to intestinal health and neurological functions. This communication also treats brain disorders, including Alzheimer's disease, Parkinson's disease, stress, autism spectrum disorder, and multiple sclerosis. These disorders are due to a bidirectional relationship that causes loss of homeostasis. The central nervous system and gut microbiota interlink with each other in terms of health and disease. Both can control each other; the brain controls the signals for gut functionality, and the gut controls the satisfaction of the mind. The autonomic nervous system stabilizes breathing rate, heartbeat rate, and central nervous system-mediated changes in the gastrointestinal tract (GIT). It also controls the functions or changes that occur in the gut, such as gut motility, digestion, gut permeability, bile secretion, carbs level, mucosa distortion, epithelial fluid maintenance, mucus production, luminal osmolality, and immunity response by mucosa production. The autonomic nervous system communicates with the gut directly through the central nervous system. The gut flora is affected by changes in metabolites that are controlled by the autonomic nervous system. By keeping in mind the gut microbiota, synbiotic relations can maintain neurological health by lowering the risk of mental problems, stress, depression, anxiety, and neurological inflammation (Suganya and Koo, 2020).

Digestion, metabolism, absorption, detoxification, and a strong immune system can be established by gut flora. It provides enzymes for the production of vitamins (B and K) and several amino acids and comprises the metabolism of polysaccharides and the synthesis of short-chain fatty acids from food. The synthesis of hormones in the gastrointestinal tract plays a crucial role in growth and body homeostasis development and proper functioning of the gut architecture, immunity, and microflora–gut–brain axis (MGBA) (Peterson, 2020). The brain and gut microbiota are linked with each other by involving immune-modulators, endocrine system modulators, and neuro-modulators. Gut flora increases immunity and strength by controlling cytokines and chemokines that have a fundamental role in the neuronal health system. The production of SCFAs is important for the healthy regulation of the central nervous system. The decline in SCFA production due to dysbiosis can lead to impaired brain functions and neuroinflammations. The imbalance in gut microflora (dysbiosis) also influences neurodegenerative disorders via immune system modulation. Overall, the importance of gut microbiota can help in maintaining brain health and offer novel therapeutic strategies. The synbiotic potential of probiotics and prebiotics offers potential benefits in managing brain degenerative disorders (Ashique et al., 2024).

18.2.5 Potential therapeutic applications

According to Yadav et al. (2022), the probiotic-prebiotic synergy helps in the modulation of the host (human) immunity and serves as a protective shield from infectious and non-infectious diseases. The synbiotic association not only increases beneficial bacteria in number, such as *Lactobacillus* and *Bifidobacterium*. It also has a significant approach to the reduction of cardiovascular diseases, the prevalence of metabolic dysfunction, and signs of insulin intolerance in elderly patients. Some synbiotic associations, like *B. animalis* and amylose corn starch in food products, improve the production of *Bifidobacteria* (probiotics). The association of *B. longum* and FOS in curd promotes the growth of *B. longum* and affects health by decreasing CVD risk factors, the rate of metabolic abnormalities, and determinants of insulin resistance. The linkage of *L. plantarum* with FOS (prebiotic) present in the oral cavity reduces sepsis and infections in the respiratory canal, which promotes the development of *L. plantarum* ATCC 202195. The connection of 5 probiotics (*L. delbrueckii* spp. *Bulgaricus, L. rhamnosus, L. acidophilus, L. plantarum*, and *Bifidobacterium bifidum*) and (prebiotic) insulin lowers the risk of intrahepatic triacylglycerol (IHTG) in nonalcoholic steatohepatitis (NASH) adults. An association of *Bifidobacterium lactis* and *L. rhamnosus* with oligofructose and inulin in hepatic conditions increases the level of protective antibodies (mucosal immunoglobulin A), reduces cholesterol levels in the blood, and lowers blood pressure. Overall, the combination of probiotics and prebiotics (synbiotics) treats and prevents a large number of fatal diseases like HIV infection, cancer, gut diseases, and numerous others (Table 18.1).

18.3 Technological challenges in synbiotic development

Symbiosis explains the interaction among living beings and is commonly used in biology to describe interdependencies between species in ecosystems. The concept of symbiosis is also applied to industrial systems, where companies exchange waste and by-products to reduce their collective amount of resource usage (Lybæk et al., 2021). The core concept is that the technique in the provisions of the development of appliances for human utilization is unbiased. However, the societal emphasis in finance on techniques and how innovations are utilized is not unbiased. The perspective for political, community-oriented, and economic strategy is established, determining choices in these domains, so human curiosity inspires these technological barriers (Blair, 2021). However, the development of synbiotic products, which combine probiotics and prebiotics to promote synergistic health benefits, faces several technological challenges that need to be addressed for their successful commercialization. These challenges span formulation, processing, and storage, as well as ensuring functional and sensory quality (Fig. 18.1).

18.4 Formulation challenges in synbiotic development

Traditional forms of nutrient supply, such as synthetic and chemical manure, have established different injurious effects on the surroundings, individuals, and cultivated plants. The usage of microbial inoculants as an organic fertilizer in farming has been certified to enhance plant development and resist pathogenic interactions. Conventional formulations of microbial inoculants offered a perspective into the positive utilization of microbes in fields. Despite its benefits to the surroundings, the efficiency of conventional organic fertilizer is ordinary compared to chemical manure. Thus a diversity of organic manure formulations has evolved to enhance the effectiveness rate of organic fertilizer in improving plant efficiency (Sarbani and Yahaya, 2022). An excellent bioformulation is supposed to be eco-friendly, efficient, and rapidly decomposable with adequate storage duration and elevated moisture retention ability. The formulation method

TABLE 18.1 Therapeutic applications of synbiotics.

Therapeutic applications	Mechanism of action	Targeted organs/ systems	Clinical implications	References
Prevention of colorectal cancer	Gut flora modulation by increasing production of SCFA	Colon	Reduce the risk of colorectal cancer via enhanced colorectal health	Kaźmierczak-Siedlecka et al. (2020)
Irritable bowel syndrome management (IBSM)	Bowel motility regulation and increase the function of the mucosal barrier	Gastrointestinal tract (GIT), enteric nervous system	Relief in boating, pain, constipation	Simon et al. (2021)
Immune system modulation	Enhanced production of cytokines and immune response	Immune cells	Strengthen immunity and reduce risks of infections	Rousseaux et al. (2023)
Prevention of diarrhea (antibiotic-associated)	Support intestinal epithelial cell integrity and restore gut microbiota	GIT (Colon)	Lowers the occurrence and intensity of antibiotic-related diarrhea	Kesavelu and Jog (2023)
Obesity & metabolic syndrome	Modulate homeostatic functions and increase SCFA synthesis	Metabolic organs	Lowers obesity risk and improves metabolic syndrome	Santos-Marcos et al. (2019)
Gut microbiota modulation	Develop probiotics growth and make sure microbial balance	GIT	Reduce dysbiosis and promote gut well-being	Liu et al. (2022a,b)
Cardiovascular health	Minimizes cholesterol absorption, boosts endothelial health, lowers oxidative damage	Cardiovascular system	Lowers blood cholesterol (LDL) and triglycerides, lower oxidative stress	Saini et al. (2024)

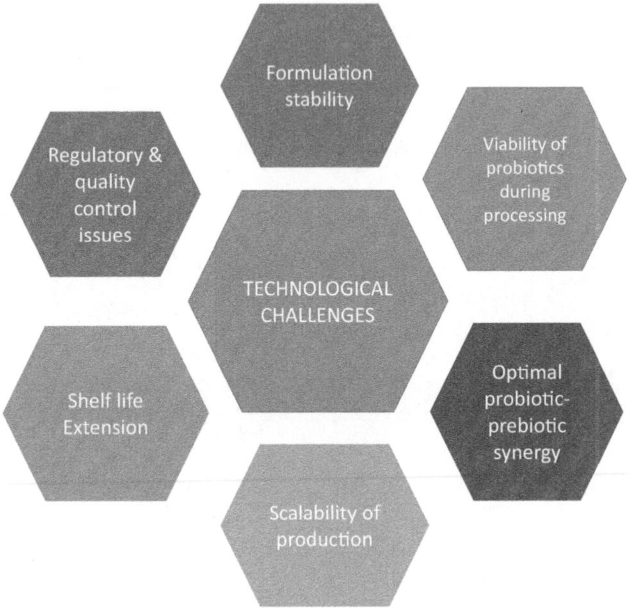

FIGURE 18.1 Technical challenges in development of synbiotic.

enhances the storage life and efficiency of organic fertilizers. Occasionally, the storage life of reproductive bodies increases in harsh environments because formulation comprises some cellular defense agents with optimal microorganisms. Various forms of formulation are generated according to their survival level and efficiency (Chaudhary et al., 2020) (Table 18.2).

18.5 Stability and viability issues in synbiotic development

The combination of probiotics and prebiotics has a crucial role in the food and health sectors in the form of synbiotic solutions. The significant potential of synbiotic association for the improvement of gut health and well-being is notable (Siddiqui et al., 2023). However, the commercialization of synbiotic solutions is facing a challenge of stability and viability during manufacturing, preservation, and utilization. The most important factors are probiotic strain viability and the systemic functionality of synbiotic products (Hathi et al., 2021).

18.5.1 Effect of environmental stress on probiotic stability

One of the fundamental issues regarding the longevity of probiotics is the survival of probiotic strains under environmental harsh conditions. The unfavorable impacts of the environment increase the survival challenges for probiotic strains. Probiotics can lose viability in the case of environmental

TABLE 18.2 Challenges and possible solutions in synbiotic survival.

Challenge	Description	Technical implication	Possible solutions	References
Probiotics stability	Probiotics are sensitive to unfavorable environmental issues	Viability Loss has negative impacts on symbiosis	Freeze-drying most effectively	Gurram et al. (2021)
Probiotics-prebiotics synergy	Compatible selection of prebiotics and probiotics strains that work effectively	The mismatch can lead to negative impacts on symbiosis and end product	In-depth microbial screening	Georgieva et al. (2022), Shinde (2019)
Viability and shelf life	Sustain the viability of probiotics under different storage conditions	Probiotic species degradation can lead to lower efficacy concerning time	Modified environment of storage, low temperatures, vacuum packaging	Jannah et al. (2022)
Regulatory compliance	Ensure formulation quality and safety in various regions (USA, China, Europe)	Failure to meet standards can lead to losses in market and quality also	Regular authority consultation and optimum or strictness during quality control processing	Sharma et al. (2021), Chaturvedi and Chakraborty (2024)
Encapsulation efficiency	Ensure the survival of probiotics and targeted release at the correct site in the gut	Low efficiency reduces bioavailability and less efficiency in targeted delivery	Nano-encapsulation, microencapsulation, and use of controlled release technologies	Xu et al. (2022), Centurion et al. (2021)

stresses such as changes in pH (acidity), temperature, exposure to air, and water content (Onwe et al., 2022). The beneficial bacteria are highly sensitive to high temperatures during manufacturing, such as *Lactobacillus* and *Bifidobacterium*. The probiotic strains survive mostly under optimal conditions, such as refrigeration and free from oxygen, to ensure preservation for a long time. This study concluded that stability and viability losses may significantly reduce the effectiveness of products containing probiotics and prebiotics, lowering the nutritional advantage for consumers (Rashidinejad et al., 2022). According to Mabrouk et al. (2021), pH sensitivity creates considerable difficulty, especially in the gastrointestinal tract (GIT). This stability and viability depend on the conditions of the gut, particularly pH.

18.5.2 Shelf life and storage conditions

The contribution of moisture levels and temperature from the environment plays a very important role in the stability of synbiotic products. The storage conditions determined that exposure to air or oxygen is the most important factor. It has been analyzed that under optimal conditions, such as very low moisture content and refrigeration, the stability of probiotics decreases with time, specifically with the high-water activity of the product (Chaturvedi and Chakraborty, 2023). The availability of free water plays a crucial role in the survival and growth of microbes. The encapsulated probiotics in an environment with conditions like dry and low moisture are better in quality and survive longer than the probiotics that are encapsulated in high or semi-moisture formats (Sharma et al., 2022). However, when synbiotic products are exposed to air or oxygen during packaging or storage, this leads to oxidative stress and lowers the shelf life due to lower stability (Hellebois et al., 2020).

18.5.3 Probiotics stability during processing

The manufacturing of nondairy synbiotic products involving such treatments that are not suitable for the stability of probiotic lives can negatively affect them (Chaturvedi et al., 2021). The heat treatments in the processing of products ensure extended shelf life and safety of the product, such as pasteurization and high temperatures. However, this can destroy the stability of microbial cells that are necessary for effective products in which probiotics are added. The association of probiotics and prebiotics is viable in high-quality products due to probiotic counts. The optimized conditions for the safety of the balanced product are dependent on the viability of probiotics (González-Herrera et al., 2021). According to Reque and Brandelli (2021), the wellness benefits of synbiotic solutions can be reduced without ensuring the durability and persistence of probiotics. It can limit the potential of nutrition in health and other related sectors.

18.5.4 Regulatory hurdles in synbiotic development

Regulatory challenges faced during development and commercialization in the probiotic field significantly affect the launch and growth of the market (Kohut et al., 2021). According to the study Speckmann et al. (2024), there are variations in regulations according to the requirements and standards for health claims across the areas. The regulatory landscapes with different standards add to the difficulties of navigating and maintaining compliance with food safety laws.

18.5.5 Diverse regulatory standards

The lack of uniform standards is one of the most fundamental challenges in regulations for product growth (Liu et al., 2022a,b). The international markets of probiotics and prebiotics (synbiotics) products require different requirements according to the regions where they are. The regulation development concerns the use of ingredients in the formulation of food. The introduction of beneficial microorganisms (probiotics) into foods is internationally gaining traditional safe profits in markets without extensive risk assessments targeting limited resources in essential areas. This is notably allowed by the Generally Recognized as Safe (GRAS) system and the European Qualified Presumption of Safety (QPS) framework (Bondar, 2020). The regulatory requirements of Europe, America (USA), India, and Japan are significantly varied from each other (Chugh et al., 2023). According to Hussian et al. (2024), the synbiotic product standards differ from one country to another. So, the EFSA (European Food Safety Authority) in Europe, the FDA (Food & Drug Administration) in America (USA) (Maftei, 2019), the FSSAI (Food Safety and Standards Authority of India) in India (Harshapriya et al., 2024), and MHLW (Ministry of Health, Labour, and Welfare) in Japan (Chugh et al., 2023). According to the study (Koirala et al., 2022), the regularities establish and regulate food safety standards within their respective regions.

18.5.6 Health claims and marketing restrictions

Regulatory authorities claim and allow premarket approvals related to the health, safety, and quality of the product. The labeling of probiotics or prebiotics used on food products is subject to a health claim approval process according to European regulations. The amount of pro-prebiotic added in the formulation must be labeled to instruct the synbiotic term usage based on scientific evidence (Swanson et al., 2020). The complexity of delivering health-conscious food products to the consumer is due to the utilization of the synbiotic association (Gomez Quintero et al., 2022). The European Union (EU) claims must be approved by EFSA on product labels (Fusco et al., 2022). Likewise, FDS in America monitors and ensures that health claims are not being misled by scientific evidence (Granato et al., 2020). Furthermore, the difference in regulations according to regions could be confusing and costly for different companies that are struggling to do business internationally (Liang et al., 2024).

18.5.7 Future directions

Collaboration of industrial stakeholders, regulators, and researchers is very important to address regulatory challenges. The regulatory communities should cooperate with technology developers towards innovative or developing strategies in the market (Anklam et al., 2022). According to the study (Mukherjee et al., 2022), the guidelines for regularities followed by a region on a commercial level should be clear for health claims that help to develop and approve products to launch in the market. de Simone (2019) reported that ensuring the standards, health claims with potent scientific evidence, and informative labeling are crucial to commercializing an innovative, successful product in the market.

18.6 Opportunities in encapsulation of probiotics

Probiotics are encapsulated using different techniques to check their effects on various foodstuffs. This increases the viability, stability, and physicochemical properties, having additional benefits for consumers. By adopting probiotic encapsulation, opportunities are building up to create innovative functional products. Strategies to make functional foods are responsible for food system advances (Agriopoulou et al., 2024). Probiotics-based functional foods deliver unique benefits to gastrointestinal tract health. Numerous probiotics are destroyed and unable to provide sufficient benefits due to improper processing, inappropriate encapsulating material selection, and unfavorable conditions in the gut. According to studies in recent years, attention toward these beneficial microorganisms has increased due to their potential applications in gut health, enhancing the immune system, and combating diseases. Traditional probiotics, classic probiotics, first-generation probiotics, and next-generation probiotics are recognized as live biotherapeutics to promote health, food development, and bioengineering research (Fentie et al., 2024). Unconventional foods like animal-sourced material (milk) and food by-products can be used as encapsulating material. Suitable material selection protects the probiotic culture and is incorporated into foodstuffs to make them more functional and improve sustainability. Encapsulation acts as a shield to safeguard probiotics (Somera et al., 2024).

Encapsulation protects probiotics that emerge as an optimistic solution for preserving viability in storage—strong barrier properties of encapsulation yield outstanding bioactive performance (Thinkohkaew et al., 2024). One of the significant and primary challenges is probiotic instability. Probiotics need to improve stability and viability by encapsulation. Due to various benefits like blood sugar control, health maintenance, and the cure of diseases, probiotics are combined with polymer coatings and become more effective in targeted delivery. Encapsulated probiotics stay longer than unencapsulated probiotics in the gastrointestinal tract (Mohamadzadeh et al., 2024). Encapsulation of beneficial microorganisms protects them against stressful environmental conditions, depending on high consistency and reliable emulsification abilities (Guo et al., 2024).

Functional foods have unique health benefits other than nutrition. Enhanced functional foods are developed by integrating encapsulated probiotics. These preserved probiotics in preparing, storing, and transporting foods face various physicochemical changes and environmental stress. Probiotic encapsulation also deals with the harsh digestive conditions in the gut. But remains intent and challengeable due to encapsulation. Encapsulation offers various opportunities for potent health applications and to improve the efficacy of probiotics (Zhao et al., 2024). Encapsulation prolongs the shelf life of probiotic-based products and also preserves them. Incorporation of probiotics into foods after encapsulation within food-grade colloidal particles to protect from the upper gastrointestinal tract, and they are released from the capsule when they reach the portion of the large intestine. However, various strategies are used to improve probiotic stability within foods, in the gut, and through encapsulation. Probiotics are trapped inside the capsule to control the release of probiotics (Nezamdoost-Sani et al., 2024).

Probiotic encapsulation enhances the efficacy and marketability of probiotic-enhanced food products. Nano-encapsulation of probiotics develops synbiotic products and makes products more potent for personalized nutrition. Limitations in encapsulating probiotics in dairy and fermentable products meet challenges. In the research field of probiotic encapsulation, next-generation probiotic products are developed that provide greater wellness advantages and therapeutic potential (Arratia-Quijada et al., 2024).

18.6.1 Enhancement of probiotic stability and shelf-life

Despite just nutritional benefits, foods are recommended that have unique health benefits and cure diseases. That is why probiotics are introduced into food composition and make them functional foods. However, in food industries, thermal processing is used for the desired quality and safety of products, and probiotics are not very stable in heat processing. So, the encapsulation technique was introduced in which probiotics were enclosed in capsules to improve their thermal stability. The potency of heat stability increases by increasing the interaction between the probiotic and the capsule in which it is enclosed, which is affected according to different encapsulation techniques (Bodbodak et al., 2024).

Probiotics' response under freeze or hot-air-drying processes and encapsulation methods was examined to check encapsulated particles' bio-potent qualities and durability. Hot-air drying increases the viability and survivability of probiotics and has more positive impacts on probiotics when compared with freeze-drying outcomes. Generally, the drying process stimulates viability in gastric and intestinal tolerance of probiotics compared with fresh cells. After drying, many probiotic attributes like antibiotic, antimicrobial, and bioactive properties (e.g., carotenoid) are investigated. Probiotics are encapsulated to form shelf stability with specific encapsulating material employing a hot-air-drying process. Most of the properties are better in contrast to fresh cells in the context of the stability of probiotics (Khullar et al., 2024).

To protect beneficial microorganisms from degradation, the requirement to shield them from acidity (e.g., in the stomach), high temperature (e.g., cooking and pasteurization), and oxygen exposure (e.g., air) is executed through encapsulation. Various methods highlighted for probiotic encapsulation include emulsion polymerization, microcapsules, coacervation process, spray-based drying, multilayer assembly, electrospinning, & hydrogel techniques. The microencapsulation technique shows Remarkable stability during storage, processing, and gastrointestinal tract digestion (Yang et al., 2024). Other encapsulation techniques are used to encapsulate probiotics, such as electrohydrodynamic, elevated surface-to-volume ratio, and reduced process temperatures. Probiotics encapsulated by electrohydrodynamic processes have greater stability in terms of air, light, thermal conditions, and pH level. This approach focuses on the viability and precise release of beneficial microorganisms on time (Vignesh et al., 2024). Encapsulated probiotics-enriched food items do not require costly storage and transportation strategies, improve convenience, extend shelf life, preserve probiotics, and are easy to uptake (Chaturvedi and Chakraborty, 2024).

Encapsulated probiotics use protective carriers to improve viability and stability. Some fibers are also incorporated with probiotics using electrospinning. The electrospinning becomes more efficient and exhibits more viability by controlling different factors such as processing conditions, solvent, solution, and environmental parameters. Types of polymers can also affect the viability of probiotics (Alizadeh et al., 2024). Microencapsulation preserves the probiotics using succinylated chickpea protein serving as an encapsulant. This process increases the solubility in alkaline conditions and reduces it in acidic conditions. Encapsulation yields better stability and viability of beneficial microorganisms. This superior protection against harsh conditions helps to stay stable due to encapsulation (Yu et al., 2024).

18.6.2 Improvement in probiotic survival rates during gastrointestinal transit

Many polymers, like alginate and chitosan, can be used as encapsulating material. Probiotic stability and shelf life are enhanced by encapsulation using poly-γ-glutamic acid. The nonencapsulated probiotics may completely decay in gastric juice within 3 hours but encapsulated with poly-γ-glutamic acid remain intact. Here, poly-γ-glutamic acid as an encapsulating material offers a greater survival rate to probiotics during different unfavorable conditions (Jang et al., 2024). The bread made up of white flour does not contain an adequate amount of valuable nutrients. The addition of some probiotics is the key solution to fulfill the nutrient gains. However, these probiotics can be lost during baking. Probiotic microorganisms (*Lactiplantibacillus plantarum*) were encapsulated with maltodextrin and sweet whey by the spray drying technique. Survival of encapsulated probiotics was evaluated by the INFOGEST protocol, and gastrointestinal transit was possible by adjusting intestinal electrolyte solution composition. Probiotics, due to encapsulation, are found after each digestive, gastric, and intestinal phase (Umaña et al., 2024).

Microencapsulation of probiotics with whey protein isolates due to their dietary advantages. The spray drying technique used during the encapsulation process exhibits higher survival rates in the gastrointestinal tract. Salts of pancreatic and bile can disturb probiotic walls in the intestinal juice and have strong interactions between probiotic membranes and encapsulant walls. Thus, maintaining viability without affecting gastrointestinal transit and spray drying (Ngamekaue et al., 2024). Both nonencapsulated and encapsulated probiotics are tested by comparison to check their viability under artificial gastrointestinal conditions. Encapsulation forms a protective barrier and helps the probiotics to survive more in gastric intestinal conditions by being the most vital element in their success. Many other factors enhance the survival rate, including acid tolerance properties by probiotics, coating material and their amounts, encapsulation techniques, and polymer varieties in composition (Nazir et al., 2024).

Probiotics are engineered by encapsulation to improve their delivery at appropriate sites in the gut. Microcapsules are formed to increase their viability and controlled release of probiotics in extreme conditions, such as simulated gastric and intestinal juices. Probiotics are encapsulated in a chitosan-coated alginate inulin structure, having several benefits like high mechanical strength, control of bioavailability, and enhanced survivability in the gastrointestinal environment (Kumar and Dhanda, 2024). Most unprotected (non-encapsulated) probiotics completely diminish under gastrointestinal conditions and show poor digestive tract delivery. Encapsulating probiotics with a chitosan-alginate inulin framework increases the viability under gastric and intestinal fluids. This modification with *Bacillus coagulans* and *Enterococcus faesium* helps to increase their shelf life, stability, and survival rates in the simulated gastrointestinal situation. Encapsulation prolongs their survival rate from one day to two days under such harsh conditions (Golnari et al., 2024).

Probiotics deliver their benefits when they survive the processing procedures, storage conditions, packaging, and the gut. Probiotics are modified by microencapsulation with spray drying to maximize the survival rates in gastrointestinal conditions (Kamble et al., 2024). Combined encapsulation of two probiotics by freeze-drying and coaxial wet electro-spraying. Probiotics are subjected to artificial gastric and intestinal fluids for four hours. The survival rates of these encapsulated bacteria (*Lactobacillus plantarum* and *Bifidobacterium animalis*) increase and are well-maintained. In combination with cryoprotectants in food industries, it has significant potential for survivability (Farahmand et al., 2024).

18.6.3 Tailored delivery systems for targeted release and site-specific action

Novel delivery systems were introduced for potential therapeutic effects on disease. Probiotic-based products challenge a lot of delivery obstacles, such as thermal stability and acidity issues. Medicinal probiotics are delivered in the form of capsules or tablets that can degrade prematurely and reduce their effectiveness. However, some probiotics are encapsulated in matrices or capsules made from polymers to increase the stability of the cell wall. Microcapsules and traditional capsule technologies are significant in improving probiotics that tolerate harsh conditions and remain stable. Multilayer encapsulation, polyelectrolyte complexation, and electrohydrodynamic technologies show great potential in probiotic delivery at specific sites where needed for specified purposes (Li and Zhang, 2024). Treatment of inflammatory bowel disease by the probiotic-based delivery system involves targeted release of drugs to specific sites where inflammation occurs; therapeutic nanostructures play a significant role in curing inflammation and tumors. Modified probiotics are used as drugs for treating this disease. The targeted site for these drug carriers is the colonic phase, which is the active enzymatic region of the gut. These encapsulated medicines have remarkable therapeutic results in clinical practices. Advanced probiotic delivery systems include microparticle-based and nanoparticle-based (Lopes et al., 2023).

Targeted delivery to the specific site and stability of probiotics in the gastrointestinal tract were challenging before encapsulation techniques emerged. Conventional technologies of encapsulation damage probiotics. Single-cell encapsulation involves the layering of biofilms or biopolymer layers around single probiotic cells and has the potential for oral delivery of probiotics. Emulsion technology, microscopic or macroscopic hydrogels develops delivery systems to enclose probiotics in encapsulating material with high mechanical strength, preserve in harsh conditions, and discharge at the specific site to perform their action (Harshapriya et al., 2024).

The magnificent yield of probiotic encapsulation by the double emulsion method depends on various factors such as the emulsification method, selection of emulsifier, probiotic response, and modifications in the emulsification technique. Not only is it helpful in the targeted delivery, but it also exhibits potential in immunomodulation, nutritional effects, clinical uses, safety for food, regulating gut microbiota, and controlled release in the intestinal tract (Han et al., 2024).

Heidarrezaei et al. (2025) reported that probiotics that are encapsulated possess improved viability and tolerance levels to tackle gastrointestinal stress. Orally consumed probiotics require the correct coating material and food matrices for proper encapsulation. Lengthening survival time during passing through the gastrointestinal tract and storage, strengthened mucosal attachment to the gut wall for extended retention, optimal transfer of probiotic cells to the designated area, rapid and efficient restoration of probiotics when released at the intended site, robust re-establishment of the probiotics, and carrier's natural decomposition to avoid particle lodging in organs and the necessity for removal ensured a potent delivery system (Table 18.3).

18.7 Comparative analysis of synbiotic versus self-probiotic or self-prebiotic approaches

18.7.1 Synbiotic approaches

A technique that combines both prebiotics and probiotics that promotes the mutualistic effect where the colonization and survival of probiotic bacteria is boosted by prebiotic bacteria, providing ameliorating benefits (Douglas, 2021).

18.7.1.1 Health benefits of synbiotics

Maintain the health of the gut and enhance gut flora composition, reduce inflammation and improve nutrient absorption, particularly of minerals like calcium and magnesium, and can even influence vitamin synthesis, prevent gastrointestinal disorders, and also have systemic effects on cardiovascular health and show antibacterial effects (Oliveira and Pamer, 2023).

18.7.1.2 Efficacy of synbiotics

This word comprises probiotics and prebiotics, and the aim is to enhance the health of the gut by leveraging the benefits of both components. Growth of favorable gut bacteria is supported by prebiotics, that is, fibers, while living microorganisms called probiotics contribute to a healthy microbiome (Rinaldi et al., 2022). The idea is that the effectiveness of probiotics can be improved by prebiotics, leading to greater overall health benefits. By enhancing bowel regularity, stimulating balanced microbiota, and reducing symptoms of digestive disorders like IBS and IBD, synbiotics have a positive impact on gut health. And they also have potential benefits for metabolic health and can

TABLE 18.3 Targeted delivery systems for targeted release.

Delivery systems	Mechanism	Advantages	Applications	Challenges	References
Enteric coating	Probiotics protection from gastric acid	Enhance survival or probiotic life, targeted delivery in the intestine	Pharmaceutical development, Colonial targeted therapeutics	Potential to increase drug bioavailability, specific formulation requirements	McCoubrey et al. (2023)
Lipid-based systems	Lipid bilayers used to coat	Higher stability and controlled release at specific site	Therapeutic formulations, value-added functional foods, health benefits	Expanding the applications on a wide range	Liu et al. (2024)
Polysaccharide-based coatings	Coating with biopolymers like chitosan to stabilize probiotics	Enhance gut health, controlled and targeted release, sustained release	Food processing, health improvements, homeostasis modulation	Complex combination with other materials	Liu et al. (2024)
Microencapsulation	Protected by microscale coats	Encapsulation enhances probiotic viability	Health care, food industry functional products	High mortality from dehydration	Vivek et al. (2023)
Hydrogel beads	Coat of gel matrix that releases probiotics at the target sites	Protection of probiotics against harsh conditions	Probiotics delivery systems, used in functional foods and medicines	Health treatments	Amiri et al. (2024)
Nanoparticles	Nano-sized carriers for probiotics encapsulation	Targeted delivery system of probiotics	Medicinal applications	Regulatory hurdles	Calinoiu et al. (2023)
Peptide-based carriers	Encapsulate using peptide barriers around probiotics	Protection and targeted delivery system	Food and biomedical industries	Expanding challenges in the future	Jiang et al. (2024)

support the functions of the immune system (Haro-Soler and Kiraly, 2019). In some cases, they can cause digestive discomfort if taken in large amounts. So, more research is needed for a complete understanding of its efficacy (Quigley, 2019).

18.7.1.3 Clinical applications of synbiotics

Help with constipation by improving stool consistency and bowel regularity. For metabolic health, due to their effects on glycemic control and weight management, synbiotics are being studied. After antibiotic use, they can reduce gut microbiota balance, also reducing the risk of disturbance, such as antibiotic-associated diarrhea (Matzaras et al., 2022).

18.7.2 Self-probiotic approaches

A technique that involves the intake of live microorganisms (probiotics) like *Bifidobacterium* and *Lactobacillus* species that has the aim to subjugate the gut, confer health benefits like increasing immunity, improve digestion, and maintain microbiota health (Scott et al., 2021).

18.7.2.1 Health benefits of self-probiotics

After antibiotic use, it is very useful for restoring microbial balance, boosting the immune system, infection prevention (*Clostridium difficile)*, reduction of pathogenic bacteria. Reduce the development of atopic eczema later in life, improve immune functions in older adults, and those with liver cirrhosis (Liu et al., 2019). It can also have a positive impact on skin health, help in the reduction of conditions such as eczema and acne by balancing the skin microbiome, and also minimize inflammation. We can ensure the safety of homemade probiotic foods and can also maximize these benefits by proper fermentation and storage practices (Skowron et al., 2021).

18.7.2.2 Efficacy of self-probiotics

Different strains of probiotics, like *Bifidobacterium* and *Lactobacillus*, have apparent benefits, and their effects are also different from person to person. Probiotics' effectiveness depends upon the consumption of the right strains in an adequate amount, but we also have to ensure that the strain of probiotic is of high viability and quality (Wallace and Milev, 2021). The overall efficacy can be impacted by homemade or self-administered probiotics that can also vary in potency and quality. It also has significant short-term benefits, especially in cases of imbalance in gut microbiota and also in cases of dysbiosis. However, their long-term efficacy may be compromised without proper nourishment (Shi et al., 2019).

18.7.2.3 Clinical applications of self-probiotics

One of its prominent uses is in Fecal Microbiota Transplantation (FMT), where healthy donors' stool is transplanted into the patient's gut. If the microbiota of a patient is healthy, their own stool could be used for similar therapeutic purposes; it can potentially aid in the treatment of conditions like recurrent Clostridium difficile infections (Mishra et al., 2021). Research is ongoing, with a focus on optimizing methods for isolating, culturing, and administering treatments, as well as understanding their long-term effects and potential risks.

18.7.3 Self-prebiotic approaches

A technique selectively promotes the growth of beneficial bacteria in the gut by intake of nondigestible fiber or compounds, that is, inulin and fructooligosaccharides (Smith, 2020).

18.7.3.1 Health benefits of self-prebiotics

By consuming food rich in prebiotics such as vegetables, fruits, grains, etc., helps in the pefect growth of beneficial gut bacteria which reduce the health related issues like bloating and constipation, by enhancing the growth of beneficial bacteria (*Bifidobacteria* and *Lactobacillus*), it improves the digestive health, improvement of calcium absorption, gut inflammation reduction, lower the risk of chronic diseases (obesity and diabetes), by improve mineral absorption, it lowers the risk of metabolic disorders (Kumari et al., 2024). Also has a good impact on mood and lowers the symptoms of anxiety and depression (Prajapati et al., 2024).

18.7.3.2 Efficacy of self-prebiotics

The promotion of the growth of beneficial gut bacteria by using some substances is known as self-prebiotics. Based on some factors, its efficacy can vary. Such as fructooligosaccharides (FOS), insulin, and resistant starch, and by providing essential nutrients, it supports beneficial microbes (Włodarczyk and Śliżewska, 2021). But due to individual variations in gut microbiomes, their effectiveness can differ from person to person. To determine their impact, there are some factors that play a very important role, like dosage, lifestyle, overall diet, and form of prebiotics. Gut health has a positive effect of prebiotics, including improved digestion and the immune system (Steels et al., 2023).

18.7.3.3 Clinical applications of self-prebiotics

They have an important role in improving the insulin sensitivity and glycemic control in type 2 diabetes. Fostering beneficial bacteria, it can support the immune system, which can reduce inflammation and enhance the body's response to infection. It also has a positive impact on mental health and give gut–brain axis connection. They can also help in the maintenance of mood disorders (Yadav et al., 2020) and can also be helpful to maintain cardiovascular problems; some studies suggest that they can also regulate blood pressure and lower the LDL-cholesterol level. It can also reduce the risk of colorectal cancer by improving gut health and has a role in the management of allergic conditions (Shamekhi et al., 2020) (Table 18.4).

18.8 Implication of synbiotic solutions in food for health care

Nowadays, the way of living has changed greatly as it has become more sedentary, because the majority of people engage in more seated activities, including work or entertainment, and consequently, there is a detrimental change in people's general health. Furthermore, accelerations in family dynamics lead to a small amount of time for food preparation and consumption, resulting in ingesting unhealthy food, which leads to ineffective completion of digestion stages. Nowadays, researchers are focused on methods to prepare healthy foods, which include the addition of bioactive compounds to food products to deliver direct beneficial effects on consumer health along with nutrients. Among the bioactive compounds, the synbiotic solution of pro and prebiotics is of great interest for food researchers (Christmann, 2020; Kristo et al., 2020; González-Herrera et al., 2021).

Krasaekoopt and Watcharapoka (2014) incorporated a commercial yogurt matrix with coencapsulated *Lactobacillus casei* 01 and *Lactobacillus acidophilus* 5 with galacto-oligosaccharides (GOS) and inulin as prebiotics in beads of alginate coated with chitosan. The encapsulated particles having galacto-oligosaccharides displayed enhanced growth in the simulated gastrointestinal tract and better protection. It is concluded that prebiotics delivers major nutrients, including carbon and nitrogen, which consequently enhances organism proliferation in the large intestine and provides some health benefits to host (Rashidinejad et al., 2022).

Peres et al. (2018) developed chocolate-flavored ice cream by mixing 6.7% inulin with all ingredients using an artisanal ice cream machine. Pasteurization and cooling of the mixture were carried out. Afterwards, probiotic *L. acidophilus* was incorporated while stirring until the required creaminess was achieved. Then, after packaging, the ice cream was stored at −18°C. The concentration of prebiotics fulfills the required doses to impart beneficial effects on the health of the consumer, whereas the concentration of probiotics in the final product was specified (González-Herrera et al., 2021).

According to Schoina et al. (2018), fortification of Mizithira cheese with encapsulated *Lactobacillus casei* ATCC 393 into *P. terebinthus* resin matrix showcased antimicrobial activity, and the viability of *Lactobacillus casei* is enhanced by *P. terebinthus* during storage (Rashidinejad et al., 2022).

A combination of *L. rhamnosus* (probiotic) with soluble corn fibers (prebiotic) synergistically enhances the immunity of elders aged 60–80 years,

TABLE 18.4 Comparative analysis of synbiotic, self-probiotic, and self-prebiotic approaches.

Category	Self-probiotic Approaches	Self-prebiotic Approaches	Synbiotic Approaches	References
Definitions	Intake of live microorganisms without prebiotics	Selective promote of beneficial bacteria	Combination of prebiotics and probiotics that benefit the host by working together	Scott et al. (2021), Desmedt et al. (2019), Douglas (2021)
Health benefits	Help with eczema, diarrhea, lactose intolerance – Strain-specific benefits	– Increase *Lactobacilli* and *Bifidobacteria* – Gut motility and calcium absorption improvement.	– Improvement in absorption of nutrients – Inflate immune system – Increased effectiveness in bloating, constipation	Skowron et al. (2021), Kumari et al. (2024), Oliveira and Pamer (2023)
Efficacy	Have mixed results depending on the strains, also effective for specific conditions like eczema and diarrhea.	Less potential in reducing inflammation, also have positive impact on microbial composition	Great reduction in IBS symptoms and also increased diversity of microbes, inflammatory reduction	Wallace and Milev (2021), Włodarczyk and Śliżewska (2021), Rinaldi et al. (2022)
Immune system impact	Specific probiotic strain-depended benefits	Enhancing beneficial microbes by indirect immune modulation	Due to combined probiotic and prebiotic actions, there is superior immune modulation	Mishra et al. (2021), Shamekhi et al. (2020), Matzaras et al. (2022)

as evidenced by enhanced natural killer cell activity relative to baseline and reduced pro-inflammatory cytokine IL-6. This synbiotic solution is also helpful in reducing LDL-cholesterol and total cholesterol, which were initially elevated (Costabile et al., 2017; Mishra et al., 2021).

Synbiotic solutions of milk-based shake with *Bifidobacterium, L. acidophilus*, and 2 g oligofructose showcased a considerable reduction in fasting glycemia and increased level of HDL cholesterol when 200 ml of synbiotic solution is administered for 30 days in a daily dose to elderly people (Zepeda-Hernández et al., 2021). Consumption of yogurt with *L. acidophilus* LA-5, *B. lactis* BB-12, inulin, and extract of monk fruit enhanced regulation of blood by a remarkable decrease in HbA1c levels and insulin resistance, whereas an increased level of short-chain fatty acids and reduced kidney and liver damage associated with an elevated fat diet in relation to conventional yogurt (Ban et al., 2020).

18.8.1 Emerging technologies and solutions

The synbiotic solutions are conventionally prepared by various methods, including ionic gelation (Garrido-Cruz et al., 2023), spray drying (Kanimozhi and Sukumar, 2024), coacervation (Rashidinejad et al., 2022), freeze drying, refractive window drying (Jiménez-Villeda et al., 2023), and microencapsulation. Nowadays, because of advancements in technology, nanotechnology (Dangi et al., 2023), 3D printing (Jiménez-Villeda et al., 2023), artificial intelligence, and machine learning (Malwe et al., 2023) are regarded as emerging technologies in synbiotic solutions.

18.8.2 Nanotechnology in synbiotic solutions development

The health benefits of probiotics and prebiotics have been verified when incorporated in diet (Lee et al., 2024; Harshapriya et al., 2024; Dangi et al., 2023) (Se). However, they may face the vitality loss during absorption along the gastrointestinal tract (GIT), which causes a loss in efficiency for probiotic active ingredients' intestinal delivery. Nanotechnology is an innovative and less employed technique to entrap the synbiotics in capsule form to protect them from detrimental gastric conditions for their survival and ample concentrations of probiotics at the target site. Furthermore, the nanosynbiotics exhibit various benefits, including enhancing probiotic coupling with intestinal mucosal tissues and safeguarding the targeted and orderly release of synbiotic solution in desired quantity (Razavi et al., 2021; Dangi et al., 2023). The nanosynbiotics exhibit various benefits such as Selenium nanoparticles in which the synbiotic solution of probiotic *L. rhamosus* and prebiotic selenium showcased UV-protective and antioxidant activity (Kaur and Rath, 2019).

Ul Ain et al. (2024) and Harshapriya et al. (2024) prepared a synbiotic solution by combining 100 ml of *Acetobacter orzoenii* cultured in the medium of MRS and inulin. After fermentation, color changes to cloudy whitish, from light brown, confirming formation of a synbiotic solution. Nano-formulation is affirmed by a color change to dark brown from cloudy white after the addition of 1 M $FeSO_4$ solution in a synbiotic solution and incubation of the synbiotic solution.

Hashem et al. (2023) prepared alginate nano-encapsulated synbiotic solution of pomegranate peel phytogenics (prebiotic) and various probiotic species, including *L. plantarum, Lactococcus lactis*, *S. cerevisiae*, and *L. paracasei*. The Alginate-$CaCl_2$ nano-encapsulated synbiotic solution demonstrated more thermal stability at elevated temperatures. Furthermore, the number of viable probiotics after exposure to gastric acidity and storage at room temperature for six months is higher in relation to free synbiotics. The nano-encapsulated probiotics exhibit antimicrobial and antifungal activity against *E. coli, Staphylococcus aureus, Pseudomonas aeruginosa, Salmonella typhi, Aspergillus flavus*, and *Aspergillus niger*.

18.8.3 Artificial intelligence and machine learning

The use of AI for the discovery and authentication of a Synbiotic solution for patients and consumers reduces time from idea to final solution (Kussmann and Cunha, 2022). Artificial intelligence and machine learning are emerging fields regarding the development and application of the synbiotic solutions in various therapies. Algorithms of artificial intelligence (AI) and machine learning (ML) exhibit potential to scan an extensive amount of data, such as Unique microbiome data of an individual, spot trends, and project results. The ability of AI and ML to scan unique microbiome data of an individual assists in the formulation of precision medicine, which is formulated according to distinctive demands and purposes of an individual, and discover probiotic strains and prebiotic material that can enhance the health results (Abdelkawi et al., 2023)

Pasinetti (2020) designed an in vitro model of the human gastrointestinal tract (GIT) employed to maximize metabolite production from synbiotic solutions tested for combinations of polyphenols and 15 combinations of 6 probiotic bacteria as synbiotics to maximize the production of metabolites employing MARS (multivariate regression algorithm) discovered that three probiotic. Together with effective grape-derived polyphenolic metabolite precursors, including 3-(3-hydroxyphenyl) propionic acid and 3-hydroxybenzoic acid, synergistically exhibit the ability to retard progression and treat Alzheimer's disease by synergistically targeting many of its neuropathologies, such as tau fibril formation, inflammation, and amyloid aggregation (Smith, 2020).

18.8.4 3-D printing

3D printing is continuously emerging and revolutionized manufacturing industry by delivering numerous advantages such as prototyping, customization of composition, low-cost complex geometrics in a short time span, various

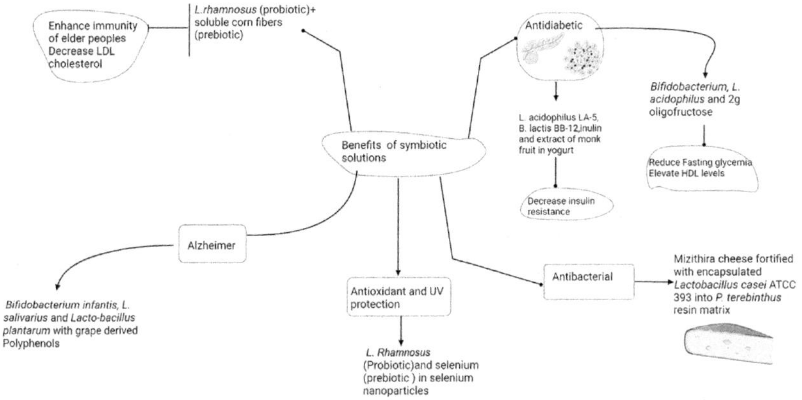

FIGURE 18.2 Various beneficial effects of synbiotic solutions on human health.

manufacturing techniques, and customized textures to fulfill taste preferences of every person (Iftekar et al., 2023; Jiménez-Villeda et al., 2023; Baiano, 2022; Zhao et al., 2021). A new wide range of healthy food products has been developed by 3D printing with added functional ingredients and nutraceuticals (Krishnaraj et al., 2019; Jiménez-Villeda et al., 2023).

Combining the synbiotics and 3D printing process on probiotics viability by formulating food with flour rich in protein and fiber. The probiotic stability was examined during storage at room temperature and 4°C. No substantial loss in the viability of the probiotic was examined during the process of 3D printing. Furthermore, the lyophilization encapsulation followed by lyophilization postprocessing showcased the best probiotic viability of 8.18 log cfu/g, along with the best survival rates of 6.43 log cfu/ml in static conditions and 7.98 log cfu/g during storage for 35 days. That's why it is summarized that encapsulated synbiotic powders incorporation in 3D foods can considerably enhance the probiotic cell stability. Whereas, incorporation of encapsulated synbiotics in 3D foods might be less complex as in 3D foods surface-to-volume ratio can be regulated, providing benefits for probiotic survival (Yoha et al., 2021; Jiménez-Villeda et al., 2023) (Fig. 18.2).

18.9 Conclusion

Synbiotics received fame due to their best role in strengthening the immune system, maintaining gut microbiome, reducing lactose intolerance, and preventing gastrointestinal problems like diarrhea and constipation. This study deals with the advancement of knowledge about therapeutic and pharmaconutritional effects. Different techniques are used for the manufacturing of synbiotic solutions, like microencapsulation (spray drying, microbeads, and extrusion) and nanoencapsulation (nano shells, electrospinning, and electrospraying). The chapter focuses on identifying and highlighting the merits of symbiosis in the context of food webs and exploring the benefits of synbiosis solutions for human health care, zero hunger, sustainable consumption, and maximizing production goals. The main purpose of this study is to explore case studies, research findings, and visualize the benefits and challenges of summarizing industrial symbiosis in the context of the food industry. It provides a novel framework that integrates symbiosis with the food industry as well as human health care. This study provides actionable perceptions for industry leaders and policymakers, advocating for symmetrical changes that line up intrinsic harmony of the natural world with food systems, as well as raise awareness among consumers.

References

Abdelkawi, A., Martinez, J.P.P., Pathak, S., Pathak, Y., 2023. Traditional approach, present status, and future outlook. Anxiety, Gut Microbiome, and Nutraceuticals: Recent Trends and Clinical Evidence 333.

Agriopoulou, S., Smaoui, S., Chaari, M., Varzakas, T., Can Karaca, A., Jafari, S.M., 2024. Encapsulation of probiotics within double/multiple layer beads/carriers: a concise review. Molecules (Basel, Switzerland) 29 (11), 2431. Available at https://doi.org/10.3390/molecules29112431.

Alizadeh, A.M., Mohseni, M., Gerami, K., Gharavi-nakhjavani, M., Aminzare, M., Rastegar, H., et al., 2024. Electrospun fibers loaded with probiotics: fundamentals, characterization, and applications. Probiotics and Antimicrobial Proteins 16 (3), 1099–1116. Available at https://doi.org/10.1007/s12602-023-10174-3, https://www.springer.com/journal/12602.

Amiri, S., Nezamdoost-Sani, N., Mostashari, P., McClements, D.J., Marszałek, K., Mousavi Khaneghah, A., 2024. Effect of the molecular structure and mechanical properties of plant-based hydrogels in food systems to deliver probiotics: an updated review. Critical Reviews in Food Science and Nutrition 64 (8), 2130–2156. Available at https://doi.org/10.1080/10408398.2022.2121260, https://www.tandfonline.com/loi/bfsn20.

Anklam, E., Bahl, M.I., Ball, R., Beger, R.D., Cohen, J., Fitzpatrick, S., et al., 2022. Emerging technologies and their impact on regulatory science. Experimental Biology and Medicine 247 (1), 1–75. Available at https://doi.org/10.1177/15353702211052280, https://journals.sagepub.com/home/EBM.

Arratia-Quijada, J., Nuño, K., Ruíz-Santoyo, V., Andrade-Espinoza, B.A., 2024. Nano-encapsulation of probiotics: need and critical considerations to design new non-dairy probiotic products. Journal of Functional Foods 116, 106192. Available at https://doi.org/10.1016/j.jff.2024.106192.

Ashique, S., Mohanto, S., Ahmed, M.G., Mishra, N., Garg, A., Chellappan, D.K., et al., 2024. Gut-brain axis: a cutting-edge approach to target neurological disorders and potential synbiotic application. Heliyon 10 (13). Available at https://doi.org/10.1016/j.heliyon.2024.e34092, https://www.sciencedirect.com/science/journal/24058440.

Baiano, A., 2022. 3D printed foods: a comprehensive review on technologies, nutritional value, safety, consumer attitude, regulatory framework, and economic and sustainability issues. Food Reviews International 38 (5), 986–1016. Available at https://doi.org/10.1080/87559129.2020.1762091.

Ban, Q., Cheng, J., Sun, X., Jiang, Y., Zhao, S., Song, X., et al., 2020. Effects of a synbiotic yogurt using monk fruit extract as sweetener on glucose regulation and gut microbiota in rats with type 2 diabetes mellitus. Journal of Dairy Science 103 (4), 2956–2968. Available at https://doi.org/10.3168/jds.2019-17700.

Bisht, D., Pal, D., Shrestha, R., 2024. Introduction to probiotics, prebiotics, and synbiotics a holistic approach. Probiotics: A Comprehensive Guide to Enhance Health and Mitigate Disease. CRC Press, India, pp. 1–28. Available at https://doi.org/10.1201/9781003452249-1, https://www.taylorfrancis.com/books/edit/10.1201/9781003452249/probiotics-gowhar-rashid-luxita-sharma-nihad-khan.

Blair, G., 2021. Technology and the environment – a framework for a symbiotic relationship. Journal of Advanced Research in Alternative Energy, Environment and Ecology 8 (2), 4–8. Available at https://doi.org/10.24321/2455.3093.202102.

Bodbodak, S., Nejatian, M., Ghandehari Yazdi, A.P., Kamali Rousta, L., Rafiee, Z., Jalali-Jivan, M., et al., 2024. Improving the thermal stability of natural bioactive ingredients via encapsulation technology. Critical Reviews in Food Science and Nutrition 64 (10), 2824–2846. Available at https://doi.org/10.1080/10408398.2022.2127145, https://www.tandfonline.com/loi/bfsn20.

Bondar, M., 2020. Regulations on functional products in different countries. Norwegian Journal of Development of the International Science (51-1), 16–23.

Calinoiu, L.F., Mitrea, L., Teleky, B.E., Szabo, K., Martău, A.G., Nemes, S.A., et al., 2023. Fruit and vegetable waste and by-products for pigments and color. Fruit and Vegetable Waste Utilization and Sustainability. Elsevier, Romania, pp. 77–100. Available at https://doi.org/10.1016/B978-0-323-91743-8.00011-3, https://www.sciencedirect.com/book/9780323917438.

Centurion, F., Basit, A.W., Liu, J., Gaisford, S., Rahim, M.A., Kalantar-Zadeh, K., 2021. Nanoencapsulation for probiotic delivery. ACS Nano 15 (12), 18653–18660. Available at https://doi.org/10.1021/acsnano.1c09951, http://pubs.acs.org/journal/ancac3.

Chaturvedi, S., Chakraborty, S., 2023. Evaluation of quality attributes and in vitro characteristics of synbiotic legume-based beverage during storage. Food Bioscience 55, 103000. Available at https://doi.org/10.1016/j.fbio.2023.103000.

Chaturvedi, S., Chakraborty, S., 2024. Effect of temperature and packaging materials on the shelf-life stability and in vitro properties of microencapsulated and spray-dried synbiotic legume-based instant beverage powder. Sustainable Food Technology 2 (1), 162–174. Available at https://doi.org/10.1039/d3fb00094j.

Chaturvedi, S., Khartad, A., Chakraborty, S., 2021. The potential of non-dairy synbiotic instant beverage powder: review on a new generation of healthy ready-to-reconstitute drinks. Food Bioscience 42, 101195. Available at https://doi.org/10.1016/j.fbio.2021.101195.

Chaudhary, T., Dixit, M., Gera, R., Shukla, A.K., Prakash, A., Gupta, G., et al., 2020. Techniques for improving formulations of bioinoculants. 3 Biotech 10 (5). Available at https://doi.org/10.1007/s13205-020-02182-9, http://www.springerlink.com/content/2190-572x/.

Christmann, M., 2020. Physical activity in the modern working world the networked health-relevant. Factors for Office Buildings: The Planned Health. Springer International Publishing, Germany, pp. 157–165. Available at https://doi.org/10.1007/978-3-030-59226-4_7, https://link.springer.com/book/10.1007/978-3-030-59226-4.

Chugh, P., Misra, S., Dhar, M.S., Raghuwanshi, S., 2023. Regulatory aspects relevant to probiotic products. Probiotics, Prebiotics, Synbiotics, and Postbiotics: Human Microbiome and Human Health. Springer Nature, India 513–234. Available at https://doi.org/10.1007/978-981-99-1463-0_25, https://link.springer.com/book/10.1007/978-981-99-1463-0.

Costabile, A., Bergillos-Meca, T., Rasinkangas, P., Korpela, K., de Vos, W.M., Gibson, G.R., 2017. Effects of soluble corn fiber alone or in synbiotic combination with *Lactobacillus rhamnosus* GG and the pilus-deficient derivative GG-PB12 on fecal microbiota, metabolism, and markers of immune function: a randomized, double-blind, placebo-controlled, crossover study in healthy elderly (saimes study). Frontiers in Immunology 8. Available at https://doi.org/10.3389/fimmu.2017.01443.

da Silva, T.F., Casarotti, S.N., de Oliveira, G.L.V., Penna, A.L.B., 2021. The impact of probiotics, prebiotics, and synbiotics on the biochemical, clinical, and immunological markers, as well as on the gut microbiota of obese hosts. Critical Reviews in Food Science and Nutrition 61 (2), 337–355. Available at https://doi.org/10.1080/10408398.2020.1733483, https://www.tandfonline.com/loi/bfsn20.

Dahiya, D., Nigam, P.S., 2022. Probiotics, prebiotics, synbiotics, and fermented foods as potential biotics in nutrition improving health via microbiome-gut-brain axis. Fermentation 8 (7). Available at https://doi.org/10.3390/fermentation8070303, https://www.mdpi.com/2311-5637/8/7/303/pdf?version=1656345023.

Dangi, P., Chaudhary, N., Chaudhary, V., Virdi, A.S., Kajla, P., Khanna, P., et al., 2023. Nanotechnology impacting probiotics and prebiotics: a paradigm shift in nutraceuticals technology. International Journal of Food Microbiology 388, 110083. Available at https://doi.org/10.1016/j.ijfoodmicro.2022.110083.

de Figueiredo, F.C., de Oliva-Neto, P., 2022. Advances and New Perspectives in Prebiotic, Probiotic and Symbiotic Products for Food Nutrition and Feed. Springer Science and Business Media LLC, pp. 311–336. Available at https://doi.org/10.1007/978-981-16-3682-0_10.

de Simone, C., 2019. The unregulated probiotic market. Clinical Gastroenterology and Hepatology 17 (5), 809–817. Available at https://doi.org/10.1016/j.cgh.2018.01.018.

Desmedt, O., Broers, V.J.V., Zamariola, G., Pachikian, B., Delzenne, N., Luminet, O., 2019. Effects of prebiotics on affect and cognition in human intervention studies. Nutrition Reviews 77 (2), 81–95. Available at https://doi.org/10.1093/nutrit/nuy052.

Dey, P., Mukherjee, S.K., Parai, D., 2023. Association of probiotics and prebiotics with human microbiome and the functioning of immune system. Probiotics, Prebiotics, Synbiotics, and Postbiotics: Human Microbiome and Human Health. Springer Nature, India, pp. 101–115. Available at https://doi.org/10.1007/978-981-99-1463-0_6, https://link.springer.com/book/10.1007/978-981-99-1463-0.

Douglas A.E., 2021. The synbiotic habit.

Farahmand, A., Ghorani, B., Emadzadeh, B., Sarabi-Jamab, M., Emadzadeh, M., Modiri, A., et al., 2024. Double protection of probiotics in alginate hydrogel through emulsification incorporated with freeze drying and coaxial wet-electrospraying: survivability and targeted delivery. LWT 204, 116459. Available at https://doi.org/10.1016/j.lwt.2024.116459.

Fekete, M., Lehoczki, A., Major, D., Fazekas-Pongor, V., Csípő, T., Tarantini, S., et al., 2024. Exploring the influence of gut–brain axis modulation on cognitive health: a comprehensive review of prebiotics, probiotics, and symbiotics. Nutrients 16 (6). Available at https://doi.org/10.3390/nu16060789, http://www.mdpi.com/journal/nutrients/.

Fentie, E.G., Lim, K., Jeong, M., Shin, J.H., 2024. A comprehensive review of the characterization, host interactions, and stabilization advancements on probiotics: addressing the challenges in functional food diversification. Comprehensive Reviews in Food Science and Food Safety 23 (5). Available at https://doi.org/10.1111/1541-4337.13424, http://onlinelibrary.wiley.com/journal/10.1111/(ISSN)1541-4337.

Fusco, V., Fanelli, F., Chieffi, D., 2022. Authenticity of probiotic foods and dietary supplements: a pivotal issue to address. Critical Reviews in Food Science and Nutrition 62 (25), 6854–6871. Available at https://doi.org/10.1080/10408398.2021.1907300.

Garrido-Cruz, J.L., Totosaus, A., Pérez-Alvarez, J.A., Cortés-Barberena, E., Pérez-Chabela, M.L., 2023. Bioactivity of pomegranate (*Punica granatum* L.) peels flour to improve probiotics viability in micro co-encapsulation by ionic gelation. Food Biotechnology 37 (4), 393–412. Available at https://doi.org/10.1080/08905436.2023.2269256, http://www.tandfonline.com/loi/lfbt20.

Georgieva, M., Georgiev, K., Hvarchanova, N., 2022. Probiotics: Past, Present, and Future Challenges. Elsevier BV, pp. 431–448. Available at https://doi.org/10.1016/b978-0-12-823733-5.00025-8.

Gillor, O., Etzion, A., Riley, M.A., 2008. The dual role of bacteriocins as anti- and probiotics. Applied Microbiology and Biotechnology 81 (4), 591–606. Available at https://doi.org/10.1007/s00253-008-1726-5.

Golnari, M., Behbahani, M., Mohabatkar, H., 2024. Comparative survival study of *Bacillus coagulans* and *Enterococcus faecium* microencapsulated in chitosan-alginate nanoparticles in simulated gastrointestinal condition. LWT 197, 115930. Available at https://doi.org/10.1016/j.lwt.2024.115930.

Gomez Quintero, D.F., Kok, C.R., Hutkins, R., 2022. The future of synbiotics: rational formulation and design. Frontiers in Microbiology 13. Available at https://doi.org/10.3389/fmicb.2022.919725, https://www.frontiersin.org/journals/microbiology#.

González-Herrera, S.M., Bermúdez-Quiñones, G., Ochoa-Martínez, L.A., Rutiaga-Quiñones, O.M., Gallegos-Infante, J.A., 2021. Synbiotics: a technological approach in food applications. Journal of Food Science and Technology 58 (3), 811–824. Available at https://doi.org/10.1007/s13197-020-04532-0, http://www.springerlink.com/content/121580/.

Granato, D., Barba, F.J., Bursać Kovačević, D., Lorenzo, J.M., Cruz, A.G., Putnik, P., 2020. Functional foods: product development, technological trends, efficacy testing, and safety. Annual Review of Food Science and Technology 11, 93–118. Available at https://doi.org/10.1146/annurev-food-032519-051708, http://www.annualreviews.org/journal/food.

Guo, Q., Cui, B., Yuan, C., Guo, L., Li, Z., Chai, Q., et al., 2024. Fabrication of dry S/O/W microcapsule and its probiotic protection against different stresses. Journal of the Science of Food and Agriculture 104 (5), 2842–2850. Available at https://doi.org/10.1002/jsfa.13175, http://onlinelibrary.wiley.com/journal/10.1002/(ISSN)1097-0010.

Gurram, S., Jha, D.K., Shah, D.S., Kshirsagar, M.M., Amin, P.D., 2021. Insights on the critical parameters affecting the probiotic viability during stabilization process and formulation development. AAPS PharmSciTech 22 (5). Available at https://doi.org/10.1208/s12249-021-02024-8, http://www.springerlink.com/content/1530-9932/.

Han, J., McClements, D.J., Liu, X., Liu, F., 2024. Oral delivery of probiotics using single-cell encapsulation. Comprehensive Reviews in Food Science and Food Safety 23 (3). Available at https://doi.org/10.1111/1541-4337.13322, http://onlinelibrary.wiley.com/journal/10.1111/(ISSN)1541-4337.

Harman, G.E., Uphoff, N., 2019. Synbiotic root-endophytic soil microbes improve crop productivity and provide environmental benefits. Scientifica 2019 (1), 9106395.

Haro-Soler, Md.M., Kiraly, D., 2019. Exploring self-efficacy beliefs in symbiotic collaboration with students: an action research project. Interpreter and Translator Trainer 13 (3), 255–270. Available at https://doi.org/10.1080/1750399X.2019.1656405, http://www.tandfonline.com/loi/ritt20.

Harshapriya, N., Manjunatha, H., Rajunaik, B., Praveen, A.R., Malashree, L., Ramachandra, B., 2024. Development and performance evaluation of synbiotic red banana yoghurt. Journal of Scientific Research and Reports 30 (6), 526–533. Available at https://doi.org/10.9734/jsrr/2024/v30i62069.

Hashem, N.M., Hosny, N.S., El-Desoky, N., Soltan, Y.A., Elolimy, A.A., Sallam, S.M.A., et al., 2023. Alginate nanoencapsulated synbiotic composite of pomegranate peel phytogenics and multi-probiotic species as a potential feed additive: physicochemical, antioxidant, and antimicrobial activities. Animals 13 (15). Available at https://doi.org/10.3390/ani13152432, http://www.mdpi.com/journal/animals.

Hassanien, A.E., EL-kaiaty, A.M., Sobhy, H.M., El-Moghazy, G.M., Gomaa, A.H., Abdel A'al, M.H., et al., 2024. Evaluation of using probiotics, prebiotics and symbiotic as growth promoters in broilers and their effects on some growth performance-related genes. Journal of the Hellenic Veterinary Medical Society 75 (2), 7621–7628. Available at https://doi.org/10.12681/jhvms.35116, https://ejournals.epublishing.ekt.gr/index.php/jhvms/issue/archive.

Hathi, Z., Mettu, S., Priya, A., Athukoralalage, S., Lam, T.N., Choudhury, N.R., et al., 2021. Methodological advances and challenges in probiotic bacteria production: ongoing strategies and future perspectives. Biochemical Engineering Journal 176. Available at https://doi.org/10.1016/j.bej.2021.108199, www.elsevier.com/locate/bej.

Heidarrezaei, M., Mauriello, G., Shokravi, H., Lau, W.J., Ismail, A.F., 2025. Delivery of probiotic-loaded microcapsules in the gastrointestinal tract: a review. Probiotics and Antimicrobial Proteins 17 (1), 193–211. Available at https://doi.org/10.1007/s12602-024-10311-6, https://www.springer.com/journal/12602.

Hellebois, T., Tsevdou, M., Soukoulis, C., 2020. Functionalizing and bio-preserving processed food products via probiotic and synbiotic edible films and coatings. Advances in Food and Nutrition Research 94, 161–221. Available at https://doi.org/10.1016/bs.afnr.2020.06.004, http://www.elsevier.com/wps/find/bookdescription.cws_home/704624/description#description.

Hussian, T., Lokesh, A., Saini, S., Tripathy, H., Dureja, 2024. Regulatory aspects of synbiotics-based delivery system. Synbiotics in Human Health: Biology to Drug Delivery. Springer Singapore, India, pp. 495–506. Available at https://doi.org/10.1007/978-981-99-5575-6_24, https://link.springer.com/book/9789819955756.

Iftekar, S.F., Aabid, A., Amir, A., Baig, M., 2023. Advancements and limitations in 3D printing materials and technologies: a critical review. Polymers 15 (11), 2519. Available at https://doi.org/10.3390/polym15112519.

Jang, W.J., Oh, D.N., Kim, D.Y., Lee, E.J., Lee, J.M., 2024. Encapsulation of probiotics with poly-γ-glutamic acid alters gut microbiota and short-chain fatty acid content by maintaining cell viability in the gastrointestinal tract. Food Hydrocolloids 147. Available at https://doi.org/10.1016/j.foodhyd.2023.109328, https://www.journals.elsevier.com/food-hydrocolloids.

Jannah, S.R., Rahayu, E.S., Yanti, R., Suroto, D.A., Wikandari, R., 2022. Study of viability, storage stability, and shelf life of probiotic instant coffee *Lactiplantibacillus plantarum* Subsp. plantarum Dad-13 in vacuum and nonvacuum packaging at different storage temperatures. International Journal of Food Science 2022. Available at https://doi.org/10.1155/2022/1663772, http://www.hindawi.com/journals/ijfs/.

Jiang, L., Zhang, Z., Qiu, C., Wen, J., 2024. A review of whey protein-based bioactive delivery systems: design, fabrication, and application. Foods 13 (15), 2453. Available at https://doi.org/10.3390/foods13152453.

Jiménez-Villeda, B.E., Falfán-Cortés, R.N., Rangel-Vargas, E., Santos-López, E.M., Gómez-Aldapa, C.A., Torres-Vitela, M.R., et al., 2023. Synbiotic encapsulation: a trend towards increasing viability and probiotic effect. Journal of Food Processing and Preservation 2023. Available at https://doi.org/10.1155/2023/7057462, https://www.hindawi.com/journals/jfpp.

Kamble, M., Singh, A., Singh, S.V., Upadhyay, A., Kondepudi, K.K., Chinchkar, A.V., 2024. Effect of gastrointestinal resistant encapsulate matrix on spray dried microencapsulated *Lacticaseibacillus rhamnosus* GG powder and its characterization. Food Research

International 192. Available at https://doi.org/10.1016/j.foodres.2024.114804, https://www.sciencedirect.com/science/journal/09639969.

Kanimozhi, N.V., Sukumar, M., 2024. Development of synbiotic juice by spray drying of *Lactobacillus paracaesii* sub sp. tolerans N2-45 along with galacto-oligosaccharides using different carrier materials. Bioactive Carbohydrates and Dietary Fibre 31. Available at https://doi.org/10.1016/j.bcdf.2024.100413, https://www.sciencedirect.com/science/journal/22126198.

Kaur, K., Rath, G., 2019. Formulation and evaluation of UV protective synbiotic skin care topical formulation. Journal of Cosmetic and Laser Therapy 21 (6), 332–342. Available at https://doi.org/10.1080/14764172.2019.1658878.

Kaźmierczak-Siedlecka, K., Daca, A., Fic, M., van de Wetering, T., Folwarski, M., Makarewicz, W., 2020. Therapeutic methods of gut microbiota modification in colorectal cancer management – fecal microbiota transplantation, prebiotics, probiotics, and synbiotics. Gut Microbes 11 (6), 1518–1530. Available at https://doi.org/10.1080/19490976.2020.1764309.

Kesavelu, D., Jog, P., 2023. Current understanding of antibiotic-associated dysbiosis and approaches for its management. Therapeutic Advances in Infectious Disease 10. Available at https://doi.org/10.1177/20499361231154443.

Khullar, G., Karami, Z., Prakitchaiwattana, C., 2024. Development of microencapsulated dried Bacillus sp. 63-11 with enhanced shelf stability and bioactivity for use as a food supplement. International Journal of Food Science and Technology 59 (3), 1291–1298. Available at https://doi.org/10.1111/ijfs.16853.

Kohut, M., Lohne, O., Leker, J., Bröring, S., 2021. Market convergence from a start-up perspective: the case of probiotics. PharmaNutrition 15, 100243. Available at https://doi.org/10.1016/j.phanu.2020.100243.

Koirala, S., Thapa, S.P., Anal, A.K., 2022. Overall safety considerations and regulatory oversight for probiotics-based foods and beverages. Probiotics, Prebiotics and Synbiotics: Technological Advancements Towards Safety and Industrial Applications. wiley, Thailand, pp. 441–461. Available at https://doi.org/10.1002/9781119702160.ch18, https://www.wiley.com/en-gb/Probiotics,+Prebiotics+and+Synbiotics:+Technological+Advancements+Towards+Safety+and+Industrial+Applications-p-9781119701200.

Krishnaraj, P., Anukiruthika, T., Choudhary, P., Moses, J.A., Anandharamakrishnan, C., 2019. 3D extrusion printing and post-processing of fibre-rich snack from indigenous composite flour. Food and Bioprocess Technology 12 (10), 1776–1786. Available at https://doi.org/10.1007/s11947-019-02336-5.

Kristo, A.S., Gültekin, B., Öztağ, M., Sikalidis, A.K., 2020. The effect of eating habits' quality on scholastic performance in Turkish adolescents. Behavioral Sciences 10 (1), 31. Available at https://doi.org/10.3390/bs10010031.

Kumar, P., Dhanda, S., 2024. Chitosan encapsulation of *Pediococcus acidilactici* NCDC 252 improved its survival in simulated gastro-intestinal conditions. Process Biochemistry 138, 130–138. Available at https://doi.org/10.1016/j.procbio.2024.01.013.

Kumari, A., K. G, R., Sudhakaran. V, A., Warrier, A.S., Singh, N.K., 2024. Unveiling the health benefits of prebiotics: a comprehensive review. Indian Journal of Microbiology 64 (2), 376–388. Available at https://doi.org/10.1007/s12088-024-01235-4, https://www.springer.com/journal/12088.

Kussmann, M., Cunha, D.H.A., 2022. Nature has the answers: discovering and validating natural bioactives for human health. Bioactive Compounds in Health and Disease 5 (10), 222–234. Available at https://doi.org/10.31989/bchd.v5i10.1000, https://www.ffhdj.com/index.php/BioactiveCompounds/article/view/1000.

Krasaekoopt, W., & Watcharapoka, S. (2014). Effect of addition of inulin and galactooligosaccharide on the survival of microencapsulated probiotics in alginate beads coated with chitosan in simulated digestive system, yogurt and fruit juice. LWT-Food Science and Technology, 57(2), 761-766..

Leach, S.T., 2024. Role of probiotics and prebiotics in gut symbiosis. Nutrients 16 (2). Available at https://doi.org/10.3390/nu16020238, http://www.mdpi.com/journal/nutrients/.

Lee, S., Choi, S.P., Choi, H.J., Jeong, H., Park, Y.S., 2024. A comprehensive review of synbiotics: an emerging paradigm in health promotion and disease management. World Journal of Microbiology and Biotechnology 40 (9). Available at https://doi.org/10.1007/s11274-024-04085-w, https://www.springer.com/journal/11274.

Lemoine, A., Tounian, P., Adel-Patient, K., Thomas, M., 2023. Pre-, pro-, syn-, and postbiotics in infant formulas: what are the immune benefits for infants? Nutrients 15 (5), 1231. Available at https://doi.org/10.3390/nu15051231.

Li, S., Zhang, Y.X., 2024. Sensitive delivery systems and novel encapsulation technologies for live biotherapeutic products and probiotics. Critical Reviews in Microbiology 50 (3), 371–384. Available at https://doi.org/10.1080/1040841X.2023.2202237, https://www.tandfonline.com/loi/imby20.

Liang, D., Wu, F., Zhou, D., Tan, B., Chen, T., 2024. Commercial probiotic products in public health: current status and potential limitations. Critical Reviews in Food Science and Nutrition 64 (19), 6455–6476. Available at https://doi.org/10.1080/10408398.2023.2169858.

Liu, R.T., Walsh, R.F.L., Sheehan, A.E., 2019. Prebiotics and probiotics for depression and anxiety: a systematic review and meta-analysis of controlled clinical trials. Neuroscience and Biobehavioral Reviews 102, 13–23. Available at https://doi.org/10.1016/j.neubiorev.2019.03.023, www.elsevier.com/locate/neubiorev.

Liu, J., Qian, Y., Song, S.F., Duan, R.R., 2022a. Industrial synbiotic agglomeration and green economic growth: a spatial difference-in-differences approach. Journal of Cleaner Production 364, 132560.

Liu, Y., Wang, J., Wu, C., 2022b. Modulation of gut microbiota and immune system by probiotics, pre-biotics, and post-biotics. Frontiers in Nutrition 8. Available at https://doi.org/10.3389/fnut.2021.634897.

Liu, X., Song, Z., Tian, W., Abdullah, Huang, Q., Chen, M., Xiao, J., 2024. Advancements in lipid-based delivery systems for functional foods: a comprehensive review of literature and patent trends. Critical Reviews in Food Science and Nutrition 1–17.

Lopes, S.A., Roque-Borda, C.A., Duarte, J.L., Di Filippo, L.D., Borges Cardoso, V.M., Pavan, F.R., et al., 2023. Delivery strategies of probiotics from nano- and microparticles: trends in the treatment of inflammatory bowel disease—an overview. Pharmaceutics 15 (11). Available at https://doi.org/10.3390/pharmaceutics15112600, http://www.mdpi.com/journal/pharmaceutics.

Lybæk, R., Christensen, T.B., Thomsen, T.P., 2021. Enhancing policies for deployment of Industrial symbiosis – what are the obstacles, drivers and future way forward? Journal of Cleaner Production 280. Available at https://doi.org/10.1016/j.jclepro.2020.124351, https://www.journals.elsevier.com/journal-of-cleaner-production.

Mabrouk, A.M., Salama, H.H., El Sayed, H.S., El Sayed, S.M., 2021. Preparation of synbiotic whey protein gel as a carrier of free and encapsulated probiotic bacteria. Journal of Food Processing and Preservation 45 (7), e15612.

Maftei, N.-M., 2019. Probiotic, Prebiotic and Synbiotic Products in Human Health. IntechOpen. Available at https://doi.org/10.5772/intechopen.81553.

Malwe, A.S., Srivastava, G.N., Sharma, V.K., 2023. GutBug: a tool for prediction of human gut bacteria mediated biotransformation of biotic and xenobiotic molecules using machine learning. Journal of Molecular Biology 435 (14). Available at https://doi.org/10.1016/j.jmb.2023.168056, https://www.journals.elsevier.com/journal-of-molecular-biology.

March, D.S., Jones, A.W., Bishop, N.C., Burton, J.O., 2020. The efficacy of prebiotic, probiotic, and synbiotic supplementation in modulating gut-derived circulatory particles associated with cardiovascular disease in individuals receiving dialysis: a systematic review and meta-analysis of randomized controlled trials. Journal of Renal Nutrition 30 (4), 347–359. Available at https://doi.org/10.1053/j.jrn.2019.07.006, http://www.sciencedirect.com/science/journal/10512276.

Matzaras, R., Nikopoulou, A., Protonotariou, E., Christaki, E., 2022. Gut microbiota modulation and prevention of dysbiosis as an alternative approach to antimicrobial resistance: a narrative review. Yale Journal of Biology and Medicine 95 (4), 479–494. Available at https://www.ncbi.nlm.nih.gov/pmc/articles/PMC9765331/pdf/yjbm_95_4_479.pdf.

McCoubrey, L.E., Favaron, A., Awad, A., Orlu, M., Gaisford, S., Basit, A.W., 2023. Colonic drug delivery: formulating the next generation of colon-targeted therapeutics. Journal of Controlled Release 353, 1107–1126. Available at https://doi.org/10.1016/j.jconrel.2022.12.029, www.elsevier.com/locate/jconrel.

McFarlane, C., Krishnasamy, R., Stanton, T., Savill, E., Snelson, M., Mihala, G., et al., 2021. Synbiotics easing renal failure by improving gut microbiology II (SYNERGY II): a feasibility randomized controlled trial. Nutrients 13 (12), 4481. Available at https://doi.org/10.3390/nu13124481.

McKenzie, V.J., Kueneman, J.G., Harris, R.N., 2018. Probiotics as a tool for disease mitigation in wildlife: insights from food production and medicine. Annals of the New York Academy of Sciences 1429 (1), 18–30. Available at https://doi.org/10.1111/nyas.13617, http://onlinelibrary.wiley.com/journal/10.1111/(ISSN)1749-6632.

Mishra, A., Chakravarty, I., Mandavgane, S., 2021. Current trends in non-dairy based synbiotics. Critical Reviews in Biotechnology 41 (6), 935–952. Available at https://doi.org/10.1080/07388551.2021.1898329.

Mohamadzadeh, M., Fazeli, A., Shojaosadati, S.A., 2024. Polysaccharides and proteins-based bionanocomposites for microencapsulation of probiotics to improve stability and viability in the gastrointestinal tract: a review. International Journal of Biological Macromolecules 259 (Pt 2), 129287. Available at https://doi.org/10.1016/j.ijbiomac.2024.129287.

Mukherjee, A., Gómez-Sala, B., O'Connor, E.M., Kenny, J.G., Cotter, P.D., 2022. Global regulatory frameworks for fermented foods: a review. Frontiers in Nutrition 9. Available at https://doi.org/10.3389/fnut.2022.902642, journal.frontiersin.org/journal/nutrition.

Nazir, S., Afzaal, M., Saeed, F., Ahmad, A., Ateeq, H., Ikram, A., et al., 2024. Survivability and behavior of probiotic bacteria encapsulated by internal gelation in non-dairy matrix and in vitro GIT conditions. PLoS One 19 (6), e0303091. Available at https://doi.org/10.1371/journal.pone.0303091.

Nezamdoost-Sani, N., Khaledabad, M.A., Amiri, S., Phimolsiripol, Y., Mousavi Khaneghah, A., 2024. A comprehensive review on the utilization of biopolymer hydrogels to encapsulate and protect probiotics in foods. International Journal of Biological Macromolecules 254. Available at https://doi.org/10.1016/j.ijbiomac.2023.127907, https://www.sciencedirect.com/journal/international-journal-of-biological-macromolecules.

Ngamekaue, N., Dumrongchai, T., Rodklongtan, A., Chitprasert, P., 2024. Improving probiotic survival through encapsulation in coconut oil in whey protein isolate emulsions during spray drying and gastrointestinal digestion. LWT 198, 116061. Available at https://doi.org/10.1016/j.lwt.2024.116061.

Olas, B., 2020. Probiotics, prebiotics and synbiotics—a promising strategy in prevention and treatment of cardiovascular diseases? International Journal of Molecular Sciences 21 (24), 9737. Available at https://doi.org/10.3390/ijms21249737.

Oliveira, R.A., Pamer, E.G., 2023. Assembling symbiotic bacterial species into live therapeutic consortia that reconstitute microbiome functions. Cell Host and Microbe 31 (4), 472–484. Available at https://doi.org/10.1016/j.chom.2023.03.002, http://www.elsevier.com.

Olvera-Rosales, L.B., Cruz-Guerrero, A.E., Ramírez-Moreno, E., Quintero-Lira, A., Contreras-López, E., Jaimez-Ordaz, J., et al., 2021. Impact of the gut microbiota balance on the health-disease relationship: the importance of consuming probiotics and prebiotics. Foods 10 (6). Available at https://doi.org/10.3390/foods10061261, https://www.mdpi.com/2304-8158/10/6/1261/pdf.

O'Malley, M.A., 2017. From endosymbiosis to holobionts: evaluating a conceptual legacy. Journal of Theoretical Biology 434, 34–41. Available at https://doi.org/10.1016/j.jtbi.2017.03.008, http://www.elsevier.com/inca/publications/store/6/2/2/9/0/4/index.htt.

Oniszczuk, A., Oniszczuk, T., Gancarz, M., Szymańska, J., 2021. Role of gut microbiota, probiotics and prebiotics in the cardiovascular diseases. Molecules (Basel, Switzerland) 26 (4), 1172. Available at https://doi.org/10.3390/molecules26041172.

Onwe, R.O., Onwosi, C.O., Ezugworie, F.N., Ekwealor, C.C., Okonkwo, C.C., 2022. Microbial trehalose boosts the ecological fitness of biocontrol agents, the viability of probiotics during long-term storage and plants tolerance to environmental-driven abiotic stress. Science of the Total Environment 806. Available at https://doi.org/10.1016/j.scitotenv.2021.150432, www.elsevier.com/locate/scitotenv.

Pasinetti, G., 2020. Synbiotic-derived metabolites reduce neuroinflammatory symptoms of Alzheimer's disease. Current Developments in Nutrition 4. Available at https://doi.org/10.1093/cdn/nzaa062_035 nzaa062_035.

Pavlidou, E., Fasoulas, A., Mantzorou, M., Giaginis, C., 2022. Clinical evidence on the potential beneficial effects of probiotics and prebiotics in cardiovascular disease. International Journal of Molecular Sciences 23 (24), 15898. Available at https://doi.org/10.3390/ijms232415898.

Peres, J., Esmerino, E., da Silva, A.L., Racowski, I., Bolini, H., 2018. Sensory profile, drivers of liking, and influence of information on the acceptance of low-calorie synbiotic and probiotic chocolate ice cream. Journal of Food Science 83 (5), 1350–1359. Available at https://doi.org/10.1111/1750-3841.14120, http://onlinelibrary.wiley.com/journal/10.1111/(ISSN)1750-3841.

Peterson, C.T., 2020. Dysfunction of the microbiota-gut-brain axis in neurodegenerative disease: the promise of therapeutic modulation with prebiotics, medicinal herbs, probiotics, and synbiotics. Journal of Evidence-Based Integrative Medicine 25. Available at https://doi.org/10.1177/2515690x20957225 2515690X2095722.

Prajapati, K., Bisani, K., Prajapati, H., Prajapati, S., Agrawal, D., Singh, S., et al., 2024. Advances in probiotics research: mechanisms of action, health benefits, and limitations in applications. Systems Microbiology and Biomanufacturing 4 (2), 386–406. Available at https://doi.org/10.1007/s43393-023-00208-w.

Quigley, E.M.M., 2019. Prebiotics and probiotics in digestive health. Clinical Gastroenterology and Hepatology 17 (2), 333–344. Available at https://doi.org/10.1016/j.cgh.2018.09.028.

Rashidinejad, A., Bahrami, A., Rehman, A., Rezaei, A., Babazadeh, A., Singh, H., et al., 2022. Co-encapsulation of probiotics with prebiotics and their application in functional/synbiotic dairy products. Critical Reviews in Food Science and Nutrition 62 (9), 2470–2494. Available at https://doi.org/10.1080/10408398.2020.1854169, https://www.tandfonline.com/loi/bfsn20.

Razavi, S., Janfaza, S., Tasnim, N., Gibson, D.L., Hoorfar, M., 2021. Nanomaterial-based encapsulation for controlled gastrointestinal delivery of viable probiotic bacteria. Nanoscale Advances 3 (10), 2699–2709. Available at https://doi.org/10.1039/d0na00952k, http://pubs.rsc.org/en/journals/journalissues/na?_ga2.190536939.1555337663.1552312502-1364180372.1550481316#!issueidna001002&typecurrent&issnonline2516-0230.

Reque, P.M., Brandelli, A., 2021. Encapsulation of probiotics and nutraceuticals: applications in functional food industry. Trends in Food Science and Technology 114, 1–10. Available at https://doi.org/10.1016/j.tifs.2021.05.022, http://www.elsevier.com/wps/find/journaldescription.cws_home/601278/description#description.

Rinaldi, F., Marotta, L., Mascolo, A., Amoruso, A., Pane, M., Giuliani, G., et al., 2022. Facial acne: a randomized, double-blind, placebo-controlled study on the clinical efficacy of a symbiotic dietary supplement. Dermatology and Therapy 12 (2), 577–589. Available at https://doi.org/10.1007/s13555-021-00664-z.

Rousseaux, A., Brosseau, C., Bodinier, M., 2023. Immunomodulation of B lymphocytes by prebiotics, probiotics and synbiotics: application in pathologies. Nutrients 15 (2), 269. Available at https://doi.org/10.3390/nu15020269.

Roy, S., Dhaneshwar, S., 2023. Role of prebiotics, probiotics, and synbiotics in management of inflammatory bowel disease: current perspectives. World Journal of Gastroenterology 29 (14), 2078–2100. Available at https://doi.org/10.3748/wjg.v29.i14.2078.

Sadiq, M.B., Azhar, F.U.A., Ahmad, I., 2022. Probiotic and prebiotic interactions and their role in maintaining host immunity. Microbiome-Gut-Brain Axis: Implications on Health. Springer Nature, Pakistan, pp. 425–443. Available at https://doi.org/10.1007/978-981-16-1626-6, https://doi.org/10.1007/978-981-16-1626-6_22.

Saini, M., Singh, S.K., Yadav, H.K., Dave, P., Gupta, M., Bhat, A.A., et al., 2024. Role of synbiotics in cardiovascular diseases. Synbiotics in Human Health: Biology to Drug Delivery. Springer Singapore, India, pp. 137–153. Available at https://doi.org/10.1007/978-981-99-5575-6_6, https://link.springer.com/book/9789819955756.

Santos-Marcos, J.A., Perez-Jimenez, F., Camargo, A., 2019. The role of diet and intestinal microbiota in the development of metabolic syndrome. Journal of Nutritional Biochemistry 70, 1–27. Available at https://doi.org/10.1016/j.jnutbio.2019.03.017, www.elsevier.com/locate/jnutbio.

Sapp, J., 2016. The symbiotic self. Evolutionary Biology 43 (4), 596–603. Available at https://doi.org/10.1007/s11692-016-9378-3.

Sarbani, N.M.M., Yahaya, N., 2022. Advanced development of bio-fertilizer formulations using microorganisms as inoculant for sustainable agriculture and environment – a review. Malaysian Journal of Science Health and Technology 8 (1), 92–101. Available at https://doi.org/10.33102/mjosht.v8i1.228.

Scarpellini, E., Rinninella, E., Basilico, M., Colomier, E., Rasetti, C., Larussa, T., et al., 2022. From pre-and probiotics to post-biotics: a narrative review. International Journal of Environmental Research and Public Health 19 (1). Available at https://doi.org/10.3390/ijerph19010037, https://www.mdpi.com/1660-4601/19/1/37/pdf.

Schoina, V., Terpou, A., Bosnea, L., Kanellaki, M., Nigam, P.S., 2018. Entrapment of *Lactobacillus casei* ATCC393 in the viscus matrix of *Pistacia terebinthus* resin for functional myzithra cheese manufacture. LWT 89, 441–448. Available at https://doi.org/10.1016/j.lwt.2017.11.015, https://www.journals.elsevier.com/lwt.

Scott, B.M., Gutiérrez-Vázquez, C., Sanmarco, L.M., da Silva Pereira, J.A., Li, Z., Plasencia, A., et al., 2021. Self-tunable engineered yeast probiotics for the treatment of inflammatory bowel disease. Nature Medicine 27 (7), 1212–1222. Available at https://doi.org/10.1038/s41591-021-01390-x, http://www.nature.com/nm/index.html.

Shamekhi, S., Lotfi, H., Abdolalizadeh, J., Bonabi, E., Zarghami, N., 2020. An overview of yeast probiotics as cancer biotherapeutics: possible clinical application in colorectal cancer. Clinical and Translational Oncology 22 (8), 1227–1239. Available at https://doi.org/10.1007/s12094-019-02270-0.

Sharma, S., Singh, A., Sharma, S., Kant, A., Sevda, S., Taherzadeh, M.J., et al., 2021. Functional foods as a formulation ingredients in beverages: technological advancements and constraints. Bioengineered 12 (2), 11055–11075. Available at https://doi.org/10.1080/21655979.2021.2005992, http://www.tandfonline.com/loi/kbie20.

Sharma, R., Rashidinejad, A., Jafari, S.M., 2022. Application of spray dried encapsulated probiotics in functional food formulations. Food and Bioprocess Technology 15 (10), 2135–2154. Available at https://doi.org/10.1007/s11947-022-02803-6, http://www.springer.com/life+sci/food+science/journal/11947.

Shen, R., Chen, S., Lei, W., Shen, J., Lv, L., Wei, T., 2023. Nonfood probiotic, prebiotic, and synbiotic use reduces all-cause and cardiovascular mortality risk in older adults: a population-based cohort study. The Journal of Nutrition, Health and Aging 27 (5), 391–397. Available at https://doi.org/10.1007/s12603-023-1921-1.

Shi, X., Zhang, J., Mo, L., Shi, J., Qin, M., Huang, X., 2019. Efficacy and safety of probiotics in eradicating *Helicobacter pylori*. Medicine 98 (15), e15180. Available at https://doi.org/10.1097/md.0000000000015180.

Shinde T.S., 2019. Synbiotic efficacy of probiotic and prebiotic food ingredients for gut health.

Siddiqui, S.A., Khan, S., Mehdizadeh, M., Nirmal, N.P., Khanashyam, A.C., Fernando, I., et al., 2023. Consumer studies focus on prebiotics, probiotics, and synbiotics in food packaging: a review. Current Food Science and Technology Reports 1 (1), 13–29. Available at https://doi.org/10.1007/s43555-023-00003-7.

Simon, E., Călinoiu, L.F., Mitrea, L., Vodnar, D.C., 2021. Probiotics, prebiotics, and synbiotics: implications and beneficial effects against irritable bowel syndrome. Nutrients 13 (6). Available at https://doi.org/10.3390/nu13062112, https://www.mdpi.com/2072-6643/13/6/2112/pdf.

Skowron, K., Bauza-kaszewska, J., Kraszewska, Z., Wiktorczyk-kapischke, N., Grudlewska-buda, K., Kwiecińska-piróg, J., Gospodarek-komkowska, E., et al., 2021. Human skin microbiome: impact of intrinsic and extrinsic factors on skin microbiota. Microorganisms 9 (3), 1–20. Available at https://doi.org/10.3390/microorganisms9030543, https://www.mdpi.com/2076-2607/9/3/543/pdf.

Smith, C., 2020. Synbiotic-derived metabolites reduce neuroinflammatory symptoms of Alzheimer's disease. Alzheimer's and Dementia 16 (S9). Available at https://doi.org/10.1002/alz.046288.

Somera, T., Mafaldo, Í.M., Bezerril, F.F., Rosset, M., Magnani, M., Pimentel, T.C., 2024. Unconventional foods and by-products as sustainable encapsulating materials carrying probiotic cultures: an overview of the encapsulation methods, and incorporation on food products. Food Bioscience 61. Available at https://doi.org/10.1016/j.fbio.2024.104892, https://www.sciencedirect.com/science/journal/22124292.

Speckmann, B., Ehring, E., Hu, J., Rodriguez Mateos, A., 2024. Exploring substrate–microbe interactions: a metabiotic approach toward developing targeted synbiotic compositions. Gut Microbes 16 (1). Available at https://doi.org/10.1080/19490976.2024.2305716.

Steels, E., Erhardt, R., Harnett, J.E., Vigar, V., Steadman, K.J., 2023. Efficacy of a prebiotic formulation for treatment of functional constipation and associated gastrointestinal symptoms in adults: a randomised controlled trial. Advances in Integrative Medicine 10 (2), 43–50.

Available at https://doi.org/10.1016/j.aimed.2023.05.001, http://www.elsevier.com/journals/advances-in-integrative-medicine/2212-9626.

Suganya, K., Koo, B.-S., 2020. Gut–brain axis: role of gut microbiota on neurological disorders and how probiotics/prebiotics beneficially modulate microbial and immune pathways to improve brain functions. International Journal of Molecular Sciences 21 (20), 7551. Available at https://doi.org/10.3390/ijms21207551.

Swanson, K.S., Gibson, G.R., Hutkins, R., Reimer, R.A., Reid, G., Verbeke, K., et al., 2020. The International Scientific Association for Probiotics and Prebiotics (ISAPP) consensus statement on the definition and scope of synbiotics. Nature Reviews Gastroenterology and Hepatology 17 (11), 687–701. Available at https://doi.org/10.1038/s41575-020-0344-2, http://www.nature.com/nrgastro/index.html.

Thakur, N., 2020. An Insight through Root-Endophytic-Mutualistic Association in Improving Crop Productivity and Sustainability. Springer Science and Business Media LLC, pp. 31–44. Available at https://doi.org/10.1007/978-3-030-51916-2_2.

Thinkohkaew, K., Jonjaroen, V., Niamsiri, N., Panya, A., Suppavorasatit, I., Potiyaraj, P., 2024. Microencapsulation of probiotics in chitosan-coated alginate/gellan gum: optimization for viability and stability enhancement. Food Hydrocolloids 151, 109788. Available at https://doi.org/10.1016/j.foodhyd.2024.109788.

Tsai, Y.L., Lin, T.L., Chang, C.J., Wu, T.R., Lai, W.F., Lu, C.C., et al., 2019. Probiotics, prebiotics and amelioration of diseases. Journal of Biomedical Science 26 (1). Available at https://doi.org/10.1186/s12929-018-0493-6, http://www.jbiomedsci.com.

Ul Ain, N., Naveed, M., Aziz, T., Shabbir, M.A., Al Asmari, F., Abdi, G., et al., 2024. Mix-match synthesis of nanosynbiotics from probiotics and prebiotics to counter gut dysbiosis via AI integrated formulation profiling. Scientific Reports 14 (1). Available at https://doi.org/10.1038/s41598-024-69515-z, https://www.nature.com/srep/.

Umaña, J.J., Bauer-Estrada, K., Filomena-Ambrosio, A., Quintanilla-Carvajal, M.X., 2024. Survival of encapsulated *Lactiplantibacillus plantarum* probiotics on soft bread throughout gastrointestinal tract transit: physicochemical characteristics, sensory profile, and functional activity. Journal of Cereal Science 118. Available at https://doi.org/10.1016/j.jcs.2024.103971, https://www.sciencedirect.com/science/journal/07335210.

Vasquez, E.C., Pereira, T.M.C., Peotta, V.A., Baldo, M.P., Campos-Toimil, M., 2019. Probiotics as beneficial dietary supplements to prevent and treat cardiovascular diseases: uncovering their impact on oxidative stress. Oxidative Medicine and Cellular Longevity 2019. Available at https://doi.org/10.1155/2019/3086270, http://www.hindawi.com/journals/oximed/.

Velayati, A., Kareem, I., Sedaghat, M., Sohrab, G., Nikpayam, O., Hedayati, M., et al., 2023. Does symbiotic supplementation which contains *Bacillus Coagulans Lactobacillus rhamnosus*, *Lactobacillus acidophilus* and fructooligosaccharide has favourite effects in patients with type-2 diabetes? A randomised, double-blind, placebo-controlled trial. Archives of Physiology and Biochemistry 129 (6), 1211–1218. Available at https://doi.org/10.1080/13813455.2021.1928225.

Vignesh, V., Seth, D., Singh, C.N., Dwivedi, M., Singh, S.K., Kumari, S., 2024. Electrohydrodynamic encapsulation: a novel technique to enhance the stability of bioactive compounds in food. Trends in Food Science and Technology 151. Available at https://doi.org/10.1016/j.tifs.2024.104626, https://www.sciencedirect.com/science/journal/09242244.

Vivek, K., Mishra, S., Pradhan, R.C., Nagarajan, M., Kumar, P.K., Singh, S.S., et al., 2023. A comprehensive review on microencapsulation of probiotics: technology, carriers and current trends. Applied Food Research 3 (1). Available at https://doi.org/10.1016/j.afres.2022.100248, https://www.journals.elsevier.com/applied-food-research.

Wallace, C.J.K., Milev, R.V., 2021. The efficacy, safety, and tolerability of probiotics on depression: clinical results from an open-label pilot study. Frontiers in Psychiatry 12. Available at https://doi.org/10.3389/fpsyt.2021.618279, http://www.frontiersin.org/Psychiatry.

Włodarczyk, M., Śliżewska, K., 2021. Efficiency of resistant starch and dextrins as prebiotics: a review of the existing evidence and clinical trials. Nutrients 13 (11), 3808. Available at https://doi.org/10.3390/nu13113808.

Xu, C., Ban, Q., Wang, W., Hou, J., Jiang, Z., 2022. Novel nano-encapsulated probiotic agents: encapsulate materials, delivery, and encapsulation systems. Journal of Controlled Release 349, 184–205. Available at https://doi.org/10.1016/j.jconrel.2022.06.061.

Yadav, M., Mandeep, Shukla, P., 2020. Probiotics of diverse origin and their therapeutic applications: a review. Journal of the American College of Nutrition 39 (5), 469–479. Available at https://doi.org/10.1080/07315724.2019.1691957.

Yadav, M.K., Kumari, I., Singh, B., Sharma, K.K., Tiwari, S.K., 2022. Probiotics, prebiotics and synbiotics: safe options for next-generation therapeutics. Applied Microbiology and Biotechnology 106 (2), 505–521. Available at https://doi.org/10.1007/s00253-021-11646-8, link.springer.de/link/service/journals/00253/index.htm.

Yang, S., Wei, S., Wu, Y., Fang, Y., Deng, Z., Xu, J., et al., 2024. Encapsulation techniques, action mechanisms, and evaluation models of probiotics: recent advances and future prospects. Food Frontiers 5 (3), 1212–1239. Available at https://doi.org/10.1002/fft2.374.

Yeşilyurt, N., Yılmaz, B., Ağagündüz, D., Capasso, R., 2021. Involvement of probiotics and postbiotics in the immune system modulation. Biologics 1 (2), 89–110. Available at https://doi.org/10.3390/biologics1020006.

Yoha, K.S., Anukiruthika, T., Anila, W., Moses, J.A., Anandharamakrishnan, C., 2021. 3D printing of encapsulated probiotics: effect of different post-processing methods on the stability of *Lactiplantibacillus plantarum* (NCIM 2083) under static in vitro digestion conditions and during storage. LWT 146. Available at https://doi.org/10.1016/j.lwt.2021.111461, https://www.journals.elsevier.com/lwt.

Yu, H., Kong, Q., Wang, M., Han, Z., Xu, J., 2024. Improved viability of probiotics by encapsulation in chickpea protein matrix during simulated gastrointestinal digestion by succinylated modification. International Journal of Biological Macromolecules 260 (Pt 2), 129614. Available at https://doi.org/10.1016/j.ijbiomac.2024.129614.

Zepeda-Hernández, A., Garcia-Amezquita, L.E., Requena, T., García-Cayuela, T., 2021. Probiotics, prebiotics, and synbiotics added to dairy products: Uses and applications to manage type 2 diabetes. Food Research International 142, 110208. Available at https://doi.org/10.1016/j.foodres.2021.110208.

Zhao, L., Zhang, M., Chitrakar, B., Adhikari, B., 2021. Recent advances in functional 3D printing of foods: a review of functions of ingredients and internal structures. Critical Reviews in Food Science and Nutrition 61 (21), 3489–3503. Available at https://doi.org/10.1080/10408398.2020.1799327.

Zhao, R., Yu, T., Li, J., Niu, R., Liu, D., Wang, W., 2024. Single-cell encapsulation systems for probiotic delivery: armor probiotics. Advances in Colloid and Interface Science 332, 103270. Available at https://doi.org/10.1016/j.cis.2024.103270.

Chapter 19

Probiotic, prebiotic, and synbiotic product development: principles, challenges, market, and trends

Hamideh Homayouni-Rad[1], Negin Hosseinzadeh[1], Saba Kamalledin Moghadam[2], Vahideh Sarabi-Aghdam[1], Nader Khani[1], Baharak Aghapour[3], and Aziz Homayouni-Rad[1]

[1]*Department of Food Science and Technology, Faculty of Nutrition and Food Sciences, Tabriz University of Medical Sciences, Tabriz, Iran,* [2]*Department of Food Science and Technology, Faculty of Nutrition Sciences and Food Technology, National Nutrition and Food Technology Research Institute, Shahid Beheshti University of Medical Sciences, Tehran, Iran,* [3]*Department of Health, Sarab Faculty of Medical Sciences, Sarab, East Azerbaijan, Iran*

19.1 Introduction to probiotics, prebiotics, and synbiotics

Probiotics, prebiotics, and synbiotics have garnered significant attention in both the scientific community and the consumer market for their beneficial effects on health, particularly related to gut health and the immune system. Each of these terms, while related, refers to different concepts within the field of microbiome research and functional foods (Asqardokht-Aliabadi et al., 2025).

Probiotics are live microorganisms that, when administered in adequate amounts, provide a health benefit to the host (Ji et al., 2023). The most common probiotic strains belong to the genera *Lactobacillus*, *Bifidobacterium*, and *Saccharomyces*, although other strains are also used. Probiotics function primarily by balancing the intestinal microbiota, enhancing immune responses, and preventing the growth of harmful pathogens in the digestive tract (Sotoudegan et al., 2019). They are often incorporated into fermented foods like yogurt, kefir, and kimchi, as well as dietary supplements (Cunningham et al., 2021).

Prebiotics are compounds that are not digestible by human enzymes but serve as food for beneficial bacteria in the gut. These are typically fibers, such as fructooligosaccharides (FOS), galactooligosaccharides (GOS), and inulin, found naturally in foods like garlic, onions, bananas, and artichokes. By nourishing

Evaluating Oral Probiotics and Prebiotics. DOI: https://doi.org/10.1016/B978-0-443-29835-6.00016-2

beneficial bacteria, prebiotics promote a healthier balance of the microbiota, improve gut barrier function, and reduce inflammation (Dong et al., 2024).

Synbiotics are a combination of probiotics and prebiotics, working synergistically to enhance the survival and colonization of probiotic strains in the gut (Sotoudegan et al., 2019). The prebiotic component in synbiotics provides the necessary sustenance for the probiotics, ensuring their survival as they pass through the digestive system and allowing them to establish a beneficial microbiota balance more effectively (Sharma et al., 2024). The development of these products is driven by a variety of principles, including the need for efficacy, safety, and innovation in formulation (Terpou et al., 2019). However, the journey from initial research to market-ready products is fraught with challenges (Grumet et al., 2020). Manufacturers must navigate regulatory landscapes, ensure scientific validation, and overcome consumer skepticism while addressing emerging trends that prioritize personalization, sustainability, and transparency (Halder et al., 2021).

This chapter aims to explore the principles underlying probiotic, prebiotic, and synbiotic product development, delve into the challenges faced by industry stakeholders, and analyze current market dynamics and trends. By examining the interplay between advancements in scientific research, consumer demand, and regulatory frameworks, this study seeks to provide a comprehensive overview of the evolving landscape of gut health products. Ultimately, understanding these elements will illuminate the opportunities for growth and innovation in this rapidly expanding sector.

19.1.1 Importance in human health

Probiotics, prebiotics, and synbiotics play critical roles in maintaining and restoring gut health, which is central to overall well-being (Homayouni Rad et al., 2023). The human gut houses trillions of microorganisms, collectively known as the gut microbiota (Gomaa, 2020). These microbes are involved in various essential processes, including digestion, nutrient absorption, and immune regulation (Brestoff and Artis, 2013). However, factors such as poor diet, stress, antibiotic use, and environmental toxins can disrupt this balance, leading to a condition known as dysbiosis (Álvarez et al., 2021). Dysbiosis is associated with a variety of health problems, including gastrointestinal disorders like irritable bowel syndrome (IBS), inflammatory bowel disease (IBD), allergies, metabolic conditions like obesity, and even mental health issues such as anxiety and depression (Volarić et al., 2024). Probiotics help restore the balance of the gut microbiota by introducing beneficial bacteria that can outcompete pathogenic microbes, enhance mucosal barrier function, and modulate immune responses (Hosseinzadeh et al., 2025) (Fig. 19.1).

They are also linked to the prevention and management of specific infections, such as Clostridium difficile-associated diarrhea and urinary tract infections.

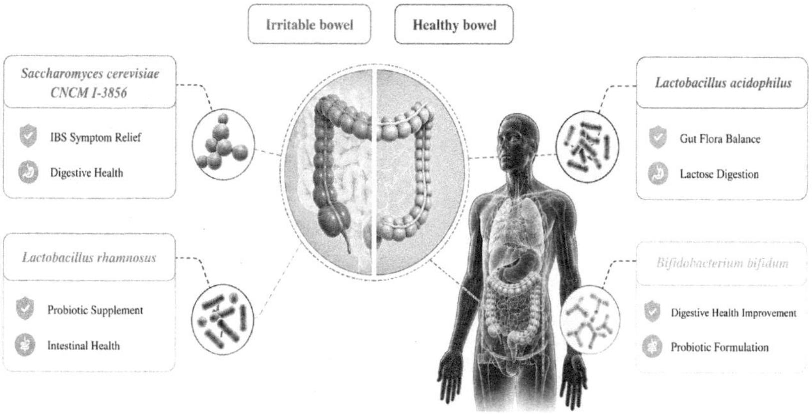

FIGURE 19.1 Probiotic strains and their role in alleviating IBS symptoms. *IBS*, Irritable bowel syndrome.

Prebiotics offer a different but complementary approach by selectively stimulating the growth of beneficial bacteria already present in the gut (Volarić et al., 2024). Prebiotics contribute to the production of short-chain fatty acids (SCFAs), which serve as fuel for the cells lining the colon and play a role in reducing inflammation, regulating metabolism, and maintaining a healthy gut environment. Synbiotics, by combining these two approaches, aim to provide a more comprehensive solution for maintaining gut health (Roy and Dhaneshwar, 2023) (Table 19.1). The prebiotic component ensures that the probiotic strains survive the acidic environment of the stomach and reach the intestines, where they can effectively colonize and perform their beneficial functions (Slavin, 2013). The benefits of probiotics, prebiotics, and synbiotics extend beyond gut health. Increasing evidence suggests that they can have systemic effects on immune function, metabolic health, and even brain function (Luo et al., 2024). This connection between the gut and other bodily systems, often referred to as the "gut-brain axis" or "gut-immune axis," underscores the importance of a balanced microbiota in maintaining overall health (Dinakis et al., 2024).

Probiotics, prebiotics, and synbiotics represent a significant advancement in understanding how to maintain and enhance health by managing the gut microbiota. Simultaneously, the market for these products is expanding rapidly, driven by key factors highlighting the growing interest in gut health and its essential role in overall well-being.

19.1.2 Consumer awareness and health trends

Consumer awareness regarding gut health has significantly evolved, particularly in the context of dietary choices. Recent research indicates that many individuals associate gut health with overall well-being, recognizing the

TABLE 19.1 Probiotic interventions across various health outcomes.

Topic	Analysis	Findings	References
Neuropsychiatric Outcomes	Using findings from 47 metaanalyses	Improvement in neuropsychological test results, depressive symptoms, mood states, and sleep quality; probiotics showed significant effects on enhancing BDNF levels	Du et al. (2024)
Probiotics' Effects on Anxiety and Depression	Clinical trials from 2014 to 2023 examining the impact of probiotics on anxiety and depression through the MGB axis.	Most recent studies indicate a beneficial role of probiotics in alleviating symptoms of anxiety and depression, though inconsistent findings exist; further targeted research is recommended to clarify mechanisms and treatment parameters	Merkouris et al. (2024)
Cardiometabolic Risk Factors	Evaluated the impact of probiotics, prebiotics, and synbiotics on cardiometabolic risk factors in children and adolescents with overweight	Multistrain probiotics reduced BMI, waist circumference, total cholesterol, triglycerides; synbiotics reduced LDL but slightly increased fasting glucose levels	Zhang et al. (2024)
Pediatric Obesity	Impact of gut microbiota on childhood obesity and associated metabolic complications	Certain probiotic strains like Lactobacillus rhamnosus and Bifidobacterium animalis reduced BMI and improved lipid profiles	Borka Balas et al. (2023)
IBD	Explores the therapeutic potential of prebiotics,	Probiotics suppressed pathogen growth and strengthened	Roy and Dhaneshwar (2023)

(Continued)

TABLE 19.1 (Continued)

Topic	Analysis	Findings	References
	probiotics, and synbiotics in managing IBD and restoring gut flora balance	intestinal barrier; prebiotics like inulin promoted beneficial gut microbiota changes	
IBS	Metaanalysis assessed the efficacy of probiotics, prebiotics, and synbiotics in treating IBS	Probiotics improved overall response rates and abdominal pain relief; specific genera like Bifidobacterium and Lactobacillus were particularly effective	Zhang et al. (2022), Zhang et al. (2023a)
ASD	This metaanalysis examined the potential benefits of probiotics, prebiotics, and synbiotics on ASD symptoms	No significant improvement in behavioral symptoms; changes in brain activity were associated with probiotic treatment	Rahim et al. (2023)
CKD	This research evaluated the effects of synbiotics, prebiotics, and probiotics on CKD patients, focusing on gut health	Interventions helped manage CKD symptoms by balancing gut microbiota, reducing uremic toxins, and improving kidney function	Zheng et al. (2021)

ASD, Autism spectrum disorder; *CKD*, chronic kidney disease; *IBD*, inflammatory bowel disease; *IBS*, irritable bowel syndrome; *MGB*, microbiota–gut–brain.

importance of a diet rich in diverse whole foods as recommended by national guidelines (Sharma et al., 2023). However, there remains a common misconception that specific foods, particularly fermented or probiotic options, are the primary means to achieve better gut health. This highlights a need for broader education on nutrition and food composition to align consumer perceptions with established dietary recommendations (Jiao et al., 2024).

The qualitative nature of studies exploring consumer perspectives reveals that while there is a general interest in gut health, this does not always translate into informed dietary behavior (Swann et al., 2020). In a study published by Williams and Tapsell in 2023, Participants often expressed divergent views on how diet influences gut health, and many acknowledged gastrointestinal symptoms as a factor in their dietary choices. Interestingly, despite a recognition of scientific evidence, this awareness did not necessarily lead to changes in behavior, suggesting a disconnect between knowledge and action (Williams et al., 2023).

Moreover, the commercialization of gut health products has contributed to consumer confusion, as mixed messages from food labels and advertising can obscure the true health benefits of these products (Eden, 2011). Participants in the study noted that while some food label claims increased their likelihood of purchase, others viewed them skeptically, perceiving them as mere marketing tactics. This underscores the importance of clear communication and education from health professionals to help consumers navigate the complexities of gut health and make informed dietary choices.

19.1.3 Types and sources of probiotics

Probiotics are primarily classified into various types based on their microbial origin, with the most common being bacteria, yeast, and molds (Latif et al., 2023). The majority of probiotics belong to the lactic acid bacteria group, particularly from the genera Lactobacillus and Bifidobacterium (Hosseinzadeh et al., 2025). These microorganisms are known for their beneficial effects on human health, especially in improving gut health and enhancing the immune response (Asqardokht-Aliabadi et al., 2025). Additionally, some yeasts and molds have also been identified to possess probiotic properties, expanding the diversity of probiotics available for health applications (Kamalledin Moghadam et al., 2024).

The sources of probiotics are diverse and can be categorized into three main groups: human and animal sources, food sources, and environmental sources (Mazziotta et al., 2023). Human gastrointestinal tracts, particularly the intestines and stomach, are significant reservoirs of probiotics, having been extensively studied for their probiotic content (George Kerry et al., 2018). Animal sources include the gastrointestinal tracts of various animals such as cows, pigs, and chickens, as well as animal milk (Bomba et al., 2012). Food sources are particularly rich in probiotics, including raw foods like milk and dairy products, fermented foods such as kimchi and pickles, and a variety of fruits and vegetables. In addition to human and animal sources, probiotics can also be found in the environment, including water, air, and soil (Rodzi and Lee, 2021). This environmental presence highlights the natural occurrence of probiotics beyond just dietary sources. Environmental probiotics and dietary probiotics differ primarily in their sources and the contexts in which they are found (George Kerry et al., 2018). Environmental probiotics are naturally occurring microorganisms found in various habitats such as soil, water, and air. For instance, specific strains like *Bacillus* spp. and *Enterobacter* sp. can be isolated from soil, while coastal waters may contain *Pseudoalteromonas mariniglutinosa* and *Bacillus albus* (Navarro-Torre et al., 2020). These probiotics are often less studied in terms of their health benefits compared to dietary sources, and their roles in human health are not as well established.

On the other hand, dietary probiotics are specifically sourced from foods and beverages that are consumed for their health benefits (Luo et al., 2024). Common dietary sources include fermented foods like kimchi, pickles, and dairy products, which are rich in beneficial bacteria such as *Lactobacillus acidophilus* and *Lactobacillus plantarum* (Behera et al., 2020). These probiotics are often selected for their ability to survive the gastrointestinal tract and confer health benefits, such as improving gut health and enhancing the immune response (Latif et al., 2023). The selection of dietary probiotics is crucial, as they are intended for direct consumption and must meet specific criteria for efficacy and safety.

In summary, while environmental probiotics are found in natural settings and may have potential health benefits, dietary probiotics are specifically cultivated and consumed for their proven health effects (Agarbati et al., 2020). The distinction between these two sources is important for understanding their applications in health and nutrition, as well as their varying levels of research and efficacy. The selection of probiotics for health benefits is crucial, as their efficacy can vary significantly depending on the strain and source (Table 19.2). Therefore understanding the types and sources of probiotics is essential for their effective application in health and nutrition.

19.1.4 Health benefits of probiotics

The effects of probiotics can be categorized into three main modes of action (Plaza-Diaz et al., 2019).

First, probiotics have the ability to modulate the host's immune system, influencing both the innate and acquired immune responses. This immunomodulatory capacity is particularly important for the prevention and treatment of infectious diseases, as well as for managing chronic digestive tract inflammation. Additionally, this probiotic mechanism may play a role in eradicating neoplastic host cells (Azad et al., 2018).

Second, probiotics can directly interact with and influence other microorganisms, both commensal and pathogenic. This principle is crucial for preventing and treating infections, as well as for restoring the microbial equilibrium within the gut (Raheem et al., 2021).

Finally, probiotic effects may stem from actions that target microbial products, such as toxins, as well as host-derived compounds like bile salts and food ingredients. These actions may result in the inactivation of toxins and the detoxification of host and food components within the gut (Yan Zhang et al., 2023b).

While all three modes of probiotic action are likely involved in defending against infections, preventing cancer, and stabilizing or reconstituting the

TABLE 19.2 Comprehensive comparison of probiotics, prebiotics, and synbiotics.

Characteristic	Probiotics	Prebiotics	Synbiotics
Definition	Live microorganisms which when administered in adequate amounts confer a health benefit on the host (Parhi et al., 2024; Venkatesh et al., 2024)	Nondigestible ingredients that promote beneficial bacteria growth in the colon, improving health (Gomez Quintero et al., 2022)	A synbiotic is a combination of probiotics and prebiotics that enhances host health by improving the survival, implantation, and metabolic activity of beneficial gut bacteria (Parhi et al., 2024; Gomez Quintero et al., 2022).
Details	Probiotic preparations often contain dead or damaged microorganisms, particularly near the end of shelf life. A minimum concentration of 8 log CFU/g or mL is recommended for effective consumption (Parhi et al., 2024; Pimentel et al., 2021).	Prebiotics are not hydrolyzed or absorbed in the upper part of the gastrointestinal tract, but they are fermented by the intestinal microflora in the colon (Bamigbade et al., 2022)	Synbiotics are fermented in the colon, producing SCFAs such as acetate, propionate, and butyrate
Examples	*Lactobacillus*, *Bifidobacterium*, *Streptococcus*, *Enterococcus*, *Bacillus*, and yeasts (Parhi et al., 2024)	Inulin, FOS, GOS, lactulose, XOS, AXOS, HMO, and resistant starch (Bamigbade et al., 2022)	*Bifidobacterium longum* with Synergy for ulcerative colitis, *Bifidobacterium infantis* EVC001 with HMO for severe acute malnutrition, and a multistrain synbiotic with inulin for type 2 diabetes
Strains	Some commercially available strains are *Lactobacillus acidophilus* LA-5, *Lactobacillus casei* Shirota, *Bifidobacterium lactis* BB-12, *Saccharomyces boulardii* (Parhi et al., 2024)	Not applicable	Selecting compatible prebiotics and probiotics with proven health benefits and ensuring synergistic effects are crucial (Gomez Quintero et al., 2022)

Selection criteria	Safety, functionality, and technological applicability. Origin, viability, stability, and compatibility with the food matrix. Stability of phenotypes and genotypes, tolerance to bile and acid, adhesion characteristics of intestinal epithelial cells, antimicrobial compound production (Venkatesh et al., 2024).	Selective stimulation of beneficial bacteria, nondigestibility, fermentation by gut bacteria, and a dose-dependent effect (Parhi et al., 2024)	Compatibility and synergism, health benefits, safety and functionality, ecological considerations, technological suitability, clinical evidence, and dosage and delivery (Kiepś and Dembczyński, 2022)
Viability	Maintaining viability during processing, storage, and passage through the digestive system is crucial for probiotic efficacy (Venkatesh et al., 2024)	Not applicable	Maintaining viability and stability during manufacturing, storage, and passage through the digestive system is critical for efficacy (Pimentel et al., 2021)
Assessment of viability	Plate count method, flow cytometry, and viability staining techniques are commonly used (Kiepś and Dembczyński, 2022)	Not applicable	Assessing the viability of the probiotic component in a synbiotic product is plate counting, flow cytometry, scanning electron microscopy, laser diffraction, two-dimensional gel electrophoresis, and gene expression analysis (Kiepś and Dembczyński, 2022)
Production methods	Fermentation is the primary method for producing probiotic cultures. Common fermentation parameters include temperature, pH, agitation, and oxygen levels. Other parameters are the media	Extraction from natural sources or enzymatic synthesis (Buitron et al., 2018)	Involve integrating the production processes of both components (probiotics and prebiotics), ensuring their compatibility and stability in the final product.

(Continued)

TABLE 19.2 (Continued)

Characteristic	Probiotics	Prebiotics	Synbiotics
	composition and the incubation time (Pimentel et al., 2021).		The specific production steps for synbiotics would depend on the chosen prebiotic and probiotic, the desired formulation (e.g., capsules, powders, food products), and the intended application (Dixit et al., 2016; González-Herrera et al., 2021).
Assessment of efficiency	In vitro and in vivo studies are conducted to evaluate the health benefits of probiotic strains. Clinical trials assess the efficacy of probiotics for specific health conditions (Cheon et al., 2023).	In vitro and in vivo studies are used to assess the prebiotic effects (Du et al.; 2023; Śliżewska and Chlebicz-Wójcik, 2020)	The efficacy of synbiotics is evaluated through clinical trials and studies focusing on the synergistic interaction between prebiotics and probiotics (Song et al., 2022; Hadi et al., 2020)
Formulation criteria	Selecting strains with proven efficacy and safety, optimizing dosage, and ensuring stability during storage and consumption are crucial	Selecting substrates that are nondigestible, fermentable by beneficial bacteria, and safe for consumption. Optimizing prebiotic concentration for desired effects (Bamigbade et al., 2022; Rahimi et al., 2022)	Selecting compatible prebiotics and probiotics with proven health benefits, ensuring a synergistic effect, and considering the ecological constraints of the gut environment (Parhi et al., 2024; Gomez Quintero et al., 2022; Sommermeyer et al., 2024)
Product development	strain selection, substrate compatibility, formulation stability, dosage optimization, safety assessments, regulatory compliance, and sensory aspects (Parhi et al., 2024)	Characterization, purification, and incorporation into food products while preserving stability and efficacy (Bamigbade et al., 2022; Buitron et al., 2018)	Optimizing the prebiotic and probiotic combination for synergistic effects. Developing products with acceptable taste, texture, and aroma (Parhi et al., 2024)

Product type	Foods: Fermented dairy products (yogurt, kefir), fermented vegetables, soy products, beverages Dietary supplements: Capsules, tablets, powders, liquids (Parhi et al., 2024; Pimentel et al., 2021)	Foods: Fruits, vegetables, grains, legumes Dietary supplements: Powders, capsules, tablets (Bamigbade et al., 2022; Cruz et al., 2010)	Functional foods: Yogurt, fermented beverages, cereal products Dietary supplements: Capsules, tablets, powders (Anđela et al., 2024; Rodionova et al., 2021)
Key challenges	Maintaining viability and stability during processing, storage, and digestion. Scientific validation and regulatory hurdles. Consumer education and awareness (Parhi et al., 2024).	Ensuring selective fermentation by beneficial bacteria, optimizing dosage, and addressing potential side effects like gas and bloating. Sensory challenges, especially related to taste and texture, can impact consumer acceptance. Standardization and quality control of prebiotic ingredients can be difficult (Bamigbade et al., 2022).	Ensuring synergistic interaction between prebiotics and probiotics, maintaining stability and viability of both components. Scientific validation of synbiotic combinations. Consumer education about the benefits and proper use of synbiotics (Parhi et al., 2024).
Market trends	Rising health consciousness and demand for functional foods. Growing interest in personalized nutrition solutions and sustainable products. Focus on scientific validation and regulatory compliance (Pimentel et al., 2021).	Increasing demand for natural and plant-based prebiotic sources. Incorporation of prebiotics into a wider range of food products. Focus on specific health benefits and targeted prebiotic development (Buitron et al., 2018).	Expanding applications in various food and beverage categories. Development of personalized synbiotic formulations. Growing research on the gut microbiome and its impact on health (Dixit et al., 2016).
Commercially available	Pure encapsulations Probiotic G.I: Shelf-stable probiotic blend, supports immune function in the G.I. tract. FLORASSIST GI with Phage Technology: Proprietary blend targeting harmful	Commonly found in dietary supplements, functional foods, and beverages Hyperbiotics organic prebiotic powder: Food-based prebiotic blend for gut health and immune support, in the form	Increasingly incorporated into functional foods, beverages, and dietary supplements DS-01 Daily Synbiotic: 53.6 billion AFU of various probiotics (e.g., *Bifidobacterium longum*,

(*Continued*)

TABLE 19.2 (Continued)

Characteristic	Probiotics	Prebiotics	Synbiotics
	bacteria, promotes healthy probiotic growth. Garden of Life Dr. Formulated Probiotics Once Daily Ultra: 90 billion CFU per capsule, 15 different strains, resistant to stomach acid. Member's Mark 10 Strain Probiotic: 10 billion active cultures, includes 10 strains of beneficial bacteria for digestive health.Thorne FloraMend Prime Probiotic: 5 billion CFUs per capsule, vegetarian, supports gut health and immune function.	of powder: Organic prebiotic blend, vegan, non-GMO Jarrow formulas prebiotic inulin FOS: Soluble fibers for gut health and metabolism support, in the form of powder: Inulin, fructooligosaccharides (FOS) Benefiber prebiotic fiber supplement powder: Daily fiber supplement to nourish gut bacteria, in the form of Powder: Wheat dextrin Dr. Tobias Prebiotics Digestive Support: Supports digestion and feeds probiotic bacteria, in the form of capsules: Advanced prebiotic technology pure encapsulations poly-prebiotic powder: Blend of prebiotic fibers and polyphenols for gut health in the form of Powder: Various prebiotic fibers, polyphenols	*Lactiplantibacillus plantarum*) and prebiotics (punicalagins from pomegranate) in the form of Capsules for Gut health, immune support, digestive regularity *Lactobacillus rhamnosus* GG + Inulin: *Lactobacillus rhamnosus* GG and inulin in the form of Capsules/Powder for Digestive health, improved gut flora *Bifidobacteria* + Fructooligosaccharides (FOS): *Bifidobacteria* and FOS in the form of capsules which supports gut microbiota, enhances digestion Synbiotic Health Products: Various combinations of probiotics and prebiotics in various forms (Capsules, Powders) for General health, digestive support DSM-Firmenich Synbiotics: Human milk oligosaccharides and specific probiotic strains in the form of various (capsules, powders) for early life nutrition, gut health

AXOS, Arabinoxylooligosaccharides; *FOS*, fructooligosaccharides; *GOS*, galactooligosaccharides; *HMO*, human milk oligosaccharides; *SCFAs*, short-chain fatty acids; *XOS*, xylooligosaccharides.

physiological balance between the intestinal microbiota and the host, it is important to note that not all probiotic strains exhibit all three principles to the same extent (Maleki et al., 2016). The specific metabolic properties, surface molecules, and secreted components of a probiotic strain will determine which of these mechanisms are its primary modes of action (Rad, et al., 2021a). The advent of "omics" technologies, such as genomics, transcriptomics, proteomics, and secretomics, has enabled the comprehensive analysis and comparison of probiotic characteristics, which can help identify and distinguish true probiotics from those that may not be as effective (O'donnell et al., 2020).

The consumption of probiotics has been associated with several health benefits, particularly in relation to digestive issues and immune function (Yousefi et al., 2019). Probiotics play a significant role in enhancing immune response (Rad et al., 2021b). Colares et al. (2024) highlighted the dual benefits of *Lactiplantibacillus plantarum* as a probiotic strain and antioxidant enhancer. The integration of this strain into a sorbet matrix opens pathways for innovative functional food development. They can stimulate the production of antibodies and modulate the activity of immune cells, which helps in reducing the risk of infections (Colares et al., 2024). This immune enhancement is crucial for maintaining overall health and preventing various diseases (Shao et al., 2021). Beyond digestive health, probiotics contribute to the prevention and treatment of infectious diseases and may even have a role in cancer prevention (Legesse Bedada et al., 2020). Probiotic supplementation has been reported to alleviate side effects of chemotherapy, such as diarrhea and mucositis, by restoring gut microbiota (Homayouni Rad et al., 2023).

19.1.5 Probiotics and immune modulation

Therefore probiotics can influence the immune system through various mechanisms, including their metabolites, cell wall components, and DNA (Hosseinzadeh et al., 2025). Even dead probiotic bacteria or their derived components, such as peptidoglycan fragments or DNA, can elicit immune-modulatory effects (Mehta et al., 2023). These probiotic products are recognized by host cells equipped with specialized receptors. The primary target cells are the gut epithelial cells and gut-associated immune cells. The direct interaction and adhesion of probiotics to host epithelial cells can trigger signaling cascades that lead to immune modulation (Gorreja and Walker, 2022). Alternatively, the release of soluble factors by probiotics can activate signaling pathways in immune cells or epithelial cells, subsequently affecting immune responses.

Probiotics, such as *Lactobacillus rhamnosus*, have been shown to induce a hyporesponsive state in T cells, characterized by decreased proliferation and cytokine production (Gao et al., 2021; Ludwig et al., 2018). This effect is mediated, in part, through the interaction of probiotics with dendritic cells, which can then modulate T cell responses. In addition, probiotics even influence the activity of natural killer cells, as shown with *Lactobacillus casei Shirota* (Seifert et al., 2011).

Specific probiotic components, like the teichoic acid of Lactobacillus plantarum, can also contribute to antiinflammatory activities in a Toll-like receptor (TLR)-dependent manner (Ren et al., 2020; Smelt et al., 2013). Additionally, probiotics protect the integrity of the mucosal gut barrier through TLR-independent mechanisms, such as modulating protein kinase C signaling and enhancing the expression of tight junction proteins (Rose et al., 2021).

Probiotics also directly influence the production of cytokines by modulating cellular signal transduction pathways, such as inhibiting the degradation of the NF-κB inhibitor IκB or interfering with proteasome function (Guo et al., 2024). Furthermore, probiotics promote intestinal epithelial cell survival and growth, as demonstrated by the two soluble proteins from *Lactobacillus rhamnosus GG* (Yan et al., 2007).

Probiotics produce a variety of antimicrobial substances that can directly inhibit the replication of pathogens. Key among these are SCFAs, such as lactic acid, and hydrogen peroxide, which exhibit broad-spectrum antimicrobial activity (Denkova et al., 2017). Probiotics also produce bacteriocins, which are low-molecular-weight antimicrobial peptides. These bacteriocins can be classified into different groups, including antibiotics, heat-stable nonantibiotics, and cyclic antimicrobial peptides (Solis-Balandra and Sanchez-Salas, 2024). The bacteriocin Abp118 produced by *Lactobacillus salivarius* has been shown to protect mice against infection by the pathogen *Listeria monocytogenes*(Riboulet-Bisson et al., 2012). Additionally, probiotics produce certain antibiotics, such as the broad-spectrum antibiotic reuterin produced by *Lactobacillus reuteri* (Sagheddu et al., 2020). Probiotics also compete with pathogens for limiting resources, such as iron, which is essential for the growth of most bacteria, but not for lactobacilli (Rabetafika et al., 2023). Probiotics like *Escherichia coli Nissle* 1917 (EcN) can effectively compete for iron by secreting siderophores and expressing multiple iron uptake systems (Tsylents et al., 2024). Their ability to balance gut microbiota is essential for maintaining a healthy digestive system and supporting immune function. In conclusion, the health benefits of probiotics are extensive, particularly in improving digestive issues and enhancing immune function, making them a valuable addition to a balanced diet.

19.2 Types, sources, and health benefits of prebiotics

The concept of prebiotics has emerged more recently, with the term first being defined about three decades ago (Gibson and Roberfroid, 1995). Prebiotics are nondigestible food ingredients that selectively stimulate the growth and activity of beneficial gut bacteria, including probiotic strains (Gibson et al., 2017). This selective promotion of desirable intestinal microbes is a key aspect of the prebiotic approach.

Prebiotics play a critical role in supporting the growth and proliferation of beneficial gut bacteria. They serve as selective nutrient sources that not only nourish favorable microorganisms but also inhibit the growth of potentially harmful bacteria. This selective stimulation leads to an optimized microbial balance characterized by an increase in beneficial strains such as *Bifidobacterium* and *Lactobacillus*, while suppressing pathogenic species like *Clostridia*, *Bacteroides*, and *Escherichia coli* (You et al., 2022). This change in microbial composition is associated with improved gut health and immunity, stress resilience, and overall metabolic well-being.

Fructans, including inulin, are among the most studied prebiotics known for their ability to enhance microbial diversity (Birkeland et al., 2020). They promote the growth of beneficial bacteria like *Bifidobacterium* and *Lactobacillus*, which are essential for a balanced gut microbiome. Galactooligosaccharides (GOS) are effective in increasing the abundance of beneficial microbes (Du et al., 2024). Studies have shown that GOS can significantly enhance the diversity of gut bacteria, contributing to improved gut health and function. Xylooligosaccharides, as a prebiotic, also play a role in promoting microbial diversity (Wang et al., 2022). They are fermented by gut bacteria, leading to the production of SCFAs that support the growth of beneficial microbial populations (Milani et al., 2023). Derived from chitin, chitooligosaccharides have been shown to positively influence gut microbiota composition, enhancing the diversity of beneficial bacteria (Chen et al., 2022). This type of prebiotic is not digested in the small intestine and reaches the colon, where it is fermented by gut bacteria (Tan et al., 2021). This fermentation process helps increase microbial diversity and supports the growth of beneficial bacteria. Gums are a type of soluble fiber that can serve as effective prebiotics, contributing to gut health by promoting the growth of beneficial bacteria (Birkeland et al., 2020). These natural polysaccharides are derived from various plant sources and are commonly used in food products for their thickening and stabilizing properties (Goh et al., 2022). Gums such as guar gum, xanthan gum, and gum arabic are notable examples that not only enhance the texture of foods but also provide health benefits by acting as prebiotics (Gajbhiye et al., 2024). Guar gum, extracted from the seeds of the guar plant, is known for its ability to improve digestion and increase the viscosity of intestinal contents (Chen et al., 2022). This can help slow down the absorption of nutrients, leading to better blood sugar control and improved satiety (Furtado et al., 2022). Xanthan gum, produced through the fermentation of sugars by the *bacterium Xanthomonas* campestris, is often used in gluten-free products and can also support gut health by fostering a favorable environment for beneficial gut bacteria (Gajbhiye et al., 2024). Gum arabic, derived from the sap of acacia trees, is another prebiotic that can enhance gut microbiota composition and improve digestive health.

Incorporating gums into the diet can provide significant health benefits, particularly in enhancing gut microbiota diversity and function. As prebiotics, they can stimulate the production of SCFAs, which are crucial for maintaining gut health and reducing inflammation. In conclusion, prebiotics such as fructans, GOS, xylooligosaccharides, chitooligosaccharides, and resistant starch are effective in enhancing microbial diversity, which is crucial for maintaining a healthy gut microbiome and overall health. In conclusion, prebiotics such as fructans, GOS, xylooligosaccharides, chitooligosaccharides, and resistant starch are effective in enhancing microbial diversity, which is crucial for maintaining a healthy gut microbiome and overall health. The fermentation of prebiotics by gut bacteria (Bozyigit et al., 2021), produce SCFAs like acetate, propionate, and butyrate. SCFAs serve as an energy source for colon cells, help maintain luminal pH, inhibit pathogen growth, and may reduce the risk of colorectal cancer (Alvandi et al., 2022). Evidence indicates that prebiotic intake may positively affect mood and cognitive function, showcasing potential benefits in conditions such as anxiety and depression. SCFAs produced from prebiotic fermentation can influence neuroinflammation, neurotransmitter signaling, and even gut-derived neuropeptide release, impacting mental health and cognitive performance (Ince Palamutoglu et al., 2024). Moreover, prebiotics have been associated with improved metabolic health, potentially aiding in weight management, insulin sensitivity, and glucose metabolism. Systemic effects through metabolic pathways may provide a deeper understanding of how gut health correlates with systemic health (Cerdó et al., 2019). Prebiotics enhance immune function by increasing the population of beneficial bacteria, which can stimulate the immune response (Ashaolu, 2020). For instance, inulin has been shown to upregulate Th2-related immune genes while suppressing pro-inflammatory genes, contributing to a balanced immune response (Hughes et al., 2022). Prebiotics help strengthen the intestinal barrier, reducing permeability and preventing the translocation of harmful bacteria and toxins into the bloodstream. This protective effect is vital for preventing inflammatory conditions. By modulating gut microbiota composition and increasing SCFA production, prebiotics can lower the levels of pro-inflammatory cytokines, thereby reducing inflammation in the gut and potentially alleviating conditions like IBD (Martyniak et al., 2021) (Fig. 19.2).

Prebiotics are not just influential in microbial dynamics; they also play a role in enhancing gut motility and promoting regular bowel movements (Homayoni Rad et al., 2012). The fermentation of prebiotics contributes to the increased production of SCFAs, which can stimulate colonic motility and fluid secretion (Ashaolu et al., 2021). This mechanism supports regularity and alleviates conditions like constipation (Homayouni Rad et al., 2013). Regular consumption of prebiotics has been linked to improved bowel function, reducing symptoms associated with constipation (Rau et al., 2024). The market for gums as prebiotics is growing, similar to other prebiotics, reflecting a rising interest in functional foods (Halder et al., 2021). However, the specific applications and consumer preferences may vary based on the unique properties of each prebiotic type.

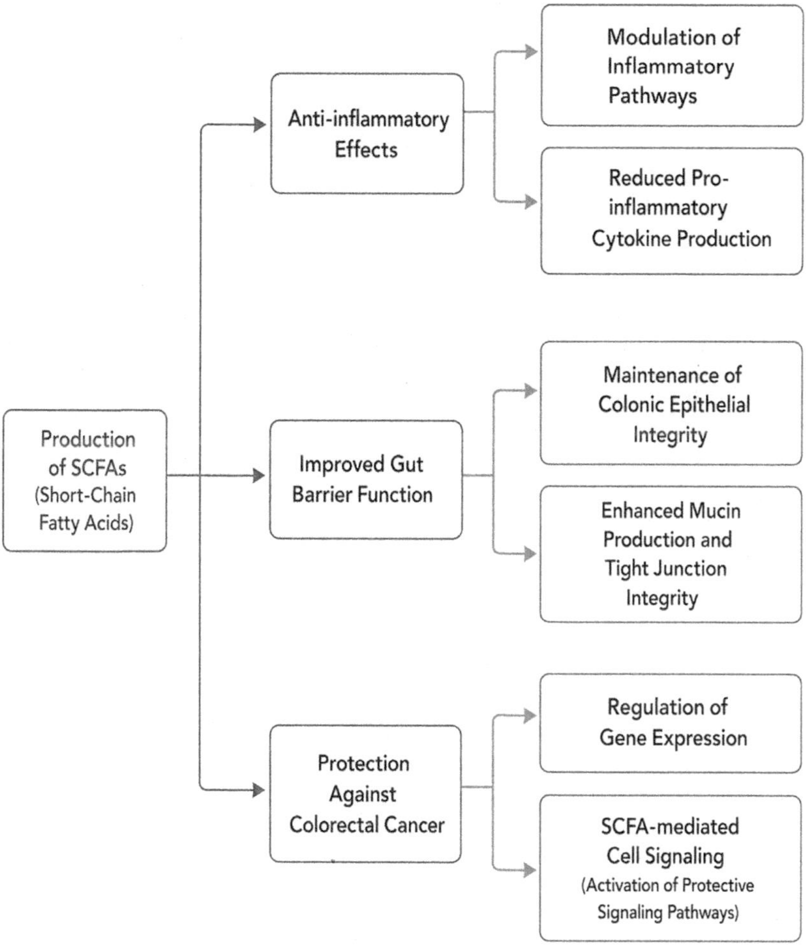

FIGURE 19.2 The multifaceted role of SCFAs in gut health and disease prevention. *SCFAs*, Short-chain fatty acids.

19.3 Synbiotics: a synergistic approach

The global synbiotics market is poised for sustained growth, driven by increasing consumer awareness of gut health, rising demand for functional foods, and significant innovation in biotechnology (Mounir et al., 2022). Synbiotics, combining probiotics and prebiotics, have demonstrated their unique ability to enhance the survival and activity of beneficial microorganisms in the digestive tract, offering targeted health benefits (Dahiya and Nigam, 2022). The synergistic effects of synbiotics are particularly relevant for clinical applications in gut disorders, including, but not limited to, inflammatory bowel disease (IBD), irritable bowel syndrome (IBS), and antibiotic-associated diarrhea. By simultaneously restoring microbial equilibrium and fostering beneficial bacteria, synbiotics offer a promising adjunctive strategy to traditional therapeutic approaches (Markowiak and Śliżewska, 2017). This convergence of factors is shaping market dynamics across key regions like North America, Europe, and the Asia-Pacific, while opening up future opportunities for innovation, personalized nutrition, and consumer education.

19.3.1 North American market

The North American synbiotics market is expected to register steady growth, fueled by rising consumer demand for functional foods and dietary supplements that support gut health and overall well-being. Changes in lifestyle, eating habits, and an increasing focus on preventive healthcare are key drivers of this trend. Major companies in North America are investing in the development of advanced nutritional supplements incorporating prebiotics, probiotics, and synbiotics, offering enhanced nutritional value to health-conscious consumers. As the prevalence of digestive disorders continues to rise, synbiotics present an attractive solution for addressing these issues, driving market growth.

19.3.2 European market

Europe accounted for the largest revenue share in the global synbiotics market in 2021, with high visibility and availability of synbiotic products playing a key role (Halder et al., 2021). Consumers in this region have become increasingly aware of the benefits of synbiotics, particularly their role in improving gut health, boosting immunity, and supporting overall wellness (Rodionova et al., 2021). Rising disposable income and a growing preference for nutritional supplements incorporated into daily diets are further boosting demand for functional foods and beverages in Europe (Raji et al., 2024). Preventive healthcare trends, coupled with a well-established retail infrastructure, make Europe a crucial player in the global synbiotics market (Kearney and Gibbons, 2018).

19.3.3 Asia-Pacific market

The Asia-Pacific region is experiencing rapid expansion in the synbiotics market, driven by the growing dairy industry in countries like China, India, and New Zealand (Bamola et al., 2013). Producers are increasingly fortifying dairy products with synbiotics to enhance their nutritional profile, catering to the rising health consciousness among consumers (Raman et al., 2019). Additionally, changing dietary habits, rising disposable incomes, and urbanization are all contributing to the growth of the synbiotics market in this region (Rodionova et al., 2021). As the middle class expands, so does spending on health and wellness products, further propelling market expansion (Mishra et al., 2024). The pet food industry in Asia-Pacific is also adopting synbiotics, promoting digestive health in pets and contributing to overall market growth.

19.3.4 Global market overview and projections

On a global scale, the synbiotics market reached USD 817.2 million in 2021 and is expected to grow to USD 1.67 billion by 2030, registering a compound annual growth rate (CAGR) of 8.3% (Huynh et al., 2017). This growth is particularly pronounced in regions with a high level of health consciousness, such as North America and Europe, where functional foods and dietary supplements account for the largest market share (Liu et al., 2024). As consumer interest in gut health and functional foods continues to rise, synbiotics are expected to play an increasingly important role in global dietary trends (Liu et al., 2024).

19.4 Future opportunities for innovation

The future of the synbiotics market offers significant opportunities for innovation, driven by advances in biotechnology and a deeper understanding of the human microbiome (Grover et al., 2012). One of the key areas of innovation lies in the development of targeted probiotic strains that address specific health conditions (Praveen and Brogi, 2025). For example, research is focusing on creating probiotics tailored for conditions like irritable bowel syndrome (IBS), inflammatory bowel disease (IBD), and even mental health issues such as anxiety and depression (Li et al., 2025). These targeted probiotics could provide more precise treatments, enhancing the therapeutic potential of synbiotics (de Leon-Derecho, 2024). The concept of personalized nutrition is also gaining traction, where synbiotics can be customized to an individual's genetic makeup and microbiome profile (Li et al., 2021). This emerging trend holds significant potential, as personalized solutions offer consumers tailored health benefits that align with their unique needs (Milani et al., 2023).

Additionally, the broader incorporation of prebiotics into everyday food products is a promising area for market expansion (Hosseinzadeh et al., 2025). Beyond traditional supplements, prebiotics are increasingly being added to common foods such as bread, cereals, and snacks, making it easier for consumers to incorporate these health-enhancing ingredients into their daily diets (Asqardokht-Aliabadi et al., 2025). The beverage industry is also exploring ways to include prebiotics in functional beverages like smoothies, juices, and even coffee, offering consumers convenient and enjoyable options to boost their gut health (Kellershohn, 2021).

19.4.1 Education and outreach

As the synbiotics market evolves, effective education and outreach will be essential for driving consumer adoption and maximizing market growth (Lieberman, 2020). Consumer awareness campaigns can highlight the scientifically proven health benefits of synbiotics, such as improved digestion, enhanced immune function, and mental health support. These campaigns can leverage various media platforms to reach a broad audience and foster greater consumer interest (Siddiqui et al., 2023). Equally important is educating consumers on the proper usage of synbiotics, including dosage, timing, and potential interactions with other supplements or medications, to ensure safety and efficacy (Pop et al., 2020). On the professional front, training healthcare providers about the benefits of synbiotics can lead to more informed recommendations and prescriptions, significantly boosting consumer confidence (Pires et al., 2024). Similarly, educating retail staff in specialty stores and pharmacies about synbiotics can improve customer service and satisfaction, as knowledgeable staff can provide valuable advice and guidance (Mcmillin, 2017).

Community engagement is another vital aspect of education. Workshops, seminars, and webinars focused on gut health and synbiotics can engage and educate communities, while collaborations with schools and universities can raise awareness among younger generations. These initiatives not only promote understanding but also cultivate long-term healthy habits that support market growth (Rausch and Kopplin, 2021). The synbiotics market is on a promising trajectory, fueled by advances in biotechnology, personalized nutrition, and the incorporation of synbiotics into a broader array of food and beverage products (Mounir et al., 2022; Zambelli and Mendonça, 2024). As the science behind gut health deepens, the potential for targeted and more effective synbiotic solutions grows (Markowiak and Śliżewska, 2017). These developments, coupled with comprehensive education and outreach, will help increase consumer awareness and trust, driving wider adoption of synbiotic products (Roy and Dhaneshwar, 2023). Companies that seize these opportunities, focusing on innovation and informed consumer engagement, will be well-positioned to meet the rising demand for gut health solutions (Zheng et al., 2021). On a broader scale, the entire probiotics, prebiotics, and synbiotics market is experiencing robust growth, propelled by rising health consciousness and scientific progress (Zambelli and Mendonça, 2024). This convergence of consumer interest and research creates significant opportunities for stakeholders, manufacturers, marketers, and healthcare professionals alike (Raji et al., 2024). Continuous research and effective consumer education will be essential in shaping perceptions and encouraging the widespread use of these products. As this field continues to evolve, those who invest in both innovation and education will lead the charge in promoting overall wellness and meeting the growing demand for gut health solutions.

19.4.2 Trend

The landscape of probiotic, prebiotic, and synbiotic product development is evolving rapidly, driven by growing consumer interest in gut health and its overall impact on well-being (Dahiya and Nigam, 2022). Recent trends reflect the convergence of scientific advancements and consumer demand, creating significant opportunities for innovation and personalized healthcare solutions (Halder et al., 2021).

One of the most transformative developments in this field is the emergence of next-generation probiotics (NGPs), which are poised to play a crucial role in personalized therapies, especially in cancer treatment and chronic disease management (Al-Akayleh et al., 2024). These engineered probiotics are designed to enhance therapeutic efficacy by modulating the gut microbiome and immune responses (Han et al., 2024). For instance, specific strains like *Akkermansia muciniphila* have demonstrated the ability to boost gastrointestinal immunity and alleviate chemotherapy complications (Al-Fakhrany and Elekhnawy, 2024; Teng et al., 2024). Furthermore, NGPs are being explored as targeted treatments for chronic diseases, where their ability to deliver bioactive compounds and reinforce gut barrier integrity has shown promise in improving patient outcomes (Murali and Mansell, 2024). The concept of precision probiotics is gaining momentum, facilitated by cutting-edge technologies like metaproteomics and artificial intelligence (AI), which enable the identification of probiotic strains tailored to individual health needs (Das et al., 2023). This opens the door to personalized healthcare solutions, where probiotics can be customized based on an individual's microbiome profile, offering more effective treatments for conditions such as irritable bowel syndrome (IBS) and inflammatory bowel disease (IBD) (Westfall et al., 2021). Alongside these scientific advancements, innovative delivery formats are reshaping the consumer experience (Murali and Mansell, 2024). Probiotics are now being integrated into a broader range of functional foods and beverages, including yogurts, smoothies, and even nondairy alternatives, making it easier for consumers to incorporate gut health solutions into their daily diets (Grumet et al., 2020; D'amico et al., 2024). Additionally, new formats like chewable tablets, gummy supplements, and ready-to-mix powders offer convenience and variety, appealing to a wide range of consumer preferences (Dixit et al., 2016; Vojvodić Cebin et al., 2024).

Another important trend is the rise of synergistic formulations, particularly the increased development of synbiotic products that combine both probiotics and prebiotics to maximize gut health benefits (Rodionova et al., 2021). Brands are increasingly recognizing the advantages of pairing these ingredients to improve digestive health, and many are incorporating whole-food sources of prebiotics, such as chicory root and apple cider vinegar, to attract consumers seeking natural solutions (Tsoupras et al., 2024). This trend toward natural, whole-food ingredients aligns with the growing demand for Sustainability and

Clean Labeling, where consumers are favoring products with minimal, recognizable ingredients and sustainable sourcing practices (Alam et al., 2025).

Incorporating functional ingredients such as microbiome-enhancing compounds like algae, seaweed, and polyphenols is another growing area of interest (Halder et al., 2021). These ingredients are being explored as novel prebiotic sources that offer additional health benefits beyond traditional prebiotics (Song et al., 2022). In tandem with this, many probiotic formulations are now being fortified with vitamins and minerals, such as vitamin D for immune support or zinc for gut barrier function, enhancing their overall health benefits (Pannerchelvan et al., 2024). However, as this market grows, so does the need for research and regulation. Companies are investing in rigorous clinical trials to validate the health claims associated with their products (Krychtiuk et al., 2024). This not only builds consumer trust but also justifies premium pricing. Regulatory compliance is also becoming more stringent, with bodies enforcing higher standards for product labeling and health claims, ensuring that consumers receive high-quality, effective products (Rausch and Kopplin, 2021).

A critical component of market expansion is Education and Awareness (Raji et al., 2024). Companies are taking active roles in educating consumers about the benefits of probiotics, prebiotics, and synbiotics through blogs, social media, and detailed product labeling (Siddiqui et al., 2023). Collaboration with healthcare professionals, such as dietitians and nutritionists, further bolsters consumer confidence by providing credible, science-backed information (Green, 2024). Additionally, community engagement efforts, such as workshops and seminars, are helping to foster a deeper understanding of gut health and its broader health implications. The role of Technological Innovations cannot be understated (Anđela et al., 2024). Advances in fermentation techniques are improving the viability and potency of probiotic strains, resulting in products that offer enhanced efficacy (Dahiya and Nigam, 2022). Technologies like microencapsulation are protecting sensitive probiotic strains during digestion, ensuring that more live cultures reach the gut intact, thereby increasing their effectiveness (Cunningham et al., 2021).

So, the trends in probiotic, prebiotic, and synbiotic product development underscore a dynamic shift toward more personalized, innovative, and research-backed solutions for gut health (Dixit et al., 2016). With the integration of next-generation probiotics into personalized therapies and the expansion of synbiotics into everyday food and beverage products, the potential for this industry is vast (Murali and Mansell, 2024). Education and scientific validation will be pivotal in driving consumer trust and adoption, while innovation in product delivery, sustainability, and functional ingredients will shape the future of this growing market (Behera et al., 2020). Companies that invest in these areas will be well-positioned to capitalize on the rising demand for gut health solutions, ensuring long-term growth and market leadership (Solis-Balandra and Sanchez-Salas, 2024). This promising trajectory highlights the critical intersection of consumer health, science, and innovation, offering exciting opportunities for stakeholders across the probiotic, prebiotic, and synbiotic spectrum to redefine wellness in the modern era.

19.5 Conclusion

In conclusion, the development of probiotic, prebiotic, and synbiotic products represents an exciting intersection of scientific research, consumer interest, and market opportunity. As awareness of gut health continues to rise, consumers are increasingly seeking evidence-based solutions that promise to enhance their well-being. The principles guiding product formulation, such as efficacy, safety, and innovation, are essential in meeting these growing expectations. However, numerous challenges persist, from navigating complex regulatory landscapes to ensuring the effectiveness and viability of formulations. As manufacturers strive to meet these demands, they must also keep pace with emerging trends, including the emphasis on personalized nutrition, sustainable practices, and transparency, which are reshaping consumer preferences. The future of the probiotic, prebiotic, and synbiotic market holds great promise. With ongoing research into the microbiome and its myriad effects on health, there is significant potential for new product innovations that address specific health concerns. Furthermore, as the industry moves toward more sustainable and ethically responsible practices, there will be opportunities for competitive differentiation. By fostering dialog across disciplines and remaining attuned to consumer needs, collective understanding of gut health and its essential role in holistic well-being can be enhanced, paving the way for a flourishing market that prioritizes health and sustainability.

References

Agarbati, A., Canonico, L., Marini, E., Zannini, E., Ciani, M., Comitini, F., 2020. Potential probiotic yeasts sourced from natural environmental and spontaneous processed foods. Foods 9 (3), 287. Available at https://doi.org/10.3390/foods9030287.

Al-Akayleh, F., Agha, A.S.A.A., Al-Remawi, M., Al-Adham, I.S.I., Daadoue, S., Alsisan, A., Khattab, D., et al., 2024. What we know about the actual role of traditional probiotics in health and disease. Probiotics and Antimicrobial Proteins 16 (5), 1836–1856. Available at https://doi.org/10.1007/s12602-024-10275-7.

Alam, M.W., Kumar, J.V., Awad, M., Saravanan, P., Al-Sowayan, N.S., Rosaiah, P., et al., 2025. Emerging trends in food process engineering: integrating sensing technologies for health, sustainability, and consumer preferences. Journal of Food Process Engineering 48, e70035. Available at https://doi.org/10.1111/jfpe.70035.

Al-Fakhrany, O.M., Elekhnawy, E., 2024. Next-generation probiotics: the upcoming biotherapeutics. Molecular Biology Reports 51, 505. Available at https://doi.org/10.1007/s11033-024-09398-5.

Alvandi, E., Wong, W.K.M., Joglekar, M.V., Spring, K.J., Hardikar, A.A., 2022. Short-chain fatty acid concentrations in the incidence and risk-stratification of colorectal cancer: a systematic review and meta-analysis. BMC Medicine 20, 323. Available at https://doi.org/10.1186/s12916-022-02529-4.

Álvarez, J., Fernández Real, J.M., Guarner, F., Gueimonde, M., Rodríguez, J.M., Saenz De Pipaon, M., et al., 2021. Microbiota intestinal y salud. Gastroenterología y Hepatología 44 (7), 519–535. Available at https://doi.org/10.1016/j.gastrohep.2021.01.009.

Anđela, M., Ivan, M., Mora, D., Arioli, S., 2024. Uncovering the probiotic supplement landscape: market offerings, sales patterns, and future forecasts using machine learning approach—a case study of Montenegro. Probiotics and Antimicrobial Proteins 17, 4635–4658. Available at https://doi.org/10.1007/s12602-024-10400-6.

Ashaolu, T.J., Ashaolu, J.O., Adeyeye, S.A.O., 2021. Fermentation of prebiotics by human colonic microbiota in vitro and short-chain fatty acids production: a critical review. Journal of Applied Microbiology 130 (3), 677–687. Available at https://doi.org/10.1111/jam.14843.

Ashaolu, T.J., 2020. Immune boosting functional foods and their mechanisms: a critical evaluation of probiotics and prebiotics. Biomedicine & Pharmacotherapy 130, 110625. Available at https://doi.org/10.1016/j.biopha.2020.110625.

Asqardokht-Aliabadi, A., Sarabi-Aghdam, V., Homayouni-Rad, A., Hosseinzadeh, N., 2025. Postbiotics in the bakery products: applications and nutritional values. Probiotics and Antimicrobial Proteins 17 (1), 292–314. Available at https://doi.org/10.1007/s12602-024-10327-y.

Azad, M.A.K., Sarker, M., Wan, D., 2018. Immunomodulatory effects of probiotics on cytokine profiles. BioMed Research International 2018. Available at https://doi.org/10.1155/2018/8063647.

Bamigbade, G.B., Subhash, A.J., Kamal-Eldin, A., Nyström, L., Ayyash, M., 2022. An updated review on prebiotics: insights on potentials of food seeds waste as source of potential prebiotics. Molecules 27 (18), 5947. Available at https://doi.org/10.3390/molecules27185947.

Bamola, V.D., Sharma, N., Chaudhry, R., 2013. Probiotics market – current trends. In: Chaudhry, R. (Ed.), Probiotics. KONTENTWORX:KWX Communications Pvt. Ltd., pp. 154–165.

Behera, S.S., El Sheikha, A.F., Hammami, R., Kumar, A., 2020. Traditionally fermented pickles: how the microbial diversity associated with their nutritional and health benefits? Journal of Functional Foods 70, 103971. Available at https://doi.org/10.1016/j.jff.2020.103971.

Birkeland, E., Gharagozlian, S., Birkeland, K.I., Valeur, J., Måge, I., Rud, I., et al., 2020. Prebiotic effect of inulin-type fructans on faecal microbiota and short-chain fatty acids in type 2 diabetes: a randomised controlled trial. European Journal of Nutrition 59 (7), 3325–3338. Available at https://doi.org/10.1007/s00394-020-02282-5.

Bomba, Nemcová, Strojný, L., Mudroňová, D., 2012. Probiotics for farm animals. Lact acid Bact Microbiol Funct Asp. Tayor Fr Gr LLC, pp. 633670 4th ed.

Borka Balas, R., Meliț, L.E., Lupu, A.ța, Lupu, V.V., Mărginean, C.O., 2023. Prebiotics, probiotics, and synbiotics—a research hotspot for pediatric obesity. Microorganisms 11 (11), 2651. Available at https://doi.org/10.3390/microorganisms11112651.

Bozyigit, I., Javadi, A., Altun, S., 2021. Strength properties of xanthan gum and guar gum treated kaolin at different water contents. Journal of Rock Mechanics and Geotechnical Engineering 13 (5), 1160–1172. Available at https://doi.org/10.1016/j.jrmge.2021.06.007.

Brestoff, J.R., Artis, D., 2013. Commensal bacteria at the interface of host metabolism and the immune system. Nature Immunology 14 (7), 676–684. Available at https://doi.org/10.1038/ni.2640.

Buitron, D.I., Sepulveda, L., Martinez, T.K.M., Aguilar, C.N., Medina, D.D., Rodriguez-Herrera, R., et al., 2018. Biotechnological approach for the production of prebiotics and search for new probiotics and their application in the food industry. Applied Food Biotechnology 5 (4), 185–192. Available at https://doi.org/10.22037/afb.v5i4.20174.

Cerdó, T., García-Santos, J., G. Bermúdez, M., Campoy, C., 2019. The role of probiotics and prebiotics in the prevention and treatment of obesity. Nutrients 11 (3), 635. Available at https://doi.org/10.3390/nu11030635.

Chen, Y., Ling, Z., Wang, X., Zong, S., Yang, J., Zhang, Q., et al., 2022. The beneficial mechanism of chitosan and chitooligosaccharides in the intestine on different health status. Journal of Functional Foods 97, 105232. Available at https://doi.org/10.1016/j.jff.2022.105232.

Cheon, S., Kim, G., Bae, J.H., Lee, D.H., Seong, H., Kim, D.H., et al., 2023. Comparative analysis of prebiotic effects of four oligosaccharides using in vitro gut model: digestibility, microbiome, and metabolome changes. FEMS Microbiology Ecology 99 (2). Available at https://doi.org/10.1093/femsec/fiad002.

Colares, H.C., Tarabal, V.S., Silva, C.I.da, Meira, H.G.R., Toledo, M.M.L.de M., Guimarães, P.P.G., et al., 2024. Optimization of bioprocesses of *Lactiplantibacillus plantarum* UFSJP2 with antioxidant activity and its viability in graviola (Annona muricata) sorbet. Food Science and Technology 44. Available at https://doi.org/10.5327/fst.00373.

Cruz, A.G., Cadena, R.S., Walter, E.H.M., Mortazavian, A.M., Granato, D., Faria, J.A.F., et al., 2010. Sensory analysis: relevance for prebiotic, probiotic, and synbiotic product development. Comprehensive Reviews in Food Science and Food Safety 9 (4), 358–373. Available at https://doi.org/10.1111/j.1541-4337.2010.00115.x.

Cunningham, M., Vinderola, G., Charalampopoulos, D., Lebeer, S., Sanders, M.E., Grimaldi, R., 2021. Applying probiotics and prebiotics in new delivery formats – is the clinical evidence transferable? Trends in Food Science and Technology 112, 495–506. Available at https://doi.org/10.1016/j.tifs.2021.04.009.

Dahiya, D., Nigam, P.S., 2022. Probiotics, prebiotics, synbiotics, and fermented foods as potential biotics in nutrition improving health via microbiome-gut-brain axis. Fermentation 8 (7), 303. Available at https://doi.org/10.3390/fermentation8070303.

D'amico, A., Buzzanca, C., Pistorio, E., Melilli, M.G., Di Stefano, V., 2024. Fruit juices as alternative to dairy products for probiotics' intake. Beverages 10 (4), 100. Available at https://doi.org/10.3390/beverages10040100.

Das, A., Behera, R.N., Kapoor, A., Ambatipudi, K., 2023. The potential of meta-proteomics and artificial intelligence to establish the next generation of probiotics for personalized healthcare. Journal of Agricultural and Food Chemistry 71 (46), 17528–17542. Available at https://doi.org/10.1021/acs.jafc.3c03834.

Denkova, R., Goranov, B., Teneva, D., Denkova, Z., Kostov, G., 2017. Antimicrobial activity of probiotic microorganisms: mechanisms of interaction and methods of examination. Antimicrobial Research: Novel Bioknowledge and Educational Programs 1, 201–2012.

Dinakis, E., O'donnell, J.A., Marques, F.Z., 2024. The gut–immune axis during hypertension and cardiovascular diseases. Acta Physiologica 240 (8). Available at https://doi.org/10.1111/apha.14193.

Dixit, Y., Wagle, A., Vakil, B., 2016. Patents in the field of probiotics, prebiotics, synbiotics: a review. Journal of Food: Microbiology, Safety & Hygiene 01 (02). Available at https://doi.org/10.4172/2476-2059.1000111.

Dong, Y., Li, M., Yue, X., 2024. Current research on probiotics and fermented products. Foods 13 (9), 1406. Available at https://doi.org/10.3390/foods13091406.

Du, Z., Li, J., Li, W., Fu, H., Ding, J., Ren, G., et al., 2023. Effects of prebiotics on the gut microbiota in vitro associated with functional diarrhea in children. Frontiers in Microbiology 14. Available at https://doi.org/10.3389/fmicb.2023.1233840.

Du, Q., Li, Q., Liu, C., Liao, G., Li, J., Yang, J., et al., 2024. Probiotics/prebiotics/synbiotics and human neuropsychiatric outcomes: an umbrella review. Beneficial Microbes 15 (6), 589–608. Available at https://doi.org/10.1163/18762891-bja00035.

Eden, S., 2011. Food labels as boundary objects: how consumers make sense of organic and functional foods. Public Understanding of Science 20 (2), 179–194. Available at https://doi.org/10.1177/0963662509336714.

Furtado, I.F.S.P.C., Sydney, E.B., Rodrigues, S.A., Sydney, A.C.N., 2022. Xanthan gum: applications, challenges, and advantages of this asset of biotechnological origin. Biotechnology Research and Innovation 6 (1), e202204. Available at https://doi.org/10.4322/biori.202205.

Gajbhiye, S., Dhoble, S., Tobin, D., 2024. Natural gum (xanthan, gellan, arabic, guar, ghatti gum, etc.)-based bio-scaffold and their application in tissue engineering. Natural Product Inspired Scaffolds: Applications in Tissue Engineering. Springer Nature, India, pp. 61–91. Available at https://doi.org/10.1007/978-981-97-3111-4_4.

Gao, D., Liu, Z., Liu, F., Chen, L., Wang, W., Ma, J., et al., 2021. Study of the immunoregulatory effect of *Lactobacillus rhamnosus* 1.0320 in immunosuppressed mice. Journal of Functional Foods 79, 104423. Available at https://doi.org/10.1016/j.jff.2021.104423.

George Kerry, R., Patra, J.K., Gouda, S., Park, Y., Shin, H.S., Das, G., 2018. Benefaction of probiotics for human health: a review. Journal of Food and Drug Analysis 26 (3), 927–939. Available at https://doi.org/10.1016/j.jfda.2018.01.002.

Gibson, G.R., Roberfroid, M.B., 1995. Dietary modulation of the human colonic microbiota: Introducing the concept of prebiotics. Journal of Nutrition 125 (6), 1401–1412. Available at https://doi.org/10.1093/jn/125.6.1401.

Gibson, G.R., Hutkins, R., Sanders, M.E., Prescott, S.L., Reimer, R.A., Salminen, S.J., et al., 2017. Expert consensus document: The International Scientific Association for Probiotics and Prebiotics (ISAPP) consensus statement on the definition and scope of prebiotics. Nature Reviews Gastroenterology and Hepatology 14 (8), 491–502. Available at https://doi.org/10.1038/nrgastro.2017.75.

Goh, J.X.H., Tan, L.T.H., Law, J.W.F., Ser, H.L., Khaw, K.Y., Letchumanan, V., et al., 2022. Harnessing the potentialities of probiotics, prebiotics, synbiotics, paraprobiotics, and postbiotics for shrimp farming. Reviews in Aquaculture 14 (3), 1478–1557. Available at https://doi.org/10.1111/raq.12659.

Gomaa, E.Z., 2020. Human gut microbiota/microbiome in health and diseases: a review. Antonie van Leeuwenhoek 113 (12), 2019–2040. Available at https://doi.org/10.1007/s10482-020-01474-7.

Gomez Quintero, D.F., Kok, C.R., Hutkins, R., 2022. The future of synbiotics: rational formulation and design. Frontiers in Microbiology 13. Available at https://doi.org/10.3389/fmicb.2022.919725.

González-Herrera, S.M., Bermúdez-Quiñones, G., Ochoa-Martínez, L.A., Rutiaga-Quiñones, O.M., Gallegos-Infante, J.A., 2021. Synbiotics: a technological approach in food applications. Journal of Food Science and Technology 58 (3), 811–824. Available at https://doi.org/10.1007/s13197-020-04532-0.

Gorreja, F., Walker, W.A., 2022. The potential role of adherence factors in probiotic function in the gastrointestinal tract of adults and pediatrics: a narrative review of experimental and human studies. Gut Microbes 14 (1). Available at https://doi.org/10.1080/19490976.2022.2149214.

Green, W.N., 2024. The anti-politics of impact investment: financial self-regulation, market competition and over-indebtedness in Cambodia. Environment and Planning A: Economy and Space 56 (6), 1668–1685. Available at https://doi.org/10.1177/0308518x241239797.

Grover, S., Rashmi, H.M., Srivastava, A.K., Batish, V.K., 2012. Probiotics for human health–new innovations and emerging trends. Gut Pathogens 4, 1–14.

Grumet, L., Tromp, Y., Stiegelbauer, V., 2020. The development of high-quality multispecies probiotic formulations: from bench to market. Nutrients 12 (8), 2453. Available at https://doi.org/10.3390/nu12082453.

Guo, Q., Jin, Y., Chen, X., Ye, X., Shen, X., Lin, M., et al., 2024. NF-κB in biology and targeted therapy: new insights and translational implications. Signal Transduction and Targeted Therapy 9 (1). Available at https://doi.org/10.1038/s41392-024-01757-9.

Hadi, A., Alizadeh, K., Hajianfar, H., Mohammadi, H., Miraghajani, M., 2020. Efficacy of synbiotic supplementation in obesity treatment: a systematic review and meta-analysis of clinical trials. Critical Reviews in Food Science and Nutrition 60 (4), 584–596. Available at https://doi.org/10.1080/10408398.2018.1545218.

Halder, T., Mehta, P., Acharya, N., 2021. Trends in the functional food market and nutraceutical product development. Nutraceuticals for Aging and Anti-Aging: Basic Understanding and Clinical Evidence. CRC Press, India, pp. 1–26. Available at https://doi.org/10.1201/9781003110866-1.

Han, K., Xu, J., Xie, F., Crowther, J., Moon, J.J., 2024. Engineering strategies to modulate the gut microbiome and immune system. Journal of Immunology 212 (2), 208–215. Available at https://doi.org/10.4049/jimmunol.2300480.

Homayoni Rad, A., Akbarzadeh, F., Mehrabany, E.V., 2012. Which are more important: prebiotics or probiotics? Nutrition 28 (11-12), 1196–1197. Available at https://doi.org/10.1016/j.nut.2012.03.017.

Homayouni Rad, A., Torab, R., Ghalibaf, M., Norouzi, S., Mehrabany, E.V., 2013. Might patients with immune-related diseases benefit from probiotics? Nutrition 29 (3), 583–586. Available at https://doi.org/10.1016/j.nut.2012.10.008.

Homayouni Rad, A., Pourjafar, H., Mirzakhani, E., 2023. A comprehensive review of the application of probiotics and postbiotics in oral health. Frontiers in Cellular and Infection Microbiology 13. Available at https://doi.org/10.3389/fcimb.2023.1120995.

Kamalledin Moghadam, S., Farhoodi, M., Mofid, V., Homayouni-Rad, A., Mortazavian-Farsani, A.-M., Milani, A., 2024. Effects of Adding Non-viable Lacticaseibacillus casei and Lactobacillus acidophilus on Physicochemical, Microbial, Chemical and Sensory Attributes of Probiotic Doogh. Applied Food Biotechnology 11 (1), e14. Available at https://doi.org/10.22037/afb.v11i1.44105.

Hosseinzadeh, N., Asqardokht-Aliabadi, A., Sarabi-Aghdam, V., Hashemi, N., Dogahi, P.R., Sarraf-Ov, N., et al., 2025. Antioxidant properties of postbiotics: an overview on the analysis and evaluation methods. Probiotics and Antimicrobial Proteins 17 (2), 606–624. Available at https://doi.org/10.1007/s12602-024-10372-7.

Hughes, R.L., Alvarado, D.A., Swanson, K.S., Holscher, H.D., 2022. The prebiotic potential of inulin-type fructans: a systematic review. Advances in Nutrition 13 (2), 492–529. Available at https://doi.org/10.1093/advances/nmab119.

Huynh, T.G., Shiu, Y.L., Nguyen, T.P., Truong, Q.P., Chen, J.C., Liu, C.H., 2017. Current applications, selection, and possible mechanisms of actions of synbiotics in improving the growth and health status in aquaculture: a review. Fish and Shellfish Immunology 64, 367–382. Available at https://doi.org/10.1016/j.fsi.2017.03.035.

Ince Palamutoglu, M., Köse, G., Bas, M., 2024. Probiotics and prebiotics affecting mental and gut health. Healthcare 12 (5), 510. Available at https://doi.org/10.3390/healthcare12050510.

Ji, J., Jin, W., Liu, S.-J., Jiao, Z., Li, X., 2023. Probiotics, prebiotics, and postbiotics in health and disease. MedComm 4 (6). Available at https://doi.org/10.1002/mco2.420.

Jiao, J., Chen, N., Yang, R., 2024. How to promote green travel effectively: a study of niche information interventions based on meta-analysis. Environment, Development and Sustainability 26 (4), 8267–8301. Available at https://doi.org/10.1007/s10668-023-03095-2.

Kearney, S.M., Gibbons, S.M., 2018. Designing synbiotics for improved human health. Microbial Biotechnology 11 (1), 141–144. Available at https://doi.org/10.1111/1751-7915.12885.

Kellershohn, J., 2021. Marketing probiotic beverages. Probiotic Beverages. Elsevier, Canada, pp. 427–440. Available at https://doi.org/10.1016/B978-0-12-818588-9.00019-X.

Kiepś, J., Dembczyński, R., 2022. Current trends in the production of probiotic formulations. Foods 11 (15), 2330. Available at https://doi.org/10.3390/foods11152330.

Krychtiuk, K.A., Andersson, T.L., Bodesheim, U., Butler, J., Curtis, L.H., Elkind, M., et al., 2024. Drug development for major chronic health conditions—aligning with growing public health needs: proceedings from a multistakeholder think tank. American Heart Journal 270, 23–43. Available at https://doi.org/10.1016/j.ahj.2024.01.004.

Latif, A., Shehzad, A., Niazi, S., Zahid, A., Ashraf, W., Iqbal, M.W., et al., 2023. Probiotics: mechanism of action, health benefits and their application in food industries. Frontiers in Microbiology 14. Available at https://doi.org/10.3389/fmicb.2023.1216674.

Legesse Bedada, T., Feto, T.K., Awoke, K.S., Garedew, A.D., Yifat, F.T., Birri, D.J., 2020. Probiotics for cancer alternative prevention and treatment. Biomedicine and Pharmacotherapy 129. Available at https://doi.org/10.1016/j.biopha.2020.110409.

Leon-Derecho, C.M.P.de, 2024. Microbiome-based therapeutics. Human Microbiome Drug Targets: Modern Approaches in Disease Management. Elsevier, Philippines, pp. 233–244. Available at https://doi.org/10.1016/B978-0-443-15435-5.00017-7.

Li, H.Y., Zhou, D.D., Gan, R.Y., Huang, S.Y., Zhao, C.N., Shang, A., et al., 2021. Effects and mechanisms of probiotics, prebiotics, synbiotics, and postbiotics on metabolic diseases targeting gut microbiota: a narrative review. Nutrients 13 (9). Available at https://doi.org/10.3390/nu13093211.

Li, Y.I., Pagulayan, K., Rau, H., Hendrickson, R., Schindler, A.G., 2025. Gut microbial composition is associated with symptom self-report in trauma-exposed Iraq and Afghanistan veterans. Neurotrauma Reports 6 (1), 1–12. Available at https://doi.org/10.1089/neur.2024.0011.

Lieberman, I.W., 2020. The growth and commercial evolution of microfinance. Futur Microfinance 9–35.

Liu, C., Yang, L., Wei, W., Fu, P., 2024. Efficacy of probiotics/synbiotics supplementation in patients with chronic kidney disease: a systematic review and meta-analysis of randomized controlled trials. Frontiers in Nutrition 11. Available at https://doi.org/10.3389/fnut.2024.1434613.

Ludwig, I.S., Broere, F., Manurung, S., Lambers, T.T., Van Der Zee, R., Van Eden, W., 2018. *Lactobacillus rhamnosus* GG-derived soluble mediators modulate adaptive immune cells. Frontiers in Immunology 9. Available at https://doi.org/10.3389/fimmu.2018.01546.

Luo, S.C., Wei, S.M., Luo, X.T., Yang, Q.Q., Wong, K.H., Cheung, P.C.K., et al., 2024. How probiotics, prebiotics, synbiotics, and postbiotics prevent dental caries: an oral microbiota perspective. npj Biofilms and Microbiomes 10 (1). Available at https://doi.org/10.1038/s41522-024-00488-7.

Maleki, D., Homayouni, A., Khalili, L., Golkhalkhali, B., 2016. P. Robiotics in Cancer Prevention, Updating the Evidence. Elsevier BV, pp. 781–791. Available at https://doi.org/10.1016/b978-0-12-802189-7.00059-9.

Markowiak, P., Śliżewska, K., 2017. Effects of probiotics, prebiotics, and synbiotics on human health. Nutrients 9 (9), 1021. Available at https://doi.org/10.3390/nu9091021.

Martyniak, A., Medyńska-Przęczek, A., Wędrychowicz, A., Skoczeń, S., Tomasik, P.J., 2021. Prebiotics, probiotics, synbiotics, paraprobiotics and postbiotic compounds in IBD. Biomolecules 11 (12), 1903. Available at https://doi.org/10.3390/biom11121903.

Mazziotta, C., Tognon, M., Martini, F., Torreggiani, E., Rotondo, J.C., 2023. Probiotics mechanism of action on immune cells and beneficial effects on human health. Cells 12 (1), 184. Available at https://doi.org/10.3390/cells12010184.

Mcmillin, K.W., 2017. Advancements in meat packaging. Meat Science 132, 153–162. Available at https://doi.org/10.1016/j.meatsci.2017.04.015.

Mehta, J.P., Ayakar, S., Singhal, R.S., 2023. The potential of paraprobiotics and postbiotics to modulate the immune system: a review. Microbiological Research 275, 127449. Available at https://doi.org/10.1016/j.micres.2023.127449.

Merkouris, E., Mavroudi, T., Miliotas, D., Tsiptsios, D., Serdari, A., Christidi, F., et al., 2024. Probiotics' effects in the treatment of anxiety and depression: a comprehensive review of 2014–2023 clinical trials. Microorganisms 12 (2), 411. Available at https://doi.org/10.3390/microorganisms12020411.

Milani, P.G., Soleimani, R.A., Khani, N., Rad, A.H., 2023. Gut microbiota: a perspective for colorectal cancer. Gastrointestinal Nursing 21 (7), 26–31. Available at https://doi.org/10.12968/gasn.2023.21.7.26.

Mishra, N., Ashique, S., Farid, A., Garg, A., 2024. Synbiotics in Metabolic Disorders: Mechanisms, Therapeutic Potential, and Future Perspectives. CRC Press. Available at https://doi.org/10.1201/9781032702438.

Mounir, M., Ibijbijen, A., Farih, K., Rabetafika, H.N., Razafindralambo, H.L., 2022. Synbiotics and their antioxidant properties, mechanisms, and benefits on human and animal health: a narrative review. Biomolecules 12 (10), 1443. Available at https://doi.org/10.3390/biom12101443.

Murali, S.K., Mansell, T.J., 2024. Next generation probiotics: engineering live biotherapeutics. Biotechnology Advances 72, 108336. Available at https://doi.org/10.1016/j.biotechadv.2024.108336.

Navarro-Torre, S., Carro, L., Rodríguez-Llorente, I.D., Pajuelo, E., Caviedes, M.Á., Igual, J.M., et al., 2020. *Pseudoalteromonas rhizosphaerae* sp. Nov., a novel plant growthpromoting bacterium with potential use in phytoremediation. International Journal of Systematic and Evolutionary Microbiology 70 (5), 3287–3294. Available at https://doi.org/10.1099/ijsem.0.004167.

O'donnell, S.T., Ross, R.P., Stanton, C., 2020. The progress of multi-omics technologies: determining function in lactic acid bacteria using a systems level approach. Frontiers in Microbiology 10. Available at https://doi.org/10.3389/fmicb.2019.03084.

Pannerchelvan, S., Rios-Solis, L., Wasoh, H., Sobri, M.Z.M., Faizal Wong, F.W., Mohamed, M.S., et al., 2024. Functional yogurt: a comprehensive review of its nutritional composition and health benefits. Food and Function 15 (22), 10927–10955. Available at https://doi.org/10.1039/d4fo03671a.

Parhi, P., Liu, S.Q., Choo, W.S., 2024. Synbiotics: effects of prebiotics on the growth and viability of probiotics in food matrices. Bioactive Carbohydrates and Dietary Fibre 32, 100462. Available at https://doi.org/10.1016/j.bcdf.2024.100462.

Pimentel, T.C., Costa, W.K.A.da, Barão, C.E., Rosset, M., Magnani, M., 2021. Vegan probiotic products: a modern tendency or the newest challenge in functional foods. Food Research International 140, 110033. Available at https://doi.org/10.1016/j.foodres.2020.110033.

Pires, L., González-Paramás, A.M., Heleno, S.A., Calhelha, R.C., 2024. Exploring therapeutic advances: a comprehensive review of intestinal microbiota modulators. Antibiotics 13 (8). Available at https://doi.org/10.3390/antibiotics13080720.

Plaza-Diaz, J., Ruiz-Ojeda, F.J., Gil-Campos, M., Gil, A., 2019. Mechanisms of action of probiotics. Advances in Nutrition 10 (1), S49–S66. Available at https://doi.org/10.1093/advances/nmy063.

Pop, O.L., Pop, C.R., Dufrechou, M., Vodnar, D.C., Socaci, S.A., Dulf, F.V., et al., 2020. Edible films and coatings functionalization by probiotic incorporation: a review. Polymers 12 (1). Available at https://doi.org/10.3390/polym12010012.

Praveen, M., Brogi, S., 2025. Microbial fermentation in food and beverage industries: innovations, challenges, and opportunities. Foods 14 (1), 114. Available at https://doi.org/10.3390/foods14010114.

Rabetafika, H.N., Razafindralambo, A., Ebenso, B., Razafindralambo, H.L., 2023. Probiotics as antibiotic alternatives for human and animal applications. Encyclopedia 3 (2), 561–581. Available at https://doi.org/10.3390/encyclopedia3020040.

Rad, A.H., Aghebati-Maleki, L., Kafil, H.S., Abbasi, A., 2021a. Molecular mechanisms of postbiotics in colorectal cancer prevention and treatment. Critical Reviews in Food Science and Nutrition 61 (11), 1787–1803. Available at https://doi.org/10.1080/10408398.2020.1765310.

Rad, A.H., Aghebati-Maleki, L., Kafil, H.S., Gilani, N., Abbasi, A., Khani, N., 2021b. Postbiotics, as dynamic biomolecules, and their promising role in promoting food safety. Biointerface Research in Applied Chemistry 11 (6), 14529–14544. Available at https://doi.org/10.33263/BRIAC116.1452914544.

Raheem, A., Liang, L., Zhang, G., Cui, S., 2021. Modulatory effects of probiotics during pathogenic infections with emphasis on immune regulation. Frontiers in Immunology 12. Available at https://doi.org/10.3389/fimmu.2021.616713.

Rahim, F., Toguzbaeva, K., Qasim, N.H., Dzhusupov, K.O., Zhumagaliuly, A., Khozhamkul, R., 2023. Probiotics, prebiotics, and synbiotics for patients with autism spectrum disorder: a meta-analysis and umbrella review. Frontiers in Nutrition 10. Available at https://doi.org/10.3389/fnut.2023.1294089.

Rahimi, F., Pasdar, Y., Kaviani, M., Abbasi, S., Fry, H., Hekmatdoost, A., et al., 2022. Efficacy of the synbiotic supplementation on the metabolic factors in patients with metabolic syndrome: a randomized, triple-blind, placebo-controlled trial. International Journal of Clinical Practice 2022, 2967977. Available at https://doi.org/10.1155/2022/2967977.

Raji, M.A., Olodo, H.B., Oke, T.T., Addy, W.A., Ofodile, O.C., Oyewole, A.T., 2024. Business strategies in virtual reality: a review of market opportunities and consumer experience. International Journal of Management & Entrepreneurship Research 6 (3), 722–736. Available at https://doi.org/10.51594/ijmer.v6i3.883.

Raman, M., Ambalam, P., Doble, M., 2019. Probiotics, prebiotics, and fibers in nutritive and functional beverages. Nutrients in Beverages: Volume 12: The Science of Beverages. Elsevier, India, pp. 315–367. Available at https://doi.org/10.1016/B978-0-12-816842-4.00009-5.

Rau, S., Gregg, A., Yaceczko, S., Limketkai, B., 2024. Prebiotics and probiotics for gastrointestinal disorders. Nutrients 16 (6), 778. Available at https://doi.org/10.3390/nu16060778.

Rausch, T.M., Kopplin, C.S., 2021. Bridge the gap: consumers' purchase intention and behavior regarding sustainable clothing. Journal of Cleaner Production 278, 123882. Available at https://doi.org/10.1016/j.jclepro.2020.123882.

Ren, C., Cheng, L., Sun, Y., Zhang, Q., De Haan, B.J., Zhang, H., et al., 2020. Lactic acid bacteria secrete toll like receptor 2 stimulating and macrophage immunomodulating bioactive factors. Journal of Functional Foods 66, 103783. Available at https://doi.org/10.1016/j.jff.2020.103783.

Riboulet-Bisson, E., Sturme, M.H.J., Jeffery, I.B., O'donnell, M.M., Neville, B.A., Forde, B.M., et al., 2012. Effect of *Lactobacillus salivarius* bacteriocin ABP118 on the mouse and pig intestinal microbiota. PLoS ONE 7 (2). Available at https://doi.org/10.1371/journal.pone.0031113.

Rodionova, N., Rodionov, A., Shchetilina, I., Manukowskaya, M., Rusanova, M., Razinkova, T., et al., 2021. Assessment of the marketing and innovative potential of synbiotic products with herbal biocorrector. BIO Web of Conferences 30, 01016. Available at https://doi.org/10.1051/bioconf/20213001016.

Rodzi, N.A.R.M., Lee, L.K., 2021. Traditional fermented foods as vehicle of non-dairy probiotics: perspectives in South East Asia countries. Food Research International 150. Available at https://doi.org/10.1016/j.foodres.2021.110814.

Rose, E.C., Odle, J., Blikslager, A.T., Ziegler, A.L., 2021. Probiotics, prebiotics and epithelial tight junctions: a promising approach to modulate intestinal barrier function. International Journal of Molecular Sciences 22 (13), 6729. Available at https://doi.org/10.3390/ijms22136729.

Roy, S., Dhaneshwar, S., 2023. Role of prebiotics, probiotics, and synbiotics in management of inflammatory bowel disease: current perspectives. World Journal of Gastroenterology 29 (14), 2078–2100. Available at https://doi.org/10.3748/wjg.v29.i14.2078.

Sagheddu, V., Uggeri, F., Belogi, L., Remollino, L., Brun, P., Bernabè, G., et al., 2020. The biotherapeutic potential of *Lactobacillus reuteri* characterized using a target-specific selection process. Frontiers in Microbiology 11. Available at https://doi.org/10.3389/fmicb.2020.00532.

Seifert, S., Bub, A., Franz, C.M.A.P., Watzl, B., 2011. Probiotic *Lactobacillus casei* shirota supplementation does not modulate immunity in healthy men with reduced natural killer cell activity. Journal of Nutrition 141 (5), 978–984. Available at https://doi.org/10.3945/jn.110.136440.

Shao, T., Verma, H.K., Pande, B., Costanzo, V., Ye, W., Cai, Y., et al., 2021. Physical activity and nutritional influence on immune function: an important strategy to improve immunity and health status. Frontiers in Physiology 12. Available at https://doi.org/10.3389/fphys.2021.751374.

Sharma, K., Aswal, C., Paul, J., 2023. Factors affecting green purchase behavior: a systematic literature review. Business Strategy and the Environment 32 (4), 2078–2092. Available at https://doi.org/10.1002/bse.3237.

Sharma, V., Mohammed, S.A., Devi, N., Vats, G., Tuli, H.S., Saini, A.K., et al., 2024. Unveiling the dynamic relationship of viruses and/or symbiotic bacteria with plant resilience in abiotic stress. Biology 4 (1). Available at https://doi.org/10.1007/s44154-023-00126-w.

Siddiqui, S.A., Khan, S., Mehdizadeh, M., Nirmal, N.P., Khanashyam, A.C., Fernando, I., et al., 2023. Consumer studies focus on prebiotics, probiotics, and synbiotics in food packaging: a review. Current Food Science and Technology Reports 1 (1), 13–29. Available at https://doi.org/10.1007/s43555-023-00003-7.

Slavin, J., 2013. Fiber and prebiotics: mechanisms and health benefits. Nutrients 5 (4), 1417–1435. Available at https://doi.org/10.3390/nu5041417.

Śliżewska, K., Chlebicz-Wójcik, A., 2020. The in vitro analysis of prebiotics to be used as a component of a synbiotic preparation. Nutrients 12 (5), 1272. Available at https://doi.org/10.3390/nu12051272.

Smelt, M.J., De Haan, B.J., Bron, P.A., Van Swam, I., Meijerink, M., Wells, J.M., et al., 2013. The impact of *Lactobacillus plantarum* WCFS1 teichoic acid D-alanylation on the generation of effector and regulatory T-cells in healthy mice. PLoS One 8 (4). Available at https://doi.org/10.1371/journal.pone.0063099.

Solis-Balandra, M.A., Sanchez-Salas, J.L., 2024. Classification and multi-functional use of bacteriocins in health, biotechnology, and food industry. Antibiotics 13 (7). Available at https://doi.org/10.3390/antibiotics13070666.

Sommermeyer, H., Chmielowiec, K., Bernatek, M., Olszewski, P., Kopczynski, J., Piątek, J., 2024. Effectiveness of a balanced nine-strain synbiotic in primary-care irritable bowel syndrome patients—a multi-center, randomized, double-blind, placebo-controlled trial. Nutrients 16 (10), 1503. Available at https://doi.org/10.3390/nu16101503.

Song, H., Jeon, D., Unno, T., 2022. Evaluation of prebiotics through an in vitro gastrointestinal digestion and fecal fermentation experiment: further idea on the implementation of machine learning technique. Foods 11 (16), 2490. Available at https://doi.org/10.3390/foods11162490.

Sotoudegan, F., Daniali, M., Hassani, S., Nikfar, S., Abdollahi, M., 2019. Reappraisal of probiotics' safety in human. Food and Chemical Toxicology 129, 22–29. Available at https://doi.org/10.1016/j.fct.2019.04.032.

Swann, J.R., Rajilic-Stojanovic, M., Salonen, A., Sakwinska, O., Gill, C., Meynier, A., et al., 2020. Considerations for the design and conduct of human gut microbiota intervention studies relating to foods. European Journal of Nutrition 59 (8), 3347–3368. Available at https://doi.org/10.1007/s00394-020-02232-1.

Tan, F.P.Y., Beltranena, E., Zijlstra, R.T., 2021. Resistant starch: implications of dietary inclusion on gut health and growth in pigs: a review. Journal of Animal Science and Biotechnology 12 (1). Available at https://doi.org/10.1186/s40104-021-00644-5.

Teng, Y., Zhao, D., Yan, C., Ren, X., Wang, A., Xia, X., 2024. Pasteurized Akkermansia muciniphila improves immunity and relieves other toxic effects in 5-fluorouracil treated BALB/c mice. Food Bioscience 59, 103918. Available at https://doi.org/10.1016/j.fbio.2024.103918.

Terpou, A., Papadaki, A., Lappa, I., Kachrimanidou, V., Bosnea, L., Kopsahelis, N., 2019. Probiotics in food systems: significance and emerging strategies towards improved viability and delivery of enhanced beneficial value. Nutrients 11 (7), 1591. Available at https://doi.org/10.3390/nu11071591.

Tsoupras, A., Gkika, D.A., Markopoulos, T., Curran, R., Scallon, C., Karali, M., et al., 2024. Apple Products (Apple Juice and Cider) and By-Products (Apple Pomace): Bioactive Compounds and Biological Properties. Springer Science and Business Media LLC, pp. 1–42. Available at https://doi.org/10.1007/978-3-031-04195-2_214-1.

Tsylents, U., Burmistrz, M., Wojciechowska, M., Stępień, J., Maj, P., Trylska, J., 2024. Iron uptake pathway of *Escherichia coli* as an entry route for peptide nucleic acids conjugated with a siderophore mimic. Frontiers in Microbiology 15. Available at https://doi.org/10.3389/fmicb.2024.1331021.

Venkatesh, G.P., Kuruvalli, G., Syed, K., Reddy, V.D., 2024. An updated review on probiotic production and applications. Gastroenterology Insights 15 (1), 221–236. Available at https://doi.org/10.3390/gastroent15010016.

Vojvodić Cebin, A., Bunić, M., Mandura Jarić, A., Šeremet, D., Komes, D., 2024. Physicochemical and sensory stability evaluation of gummy candies fortified with mountain germander extract and prebiotics. Polymers 16 (2), 259. Available at https://doi.org/10.3390/polym16020259.

Volarić, M., Šojat, D., Majnarić, L.T., Vučić, D., 2024. The association between functional dyspepsia and metabolic syndrome—the state of the art. International Journal of Environmental Research and Public Health 21 (2), 237. Available at https://doi.org/10.3390/ijerph21020237.

Wang, Y., Wang, B., Zhan, X., Wang, Y., Li, W., 2022. Effects of glucose oxidase and its combination with *B. amyloliquefaciens* SC06 on intestinal microbiota, immune response and antioxidative capacity in broilers. Animal 16 (3), 100473. Available at https://doi.org/10.1016/j.animal.2022.100473.

Westfall, S., Carracci, F., Estill, M., Zhao, D., Wu, Q.-li, Shen, L., et al., 2021. Optimization of probiotic therapeutics using machine learning in an artificial human gastrointestinal tract. Scientific Reports 11 (1). Available at https://doi.org/10.1038/s41598-020-79947-y.

Williams, G.M., Tapsell, L.C., Beck, E.J., 2023. Gut health, the microbiome and dietary choices: an exploration of consumer perspectives. Nutrition and Dietetics 80 (1), 85–94. Available at https://doi.org/10.1111/1747-0080.12769.

Yan, F., Cao, H., Cover, T.L., Whitehead, R., Washington, M.K., Polk, D.B., 2007. Soluble proteins produced by probiotic bacteria regulate intestinal epithelial cell survival and growth. Gastroenterology 132 (2), 562–575. Available at https://doi.org/10.1053/j.gastro.2006.11.022.

You, S., Ma, Y., Yan, B., Pei, W., Wu, Q., Ding, C., et al., 2022. The promotion mechanism of prebiotics for probiotics: a review. Frontiers in Nutrition 9. Available at https://doi.org/10.3389/fnut.2022.1000517.

Yousefi, B., Eslami, M., Ghasemian, A., Kokhaei, P., Salek Farrokhi, A., Darabi, N., 2019. Probiotics importance and their immunomodulatory properties. Journal of Cellular Physiology 234 (6), 8008–8018. Available at https://doi.org/10.1002/jcp.27559.

Zambelli, R.A., Mendonça, L.G.D., 2024. Application of probiotics in food industry a promising scope for improving food quality. Probiotics: A Comprehensive Guide to Enhance Health and Mitigate Disease. CRC Press, Brazil, pp. 282–304. Available at https://doi.org/10.1201/9781003452249-13.

Zhang, T., Zhang, C., Zhang, J., Sun, F., Duan, L., 2022. Efficacy of probiotics for irritable bowel syndrome: a systematic review and network meta-analysis. Frontiers in Cellular and Infection Microbiology 12. Available at https://doi.org/10.3389/fcimb.2022.859967.

Zhang, W.X., Shi, L.B., Zhou, M.S., Wu, J., Shi, H.Y., 2023a. Efficacy of probiotics, prebiotics and synbiotics in irritable bowel syndrome: a systematic review and meta-analysis of randomized, double-blind, placebo-controlled trials. Journal of Medical Microbiology 72 (9). Available at https://doi.org/10.1099/jmm.0.001758.

Zhang, Y., Chen, R., Zhang, D.D., Qi, S., Liu, Y., 2023b. Metabolite interactions between host and microbiota during health and disease: which feeds the other? Biomedicine & Pharmacotherapy 160, 114295. Available at https://doi.org/10.1016/j.biopha.2023.114295.

Zhang, L., Wang, F., Wang, R., Sun, B., Liu, P.J., 2024. Effects of probiotics, prebiotics, and synbiotics on cardiometabolic risk factors in children and adolescents with overweight or obesity: a systematic review and Bayesian network meta-analysis. Critical Reviews in Food Science and Nutrition 1–15. Available at https://doi.org/10.1080/10408398.2024.2409956.

Zheng, H.J., Guo, J., Wang, Q., Wang, L., Wang, Y., Zhang, F., et al., 2021. Probiotics, prebiotics, and synbiotics for the improvement of metabolic profiles in patients with chronic kidney disease: a systematic review and meta-analysis of randomized controlled trials. Critical Reviews in Food Science and Nutrition 61 (4), 577–598. Available at https://doi.org/10.1080/10408398.2020.1740645.

Index

Note: Page numbers followed by "*f*" and "*t*" refer to figures and tables, respectively.

N

O

T

9780443298356